Universitext

Universitext

Universitext is a series of textbooks that presents material from a wide variety of mathematical disciplines at master's level and beyond. The books, often well class-tested by their author, may have an informal, personal even experimental approach to their subject matter. Some of the most successful and established books in the series have evolved through several editions, always following the evolution of teaching curricula, into very polished texts.

Thus as research topics trickle down into graduate-level teaching, first textbooks written for new, cutting-edge courses may make their way into *Universitext*.

More information about this series at http://www.springer.com/series/223

Achim Klenke

Probability Theory

A Comprehensive Course

Third Edition

 Springer

Achim Klenke
Institut für Mathematik
Johannes Gutenberg-Universität Mainz
Mainz, Germany

ISSN 0172-5939   ISSN 2191-6675   (electronic)
Universitext
ISBN 978-3-030-56401-8   ISBN 978-3-030-56402-5   (eBook)
https://doi.org/10.1007/978-3-030-56402-5

Mathematics Subject Classification: 60-01, 60B10, 60G42, 60G55, 60H05, 60H10, 60J10, 37-01, 28-01, 82-00

This Springer imprint is published by the registered company Springer Nature Switzerland AG.
The registered company address is: Gewerbestrasse 11, 6330 Cham, Switzerland

Preface to the Third Edition

New in the third edition: the sections close with a short "takeaways" block where highlights of the section are summarized sometimes on an informal level without full rigor. Furthermore, in some places "reflection" blocks have been added. They are of different levels of difficulty indicated by the number of clubsuits. Finally, there are more exercises and some new illustrations.

Many people have helped in correcting errors or improving the exposition by asking questions and I thank all of them. In particular, I would like to thank Philipp Neumann for many helpful comments.

Mainz, Germany
June 2020

Achim Klenke

Preface to the Second Edition

In the second edition of this book, many errors have been corrected. Furthermore, the text has been extended carefully in many places. In particular, there are more exercises and a lot more illustrations.

I would like to take the opportunity to thank all of those who helped in improving the first edition of this book, in particular: Michael Diether, Maren Eckhoff, Christopher Grant, Matthias Hammer, Heiko Hoffmann, Martin Hutzenthaler, Martin Kolb, Manuel Mergens, Thal Nowik, Felix Schneider, Wolfgang Schwarz, and Stephan Tolksdorf.

A constantly updated list of errors can be found at www.aklenke.de.

Mainz
Achim Klenke
March 2013

Preface to the First Edition

This book is based on two four-hour courses on advanced probability theory that I have held in recent years at the universities of Cologne and Mainz. It is implicitly assumed that the reader has a certain familiarity with the basic concepts of probability theory, although the formal framework will be fully developed in this book.

The aim of this book is to present the central objects and concepts of probability theory: random variables, independence, laws of large numbers and central limit theorems, martingales, exchangeability and infinite divisibility, Markov chains and Markov processes, as well as their connection with discrete potential theory, coupling, ergodic theory, Brownian motion and the Itô integral (including stochastic differential equations), the Poisson point process, percolation, and the theory of large deviations.

Measure theory and integration are necessary prerequisites for a systematic probability theory. We develop it only to the point to which it is needed for our purposes: construction of measures and integrals, the Radon–Nikodym theorem and regular conditional distributions, convergence theorems for functions (Lebesgue) and measures (Prohorov), and construction of measures in product spaces. The chapters on measure theory do not come as a block at the beginning (although they are written such that this would be possible; that is, independent of the probabilistic chapters) but are rather interlaced with probabilistic chapters that are designed to display the power of the abstract concepts in the more intuitive world of probability theory. For example, we study percolation theory at the point where we barely have measures, random variables, and independence; not even the integral is needed. As the only exception, the *systematic* construction of independent random variables is deferred to Chap. 14. Although it is rather a matter of taste, I hope that this setup helps to motivate the reader throughout the measure-theoretical chapters.

Those readers with a solid measure-theoretical education can skip in particular the first and fourth chapters and might wish only to look up this or that.

In the first eight chapters, we lay the foundations that will be needed in all the subsequent chapters. After that, there are seven more or less independent parts, consisting of Chaps. 9–20, and 23. The chapter on Brownian motion (21) makes reference to Chaps. 9–15. Again, after that, the three blocks consisting of Chaps. 22, 24, and 25, 26 can be read independently.

I should like to thank all those who read the manuscript and the German original version of this book and gave numerous hints for improvements: Roland Alkemper, René Billing, Dirk Brüggemann, Anne Eisenbürger, Patrick Jahn, Arnulf Jentzen, Ortwin Lorenz, L. Mayer, Mario Oeler, Marcus Schölpen, my colleagues Ehrhard Behrends, Wolfgang Bühler, Nina Gantert, Rudolf Grübel, Wolfgang König, Peter Mörters, and Ralph Neininger, and in particular my colleague from Munich Hans-Otto Georgii. Dr John Preater did a great job language editing the English manuscript and also pointing out numerous mathematical flaws.

I am especially indebted to my wife Katrin for proofreading the English manuscript and for her patience and support.

I would be grateful for further suggestions, errors, etc. to be sent by e-mail to math@aklenke.de.

Mainz   Achim Klenke
October 2007

Contents

Chapter 1
Basic Measure Theory

In this chapter, we introduce the classes of sets that allow for a systematic treatment of events and random observations in the framework of probability theory. Furthermore, we construct measures, in particular probability measures, on such classes of sets. Finally, we define random variables as measurable maps.

1.1 Classes of Sets

In the following, let $\Omega \neq \emptyset$ be a nonempty set and let $\mathcal{A} \subset 2^{\Omega}$ (set of all subsets of Ω) be a class of subsets of Ω. Later, Ω will be interpreted as the space of elementary events and $\mathcal{A}$ will be the system of observable events. In this section, we introduce names for classes of subsets of Ω that are stable under certain set operations and we establish simple relations between such classes.

Definition 1.1 *A class of sets $\mathcal{A}$ is called*

- *∩-closed (closed under intersections) or a π-**system** if $A \cap B \in \mathcal{A}$ whenever $A, B \in \mathcal{A}$,*
- *σ-∩-closed (closed under countable[1] intersections) if $\bigcap_{n=1}^{\infty} A_n \in \mathcal{A}$ for any choice of countably many sets $A_1, A_2, \ldots \in \mathcal{A}$,*
- *∪-closed (closed under unions) if $A \cup B \in \mathcal{A}$ whenever $A, B \in \mathcal{A}$,*
- *σ-∪-closed (closed under countable unions) if $\bigcup_{n=1}^{\infty} A_n \in \mathcal{A}$ for any choice of countably many sets $A_1, A_2, \ldots \in \mathcal{A}$,*
- *\-closed (closed under differences) if $A \setminus B \in \mathcal{A}$ whenever $A, B \in \mathcal{A}$, and*
- *closed under complements if $A^c := \Omega \setminus A \in \mathcal{A}$ for any set $A \in \mathcal{A}$.*

---

[1] By "countable" we always mean either finite or countably infinite.

© The Editor(s) (if applicable) and The Author(s), under exclusive license to Springer Nature Switzerland AG 2020
A. Klenke, *Probability Theory*, Universitext,
https://doi.org/10.1007/978-3-030-56402-5_1

Definition 1.2 (σ-algebra) *A class of sets $\mathcal{A} \subset 2^{\Omega}$ is called a σ-**algebra** if it fulfills the following three conditions:*

(i) $\Omega \in \mathcal{A}$.
(ii) $\mathcal{A}$ is closed under complements.
(iii) $\mathcal{A}$ is closed under countable unions.

Sometimes a σ-algebra is also named a σ-field. As we will see, we can define probabilities on σ-algebras in a consistent way. Hence these are the natural classes of sets to be considered as *events* in probability theory.

Theorem 1.3 *If $\mathcal{A}$ is closed under complements, then we have the equivalences*

$$\mathcal{A} \text{ is } \cap\text{-closed} \quad \Longleftrightarrow \quad \mathcal{A} \text{ is } \cup\text{-closed,}$$

$$\mathcal{A} \text{ is } \sigma\text{-}\cap\text{-closed} \quad \Longleftrightarrow \quad \mathcal{A} \text{ is } \sigma\text{-}\cup\text{-closed.}$$

Proof The two statements are immediate consequences of de Morgan's rule (reminder: $(\bigcup A_i)^c = \bigcap A_i^c$). For example, let $\mathcal{A}$ be σ-$\cap$-closed and let $A_1, A_2, \ldots \in \mathcal{A}$. Hence

$$\bigcup_{n=1}^{\infty} A_n = \left(\bigcap_{n=1}^{\infty} A_n^c \right)^c \in \mathcal{A}.$$

Thus $\mathcal{A}$ is σ-$\cup$-closed. The other cases can be proved similarly. □

Theorem 1.4 *Assume that $\mathcal{A}$ is $\backslash$-closed. Then the following statements hold:*

(i) $\mathcal{A}$ is $\cap$-closed.
(ii) If in addition $\mathcal{A}$ is σ-$\cup$-closed, then $\mathcal{A}$ is σ-$\cap$-closed.
(iii) Any countable (respectively finite) union of sets in $\mathcal{A}$ can be expressed as a countable (respectively finite) disjoint union of sets in $\mathcal{A}$.

Proof

(i) Assume that $A, B \in \mathcal{A}$. Hence also $A \cap B = A \setminus (A \setminus B) \in \mathcal{A}$.
(ii) Assume that $A_1, A_2, \ldots \in \mathcal{A}$. Hence

$$\bigcap_{n=1}^{\infty} A_n = \bigcap_{n=2}^{\infty} (A_1 \cap A_n) = \bigcap_{n=2}^{\infty} A_1 \setminus (A_1 \setminus A_n) = A_1 \setminus \bigcup_{n=2}^{\infty} (A_1 \setminus A_n) \in \mathcal{A}.$$

(iii) Assume that $A_1, A_2, \ldots \in \mathcal{A}$. Hence a representation of $\bigcup_{n=1}^{\infty} A_n$ as a countable disjoint union of sets in $\mathcal{A}$ is

$$\bigcup_{n=1}^{\infty} A_n = A_1 \uplus (A_2 \setminus A_1) \uplus ((A_3 \setminus A_1) \setminus A_2) \uplus (((A_4 \setminus A_1) \setminus A_2) \setminus A_3) \uplus \ldots. \quad \square$$

Remark 1.5 Sometimes the disjoint union of sets is denoted by the symbol $\uplus$. Note that this is not a new operation but only stresses the fact that the sets involved are mutually disjoint. ◊

Definition 1.6 *A class of sets $\mathcal{A} \subset 2^{\Omega}$ is called an **algebra** if the following three conditions are fulfilled:*

(i) $\Omega \in \mathcal{A}$.
(ii) $\mathcal{A}$ *is* $\setminus$*-closed.*
(iii) $\mathcal{A}$ *is* $\cup$*-closed.*

If $\mathcal{A}$ is an algebra, then obviously $\emptyset = \Omega \setminus \Omega$ is in $\mathcal{A}$. However, in general, this property is weaker than (i) in Definition 1.6.

Theorem 1.7 *A class of sets $\mathcal{A} \subset 2^{\Omega}$ is an algebra if and only if the following three properties hold:*

(i) $\Omega \in \mathcal{A}$.
(ii) $\mathcal{A}$ *is closed under complements.*
(iii) $\mathcal{A}$ *is closed under intersections.*

Proof This is left as an exercise. □

Definition 1.8 *A class of sets $\mathcal{A} \subset 2^{\Omega}$ is called a **ring** if the following three conditions hold:*

(i) $\emptyset \in \mathcal{A}$.
(ii) $\mathcal{A}$ *is* $\setminus$*-closed.*
(iii) $\mathcal{A}$ *is* $\cup$*-closed.*

*A ring is called a σ-**ring** if it is also σ-$\cup$-closed.*

Definition 1.9 *A class of sets $\mathcal{A} \subset 2^{\Omega}$ is called a **semiring** if*

(i) $\emptyset \in \mathcal{A}$,
(ii) *for any two sets $A, B \in \mathcal{A}$ the difference set $B \setminus A$ is a finite union of mutually disjoint sets in $\mathcal{A}$,*
(iii) $\mathcal{A}$ *is* $\cap$*-closed.*

Definition 1.10 *A class of sets $\mathcal{A} \subset 2^{\Omega}$ is called a λ-**system** (or Dynkin's λ-system) if*

(i) $\Omega \in \mathcal{A}$,
(ii) *for any two sets $A, B \in \mathcal{A}$ with $A \subset B$, the difference set $B \setminus A$ is in $\mathcal{A}$, and*
(iii) $\uplus_{n=1}^{\infty} A_n \in \mathcal{A}$ *for any choice of countably many pairwise disjoint sets $A_1, A_2, \ldots \in \mathcal{A}$.*

Example 1.11

(i) For any nonempty set Ω, the classes $\mathcal{A} = \{\emptyset, \Omega\}$ and $\mathcal{A} = 2^{\Omega}$ are the trivial examples of algebras, σ-algebras and λ-systems. On the other hand, $\mathcal{A} = \{\emptyset\}$ and $\mathcal{A} = 2^{\Omega}$ are the trivial examples of semirings, rings and σ-rings.

(ii) Let $\Omega = \mathbb{R}$. Then $\mathcal{A} = \{A \subset \mathbb{R} : A \text{ is countable}\}$ is a σ-ring.

(iii) $\mathcal{A} = \{(a, b] : a, b \in \mathbb{R}, \ a \leq b\}$ is a semiring on $\Omega = \mathbb{R}$ (but is not a ring).

(iv) The class of finite unions of bounded intervals is a ring on $\Omega = \mathbb{R}$ (but is not an algebra).

(v) The class of finite unions of arbitrary (also unbounded) intervals is an algebra on $\Omega = \mathbb{R}$ (but is not a σ-algebra).

(vi) Let E be a finite nonempty set and let $\Omega := E^{\mathbb{N}}$ be the set of all E-valued sequences $\omega = (\omega_n)_{n \in \mathbb{N}}$. For any $\omega_1, \ldots, \omega_n \in E$, let

$$[\omega_1, \ldots, \omega_n] := \{\omega' \in \Omega : \omega_i' = \omega_i \text{ for all } i = 1, \ldots, n\}$$

be the set of all sequences whose first n values are $\omega_1, \ldots, \omega_n$. Let $\mathcal{A}_0 = \{\emptyset\}$. For $n \in \mathbb{N}$, define

$$\mathcal{A}_n := \{[\omega_1, \ldots, \omega_n] : \omega_1, \ldots, \omega_n \in E\}. \tag{1.1}$$

Hence $\mathcal{A} := \bigcup_{n=0}^{\infty} \mathcal{A}_n$ is a semiring but is not a ring (if $\#E > 1$).

(vii) Let Ω be an arbitrary nonempty set. Then

$$\mathcal{A} := \{A \subset \Omega : A \text{ or } A^c \text{ is finite}\}$$

is an algebra. However, if $\#\Omega = \infty$, then $\mathcal{A}$ is not a σ-algebra.

(viii) Let Ω be an arbitrary nonempty set. Then

$$\mathcal{A} := \{A \subset \Omega : A \text{ or } A^c \text{ is countable}\}$$

is a σ-algebra.

(ix) Every σ-algebra is a λ-system.

(x) Let $\Omega = \{1, 2, 3, 4\}$ and $\mathcal{A} = \{\emptyset, \{1, 2\}, \{1, 4\}, \{2, 3\}, \{3, 4\}, \{1, 2, 3, 4\}\}$. Hence $\mathcal{A}$ is a λ-system but is not an algebra. $\Diamond$

Theorem 1.12 (Relations between classes of sets)

(i) *Every σ-algebra also is a λ-system, an algebra and a σ-ring.*

(ii) *Every σ-ring is a ring, and every ring is a semiring.*

(iii) *Every algebra is a ring. An algebra on a finite set Ω is a σ-algebra.*

Proof

(i) This is obvious.

(ii) Let $\mathcal{A}$ be a ring. By Theorem 1.4, $\mathcal{A}$ is closed under intersections and is hence a semiring.

(iii) Let $\mathcal{A}$ be an algebra. Then $\emptyset = \Omega \setminus \Omega \in \mathcal{A}$, and hence $\mathcal{A}$ is a ring. If in addition Ω is finite, then $\mathcal{A}$ is finite. Hence any countable union of sets in $\mathcal{A}$ is a finite union of sets. $\square$

Definition 1.13 (liminf and limsup) *Let $A_1, A_2, \ldots$ be subsets of Ω. The sets*

$$\liminf_{n\to\infty} A_n := \bigcup_{n=1}^{\infty} \bigcap_{m=n}^{\infty} A_m \quad and \quad \limsup_{n\to\infty} A_n := \bigcap_{n=1}^{\infty} \bigcup_{m=n}^{\infty} A_m$$

*are called **limes inferior** and **limes superior**, respectively, of the sequence $(A_n)_{n\in\mathbb{N}}$.*

Remark 1.14

(i) lim inf and lim sup can be rewritten as

$$\liminf_{n\to\infty} A_n = \big\{\omega \in \Omega : \#\{n \in \mathbb{N} : \omega \notin A_n\} < \infty\big\},$$

$$\limsup_{n\to\infty} A_n = \big\{\omega \in \Omega : \#\{n \in \mathbb{N} : \omega \in A_n\} = \infty\big\}.$$

In other words, limes inferior is the event where *eventually all* of the A_n occur. On the other hand, limes superior is the event where *infinitely many* of the A_n occur. In particular, $A_* := \liminf_{n\to\infty} A_n \subset A^* := \limsup_{n\to\infty} A_n$.

(ii) We define the **indicator function** on the set A by

$$\mathbb{1}_A(x) := \begin{cases} 1, & \text{if } x \in A, \\ 0, & \text{if } x \notin A. \end{cases} \tag{1.2}$$

With this notation,

$$\mathbb{1}_{A_*} = \liminf_{n\to\infty} \mathbb{1}_{A_n} \quad and \quad \mathbb{1}_{A^*} = \limsup_{n\to\infty} \mathbb{1}_{A_n}.$$

(iii) If $\mathcal{A} \subset 2^{\Omega}$ is a σ-algebra and if $A_n \in \mathcal{A}$ for every $n \in \mathbb{N}$, then $A_* \in \mathcal{A}$ and $A^* \in \mathcal{A}$. $\Diamond$

Proof This is left as an exercise. $\qquad\square$

Theorem 1.15 (Intersection of classes of sets) *Let I be an arbitrary index set, and assume that $\mathcal{A}_i$ is a σ-algebra for every $i \in I$. Hence the intersection*

$$\mathcal{A}_I := \big\{A \subset \Omega : A \in \mathcal{A}_i \text{ for every } i \in I\big\} = \bigcap_{i\in I} \mathcal{A}_i$$

is a σ-algebra. The analogous statement holds for rings, σ-rings, algebras and λ-systems. However, it fails for semirings.

Proof We give the proof for σ-algebras only. To this end, we check (i)–(iii) of Definition 1.2.

(i) Clearly, $\Omega \in \mathcal{A}_i$ for every $i \in I$, and hence $\Omega \in \mathcal{A}_I$.

(ii) Assume $A \in \mathcal{A}_I$. Hence $A \in \mathcal{A}_i$ for any $i \in I$. Thus also $A^c \in \mathcal{A}_i$ for any $i \in I$. We conclude that $A^c \in \mathcal{A}_I$.

(iii) Assume $A_1, A_2, \ldots \in \mathcal{A}_I$. Hence $A_n \in \mathcal{A}_i$ for every $n \in \mathbb{N}$ and $i \in I$. Thus $A := \bigcup_{n=1}^{\infty} A_n \in \mathcal{A}_i$ for every $i \in I$. We conclude $A \in \mathcal{A}_I$.

Counterexample for semirings: Let $\Omega = \{1, 2, 3, 4\}$, $\mathcal{A}_1 = \{\emptyset, \Omega, \{1\}, \{2, 3\}, \{4\}\}$ and $\mathcal{A}_2 = \{\emptyset, \Omega, \{1\}, \{2\}, \{3, 4\}\}$. Then $\mathcal{A}_1$ and $\mathcal{A}_2$ are semirings but $\mathcal{A}_1 \cap \mathcal{A}_2 = \{\emptyset, \Omega, \{1\}\}$ is not.   □

Theorem 1.16 (Generated σ-algebra) *Let $\mathcal{E} \subset 2^{\Omega}$. Then there exists a smallest σ-algebra $\sigma(\mathcal{E})$ with $\mathcal{E} \subset \sigma(\mathcal{E})$:*

$$\sigma(\mathcal{E}) := \bigcap_{\substack{\mathcal{A} \subset 2^{\Omega} \text{ is a } \sigma\text{-algebra} \\ \mathcal{A} \supset \mathcal{E}}} \mathcal{A}.$$

*$\sigma(\mathcal{E})$ is called the σ-algebra **generated by** $\mathcal{E}$. $\mathcal{E}$ is called a **generator** of $\sigma(\mathcal{E})$. Similarly, we define $\delta(\mathcal{E})$ as the λ-system generated by $\mathcal{E}$.*

Proof $\mathcal{A} = 2^{\Omega}$ is a σ-algebra with $\mathcal{E} \subset \mathcal{A}$. Hence the intersection is nonempty. By Theorem 1.15, $\sigma(\mathcal{E})$ is a σ-algebra. Clearly, it is the smallest σ-algebra that contains $\mathcal{E}$. For λ-systems the proof is similar.   □

Remark 1.17 The following three statements hold:

(i) $\mathcal{E} \subset \sigma(\mathcal{E})$.
(ii) If $\mathcal{E}_1 \subset \mathcal{E}_2$, then $\sigma(\mathcal{E}_1) \subset \sigma(\mathcal{E}_2)$.
(iii) $\mathcal{A}$ is a σ-algebra if and only if $\sigma(\mathcal{A}) = \mathcal{A}$.

The same statements hold for λ-systems. Furthermore, $\delta(\mathcal{E}) \subset \sigma(\mathcal{E})$. ◊

Theorem 1.18 ($\cap$-closed λ-system) *Let $\mathcal{D} \subset 2^{\Omega}$ be a λ-system. Then*

$$\mathcal{D} \text{ is a } \pi\text{-system} \quad \Longleftrightarrow \quad \mathcal{D} \text{ is a } \sigma\text{-algebra}.$$

Proof "$\Longleftarrow$"   This is obvious.
"$\Longrightarrow$"   We check (i)–(iii) of Definition 1.2.

(i) Clearly, $\Omega \in \mathcal{D}$.
(ii) (Closedness under complements) Let $A \in \mathcal{D}$. Since $\Omega \in \mathcal{D}$ and by property (ii) of the λ-system, we get that $A^c = \Omega \setminus A \in \mathcal{D}$.
(iii) (σ-$\cup$-closedness) Let $A, B \in \mathcal{D}$. By assumption, $A \cap B \in \mathcal{D}$, and trivially $A \cap B \subset A$. Thus $A \setminus B = A \setminus (A \cap B) \in \mathcal{D}$. This implies that $\mathcal{D}$ is $\setminus$-closed. Now let $A_1, A_2, \ldots \in \mathcal{D}$. By Theorem 1.4(iii), there exist mutually disjoint sets $B_1, B_2, \ldots \in \mathcal{D}$ with $\bigcup_{n=1}^{\infty} A_n = \biguplus_{n=1}^{\infty} B_n \in \mathcal{D}$.   □

Theorem 1.19 (Dynkin's π-λ theorem) *If $\mathcal{E} \subset 2^{\Omega}$ is a π-system, then*

$$\sigma(\mathcal{E}) = \delta(\mathcal{E}).$$

Proof "⊃" This follows from Remark 1.17.

"⊂" We have to show that $\delta(\mathcal{E})$ is a σ-algebra. By Theorem 1.18, it is enough to show that $\delta(\mathcal{E})$ is a π-system. For any $B \in \delta(\mathcal{E})$ define

$$\mathcal{D}_B := \{A \in \delta(\mathcal{E}) : A \cap B \in \delta(\mathcal{E})\}.$$

In order to show that $\delta(\mathcal{E})$ is a π-system, it is enough to show that

$$\delta(\mathcal{E}) \subset \mathcal{D}_B \quad \text{for any } B \in \delta(\mathcal{E}). \tag{1.3}$$

In order to show that $\mathcal{D}_E$ is a λ-system for any $E \in \delta(\mathcal{E})$, we check (i)–(iii) of Definition 1.10:

(i) Clearly, $\Omega \cap E = E \in \delta(\mathcal{E})$; hence $\Omega \in \mathcal{D}_E$.
(ii) For any $A, B \in \mathcal{D}_E$ with $A \subset B$, we have $(B \setminus A) \cap E = (B \cap E) \setminus (A \cap E) \in \delta(\mathcal{E})$.
(iii) Assume that $A_1, A_2, \dots \in \mathcal{D}_E$ are mutually disjoint. Hence

$$\left(\bigcup_{n=1}^{\infty} A_n \right) \cap E = \biguplus_{n=1}^{\infty} (A_n \cap E) \in \delta(\mathcal{E}).$$

By assumption, $A \cap E \in \mathcal{E}$ if $A, E \in \mathcal{E}$; thus $\mathcal{E} \subset \mathcal{D}_E$ if $E \in \mathcal{E}$. By Remark 1.17(ii), we conclude that $\delta(\mathcal{E}) \subset \mathcal{D}_E$ for any $E \in \mathcal{E}$. Hence we get that $B \cap E \in \delta(\mathcal{E})$ for any $B \in \delta(\mathcal{E})$ and $E \in \mathcal{E}$. This implies that $E \in \mathcal{D}_B$ for any $B \in \delta(\mathcal{E})$. Thus $\mathcal{E} \subset \mathcal{D}_B$ for any $B \in \delta(\mathcal{E})$, and hence (1.3) follows. $\square$

For an illustration of the inclusions between the classes of sets, see Fig. 1.1.

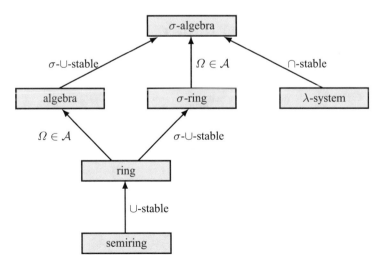

Fig. 1.1 Inclusions between classes of sets $\mathcal{A} \subset 2^{\Omega}$.

Reflection Where does the proof of Theorem 1.19 fail if $\mathcal{E}$ is not $\cap$-stable? Find an example of a class of sets $\mathcal{E}$ that is not $\cap$-stable and such that $\sigma(\mathcal{E}) \neq \delta(\mathcal{E})$. ♠

We are particularly interested in σ-algebras that are generated by topologies. The most prominent role is played by the Euclidean space $\mathbb{R}^n$; however, we will also consider the (infinite-dimensional) space $C([0, 1])$ of continuous functions $[0, 1] \rightarrow \mathbb{R}$. On $C([0, 1])$ the norm $\|f\|_\infty = \sup_{x \in [0,1]} |f(x)|$ induces a topology. For the convenience of the reader, we recall the definition of a topology.

Definition 1.20 (Topology) *Let $\Omega \neq \emptyset$ be an arbitrary set. A class of sets $\tau \subset 2^\Omega$ is called a **topology** on Ω if it has the following three properties:*

(i) $\emptyset, \Omega \in \tau$.
(ii) $A \cap B \in \tau$ for any $A, B \in \tau$.
(iii) $\left(\bigcup_{A \in \mathcal{F}} A \right) \in \tau$ for any $\mathcal{F} \subset \tau$.

*The pair (Ω, τ) is called a **topological space**. The sets $A \in \tau$ are called **open**, and the sets $A \subset \Omega$ with $A^c \in \tau$ are called **closed**.*

In contrast with σ-algebras, topologies are closed under finite intersections only, but they are also closed under arbitrary unions.

Let d be a metric on Ω, and denote the open ball with radius $r > 0$ centered at $x \in \Omega$ by

$$B_r(x) = \{y \in \Omega : d(x, y) < r\}.$$

Then the usual class of open sets is the topology

$$\tau = \left\{ \bigcup_{(x,r) \in F} B_r(x) : F \subset \Omega \times (0, \infty) \right\}.$$

Definition 1.21 (Borel σ-algebra) *Let (Ω, τ) be a topological space. The σ-algebra*

$$\mathcal{B}(\Omega) := \mathcal{B}(\Omega, \tau) := \sigma(\tau)$$

*that is generated by the open sets is called the **Borel σ-algebra** on Ω. The elements $A \in \mathcal{B}(\Omega, \tau)$ are called **Borel sets** or **Borel measurable sets**.*

Remark 1.22 In many cases, we are interested in $\mathcal{B}(\mathbb{R}^n)$, where $\mathbb{R}^n$ is equipped with the Euclidean distance

$$d(x, y) = \|x - y\|_2 = \sqrt{\sum_{i=1}^n (x_i - y_i)^2}.$$

(i) There are subsets of $\mathbb{R}^n$ that are not Borel sets. These sets are not easy to construct like, for example, **Vitali sets** that can be found in calculus books (see also [37, Theorem 3.4.4]). Here we do not want to stress this point but state that, vaguely speaking, all sets that can be constructed explicitly are Borel sets.

(ii) If $C \subset \mathbb{R}^n$ is a closed set, then $C^c \in \tau$ is in $\mathcal{B}(\mathbb{R}^n)$ and hence C is a Borel set. In particular, $\{x\} \in \mathcal{B}(\mathbb{R}^n)$ for every $x \in \mathbb{R}^n$.

(iii) $\mathcal{B}(\mathbb{R}^n)$ is not a topology. To show this, let $V \subset \mathbb{R}^n$ such that $V \notin \mathcal{B}(\mathbb{R}^n)$. If $\mathcal{B}(\mathbb{R}^n)$ were a topology, then it would be closed under arbitrary unions. As $\{x\} \in \mathcal{B}(\mathbb{R}^n)$ for all $x \in \mathbb{R}^n$, we would get the contradiction $V = \bigcup_{x \in V}\{x\} \in \mathcal{B}(\mathbb{R}^n)$. ◊

In most cases the class of open sets that generates the Borel σ-algebra is too big to work with efficiently. Hence we aim at finding smaller (in particular, countable) classes of sets that generate the Borel σ-algebra and that are more amenable. In some of the examples, the elements of the generating class are simpler sets such as rectangles or compact sets.

We introduce the following notation. We denote by $\mathbb{Q}$ the set of rational numbers and by $\mathbb{Q}^+$ the set of strictly positive rational numbers. For $a, b \in \mathbb{R}^n$, we write

$$a < b \quad \text{if } a_i < b_i \quad \text{for all } i = 1, \ldots, n. \tag{1.4}$$

For $a < b$, we define the open **rectangle** as the Cartesian product

$$(a, b) := \bigtimes_{i=1}^{n} (a_i, b_i) := (a_1, b_1) \times (a_2, b_2) \times \cdots \times (a_n, b_n). \tag{1.5}$$

Analogously, we define $[a, b]$, $(a, b]$ and $[a, b)$. Furthermore, we define $(-\infty, b) := \bigtimes_{i=1}^{n}(-\infty, b_i)$, and use an analogous definition for $(-\infty, b]$ and so on. We introduce the following classes of sets:

$$\mathcal{E}_1 := \{A \subset \mathbb{R}^n : A \text{ is open}\}, \qquad \mathcal{E}_2 := \{A \subset \mathbb{R}^n : A \text{ is closed}\},$$

$$\mathcal{E}_3 := \{A \subset \mathbb{R}^n : A \text{ is compact}\}, \qquad \mathcal{E}_4 := \{B_r(x) : x \in \mathbb{Q}^n, r \in \mathbb{Q}^+\},$$

$$\mathcal{E}_5 := \{(a, b) : a, b \in \mathbb{Q}^n, a < b\}, \qquad \mathcal{E}_6 := \{[a, b) : a, b \in \mathbb{Q}^n, a < b\},$$

$$\mathcal{E}_7 := \{(a, b] : a, b \in \mathbb{Q}^n, a < b\}, \qquad \mathcal{E}_8 := \{[a, b] : a, b \in \mathbb{Q}^n, a < b\},$$

$$\mathcal{E}_9 := \{(-\infty, b) : b \in \mathbb{Q}^n\}, \qquad \mathcal{E}_{10} := \{(-\infty, b] : b \in \mathbb{Q}^n\},$$

$$\mathcal{E}_{11} := \{(a, \infty) : a \in \mathbb{Q}^n\}, \qquad \mathcal{E}_{12} := \{[a, \infty) : a \in \mathbb{Q}^n\}.$$

Theorem 1.23 *The Borel σ-algebra $\mathcal{B}(\mathbb{R}^n)$ is generated by any of the classes of sets $\mathcal{E}_1, \ldots, \mathcal{E}_{12}$, that is, $\mathcal{B}(\mathbb{R}^n) = \sigma(\mathcal{E}_i)$ for any $i = 1, \ldots, 12$.*

Proof We show only some of the identities.

(1) By definition, $\mathcal{B}(\mathbb{R}^n) = \sigma(\mathcal{E}_1)$.

(2) Let $A \in \mathcal{E}_1$. Then $A^c \in \mathcal{E}_2$, and hence $A = (A^c)^c \in \sigma(\mathcal{E}_2)$. It follows that $\mathcal{E}_1 \subset \sigma(\mathcal{E}_2)$. By Remark 1.17, this implies $\sigma(\mathcal{E}_1) \subset \sigma(\mathcal{E}_2)$. Similarly, we obtain $\sigma(\mathcal{E}_2) \subset \sigma(\mathcal{E}_1)$ and hence equality.

(3) Any compact set is closed; hence $\sigma(\mathcal{E}_3) \subset \sigma(\mathcal{E}_2)$. Now let $A \in \mathcal{E}_2$. The sets $A_K := A \cap [-K, K]^n$, $K \in \mathbb{N}$, are compact; hence the countable union $A = \bigcup_{K=1}^{\infty} A_K$ is in $\sigma(\mathcal{E}_3)$. It follows that $\mathcal{E}_2 \subset \sigma(\mathcal{E}_3)$ and thus $\sigma(\mathcal{E}_2) = \sigma(\mathcal{E}_3)$.

(4) Clearly, $\mathcal{E}_4 \subset \mathcal{E}_1$; hence $\sigma(\mathcal{E}_4) \subset \sigma(\mathcal{E}_1)$. Now let $A \subset \mathbb{R}^n$ be an open set. For any $x \in A$, define $R(x) = \min(1, \sup\{r > 0 : B_r(x) \subset A\})$. Note that $R(x) > 0$, as A is open. Let $r(x) \in (R(x)/2, R(x)) \cap \mathbb{Q}$. For any $y \in A$ and $x \in (B_{R(y)/3}(y)) \cap \mathbb{Q}^n$, we have $R(x) \geq R(y) - \|x - y\|_2 > \frac{2}{3}R(y)$, and hence $r(x) > \frac{1}{3}R(y)$ and thus $y \in B_{r(x)}(x)$. It follows that $A = \bigcup_{x \in A \cap \mathbb{Q}^n} B_{r(x)}(x)$ is a countable union of sets from $\mathcal{E}_4$ and is hence in $\sigma(\mathcal{E}_4)$. We have shown that $\mathcal{E}_1 \subset \sigma(\mathcal{E}_4)$. By Remark 1.17, this implies $\sigma(\mathcal{E}_1) \subset \sigma(\mathcal{E}_4)$.

(5–12) Exhaustion arguments similar to that in (4) also work for rectangles. If in (4) we take open rectangles instead of open balls $B_r(x)$, we get $\mathcal{B}(\mathbb{R}^n) = \sigma(\mathcal{E}_5)$. For example, we have

$$\underset{i=1}{\overset{n}{\bigtimes}} [a_i, b_i) = \bigcap_{k=1}^{\infty} \underset{i=1}{\overset{n}{\bigtimes}} \left(a_i - \frac{1}{k}, b_i\right) \in \sigma(\mathcal{E}_5).$$

The other inclusions $\mathcal{E}_i \subset \sigma(\mathcal{E}_j)$ can be shown similarly. $\qquad \square$

Remark 1.24 Any of the classes $\mathcal{E}_1, \mathcal{E}_2, \mathcal{E}_3, \mathcal{E}_5, \ldots, \mathcal{E}_{12}$ (but not $\mathcal{E}_4$) is a π-system. Hence, the Borel σ-algebra equals the generated λ-system: $\mathcal{B}(\mathbb{R}^n) = \delta(\mathcal{E}_i)$ for $i = 1, 2, 3, 5, \ldots, 12$. In addition, the classes $\mathcal{E}_4, \ldots, \mathcal{E}_{12}$ are countable. This is a crucial property that will be needed later. $\lozenge$

Definition 1.25 (Trace of a class of sets) *Let $\mathcal{A} \subset 2^{\Omega}$ be an arbitrary class of subsets of Ω and let $A \in 2^{\Omega} \setminus \{\emptyset\}$. The class*

$$\mathcal{A}\big|_A := \{A \cap B : B \in \mathcal{A}\} \subset 2^A \tag{1.6}$$

*is called the **trace** of $\mathcal{A}$ on A or the **restriction** of $\mathcal{A}$ to A.*

Theorem 1.26 *Let $A \subset \Omega$ be a nonempty set and let $\mathcal{A}$ be a σ-algebra on Ω or any of the classes of Definitions 1.6–1.9. Then $\mathcal{A}\big|_A$ is a class of sets of the same type as $\mathcal{A}$; however, on A instead of Ω. For λ-systems this is not true in general.*

Proof This is left as an exercise. $\qquad \square$

> **Takeaways** σ-algebras are classes of sets that are stable under countable intersections and unions. They can be generated by classes with less structure (algebras, rings, semirings), but also by classes with a different structure (e.g., a topology). In the case of a topology we get a Borel σ-algebra, that can also be generated using simple sets such as rectangles.

Exercise 1.1.1 Let $\mathcal{A}$ be a semiring. Show that any countable (respectively finite) union of sets in $\mathcal{A}$ can be written as a countable (respectively finite) *disjoint* union of sets in $\mathcal{A}$. ♣

Exercise 1.1.2 Give a counterexample that shows that, in general, the union $\mathcal{A} \cup \mathcal{A}'$ of two σ-algebras need not be a σ-algebra. ♣

Exercise 1.1.3 Let (Ω_1, d_1) and (Ω_2, d_2) be metric spaces and let $f : \Omega_1 \to \Omega_2$ be an arbitrary map. Denote by $U_f = \{x \in \Omega_1 : f \text{ is discontinuous at } x\}$ the set of points of discontinuity of f. Show that $U_f \in \mathcal{B}(\Omega_1)$.
Hint: First show that for any $\varepsilon > 0$ and $\delta > 0$ the set

$$U_f^{\delta,\varepsilon} := \{x \in \Omega_1 : \text{ there are } y, z \in B_\varepsilon(x) \text{ with } d_2(f(y), f(z)) > \delta\}$$

is open (where $B_\varepsilon(x) = \{y \in \Omega_1 : d_1(x, y) < \varepsilon\}$). Then construct U_f from such $U_f^{\delta,\varepsilon}$. ♣

Exercise 1.1.4 Let Ω be an uncountably infinite set and $\mathcal{A} = \sigma(\{\omega\} : \omega \in \Omega)$. Show that

$$\mathcal{A} = \{A \subset \Omega : A \text{ is countable or } A^c \text{ is countable}\}. \quad ♣$$

Exercise 1.1.5 Let $\mathcal{A}$ be a ring on the set Ω. Show that $\mathcal{A}$ is an Abelian algebraic ring with multiplication "$\cap$" and addition "$\triangle$". ♣

1.2 Set Functions

We aim at assigning to each "event" (which will be formalised later) a number that can be interpreted as the probability for the event to occur. To this end, we first study more general set functions that assign nonnegative numbers to subsets. Then we describe those properties necessary for such a function to qualify as a probability assignment.

Definition 1.27 *Let $\mathcal{A} \subset 2^\Omega$ and let $\mu : \mathcal{A} \to [0, \infty]$ be a set function. We say that μ is*

(i) **monotone** *if $\mu(A) \leq \mu(B)$ for any two sets $A, B \in \mathcal{A}$ with $A \subset B$,* *(ii)* **additive** *if* $\mu\left(\biguplus_{i=1}^{n} A_i\right) = \sum_{i=1}^{n} \mu(A_i)$ *for any choice of finitely many mutually*

disjoint sets $A_1, \ldots, A_n \in \mathcal{A}$ with $\bigcup_{i=1}^{n} A_i \in \mathcal{A}$,

(iii) **σ-additive** *if* $\mu\left(\biguplus_{i=1}^{\infty} A_i\right) = \sum_{i=1}^{\infty} \mu(A_i)$ *for any choice of countably many*

mutually disjoint sets $A_1, A_2, \ldots \in \mathcal{A}$ with $\bigcup_{i=1}^{\infty} A_i \in \mathcal{A}$,

(iv) **subadditive** *if for any choice of finitely many sets* $A, A_1, \ldots, A_n \in \mathcal{A}$ *with*
   $A \subset \bigcup_{i=1}^{n} A_i$, we have $\mu(A) \le \sum_{i=1}^{n} \mu(A_i)$, and

(v) **σ-subadditive** *if for any choice of countably many sets* $A, A_1, A_2, \ldots \in \mathcal{A}$
   with $A \subset \bigcup_{i=1}^{\infty} A_i$, we have $\mu(A) \le \sum_{i=1}^{\infty} \mu(A_i)$.

Definition 1.28 *Let $\mathcal{A}$ be a semiring and let $\mu : \mathcal{A} \to [0, \infty]$ be a set function with $\mu(\emptyset) = 0$. μ is called a*

- **content** *if μ is additive,*
- **premeasure** *if μ is σ-additive,*
- **measure** *if μ is a premeasure and $\mathcal{A}$ is a σ-algebra, and*
- **probability measure** *if μ is a measure and $\mu(\Omega) = 1$.*

Definition 1.29 *Let $\mathcal{A}$ be a semiring. A content μ on $\mathcal{A}$ is called*

(i) **finite** *if $\mu(A) < \infty$ for every $A \in \mathcal{A}$ and*
(ii) **σ-finite** *if there exists a sequence of sets $\Omega_1, \Omega_2, \ldots \in \mathcal{A}$ such that $\Omega = \bigcup_{n=1}^{\infty} \Omega_n$ and such that $\mu(\Omega_n) < \infty$ for all $n \in \mathbb{N}$.*

Example 1.30 (Contents, measures)

(i) Let $\omega \in \Omega$ and $\delta_\omega(A) = \mathbb{1}_A(\omega)$ (see (1.2)). Then δ_ω is a probability measure on any σ-algebra $\mathcal{A} \subset 2^{\Omega}$. δ_ω is called the **Dirac measure** for the point ω.

(ii) Let Ω be a finite nonempty set. By

$$\mu(A) := \frac{\#A}{\#\Omega} \quad \text{for } A \subset \Omega,$$

we define a probability measure on $\mathcal{A} = 2^{\Omega}$. This μ is called the **uniform distribution** on Ω. For this distribution, we introduce the symbol $\mathcal{U}_\Omega := \mu$. The resulting triple $(\Omega, \mathcal{A}, \mathcal{U}_\Omega)$ is called a **Laplace space**.

(iii) Let Ω be countably infinite and let

$$\mathcal{A} := \{A \subset \Omega : \#A < \infty \text{ or } \#A^c < \infty\}.$$

Then $\mathcal{A}$ is an algebra. The set function μ on $\mathcal{A}$ defined by

$$\mu(A) = \begin{cases} 0, & \text{if } A \text{ is finite,} \\ \infty, & \text{if } A^c \text{ is finite,} \end{cases}$$

is a content but is not a premeasure. Indeed, $\mu\left(\bigcup_{\omega\in\Omega}\{\omega\}\right) = \mu(\Omega) = \infty$, but $\sum_{\omega\in\Omega}\mu(\{\omega\}) = 0$.

(iv) Let $(\mu_n)_{n\in\mathbb{N}}$ be a sequence of measures (premeasures, contents) and let $(\alpha_n)_{n\in\mathbb{N}}$ be a sequence of nonnegative numbers. Then also $\mu := \sum_{n=1}^{\infty}\alpha_n\mu_n$ is a measure (premeasure, content).

(v) Let Ω be an (at most) countable nonempty set and let $\mathcal{A} = 2^{\Omega}$. Further, let $(p_\omega)_{\omega\in\Omega}$ be nonnegative numbers. Then $A \mapsto \mu(A) := \sum_{\omega\in A} p_\omega$ defines a σ-finite measure on 2^Ω. We call $p = (p_\omega)_{\omega\in\Omega}$ the **weight function** of μ. The number p_ω is called the weight of μ at point ω.

(vi) If in (v) the sum $\sum_{\omega\in\Omega} p_\omega$ equals one, then μ is a probability measure. In this case, we interpret p_ω as the probability of the elementary event ω. The vector $p = (p_\omega)_{\omega\in\Omega}$ is called a **probability vector**.

(vii) If in (v) $p_\omega = 1$ for every $\omega \in \Omega$, then μ is called **counting measure** on Ω. If Ω is finite, then so is μ.

(viii) Let $\mathcal{A}$ be the ring of finite unions of intervals $(a, b] \subset \mathbb{R}$. For $a_1 < b_1 < a_2 < b_2 < \ldots < b_n$ and $A = \biguplus_{i=1}^{n}(a_i, b_i]$, define

$$\mu(A) = \sum_{i=1}^{n}(b_i - a_i).$$

Then μ is a σ-finite content on $\mathcal{A}$ (even a premeasure) since $\bigcup_{n=1}^{\infty}(-n, n] = \mathbb{R}$ and $\mu((-n, n]) = 2n < \infty$ for all $n \in \mathbb{N}$.

(ix) Let $f : \mathbb{R} \to [0, \infty)$ be continuous. In a similar way to (viii), we define

$$\mu_f(A) = \sum_{i=1}^{n}\int_{a_i}^{b_i} f(x)\,dx.$$

Then μ_f is a σ-finite content on $\mathcal{A}$ (even a premeasure). The function f is called the **density** of μ and plays a role similar to the weight function p in (v). $\Diamond$

Lemma 1.31 (Properties of contents) *Let $\mathcal{A}$ be a semiring and let μ be a content on $\mathcal{A}$. Then the following statements hold.*

(i) *If $\mathcal{A}$ is a ring, then $\mu(A \cup B) + \mu(A \cap B) = \mu(A) + \mu(B)$ for any two sets $A, B \in \mathcal{A}$.*

(ii) *μ is monotone. If $\mathcal{A}$ is a ring, then $\mu(B) = \mu(A) + \mu(B \setminus A)$ for any two sets $A, B \in \mathcal{A}$ with $A \subset B$.*

(iii) μ *is subadditive. If μ is σ-additive, then μ is also σ-subadditive.*

(iv) *If $\mathcal{A}$ is a ring, then $\sum\limits_{n=1}^{\infty} \mu(A_n) \leq \mu\left(\bigcup\limits_{n=1}^{\infty} A_n\right)$ for any choice of countably many*

mutually disjoint sets $A_1, A_2, \ldots \in \mathcal{A}$ with $\bigcup\limits_{n=1}^{\infty} A_n \in \mathcal{A}$.

Proof

(i) Note that $A \cup B = A \uplus (B \setminus A)$ and $B = (A \cap B) \uplus (B \setminus A)$. As μ is additive, we obtain

$$\mu(A \cup B) = \mu(A) + \mu(B \setminus A) \quad \text{and} \quad \mu(B) = \mu(A \cap B) + \mu(B \setminus A).$$

This implies (i).

(ii) Let $A \subset B$. Since $A \cap B = A$, we obtain $\mu(B) = \mu(A \uplus (B \setminus A)) = \mu(A) + \mu(B \setminus A)$ if $B \setminus A \in \mathcal{A}$. In particular, this is true if $\mathcal{A}$ is a ring. If $\mathcal{A}$ is only a semiring, then there exists an $n \in \mathbb{N}$ and mutually disjoint sets $C_1, \ldots, C_n \in \mathcal{A}$ such that $B \setminus A = \uplus_{i=1}^{n} C_i$. Hence $\mu(B) = \mu(A) + \sum_{i=1}^{n} \mu(C_i) \geq \mu(A)$ and thus μ is monotone.

(iii) Let $n \in \mathbb{N}$ and $A, A_1, \ldots, A_n \in \mathcal{A}$ with $A \subset \bigcup_{i=1}^{n} A_i$. Define $B_1 = A_1$ and

$$B_k = A_k \setminus \bigcup_{i=1}^{k-1} A_i = \bigcap_{i=1}^{k-1}(A_k \setminus (A_k \cap A_i)) \quad \text{for } k = 2, \ldots, n.$$

By the definition of a semiring, any $A_k \setminus (A_k \cap A_i)$ is a finite disjoint union of sets in $\mathcal{A}$. Hence there exists a $c_k \in \mathbb{N}$ and sets $C_{k,1}, \ldots, C_{k,c_k} \in \mathcal{A}$ such that $\uplus_{i=1}^{c_k} C_{k,i} = B_k \subset A_k$. Similarly, there exist $d_k \in \mathbb{N}$ and $D_{k,1}, \ldots, D_{k,d_k} \in \mathcal{A}$ such that $A_k \setminus B_k = \uplus_{i=1}^{d_k} D_{k,i}$. Since μ is additive, we have

$$\mu(A_k) = \sum_{i=1}^{c_k} \mu(C_{k,i}) + \sum_{i=1}^{d_k} \mu(D_{k,i}) \geq \sum_{i=1}^{c_k} \mu(C_{k,i}).$$

Again due to additivity and monotonicity, we get

$$\mu(A) = \mu\left(\biguplus_{k=1}^{n} \biguplus_{i=1}^{c_k}(C_{k,i} \cap A)\right) = \sum_{k=1}^{n} \sum_{i=1}^{c_k} \mu(C_{k,i} \cap A)$$

$$\leq \sum_{k=1}^{n} \sum_{i=1}^{c_k} \mu(C_{k,i}) \leq \sum_{k=1}^{n} \mu(A_k).$$

Hence μ is subadditive. By a similar argument, σ-subadditivity follows from σ-additivity.

(iv) Let $\mathcal{A}$ be a ring and let $A = \bigcup_{n=1}^{\infty} A_n \in \mathcal{A}$. Since μ is additive (and thus monotone), we have by (ii)

$$\sum_{n=1}^{m} \mu(A_n) = \mu\left(\biguplus_{n=1}^{m} A_n\right) \le \mu(A) \quad \text{for any } m \in \mathbb{N}.$$

It follows that $\sum_{n=1}^{\infty} \mu(A_n) \le \mu(A)$. $\quad\Box$

Remark 1.32 The inequality in (iv) can be strict (see Example 1.30(iii)). In other words, there are contents that are not premeasures. $\Diamond$

If $\mathcal{A}$ is a ring an μ is a content on $\mathcal{A}$, then by Lemma 1.31, for $A, B \in \mathcal{E}$ such that $\mu(A), \mu(B) < \infty$, we have

$$\mu(A \cup B) = \mu(A) + \mu(B) - \mu(A \cap B).$$

Similarly, for three sets $A, B, C \in \mathcal{A}$ with finite content, we have

$$
\begin{aligned}
\mu(A \cup B \cup C) =\,& \mu(A \cup B) + \mu(C) - \mu((A \cap C) \cup (B \cap C)) \\
=\,& \mu(A) + \mu(B) + \mu(C) \\
& - \mu(A \cap B) - \mu(A \cap C) - \mu(B \cap C) + \mu(A \cap B \cap C).
\end{aligned}
$$

Note that the sign of each expression changes with the number of sets that are cut. This statement will now be generalised to an arbitrary finite number of sets.

Theorem 1.33 (Inclusion–exclusion formula) *Let $\mathcal{A}$ be a ring and let μ be a content on $\mathcal{A}$. Let $n \in \mathbb{N}$ and $A_1, \ldots, A_n \in \mathcal{A}$ such that $\mu(A_1 \cup \ldots \cup A_n) < \infty$. Then the following inclusion and exclusion formulas hold:*

$$\mu(A_1 \cup \ldots \cup A_n) = \sum_{k=1}^{n} (-1)^{k-1} \sum_{\{i_1,\ldots,i_k\} \subset \{1,\ldots,n\}} \mu(A_{i_1} \cap \ldots \cap A_{i_k}),$$

$$\mu(A_1 \cap \ldots \cap A_n) = \sum_{k=1}^{n} (-1)^{k-1} \sum_{\{i_1,\ldots,i_k\} \subset \{1,\ldots,n\}} \mu(A_{i_1} \cup \ldots \cup A_{i_k}).$$

Here summation is over all subsets of $\{1, \ldots, n\}$ with k elements.

Proof This is left as an exercise. *Hint:* Use induction on n. $\quad\Box$

The next goal is to characterize σ-subadditivity by a certain continuity property (Theorem 1.36). To this end, we agree on the following conventions.

Definition 1.34 *Let* $A, A_1, A_2, \ldots$ *be sets. We write*

- $A_n \uparrow A$ *and say that* $(A_n)_{n\in\mathbb{N}}$ *increases to* A *if* $A_1 \subset A_2 \subset \ldots$ *and* $\bigcup_{n=1}^{\infty} A_n = A$, *and*
- $A_n \downarrow A$ *and say that* $(A_n)_{n\in\mathbb{N}}$ *decreases to* A *if* $A_1 \supset A_2 \supset A_3 \supset \ldots$ *and* $\bigcap_{n=1}^{\infty} A_n = A$.

Assume that we have a sequence of events $A_1, A_2, \ldots$ that cannot *all* occur jointly. Then we should have that the probability for $A_1, \ldots, A_n$ to occur jointly vanishes as $n \to \infty$. This is a property of continuity that cannot be deduced from the axioms of a content and thus must be postulated separately.

Definition 1.35 (Continuity of contents) *Let* μ *be a content on the ring* $\mathcal{A}$.

(i) μ *is called* **lower semicontinuous** *if* $\mu(A_n) \xrightarrow{n\to\infty} \mu(A)$ *for any* $A \in \mathcal{A}$ *and any sequence* $(A_n)_{n\in\mathbb{N}}$ *in* $\mathcal{A}$ *with* $A_n \uparrow A$.

(ii) μ *is called* **upper semicontinuous** *if* $\mu(A_n) \xrightarrow{n\to\infty} \mu(A)$ *for any* $A \in \mathcal{A}$ *and any sequence* $(A_n)_{n\in\mathbb{N}}$ *in* $\mathcal{A}$ *with* $\mu(A_n) < \infty$ *for some (and then eventually all)* $n \in \mathbb{N}$ *and* $A_n \downarrow A$.

(iii) μ *is called* $\emptyset$**-continuous** *if (ii) holds for* $A = \emptyset$.

In the definition of upper semicontinuity, we needed the assumption $\mu(A_n) < \infty$ since otherwise we would not even have $\emptyset$-continuity for an example as simple as the counting measure μ on $(\mathbb{N}, 2^{\mathbb{N}})$. Indeed, $A_n := \{n, n+1, \ldots\} \downarrow \emptyset$ but $\mu(A_n) = \infty$ for all $n \in \mathbb{N}$.

Theorem 1.36 (Continuity and premeasure) *Let* μ *be a content on the ring* $\mathcal{A}$. *Consider the following five properties.*

(i) μ *is* σ-additive (and hence a premeasure).
(ii) μ *is* σ-subadditive.
(iii) μ *is lower semicontinuous.*
(iv) μ *is* $\emptyset$-continuous.
(v) μ *is upper semicontinuous.*

   Then the following implications hold:

$$(i) \iff (ii) \iff (iii) \implies (iv) \iff (v).$$

If μ *is finite, then we also have* $(iv) \implies (iii)$.

Proof "(i) $\implies$ (ii)"   Let $A, A_1, A_2, \ldots \in \mathcal{A}$ with $A \subset \bigcup_{i=1}^{\infty} A_i$. Define $B_1 = A_1$ and $B_n = A_n \setminus \bigcup_{i=1}^{n-1} A_i \in \mathcal{A}$ for $n = 2, 3, \ldots$. Then $A = \biguplus_{n=1}^{\infty}(A \cap B_n)$. Since μ is monotone and σ-additive, we infer

$$\mu(A) = \sum_{n=1}^{\infty} \mu(A \cap B_n) \le \sum_{n=1}^{\infty} \mu(A_n).$$

Hence μ is σ-subadditive.

"(ii) $\implies$ (i)" This follows from Lemma 1.31(iv).

"(i) $\implies$ (iii)" Let μ be a premeasure and $A \in \mathcal{A}$. Let $(A_n)_{n \in \mathbb{N}}$ be a sequence in $\mathcal{A}$ such that $A_n \uparrow A$ and let $A_0 = \emptyset$. Then

$$\mu(A) = \sum_{i=1}^{\infty} \mu(A_i \setminus A_{i-1}) = \lim_{n \to \infty} \sum_{i=1}^{n} \mu(A_i \setminus A_{i-1}) = \lim_{n \to \infty} \mu(A_n).$$

"(iii) $\implies$ (i)" Assume now that (iii) holds. Let $B_1, B_2, \ldots \in \mathcal{A}$ be mutually disjoint, and assume that $B = \biguplus_{n=1}^{\infty} B_n \in \mathcal{A}$. Define $A_n = \bigcup_{i=1}^{n} B_i$ for all $n \in \mathbb{N}$. Then it follows from (iii) that

$$\mu(B) = \lim_{n \to \infty} \mu(A_n) = \sum_{i=1}^{\infty} \mu(B_i).$$

Hence μ is σ-additive and therefore a premeasure.

"(iv) $\implies$ (v)" Let $A, A_1, A_2, \ldots \in \mathcal{A}$ with $A_n \downarrow A$ and $\mu(A_1) < \infty$. Define $B_n = A_n \setminus A \in \mathcal{A}$ for all $n \in \mathbb{N}$. Then $B_n \downarrow \emptyset$. This implies $\mu(A_n) - \mu(A) = \mu(B_n) \xrightarrow{n \to \infty} 0$.

"(v) $\implies$ (iv)" This is evident.

"(iii) $\implies$ (iv)" Let $A_1, A_2, \ldots \in \mathcal{A}$ with $A_n \downarrow \emptyset$ and $\mu(A_1) < \infty$. Then $A_1 \setminus A_n \in \mathcal{A}$ for any $n \in \mathbb{N}$ and $A_1 \setminus A_n \uparrow A_1$. Hence

$$\mu(A_1) = \lim_{n \to \infty} \mu(A_1 \setminus A_n) = \mu(A_1) - \lim_{n \to \infty} \mu(A_n).$$

Since $\mu(A_1) < \infty$, we have $\lim_{n \to \infty} \mu(A_n) = 0$.

"(iv) $\implies$ (iii)" (for finite μ) Assume that $\mu(A) < \infty$ for every $A \in \mathcal{A}$ and that μ is $\emptyset$-continuous. Let $A, A_1, A_2, \ldots \in \mathcal{A}$ with $A_n \uparrow A$. Then we have $A \setminus A_n \downarrow \emptyset$ and

$$\mu(A) - \mu(A_n) = \mu(A \setminus A_n) \xrightarrow{n \to \infty} 0.$$

Hence (iii) follows. $\qquad\square$

Example 1.37 (Compare Example 1.30(iii).) Let Ω be a countable set, and define

$$\mathcal{A} = \{A \subset \Omega : \#A < \infty \text{ or } \#A^c < \infty\},$$

$$\mu(A) = \begin{cases} 0, & \text{if } A \text{ is finite,} \\ \infty, & \text{if } A \text{ is infinite.} \end{cases}$$

Then μ is an $\emptyset$-continuous content but not a premeasure. $\lozenge$

Definition 1.38

(i) *A pair* $(\Omega, \mathcal{A})$ *consisting of a nonempty set* Ω *and a* σ*-algebra* $\mathcal{A} \subset 2^\Omega$ *is called a* ***measurable space***. *The sets* $A \in \mathcal{A}$ *are called* ***measurable sets***. *If* Ω *is at most countably infinite and if* $\mathcal{A} = 2^\Omega$, *then the measurable space* $(\Omega, 2^\Omega)$ *is called* ***discrete***.

(ii) *A triple* $(\Omega, \mathcal{A}, \mu)$ *is called a* ***measure space*** *if* $(\Omega, \mathcal{A})$ *is a measurable space and if* μ *is a measure on* $\mathcal{A}$.

(iii) *If in addition* $\mu(\Omega) = 1$, *then* $(\Omega, \mathcal{A}, \mu)$ *is called a* ***probability space***. *In this case, the sets* $A \in \mathcal{A}$ *are called* ***events***.

(iv) *The set of all finite measures on* $(\Omega, \mathcal{A})$ *is denoted by* $\mathcal{M}_f(\Omega) := \mathcal{M}_f(\Omega, \mathcal{A})$. *The subset of probability measures is denoted by* $\mathcal{M}_1(\Omega) := \mathcal{M}_1(\Omega, \mathcal{A})$. *Finally, the set of* σ*-finite measures on* $(\Omega, \mathcal{A})$ *is denoted by* $\mathcal{M}_\sigma(\Omega, \mathcal{A})$.

Takeaways In this section, we have compiled a wish list of the properties that a probability assignment should have: σ-additivity and normalization (Definition 1.28). We have seen how σ-additivity follows from additivity (which is easier to check) and continuity (Theorem 1.36). In order for the notion of σ-additivity to make sense, the underlying class of sets must be closed under countable set operations; that is, it must be a σ-algebra. This shows that the concepts formed in Sect. 1.1 are sensible.

Exercise 1.2.1 Let $\mathcal{A} = \{(a, b] \cap \mathbb{Q} : a, b \in \mathbb{R}, a \leq b\}$. Define $\mu : \mathcal{A} \to [0, \infty)$ by $\mu\big((a, b] \cap \mathbb{Q}\big) = b - a$. Show that $\mathcal{A}$ is a semiring and μ is a content on $\mathcal{A}$ that is lower and upper semicontinuous but is not σ-additive. ♣

1.3   The Measure Extension Theorem

In this section, we construct measures μ on σ-algebras. The starting point will be to define the values of μ on a smaller class of sets; that is, on a semiring. Under a mild consistency condition, the resulting set function can be extended to the whole σ-algebra.

Before we develop the complete theory, we begin with two examples: The Lebesgue measure and the infinite product measure. While the Lebesgue measure is ubiquitous in analysis, the infinite product measure plays an important role in probability theory for modelling infinitely many independent events.

Example 1.39 (Lebesgue measure) Let $n \in \mathbb{N}$ and let

$$\mathcal{A} = \{(a, b] : a, b \in \mathbb{R}^n, a \leq b\}$$

be the semiring of half open rectangles $(a, b] \subset \mathbb{R}^n$ (see (1.5)). The n-dimensional volume of such a rectangle is

$$\mu((a, b]) = \prod_{i=1}^{n} (b_i - a_i).$$

Can we extend the set function μ to a (uniquely determined) measure on the Borel σ-algebra $\mathcal{B}(\mathbb{R}^n) = \sigma(\mathcal{A})$? We will see that this is indeed possible. The resulting measure is called Lebesgue measure (or sometimes Lebesgue–Borel measure) λ on $(\mathbb{R}^n, \mathcal{B}(\mathbb{R}^n))$. $\Diamond$

Example 1.40 (Product measure, Bernoulli measure) We construct a measure for an infinitely often repeated random experiment with finitely many possible outcomes. Let E be the set of possible outcomes. For $e \in E$, let $p_e \geq 0$ be the probability that e occurs. Hence $\sum_{e \in E} p_e = 1$. For a fixed realization of the repeated experiment, let $\omega_1, \omega_2, \ldots \in E$ be the observed outcomes. Hence the space of *all* possible outcomes of the repeated experiment is $\Omega = E^{\mathbb{N}}$. As in Example 1.11(vi), we define the set of all sequences whose first n values are $\omega_1, \ldots, \omega_n$:

$$[\omega_1, \ldots, \omega_n] := \{\omega' \in \Omega : \omega'_i = \omega_i \text{ for any } i = 1, \ldots, n\}. \tag{1.7}$$

Let $\mathcal{A}_0 = \{\emptyset\}$. For $n \in \mathbb{N}$, define the class of cylinder sets that depend only on the first n coordinates

$$\mathcal{A}_n := \{[\omega_1, \ldots, \omega_n] : \omega_1, \ldots, \omega_n \in E\}, \tag{1.8}$$

and let $\mathcal{A} := \bigcup_{n=0}^{\infty} \mathcal{A}_n$.

We interpret $[\omega_1, \ldots, \omega_n]$ as the event where the outcome of the first experiment is ω_1, the outcome of the second experiment is ω_2 and finally the outcome of the nth experiment is ω_n. The outcomes of the other experiments do not play a role for the occurrence of this event. As the individual experiments ought to be independent, we should have for any choice $\omega_1, \ldots, \omega_n \in E$ that the probability of the event $[\omega_1, \ldots, \omega_n]$ is the product of the probabilities of the individual events; that is,

$$\mu([\omega_1, \ldots, \omega_n]) = \prod_{i=1}^{n} p_{\omega_i}.$$

This formula defines a content μ on the semiring $\mathcal{A}$, and our aim is to extend μ in a unique way to a probability measure on the σ-algebra $\sigma(\mathcal{A})$ that is generated by $\mathcal{A}$.

Before we do so, we make the following definition. Define the (ultra-)metric d on Ω by

$$d(\omega, \omega') = \begin{cases} 2^{-\inf\{n \in \mathbb{N} : \omega_n \neq \omega'_n\}}, & \text{if } \omega \neq \omega', \\ 0, & \text{if } \omega = \omega'. \end{cases} \tag{1.9}$$

Hence (Ω, d) is a compact metric space. Clearly,

$$[\omega_1, \ldots, \omega_n] = B_{2^{-n}}(\omega) = \{\omega' \in \Omega : d(\omega, \omega') < 2^{-n}\}.$$

The complement of $[\omega_1, \ldots, \omega_n]$ is an open set, as it is the union of $(\#E)^n - 1$ open balls

$$[\omega_1, \ldots, \omega_n]^c = \bigcup_{(\omega'_1, \ldots, \omega'_n) \neq (\omega_1, \ldots, \omega_n)} [\omega'_1, \ldots, \omega'_n].$$

Since Ω is compact, the closed subset $[\omega_1, \ldots, \omega_n]$ is compact. As in Theorem 1.23, it can be shown that $\sigma(\mathcal{A}) = \mathcal{B}(\Omega, d)$.

Exercise: Prove the statements made above. $\Diamond$

Reflection Why is there no infinite product measure if $\sum_e p(e) \in (0, \infty) \setminus \{1\}$? ♠

The main result of this chapter is Carathéodory's measure extension theorem.

Theorem 1.41 (Carathéodory) *Let $\mathcal{A} \subset 2^{\Omega}$ be a ring and let μ be a σ-finite premeasure on $\mathcal{A}$. There exists a unique measure $\widetilde{\mu}$ on $\sigma(\mathcal{A})$ such that $\widetilde{\mu}(A) = \mu(A)$ for all $A \in \mathcal{A}$. Furthermore, $\widetilde{\mu}$ is σ-finite.*

We prepare for the proof of this theorem with a couple of lemmas. In fact, we will show a slightly stronger statement in Theorem 1.53.

Lemma 1.42 (Uniqueness by an ∩-closed generator) *Let $(\Omega, \mathcal{A}, \mu)$ be a σ-finite measure space and let $\mathcal{E} \subset \mathcal{A}$ be a π-system that generates $\mathcal{A}$. Assume that there exist sets $\Omega_1, \Omega_2, \ldots \in \mathcal{E}$ such that $\bigcup_{n=1}^{\infty} \Omega_n = \Omega$ and $\mu(\Omega_n) < \infty$ for all $n \in \mathbb{N}$. Then μ is uniquely determined by the values $\mu(E)$, $E \in \mathcal{E}$.*
If μ is a probability measure, the existence of the sequence $(\Omega_n)_{n \in \mathbb{N}}$ is not needed.

Proof Let ν be a (possibly different) σ-finite measure on $(\Omega, \mathcal{A})$ such that

$$\mu(E) = \nu(E) \quad \text{for every } E \in \mathcal{E}.$$

Let $E \in \mathcal{E}$ with $\mu(E) < \infty$. Consider the class of sets

$$\mathcal{D}_E = \{A \in \mathcal{A} : \mu(A \cap E) = \nu(A \cap E)\}.$$

In order to show that $\mathcal{D}_E$ is a λ-system, we check the properties of Definition 1.10:

(i) Clearly, $\Omega \in \mathcal{D}_E$.
(ii) Let $A, B \in \mathcal{D}_E$ with $A \supset B$. Then

$$\mu((A \setminus B) \cap E) = \mu(A \cap E) - \mu(B \cap E)$$
$$= \nu(A \cap E) - \nu(B \cap E) = \nu((A \setminus B) \cap E).$$

Hence $A \setminus B \in \mathcal{D}_E$.

(iii) Let $A_1, A_2, \ldots \in \mathcal{D}_E$ be mutually disjoint and $A = \bigcup_{n=1}^{\infty} A_n$. Then

$$\mu(A \cap E) = \sum_{n=1}^{\infty} \mu(A_n \cap E) = \sum_{n=1}^{\infty} \nu(A_n \cap E) = \nu(A \cap E).$$

Hence $A \in \mathcal{D}_E$.

Clearly, $\mathcal{E} \subset \mathcal{D}_E$; hence $\delta(\mathcal{E}) \subset \mathcal{D}_E$. Since $\mathcal{E}$ is a π-system, Theorem 1.19 yields

$$\mathcal{A} \supset \mathcal{D}_E \supset \delta(\mathcal{E}) = \sigma(\mathcal{E}) = \mathcal{A}.$$

Hence $\mathcal{D}_E = \mathcal{A}$.

This implies $\mu(A \cap E) = \nu(A \cap E)$ for any $A \in \mathcal{A}$ and $E \in \mathcal{E}$ with $\mu(E) < \infty$. Now let $\Omega_1, \Omega_2, \ldots \in \mathcal{E}$ be a sequence such that $\bigcup_{n=1}^{\infty} \Omega_n = \Omega$ and $\mu(\Omega_n) < \infty$ for all $n \in \mathbb{N}$. Let $E_n := \bigcup_{i=1}^{n} \Omega_i$, $n \in \mathbb{N}$, and $E_0 = \emptyset$. Hence $E_n = \biguplus_{i=1}^{n} (E_{i-1}^c \cap \Omega_i)$. For any $A \in \mathcal{A}$ and $n \in \mathbb{N}$, we thus get

$$\mu(A \cap E_n) = \sum_{i=1}^{n} \mu\big((A \cap E_{i-1}^c) \cap \Omega_i\big) = \sum_{i=1}^{n} \nu\big((A \cap E_{i-1}^c) \cap \Omega_i\big) = \nu(A \cap E_n).$$

Since $E_n \uparrow \Omega$ and since μ and ν are lower semicontinuous, we infer

$$\mu(A) = \lim_{n \to \infty} \mu(A \cap E_n) = \lim_{n \to \infty} \nu(A \cap E_n) = \nu(A).$$

The additional statement is trivial as $\tilde{\mathcal{E}} := \mathcal{E} \cup \{\Omega\}$ is a π-system that generates $\mathcal{A}$, and the value $\mu(\Omega) = 1$ is given. Hence one can choose the constant sequence $E_n = \Omega$, $n \in \mathbb{N}$. However, note that it is not enough to assume that μ is finite. In this case, in general, the total mass $\mu(\Omega)$ is not uniquely determined by the values $\mu(E)$, $E \in \mathcal{E}$; see Example 1.45(ii). □

Reflection Where in the previous proof did we exploit the $\cap$-stability? What goes wrong if $\cap$-stability is missing? Compare Example 1.45. ♠

Example 1.43 Let $\Omega = \mathbb{Z}$ and $\mathcal{E} = \{E_n : n \in \mathbb{Z}\}$ where $E_n = (-\infty, n] \cap \mathbb{Z}$. Then $\mathcal{E}$ is a π-system and $\sigma(\mathcal{E}) = 2^{\Omega}$. Hence a finite measure μ on $(\Omega, 2^{\Omega})$ is uniquely determined by the values $\mu(E_n)$, $n \in \mathbb{Z}$.

However, a σ-finite measure on $\mathbb{Z}$ is not uniquely determined by the values on $\mathcal{E}$: Let μ be the counting measure on $\mathbb{Z}$ and let $\nu = 2\mu$. Hence $\mu(E) = \infty = \nu(E)$ for all $E \in \mathcal{E}$. In order to distinguish μ and ν one needs a generator that contains sets of finite measure (of μ). Do the sets $\tilde{F}_n = [-n, n] \cap \mathbb{Z}$, $n \in \mathbb{N}$ do the trick? Indeed, for any σ-finite measure μ, we have $\mu(\tilde{F}_n) < \infty$ for all $n \in \mathbb{N}$. However, the sets $\tilde{F}_n$ do not generate 2^{Ω} (but which σ-algebra?). We get things to work out better if we modify the definition: $F_n = [-n/2, (n+1)/2] \cap \mathbb{Z}$. Now $\sigma(\{F_n, n \in \mathbb{N}\}) = 2^{\Omega}$, and hence $\mathcal{E} = \{F_n, \ n \in \mathbb{N}\}$ is a π-system that generates 2^Ω and such that $\mu(F_n) < \infty$ for all $n \in \mathbb{N}$. The conditions of the theorem are fulfilled as $F_n \uparrow \Omega$. $\Diamond$

Example 1.44 (Distribution function) A probability measure μ on the space $\left(\mathbb{R}^n, \mathcal{B}(\mathbb{R}^n)\right)$ is uniquely determined by the values $\mu((-\infty, b])$ (where $(-\infty, b] = \bigtimes_{i=1}^n (-\infty, b_i]$, $b \in \mathbb{R}^n$). In fact, these sets form a π-system that generates $\mathcal{B}(\mathbb{R}^n)$ (see Theorem 1.23). In particular, a probability measure μ on $\mathbb{R}$ is uniquely determined by its **distribution function** $F : \mathbb{R} \to [0, 1]$, $x \mapsto \mu((-\infty, x])$. $\Diamond$

Example 1.45

(i) Let $\Omega = \{1, 2, 3, 4\}$ and $\mathcal{E} = \{\{1, 2\}, \{2, 3\}\}$. Clearly, $\sigma(\mathcal{E}) = 2^\Omega$ but $\mathcal{E}$ is not a π-system. In fact, here a probability measure μ is not uniquely determined by the values, say $\mu(\{1, 2\}) = \mu(\{2, 3\}) = \frac{1}{2}$. We give just two different possibilities: $\mu = \frac{1}{2}\delta_1 + \frac{1}{2}\delta_3$ and $\mu' = \frac{1}{2}\delta_2 + \frac{1}{2}\delta_4$.

(ii) Let $\Omega = \{1, 2\}$ and $\mathcal{E} = \{\{1\}\}$. Then $\mathcal{E}$ is a π-system that generates 2^Ω. Hence a probability measure μ is uniquely determined by the value $\mu(\{1\})$. However, a *finite* measure is not determined by its value on $\{1\}$, as $\mu = 0$ and $\nu = \delta_2$ are different finite measures that agree on $\mathcal{E}$. $\Diamond$

Lemma 1.42 yields uniqueness in Carathéodory's theorem. The more challenging part is to come up with a candidate $\widetilde{\mu}$ for the extension of the pre-measure in the first place. The strategy is to define a number $\mu^*(E)$ for each $E \in 2^\Omega$ by covering E with elements of $\mathcal{E}$ and then determine the total content. The smallest value $\mu^*(E)$ that can be obtained by such an approximation is called the outer measure of E. The second step is to check that μ^* is a measure at least on $\sigma(\mathcal{E})$. This gives a good candidate for $\widetilde{\mu}$.

Definition 1.46 (Outer measure) *A set function $\mu^* : 2^\Omega \to [0, \infty]$ is called an* **outer measure** *if*

(i) $\mu^*(\emptyset) = 0$, and
(ii) μ^* is monotone,
(iii) μ^* is σ-subadditive.

Lemma 1.47 *Let $\mathcal{A} \subset 2^\Omega$ be an arbitrary class of sets with $\emptyset \in \mathcal{A}$ and let μ be a nonnegative set function on $\mathcal{A}$ with $\mu(\emptyset) = 0$. For $A \subset \Omega$, define the set of countable coverings $\mathcal{F}$ with sets $F \in \mathcal{A}$:*

$$\mathcal{U}(A) = \left\{\mathcal{F} \subset \mathcal{A} : \mathcal{F} \text{ is at most countable and } A \subset \bigcup_{F \in \mathcal{F}} F\right\}.$$

Define

$$\mu^*(A) := \inf\left\{\sum_{F \in \mathcal{F}} \mu(F) : \mathcal{F} \in \mathcal{U}(A)\right\},$$

where $\inf \emptyset = \infty$. *Then* μ^* *is an outer measure. If in addition* μ *is* σ-*subadditive, then* $\mu^*(A) = \mu(A)$ *for all* $A \in \mathcal{A}$.

Proof We check properties (i)–(iii) of an outer measure.

(i) Since $\emptyset \in \mathcal{A}$, we have $\{\emptyset\} \in \mathcal{U}(\emptyset)$; hence $\mu^*(\emptyset) = 0$.

(ii) If $A \subset B$, then $\mathcal{U}(A) \supset \mathcal{U}(B)$; hence $\mu^*(A) \leq \mu^*(B)$.

(iii) Let $A_n \subset \Omega$ for any $n \in \mathbb{N}$ and let $A \subset \bigcup_{n=1}^{\infty} A_n$. We show that $\mu^*(A) \leq \sum_{n=1}^{\infty} \mu^*(A_n)$. Without loss of generality, assume $\mu^*(A_n) < \infty$ and hence $\mathcal{U}(A_n) \neq \emptyset$ for all $n \in \mathbb{N}$. Fix $\varepsilon > 0$. For every $n \in \mathbb{N}$, choose a covering $\mathcal{F}_n \in \mathcal{U}(A_n)$ such that

$$\sum_{F \in \mathcal{F}_n} \mu(F) \leq \mu^*(A_n) + \varepsilon \, 2^{-n}.$$

Then $\mathcal{F} := \bigcup_{n=1}^{\infty} \mathcal{F}_n \in \mathcal{U}(A)$ and

$$\mu^*(A) \leq \sum_{F \in \mathcal{F}} \mu(F) \leq \sum_{n=1}^{\infty} \sum_{F \in \mathcal{F}_n} \mu(F) \leq \sum_{n=1}^{\infty} \mu^*(A_n) + \varepsilon.$$

Let $A \in \mathcal{A}$. Since $\{A\} \in \mathcal{U}(A)$, we have $\mu^*(A) \leq \mu(A)$. If μ is σ-subadditive, then for any $\mathcal{F} \in \mathcal{U}(A)$, we have $\sum_{F \in \mathcal{F}} \mu(F) \geq \mu(A)$; hence $\mu^*(A) \geq \mu(A)$. $\quad\square$

Definition 1.48 (μ^*-measurable sets) *Let* μ^* *be an outer measure. A set* $A \in 2^{\Omega}$ *is called* μ^*-***measurable*** *if*

$$\mu^*(A \cap E) + \mu^*(A^c \cap E) = \mu^*(E) \quad \text{for any } E \in 2^{\Omega}. \tag{1.10}$$

We write $\mathcal{M}(\mu^*) = \{A \in 2^{\Omega} : A \text{ is } \mu^*\text{-measurable}\}$.

Lemma 1.49 $A \in \mathcal{M}(\mu^*)$ *if and only if*

$$\mu^*(A \cap E) + \mu^*(A^c \cap E) \leq \mu^*(E) \quad \text{for any } E \in 2^{\Omega}.$$

Proof As μ^* is subadditive, the other inequality is trivial. $\quad\square$

Lemma 1.50 $\mathcal{M}(\mu^*)$ *is an algebra.*

Proof We check properties (i)–(iii) of an algebra from Theorem 1.7.

(i) $\Omega \in \mathcal{M}(\mu^*)$ is evident.

(ii) (Closedness under complements) By definition, $A \in \mathcal{M}(\mu^*) \iff A^c \in \mathcal{M}(\mu^*)$.

(iii) (π-system)   Let $A, B \in \mathcal{M}(\mu^*)$ and $E \in 2^\Omega$. Then

$$\mu^*\big((A \cap B) \cap E\big) + \mu^*\big((A \cap B)^c \cap E\big)$$
$$= \mu^*(A \cap B \cap E) + \mu^*\big((A^c \cap B \cap E) \cup (A^c \cap B^c \cap E) \cup (A \cap B^c \cap E)\big)$$
$$\le \mu^*(A \cap B \cap E) + \mu^*(A^c \cap B \cap E)$$
$$+ \mu^*(A^c \cap B^c \cap E) + \mu^*(A \cap B^c \cap E)$$
$$= \mu^*(B \cap E) + \mu^*(B^c \cap E)$$
$$= \mu^*(E).$$

Here we used $A \in \mathcal{M}(\mu^*)$ in the last but one equality and $B \in \mathcal{M}(\mu^*)$ in the last equality. $\square$

Lemma 1.51   *An outer measure μ^* is σ-additive on $\mathcal{M}(\mu^*)$.*

Proof   Let $A, B \in \mathcal{M}(\mu^*)$ with $A \cap B = \emptyset$. Then

$$\mu^*(A \cup B) = \mu^*(A \cap (A \cup B)) + \mu^*(A^c \cap (A \cup B)) = \mu^*(A) + \mu^*(B).$$

Inductively, we get (finite) additivity. By definition, μ^* is σ-subadditive; hence we conclude by Theorem 1.36 that μ^* is also σ-additive. $\square$

Lemma 1.52   *If μ^* is an outer measure, then $\mathcal{M}(\mu^*)$ is a σ-algebra. In particular, μ^* is a measure on $\mathcal{M}(\mu^*)$.*

Proof   By Lemma 1.50, $\mathcal{M}(\mu^*)$ is an algebra and hence a π-system. By Theorem 1.18, it is sufficient to show that $\mathcal{M}(\mu^*)$ is a λ-system.

Hence, let $A_1, A_2, \ldots \in \mathcal{M}(\mu^*)$ be mutually disjoint, and define $A := \biguplus_{n=1}^{\infty} A_n$. We have to show $A \in \mathcal{M}(\mu^*)$; that is,

$$\mu^*(A \cap E) + \mu^*(A^c \cap E) \le \mu^*(E) \quad \text{for any } E \in 2^\Omega. \tag{1.11}$$

Let $B_n = \bigcup_{i=1}^{n} A_i$ for all $n \in \mathbb{N}$. For all $n \in \mathbb{N}$, we have

$$\mu^*(E \cap B_{n+1}) = \mu^*\big((E \cap B_{n+1}) \cap B_n\big) + \mu^*\big((E \cap B_{n+1}) \cap B_n^c\big)$$
$$= \mu^*(E \cap B_n) + \mu^*(E \cap A_{n+1}).$$

Inductively, we get $\mu^*(E \cap B_n) = \sum_{i=1}^{n} \mu^*(E \cap A_i)$. The monotonicity of μ^* now implies that

$$\mu^*(E) = \mu^*(E \cap B_n) + \mu^*(E \cap B_n^c) \geq \mu^*(E \cap B_n) + \mu^*(E \cap A^c)$$

$$= \sum_{i=1}^{n} \mu^*(E \cap A_i) + \mu^*(E \cap A^c).$$

Letting $n \to \infty$ and using the σ-subadditivity of μ^*, we conclude

$$\mu^*(E) \geq \sum_{i=1}^{\infty} \mu^*(E \cap A_i) + \mu^*(E \cap A^c) \geq \mu^*(E \cap A) + \mu^*(E \cap A^c).$$

Hence (1.11) holds and the proof is complete. $\square$

We come to an extension theorem for measures that makes slightly weaker assumptions than Carathéodory's theorem (Theorem 1.41).

Theorem 1.53 (Extension theorem for measures) *Let $\mathcal{A}$ be a semiring and let $\mu :$
$\mathcal{A} \to [0, \infty]$ be an additive, σ-subadditive and σ-finite set function with $\mu(\emptyset) = 0$.
   Then there is a unique σ-finite measure $\tilde{\mu} : \sigma(\mathcal{A}) \to [0, \infty]$ such that $\tilde{\mu}(A) = \mu(A)$ for all $A \in \mathcal{A}$.*

Proof As $\mathcal{A}$ is a π-system, uniqueness follows by Lemma 1.42.
   In order to establish the existence of $\tilde{\mu}$, we define as in Lemma 1.47

$$\mu^*(A) := \inf \left\{ \sum_{F \in \mathcal{F}} \mu(F) : \mathcal{F} \in \mathcal{U}(A) \right\} \quad \text{for any } A \in 2^{\Omega}.$$

By Lemma 1.47, μ^* is an outer measure and $\mu^*(A) = \mu(A)$ for any $A \in \mathcal{A}$. We have to show that $\mathcal{M}(\mu^*) \supset \sigma(\mathcal{A})$. Since $\mathcal{M}(\mu^*)$ is a σ-algebra (Lemma 1.52), it is enough to show $\mathcal{A} \subset \mathcal{M}(\mu^*)$.
To this end, let $A \in \mathcal{A}$ and $E \in 2^{\Omega}$ with $\mu^*(E) < \infty$. Fix $\varepsilon > 0$. Then there is a sequence $E_1, E_2, \ldots \in \mathcal{A}$ such that

$$E \subset \bigcup_{n=1}^{\infty} E_n \quad \text{and} \quad \sum_{n=1}^{\infty} \mu(E_n) \leq \mu^*(E) + \varepsilon.$$

Define $B_n := E_n \cap A \in \mathcal{A}$. Since $\mathcal{A}$ is a semiring, for every $n \in \mathbb{N}$ there is an $m_n \in \mathbb{N}$ and sets $C_n^1, \ldots, C_n^{m_n} \in \mathcal{A}$ such that $E_n \setminus A = E_n \setminus B_n = \biguplus_{k=1}^{m_n} C_n^k$. Hence

$$E \cap A \subset \bigcup_{n=1}^{\infty} B_n, \quad E \cap A^c \subset \bigcup_{n=1}^{\infty} \bigcup_{k=1}^{m_n} C_n^k \quad \text{and} \quad E_n = B_n \uplus \biguplus_{k=1}^{m_n} C_n^k.$$

By the definition of the outer measure and since μ is assumed to be (finitely) additive, we get

$$\mu^*(E \cap A) + \mu^*(E \cap A^c) \leq \sum_{n=1}^{\infty} \mu(B_n) + \sum_{n=1}^{\infty} \sum_{k=1}^{m_n} \mu(C_n^k)$$

$$= \sum_{n=1}^{\infty} \left(\mu(B_n) + \sum_{k=1}^{m_n} \mu(C_n^k) \right)$$

$$= \sum_{n=1}^{\infty} \mu(E_n)$$

$$\leq \mu^*(E) + \varepsilon.$$

Hence $\mu^*(E \cap A) + \mu^*(E \cap A^c) \leq \mu^*(E)$ and thus $A \in \mathcal{M}(\mu^*)$, which implies $\mathcal{A} \subset \mathcal{M}(\mu^*)$. Now define $\widetilde{\mu} : \sigma(\mathcal{A}) \to [0, \infty]$, $A \mapsto \mu^*(A)$. By Lemma 1.51, $\widetilde{\mu}$ is a measure and $\widetilde{\mu}$ is σ-finite since μ is σ-finite. $\square$

Reflection In Theorem 1.53, in general, μ cannot be extended to a measure on all of 2^Ω. Why? At which point would the proof fail? Usually it is difficult to show in a specific situation that the extension to 2^Ω is impossible. We refer to analysis books like [37] where Vitali sets are used in order to show that the Lebesgue measure cannot be defined on $2^\mathbb{R}$. ♠♠

Example 1.54 (Lebesgue measure, continuation of Example 1.39) We aim at extending the volume $\mu((a, b]) = \prod_{i=1}^{n}(b_i - a_i)$ that was defined on the class of rectangles $\mathcal{A} = \{(a, b] : a, b \in \mathbb{R}^n, a \leq b\}$ to the Borel σ-algebra $\mathcal{B}(\mathbb{R}^n)$. In order to check the assumptions of Theorem 1.53, we only have to check that μ is σ-subadditive. To this end, let $(a, b]$, $(a(1), b(1)]$, $(a(2), b(2)], \ldots \in \mathcal{A}$ with

$$(a, b] \subset \bigcup_{k=1}^{\infty} (a(k), b(k)].$$

We show that

$$\mu((a, b]) \leq \sum_{k=1}^{\infty} \mu\big((a(k), b(k)]\big). \tag{1.12}$$

For this purpose we use a compactness argument to reduce (1.12) to finite additivity. Fix $\varepsilon > 0$. For any $k \in \mathbb{N}$, choose $b_\varepsilon(k) > b(k)$ such that

$$\mu\big((a(k), b_\varepsilon(k)]\big) \leq \mu\big((a(k), b(k)]\big) + \varepsilon \, 2^{-k-1}.$$

Further choose $a_\varepsilon \in (a, b)$ such that $\mu((a_\varepsilon, b]) \geq \mu((a, b]) - \frac{\varepsilon}{2}$. Now $[a_\varepsilon, b]$ is compact and

$$\bigcup_{k=1}^{\infty} (a(k), b_\varepsilon(k)) \supset \bigcup_{k=1}^{\infty} (a(k), b(k)] \supset (a, b] \supset [a_\varepsilon, b],$$

whence there exists a K_0 such that $\bigcup_{k=1}^{K_0} (a(k), b_\varepsilon(k)) \supset (a_\varepsilon, b]$. As μ is (finitely) subadditive (see Lemma 1.31(iii)), we obtain

$$\mu((a, b]) \leq \frac{\varepsilon}{2} + \mu((a_\varepsilon, b]) \leq \frac{\varepsilon}{2} + \sum_{k=1}^{K_0} \mu((a(k), b_\varepsilon(k)])$$

$$\leq \frac{\varepsilon}{2} + \sum_{k=1}^{K_0} \left(\varepsilon \, 2^{-k-1} + \mu((a(k), b(k)]) \right) \leq \varepsilon + \sum_{k=1}^{\infty} \mu((a(k), b(k)]).$$

Letting $\varepsilon \downarrow 0$ yields (1.12); hence μ is σ-subadditive. $\lozenge$

Combining the last example with Theorem 1.53, we have shown the following theorem.

Theorem 1.55 (Lebesgue measure) *There exists a uniquely determined measure* λ^n *on* $(\mathbb{R}^n, \mathcal{B}(\mathbb{R}^n))$ *with the property that*

$$\lambda^n((a, b]) = \prod_{i=1}^{n} (b_i - a_i) \quad \text{for all } a, b \in \mathbb{R}^n \text{ with } a < b.$$

λ^n *is called the* **Lebesgue measure** *on* $(\mathbb{R}^n, \mathcal{B}(\mathbb{R}^n))$ *or* **Lebesgue–Borel measure**.

Example 1.56 (Lebesgue–Stieltjes measure) Let $\Omega = \mathbb{R}$ and $\mathcal{A} = \{(a, b] : a, b \in \mathbb{R}, \, a \leq b\}$. $\mathcal{A}$ is a semiring and $\sigma(\mathcal{A}) = \mathcal{B}(\mathbb{R})$, where $\mathcal{B}(\mathbb{R})$ is the Borel σ-algebra on $\mathbb{R}$. Furthermore, let $F : \mathbb{R} \to \mathbb{R}$ be monotone increasing and right continuous. We define a set function

$$\tilde{\mu}_F : \mathcal{A} \to [0, \infty), \qquad (a, b] \mapsto F(b) - F(a).$$

Clearly, $\tilde{\mu}_F(\emptyset) = 0$ and $\tilde{\mu}_F$ is additive.

Let $(a, b], (a(1), b(1)], (a(2), b(2)], \ldots \in \mathcal{A}$ such that $(a, b] \subset \bigcup_{n=1}^{\infty} (a(n), b(n)]$ and $a < b$. Fix $\varepsilon > 0$ and choose $a_\varepsilon \in (a, b)$ such that $F(a_\varepsilon) - F(a) < \varepsilon/2$. This is possible, as F is right continuous. For any $k \in \mathbb{N}$, choose $b_\varepsilon(k) > b(k)$ such that

$$F(b_\varepsilon(k)) - F(b(k)) < \varepsilon \, 2^{-k-1}.$$

As in Example 1.54, it can be shown that $\tilde{\mu}_F((a, b]) \leq \varepsilon + \sum_{k=1}^{\infty} \tilde{\mu}_F((a(k), b(k)])$. This implies that $\tilde{\mu}_F$ is σ-subadditive. By Theorem 1.53, we can extend $\tilde{\mu}_F$ uniquely to a σ-finite measure μ_F on $\mathcal{B}(\mathbb{R})$. $\Diamond$

Definition 1.57 (Lebesgue–Stieltjes measure) *The measure* μ_F *on* $\left(\mathbb{R}, \mathcal{B}(\mathbb{R})\right)$ *defined by*

$$\mu_F((a, b]) = F(b) - F(a) \quad \text{for all } a, b \in \mathbb{R} \text{ with } a < b$$

*is called the **Lebesgue–Stieltjes measure** with distribution function F.*

Example 1.58 Important special cases for the Lebesgue–Stieltjes measure are the following:

 (i) If $F(x) = x$, then $\mu_F = \lambda^1$ is the Lebesgue measure on $\mathbb{R}$.
 (ii) Let $f : \mathbb{R} \to [0, \infty)$ be continuous and let $F(x) = \int_0^x f(t)\, dt$ for all $x \in \mathbb{R}$. Then μ_F is the extension of the premeasure with **density** f that was defined in Example 1.30(ix).
 (iii) Let $x_1, x_2, \ldots \in \mathbb{R}$ and $\alpha_n \geq 0$ for all $n \in \mathbb{N}$ such that $\sum_{n=1}^{\infty} \alpha_n < \infty$. Then $F = \sum_{n=1}^{\infty} \alpha_n \mathbb{1}_{[x_n, \infty)}$ is the distribution function of the finite measure $\mu_F = \sum_{n=1}^{\infty} \alpha_n \delta_{x_n}$.
 (iv) Let $x_1, x_2, \ldots \in \mathbb{R}$ such that $\mu = \sum_{n=1}^{\infty} \delta_{x_n}$ is a σ-finite measure. Then μ is a Lebesgue–Stieltjes measure if and only if the sequence $(x_n)_{n \in \mathbb{N}}$ does not have a limit point. Indeed, if $(x_n)_{n \in \mathbb{N}}$ does not have a limit point, then by the Bolzano–Weierstraß theorem, $\#\{n \in \mathbb{N} : x_n \in [-K, K]\} < \infty$ for every $K > 0$. If we let $F(x) = \#\{n \in \mathbb{N} : x_n \in [0, x]\}$ for $x \geq 0$ and $F(x) = -\#\{n \in \mathbb{N} : x_n \in (x, 0)\}$ for $x < 0$, then $\mu = \mu_F$. On the other hand, if μ is a Lebesgue–Stieltjes measure, this is $\mu = \mu_F$ for some F, then $\#\{n \in \mathbb{N} : x_n \in (-K, K]\} = F(K) - F(-K) < \infty$ for all $K > 0$; hence $(x_n)_{n \in \mathbb{N}}$ does not have a limit point.
 (v) If $\lim_{x \to \infty} \left(F(x) - F(-x)\right) = 1$, then μ_F is a probability measure. $\Diamond$

We will now have a closer look at the case where μ_F is a probability measure.

Definition 1.59 (Distribution function) *A right continuous monotone increasing function* $F : \mathbb{R} \to [0, 1]$ *with* $F(-\infty) := \lim_{x \to -\infty} F(x) = 0$ *and* $F(\infty) := \lim_{x \to \infty} F(x) = 1$ *is called a (proper)* **probability distribution function** *(p.d.f.). If we only have* $F(\infty) \leq 1$ *instead of* $F(\infty) = 1$, *then F is called a (possibly) defective p.d.f. If* μ *is a (sub-) probability measure on* $\left(\mathbb{R}, \mathcal{B}(\mathbb{R})\right)$, *then* $F_\mu : x \mapsto \mu((-\infty, x])$ *is called the distribution function of* μ.

Clearly, F_μ is right continuous and $F_\mu(-\infty) = 0$, since μ is upper semicontinuous and finite (Theorem 1.36). Since μ is lower semicontinuous, we have $F_\mu(\infty) = \mu(\mathbb{R})$; hence F_μ is indeed a (possibly defective) distribution function if μ is a (sub-) probability measure.

The argument of Example 1.56 yields the following theorem.

Theorem 1.60 *The map $\mu \mapsto F_\mu$ is a bijection from the set of probability measures on $(\mathbb{R}, \mathcal{B}(\mathbb{R}))$ to the set of probability distribution functions, respectively from the set of sub-probability measures to the set of defective distribution functions.*

We have established that every finite measure on $(\mathbb{R}, \mathcal{B}(\mathbb{R}))$ is a Lebesgue–Stieltjes measure for some function F. For σ-finite measures, the corresponding statement does not hold in this generality as we saw in Example 1.58(iv).

We come now to a theorem that combines Theorem 1.55 with the idea of Lebesgue–Stieltjes measures. Later we will see that the following theorem is valid in greater generality. In particular, the assumption that the factors are of Lebesgue–Stieltjes type can be dropped.

Theorem 1.61 (Finite products of measures) *Let $n \in \mathbb{N}$ and let $\mu_1, \ldots, \mu_n$ be finite measures or, more generally, Lebesgue–Stieltjes measures on $(\mathbb{R}, \mathcal{B}(\mathbb{R}))$. Then there exists a unique σ-finite measure μ on $(\mathbb{R}^n, \mathcal{B}(\mathbb{R}^n))$ such that*

$$\mu((a, b]) = \prod_{i=1}^n \mu_i((a_i, b_i]) \quad \text{for all } a, b \in \mathbb{R}^n \text{ with } a < b.$$

*We call $\mu =: \displaystyle\bigotimes_{i=1}^n \mu_i$ the **product measure** of the measures $\mu_1, \ldots, \mu_n$.*

Proof The proof is the same as for Theorem 1.55. One has to check that the intervals $(a, b_\varepsilon]$ and so on can be chosen such that $\mu((a, b_\varepsilon]) < \mu((a, b]) + \varepsilon$. Here we employ the right continuity of the increasing function F_i that belongs to μ_i. The details are left as an exercise.   □

Remark 1.62 Later we will see in Theorem 14.14 that the statement holds even for arbitrary σ-finite measures $\mu_1, \ldots, \mu_n$ on arbitrary (even different) measurable spaces. One can even construct infinite products if all factors are probability spaces (Theorem 14.39). ◊

Example 1.63 (Infinite product measure, continuation of Example 1.40) Let E be a finite set and let $\Omega = E^{\mathbb{N}}$ be the space of E-valued sequences. Further, let $(p_e)_{e \in E}$ be a probability vector. Define a content μ on $\mathcal{A} = \{[\omega_1, \ldots, \omega_n] : \omega_1, \ldots, \omega_n \in E, n \in \mathbb{N}\}$ by

$$\mu([\omega_1, \ldots, \omega_n]) = \prod_{i=1}^n p_{\omega_i}.$$

We aim at extending μ to a measure on $\sigma(\mathcal{A})$. In order to check the assumptions of Theorem 1.53, we have to show that μ is σ-subadditive. As in the preceding example, we use a compactness argument.

Let $A, A_1, A_2, \ldots \in \mathcal{A}$ and $A \subset \bigcup_{n=1}^{\infty} A_n$. We are done if we can show that there exists an $N \in \mathbb{N}$ such that

$$A \subset \bigcup_{n=1}^{N} A_n. \tag{1.13}$$

Indeed, due to the (finite) subadditivity of μ (see Lemma 1.31(iii)), this implies $\mu(A) \leq \sum_{n=1}^{N} \mu(A_n) \leq \sum_{n=1}^{\infty} \mu(A_n)$; hence μ is σ-subadditive.

We now give two different proofs for (1.13).

1. Proof.   The metric d from (1.9) induces the product topology on Ω; hence, as remarked in Example 1.40, (Ω, d) is a compact metric space. Every $A \in \mathcal{A}$ is closed and thus compact. Since every A_n is also open, A can be covered by finitely many A_n; hence (1.13) holds.

2. Proof.   We now show by *elementary* means the validity of (1.13). The procedure imitates the proof that Ω is compact. Let $B_n := A \setminus \bigcup_{i=1}^{n} A_i$. We assume $B_n \neq \emptyset$ for all $n \in \mathbb{N}$ in order to get a contradiction. By Dirichlet's pigeonhole principle (recall that E is finite), we can choose $\omega_1 \in E$ such that $[\omega_1] \cap B_n \neq \emptyset$ for infinitely many $n \in \mathbb{N}$. Since $B_1 \supset B_2 \supset \ldots$, we obtain

$$[\omega_1] \cap B_n \neq \emptyset \quad \text{for all } n \in \mathbb{N}.$$

Successively choose $\omega_2, \omega_3, \ldots \in E$ in such a way that

$$[\omega_1, \ldots, \omega_k] \cap B_n \neq \emptyset \quad \text{for all } k, n \in \mathbb{N}.$$

B_n is a disjoint union of certain sets $C_{n,1}, \ldots, C_{n,m_n} \in \mathcal{A}$. Hence, for every $n \in \mathbb{N}$ there is an $i_n \in \{1, \ldots, m_n\}$ such that $[\omega_1, \ldots, \omega_k] \cap C_{n,i_n} \neq \emptyset$ for infinitely many $k \in \mathbb{N}$. Since $[\omega_1] \supset [\omega_1, \omega_2] \supset \ldots$, we obtain

$$[\omega_1, \ldots, \omega_k] \cap C_{n,i_n} \neq \emptyset \quad \text{for all } k, n \in \mathbb{N}.$$

For fixed $n \in \mathbb{N}$ and large k, we have $[\omega_1, \ldots, \omega_k] \subset C_{n,i_n}$. Hence $\omega = (\omega_1, \omega_2, \ldots) \in C_{n,i_n} \subset B_n$. This implies $\bigcap_{n=1}^{\infty} B_n \neq \emptyset$, contradicting the assumption. $\Diamond$

Combining the last example with Theorem 1.53, we have shown the following theorem.

Theorem 1.64 (Product measure, Bernoulli measure) *Let E be a finite nonempty set and $\Omega = E^{\mathbb{N}}$. Let $(p_e)_{e \in E}$ be a probability vector. Then there exists a unique probability measure μ on $\sigma(\mathcal{A}) = \mathcal{B}(\Omega)$ such that*

$$\mu([\omega_1, \ldots, \omega_n]) = \prod_{i=1}^{n} p_{\omega_i} \quad \text{for all } \omega_1, \ldots, \omega_n \in E \text{ and } n \in \mathbb{N}.$$

μ is called the **product measure** or **Bernoulli measure** on Ω with weights $(p_e)_{e\in E}$. We write $\left(\sum_{e\in E} p_e \delta_e\right)^{\otimes \mathbb{N}} := \mu$. The σ-algebra $(2^E)^{\otimes \mathbb{N}} := \sigma(\mathcal{A})$ is called the product σ-algebra on Ω.

We will study product measures in a systematic way in Chap. 14.

The measure extension theorem yields an abstract statement of existence and uniqueness for measures on $\sigma(\mathcal{A})$ that were first defined on a semiring $\mathcal{A}$ only. The following theorem, however, shows that the measure of a set from $\sigma(\mathcal{A})$ can be well approximated by finite and countable operations with sets from $\mathcal{A}$.

Denote by

$$A \vartriangle B := (A \setminus B) \cup (B \setminus A) \quad \text{for } A, B \subset \Omega \tag{1.14}$$

the **symmetric difference** of the two sets A and B.

Theorem 1.65 (Approximation theorem for measures) *Let $\mathcal{A} \subset 2^\Omega$ be a semiring and let μ be a measure on $\sigma(\mathcal{A})$ that is σ-finite on $\mathcal{A}$.*

(i) *For any $A \in \sigma(\mathcal{A})$ and $\varepsilon > 0$, there exist mutually disjoint sets $A_1, A_2, \ldots \in \mathcal{A}$ such that $A \subset \bigcup_{n=1}^{\infty} A_n$ and $\mu\left(\bigcup_{n=1}^{\infty} A_n \setminus A\right) < \varepsilon$.*

(ii) *For any $A \in \sigma(\mathcal{A})$ with $\mu(A) < \infty$ and any $\varepsilon > 0$, there exists an $n \in \mathbb{N}$ and mutually disjoint sets $A_1, \ldots, A_n \in \mathcal{A}$ such that $\mu\left(A \vartriangle \bigcup_{k=1}^{n} A_k\right) < \varepsilon$.*

(iii) *For any $A \in \mathcal{M}(\mu^*)$, there are sets $A_-, A_+ \in \sigma(\mathcal{A})$ with $A_- \subset A \subset A_+$ and $\mu(A_+ \setminus A_-) = 0$.*

Remark 1.66 (iii) implies that (i) and (ii) also hold for $A \in \mathcal{M}(\mu^*)$ (with μ^* instead of μ). If $\mathcal{A}$ is an algebra, then in (ii) for any $A \in \sigma(\mathcal{A})$, we even have $\inf_{B\in\mathcal{A}} \mu(A \vartriangle B) = 0$. ◊

Proof (ii) As μ and the outer measure μ^* coincide on $\sigma(\mathcal{A})$ and since $\mu(A)$ is finite, by the very definition of μ^* (see Lemma 1.47) there exists a covering $B_1, B_2, \ldots \in \mathcal{A}$ of A such that

$$\mu(A) \geq \sum_{i=1}^{\infty} \mu(B_i) - \varepsilon/2.$$

Let $n \in \mathbb{N}$ with $\sum_{i=n+1}^{\infty} \mu(B_i) < \frac{\varepsilon}{2}$ (such an n exists since $\mu(A) < \infty$). For any three sets C, D, E, we have

$$C \vartriangle D = (D\setminus C)\cup(C\setminus D) \subset (D\setminus C)\cup(C\setminus(D\cup E))\cup E \subset (C \vartriangle (D\cup E))\cup E.$$

Choosing $C = A$, $D = \bigcup_{i=1}^{n} B_i$ and $E = \bigcup_{i=n+1}^{\infty} B_i$, this yields

$$\mu\left(A \bigtriangleup \bigcup_{i=1}^{n} B_i\right) \leq \mu\left(A \bigtriangleup \bigcup_{i=1}^{\infty} B_i\right) + \mu\left(\bigcup_{i=n+1}^{\infty} B_i\right)$$

$$\leq \mu\left(\bigcup_{i=1}^{\infty} B_i\right) - \mu(A) + \frac{\varepsilon}{2} \leq \varepsilon.$$

As $\mathcal{A}$ is a semiring, there exist a $k \in \mathbb{N}$ and $A_1, \ldots, A_k \in \mathcal{A}$ such that

$$\bigcup_{i=1}^{n} B_i = B_1 \uplus \biguplus_{i=2}^{n} \bigcap_{j=1}^{i-1}(B_i \setminus B_j) =: \biguplus_{i=1}^{k} A_i.$$

(i) Let $A \in \sigma(\mathcal{A})$ and $E_n \uparrow \Omega$, $E_n \in \sigma(\mathcal{A})$ with $\mu(E_n) < \infty$ for any $n \in \mathbb{N}$. For every $n \in \mathbb{N}$, choose a covering $(B_{n,m})_{m \in \mathbb{N}}$ of $A \cap E_n$ with

$$\mu(A \cap E_n) \geq \sum_{m=1}^{\infty} \mu(B_{n,m}) - 2^{-n}\varepsilon.$$

(This is possible due to the definition of the outer measure μ^*, which coincides with μ on $\mathcal{A}$.) Let $\bigcup_{m,n=1}^{\infty} B_{n,m} = \biguplus_{n=1}^{\infty} A_n$ for certain $A_n \in \mathcal{A}$, $n \in \mathbb{N}$ (Exercise 1.1.1). Then

$$\mu\left(\biguplus_{n=1}^{\infty} A_n \setminus A\right) = \mu\left(\bigcup_{n=1}^{\infty} \bigcup_{m=1}^{\infty} B_{n,m} \setminus A\right)$$

$$\leq \mu\left(\bigcup_{n=1}^{\infty} \bigcup_{m=1}^{\infty} \left(B_{n,m} \setminus (A \cap E_n)\right)\right)$$

$$\leq \sum_{n=1}^{\infty}\left(\left(\sum_{m=1}^{\infty} \mu(B_{n,m})\right) - \mu(A \cap E_n)\right) \leq \varepsilon.$$

(iii) Let $A \in \mathcal{M}(\mu^*)$ and $(E_n)_{n \in \mathbb{N}}$ as above. For any $m, n \in \mathbb{N}$, choose $A_{n,m} \in \sigma(\mathcal{A})$ such that $A_{n,m} \supset A \cap E_n$ and $\mu^*(A_{n,m}) \leq \mu^*(A \cap E_n) + \frac{2^{-n}}{m}$.

Define $A_m := \bigcup_{n=1}^{\infty} A_{n,m} \in \sigma(\mathcal{A})$. Then $A_m \supset A$ and $\mu^*(A_m \setminus A) \le \frac{1}{m}$. Define
$A_+ := \bigcap_{m=1}^{\infty} A_m$. Then $\sigma(\mathcal{A}) \ni A_+ \supset A$ and $\mu^*(A_+ \setminus A) = 0$. Similarly, choose
$(A_-)^c \in \sigma(\mathcal{A})$ with $(A_-)^c \supset A^c$ and $\mu^*((A_-)^c \setminus A^c) = 0$. Then $A_+ \supset A \supset A_-$ and $\mu(A_+ \setminus A_-) = \mu^*(A_+ \setminus A_-) = \mu^*(A_+ \setminus A) + \mu^*(A \setminus A_-) = 0$.   □

Remark 1.67 (Regularity of measures)   (Compare with Theorem 13.6.) Let λ^n be the Lebesgue measure on $(\mathbb{R}^n, \mathcal{B}(\mathbb{R}^n))$. Let $\mathcal{A}$ be the semiring of rectangles of the form $(a, b] \subset \mathbb{R}^n$; hence $\mathcal{B}(\mathbb{R}^n) = \sigma(\mathcal{A})$ by Theorem 1.23. By the approximation theorem, for any $A \in \mathcal{B}(\mathbb{R}^n)$ and $\varepsilon > 0$, there exist countably many $A_1, A_2, \ldots \in \mathcal{A}$ with $A \subset \bigcup_{i=1}^{\infty} A_i$ and

$$\lambda^n \left(\bigcup_{i=1}^{\infty} A_i \setminus A \right) < \varepsilon/2.$$

For any A_i, there exists an *open* rectangle $B_i \supset A_i$ with $\lambda^n(B_i \setminus A_i) < \varepsilon 2^{-i-1}$ (upper semicontinuity of λ^n). Hence $U = \bigcup_{i=1}^{\infty} B_i$ is an open set $U \supset A$ with

$$\lambda^n(U \setminus A) < \varepsilon.$$

This property of λ^n is called **outer regularity**.
If $\lambda^n(A)$ is finite, then for any $\varepsilon > 0$ there exists a compact $K \subset A$ such that

$$\lambda^n(A \setminus K) < \varepsilon.$$

This property of λ^n is called **inner regularity**. Indeed, let $N > 0$ be such that $\lambda^n(A) - \lambda^n(A \cap [-N, N]^n) < \varepsilon/2$. Choose an open set $U \supset (A \cap [-N, N]^n)^c$ such that $\lambda^n(U \setminus (A \cap [-N, N]^n)^c) < \varepsilon/2$, and let $K := [-N, N]^n \setminus U \subset A$. ◊

Definition 1.68 (Null set)   *Let $(\Omega, \mathcal{A}, \mu)$ be a measure space.*

   (i) *A set $A \in \mathcal{A}$ is called a μ-**null set**, or briefly a null set, if $\mu(A) = 0$. By $\mathcal{N}_\mu$ we denote the class of all subsets of μ-null sets.*
  (ii) *Let $E(\omega)$ be a property that a point $\omega \in \Omega$ can have or not have. We say that E holds μ-**almost everywhere** (a.e.) or for **almost all** (a.a.) ω if there exists a null set N such that $E(\omega)$ holds for every $\omega \in \Omega \setminus N$. If $A \in \mathcal{A}$ and if there exists a null set N such that $E(\omega)$ holds for every $\omega \in A \setminus N$, then we say that E holds almost everywhere on A.*
       *If $\mu = P$ is a probability measure, then we say that E holds P-**almost surely** (a.s.), respectively almost surely on A.*
 (iii) *Let $A, B \in \mathcal{A}$ be such that $\mu(A \triangle B) = 0$. Then we write $A = B \pmod{\mu}$.*

Definition 1.69   *A measure space $(\Omega, \mathcal{A}, \mu)$ is called **complete** if $\mathcal{N}_\mu \subset \mathcal{A}$.*

Remark 1.70 (Completion of a measure space) Let $(\Omega, \mathcal{A}, \mu)$ be a σ-finite measure space. There exists a unique smallest σ-algebra $\mathcal{A}^* \supset \mathcal{A}$ and an extension μ^* of μ to $\mathcal{A}^*$ such that $(\Omega, \mathcal{A}^*, \mu^*)$ is complete. $(\Omega, \mathcal{A}^*, \mu^*)$ is called the **completion** of $(\Omega, \mathcal{A}, \mu)$. With the notation of Theorem 1.53, this completion is

$$\left(\Omega, \mathcal{M}(\mu^*), \mu^* \big|_{\mathcal{M}(\mu^*)} \right).$$

Furthermore,

$$\mathcal{M}(\mu^*) = \sigma(\mathcal{A} \cup \mathcal{N}_\mu) = \{A \cup N : A \in \mathcal{A}, \ N \in \mathcal{N}_\mu\}$$

and $\mu^*(A \cup N) = \mu(A)$ for any $A \in \mathcal{A}$ and $N \in \mathcal{N}_\mu$.

In the following, we will not need these statements. Hence, instead of giving a proof, we refer to the textbooks on measure theory (e.g., [37]). $\Diamond$

Example 1.71 Let λ be the Lebesgue measure (more accurately, the Lebesgue–Borel measure) on $\big(\mathbb{R}^n, \mathcal{B}(\mathbb{R}^n)\big)$. Then λ can be extended uniquely to a measure λ^* on

$$\mathcal{B}^*(\mathbb{R}^n) = \sigma\big(\mathcal{B}(\mathbb{R}^n) \cup \mathcal{N}\big),$$

where $\mathcal{N}$ is the class of subsets of Lebesgue–Borel null sets. $\mathcal{B}^*(\mathbb{R}^n)$ is called the σ-algebra of Lebesgue measurable sets. For the sake of distinction, we sometimes call λ the **Lebesgue–Borel measure** and λ^* the **Lebesgue measure**. However, in practice, this distinction will not be needed in this book. $\Diamond$

Example 1.72 Let $\mu = \delta_\omega$ be the Dirac measure for the point $\omega \in \Omega$ on some measurable space $(\Omega, \mathcal{A})$. If $\{\omega\} \in \mathcal{A}$, then the completion is $\mathcal{A}^* = 2^\Omega$, $\mu^* = \delta_\omega$. In the extreme case of a trivial σ-algebra $\mathcal{A} = \{\emptyset, \Omega\}$, however, the empty set is the only null set, $\mathcal{N}_\mu = \{\emptyset\}$; hence $\mathcal{A}^* = \{\emptyset, \Omega\}$, $\mu^* = \delta_\omega$. Note that, on the trivial σ-algebra, Dirac measures for different points $\omega \in \Omega$ cannot be distinguished. $\Diamond$

Definition 1.73 Let $(\Omega, \mathcal{A}, \mu)$ be a measure space and $\Omega' \in \mathcal{A}$. On the trace σ-algebra $\mathcal{A}\big|_{\Omega'}$, we define a measure by

$$\mu\big|_{\Omega'}(A) := \mu(A) \quad \text{for } A \in \mathcal{A} \text{ with } A \subset \Omega'.$$

*This measure is called the **restriction** of μ to Ω'.*

Example 1.74 The restriction of the Lebesgue–Borel measure λ on $\big(\mathbb{R}, \mathcal{B}(\mathbb{R})\big)$ to $[0, 1]$ is a probability measure on $([0, 1], \mathcal{B}(\mathbb{R})\big|_{[0,1]})$. More generally, for a measurable $A \in \mathcal{B}(\mathbb{R})$, we call the restriction $\lambda\big|_A$ the *Lebesgue measure* on A. Often this measure will be denoted by the same symbol λ when there is no danger of ambiguity.

Later we will see (Corollary 1.84) that $\mathcal{B}(\mathbb{R})\big|_A = \mathcal{B}(A)$, where $\mathcal{B}(A)$ is the Borel σ-algebra on A that is generated by the (relatively) open subsets of A. ◊

Example 1.75 (Uniform distribution) Let $A \in \mathcal{B}(\mathbb{R}^n)$ be a measurable set with n-dimensional Lebesgue measure $\lambda^n(A) \in (0, \infty)$. Then we can define a probability measure on $\mathcal{B}(\mathbb{R}^n)\big|_A$ by

$$\mu(B) := \frac{\lambda^n(B)}{\lambda^n(A)} \quad \text{for } B \in \mathcal{B}(\mathbb{R}^n) \text{ with } B \subset A.$$

This measure μ is called the **uniform distribution** on A and will be denoted by $\mathcal{U}_A := \mu$. ◊

Takeaways The measure extension theorem shows how to extend contents from semirings to σ-algebras but usually does not give a concrete construction. However, in the special case where the content was defined on an algebra in the first place, the measure on sets of the σ-algebra can be approximated arbitrarily well by sets from the algebra. This will be helpful in many places. In order for the measure extension to work, σ-additivity is decisive. It is an interesting finding that in two important examples we could check σ-subadditivity using topological properties. More specifically, we used compactness arguments. At that point it was only a small step to show that the Lebesgue measure is regular in the sense that the measure of an arbitrary measurable set can be approximated by compact subsets as well as by open supersets.

Exercise 1.3.1 Show the following generalization of Example 1.58(iv): A measure $\sum_{n=1}^{\infty} \alpha_n \delta_{x_n}$ is a Lebesgue–Stieltjes measure for a suitable function F if and only if $\sum_{n:\, |x_n| \leq K} \alpha_n < \infty$ for all $K > 0$. ♣

Exercise 1.3.2 Let Ω be an uncountably infinite set and let $\omega_0 \in \Omega$ be an arbitrary element. Let $\mathcal{A} = \sigma(\{\omega\} : \omega \in \Omega \setminus \{\omega_0\})$.

(i) Give a characterization of $\mathcal{A}$ as in Exercise 1.1.4 (page 11).
(ii) Show that $(\Omega, \mathcal{A}, \delta_{\omega_0})$ is complete. ♣

Exercise 1.3.3 Let $(\mu_n)_{n \in \mathbb{N}}$ be a sequence of finite measures on the measurable space $(\Omega, \mathcal{A})$. Assume that for any $A \in \mathcal{A}$ there exists the limit $\mu(A) := \lim_{n \to \infty} \mu_n(A)$.
Show that μ is a measure on $(\Omega, \mathcal{A})$.
Hint: In particular, one has to show that μ is $\emptyset$-continuous. ♣

1.4   Measurable Maps

A major task of mathematics is to study homomorphisms between objects; that is, structure-preserving maps. For topological spaces, these are the continuous maps, and for measurable spaces, these are the measurable maps.

In the rest of this chapter, we let $(\Omega, \mathcal{A})$ and $(\Omega', \mathcal{A}')$ be measurable spaces.

Definition 1.76 (Measurable maps)

(i) A map $X : \Omega \to \Omega'$ is called $\mathcal{A} - \mathcal{A}'$-measurable (or, briefly, measurable) if $X^{-1}(\mathcal{A}') := \{X^{-1}(A') : A' \in \mathcal{A}'\} \subset \mathcal{A}$; that is, if

$$X^{-1}(A') \in \mathcal{A} \quad \text{for any } A' \in \mathcal{A}'.$$

If X is measurable, we write $X : (\Omega, \mathcal{A}) \to (\Omega', \mathcal{A}')$.
(ii) If $\Omega' = \mathbb{R}$ and $\mathcal{A}' = \mathcal{B}(\mathbb{R})$ is the Borel σ-algebra on $\mathbb{R}$, then $X : (\Omega, \mathcal{A}) \to (\mathbb{R}, \mathcal{B}(\mathbb{R}))$ is called an $\mathcal{A}$-measurable real map.

Example 1.77

(i) The identity map id $: \Omega \to \Omega$ is $\mathcal{A} - \mathcal{A}$-measurable.
(ii) If $\mathcal{A} = 2^{\Omega}$ or $\mathcal{A}' = \{\emptyset, \Omega'\}$, then any map $X : \Omega \to \Omega'$ is $\mathcal{A} - \mathcal{A}'$-measurable.
(iii) Let $A \subset \Omega$. The indicator function $\mathbb{1}_A : \Omega \to \{0, 1\}$ is $\mathcal{A} - 2^{\{0,1\}}$-measurable if and only if $A \in \mathcal{A}$. $\lozenge$

Theorem 1.78 (Generated σ-algebra) *Let $(\Omega', \mathcal{A}')$ be a measurable space and let Ω be a nonempty set. Let $X : \Omega \to \Omega'$ be a map. The preimage*

$$X^{-1}(\mathcal{A}') := \{X^{-1}(A') : A' \in \mathcal{A}'\} \tag{1.15}$$

*is the smallest σ-algebra with respect to which X is measurable. We say that $\sigma(X) := X^{-1}(\mathcal{A}')$ is the σ-algebra on Ω that is **generated** by X.*

Proof This is left as an exercise.   □

We now consider σ-algebras that are generated by more than one map.

Definition 1.79 (Generated σ-algebra) *Let Ω be a nonempty set. Let I be an arbitrary index set. For any $i \in I$, let $(\Omega_i, \mathcal{A}_i)$ be a measurable space and let $X_i : \Omega \to \Omega_i$ be an arbitrary map. Then*

$$\sigma(X_i, \, i \in I) := \sigma\left(\bigcup_{i \in I} \sigma(X_i)\right) = \sigma\left(\bigcup_{i \in I} X_i^{-1}(\mathcal{A}_i)\right)$$

*is called the σ-algebra on Ω that is **generated** by $(X_i, \, i \in I)$. This is the smallest σ-algebra with respect to which all X_i are measurable.*

As with continuous maps, the composition of measurable maps is again measurable.

Theorem 1.80 (Composition of maps) *Let $(\Omega, \mathcal{A})$, $(\Omega', \mathcal{A}')$ and $(\Omega'', \mathcal{A}'')$ be measurable spaces and let $X : \Omega \to \Omega'$ and $X' : \Omega' \to \Omega''$ be measurable maps. Then the map $Y := X' \circ X : \Omega \to \Omega''$, $\omega \mapsto X'(X(\omega))$ is $\mathcal{A} - \mathcal{A}''$-measurable.*

Proof Obvious, since $Y^{-1}(\mathcal{A}'') = X^{-1}((X')^{-1}(\mathcal{A}'')) \subset X^{-1}(\mathcal{A}') \subset \mathcal{A}$.   □

In practice, it is often not possible to check if a map X is measurable by checking if all preimages $X^{-1}(A')$, $A' \in \mathcal{A}'$ are measurable. Most σ-algebras $\mathcal{A}'$ are simply too large. Thus it comes in very handy that it is sufficient to check measurability on a generator of $\mathcal{A}'$ by the following theorem.

Theorem 1.81 (Measurability on a generator) *Let $\mathcal{E}' \subset \mathcal{A}'$ be a class of $\mathcal{A}'$-measurable sets. Then $\sigma(X^{-1}(\mathcal{E}')) = X^{-1}(\sigma(\mathcal{E}'))$ and hence*

$$X \text{ is } \mathcal{A} \text{-} \sigma(\mathcal{E}')\text{-measurable} \iff X^{-1}(E') \in \mathcal{A} \quad \text{for all } E' \in \mathcal{E}'.$$

If in particular $\sigma(\mathcal{E}') = \mathcal{A}'$, then

$$X \text{ is } \mathcal{A} - -\mathcal{A}'\text{-measurable} \iff X^{-1}(\mathcal{E}') \subset \mathcal{A}.$$

Proof Clearly, $X^{-1}(\mathcal{E}') \subset X^{-1}(\sigma(\mathcal{E}')) = \sigma(X^{-1}(\sigma(\mathcal{E}')))$. Hence also

$$\sigma(X^{-1}(\mathcal{E}')) \subset X^{-1}(\sigma(\mathcal{E}')).$$

For the other inclusion, consider the class of sets

$$\mathcal{A}_0' := \{A' \in \sigma(\mathcal{E}') : X^{-1}(A') \in \sigma(X^{-1}(\mathcal{E}'))\}.$$

We first show that $\mathcal{A}_0'$ is a σ-algebra by checking (i)–(iii) of Definition 1.2:

(i) Clearly, $\Omega' \in \mathcal{A}_0'$.
(ii) (Stability under complements)   If $A' \in \mathcal{A}_0'$, then

$$X^{-1}((A')^c) = (X^{-1}(A'))^c \in \sigma(X^{-1}(\mathcal{E}'));$$

hence $(A')^c \in \mathcal{A}_0'$.
(iii) (σ-∪-stability)   Let $A_1', A_2', \ldots \in \mathcal{A}_0'$. Then

$$X^{-1}\left(\bigcup_{n=1}^{\infty} A_n'\right) = \bigcup_{n=1}^{\infty} X^{-1}(A_n') \in \sigma(X^{-1}(\mathcal{E}'));$$

hence $\bigcup_{n=1}^{\infty} A_n' \in \mathcal{A}_0'$.

Now $\mathcal{A}_0' = \sigma(\mathcal{E}')$ since $\mathcal{E}' \subset \mathcal{A}_0'$. Hence $X^{-1}(A') \in \sigma(X^{-1}(\mathcal{E}'))$ for any $A' \in \sigma(\mathcal{E}')$ and thus $X^{-1}(\sigma(\mathcal{E}')) \subset \sigma(X^{-1}(\mathcal{E}'))$.   □

Corollary 1.82 (Measurability of composed maps) *Let I be a nonempty index set and let $(\Omega, \mathcal{A})$, $(\Omega', \mathcal{A}')$ and $(\Omega_i, \mathcal{A}_i)$ be measurable spaces for any $i \in I$. Further, let $(X_i : i \in I)$ be a family of measurable maps $X_i : \Omega' \to \Omega_i$ with $\mathcal{A}' = \sigma(X_i : i \in I)$. Then the following holds: A map $Y : \Omega \to \Omega'$ is $\mathcal{A} - \mathcal{A}'$-measurable if and only if $X_i \circ Y$ is $\mathcal{A} - \mathcal{A}_i$-measurable for all $i \in I$.*

Proof If Y is measurable, then by Theorem 1.80 every $X_i \circ Y$ is measurable. Now assume that all of the composed maps $X_i \circ Y$ are $\mathcal{A} - \mathcal{A}_i$-measurable. By assumption, the set $\mathcal{E}' := \{X_i^{-1}(A'') : A'' \in \mathcal{A}_i, i \in I\}$ is a generator of $\mathcal{A}'$. Since all $X_i \circ Y$ are measurable, we have $Y^{-1}(A') \in \mathcal{A}$ for any $A' \in \mathcal{E}'$. Hence Theorem 1.81 yields that Y is measurable. $\square$

Recall the definition of the trace of a class of sets from Definition 1.25.

Corollary 1.83 (Trace of a generated σ-algebra) *Let $\mathcal{E} \subset 2^\Omega$ and assume that $A \subset \Omega$ is nonempty. Then $\sigma\left(\mathcal{E}\big|_A\right) = \sigma(\mathcal{E})\big|_A$.*

Proof Let $X : A \hookrightarrow \Omega, \omega \mapsto \omega$ be the canonical inclusion; hence $X^{-1}(B) = A \cap B$ for all $B \subset \Omega$. By Theorem 1.81, we have

$$\sigma\left(\mathcal{E}\big|_A\right) = \sigma(\{E \cap A : E \in \mathcal{E}\})$$

$$= \sigma(\{X^{-1}(E) : E \in \mathcal{E}\}) = \sigma(X^{-1}(\mathcal{E}))$$

$$= X^{-1}(\sigma(\mathcal{E})) = \{A \cap B : B \in \sigma(\mathcal{E})\} = \sigma(\mathcal{E})\big|_A.$$

$\square$

Recall that, for any subset $A \subset \Omega$ of a topological space (Ω, τ), the class $\tau\big|_A$ is the topology of relatively open sets (in A). We denote by $\mathcal{B}(\Omega, \tau) = \sigma(\tau)$ the Borel σ-algebra on (Ω, τ).

Corollary 1.84 (Trace of the Borel σ-algebra) *Let (Ω, τ) be a topological space and let $A \subset \Omega$ be a nonempty subset of Ω. Then*

$$\mathcal{B}(\Omega, \tau)\big|_A = \mathcal{B}\left(A, \tau\big|_A\right).$$

Example 1.85

(i) Let Ω' be countable. Then $X : \Omega \to \Omega'$ is $\mathcal{A} - 2^{\Omega'}$-measurable if and only if $X^{-1}(\{\omega'\}) \in \mathcal{A}$ for all $\omega' \in \Omega'$. If Ω' is uncountably infinite, this is wrong in general. (For example, consider $\Omega = \Omega' = \mathbb{R}$, $\mathcal{A} = \mathcal{B}(\mathbb{R})$, and $X(\omega) = \omega$ for all $\omega \in \Omega$. Clearly, $X^{-1}(\{\omega\}) = \{\omega\} \in \mathcal{B}(\mathbb{R})$. If, on the other hand, $A \subset \mathbb{R}$ is not in $\mathcal{B}(\mathbb{R})$, then $A \in 2^{\mathbb{R}}$, but $X^{-1}(A) \notin \mathcal{B}(\mathbb{R})$.)

(ii) For $x \in \mathbb{R}$, we agree on the following notation for rounding:

$$\lfloor x \rfloor := \max\{k \in \mathbb{Z} : k \leq x\} \quad \text{and} \quad \lceil x \rceil := \min\{k \in \mathbb{Z} : k \geq x\}. \quad (1.16)$$

The maps $\mathbb{R} \to \mathbb{Z}$, $x \mapsto \lfloor x \rfloor$ and $x \mapsto \lceil x \rceil$ are $\mathcal{B}(\mathbb{R}) - 2^{\mathbb{Z}}$-measurable since for all $k \in \mathbb{Z}$ the preimages $\{x \in \mathbb{R} : \lfloor x \rfloor = k\} = [k, k+1)$ and $\{x \in \mathbb{R} : \lceil x \rceil = k\} = (k-1, k]$ are in $\mathcal{B}(\mathbb{R})$. By the composition theorem (Theorem 1.80), for any measurable map $f : (\Omega, \mathcal{A}) \to (\mathbb{R}, \mathcal{B}(\mathbb{R}))$ the maps $\lfloor f \rfloor$ and $\lceil f \rceil$ are also $\mathcal{A} - 2^{\mathbb{Z}}$-measurable.

(iii) A map $X : \Omega \to \mathbb{R}^d$ is $\mathcal{A} - \mathcal{B}(\mathbb{R}^d)$-measurable if and only if

$$X^{-1}((-\infty, a]) \in \mathcal{A} \quad \text{for any } a \in \mathbb{R}^d.$$

In fact $\sigma((-\infty, a], a \in \mathbb{R}^d) = \mathcal{B}(\mathbb{R}^d)$ by Theorem 1.23. The analogous statement holds for any of the classes $\mathcal{E}_1, \ldots, \mathcal{E}_{12}$ from Theorem 1.23. $\Diamond$

Example 1.86 Let $d(x, y) = \|x - y\|_2$ be the usual Euclidean distance on $\mathbb{R}^n$ and let $\mathcal{B}(\mathbb{R}^n, d) = \mathcal{B}(\mathbb{R}^n)$ be the Borel σ-algebra with respect to the topology generated by d. For any subset A of $\mathbb{R}^n$, we have $\mathcal{B}(A, d) = \mathcal{B}(\mathbb{R}^n, d)\big|_A$. $\Diamond$

We want to extend the real line by the points $-\infty$ and $+\infty$. Thus we define

$$\overline{\mathbb{R}} := \mathbb{R} \cup \{-\infty, +\infty\}.$$

From a topological point of view, $\overline{\mathbb{R}}$ will be considered as the so-called two point compactification by considering $\overline{\mathbb{R}}$ as topologically isomorphic to $[-1, 1]$ via the map

$$\varphi : [-1, 1] \to \overline{\mathbb{R}}, \qquad x \mapsto \begin{cases} \tan(\pi x/2), & \text{if } x \in (-1, 1), \\ -\infty, & \text{if } x = -1, \\ \infty, & \text{if } x = +1. \end{cases}$$

In fact, $\bar{d}(x, y) = \left| \varphi^{-1}(x) - \varphi^{-1}(y) \right|$ for $x, y \in \overline{\mathbb{R}}$ defines a metric on $\overline{\mathbb{R}}$ such that φ and φ^{-1} are continuous. Hence φ is a topological isomorphism. We denote by $\bar{\tau}$ the corresponding topology induced on $\overline{\mathbb{R}}$ and by τ the usual topology on $\mathbb{R}$.

Corollary 1.87 *With the above notation, $\bar{\tau}\big|_{\mathbb{R}} = \tau$ and hence $\mathcal{B}(\overline{\mathbb{R}})\big|_{\mathbb{R}} = \mathcal{B}(\mathbb{R})$.*

In particular, if $X : (\Omega, \mathcal{A}) \to (\mathbb{R}, \mathcal{B}(\mathbb{R}))$ is measurable, then in a canonical way X is also an $\overline{\mathbb{R}}$-valued measurable map.

Thus $\overline{\mathbb{R}}$ is really an extension of the real line, and the inclusion $\mathbb{R} \hookrightarrow \overline{\mathbb{R}}$ is measurable.

Reflection Check that each of the families $\mathcal{E}_1, \ldots, \mathcal{E}_{12}$ from Theorem 1.23 (with $n = 1$) is a generator for $\mathcal{B}(\overline{\mathbb{R}})$. Also check that the families $\{[-\infty, a], a \in \mathbb{Q}\}$, $\{[-\infty, a), a \in \mathbb{Q}\}$, $\{[b, \infty], b \in \mathbb{Q}\}$ and $\{(b, \infty], b \in \mathbb{Q}\}$ are generators of $\mathcal{B}(\overline{\mathbb{R}})$.
♠

Theorem 1.88 (Measurability of continuous maps) *Let (Ω, τ) and (Ω', τ') be topological spaces and let $f : \Omega \to \Omega'$ be a continuous map. Then f is $\mathcal{B}(\Omega) - \mathcal{B}(\Omega')$-measurable.*

Proof As $\mathcal{B}(\Omega') = \sigma(\tau')$ and by Theorem 1.81, it is sufficient to show that $f^{-1}(A') \in \sigma(\tau)$ for all $A' \in \tau'$. However, since f is continuous, we even have $f^{-1}(A') \in \tau$ for all $A' \in \tau'$. $\square$

For $x, y \in \overline{\mathbb{R}}$, we agree on the following notation.

$$
\begin{aligned}
x \vee y &= \max(x, y) & \text{(maximum)}, \\
x \wedge y &= \min(x, y) & \text{(minimum)}, \\
x^{+} &= \max(x, 0) & \text{(positive part)}, \\
x^{-} &= \max(-x, 0) & \text{(negative part)}, \\
|x| &= \max(x, -x) = x^{-} + x^{+} & \text{(modulus)}, \\
\operatorname{sign}(x) &= \mathbb{1}_{\{x > 0\}} - \mathbb{1}_{\{x < 0\}} & \text{(sign function)}.
\end{aligned}
$$

Analogously, for measurable real maps we write, for example, $X^{+} = \max(X, 0)$. The maps $x \mapsto x^{+}$, $x \mapsto x^{-}$ and $x \mapsto |x|$ are continuous (and hence measurable by the preceding theorem). Clearly, the map $x \mapsto \operatorname{sign}(x)$ also is measurable. Using Corollary 1.82, we thus get the following corollary.

Corollary 1.89 *If X is a real or $\overline{\mathbb{R}}$-valued measurable map, then the maps X^{-}, X^{+}, $|X|$ and $\operatorname{sign}(X)$ also are measurable.*

Theorem 1.90 (Coordinate maps are measurable) *Let $(\Omega, \mathcal{A})$ be a measurable space and let $f_1, \ldots, f_n : \Omega \to \mathbb{R}$ be maps. Define $f := (f_1, \ldots, f_n) : \Omega \to \mathbb{R}^n$. Then*

$$f \text{ is } \mathcal{A} - \mathcal{B}(\mathbb{R}^n)\text{-measurable} \quad \Longleftrightarrow \quad \text{each } f_i \text{ is } \mathcal{A} - \mathcal{B}(\mathbb{R})\text{-measurable}.$$

The analogous statement holds for $f_i : \Omega \to \overline{\mathbb{R}} := \mathbb{R} \cup \{\pm\infty\}$.

Proof For $b \in \mathbb{R}^n$, we have $f^{-1}((-\infty, b)) = \bigcap_{i=1}^{n} f_i^{-1}((-\infty, b_i))$. If each f_i is measurable, then $f^{-1}((-\infty, b)) \in \mathcal{A}$. However, the rectangles $(-\infty, b)$, $b \in \mathbb{R}^n$, generate $\mathcal{B}(\mathbb{R}^n)$, and hence f is measurable. Now assume that f is measurable. For $i = 1, \ldots, n$, let $\pi_i : \mathbb{R}^n \to \mathbb{R}$, $x \mapsto x_i$ be the projection on the ith coordinate. Clearly, π_i is continuous and thus $\mathcal{B}(\mathbb{R}^n) - \mathcal{B}(\mathbb{R})$-measurable. Hence $f_i = \pi_i \circ f$ is measurable by Theorem 1.80. $\square$

In the following theorem, we agree that $\frac{x}{0} := 0$ for all $x \in \mathbb{R}$.

Theorem 1.91 *Let $(\Omega, \mathcal{A})$ be a measurable space. Let $h : (\Omega, \mathcal{A}) \to (\mathbb{R}, \mathcal{B}(\mathbb{R}))$ and $f, g : (\Omega, \mathcal{A}) \to (\mathbb{R}^n, \mathcal{B}(\mathbb{R}^n))$ be measurable maps. Then also the maps $f + g$, $f - g$, $f \cdot h$ and f/h are measurable.*

Proof The map $\pi : \mathbb{R}^n \times \mathbb{R} \to \mathbb{R}^n, (x, \alpha) \mapsto \alpha \cdot x$ is continuous and thus measurable. By Theorem 1.90, $(f, h) : \Omega \to \mathbb{R}^n \times \mathbb{R}$ is measurable. Hence also the composed map $f \cdot h = \pi \circ (f, h)$ is measurable. Similarly, we obtain the measurability of $f + g$ and $f - g$.

In order to show measurability of f/h, we define the map $H : \mathbb{R} \to \mathbb{R}, x \mapsto 1/x$. Note that by our convention $H(0) = 0$. Hence $f/h = f \cdot H \circ h$. Thus it is enough to show that H is measurable. Clearly, $H\big|_{\mathbb{R}\setminus\{0\}}$ is continuous. For any open set $U \subset \mathbb{R}$, $U \setminus \{0\}$ is also open and hence $H^{-1}(U \setminus \{0\}) \in \mathcal{B}(\mathbb{R})$. Furthermore, $H^{-1}(\{0\}) = \{0\}$. Concluding, we get $H^{-1}(U) = H^{-1}(U \setminus \{0\}) \cup (U \cap \{0\}) \in \mathcal{B}(\mathbb{R})$. □

Theorem 1.92 *Let* $X_1, X_2, \ldots$ *be measurable maps* $(\Omega, \mathcal{A}) \to (\overline{\mathbb{R}}, \mathcal{B}(\overline{\mathbb{R}}))$. *Then the following maps are also measurable:*

$$\inf_{n \in \mathbb{N}} X_n, \qquad \sup_{n \in \mathbb{N}} X_n, \qquad \liminf_{n \to \infty} X_n, \qquad \limsup_{n \to \infty} X_n.$$

Proof For any $a \in \overline{\mathbb{R}}$, we have

$$\left(\inf_{n \in \mathbb{N}} X_n\right)^{-1} ([-\infty, a)) = \bigcup_{n=1}^{\infty} X_n^{-1}([-\infty, a)) \in \mathcal{A}.$$

By Theorem 1.81, this implies that $\inf_{n \in \mathbb{N}} X_n$ is measurable. The proof for $\sup_{n \in \mathbb{N}} X_n$ is similar.

For any $n \in \mathbb{N}$, we define $Y_n := \inf_{m \geq n} X_m$. Note that Y_n is measurable and hence $\liminf_{n \to \infty} X_n := \sup_{n \in \mathbb{N}} Y_n$ also is measurable. The proof for the limes superior is similar. □

We come to an important example of measurable maps $(\Omega, \mathcal{A}) \to (\mathbb{R}, \mathcal{B}(\mathbb{R}))$, the so-called simple functions.

Definition 1.93 (Simple function) *Let* $(\Omega, \mathcal{A})$ *be a measurable space. A map* $f : \Omega \to \mathbb{R}$ *is called a **simple function** if there is an* $n \in \mathbb{N}$ *and mutually disjoint measurable sets* $A_1, \ldots, A_n \in \mathcal{A}$, *as well as numbers* $\alpha_1, \ldots, \alpha_n \in \mathbb{R}$, *such that*

$$f = \sum_{i=1}^{n} \alpha_i \mathbb{1}_{A_i}.$$

Remark 1.94 A measurable map that assumes only finitely many values is a simple function. (Exercise: Show this!) ◊

Definition 1.95 *Assume that* $f, f_1, f_2, \ldots$ *are maps* $\Omega \to \overline{\mathbb{R}}$ *such that*

$$f_1(\omega) \leq f_2(\omega) \leq \ldots \quad \text{and} \quad \lim_{n \to \infty} f_n(\omega) = f(\omega) \quad \text{for any } \omega \in \Omega.$$

Then we write $f_n \uparrow f$ and say that $(f_n)_{n\in\mathbb{N}}$ increases (pointwise) to f. Analogously, we write $f_n \downarrow f$ if $(-f_n) \uparrow (-f)$.

Theorem 1.96 *Let $(\Omega, \mathcal{A})$ be a measurable space and let $f : \Omega \to [0, \infty]$ be measurable. Then the following statements hold.*

(i) *There exists a sequence $(f_n)_{n\in\mathbb{N}}$ of nonnegative simple functions such that $f_n \uparrow f$.*

(ii) *There are measurable sets $A_1, A_2, \ldots \in \mathcal{A}$ and numbers $\alpha_1, \alpha_2, \ldots \geq 0$ such that $f = \sum_{n=1}^{\infty} \alpha_n \mathbb{1}_{A_n}$.*

Proof

(i) For $n \in \mathbb{N}_0$, define $f_n = \left(2^{-n}\lfloor 2^n f\rfloor\right) \wedge n$. Then f_n is measurable (by Theorem 1.92 and Example 1.85(ii)) and assumes at most $n2^n + 1$ different values. Hence it is a simple function. Clearly, $f_n \uparrow f$.

(ii) Let f_n be as above. Let $B_{n,i} := \{\omega : f_n(\omega) - f_{n-1}(\omega) = i\, 2^{-n}\}$ and $\beta_{n,i} = i\, 2^{-n}$ for $n \in \mathbb{N}$ and $i = 1, \ldots, 2^n$. Hence $f_n - f_{n-1} = \sum_{i=1}^{2^n} \beta_{n,i} \mathbb{1}_{B_{n,i}}$. By changing the numeration $(n, i) \mapsto m$, we get $(\alpha_m)_{m\in\mathbb{N}}$ and $(A_m)_{m\in\mathbb{N}}$ such that

$$f = f_0 + \sum_{n=1}^{\infty}(f_n - f_{n-1}) = \sum_{m=1}^{\infty} \alpha_m \mathbb{1}_{A_m}. \qquad \square$$

As a corollary to this statement on the structure of $[0, \infty]$-valued measurable maps, we show the following factorization lemma.

Corollary 1.97 (Factorization lemma) *Let $(\Omega', \mathcal{A}')$ be a measurable space and let Ω be a nonempty set. Let $f : \Omega \to \Omega'$ be a map. A map $g : \Omega \to \overline{\mathbb{R}}$ is $\sigma(f) - \mathcal{B}(\overline{\mathbb{R}})$-measurable if and only if there is a measurable map $\varphi : (\Omega', \mathcal{A}') \to (\overline{\mathbb{R}}, \mathcal{B}(\overline{\mathbb{R}}))$ such that $g = \varphi \circ f$.*

Proof " $\Longleftarrow$ " If φ is measurable and $g = \varphi \circ f$, then g is measurable by Theorem 1.80.

" $\Longrightarrow$ " Now assume that g is $\sigma(f) - \mathcal{B}(\overline{\mathbb{R}})$-measurable. First consider the case $g \geq 0$. Then there exist measurable sets $A_1, A_2 \ldots \in \sigma(f)$ as well as numbers $\alpha_1, \alpha_2, \ldots, \in [0, \infty)$ such that $g = \sum_{n=1}^{\infty} \alpha_n \mathbb{1}_{A_n}$. By the definition of $\sigma(f)$, for any $n \in \mathbb{N}$ there is a set $B_n \in \mathcal{A}'$ such that $f^{-1}(B_n) = A_n$; that is, such that $\mathbb{1}_{A_n} = \mathbb{1}_{B_n} \circ f$. Define $\varphi : \Omega' \to \overline{\mathbb{R}}$ by

$$\varphi = \sum_{n=1}^{\infty} \alpha_n \mathbb{1}_{B_n}.$$

Clearly, φ is $\mathcal{A}' - \mathcal{B}(\overline{\mathbb{R}})$-measurable and $g = \varphi \circ f$.

Now drop the assumption that g is nonnegative. Then there exist measurable maps φ^- and φ^+ such that $g^- = \varphi^- \circ f$ and $g^+ = \varphi^+ \circ f$. Note that $g^+(\omega) \wedge g^-(\omega) = 0$ for all $\omega \in \Omega$. Hence

$$\omega \mapsto \varphi(\omega) := \begin{cases} \varphi^+(\omega) - \varphi^-(\omega), & \text{if } \varphi^+(\omega) < \infty \text{ or } \varphi^-(\omega) < \infty, \\ 0, & \text{else} \end{cases}$$

does the trick. $\square$

A measurable map transports a measure from one space to another.

Definition 1.98 (Image measure) *Let $(\Omega, \mathcal{A})$ and $(\Omega', \mathcal{A}')$ be measurable spaces and let μ be a measure on $(\Omega, \mathcal{A})$. Further, let $X : (\Omega, \mathcal{A}) \to (\Omega', \mathcal{A}')$ be measurable. The **image measure** of μ under the map X is the measure $\mu \circ X^{-1}$ on $(\Omega', \mathcal{A}')$ that is defined by*

$$\mu \circ X^{-1} : \mathcal{A}' \to [0, \infty], \quad A' \mapsto \mu(X^{-1}(A')).$$

Example 1.99 Let μ be a measure on $\mathbb{Z}^2$ and let $X : \mathbb{Z}^2 \to \mathbb{Z}$, $(x, y) \mapsto x + y$. Then

$$\mu \circ X^{-1}(\{x\}) = \sum_{y \in \mathbb{Z}} \mu(\{(x - y, y)\}). \quad \Diamond$$

Example 1.100 Let $L : \mathbb{R}^n \to \mathbb{R}^n$ be a linear bijection and let λ be the Lebesgue measure on $(\mathbb{R}^n, \mathcal{B}(\mathbb{R}^n))$. Then $\lambda \circ L^{-1} = |\det(L)|^{-1}\lambda$. This is clear since for any $a, b \in \mathbb{R}^n$ with $a < b$, the parallelepiped $L^{-1}((a, b])$ has volume $|\det(L^{-1})| \prod_{i=1}^n (b_i - a_i)$. $\Diamond$

As a generalization of the last example, we state without proof the transformation formula for measures with continuous densities under differentiable maps. The proof can be found in textbooks on calculus.

Theorem 1.101 (Transformation formula in $\mathbb{R}^n$) *Let μ be a measure on $\mathbb{R}^n$ that has a continuous (or piecewise continuous) density $f : \mathbb{R}^n \to [0, \infty)$. That is,*

$$\mu((x, y]) = \int_{x_1}^{y_1} dt_1 \cdots \int_{x_n}^{y_n} dt_n \, f(t_1, \ldots, t_n) \quad \text{for all } x, y \in \mathbb{R}^n, \, x \le y.$$

Let $A \subset \mathbb{R}^n$ be an open or a closed subset of $\mathbb{R}^n$ with $\mu(\mathbb{R}^n \setminus A) = 0$. Further, let $B \subset \mathbb{R}^n$ be open or closed. Finally, assume that $\varphi : A \to B$ is a continuously differentiable bijection with derivative φ'. Then the image measure $\mu \circ \varphi^{-1}$ has the density

$$
f_\varphi(x) = \begin{cases} \dfrac{f(\varphi^{-1}(x))}{|\det(\varphi'(\varphi^{-1}(x)))|}, & \text{if } x \in B \ \text{mit} \ \det(\varphi'(\varphi^{-1}(x))) \neq 0, \\[3mm] 0, & \text{else.} \end{cases}
$$

> **Takeaways** Measurable maps are the natural maps between two measurable spaces as they are structure preserving. We have encountered conditions for measurability that are easy to check, e.g., continuity and measurability on a generator. Furthermore, we have seen that compositions of measurable maps, vectors of measurable maps and so on are again measurable.

Exercise 1.4.1 Let $f : \mathbb{R} \to \mathbb{R}$, $x \mapsto |x|$. Show that a Borel measurable map $g : \mathbb{R} \to \mathbb{R}$ is $\sigma(f) = f^{-1}(\mathcal{B}(\mathbb{R}))$-measurable if and only if g is even. ♣

Exercise 1.4.2 Let $(\Omega, \mathcal{A}, \mu)$ be a measure space and let $f : \Omega \to \mathbb{R}$ be measurable. Assume that $g : \Omega \to \mathbb{R}$ fulfills $g = f$ μ-almost everywhere. Show that g need not be measurable. ♣

Exercise 1.4.3 Let $f : \mathbb{R} \to \mathbb{R}$ be differentiable with derivative f'. Show that f' is $\mathcal{B}(\mathbb{R}) - \mathcal{B}(\mathbb{R})$-measurable. ♣

Exercise 1.4.4 (Compare Examples 1.40 and 1.63) Let $\Omega = \{0, 1\}^{\mathbb{N}}$ and let $\mathcal{A} = (2^{\{0,1\}})^{\otimes \mathbb{N}}$ be the σ-algebra generated by the cylinder sets

$$
\big\{ [\omega_1, \ldots, \omega_n] : n \in \mathbb{N}, \ \omega_1, \ldots, \omega_n \in \{0, 1\} \big\}.
$$

Further, let $\mu = (\frac{1}{2}\delta_0 + \frac{1}{2}\delta_1)^{\otimes \mathbb{N}}$ be the Bernoulli measure on Ω with equal weights on 0 and 1. For all $n \in \mathbb{N}$, let $X_n : \Omega \to \{0, 1\}$, $\omega \mapsto \omega_n$ be the nth coordinate map. Finally, let

$$
U(\omega) = \sum_{n=1}^{\infty} X_n(\omega)\, 2^{-n} \quad \text{for } \omega \in \Omega.
$$

(i) Show that $\mathcal{A} = \sigma(X_n : n \in \mathbb{N})$.
(ii) Show that U is $\mathcal{A} - \mathcal{B}([0, 1])$-measurable.
(iii) Determine the image measure $\mu \circ U^{-1}$ on $([0, 1], \mathcal{B}([0, 1]))$.
(iv) Determine an $\Omega_0 \in \mathcal{A}$ such that $\tilde{U} := U\big|_{\Omega_0}$ is bijective.

(v) Show that $\tilde{U}^{-1}$ is $\mathcal{B}([0, 1]) - \mathcal{A}\big|_{\Omega_0}$-measurable.

(vi) Give an interpretation of the map $X_n \circ \tilde{U}^{-1}$. ♣

Exercise 1.4.5 (Lusin's theorem) Let $f : \mathbb{R} \to \mathbb{R}$ be a Borel measurable map. Show that for any $\varepsilon > 0$, there exists a closed set $C \subset \mathbb{R}$ with $\lambda(\mathbb{R} \setminus C) < \varepsilon$ such that the restriction $f\big|_C$ of f to C is continuous. (Note: Clearly, this does not mean that f would be continuous in every point $x \in C$.)

Hint: Use the inner regularity of Lebesgue measure λ (Remark 1.67) to show the assertion first for indicator functions. Next construct a sequence of maps that approximates f uniformly on a suitable set C. ♣

1.5 Random Variables

The fundamental idea of modern probability theory is to model one or more random experiments as a probability space $(\Omega, \mathcal{A}, \mathbf{P})$. The sets $A \in \mathcal{A}$ are called **events**. In most cases, the events of Ω are not observed directly. Rather, the observations are aspects of the single experiments that are coded as measurable maps from Ω to a space of possible observations. In short, every random observation (the technical term is *random variable*) is a measurable map. The probabilities of the possible random observations will be described in terms of the distribution of the corresponding random variable, which is the image measure of $\mathbf{P}$ under X. Hence we have to develop a calculus to determine the distributions of, for example, sums of random variables.

Definition 1.102 (Random variables) *Let $(\Omega', \mathcal{A}')$ be a measurable space and let $X : \Omega \to \Omega'$ be measurable.*

(i) *X is called a **random variable** with values in $(\Omega', \mathcal{A}')$. If $(\Omega', \mathcal{A}') = (\mathbb{R}, \mathcal{B}(\mathbb{R}))$, then X is called a real random variable or simply a random variable.*

(ii) *For $A' \in \mathcal{A}'$, we denote $\{X \in A'\} := X^{-1}(A')$ and $\mathbf{P}[X \in A'] := \mathbf{P}[X^{-1}(A')]$. In particular, we let $\{X \geq 0\} := X^{-1}([0, \infty))$ and define $\{X \leq b\}$ similarly and so on.*

Definition 1.103 (Distributions) *Let X be a random variable.*

(i) *The probability measure $\mathbf{P}_X := \mathbf{P} \circ X^{-1}$ is called the **distribution** of X.*

(ii) *For a real random variable X, the map $F_X : x \mapsto \mathbf{P}[X \leq x]$ is called the **distribution function** of X (or, more accurately, of $\mathbf{P}_X$). We write $X \sim \mu$ if $\mu = \mathbf{P}_X$ and say that X has distribution μ.*

(iii) *A family $(X_i)_{i \in I}$ of random variables is called **identically distributed** if $\mathbf{P}_{X_i} = \mathbf{P}_{X_j}$ for all $i, j \in I$. We write $X \overset{\mathcal{D}}{=} Y$ if $\mathbf{P}_X = \mathbf{P}_Y$ ($\mathcal{D}$ for distribution).*

Theorem 1.104 *For any distribution function* F, *there exists a real random variable* X *with* $F_X = F$.

Proof We explicitly construct a probability space $(\Omega, \mathcal{A}, \mathbf{P})$ and a random variable $X : \Omega \to \mathbb{R}$ such that $F_X = F$.

The simplest choice would be $(\Omega, \mathcal{A}) = (\mathbb{R}, \mathcal{B}(\mathbb{R}))$, $X : \mathbb{R} \to \mathbb{R}$ the identity map and $\mathbf{P}$ the Lebesgue–Stieltjes measure with distribution function F (see Example 1.56).

A more instructive approach is based on first constructing, independently of F, a sort of standard probability space on which we define a random variable with uniform distribution on $(0, 1)$. In a second step, this random variable will be transformed by applying the inverse map F^{-1}: Let $\Omega := (0, 1)$, $\mathcal{A} := \mathcal{B}(\mathbb{R})\big|_{\Omega}$ and let $\mathbf{P}$ be the Lebesgue measure on $(\Omega, \mathcal{A})$ (see Example 1.74). Define the left continuous inverse of F:

$$F^{-1}(t) := \inf\{x \in \mathbb{R} : F(x) \geq t\} \quad \text{for } t \in (0, 1).$$

Then

$$F^{-1}(t) \leq x \iff t \leq F(x).$$

In particular, $\{t : F^{-1}(t) \leq x\} = (0, F(x)] \cap (0, 1)$; hence $F^{-1} : (\Omega, \mathcal{A}) \to (\mathbb{R}, \mathcal{B}(\mathbb{R}))$ is measurable and

$$\mathbf{P}[\{t : F^{-1}(t) \leq x\}] = F(x).$$

Concluding, $X := F^{-1}$ is the random variable that we wanted to construct. □

Example 1.105 We present some prominent distributions of real random variables X. These are some of the most important distributions in probability theory, and we will come back to these examples in many places.

(i) Let $p \in [0, 1]$ and $\mathbf{P}[X = 1] = p$, $\mathbf{P}[X = 0] = 1 - p$. Then $\mathbf{P}_X =: \text{Ber}_p$ is called the **Bernoulli distribution** with parameter p; formally

$$\text{Ber}_p = (1 - p)\,\delta_0 + p\,\delta_1.$$

Its distribution function is

$$F_X(x) = \begin{cases} 0, & \text{if } x < 0, \\ 1 - p, & \text{if } x \in [0, 1), \\ 1, & \text{if } x \geq 1. \end{cases}$$

The distribution $\mathbf{P}_Y$ of $Y := 2X - 1$ is sometimes called the **Rademacher distribution** with parameter p; formally $\text{Rad}_p = (1 - p)\,\delta_{-1} + p\,\delta_{+1}$. In particular, $\text{Rad}_{1/2}$ is called *the* Rademacher distribution.

(ii) Let $p \in [0, 1]$ and $n \in \mathbb{N}$, and let $X : \Omega \to \{0, \ldots, n\}$ be such that

$$\mathbf{P}[X = k] = \binom{n}{k} p^k (1 - p)^{n-k}.$$

Then $\mathbf{P}_X =: b_{n,p}$ is called the **binomial distribution** with parameters n and p; formally

$$b_{n,p} = \sum_{k=0}^{n} \binom{n}{k} p^k (1 - p)^{n-k} \, \delta_k.$$

(iii) Let $p \in (0, 1]$ and $X : \Omega \to \mathbb{N}_0$ with

$$\mathbf{P}[X = n] = p \, (1 - p)^n \quad \text{for any } n \in \mathbb{N}_0.$$

Then $\gamma_p := b_{1,p}^- := \mathbf{P}_X$ is called the **geometric distribution**[2] with parameter p; formally

$$\gamma_p = \sum_{n=0}^{\infty} p \, (1 - p)^n \, \delta_n.$$

Its distribution function is $F(x) = 1 - (1 - p)^{\lfloor x+1 \rfloor \vee 0}$ for $x \in \mathbb{R}$.

We can interpret $X + 1$ as the waiting time for the first success in a series of independent random experiments, any of which yields a success with probability p. Indeed, let $\Omega = \{0, 1\}^{\mathbb{N}}$ and let $\mathbf{P}$ be the product measure $\left((1 - p)\delta_0 + p \, \delta_1\right)^{\otimes \mathbb{N}}$ (Theorem 1.64), as well as $\mathcal{A} = \sigma([\omega_1, \ldots, \omega_n] : \omega_1, \ldots, \omega_n \in \{0, 1\}, n \in \mathbb{N})$. Define

$$X(\omega) := \inf\{n \in \mathbb{N} : \omega_n = 1\} - 1,$$

where $\inf \emptyset = \infty$. Clearly, any map

$$X_n : \Omega \to \mathbb{R}, \quad \omega \mapsto \begin{cases} n - 1, & \text{if } \omega_n = 1, \\ \infty, & \text{if } \omega_n = 0, \end{cases}$$

---

[2] Warning: For some authors, the geometric distribution is shifted by one to the right; that is, it is a distribution on $\mathbb{N}$.

is $\mathcal{A} - \mathcal{B}(\overline{\mathbb{R}})$-measurable. Thus also $X = \inf_{n \in \mathbb{N}} X_n$ is $\mathcal{A} - \mathcal{B}(\overline{\mathbb{R}})$-measurable and is hence a random variable. Let $\omega^0 := (0, 0, \ldots) \in \Omega$. Then $\mathbf{P}[X \geq n] = \mathbf{P}[[\omega_1^0, \ldots, \omega_n^0]] = (1 - p)^n$. Hence

$$\mathbf{P}[X = n] = \mathbf{P}[X \geq n] - \mathbf{P}[X \geq n+1] = (1-p)^n - (1-p)^{n+1} = p\,(1-p)^n.$$

(iv) Let $r > 0$ (note that r need not be an integer) and let $p \in (0, 1]$. We denote by

$$b_{r,p}^{-} := \sum_{k=0}^{\infty} \binom{-r}{k} (-1)^k\, p^r (1 - p)^k\, \delta_k \tag{1.17}$$

the **negative binomial distribution** or **Pascal distribution** with parameters r and p. (Here $\binom{x}{k} = \frac{x(x-1)\cdots(x-k+1)}{k!}$ for $x \in \mathbb{R}$ and $k \in \mathbb{N}$ is the generalized binomial coefficient.) If $r \in \mathbb{N}$, then one can show as in the preceding example that $b_{r,p}^{-}$ is the distribution of the waiting time for the rth success in a series of random experiments. We come back to this in Example 3.4(iv).

(v) Let $\lambda \in [0, \infty)$ and let $X : \Omega \to \mathbb{N}_0$ be such that

$$\mathbf{P}[X = n] = e^{-\lambda}\, \frac{\lambda^n}{n!} \quad \text{for any } n \in \mathbb{N}_0.$$

Then $\mathbf{P}_X =: \mathrm{Poi}_\lambda$ is called the **Poisson distribution** with parameter λ.

(vi) Consider an urn with $B \in \mathbb{N}$ black balls and $W \in \mathbb{N}$ white balls. Draw $n \in \mathbb{N}$ balls from the urn without replacement. A little bit of combinatorics shows that the probability of drawing exactly $b \in \{0, \ldots, n\}$ black balls is given by the **hypergeometric distribution** with parameters $B, W, n \in \mathbb{N}$:

$$\mathrm{Hyp}_{B,W;n}(\{b\}) = \frac{\dbinom{B}{b}\dbinom{W}{n-b}}{\dbinom{B+W}{n}} \quad \text{for } b \in \{0, \ldots, n\}. \tag{1.18}$$

This generalizes easily to the situation of k colors and B_i balls of color $i = 1, \ldots, k$. As above, we get that the probability of drawing out of n balls exactly b_i balls of color i for each $i = 1, \ldots, k$ (with the restriction $b_1 + \ldots + b_k = n$ and $b_i \leq B_i$ for all i) is given by the **generalized hypergeometric distribution**

$$\mathrm{Hyp}_{B_1,\ldots,B_k;n}(\{(b_1, \ldots, b_k)\}) = \frac{\dbinom{B_1}{b_1} \cdots \dbinom{B_k}{b_k}}{\dbinom{B_1 + \ldots + B_k}{n}}. \tag{1.19}$$

(vii) Let $\mu \in \mathbb{R}$, $\sigma^2 > 0$ and let X be a real random variable with

$$\mathbf{P}[X \leq x] = \frac{1}{\sqrt{2\pi\sigma^2}} \int_{-\infty}^{x} \exp\left(-\frac{(t-\mu)^2}{2\sigma^2}\right) dt \quad \text{for } x \in \mathbb{R}.$$

Then $\mathbf{P}_X =: \mathcal{N}_{\mu,\sigma^2}$ is called the Gaussian **normal distribution** with parameters μ and σ^2. In particular, $\mathcal{N}_{0,1}$ is called the standard normal distribution.

(viii) Let $\theta > 0$ and let X be a nonnegative random variable such that

$$\mathbf{P}[X \leq x] = \mathbf{P}[X \in [0, x]] = \int_0^x \theta\, e^{-\theta t}\, dt \quad \text{for } x \geq 0.$$

Then $\mathbf{P}_X$ is called the **exponential distribution** with parameter θ (in shorthand, $\exp_\theta$).

(ix) Let $\mu \in \mathbb{R}^d$ and let Σ be a positive definite symmetric $d \times d$ matrix. Let X be an $\mathbb{R}^d$-valued random variable such that

$$\mathbf{P}[X \leq x] = \det(2\pi\,\Sigma)^{-1/2} \int_{(-\infty,x]} \exp\left(-\frac{1}{2}\langle t - \mu,\ \Sigma^{-1}(t-\mu)\rangle\right) \lambda^d(dt)$$

for $x \in \mathbb{R}^d$ (where $\langle\,\cdot\,,\,\cdot\,\rangle$ denotes the inner product in $\mathbb{R}^d$). Then $\mathbf{P}_X =: \mathcal{N}_{\mu,\Sigma}$ is the d-dimensional normal distribution with parameters μ and Σ. ◊

Definition 1.106 *If the distribution function $F : \mathbb{R}^n \to [0, 1]$ is of the form*

$$F(x) = \int_{-\infty}^{x_1} dt_1 \cdots \int_{-\infty}^{x_n} dt_n\ f(t_1, \ldots, t_n) \quad \text{for } x = (x_1, \ldots, x_n) \in \mathbb{R}^n,$$

*for some integrable function $f : \mathbb{R}^n \to [0, \infty)$, then f is called the **density** of the distribution.*

Example 1.107

(i) Let $\theta, r > 0$ and let $\Gamma_{\theta,r}$ be the distribution on $[0, \infty)$ with density

$$x \mapsto \frac{\theta^r}{\Gamma(r)}\, x^{r-1} e^{-\theta x}.$$

(Here Γ denotes the gamma function.) Then $\Gamma_{\theta,r}$ is called the **Gamma distribution** with scale parameter θ and shape parameter r.

(ii) Let $r, s > 0$ and let $\beta_{r,s}$ be the distribution on $[0, 1]$ with density

$$x \mapsto \frac{\Gamma(r+s)}{\Gamma(r)\Gamma(s)}\, x^{r-1}(1-x)^{s-1}.$$

Then $\beta_{r,s}$ is called the **Beta distribution** with parameters r and s.

(iii) Let $a > 0$ and let Cau_a be the distribution on $\mathbb{R}$ with density

$$x \mapsto \frac{1}{a\pi} \frac{1}{1 + (x/a)^2}.$$

Then Cau_a is called the **Cauchy distribution** with parameter a. $\Diamond$

Takeaways A random variable X is a measurable map from a probability space to some measurable space. The image measure $\mathbf{P}_X$ describes the distribution of X. In most cases, only the distribution of a random variable is of interest but not the underlying probability space. We have got acquainted with some fundamental probability distributions:

- Bernoulli-distribution Ber_p on $\{0, 1\}$
- binomial distribution $b_{n,p}$ on $\{0, \ldots, n\}$
- geometric distribution γ_p on $\mathbb{N}_0$
- negative binomial distribution (or Pascal distribution) $b^-_{r,p}$ on $\mathbb{N}_0$
- Poisson distribution Poi_λ on $\mathbb{N}_0$
- hypergeometric distribution $\text{Hyp}_{B,W;n}$ on $\{0, \ldots, n\}$
- normal distribution $\mathcal{N}_{\mu,\sigma^2}$ on $\mathbb{R}$
- exponential distribution $\exp_\theta$ on $[0, \infty)$
- Gamma distribution $\Gamma_{\theta,r}$ on $[0, \infty)$
- Beta distribution $\beta_{r,s}$ on $[0, 1]$
- Cauchy distribution Cau_a on $\mathbb{R}$

Exercise 1.5.1 Use the identity $\binom{-n}{k}(-1)^k = \binom{n+k-1}{k}$ to deduce (1.17) by combinatorial means from its interpretation as a waiting time. ♣

Exercise 1.5.2 Give an example of two normally distributed random variables X and Y such that (X, Y) is not (two-dimensional) normally distributed. ♣

Exercise 1.5.3 Use the transformation formula (Theorem 1.101) to show the following statements.

(i) Let $X \sim \mathcal{N}_{\mu,\sigma^2}$ and let $a \in \mathbb{R} \setminus \{0\}$ and $b \in \mathbb{R}$. Then $(aX + b) \sim \mathcal{N}_{a\mu+b,a^2\sigma^2}$.
(ii) Let $X \sim \exp_\theta$ and $a > 0$. Then $aX \sim \exp_{\theta/a}$. ♣

Exercise 1.5.4 Show that $F : \mathbb{R}^2 \to [0, 1]$ is the distribution function of a (uniquely determined) probability measure μ on $(\mathbb{R}^2, \mathcal{B}(\mathbb{R}^2))$ if and only if

(i) F is monotone increasing and right continuous
(ii) $F((-x_1, y_2)) + F((y_1, -x_2)) \to 0$ and $F(x) \to 1$ for all $y_1, y_2 \in \mathbb{R}$ and for $x = (x_1, x_2) \to \infty$,
(iii) $F((y_1, y_2)) - F((y_1, x_2)) - F((x_1, y_2)) + F((x_1, x_2)) \geq 0$ for all $x_1 \leq y_1$ and $x_2 \leq y_2$. ♣

Exercise 1.5.5

(i) Let F and G be distribution functions on $\mathbb{R}$. Use Exercise 1.5.4 to show that $(x, y) \mapsto F(x) \wedge G(y)$ is a distribution function on $\mathbb{R}^2$.

(ii) Give an example of two distribution functions F and G on $\mathbb{R}^2$ such that $(x, y) \mapsto F(x) \wedge G(y)$ is *not* a distribution function on $\mathbb{R}^4$.

  Hint: First use the inclusion-exclusion formula (Theorem 1.33) to derive a criterion similar to that in Exercise 1.5.4(iii). ♣

Chapter 2
Independence

The measure theory from the preceding chapter is a linear theory that could not describe the dependence structure of events or random variables. We enter the realm of probability theory exactly at this point, where we define independence of events and random variables. Independence is a pivotal notion of probability theory, and the computation of dependencies is one of the theory's major tasks.

In the following, $(\Omega, \mathcal{A}, \mathbf{P})$ is a probability space and the sets $A \in \mathcal{A}$ are the events. As soon as constructing probability spaces has become routine, the concrete probability space will lose its importance and it will be only the random variables that will interest us. The bold font symbol $\mathbf{P}$ will then denote the universal object of a probability measure, and the probabilities $\mathbf{P}[\,\cdot\,]$ with respect to it will always be written in square brackets.

2.1 Independence of Events

We consider two events A and B as (stochastically) independent if the occurrence of A does not change the probability that B also occurs. Formally, we say that A and B are independent if

$$\mathbf{P}[A \cap B] = \mathbf{P}[A] \cdot \mathbf{P}[B]. \tag{2.1}$$

Example 2.1 (Rolling a die twice) Consider the random experiment of rolling a die twice. Hence $\Omega = \{1, \ldots, 6\}^2$ endowed with the σ-algebra $\mathcal{A} = 2^\Omega$ and the uniform distribution $\mathbf{P} = \mathcal{U}_\Omega$ (see Example 1.30(ii)).

A. Klenke, *Probability Theory*, Universitext, https://doi.org/10.1007/978-3-030-56402-5_2

(i) Two events A and B should be independent, e.g., if A depends only on the outcome of the first roll and B depends only on the outcome of the second roll. Formally, we assume that there are sets $\tilde{A}, \tilde{B} \subset \{1, \ldots, 6\}$ such that

$$A = \tilde{A} \times \{1, \ldots, 6\} \quad \text{and} \quad B = \{1, \ldots, 6\} \times \tilde{B}.$$

Now we check that A and B indeed fulfill (2.1). To this end, we compute $P[A] = \frac{\#A}{36} = \frac{\#\tilde{A}}{6}$ and $P[B] = \frac{\#B}{36} = \frac{\#\tilde{B}}{6}$. Furthermore,

$$P[A \cap B] = \frac{\#(\tilde{A} \times \tilde{B})}{36} = \frac{\#\tilde{A}}{6} \cdot \frac{\#\tilde{B}}{6} = P[A] \cdot P[B].$$

(ii) Stochastic independence can occur also in less obvious situations. For instance, let A be the event where the sum of the two rolls is odd,

$$A = \big\{(\omega_1, \omega_2) \in \Omega : \omega_1 + \omega_2 \in \{3, 5, 7, 9, 11\}\big\},$$

and let B be the event where the first roll gives at most a three

$$B = \{(\omega_1, \omega_2) \in \Omega : \omega_1 \in \{1, 2, 3\}\}.$$

Although it might seem that these two events are entangled in some way, they are stochastically independent. Indeed, it is easy to check that $P[A] = P[B] = \frac{1}{2}$ and $P[A \cap B] = \frac{1}{4}$. ◊

What is the condition for *three* events A_1, A_2, A_3 to be independent? Of course, any of the pairs (A_1, A_2), (A_1, A_3) and (A_2, A_3) has to be independent. However, we have to make sure also that the simultaneous occurrence of A_1 *and* A_2 does not change the probability that A_3 occurs. Hence, it is not enough to consider pairs only.

Formally, we call three events A_1, A_2 and A_3 (stochastically) independent if

$$P[A_i \cap A_j] = P[A_i] \cdot P[A_j] \quad \text{for all } i, j \in \{1, 2, 3\}, i \neq j, \tag{2.2}$$

and

$$P[A_1 \cap A_2 \cap A_3] = P[A_1] \cdot P[A_2] \cdot P[A_3]. \tag{2.3}$$

Note that (2.2) does not imply (2.3) (and (2.3) does not imply (2.2)).

Example 2.2 (Rolling a die three times) We roll a die three times. Hence $\Omega = \{1, \ldots, 6\}^3$ endowed with the discrete σ-algebra $\mathcal{A} = 2^\Omega$ and the uniform distribution $P = \mathcal{U}_\Omega$ (see Example 1.30(ii)).

(i) If we assume that for any $i = 1, 2, 3$ the event A_i depends only on the outcome of the ith roll, then the events A_1, A_2 and A_3 are independent. Indeed, as in the preceding example, there are sets $\tilde{A}_1, \tilde{A}_2, \tilde{A}_3 \subset \{1, \ldots 6\}$ such that

$$A_1 = \tilde{A}_1 \times \{1, \ldots, 6\}^2,$$
$$A_2 = \{1, \ldots, 6\} \times \tilde{A}_2 \times \{1, \ldots, 6\},$$
$$A_3 = \{1, \ldots, 6\}^2 \times \tilde{A}_3.$$

The validity of (2.2) follows as in Example 2.1(i). In order to show (2.3), we compute

$$\mathbf{P}[A_1 \cap A_2 \cap A_3] = \frac{\#(\tilde{A}_1 \times \tilde{A}_2 \times \tilde{A}_3)}{216} = \prod_{i=1}^{3} \frac{\#\tilde{A}_i}{6} = \prod_{i=1}^{3} \mathbf{P}[A_i].$$

(ii) Consider now the events

$$A_1 := \{\omega \in \Omega : \omega_1 = \omega_2\},$$
$$A_2 := \{\omega \in \Omega : \omega_2 = \omega_3\},$$
$$A_3 := \{\omega \in \Omega : \omega_1 = \omega_3\}.$$

Then $\#A_1 = \#A_2 = \#A_3 = 36$; hence $\mathbf{P}[A_1] = \mathbf{P}[A_2] = \mathbf{P}[A_3] = \frac{1}{6}$. Furthermore, $\#(A_i \cap A_j) = 6$ if $i \neq j$; hence $\mathbf{P}[A_i \cap A_j] = \frac{1}{36}$. Hence (2.2) holds. On the other hand, we have $\#(A_1 \cap A_2 \cap A_3) = 6$, thus $\mathbf{P}[A_1 \cap A_2 \cap A_3] = \frac{1}{36} \neq \frac{1}{6} \cdot \frac{1}{6} \cdot \frac{1}{6}$. Thus (2.3) does not hold and so the events A_1, A_2, A_3 are not independent. $\Diamond$

In order to define independence of larger families of events, we have to request the validity of product formulas, such as (2.2) and (2.3), not only for pairs and triples but for all finite subfamilies of events. We thus make the following definition.

Definition 2.3 (Independence of events) *Let I be an arbitrary index set and let $(A_i)_{i \in I}$ be an arbitrary family of events. The family $(A_i)_{i \in I}$ is called **independent** if for any finite subset $J \subset I$ the product formula holds:*

$$\mathbf{P}\left[\bigcap_{j \in J} A_j \right] = \prod_{j \in J} \mathbf{P}[A_j].$$

Reflection How do you choose four events A_1, A_2, A_3, A_4 such that each pair A_i, A_j, $i \neq j$, and each triple A_i, A_j, A_k, $\#\{i, j, k\} = 3$, is independent, but A_1, A_2, A_3, A_4 is not? ♠

Reflection An event is independent of itself if A and B are independent for $B = A$. Check that in this case A has either probability 0 or 1. ♠

The most prominent example of an independent family of infinitely many events is given by the perpetuated independent repetition of a random experiment.

Example 2.4 Let E be a finite set (the set of possible outcomes of the individual experiment) and let $(p_e)_{e \in E}$ be a probability vector on E. Equip (as in Theorem 1.64) the probability space $\Omega = E^{\mathbb{N}}$ with the σ-algebra $\mathcal{A} = \sigma(\{[\omega_1, \ldots, \omega_n] : \omega_1, \ldots, \omega_n \in E, n \in \mathbb{N}\})$ and with the product measure (or Bernoulli measure)

$$\mathbf{P} = \left(\textstyle\sum_{e \in E} p_e \delta_e\right)^{\otimes \mathbb{N}}; \text{ that is where } \mathbf{P}\big[[\omega_1, \ldots, \omega_n]\big] = \prod_{i=1}^{n} p_{\omega_i}. \text{ Let } \tilde{A}_i \subset E \text{ for any}$$

$i \in \mathbb{N}$, and let A_i be the event where $\tilde{A}_i$ occurs in the ith experiment; that is,

$$A_i = \{\omega \in \Omega : \omega_i \in \tilde{A}_i\} \quad = \quad \biguplus_{(\omega_1, \ldots, \omega_i) \in E^{i-1} \times \tilde{A}_i} [\omega_1, \ldots, \omega_i].$$

Intuitively, the family $(A_i)_{i \in \mathbb{N}}$ should be independent if the definition of independence makes any sense at all. We check that this is indeed the case. Let $J \subset \mathbb{N}$ be finite and $n := \max J$. Formally, we define $B_j = A_j$ and $\tilde{B}_j = \tilde{A}_j$ for $j \in J$ and $B_j = \Omega$ and $\tilde{B}_j = E$ for $j \in \{1, \ldots, n\} \setminus J$. Then

$$\mathbf{P}\left[\bigcap_{j \in J} A_j\right] = \mathbf{P}\left[\bigcap_{j \in J} B_j\right] = \mathbf{P}\left[\bigcap_{j=1}^{n} B_j\right]$$

$$= \sum_{e_1 \in \tilde{B}_1} \cdots \sum_{e_n \in \tilde{B}_n} \prod_{j=1}^{n} p_{e_j} = \prod_{j=1}^{n}\left(\sum_{e \in \tilde{B}_j} p_e\right) = \prod_{j \in J}\left(\sum_{e \in \tilde{A}_j} p_e\right).$$

This is true in particular for $\#J = 1$. Hence $\mathbf{P}[A_i] = \sum_{e \in \tilde{A}_i} p_e$ for all $i \in \mathbb{N}$, whence

$$\mathbf{P}\left[\bigcap_{j \in J} A_j\right] = \prod_{j \in J} \mathbf{P}[A_j]. \tag{2.4}$$

Since this holds for all finite $J \subset \mathbb{N}$, the family $(A_i)_{i \in \mathbb{N}}$ is independent. $\Diamond$

If A and B are independent, then A^c and B also are independent since $\mathbf{P}[A^c \cap B] = \mathbf{P}[B] - \mathbf{P}[A \cap B] = \mathbf{P}[B] - \mathbf{P}[A]\mathbf{P}[B] = (1 - \mathbf{P}[A])\mathbf{P}[B] = \mathbf{P}[A^c]\mathbf{P}[B]$. We generalize this observation in the following theorem.

Theorem 2.5 *Let I be an arbitrary index set and let $(A_i)_{i \in I}$ be a family of events. Define $B_i^0 = A_i$ and $B_i^1 = A_i^c$ for $i \in I$. Then the following three statements are equivalent.*

  (i) *The family $(A_i)_{i \in I}$ is independent.*
  (ii) *There is an $\alpha \in \{0, 1\}^I$ such that the family $(B_i^{\alpha_i})_{i \in I}$ is independent.*
  (iii) *For any $\alpha \in \{0, 1\}^I$, the family $(B_i^{\alpha_i})_{i \in I}$ is independent.*

Proof This is left as an exercise. □

Example 2.6 (Euler's prime number formula) The **Riemann zeta function** is defined by the Dirichlet series

$$\zeta(s) := \sum_{n=1}^{\infty} n^{-s} \quad \text{for } s \in (1, \infty).$$

Euler's prime number formula is a representation of the Riemann zeta function as an infinite product

$$\zeta(s) = \prod_{p \in \mathcal{P}} \left(1 - p^{-s}\right)^{-1}, \tag{2.5}$$

where $\mathcal{P} := \{p \in \mathbb{N} : p \text{ is prime}\}$.

We give a probabilistic proof for this formula. Let $\Omega = \mathbb{N}$, and for fixed $s > 1$ define $\mathbf{P}$ on 2^{Ω} by

$$\mathbf{P}[\{n\}] = \zeta(s)^{-1} n^{-s} \quad \text{for } n \in \mathbb{N}.$$

Let $p\mathbb{N} = \{pn : n \in \mathbb{N}\}$ and $\mathcal{P}_n = \{p \in \mathcal{P} : p \leq n\}$. We consider $p\mathbb{N} \subset \Omega$ as an event. Note that $\mathbf{P}[p\mathbb{N}] = p^{-s}$ and that $(p\mathbb{N}, \ p \in \mathcal{P})$ is independent. Indeed, for $k \in \mathbb{N}$ and mutually distinct $p_1, \ldots, p_k \in \mathcal{P}$, we have $\bigcap_{i=1}^{k}(p_i\mathbb{N}) = (p_1 \cdots p_k)\mathbb{N}$. Thus

$$\mathbf{P}\left[\bigcap_{i=1}^{k}(p_i\mathbb{N})\right] = \sum_{n=1}^{\infty} \mathbf{P}[\{p_1 \cdots p_k n\}]$$

$$= \zeta(s)^{-1} (p_1 \cdots p_k)^{-s} \sum_{n=1}^{\infty} n^{-s}$$

$$= (p_1 \cdots p_k)^{-s} = \prod_{i=1}^{k} \mathbf{P}[\, p_i\mathbb{N}\,].$$

By Theorem 2.5, the family $((p\mathbb{N})^c, \ p \in \mathcal{P})$ is also independent, whence

$$\zeta(s)^{-1} = \mathbf{P}[\{1\}] = \mathbf{P}\left[\bigcap_{p \in \mathcal{P}} (p\mathbb{N})^c\right]$$

$$= \lim_{n \to \infty} \mathbf{P}\left[\bigcap_{p \in \mathcal{P}_n} (p\mathbb{N})^c\right]$$

$$= \lim_{n \to \infty} \prod_{p \in \mathcal{P}_n} \left(1 - \mathbf{P}[\, p\mathbb{N}\,]\right) = \prod_{p \in \mathcal{P}} \left(1 - p^{-s}\right).$$

This shows (2.5). ◊

If we roll a die infinitely often, what is the chance that the face shows a six infinitely often? This probability should equal one. Otherwise there would be a last point in time when we see a six and after which the face only shows a number one to five. However, this is not very plausible.

Recall that we formalized the event where infinitely many of a series of events occur by means of the limes superior (see Definition 1.13). The following theorem confirms the conjecture mentioned above and also gives conditions under which we *cannot* expect that infinitely many of the events occur.

Theorem 2.7 (Borel–Cantelli lemma) *Let $A_1, A_2, \ldots$ be events and define $A^* = \limsup\limits_{n \to \infty} A_n$.*

(i) *If $\sum_{n=1}^{\infty} \mathbf{P}[A_n] < \infty$, then $\mathbf{P}[A^*] = 0$. (Here $\mathbf{P}$ could be an arbitrary measure on $(\Omega, \mathcal{A})$.)*

(ii) *If $(A_n)_{n \in \mathbb{N}}$ is independent and $\sum_{n=1}^{\infty} \mathbf{P}[A_n] = \infty$, then $\mathbf{P}[A^*] = 1$.*

Proof

(i) $\mathbf{P}$ is upper semicontinuous and σ-subadditive; hence, by assumption,

$$\mathbf{P}[A^*] = \lim_{n \to \infty} \mathbf{P}\left[\bigcup_{m=n}^{\infty} A_m\right] \leq \lim_{n \to \infty} \sum_{m=n}^{\infty} \mathbf{P}[A_m] = 0.$$

(ii) De Morgan's rule and the lower semicontinuity of $\mathbf{P}$ yield

$$\mathbf{P}\left[(A^*)^c\right] = \mathbf{P}\left[\bigcup_{m=1}^{\infty} \bigcap_{n=m}^{\infty} A_n^c\right] = \lim_{m \to \infty} \mathbf{P}\left[\bigcap_{n=m}^{\infty} A_n^c\right].$$

However, for every $m \in \mathbb{N}$ (since $\log(1 - x) \leq -x$ for $x \in [0, 1]$), by upper continuity of $\mathbf{P}$

$$\mathbf{P}\left[\bigcap_{n=m}^{\infty} A_n^c\right] = \lim_{N \to \infty} \mathbf{P}\left[\bigcap_{n=m}^{N} A_n^c\right] = \prod_{n=m}^{\infty} (1 - \mathbf{P}[A_n])$$

$$= \exp\left(\sum_{n=m}^{\infty} \log(1 - \mathbf{P}[A_n])\right) \leq \exp\left(-\sum_{n=m}^{\infty} \mathbf{P}[A_n]\right) = 0. \qquad \square$$

Example 2.8 We throw a die again and again and ask for the probability of seeing a six infinitely often. Hence $\Omega = \{1, \ldots, 6\}^{\mathbb{N}}$, $\mathcal{A} = (2^{\{1,\ldots,6\}})^{\otimes \mathbb{N}}$ is the product σ-algebra and $\mathbf{P} = \left(\sum_{e \in \{1,\ldots,6\}} \frac{1}{6}\delta_e\right)^{\otimes \mathbb{N}}$ is the Bernoulli measure (see Theorem 1.64). Furthermore, let $A_n = \{\omega \in \Omega : \omega_n = 6\}$ be the event where the nth roll shows a six. Then $A^* = \limsup\limits_{n \to \infty} A_n$ is the event where we see a six infinitely often (see

Remark 1.14). Furthermore, $(A_n)_{n \in \mathbb{N}}$ is an independent family with the property $\sum_{n=1}^{\infty} P[A_n] = \sum_{n=1}^{\infty} \frac{1}{6} = \infty$. Hence the Borel–Cantelli lemma yields $P[A^*] = 1$. ◊

Example 2.9 We roll a die only once and define A_n for any $n \in \mathbb{N}$ as the event where in this one roll the face showed a six. Note that $A_1 = A_2 = A_3 = \ldots$. Then $\sum_{n \in \mathbb{N}} P[A_n] = \infty$; however, $P[A^*] = P[A_1] = \frac{1}{6}$. This shows that in Part (ii) of the Borel–Cantelli lemma, the assumption of independence is indispensable. ◊

Example 2.10 Let $\Lambda \in (0, \infty)$ and $0 \leq \lambda_n \leq \Lambda$ for $n \in \mathbb{N}$. Let X_n, $n \in \mathbb{N}$, be Poisson random variables with parameters λ_n. Then

$$P\big[X_n \geq n \text{ for infinitely many } n\big] = 0.$$

Indeed,

$$\sum_{n=1}^{\infty} P[X_n \geq n] = \sum_{n=1}^{\infty} \sum_{m=n}^{\infty} P[X_n = m] = \sum_{m=1}^{\infty} \sum_{n=1}^{m} P[X_n = m]$$

$$= \sum_{m=1}^{\infty} \sum_{n=1}^{m} e^{-\lambda_n} \frac{\lambda_n^m}{m!} \leq \sum_{m=1}^{\infty} m \frac{\Lambda^m}{m!} = \Lambda e^{\Lambda} < \infty. \quad ◊$$

(2.6)

Note that in Theorem 2.7 in the case of independent events, only the probabilities $P[A^*] = 0$ and $P[A^*] = 1$ could show up. Thus the Borel–Cantelli lemma belongs to the class of so-called 0–1 laws. Later we will encounter more 0–1 laws (see, for example, Theorem 2.37).

Now we extend the notion of independence from families of events to families of classes of events.

Definition 2.11 (Independence of classes of events) *Let I be an arbitrary index set and let $\mathcal{E}_i \subset \mathcal{A}$ for all $i \in I$. The family $(\mathcal{E}_i)_{i \in I}$ is called **independent** if, for any finite subset $J \subset I$ and any choice of $E_j \in \mathcal{E}_j$, $j \in J$, we have*

$$P\bigg[\bigcap_{j \in J} E_j\bigg] = \prod_{j \in J} P[E_j]. \tag{2.7}$$

Example 2.12 As in Example 2.4, let $(\Omega, \mathcal{A}, P)$ be the product space of infinitely many repetitions of a random experiment whose possible outcomes e are the elements of the finite set E and have probabilities $p = (p_e)_{e \in E}$. For $i \in \mathbb{N}$, define

$$\mathcal{E}_i = \big\{\{\omega \in \Omega : \omega_i \in A\} : A \subset E\big\}.$$

For any choice of sets $A_i \in \mathcal{E}_i$, $i \in \mathbb{N}$, the family $(A_i)_{i \in \mathbb{N}}$ is independent; hence $(\mathcal{E}_i)_{i \in \mathbb{N}}$ is independent. ◊

Theorem 2.13

(i) *Let I be finite, and for any $i \in I$ let $\mathcal{E}_i \subset \mathcal{A}$ with $\Omega \in \mathcal{E}_i$. Then*

$$(\mathcal{E}_i)_{i \in I} \text{ is independent} \iff (2.7) \text{ holds for } J = I.$$

(ii) *$(\mathcal{E}_i)_{i \in I}$ is independent $\iff$ $((\mathcal{E}_j)_{j \in J}$ is independent for all finite $J \subset I)$.*

(iii) *If $(\mathcal{E}_i \cup \{\emptyset\})$ is $\cap$-stable, then*

$$(\mathcal{E}_i)_{i \in I} \text{ is independent} \iff (\sigma(\mathcal{E}_i))_{i \in I} \text{ is independent.}$$

(iv) *Let K be an arbitrary set and let $(I_k)_{k \in K}$ be mutually disjoint subsets of I. If $(\mathcal{E}_i)_{i \in I}$ is independent, then $\left(\bigcup_{i \in I_k} \mathcal{E}_i \right)_{k \in K}$ is also independent.*

Proof

(i) " $\Longrightarrow$ "   This is trivial.

(i) " $\Longleftarrow$ "   For $J \subset I$ and $j \in I \setminus J$, choose $E_j = \Omega$.

(ii)   This is trivial.

(iii) " $\Longleftarrow$ "   This is trivial.

(iii) " $\Longrightarrow$ "   Let $J \subset I$ be finite. We will show that for any two finite sets J and J' with $J \subset J' \subset I$,

$$\mathbf{P}\left[\bigcap_{i \in J'} E_i \right] = \prod_{i \in J'} \mathbf{P}[E_i] \text{ for any choice } \begin{cases} E_i \in \sigma(\mathcal{E}_i), & \text{if } i \in J, \\ E_i \in \mathcal{E}_i, & \text{if } i \in J' \setminus J. \end{cases} \quad (2.8)$$

The case $J' = J$ is exactly the claim we have to show.

We carry out the proof of (2.8) by induction on $\#J$. For $\#J = 0$, the statement (2.8) holds by assumption of this theorem.

Now assume that (2.8) holds for every J with $\#J = n$ and for every finite $J' \supset J$. Fix such a J and let $j \in I \setminus J$. Choose $J' \supset \tilde{J} := J \cup \{j\}$. We show the validity of (2.8) with J replaced by $\tilde{J}$. Since $\#\tilde{J} = n + 1$, this verifies the induction step.

Let $E_i \in \sigma(\mathcal{E}_i)$ for any $i \in J$, and let $E_i \in \mathcal{E}_i$ for any $i \in J' \setminus (J \cup \{j\})$. Define two measures μ and ν on $(\Omega, \mathcal{A})$ by

$$\mu : E_j \mapsto \mathbf{P}\left[\bigcap_{i \in J'} E_i \right] \quad \text{and} \quad \nu : E_j \mapsto \prod_{i \in J'} \mathbf{P}[E_i].$$

By the induction hypothesis (2.8), we have $\mu(E_j) = \nu(E_j)$ for every $E_j \in \mathcal{E}_j \cup \{\emptyset, \Omega\}$. Since $\mathcal{E}_j \cup \{\emptyset\}$ is a π-system, Lemma 1.42 yields that $\mu(E_j) = \nu(E_j)$ for all $E_j \in \sigma(\mathcal{E}_j)$. That is, (2.8) holds with J replaced by $J \cup \{j\}$.

(iv) This is trivial, as (2.7) has to be checked only for $J \subset I$ with

$$\#(J \cap I_k) \leq 1 \quad \text{for any } k \in K. \qquad \Box$$

> **Takeaways** Two events A and B are independent if $\mathbf{P}[A \cap B] = \mathbf{P}[A] \cdot \mathbf{P}[B]$. We have extended this notion to families of events and even to families of classes of sets. The Borel-Cantelli lemma shows that infinitely many of countably many independent events occur jointly with probability either 0 or 1 depending on the summability of the probabilities of the single events.

Exercise 2.1.1 In a queue each new arriving person chooses independently a random waiting position. What is the probability that the first position changes infinitely often? ♣

Exercise 2.1.2 Show that the conclusion of the interesting part of the Borel-Cantelli lemma (Theorem 2.7(ii)) still holds under the following weaker condition: Each of the families $(A_1, A_3, A_5, \ldots)$ and $(A_2, A_4, A_6, \ldots)$ is independent (but not necessarily independent of each other). ♣

2.2 Independent Random Variables

Now that we have studied independence of events, we want to study independence of random variables. Here also the definition ends up with a product formula. Formally, however, we can also define independence of random variables via independence of the σ-algebras they generate. This is the reason why we studied independence of classes of events in the last section.

Independent random variables allow for a rich calculus. For example, we can compute the distribution of a sum of two independent random variables by a simple convolution formula. Since we do not have a general notion of an integral at hand at this point, for the time being we restrict ourselves to presenting the convolution formula for integer-valued random variables only.

Let I be an arbitrary index set. For each $i \in I$, let $(\Omega_i, \mathcal{A}_i)$ be a measurable space and let $X_i : (\Omega, \mathcal{A}) \to (\Omega_i, \mathcal{A}_i)$ be a random variable with generated σ-algebra $\sigma(X_i) = X_i^{-1}(\mathcal{A}_i)$.

Definition 2.14 (Independent random variables) *The family* $(X_i)_{i \in I}$ *of random variables is called* ***independent*** *if the family* $(\sigma(X_i))_{i \in I}$ *of σ-algebras is independent.*

As a shorthand, we say that a family $(X_i)_{i \in I}$ is "i.i.d." (for "independent and identically distributed") if $(X_i)_{i \in I}$ is independent and if $\mathbf{P}_{X_i} = \mathbf{P}_{X_j}$ for all $i, j \in I$.

Remark 2.15

(i) Clearly, the family $(X_i)_{i \in I}$ is independent if and only if, for any finite set $J \subset I$ and any choice of $A_j \in \mathcal{A}_j$, $j \in J$, we have

$$\mathbf{P}\Big[\bigcap_{j \in J} \{X_j \in A_j\}\Big] = \prod_{j \in J} \mathbf{P}[X_j \in A_j].$$

The next theorem will show that it is enough to request the validity of such a product formula for A_j from an $\cap$-stable generator of $\mathcal{A}_j$ only.

(ii) If $(\tilde{\mathcal{A}}_i)_{i \in I}$ is an independent family of σ-algebras and if each X_i is $\tilde{\mathcal{A}}_i - \mathcal{A}_i$-measurable, then $(X_i)_{i \in I}$ is independent. This is a direct consequence of the fact that $\sigma(X_i) \subset \tilde{\mathcal{A}}_i$.

(iii) For each $i \in I$, let $(\Omega'_i, \mathcal{A}'_i)$ be another measurable space and assume that $f_i : (\Omega_i, \mathcal{A}_i) \rightarrow (\Omega'_i, \mathcal{A}'_i)$ is a measurable map. If $(X_i)_{i \in I}$ is independent, then $(f_i \circ X_i)_{i \in I}$ is independent. This statement is a special case of (ii) since $f_i \circ X_i$ is $\sigma(X_i) - \mathcal{A}'_i$-measurable (see Theorem 1.80).$\Diamond$

Theorem 2.16 (Independent generators) *For any $i \in I$, let $\mathcal{E}_i \subset \mathcal{A}_i$ be a π-system that generates $\mathcal{A}_i$. If $(X_i^{-1}(\mathcal{E}_i))_{i \in I}$ is independent, then $(X_i)_{i \in I}$ is independent.*

Proof By Theorem 1.81, $X_i^{-1}(\mathcal{E}_i)$ is a π-system that generates the σ-algebra $X_i^{-1}(\mathcal{A}_i) = \sigma(X_i)$. Hence the statement follows from Theorem 2.13. $\square$

Example 2.17 Let E be a countable set and let $(X_i)_{i \in I}$ be random variables with values in $(E, 2^E)$. In this case, $(X_i)_{i \in I}$ is independent if and only if, for any finite $J \subset I$ and any choice of $x_j \in E$, $j \in J$,

$$\mathbf{P}\big[X_j = x_j \text{ for all } j \in J\big] = \prod_{j \in J} \mathbf{P}[X_j = x_j].$$

This is obvious since $\big\{\{x\} : x \in E\big\} \cup \{\emptyset\}$ is a π-system that generates 2^E, thus $\big\{X_i^{-1}(\{x_i\}) : x_i \in E\big\} \cup \{\emptyset\}$ is a π-system that generates $\sigma(X_i)$ (Theorem 1.81). $\Diamond$

Example 2.18 Let E be a finite set and let $p = (p_e)_{e \in E}$ be a probability vector. Repeat a random experiment with possible outcomes $e \in E$ and probabilities p_e for $e \in E$ infinitely often (see Example 1.40 and Theorem 1.64). Let $\Omega = E^{\mathbb{N}}$ be the infinite product space and let $\mathcal{A}$ be the σ-algebra generated by the cylinder sets (see (1.8)). Let $\mathbf{P} = \big(\sum_{e \in E} p_e \delta_e\big)^{\otimes \mathbb{N}}$ be the Bernoulli measure. Further, for any $n \in \mathbb{N}$, let

$$X_n : \Omega \rightarrow E, \qquad (\omega_m)_{m \in \mathbb{N}} \mapsto \omega_n,$$

be the projection on the nth coordinate. In other words: For any simple event $\omega \in \Omega$, $X_n(\omega)$ yields the result of the nth experiment. Then, by (2.4) (in Example 2.4), for

$n \in \mathbb{N}$ and $x \in E^n$, we have

$$\mathbf{P}\big[X_j = x_j \text{ for all } j = 1, \ldots, n\big] = \mathbf{P}\big[[x_1, \ldots, x_n]\big] = \mathbf{P}\left[\bigcap_{j=1}^{n} X_j^{-1}(\{x_j\})\right]$$

$$= \prod_{j=1}^{n} \mathbf{P}\big[X_j^{-1}(\{x_j\})\big] = \prod_{j=1}^{n} \mathbf{P}[X_j = x_j],$$

and $\mathbf{P}[X_j = x_j] = p_{x_j}$. By virtue of Theorem 2.13(i), this implies that the family $(X_1, \ldots, X_n)$ is independent and hence, by Theorem 2.13(ii), $(X_n)_{n \in \mathbb{N}}$ is independent as well. ◇

In particular, we have shown the following theorem.

Theorem 2.19 *Let E be a finite set and let $(p_e)_{e \in E}$ be a probability vector on E. Then there exists a probability space $(\Omega, \mathcal{A}, \mathbf{P})$ and an independent family $(X_n)_{n \in \mathbb{N}}$ of E-valued random variables on $(\Omega, \mathcal{A}, \mathbf{P})$ such that $\mathbf{P}[X_n = e] = p_e$ for any $e \in E$.*

Later we will see that the assumption that E is finite can be dropped. Also one can allow for different distributions in the respective factors. For the time being, however, this theorem gives us enough examples of interesting families of independent random variables.

Our next goal is to deduce simple criteria in terms of distribution functions and densities for checking whether a family of random variables is independent or not.

Definition 2.20 *For any $i \in I$, let X_i be a real random variable. For any finite subset $J \subset I$, let*

$$F_J := F_{(X_j)_{j \in J}} : \mathbb{R}^J \to [0, 1],$$

$$x \mapsto \mathbf{P}\big[X_j \leq x_j \text{ for all } j \in J\big] = \mathbf{P}\left[\bigcap_{j \in J} X_j^{-1}((-\infty, x_j])\right].$$

*Then F_J is called the **joint distribution function** of $(X_j)_{j \in J}$. The probability measure $\mathbf{P}_{(X_j)_{j \in J}}$ on $\mathbb{R}^J$ is called the **joint distribution** of $(X_j)_{j \in J}$.*

Theorem 2.21 *A family $(X_i)_{i \in I}$ of real random variables is independent if and only if, for every finite $J \subset I$ and every $x = (x_j)_{j \in J} \in \mathbb{R}^J$,*

$$F_J(x) = \prod_{j \in J} F_{\{j\}}(x_j). \tag{2.9}$$

Proof The class of sets $\{(-\infty, b], \ b \in \mathbb{R}\}$ is an $\cap$-stable generator of the Borel σ-algebra $\mathcal{B}(\mathbb{R})$ (see Theorem 1.23). Equation (2.9) says that, for any choice of real numbers $(x_i)_{i \in I}$, the events $(X_i^{-1}((-\infty, x_i]))_{i \in I}$ are independent. Hence Theorem 2.16 yields the claim.   □

Corollary 2.22 *In addition to the assumptions of Theorem 2.21, we assume that any F_J has a continuous **density** $f_J = f_{(X_j)_{j \in J}}$ (the **joint density** of $(X_j)_{j \in J}$). That is, there exists a continuous map $f_J : \mathbb{R}^J \to [0, \infty)$ such that*

$$F_J(x) = \int_{-\infty}^{x_{j_1}} dt_1 \cdots \int_{-\infty}^{x_{j_n}} dt_n \, f_J(t_1, \ldots, t_n) \quad \text{for all } x \in \mathbb{R}^J$$

(where $J = \{j_1, \ldots, j_n\}$). In this case, the family $(X_i)_{i \in I}$ is independent if and only if, for any finite $J \subset I$

$$f_J(x) = \prod_{j \in J} f_j(x_j) \quad \text{for all } x \in \mathbb{R}^J. \tag{2.10}$$

Corollary 2.23 *Let $n \in \mathbb{N}$ and let $\mu_1, \ldots, \mu_n$ be probability measures on $(\mathbb{R}, \mathcal{B}(\mathbb{R}))$. Then there exists a probability space $(\Omega, \mathcal{A}, \mathbf{P})$ and an independent family of random variables $(X_i)_{i=1,\ldots,n}$ on $(\Omega, \mathcal{A}, \mathbf{P})$ with $\mathbf{P}_{X_i} = \mu_i$ for each $i = 1, \ldots, n$.*

Proof Let $\Omega = \mathbb{R}^n$ and $\mathcal{A} = \mathcal{B}(\mathbb{R}^n)$. Let $\mathbf{P} = \bigotimes_{i=1}^n \mu_i$ be the product measure of the μ_i (see Theorem 1.61). Further, let $X_i : \mathbb{R}^n \to \mathbb{R}$, $(x_1, \ldots, x_n) \mapsto x_i$ be the projection on the ith coordinate for each $i = 1, \ldots, n$. Then, for any $i = 1, \ldots, n$,

$$F_{\{i\}}(x) = \mathbf{P}[X_i \leq x] = \mathbf{P}[\mathbb{R}^{i-1} \times (-\infty, x] \times \mathbb{R}^{n-i}]$$
$$= \mu_i((-\infty, x]) \cdot \prod_{j \neq i} \mu_j(\mathbb{R}) = \mu_i((-\infty, x]).$$

Hence indeed $\mathbf{P}_{X_i} = \mu_i$. Furthermore, for all $x_1, \ldots, x_n \in \mathbb{R}$,

$$F_{\{1,\ldots,n\}}((x_1, \ldots, x_n)) = \mathbf{P}\left[\underset{i=1}{\overset{n}{\times}} (-\infty, x_i]\right] = \prod_{i=1}^n \mu_i((-\infty, x_i]) = \prod_{i=1}^n F_{\{i\}}(x_i).$$

Hence Theorem 2.21 (and Theorem 2.13(i)) yields the independence of $(X_i)_{i=1,\ldots,n}$.   □

Example 2.24 Let $X_1, \ldots, X_n$ be independent exponentially distributed random variables with parameters $\theta_1, \ldots, \theta_n \in (0, \infty)$. Then

$$F_{\{i\}}(x) = \int_0^x \theta_i e^{-\theta_i t} \, dt = 1 - e^{-\theta_i x} \quad \text{for } x \geq 0$$

and hence

$$F_{\{1,\dots,n\}}\big((x_1,\dots,x_n)\big) = \prod_{i=1}^{n}\big(1 - e^{-\theta_i x_i}\big).$$

Consider now the random variable $Y = \max(X_1,\dots,X_n)$. Then

$$\begin{aligned}F_Y(x) &= \mathbf{P}\big[X_i \le x \ \text{ for all } \ i = 1,\dots,n\big]\\[2mm]&= F_{\{1,\dots,n\}}\big((x,\dots,x)\big) = \prod_{i=1}^{n}\big(1 - e^{-\theta_i x}\big).\end{aligned}$$

The distribution function of the random variable $Z := \min(X_1,\dots,X_n)$ has a nice closed form:

$$\begin{aligned}F_Z(x) &= 1 - \mathbf{P}[Z > x]\\[2mm]&= 1 - \mathbf{P}\big[X_i > x \ \text{ for all } \ i = 1,\dots,n\big]\\[2mm]&= 1 - \prod_{i=1}^{n} e^{-\theta_i x} \;=\; 1 - \exp\big(-(\theta_1 + \dots + \theta_n)\,x\big).\end{aligned}$$

In other words, Z is exponentially distributed with parameter $\theta_1 + \dots + \theta_n$. $\Diamond$

Example 2.25 Let $\mu_i \in \mathbb{R}$ and $\sigma_i^2 > 0$ for $i \in I$. Let $(X_i)_{i\in I}$ be real random variables with joint density functions (for finite $J \subset I$)

$$f_J(x) = \prod_{j\in J}\big(2\pi\sigma_j^2\big)^{-\frac{1}{2}} \exp\Big(-\sum_{j\in J}\frac{(x_j - \mu_j)^2}{2\sigma_j^2}\Big) \qquad \text{for } x \in \mathbb{R}^J.$$

Then $(X_i)_{i\in I}$ is independent and X_i is normally distributed with parameters (μ_i, σ_i^2).

For any finite $I = \{i_1,\dots,i_n\}$ (with mutually distinct $i_1,\dots,i_n$), the vector $Y = (X_{i_1},\dots,X_{i_n})$ has the n-dimensional normal distribution with $\mu = \mu^I := (\mu_{i_1},\dots,\mu_{i_n})$ and with $\Sigma = \Sigma^I$ the diagonal matrix with entries $\sigma_{i_1}^2,\dots,\sigma_{i_n}^2$ (see Example 1.105(ix)). $\Diamond$

Theorem 2.26 *Let K be an arbitrary set and I_k, $k \in K$, arbitrary mutually disjoint index sets. Define $I = \bigcup_{k\in K} I_k$.*

If the family $(X_i)_{i\in I}$ is independent, then the family of σ-algebras $(\sigma(X_j,\, j \in I_k))_{k\in K}$ is independent.

Proof For $k \in K$, let

$$Z_k = \left\{ \bigcap_{j \in I_k} A_j : A_j \in \sigma(X_j), \ \#\{j \in I_k : A_j \neq \Omega\} < \infty \right\}$$

be the semiring of finite-dimensional rectangular cylinder sets. Clearly, Z_k is a π-system and $\sigma(Z_k) = \sigma(X_j, \ j \in I_k)$. Hence, by Theorem 2.13(iii), it is enough to show that $(Z_k)_{k \in K}$ is independent. By Theorem 2.13(ii), we can even assume that K is finite.

For $k \in K$, let $B_k \in Z_k$ and $J_k \subset I_k$ be finite with $B_k = \bigcap_{j \in J_k} A_j$ for certain $A_j \in \sigma(X_j)$. Define $J = \bigcup_{k \in K} J_k$. Then

$$\mathbf{P}\left[\bigcap_{k \in K} B_k \right] = \mathbf{P}\left[\bigcap_{j \in J} A_j \right] = \prod_{j \in J} \mathbf{P}[A_j] = \prod_{k \in K} \prod_{j \in J_k} \mathbf{P}[A_j] = \prod_{k \in K} \mathbf{P}[B_k].$$

$\square$

Example 2.27 If $(X_n)_{n \in \mathbb{N}}$ is an independent family of real random variables, then also $(Y_n)_{n \in \mathbb{N}} = (X_{2n} - X_{2n-1})_{n \in \mathbb{N}}$ is independent. Indeed, for any $n \in \mathbb{N}$, the random variable Y_n is $\sigma(X_{2n}, X_{2n-1})$-measurable by Theorem 1.91, and $(\sigma(X_{2n}, X_{2n-1}))_{n \in \mathbb{N}}$ is independent by Theorem 2.26. $\lozenge$

Example 2.28 Let $(X_{m,n})_{(m,n) \in \mathbb{N}^2}$ be an independent family of Bernoulli random variables with parameter $p \in (0, 1)$. Define the waiting time for the first "success" in the mth row of the matrix $(X_{m,n})_{m,n}$ by

$$Y_m := \inf \{n \in \mathbb{N} : X_{m,n} = 1\} - 1.$$

Then $(Y_m)_{m \in \mathbb{N}}$ are independent geometric random variables with parameter p (see Example 1.105(iii)). Indeed,

$$\{Y_m \leq k\} = \bigcup_{l=1}^{k+1} \{X_{m,l} = 1\} \in \sigma(X_{m,l}, \ l = 1, \ldots, k+1) \subset \sigma(X_{m,l}, \ l \in \mathbb{N}).$$

Hence Y_m is $\sigma(X_{m,l}, l \in \mathbb{N})$-measurable and thus $(Y_m)_{m \in \mathbb{N}}$ is independent. Furthermore,

$$\mathbf{P}[Y_m > k] = \mathbf{P}[X_{m,l} = 0, \ l = 1, \ldots, k+1] = \prod_{l=1}^{k+1} \mathbf{P}[X_{m,l} = 0] = (1 - p)^{k+1}.$$

Concluding, we get $\mathbf{P}[Y_m = k] = \mathbf{P}[Y_m > k - 1] - \mathbf{P}[Y_m > k] = p(1 - p)^k$. $\lozenge$

Definition 2.29 (Convolution) *Let μ and v be probability measures on $(\mathbb{Z}, 2^{\mathbb{Z}})$. The **convolution** $\mu * v$ is defined as the probability measure on $(\mathbb{Z}, 2^{\mathbb{Z}})$ such that*

$$(\mu * v)(\{n\}) = \sum_{m=-\infty}^{\infty} \mu(\{m\})\, v(\{n - m\}).$$

We define the nth convolution power recursively by $\mu^{*1} = \mu$ and

$$\mu^{*(n+1)} = \mu^{*n} * \mu.$$

Remark 2.30 The convolution is a symmetric operation: $\mu * v = v * \mu$. $\Diamond$

Theorem 2.31 *If X and Y are independent $\mathbb{Z}$-valued random variables, then $\mathbf{P}_{X+Y} = \mathbf{P}_X * \mathbf{P}_Y$.*

Proof For any $n \in \mathbb{Z}$,

$$\mathbf{P}_{X+Y}(\{n\}) = \mathbf{P}[X + Y = n]$$

$$= \mathbf{P}\left[\biguplus_{m \in \mathbb{Z}} \left(\{X = m\} \cap \{Y = n - m\} \right) \right]$$

$$= \sum_{m \in \mathbb{Z}} \mathbf{P}\left[\{X = m\} \cap \{Y = n - m\} \right]$$

$$= \sum_{m \in \mathbb{Z}} \mathbf{P}_X[\{m\}]\, \mathbf{P}_Y[\{n - m\}] \; = \; (\mathbf{P}_X * \mathbf{P}_Y)[\{n\}].$$

$\square$

Reflection Check that $\mathbf{P}_{X+Y} = \mathbf{P}_X * \mathbf{P}_Y$ does not imply that X and Y be independent. ♠♠

Owing to the last theorem, it is natural to define the convolution of two probability measures on $\mathbb{R}^n$ (or more generally on an Abelian group) as the distribution of the sum of two independent random variables with the corresponding distributions. Later we will encounter a different (but equivalent) definition that will, however, rely on the notion of an integral that is not yet available to us at this point (see Definition 14.20).

Definition 2.32 (Convolution of measures) *Let μ and v be probability measures on $\mathbb{R}^n$ and let X and Y be independent random variables with $\mathbf{P}_X = \mu$ and $\mathbf{P}_Y = v$. We define the **convolution** of μ and v as $\mu * v = \mathbf{P}_{X+Y}$.*
Recursively, we define the convolution powers μ^{*k} for all $k \in \mathbb{N}$ and let $\mu^{*0} = \delta_0$.

Example 2.33 Let X and Y be independent Poisson random variables with parameters μ and $\lambda \geq 0$. Then

$$\mathbf{P}[X + Y = n] = e^{-\mu} e^{-\lambda} \sum_{m=0}^{n} \frac{\mu^m}{m!} \frac{\lambda^{n-m}}{(n - m)!}$$

$$= e^{-(\mu+\lambda)} \frac{1}{n!} \sum_{m=0}^{n} \binom{n}{m} \mu^m \lambda^{n-m} = e^{-(\mu+\lambda)} \frac{(\mu + \lambda)^n}{n!}.$$

Hence $\mathrm{Poi}_\mu * \mathrm{Poi}_\lambda = \mathrm{Poi}_{\mu+\lambda}$. $\Diamond$

Takeaways Families of random variables are independent if the events they describe are independent (Definition 2.14). In order to check independence, it is enough to check it on a generator of the σ-algebra (Theorem 2.16). For example, it can be enough to check independence for intervals, rectangles or, in the discrete case, for single points. Independence of random variables can be characterised via product formulas for their joint distribution functions (Theorem 2.21), densities (Corollary 2.22) or weight functions (exercise!). The distribution of a sum of independent random variables can be computed using a convolution formula (Theorem 2.31).

Exercise 2.2.1 Let X and Y be independent random variables with $X \sim \exp_\theta$ and $Y \sim \exp_\rho$ for certain $\theta, \rho > 0$. Show that

$$\mathbf{P}[X < Y] = \frac{\theta}{\theta + \rho}. \quad \clubsuit$$

Exercise 2.2.2 (Box–Muller method) Let U and V be independent random variables that are uniformly distributed on $[0, 1]$. Define

$$X := \sqrt{-2\log(U)}\ \cos(2\pi V) \quad \text{and} \quad Y := \sqrt{-2\log(U)}\ \sin(2\pi V).$$

Show that X and Y are independent and $\mathcal{N}_{0,1}$-distributed.
Hint: First compute the distribution of $\sqrt{-2\log(U)}$ and then use the transformation formula (Theorem 1.101) as well as polar coordinates. $\clubsuit$

Exercise 2.2.3 (Multinomial distribution) Let $m \in \mathbb{N}$ and let $p = (p_1, \ldots, p_m)$ be a probability vector on $\{1, \ldots, m\}$. Let $X_1, \ldots, X_n$ be independent random variables with values in $1, \ldots, m$ and distribution p. We define an $\mathbb{N}_0^m$-valued random variable $Y = (Y_1, \ldots, Y_m)$ by

$$Y_i := \#\{k = 1, \ldots, n : X_k = i\} \quad \text{for } i = 1, \ldots, m.$$

Show that for $k = (k_1, \ldots, k_m) \in \mathbb{N}_0^m$ with $k_1 + \ldots + k_m = n$, we have

$$\mathbf{P}[Y = k] = \mathrm{Mul}_{n,p}(\{k\}) := \binom{n}{k} p^k. \tag{2.11}$$

Here

$$\binom{n}{k} = \binom{n}{k_1, \ldots, k_m} = \frac{n!}{k_1! \cdots k_m!}$$

is the **multinomial coefficient** and $p^k = p_1^{k_1} \cdots p_m^{k_m}$. The distribution $\mathrm{Mul}_{n,p}$ on $\mathbb{N}_0^m$ is called multinomial distribution with parameters n and p. ♣

2.3 Kolmogorov's 0–1 Law

With the Borel–Cantelli lemma, we have seen a first 0–1 law for independent events. We now come to another 0–1 law for independent events and for independent σ-algebras. To this end, we first introduce the notion of the tail σ-algebra.

Definition 2.34 (Tail σ-algebra) *Let I be a countably infinite index set and let $(\mathcal{A}_i)_{i \in I}$ be a family of σ-algebras. Then*

$$\mathcal{T}\big((\mathcal{A}_i)_{i \in I}\big) := \bigcap_{\substack{J \subset I \\ \#J < \infty}} \sigma\bigg(\bigcup_{j \in I \setminus J} \mathcal{A}_j\bigg)$$

*is called the **tail σ-algebra** of $(\mathcal{A}_i)_{i \in I}$. If $(A_i)_{i \in I}$ is a family of events, then we define*

$$\mathcal{T}\big((A_i)_{i \in I}\big) := \mathcal{T}\big((\{\emptyset, A_i, A_i^c, \Omega\})_{i \in I}\big).$$

If $(X_i)_{i \in I}$ is a family of random variables, then we define $\mathcal{T}\big((X_i)_{i \in I}\big) := \mathcal{T}\big((\sigma(X_i))_{i \in I}\big)$.

occurrence is independent of any fixed finite subfamily of the X_i. To put it differently, for any finite subfamily of the X_i, we can change the values of the X_i arbitrarily without changing whether A occurs or not.

Theorem 2.35 *Let $J_1, J_2, \ldots$ be finite sets with $J_n \uparrow I$. Then*

$$\mathcal{T}\big((\mathcal{A}_i)_{i \in I}\big) = \bigcap_{n=1}^{\infty} \sigma\bigg(\bigcup_{m \in I \setminus J_n} \mathcal{A}_m\bigg).$$

In the particular case $I = \mathbb{N}$, this reads $\mathcal{T}\big((\mathcal{A}_n)_{n \in \mathbb{N}}\big) = \bigcap_{n=1}^{\infty} \sigma\bigg(\bigcup_{m=n}^{\infty} \mathcal{A}_m\bigg).$

The name "tail σ-algebra" is due to the interpretation of $I = \mathbb{N}$ as a set of times. As is made clear in the theorem, any event in $\mathcal{T}$ does not depend on the first finitely many time points.

Proof "$\subset$"　This is clear.
"$\supset$"　Let $J_n \subset I$, $n \in \mathbb{N}$, be finite sets with $J_n \uparrow I$. Let $J \subset I$ be finite. Then there exists an $N \in \mathbb{N}$ with $J \subset J_N$ and

$$\bigcap_{n=1}^{\infty} \sigma\left(\bigcup_{m \in I \setminus J_n} A_m\right) \subset \bigcap_{n=1}^{N} \sigma\left(\bigcup_{m \in I \setminus J_n} A_m\right)$$

$$= \sigma\left(\bigcup_{m \in I \setminus J_N} A_m\right) \subset \sigma\left(\bigcup_{m \in I \setminus J} A_m\right).$$

The left-hand side does not depend on J. Hence we can form the intersection over all finite J and obtain

$$\bigcap_{n=1}^{\infty} \sigma\left(\bigcup_{m \in I \setminus J_n} A_m\right) \subset \mathcal{T}((A_i)_{i \in I}).$$

$\square$

Maybe at first glance it is not evident that there are any interesting events in the tail σ-algebra at all. It might not even be clear that we do not have $\mathcal{T} = \{\emptyset, \Omega\}$. Hence we now present simple examples of tail events and tail σ-algebra measurable random variables. In Sect. 2.4, we will study a more complex example.

Example 2.36

(i) Let $A_1, A_2, \ldots$ be events. Then the events $A_* := \liminf_{n \to \infty} A_n$ and $A^* := \limsup_{n \to \infty} A_n$ are in $\mathcal{T}((A_n)_{n \in \mathbb{N}})$. Indeed, if we define $B_n := \bigcap_{m=n}^{\infty} A_m$ for $n \in \mathbb{N}$, then $B_n \uparrow A_*$ and $B_n \in \sigma((A_m)_{m \geq N})$ for any $n \geq N$. Thus $A_* \in \sigma((A_m)_{m \geq N})$ for any $N \in \mathbb{N}$ and hence $A_* \in \mathcal{T}((A_n)_{n \in \mathbb{N}})$. The case A^* is similar.

(ii) Let $(X_n)_{n \in \mathbb{N}}$ be a family of $\overline{\mathbb{R}}$-valued random variables. Then the maps $X_* := \liminf_{n \to \infty} X_n$ and $X^* := \limsup_{n \to \infty} X_n$ are $\mathcal{T}((X_n)_{n \in \mathbb{N}})$-measurable. Indeed, if we define $Y_n := \sup_{m \geq n} X_m$, then for any $N \in \mathbb{N}$, the random variable $X^* = \inf_{n \geq 1} Y_n = \inf_{n \geq N} Y_n$ is $\mathcal{T}_N := \sigma(X_n, n \geq N)$-measurable and hence also measurable with respect to $\mathcal{T}((X_n)_{n \in \mathbb{N}}) = \bigcap_{n=1}^{\infty} \mathcal{T}_n$.
　　The case X_* is similar.

(iii) Let $(X_n)_{n \in \mathbb{N}}$ be real random variables. Then the **Cesàro limits**

$$\liminf_{n \to \infty} \frac{1}{n} \sum_{i=1}^{n} X_i \quad \text{and} \quad \limsup_{n \to \infty} \frac{1}{n} \sum_{i=1}^{n} X_i$$

are $\mathcal{T}((X_n)_{n \in \mathbb{N}})$-measurable. In order to show this, choose $N \in \mathbb{N}$ and note that

$$X_* := \liminf_{n \to \infty} \frac{1}{n} \sum_{i=1}^{n} X_i = \liminf_{n \to \infty} \frac{1}{n} \sum_{i=N}^{n} X_i$$

is $\sigma((X_n)_{n \geq N})$-measurable. Since this holds for any N, X_* is $\mathcal{T}((X_n)_{n \in \mathbb{N}})$-measurable. The case of the limes superior is similar. $\Diamond$

Theorem 2.37 (Kolmogorov's 0–1 Law) *Let I be a countably infinite index set and let $(\mathcal{A}_i)_{i \in I}$ be an independent family of σ-algebras. Then the tail σ-algebra is **P**-trivial, that is,*

$$\mathbf{P}[A] \in \{0, 1\} \quad \text{for any } A \in \mathcal{T}((\mathcal{A}_i)_{i \in I}).$$

Proof It is enough to consider the case $I = \mathbb{N}$. For $n \in \mathbb{N}$, let

$$\mathcal{F}_n := \left\{ \bigcap_{k=1}^{n} A_k : A_1 \in \mathcal{A}_1, \dots, A_n \in \mathcal{A}_n \right\}.$$

Then $\mathcal{F} := \bigcup_{n=1}^{\infty} \mathcal{F}_n$ is a semiring and $\sigma(\mathcal{F}) = \sigma(\bigcup_{n \in \mathbb{N}} \mathcal{A}_n)$. Indeed, for any $n \in \mathbb{N}$ and $A_n \in \mathcal{A}_n$, we have $A_n \in \mathcal{F}$; hence $\sigma(\bigcup_{n \in \mathbb{N}} \mathcal{A}_n) \subset \sigma(\mathcal{F})$. On the other hand, we have $\mathcal{F}_m \subset \sigma(\bigcup_{n=1}^{m} \mathcal{A}_n) \subset \sigma(\bigcup_{n \in \mathbb{N}} \mathcal{A}_n)$ for any $m \in \mathbb{N}$; hence $\mathcal{F} \subset \sigma(\bigcup_{n \in \mathbb{N}} \mathcal{A}_n)$.

Let $A \in \mathcal{T}((\mathcal{A}_n)_{n \in \mathbb{N}})$ and $\varepsilon > 0$. By the approximation theorem for measures (Theorem 1.65), there exists an $N \in \mathbb{N}$ and mutually disjoint sets $F_1, \dots, F_N \in \mathcal{F}$ such that $\mathbf{P}[A \triangle (F_1 \cup \dots \cup F_N)] < \varepsilon$. Clearly, there is an $n \in \mathbb{N}$ such that $F_1, \dots, F_N \in \mathcal{F}_n$ and thus $F := F_1 \cup \dots \cup F_N \in \sigma(\mathcal{A}_1 \cup \dots \cup \mathcal{A}_n)$. Obviously, $A \in \sigma(\bigcup_{m=n+1}^{\infty} \mathcal{A}_m)$; hence A is independent of F. Thus

$$\varepsilon > \mathbf{P}[A \setminus F] = \mathbf{P}[A \cap (\Omega \setminus F)] = \mathbf{P}[A](1 - \mathbf{P}[F]) \geq \mathbf{P}[A](1 - \mathbf{P}[A] - \varepsilon).$$

Letting $\varepsilon \downarrow 0$ yields $0 = \mathbf{P}[A](1 - \mathbf{P}[A])$.   □

Corollary 2.38 *Let $(A_n)_{n \in \mathbb{N}}$ be a sequence of independent events. Then*

$$\mathbf{P}\left[\limsup_{n \to \infty} A_n \right] \in \{0, 1\} \quad \text{and} \quad \mathbf{P}\left[\liminf_{n \to \infty} A_n \right] \in \{0, 1\}.$$

Proof Essentially this is a simple conclusion of the Borel–Cantelli lemma. However, the statement can also be deduced from Kolmogorov's 0–1 law as limes superior and limes inferior are in the tail σ-algebra.   □

Corollary 2.39 *Let $(X_n)_{n\in\mathbb{N}}$ be an independent family of $\overline{\mathbb{R}}$-valued random variables. Then $X_* := \liminf_{n\to\infty} X_n$ and $X^* := \limsup_{n\to\infty} X_n$ are almost surely constant. That is, there exist $x_*, x^* \in \overline{\mathbb{R}}$ such that $\mathbf{P}[X_* = x_*] = 1$ and $\mathbf{P}[X^* = x^*] = 1$.*

If all X_i are real-valued, then the Cesàro limits

$$\liminf_{n\to\infty} \frac{1}{n} \sum_{i=1}^{n} X_i \quad and \quad \limsup_{n\to\infty} \frac{1}{n} \sum_{i=1}^{n} X_i$$

are also almost surely constant.

Proof Let $X_* := \liminf_{n\to\infty} X_n$. For any $x \in \overline{\mathbb{R}}$, we have $\{X_* \leq x\} \in \mathcal{T}((X_n)_{n\in\mathbb{N}})$; hence $\mathbf{P}[X_* \leq x] \in \{0, 1\}$. Define

$$x_* := \inf\{x \in \mathbb{R} : \mathbf{P}[X_* \leq x] = 1\} \in \overline{\mathbb{R}}.$$

If $x_* = \infty$, then evidently

$$\mathbf{P}[X_* < \infty] = \lim_{n\to\infty} \mathbf{P}[X_* \leq n] = 0.$$

If $x_* \in \mathbb{R}$, then

$$\mathbf{P}[X_* \leq x_*] = \lim_{n\to\infty} \mathbf{P}\left[X_* \leq x_* + \frac{1}{n}\right] = 1$$

and

$$\mathbf{P}[X_* < x_*] = \lim_{n\to\infty} \mathbf{P}\left[X_* \leq x_* - \frac{1}{n}\right] = 0.$$

If $x_* = -\infty$, then

$$\mathbf{P}[X_* > -\infty] = \lim_{n\to\infty} \mathbf{P}[X_* > -n] = 0.$$

The cases of the limes superior and the Cesàro limits are similar.   □

Takeaways Consider an event that is described by the values of infinitely many random variables. If the occurrence of the event does not change when we change finitely many values of the random variables, then the event is called *terminal*. If the random variables are independent, then terminal events either have probability 0 or 1 (Kolmogorov's 0–1 law). In the Borel-Cantelli lemma we have encountered a special case.

Exercise 2.3.1 Let $(X_n)_{n \in \mathbb{N}}$ be an independent family of $\mathrm{Rad}_{1/2}$ random variables (i.e., $\mathbf{P}[X_n = -1] = \mathbf{P}[X_n = +1] = \frac{1}{2}$) and let $S_n = X_1 + \ldots + X_n$ for any $n \in \mathbb{N}$. Show that $\limsup_{n \to \infty} S_n = \infty$ almost surely. ♣

Exercise 2.3.2 Consider two families $(A_1, A_3, A_5, \ldots)$ and $(A_2, A_4, A_6, \ldots)$ of events and assume that each family is independent but they are not necessarily independent of each other. In Exercise 2.1.2 it was shown that the conclusion of the Borel-Cantelli lemma still holds under this weaker assumption. Now find an example that shows that the conclusion of Kolmogorov's 0–1 law need not hold under this assumption. ♣

2.4 Example: Percolation

Consider the d-dimensional integer lattice $\mathbb{Z}^d$, where any point is connected to any of its $2d$ nearest neighbors by an edge. If $x, y \in \mathbb{Z}^d$ are nearest neighbors (that is, $\|x - y\|_2 = 1$), then we denote by $e = \langle x, y \rangle = \langle y, x \rangle$ the edge that connects x and y. Formally, the set of edges is a subset of the set of subsets of $\mathbb{Z}^d$ with two elements:

$$E = \big\{ \{x, y\} : x, y \in \mathbb{Z}^d \text{ with } \|x - y\|_2 = 1 \big\}.$$

Somewhat more generally, an undirected **graph** G is a pair $G = (V, E)$, where V is a set (the set of "vertices" or nodes) and $E \subset \{\{x, y\} : x, y \in V, x \neq y\}$ is a subset of the set of subsets of V of cardinality two (the set of **edges** or **bonds**).

Our intuitive understanding of an edge is a connection between two points x and y and not an (unordered) pair $\{x, y\}$. To stress this notion of a connection, we use a different symbol from the set brackets. That is, we denote the edge that connects x and y by $\langle x, y \rangle = \langle y, x \rangle$ instead of $\{x, y\}$.

Our graph (V, E) is the starting point for a stochastic model of a porous medium. We interpret the edges as tubes along which water can flow. However, we want the medium not to have a homogeneous structure, such as $\mathbb{Z}^d$, but an amorphous structure. In order to model this, we randomly destroy a certain fraction $1 - p$ of the tubes (with $p \in [0, 1]$ a parameter) and keep the others. Water can flow only through the remaining tubes. The destroyed tubes will be called "closed", the others "open". The fundamental question is: For which values of p is there a connected infinite system of tubes along which water can flow? The physical interpretation is that if we throw a block of the considered material into a bathtub, then the block will soak up water; that is, it will be wetted inside. If there is no infinite open component, then the water may wet only a thin layer at the surface. See Fig. 2.1 for a computer simulation of the percolation model.

We now come to a formal description of the model. Choose a parameter $p \in [0, 1]$ and an independent family of identically distributed random variables $(X_e^p)_{e \in E}$ with $X_e^p \sim \mathrm{Ber}_p$; that is, $\mathbf{P}[X_e^p = 1] = 1 - \mathbf{P}[X_e^p = 0] = p$ for any

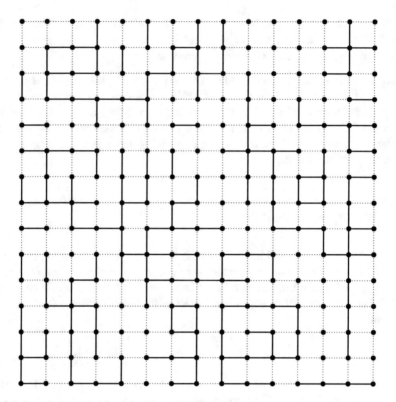

Fig. 2.1 Percolation on a 15×15 grid, $p = 0.42$.

$e \in E$. We define the set of *open* edges as

$$E^p := \{e \in E : X_e^p = 1\}. \tag{2.12}$$

Consequently, the edges in $E \setminus E^p$ are called *closed*. Hence we have constructed a (random) subgraph $(\mathbb{Z}^d, E^p)$ of $(\mathbb{Z}^d, E)$. We call $(\mathbb{Z}^d, E^p)$ a percolation model (more precisely, a model for **bond percolation**, in contrast to **site percolation**, where vertices can be open or closed). An (open) path (of length n) in this subgraph is a sequence $\pi = (x_0, x_1, \ldots, x_n)$ of points in $\mathbb{Z}^d$ with $\langle x_{i-1}, x_i \rangle \in E^p$ for all $i = 1, \ldots, n$. We say that two points $x, y \in \mathbb{Z}^d$ are connected by an open path if there is an $n \in \mathbb{N}$ and an open path $(x_0, x_1, \ldots, x_n)$ with $x_0 = x$ and $x_n = y$. In this case, we write $x \longleftrightarrow_p y$. Note that "$\longleftrightarrow_p$" is an equivalence relation; however, a random one, as it depends on the values of the random variables $(X_e^p)_{e \in E}$. For every $x \in \mathbb{Z}^d$, we define the (random) open cluster of x; that is, the connected component of x in the graph $(\mathbb{Z}^d, E^p)$:

$$C^p(x) := \{y \in \mathbb{Z}^d : x \longleftrightarrow_p y\}. \tag{2.13}$$

Lemma 2.40 *Let* $x, y \in \mathbb{Z}^d$. *Then* $\mathbb{1}_{\{x \longleftrightarrow_p y\}}$ *is a random variable. In particular,* $\#C^p(x)$ *is a random variable for any* $x \in \mathbb{Z}^d$.

Proof We may assume $x = 0$. Let $f_{y,n} = 1$ if there exists an open path of length at most n that connects 0 to y, and $f_{y,n} = 0$ otherwise. Clearly, $f_{y,n} \uparrow \mathbb{1}_{\{0 \longleftrightarrow_p y\}}$ for $n \to \infty$; hence it suffices to show that each $f_{y,n}$ is measurable. Let $B_n := \{-n, -n + 1, \ldots, n - 1, n\}^d$ and $E_n := \{e \in E : e \cap B_n \neq \emptyset\}$. Then $Y_n := (X_e^p : e \in E_n) : \Omega \to \{0, 1\}^{E_n}$ is measurable (with respect to $2^{(\{0,1\}^{E_n})}$) by Theorem 1.90. However, $f_{y,n}$ is a function of Y_n, say $f_{y,n} = g_{y,n} \circ Y_n$ for some map $g_{y,n} : \{0, 1\}^{E_n} \to \{0, 1\}$. By the composition theorem for maps (Theorem 1.80), $f_{y,n}$ is measurable.

The additional statement holds since $\#C^p(x) = \sum_{y \in \mathbb{Z}^d} \mathbb{1}_{\{x \longleftrightarrow_p y\}}$. □

Definition 2.41 *We say that percolation occurs if there exists an infinitely large open cluster. We call*

$$\psi(p) := \mathbf{P}[\textit{there exists an infinite open cluster}]$$

$$= \mathbf{P}\left[\bigcup_{x \in \mathbb{Z}^d} \{\#C^p(x) = \infty\} \right]$$

the probability of percolation. We define

$$\theta(p) := \mathbf{P}[\#C^p(0) = \infty]$$

as the probability that the origin is in an infinite open cluster.

By the translation invariance of the lattice, we have

$$\theta(p) = \mathbf{P}[\#C^p(y) = \infty] \quad \text{for any } y \in \mathbb{Z}^d. \tag{2.14}$$

The fundamental question is: How large are $\theta(p)$ and $\psi(p)$ depending on p?

We make the following simple observation.

Theorem 2.42 *The map* $[0, 1] \to [0, 1]$, $p \mapsto \theta(p)$ *is monotone increasing.*

Proof Although the statement is intuitively so clear that it might not need a proof, we give a formal proof in order to introduce a technique called **coupling**. Let $p, p' \in [0, 1]$ with $p < p'$. Let $(Y_e)_{e \in E}$ be an independent family of random variables with $\mathbf{P}[Y_e \leq q] = q$ for any $e \in E$ and $q \in \{p, p', 1\}$. At this point, we could, for example, assume that $Y_e \sim \mathcal{U}_{[0,1]}$ is uniformly distributed on $[0, 1]$. Since we have not yet shown the existence of an independent family with this distribution, we content ourselves with Y_e that assume only three values $\{p, p', 1\}$. Hence

$$\mathbf{P}[Y_e = q] = \begin{cases} p, & \text{if } q = p, \\ p' - p, & \text{if } q = p', \\ 1 - p', & \text{if } q = 1. \end{cases}$$

Such a family $(Y_e)_{e \in E}$ exists by Theorem 2.19. For $q \in \{p, p'\}$ and $e \in E$, we define

$$X_e^q := \begin{cases} 1, & \text{if } Y_e \leq q, \\ 0, & \text{else.} \end{cases}$$

Clearly, for any $q \in \{p, p'\}$, the family $(X_e^q)_{e \in E}$ of random variables is independent (see Remark 2.15(iii)) and $X_e^q \sim \mathrm{Ber}_q$. Furthermore, $X_e^p \leq X_e^{p'}$ for any $e \in E$. The procedure of defining two families of random variables that are related in a specific way (here "$\leq$") on one probability space is called a *coupling*.

Clearly, $C^p(x) \subset C^{p'}(x)$ for any $x \in \mathbb{Z}^d$; hence $\theta(p) \leq \theta(p')$.   □

With the aid of Kolmogorov's 0–1 law, we can infer the following theorem.

Theorem 2.43 *For any $p \in [0, 1]$, we have* $\psi(p) = \begin{cases} 0, & \text{if } \theta(p) = 0, \\ 1, & \text{if } \theta(p) > 0. \end{cases}$

Proof If $\theta(p) = 0$, then by (2.14)

$$\psi(p) \leq \sum_{y \in \mathbb{Z}^d} \mathbf{P}[\#C^p(y) = \infty] = \sum_{y \in \mathbb{Z}^d} \theta(p) = 0.$$

Now let $A = \bigcup_{y \in \mathbb{Z}^d} \{\#C^p(y) = \infty\}$. Clearly, A remains unchanged if we change the state of finitely many edges. That is, $A \in \sigma((X_e^p)_{e \in E \setminus F})$ for every finite $F \subset E$. Hence A is in the tail σ-algebra $\mathcal{T}((X_e^p)_{e \in E})$ by Theorem 2.35. Kolmogorov's 0–1 law (Theorem 2.37) implies that $\psi(p) = \mathbf{P}[A] \in \{0, 1\}$. If $\theta(p) > 0$, then $\psi(p) \geq \theta(p)$ implies $\psi(p) = 1$.   □

Due to the monotonicity, we can make the following definition.

Definition 2.44 *The critical value p_c for percolation is defined as*

$$\begin{aligned} p_c &= \inf\{p \in [0, 1] : \theta(p) > 0\} = \sup\{p \in [0, 1] : \theta(p) = 0\} \\ &= \inf\{p \in [0, 1] : \psi(p) = 1\} = \sup\{p \in [0, 1] : \psi(p) = 0\}. \end{aligned}$$

We come to the main theorem of this section.

Theorem 2.45 *For $d = 1$, we have $p_c = 1$. For $d \geq 2$, we have $p_c(d) \in \left[\frac{1}{2d-1}, \frac{2}{3}\right]$.*

Proof First consider $d = 1$ and $p < 1$. Let $A^- := \{X_{\langle n,n+1\rangle}^p = 0 \text{ for some } n < 0\}$ and $A^+ := \{X_{\langle n,n+1\rangle}^p = 0 \text{ for some } n > 0\}$. Let $A = A^- \cap A^+$. By the Borel–Cantelli lemma, we get $\mathbf{P}[A^-] = \mathbf{P}[A^+] = 1$. Hence $\theta(p) = \mathbf{P}[A^c] = 0$.

Now assume $d \geq 2$.

Lower bound First we show $p_c \geq \frac{1}{2d-1}$. Clearly, for any $n \in \mathbb{N}$,

$$\mathbf{P}[\#C^p(0) = \infty] \leq \mathbf{P}\big[\text{there is an } x \in C^p(0) \text{ with } \|x\|_\infty = n\big].$$

We want to estimate the probability that there exists a point $x \in C^p(0)$ with distance n from the origin. Any such point is connected to the origin by a path without self-intersections π that starts at 0 and has length $m \geq n$. Let $\Pi_{0,m}$ be the set of such paths. Clearly, $\#\Pi_{0,m} \leq 2d \cdot (2d-1)^{m-1}$ since there are $2d$ choices for the first step and at most $2d - 1$ choices for any further step. For any $\pi \in \Pi_{0,m}$, the probability that π uses only open edges is

$$\mathbf{P}[\pi \text{ is open}] = p^m.$$

Hence, for $p < \frac{1}{2d-1}$,

$$\theta(p) \leq \sum_{m=n}^{\infty} \sum_{\pi \in \Pi_{0,m}} \mathbf{P}[\pi \text{ is open}] \leq \frac{2d}{2d-1} \sum_{m=n}^{\infty} \big((2d-1)p\big)^m$$

$$= \frac{2d}{(2d-1)(1-(2d-1)p)}\big((2d-1)p\big)^n \xrightarrow{n\to\infty} 0.$$

We conclude that $p_c \geq \frac{1}{2d-1}$.

Upper bound We can consider $\mathbb{Z}^d$ as a subset of $\mathbb{Z}^d \times \{0\} \subset \mathbb{Z}^{d+1}$. Hence, if percolation occurs for p in $\mathbb{Z}^d$, then it also occurs for p in $\mathbb{Z}^{d+1}$. Hence the corresponding critical values are ordered $p_c(d+1) \leq p_c(d)$.

Thus, it is enough to consider the case $d = 2$. Here we show $p_c \leq \frac{2}{3}$ by using a contour argument due to Peierls [127], originally designed for the Ising model of a ferromagnet, see Example 18.16 and (18.9).

For $N \in \mathbb{N}$, we define (compare (2.13) with $x = (i, 0)$)

$$C_N := \bigcup_{i=0}^{N} C^p\big((i,0)\big)$$

as the set of points that are connected (along open edges) to at least one of the points in $\{0, \ldots, N\} \times \{0\}$. Due to the subadditivity of probability (and since $\mathbf{P}[\#C^p((i,0)) = \infty] = \theta(p)$ for any $i \in \mathbb{Z}$), we have

$$\theta(p) = \frac{1}{N+1}\sum_{i=0}^{N}\mathbf{P}\big[\#C^p\big((i,0)\big) = \infty\big] \geq \frac{1}{N+1}\mathbf{P}[\#C_N = \infty].$$

Now consider those closed contours in the dual graph $(\tilde{\mathbb{Z}}^2, \tilde{E})$ that surrounds C_N if $\#C_N < \infty$. Here the dual graph is defined by

$$\tilde{\mathbb{Z}}^2 = \left(\frac{1}{2}, \frac{1}{2}\right) + \mathbb{Z}^2,$$

$$\tilde{E} = \left\{ \{x, y\} : x, y \in \tilde{\mathbb{Z}}^2, \ \|x - y\|_2 = 1 \right\}.$$

An edge $\tilde{e}$ in the dual graph $(\tilde{\mathbb{Z}}^2, \tilde{E})$ crosses exactly one edge e in $(\mathbb{Z}^2, E)$. We call $\tilde{e}$ open if e is open and closed otherwise. A circle γ is a self-intersection free path in $(\tilde{\mathbb{Z}}^2, \tilde{E})$ that starts and ends at the same point. A contour of the set C_N is a minimal circle that surrounds C_N. Minimal means that the enclosed area is minimal (see Fig. 2.2). For $n \geq 2N$, let

$$\Gamma_n = \left\{ \gamma : \gamma \text{ is a circle of length } n \text{ that surrounds } \{0, \ldots, N\} \times \{0\} \right\}.$$

We want to deduce an upper bound for $\#\Gamma_n$. Let $\gamma \in \Gamma_n$ and fix one point of γ. For definiteness, choose the upper point $\left(m + \frac{1}{2}, \frac{1}{2}\right)$ of the rightmost edge of γ that crosses the horizontal axis (in Fig. 2.2 this is the point $\left(5 + \frac{1}{2}, \frac{1}{2}\right)$). Clearly, $m \geq N$ and $m \leq n$ since γ surrounds the origin. Starting from $\left(m + \frac{1}{2}, \frac{1}{2}\right)$, for any further

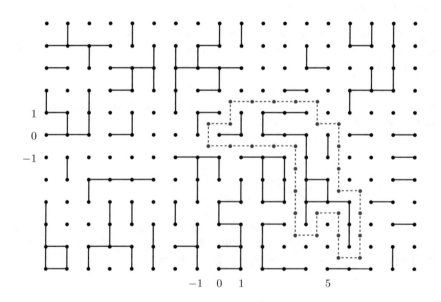

Fig. 2.2 Contour of the cluster C_5.

edge of γ, there are at most three possibilities. Hence

$$\#\Gamma_n \leq n \cdot 3^n.$$

We say that γ is closed if it uses only closed edges (in $\tilde{E}$). A contour of C_N is automatically closed and has a length of at least $2N$. Hence for $p > \frac{2}{3}$

$$\mathbf{P}[\#C_N < \infty] = \sum_{n=2N}^{\infty} \mathbf{P}\big[\text{there is a closed circle } \gamma \in \Gamma_n\big]$$

$$\leq \sum_{n=2N}^{\infty} n \cdot \big(3(1-p)\big)^n \overset{N\to\infty}{\longrightarrow} 0.$$

We conclude $p_c \leq \frac{2}{3}$.   □

In general, the value of p_c is not known and is extremely hard to determine. In the case of bond percolation on $\mathbb{Z}^2$, however, the exact value of p_c can be determined due to the self-duality of the planar graph $(\mathbb{Z}^2, E)$. (If $G = (V, E)$ is a planar graph; that is, a graph that can be embedded into $\mathbb{R}^2$ without self-intersections, then the vertex set of the dual graph is the set of faces of G. Two such vertices are connected by exactly one edge; that is, by the edge in E that separates the two faces. Evidently, the two-dimensional integer lattice is isomorphic to its dual graph. Note that the contour in Fig. 2.2 can be considered as a closed path in the dual graph.) We cite a theorem of Kesten [94].

Theorem 2.46 (Kesten [94]) *For bond percolation in $\mathbb{Z}^2$, the critical value is $p_c = \frac{1}{2}$ and $\theta(p_c) = 0$.*

Proof See, for example, the book of Grimmett [63, pages 287ff].   □

It is conjectured that $\theta(p_c) = 0$ holds in any dimension $d \geq 2$. However, rigorous proofs are known only for $d = 2$ and $d \geq 19$ (see [67]).

Uniqueness of the Infinite Open Cluster*

Fix a p such that $\theta(p) > 0$. We saw that with probability one there is *at least* one infinite open cluster. Now we want to show that there is *exactly* one.

Denote by $N \in \{0, 1, \ldots, \infty\}$ the (random) number of infinite open clusters.

Theorem 2.47 (Uniqueness of the infinite open cluster) *For any $p \in [0, 1]$, we have $\mathbf{P}_p[N \leq 1] = 1$.*

Proof This theorem was first proved by Aizenman, Kesten and Newman [2, 3]. Here we follow the proof of Burton and Keane [23] as described in [63, Section 8.2].

The cases $p = 1$ and $\theta(p) = 0$ (hence in particular the case $p = 0$) are trivial. Hence we assume now that $p \in (0, 1)$ and $\theta(p) > 0$.

Step 1.   We first show that

$$\mathbf{P}_p[N = m] = 1 \qquad \text{for some}\ \ m = 0, 1, \ldots, \infty. \tag{2.15}$$

We need a 0–1 law similar to that of Kolmogorov. However, N is not measurable with respect to the tail σ-algebra. Hence we have to find a more subtle argument. Let $u_1 = (1, 0, \ldots, 0)$ be the first unit vector in $\mathbb{Z}^d$. On the edge set E, define the translation $\tau : E \to E$ by $\tau(\langle x, y \rangle) = \langle x + u_1, y + u_1 \rangle$. Let

$$E_0 := \big\{ \langle (x_1, \ldots, x_d), (y_1, \ldots, y_d) \rangle \in E : x_1 = 0, \ y_1 \geq 0 \big\}$$

be the set of all edges in $\mathbb{Z}^d$ that either connect two points from $\{0\} \times \mathbb{Z}^{d-1}$ or one point of $\{0\} \times \mathbb{Z}^{d-1}$ with one point of $\{1\} \times \mathbb{Z}^{d-1}$. Clearly, the sets $(\tau^n(E_0), n \in \mathbb{Z})$ are disjoint and $E = \biguplus_{n \in \mathbb{Z}} \tau^n(E_0)$. Hence the random variables $Y_n := (X^p_{\tau^n(e)})_{e \in E_0}, n \in \mathbb{Z}$, are independent and identically distributed (with values in $\{0, 1\}^{E_0}$).Define $Y = (Y_n)_{n \in \mathbb{Z}}$ and $\tau(Y) = (Y_{n+1})_{n \in \mathbb{Z}}$. Define $A_m \in \{0, 1\}^E$ by

$$\{Y \in A_m\} = \{N = m\}.$$

Clearly, the value of N does not change if we shift *all* edges simultaneously. That is, $\{Y \in A_m\} = \{\tau(Y) \in A_m\}$. An event with this property is called *invariant* or *shift invariant*. Using an argument similar to that in the proof of Kolmogorov's 0–1 law, one can show that invariant events (defined by i.i.d. random variables) have probability either 0 or 1 (see Example 20.26 for a proof).

Step 2.   We will show that

$$\mathbf{P}_p[N = m] = 0 \qquad \text{for any}\ \ m \in \mathbb{N} \setminus \{1\}. \tag{2.16}$$

Accordingly, let $m = 2, 3, \ldots$. We assume that $\mathbf{P}[N = m] = 1$ and show that this leads to a contradiction.

For $L \in \mathbb{N}$, let $B_L := \{-L, \ldots, L\}^d$ and denote by $E_L = \{e = \langle x, y \rangle \in E : x, y \in B_L\}$ the set of those edges with both vertices lying in B_L. For $i = 0, 1$, let $D^i_L := \{X^p_e = i \ \text{for all}\ e \in E_L\}$. Let N^1_L be the number of infinite open clusters if we consider all edges e in E_L as open (independently of the value of X^p_e). Similarly define N^0_L where we consider all edges in E_L as closed. Since $\mathbf{P}_p[D^i_L] > 0$, and since $N = m$ almost surely, we have $N^i_L = m$ almost surely for $i = 0, 1$. Let

$$A^2_L := \bigcup_{x^1, x^2 \in B_L \setminus B_{L-1}} \big\{ C^p(x^1) \cap C^p(x^2) = \emptyset \big\} \cap \big\{ \#C^p(x^1) = \#C_p(x^2) = \infty \big\}$$

be the event where there exist two points on the boundary of B_L that lie in different infinite open clusters. Clearly, $A_L^2 \uparrow \{N \geq 2\}$ for $L \to \infty$.

Define $A_{L,0}^2$ in a similarly way to A_L^2; however, we now consider all edges $e \in E_L$ as closed, irrespective of whether $X_e^p = 1$ or $X_e^p = 0$. If A_L^2 occurs, then there are two points x^1, x^2 on the boundary of B_L such that for any $i = 1, 2$, there is an infinite self-intersection free open path π_{x^i} starting at x^i that avoids x^{3-i}. Hence $A_L^2 \subset A_{L,0}^2$. Now choose L large enough for $\mathbf{P}[A_{L,0}^2] > 0$.

If $A_{L,0}^2$ occurs and if we open all edges in B_L, then at least two of the infinite open clusters get connected by edges in B_L. Hence the total number of infinite open clusters decreases by at least one. We infer $\mathbf{P}_p[N_L^1 \leq N_L^0 - 1] \geq \mathbf{P}_p[A_{L,0}^2] > 0$, which leads to a contradiction.

Step 3. In Step 2, we have shown already that N does not assume a *finite* value larger than 1. Hence it remains to show that almost surely N does not assume the value ∞. Indeed, we show that

$$\mathbf{P}_p[N \geq 3] = 0. \tag{2.17}$$

This part of the proof is the most difficult one. We assume that $\mathbf{P}_p[N \geq 3] > 0$ and show that this leads to a contradiction.

We say that a point $x \in \mathbb{Z}^d$ is a **trifurcation point** if

- x is in an infinite open cluster $C^p(x)$,
- there are exactly three open edges with endpoint x, and
- removing all of these three edges splits $C^p(x)$ into three mutually disjoint infinite open clusters.

By T we denote the set of trifurcation points, and let $T_L := T \cap B_L$. Let $r := \mathbf{P}_p[0 \in T]$. Due to translation invariance, we have $(\#B_L)^{-1}\mathbf{E}_p[\#T_L] = r$ for any L. (Here $\mathbf{E}_p[\#T_L]$ denotes the expected value of $\#T_L$, which we define formally in Chap. 5.) Let

$$A_L^3 := \bigcup_{x^1,x^2,x^3 \in B_L \setminus B_{L-1}} \left(\bigcap_{i \neq j}\{C^p(x^i) \cap C^p(x^j) = \emptyset\} \right) \cap \left(\bigcap_{i=1}^{3}\{\#C^p(x^i) = \infty\} \right)$$

be the event where there are three points on the boundary of B_L that lie in different infinite open clusters. Clearly, $A_L^3 \uparrow \{N \geq 3\}$ for $L \to \infty$.

As for $A_{L,0}^2$, we define $A_{L,0}^3$ as the event where there are three distinct points on the boundary of B_L that lie in different infinite open clusters if we consider all edges in E_L as closed. As above, we have $A_L^3 \subset A_{L,0}^3$.

For three distinct points $x^1, x^2, x^3 \in B_L \setminus B_{L-1}$, let F_{x^1,x^2,x^3} be the event where for any $i = 1, 2, 3$, there exists an infinite self-intersection free open path π_{x^i} starting at x^i that uses only edges in $E^p \setminus E_L$ and that avoids the points x^j, $j \neq i$. Then

$$A^3_{L,0} \subset \bigcup_{\substack{x^1,x^2,x^3 \in B_L \setminus B_{L-1} \\ \text{mutually distinct}}} F_{x^1,x^2,x^3}.$$

Let L be large enough for $\mathbf{P}_p[A^3_{L,0}] \geq \mathbf{P}_p[N \geq 3]/2 > 0$. Choose three pairwise distinct points $x^1, x^2, x^3 \in B_L \setminus B_{L-1}$ with $\mathbf{P}_p[F_{x^1,x^2,x^3}] > 0$.

If F_{x^1,x^2,x^3} occurs, then we can find a point $y \in B_L$ that is the starting point of three mutually disjoint (not necessarily open) paths π_1, π_2 and π_3 that end at x^1, x^2 and x^3. Let G_{y,x^1,x^2,x^3} be the event where in E_L exactly those edges are open that belong to these three paths (that is, all other edges in E_L are closed). The events F_{x^1,x^2,x^3} and G_{y,x^1,x^2,x^3} are independent, and if both of them occur, then y is a trifurcation point. Hence

$$r = \mathbf{P}_p[y \in T] \geq \mathbf{P}_p[F_{x^1,x^2,x^3}] \cdot \left(p \wedge (1-p)\right)^{\#E_L} > 0.$$

Now we show that r must equal 0, which contradicts the assumption $\mathbf{P}_p[N \geq 3] > 0$. Let K_L be the set of all edges which have at least one endpoint in B_L. We consider two edges in K_L as equivalent if there exists a path in B_L along open edges that does not hit any trifurcation point and which joins at least one endpoint of each of the two edges. We denote the equivalence relation by R and let $U_L = K_L/R$ be the set of equivalence classes. (Note that the three neighboring edges of a trifurcation point are in different equivalence classes.) We turn the set $H_L := U_L \cup T_L$ into a graph by considering two points $x \in T_L$ and $u \in U_L$ as neighbors if there exists an edge $k \in u$ which is incident to x. Note that each point $x \in T_L$ has exactly three neighbors which are in U_L. The points in U_L can be isolated (that is, without neighbors) or can be joined to arbitrarily many points in T_L but not in U_L.

A circle is a self-avoiding (finite) path that ends at its starting point. Note that the graph H_L has no circles. To show this assume there was a self-avoiding path $(h_0, h_1, \dots, h_n)$ starting and ending in some point $h_0 = h_n = x \in T_L$. Then $h_1, h_{n-1} \in U_L$ are distinct but connected in K^p even if we remove x. However, by the definition of the trifurcation point x, this is impossible. On the other hand, if there was a self-avoiding path $(g_0, \dots, g_m)$ starting and ending in some point $g_0 = g_m = u \in U_L$, then $(g_1, g_2, \dots, g_m, g_1)$ is a self-avoiding path starting and ending in $g_1 \in T_L$. However, we have just shown that such a path could not exist.

Write $\deg_{H_L}(h)$ for the degree of $h \in H_L$; that is, the number of neighbors of h in H_L. A point h with $\deg_{H_L}(h) = 1$ is called a *leaf* of H_L. Obviously, only points of U_L can be leaves. Let Z be a connected component of H_L that contains at least one point $x \in T_L$. Since Z is a tree (that is, it is connected and contains no circles), we have

$$\#Z - 1 = \frac{1}{2} \sum_{h \in Z} \deg_{H_L}(h).$$

Rearranging this formula yields an expression for the number of leaves:

$$\#\{u \in Z : \deg_{H_L}(u) = 1\} = 2 + \sum_{h \in Z} \left(\deg_{H_L}(h) - 2 \right)^+$$

$$\geq 2 + \#\{h \in Z : \deg_{H_L}(h) \geq 3\}$$

$$\geq 2 + \#(Z \cap T_L).$$

Summing over the connected components Z of H_L with at least one point in T_L, we obtain

$$\#\{u \in H_L : \deg_{H_L}(u) = 1\} \geq \#T_L.$$

Observe that any leaf $u \in H_L$ contains an edge that is incident to a point $x \in T_L$. Hence the edges of u lie in an infinite open cluster of K^p and there is at least one edge $k \in u$ incident to a point at the boundary $B_L \setminus B_{L-1}$ of B_L. For distinct leaves these are distinct points since the leaves belong to *disjoint* open clusters. Hence we get the bound

$$\#T_L \leq \#(B_L \setminus B_{L-1})$$

and thus

$$\frac{\#T_L}{\#B_L} \leq \frac{\#(B_L \setminus B_{L-1})}{\#B_L} \leq \frac{d}{L} \xrightarrow{L \to \infty} 0.$$

Now $r = (\#B_L)^{-1} \mathbf{E}_p[\#T_L] \leq d/L$ implies $r = 0$. (Note that in the argument we used the notion of the expected value $\mathbf{E}_p[\#T_L]$ that will be formally introduced only in Chap. 5.)   □

Takeaways Independent coin tosses decide if an edge of $\mathbb{Z}^d$ is retained (probability p) or removed. The remaining random graph almost surely contains a (unique) infinite connected component if p is larger than a critical value p_c. For $d \geq 2$, we have $0 \leq \frac{1}{2d-1} \leq p_c \leq \frac{2}{3}$ (Theorem 2.45). Starting with a graph other than $\mathbb{Z}^d$, for example an infinite binary tree, can result in multiple infinite connected components (Exercise 2.4.1).

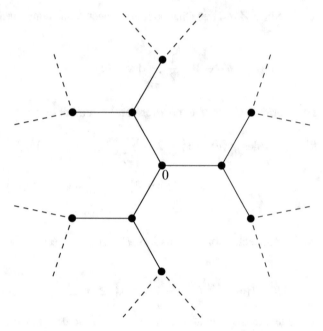

Fig. 2.3 Binary tree.

Exercise 2.4.1 Let T be the infinite binary tree (Fig. 2.3). That is, each point in T has exactly three neighbours. We single out an arbitrary point of T and name it 0. Now consider bond percolation on T with probability p.

(i) Show that $p_c \geq 1/2$. Hint: Use a similar argument as in the proof of Theorem 2.45.

(ii) Let J_n be the number of connected subgraphs of T that contain 0. Show that $J_n \leq 4^{n+1}$. Use a contour argument similarly as in Theorem 2.45 to show that $p_c \leq \frac{3}{4}$. (In fact, we could use the theory of branching processes to show that $p_c = \frac{1}{2}$.)

(iii) For $p \in (p_c, 1)$, show that with positive probability there are at least two infinite connected components.

In (iii) one can even show that almost surely there are infinitely many infinite connected components. ♣

Chapter 3
Generating Functions

It is a fundamental principle of mathematics to map a class of objects that are of interest into a class of objects where computations are easier. This map can be one to one, as with linear maps and matrices, or it may map only some properties uniquely, as with matrices and determinants.

In probability theory, in the second category fall quantities such as the median, mean and variance of random variables. In the first category, we have characteristic functions, Laplace transforms and probability generating functions. These are useful mostly because addition of independent random variables leads to multiplication of the transforms. Before we introduce characteristic functions (and Laplace transforms) later in the book, we want to illustrate the basic idea with probability generating functions that are designed for $\mathbb{N}_0$-valued random variables.

In the first section, we give the basic definitions and derive simple properties. The next two sections are devoted to two applications: The Poisson approximation theorem and a simple investigation of Galton–Watson branching processes.

3.1 Definition and Examples

Definition 3.1 (Probability generating function) *Let X be an $\mathbb{N}_0$-valued random variable. The (probability) **generating function** (p.g.f.) of $\mathbf{P}_X$ (or, loosely speaking, of X) is the map $\psi_{\mathbf{P}_X} = \psi_X$ defined by (with the understanding that $0^0 = 1$)*

$$\psi_X : [0, 1] \to [0, 1], \qquad z \mapsto \sum_{n=0}^{\infty} \mathbf{P}[X = n] \, z^n. \qquad (3.1)$$

© The Editor(s) (if applicable) and The Author(s), under exclusive license
to Springer Nature Switzerland AG 2020
A. Klenke, *Probability Theory*, Universitext,
https://doi.org/10.1007/978-3-030-56402-5_3

Theorem 3.2

(i) ψ_X is continuous on $[0, 1]$ and infinitely often continuously differentiable on $(0, 1)$. For $n \in \mathbb{N}$, the nth derivative $\psi_X^{(n)}$ fulfills

$$\lim_{z \uparrow 1} \psi_X^{(n)}(z) = \sum_{k=n}^{\infty} \mathbf{P}[X = k] \cdot k(k-1) \cdots (k-n+1), \tag{3.2}$$

where both sides can equal ∞.

(ii) The distribution $\mathbf{P}_X$ of X is uniquely determined by ψ_X.

(iii) For any $r \in (0, 1)$, ψ_X is uniquely determined by countably many values $\psi_X(x_i)$, $x_i \in [0, r]$, $i \in \mathbb{N}$. If the series in (3.1) converges for some $z > 1$, then the statement is also true for any $r \in (0, z)$ and we have

$$\lim_{x \uparrow 1} \psi_X^{(n)}(x) = \psi_X^{(n)}(1) < \infty \quad for \ n \in \mathbb{N}.$$

In this case, ψ_X is uniquely determined by the derivatives $\psi_X^{(n)}(1)$, $n \in \mathbb{N}$.

Proof The statements follow from the elementary theory of power series. For the first part of (iii), see, eg. [149, Theorem 8.5]. □

Reflection Come up with an example for X such that the series in (3.1) does not converge for any $z > 1$ but $\lim_{x \uparrow 1} \psi_X'(x)$ exists and is finite. ♠♠

Theorem 3.3 (Multiplicativity of generating functions) If $X_1, \ldots, X_n$ are independent and $\mathbb{N}_0$-valued random variables, then

$$\psi_{X_1 + \ldots + X_n} = \prod_{i=1}^{n} \psi_{X_i}.$$

Proof Let $z \in [0, 1)$ and write $\psi_{X_1}(z) \psi_{X_2}(z)$ as a Cauchy product

$$\psi_{X_1}(z) \psi_{X_2}(z) = \left(\sum_{n=0}^{\infty} \mathbf{P}[X_1 = n] z^n \right) \left(\sum_{n=0}^{\infty} \mathbf{P}[X_2 = n] z^n \right)$$

$$= \sum_{n=0}^{\infty} z^n \left(\sum_{m=0}^{n} \mathbf{P}[X_1 = m] \mathbf{P}[X_2 = n - m] \right)$$

$$= \sum_{n=0}^{\infty} z^n \sum_{m=0}^{n} \mathbf{P}[X_1 = m, \ X_2 = n - m]$$

$$= \sum_{n=0}^{\infty} \mathbf{P}[X_1 + X_2 = n] z^n = \psi_{X_1 + X_2}(z).$$

Inductively, the claim follows for all $n \geq 2$. □

Example 3.4

(i) Let X be $b_{n,p}$-distributed for some $n \in \mathbb{N}$ and let $p \in [0, 1]$. Then

$$\psi_X(z) = \sum_{m=0}^{n} \binom{n}{m} p^m (1-p)^{n-m} z^m = (pz + (1-p))^n. \tag{3.3}$$

(ii) If X, Y are independent, $X \sim b_{m,p}$ and $Y \sim b_{n,p}$, then, by Theorem 3.3,

$$\psi_{X+Y}(z) = (pz + (1-p))^m (pz + (1-p))^n = (pz + (1-p))^{m+n}.$$

Hence, by Theorem 3.2(ii), $X + Y$ is $b_{m+n,p}$-distributed and thus (by Theorem 2.31)

$$b_{m,p} * b_{n,p} = b_{m+n,p}.$$

(iii) Let X and Y be independent Poisson random variables with parameters $\lambda \geq 0$ and $\mu \geq 0$, respectively. That is, $\mathbf{P}[X = n] = e^{-\lambda} \lambda^n / n!$ for $n \in \mathbb{N}_0$. Then

$$\psi_{\mathrm{Poi}_\lambda}(z) = \sum_{n=0}^{\infty} e^{-\lambda} \frac{(\lambda z)^n}{n!} = e^{\lambda(z-1)}. \tag{3.4}$$

Hence $X + Y$ has probability generating function

$$\psi_{\mathrm{Poi}_\lambda}(z) \cdot \psi_{\mathrm{Poi}_\mu}(z) = e^{\lambda(z-1)} e^{\mu(z-1)} = \psi_{\mathrm{Poi}_{\lambda+\mu}}(z).$$

Thus $X + Y \sim \mathrm{Poi}_{\lambda+\mu}$. We conclude that

$$\mathrm{Poi}_\lambda * \mathrm{Poi}_\mu = \mathrm{Poi}_{\lambda+\mu}. \tag{3.5}$$

(iv) Let $X_1, \ldots, X_n \sim \gamma_p$ be independent geometrically distributed random variables with parameter $p \in (0, 1)$. Define $Y = X_1 + \ldots + X_n$. Then, for any $z \in [0, 1]$,

$$\psi_{X_1}(z) = \sum_{k=0}^{\infty} p(1-p)^k z^k = \frac{p}{1 - (1-p)z}. \tag{3.6}$$

By the generalized binomial theorem (see Lemma 3.5 with $\alpha = -n$), Theorem 3.3 and (3.6), we have

$$\psi_Y(z) = \psi_{X_1}(z)^n = \frac{p^n}{(1 - (1-p)z)^n}$$

$$= \sum_{k=0}^{\infty} p^n \binom{-n}{k} (-1)^k (1-p)^k z^k$$

$$= \sum_{k=0}^{\infty} b_{n,p}^-(\{k\}) z^k.$$

Here, for $r \in (0, \infty)$ and $p \in (0, 1]$,

$$b_{r,p}^- = \sum_{k=0}^{\infty} \binom{-r}{k} (-1)^k p^r (1-p)^k \delta_k \tag{3.7}$$

is the negative binomial distribution with parameters r and p. By the uniqueness theorem for probability generating functions, we get $Y \sim b_{n,p}^-$; hence (see Definition 2.29 for the nth convolution power) $b_{n,p}^- = \gamma_p^{*n}$. ◊

Lemma 3.5 (Generalized binomial theorem) *For $\alpha \in \mathbb{R}$ and $k \in \mathbb{N}_0$, we define the binomial coefficient*

$$\binom{\alpha}{k} := \frac{\alpha \cdot (\alpha - 1) \cdots (\alpha - k + 1)}{k!}. \tag{3.8}$$

Then the generalized binomial theorem holds:

$$(1 + x)^{\alpha} = \sum_{k=0}^{\infty} \binom{\alpha}{k} x^k \quad \text{for all } x \in \mathbb{C} \text{ with } |x| < 1. \tag{3.9}$$

In particular, we have

$$\frac{1}{\sqrt{1-x}} = \sum_{n=0}^{\infty} \binom{2n}{n} 4^{-n} x^n \quad \text{for all } x \in \mathbb{C} \text{ with } |x| < 1. \tag{3.10}$$

Proof The map $f : x \mapsto (1 + x)^{\alpha}$ is holomorphic up to possibly a singularity at $x = -1$. Hence it can be developed in a power series about 0 with radius of convergence at least 1:

$$f(x) = \sum_{k=0}^{\infty} \frac{f^{(k)}(0)}{k!} x^k \quad \text{for } |x| < 1.$$

For $k \in \mathbb{N}_0$, the kth derivative is $f^{(k)}(0) = \alpha(\alpha - 1) \cdots (\alpha - k + 1)$. Hence (3.9) holds.

The additional claim follows by the observation that (for $\alpha = -1/2$) we have $\binom{-1/2}{n} = \binom{2n}{n}(-4)^{-n}$.
                                                                                    □

> **Takeaways** Generating functions determine a probability distribution on $\mathbb{N}_0$. They are the perfect analytic tool for studying sums of independent random variables (on $\mathbb{N}_0$) as these sums translate into products of the generating functions.

Exercise 3.1.1 Show that $b_{r,p}^- * b_{s,p}^- = b_{r+s,p}^-$ for $r, s \in (0, \infty)$ and $p \in (0, 1]$. ♣

Exercise 3.1.2 Give an example for two different probability generating functions that coincide at countably many points $x_i \in (0, 1)$, $i \in \mathbb{N}$. (That is, in Theorem 3.2(iii), the assumption $\psi(z) < \infty$ for some $z > 1$ cannot be dropped.) ♣

3.2 Poisson Approximation

Lemma 3.6 *Let μ and $(\mu_n)_{n\in\mathbb{N}}$ be probability measures on $(\mathbb{N}_0, 2^{\mathbb{N}_0})$ with generating functions ψ and ψ_n, $n \in \mathbb{N}$. Then the following statements are equivalent.*

(i) $\mu_n(\{k\}) \xrightarrow{n\to\infty} \mu(\{k\})$ *for all $k \in \mathbb{N}_0$.*

(ii) $\mu_n(A) \xrightarrow{n\to\infty} \mu(A)$ *for all $A \subset \mathbb{N}_0$.*

(iii) $\psi_n(z) \xrightarrow{n\to\infty} \psi(z)$ *for all $z \in [0, 1]$.*

(iv) $\psi_n(z) \xrightarrow{n\to\infty} \psi(z)$ *for all $z \in [0, \eta)$ for some $\eta \in (0, 1)$.*

We write $\mu_n \xrightarrow{n\to\infty} \mu$ if any of the four conditions holds and say that $(\mu_n)_{n\in\mathbb{N}}$ converges weakly to μ.

Proof **(i)** $\Longrightarrow$ **(ii)** Fix $\varepsilon > 0$ and choose $N \in \mathbb{N}$ such that $\mu(\{N + 1, N + 2, \ldots\}) < \frac{\varepsilon}{4}$. For sufficiently large $n_0 \in \mathbb{N}$, we have

$$\sum_{k=0}^{N} |\mu_n(\{k\}) - \mu(\{k\})| < \frac{\varepsilon}{4} \quad \text{for all } n \geq n_0.$$

In particular, for any $n \geq n_0$, we have $\mu_n(\{N+1, N+2, \ldots\}) < \frac{\varepsilon}{2}$. Hence, for $n \geq n_0$,

$$\left|\mu_n(A) - \mu(A)\right| \leq \mu_n(\{N+1, N+2, \ldots\}) + \mu(\{N+1, N+2, \ldots\})$$

$$+ \sum_{k \in A \cap \{0, \ldots, N\}} \left|\mu_n(\{k\}) - \mu(\{k\})\right|$$

$$< \varepsilon.$$

(ii) $\Longrightarrow$ **(i)**   This is trivial.
(i) $\Longleftrightarrow$ **(iii)** $\Longleftrightarrow$ **(iv)**   This follows from the elementary theory of power series.
$\square$

Reflection If instead of $\mu(\mathbb{N}_0) = 1$, in the previous lemma we only assume $\mu(\mathbb{N}_0) \in [0, 1]$, then we still have $(i) \Longleftarrow (ii) \Longleftrightarrow (iii) \Longleftrightarrow (iv)$, but not $(i) \Longrightarrow (ii)$. Why? ♠

Let $(p_{n,k})_{n,k \in \mathbb{N}}$ be numbers with $p_{n,k} \in [0, 1]$ such that the limit

$$\lambda := \lim_{n \to \infty} \sum_{k=1}^{\infty} p_{n,k} \in (0, \infty) \tag{3.11}$$

exists and such that $\lim_{n \to \infty} \sum_{k=1}^{\infty} p_{n,k}^2 = 0$ (e.g., $p_{n,k} = \lambda/n$ for $k \leq n$ and $p_{n,k} = 0$ for $k > n$). For each $n \in \mathbb{N}$, let $(X_{n,k})_{k \in \mathbb{N}}$ be an independent family of random variables with $X_{n,k} \sim \mathrm{Ber}_{p_{n,k}}$.
Define

$$S^n := \sum_{l=1}^{\infty} X_{n,l} \quad \text{and} \quad S_k^n := \sum_{l=1}^{k} X_{n,l} \quad \text{for } k \in \mathbb{N}.$$

Theorem 3.7 (Poisson approximation) *Under the above assumptions, the distributions* $(\mathbf{P}_{S^n})_{n \in \mathbb{N}}$ *converge weakly to the Poisson distribution* Poi_λ.

Proof The p.g.f. of the Poisson distribution is $\psi(z) = e^{\lambda(z-1)}$ (see (3.4)). On the other hand, $S^n - S_k^n$ and S_k^n are independent for any $k \in \mathbb{N}$; hence $\psi_{S^n} = \psi_{S_k^n} \cdot \psi_{S^n - S_k^n}$. Now, for any $z \in [0, 1]$,

$$1 \geq \frac{\psi_{S^n}(z)}{\psi_{S_k^n}(z)} = \psi_{S^n - S_k^n}(z) \geq 1 - \mathbf{P}[S^n - S_k^n \geq 1] \geq 1 - \sum_{l=k+1}^{\infty} p_{n,l} \xrightarrow{k \to \infty} 1,$$

hence

$$\psi_{S^n}(z) = \lim_{k \to \infty} \psi_{S_k^n}(z) = \prod_{l=1}^{\infty} (p_{n,l} z + (1 - p_{n,l}))$$

$$= \exp\left(\sum_{l=1}^{\infty} \log\left(1 + p_{n,l}(z - 1)\right)\right).$$

Note that $|\log(1 + x) - x| \le x^2$ for $|x| < \frac{1}{2}$. By assumption, $\max_{l \in \mathbb{N}} p_{n,l} \to 0$ for $n \to \infty$; hence, for sufficiently large n,

$$\left| \left(\sum_{l=1}^{\infty} \log\left(1 + p_{n,l}(z - 1)\right)\right) - \left((z - 1)\sum_{l=1}^{\infty} p_{n,l}\right) \right|$$

$$\le \sum_{l=1}^{\infty} p_{n,l}^2 \le \left(\sum_{l=1}^{\infty} p_{n,l}\right) \max_{l \in \mathbb{N}} p_{n,l} \xrightarrow{n \to \infty} 0.$$

Together with (3.11), we infer

$$\lim_{n \to \infty} \psi_{S^n}(z) = \lim_{n \to \infty} \exp\left((z - 1)\sum_{l=1}^{\infty} p_{n,l}\right) = e^{\lambda(z-1)}. \qquad \Box$$

Takeaways The number of successes of a large number of improbable independent events is approximately Poisson distributed. Hence the Poisson distribution is used to model the number of rare successes in a large number of trials.

Exercise 3.2.1 Let $\lambda > 0$ and $p_n = \lambda/n$, $n \in \mathbb{N}$. In Theorem 3.7, it was shown that the binomial distribution $b_{n,\lambda/n}$ converges to the Poisson distribution Poi_λ. Show this with a different approach by checking condition (i) from Lemma 3.6. ♣

3.3 Branching Processes

Branching processes are models for the random development of the size of a population. Generating functions are the ideal tool for the analysis of such processes.

Let $T, X_1, X_2, \ldots$ be independent $\mathbb{N}_0$-valued random variables. What is the distribution of $S := \sum_{n=1}^{T} X_n$? First of all, note that S is measurable since

$$\{S = k\} = \bigcup_{n=0}^{\infty} \{T = n\} \cap \{X_1 + \ldots + X_n = k\}.$$

Theorem 3.8 *If the random variables $X_1, X_2, \ldots$ are also identically distributed, then the probability generating function of S is given by $\psi_S(z) = \psi_T(\psi_{X_1}(z))$.*

Proof We compute

$$\psi_S(z) = \sum_{k=0}^{\infty} \mathbf{P}[S = k] z^k$$

$$= \sum_{k=0}^{\infty} \sum_{n=0}^{\infty} \mathbf{P}[T = n] \mathbf{P}[X_1 + \ldots + X_n = k] z^k$$

$$= \sum_{n=0}^{\infty} \mathbf{P}[T = n] \psi_{X_1}(z)^n = \psi_T(\psi_{X_1}(z)). \qquad \square$$

Now assume that $p_0, p_1, p_2, \ldots \in [0, 1]$ are such that $\sum_{k=0}^{\infty} p_k = 1$. Let $(X_{n,i})_{n,i \in \mathbb{N}_0}$ be an independent family of random variables with $\mathbf{P}[X_{n,i} = k] = p_k$ for all $i, k, n \in \mathbb{N}_0$.

Let $Z_0 = 1$ and

$$Z_n = \sum_{i=1}^{Z_{n-1}} X_{n-1,i} \quad \text{for } n \in \mathbb{N}.$$

Z_n can be interpreted as the number of individuals in the nth generation of a randomly developing population. The ith individual in the nth generation has $X_{n,i}$ offspring (in the $(n + 1)$th generation).

Definition 3.9 $(Z_n)_{n \in \mathbb{N}_0}$ *is called a **Galton–Watson process** or **branching process** with offspring distribution $(p_k)_{k \in \mathbb{N}_0}$.*

Probability generating functions are an important tool for the investigation of branching processes. Hence, let

$$\psi(z) = \sum_{k=0}^{\infty} p_k z^k$$

be the p.g.f. of the offspring distribution and let ψ' be its derivative. Recursively, define the nth iterate of ψ by

$$\psi_1 := \psi \quad \text{and} \quad \psi_n := \psi \circ \psi_{n-1} \ \text{ for } n = 2, 3, \ldots.$$

Finally, let ψ_{Z_n} be the p.g.f. of Z_n.

Lemma 3.10 $\psi_n = \psi_{Z_n}$ for all $n \in \mathbb{N}$.

Proof For $n = 1$, the statement is true by definition. For $n \in \mathbb{N}$, we conclude inductively by Theorem 3.8 that $\psi_{Z_{n+1}} = \psi \circ \psi_{Z_n} = \psi \circ \psi_n = \psi_{n+1}$. $\qquad\square$

Clearly, the probability $q_n := \mathbf{P}[Z_n = 0]$ that Z is extinct by time n is monotone increasing in n. We denote by

$$q := \lim_{n \to \infty} \mathbf{P}[Z_n = 0]$$

the *extinction probability*; that is, the probability that the population will *eventually* die out.

Under what conditions do we have $q = 0$, $q = 1$, or $q \in (0, 1)$? Clearly, $q \geq p_0$. On the other hand, if $p_0 = 0$, then Z_n is monotone in n; hence $q = 0$.

Theorem 3.11 (Extinction probability of the Galton–Watson process)
Assume $p_1 \neq 1$. Then:

(i) $F := \{r \in [0, 1] : \psi(r) = r\} = \{q, 1\}$.
(ii) *The following equivalences hold:*

$$q < 1 \quad \Longleftrightarrow \quad \lim_{z \uparrow 1} \psi'(z) > 1 \quad \Longleftrightarrow \quad \sum_{k=1}^{\infty} k p_k > 1.$$

Proof

(i) We have $\psi(1) = 1$; hence $1 \in F$. Note that

$$q_n = \psi_n(0) = \psi(q_{n-1}) \quad \text{for all } n \in \mathbb{N}$$

and $q_n \uparrow q$. Since ψ is continuous, we infer

$$\psi(q) = \psi\left(\lim_{n \to \infty} q_n\right) = \lim_{n \to \infty} \psi(q_n) = \lim_{n \to \infty} q_{n+1} = q.$$

Thus $q \in F$. If $r \in F$ is an arbitrary fixed point of ψ, then $r \geq 0 = q_0$. Since ψ is monotone increasing, it follows that $r = \psi(r) \geq \psi(q_0) = q_1$. Inductively, we get $r \geq q_n$ for all $n \in \mathbb{N}_0$; that is, $r \geq q$. We conclude $q = \min F$.

(ii) If $p_0 + p_1 = 1$, then all of the statements are obvious. Now assume $p_0 + p_1 < 1$. For the first equivalence, we distinguish two cases.

Case 1: $\lim_{z \uparrow 1} \psi'(z) \le 1$. Since ψ is strictly convex, in this case, we have $\psi(z) >$ z for all $z \in [0, 1)$; hence $F = \{1\}$. We conclude $q = 1$.

Case 2: $\lim_{z \uparrow 1} \psi'(z) > 1$. As ψ is strictly convex and since $\psi(0) \ge 0$, there is a unique $r \in [0, 1)$ such that $\psi(r) = r$. Hence $F = \{r, 1\}$ and $q = \min F = r$. The second equivalence in (ii) follows by (3.2).   □

For further reading, we refer to [5].

Takeaways A branching process dies out eventually if the mean number of offspring is no larger than 1. More generally, the extinction probability is the smallest fixed point of the generating function of the offspring distribution.

Exercise 3.3.1 Assume that we have a branching process $Z = (Z_n)_{n \in \mathbb{N}_0}$ with $Z_0 = 1$ whose offspring distribution is given by $p_0 = 1/3$ and $p_2 = 2/3$. Compute $\psi'(1)$ and the extinction probability. ♣

Exercise 3.3.2 Assume that we have a branching process $Z = (Z_n)_{n \in \mathbb{N}_0}$ with $Z_0 = 1$ whose offspring distribution is given by $p_k = \frac{1}{3} \cdot (2/3)^k$, $k \in \mathbb{N}_0$.

 (i) Compute the generating function ψ and the extinction probability q.
 (ii) For this particular ψ, all the iterates are of a special form and can be computed explicitly. Do it!
 (iii) Compute $\lim_{n \to \infty} \psi_n(z)$, $z \in [0, 1]$. What does the result imply for the convergence of $\mathbf{P}_{Z_n}$? (Compare Lemma 3.6 and the comment below it.) ♣

Chapter 4
The Integral

Based on the notions of measure spaces and measurable maps, we introduce the integral of a measurable map with respect to a general measure. This generalizes the Lebesgue integral that can be found in textbooks on calculus. Furthermore, the integral is a cornerstone in a systematic theory of probability that allows for the definition and investigation of expected values and higher moments of random variables.

In this chapter, we define the integral by an approximation scheme with simple functions. Then we deduce basic statements such as Fatou's lemma. Other important convergence theorems for integrals follow in Chaps. 6 and 7.

4.1 Construction and Simple Properties

In the following, $(\Omega, \mathcal{A}, \mu)$ will always be a measure space. We denote by $\mathbb{E}$ the vector space of simple functions (see Definition 1.93) on $(\Omega, \mathcal{A})$ and by

$$\mathbb{E}^+ := \{f \in \mathbb{E} : f \geq 0\}$$

the cone (why this name?) of nonnegative simple functions. If

$$f = \sum_{i=1}^{m} \alpha_i \mathbb{1}_{A_i} \tag{4.1}$$

for some $m \in \mathbb{N}$ and for $\alpha_1, \ldots, \alpha_m \in (0, \infty)$, and for mutually disjoint sets $A_1, \ldots, A_m \in \mathcal{A}$, then (4.1) is said to be a *normal representation* of f.

© The Editor(s) (if applicable) and The Author(s), under exclusive license
to Springer Nature Switzerland AG 2020
A. Klenke, *Probability Theory*, Universitext,
https://doi.org/10.1007/978-3-030-56402-5_4

Lemma 4.1 If $f = \sum_{i=1}^{m} \alpha_i \mathbb{1}_{A_i}$ and $f = \sum_{j=1}^{n} \beta_j \mathbb{1}_{B_j}$ are two normal representations of $f \in E^+$, then

$$\sum_{i=1}^{m} \alpha_i \, \mu(A_i) = \sum_{j=1}^{n} \beta_j \, \mu(B_j).$$

Proof If $\mu(A_i \cap B_j) > 0$ for some i and j, then $A_i \cap B_j \neq \emptyset$, and $f(\omega) = \alpha_i = \beta_j$ for any $\omega \in A_i \cap B_j$. Furthermore, clearly $A_i \subset \bigcup_{j=1}^{n} B_j$ if $\alpha_i \neq 0$, and $B_j \subset \bigcup_{i=1}^{m} A_i$ if $\beta_j \neq 0$. We conclude that

$$\sum_{i=1}^{m} \alpha_i \, \mu(A_i) = \sum_{i=1}^{m} \sum_{j=1}^{n} \alpha_i \, \mu(A_i \cap B_j)$$

$$= \sum_{i=1}^{m} \sum_{j=1}^{n} \beta_j \, \mu(A_i \cap B_j) = \sum_{j=1}^{n} \beta_j \, \mu(B_j). \qquad \square$$

This lemma allows us to make the following definition (since the value of $I(f)$ does not depend on the choice of the normal representation).

Definition 4.2 Define the map $I : E^+ \to [0, \infty]$ by

$$I(f) = \sum_{i=1}^{m} \alpha_i \, \mu(A_i)$$

if f has the normal representation $f = \sum_{i=1}^{m} \alpha_i \mathbb{1}_{A_i}$.

Lemma 4.3 The map I is positive linear and monotone increasing: Let $f, g \in E^+$ and $\alpha \geq 0$. Then the following statements hold.

  (i) $I(\alpha f) = \alpha \, I(f)$.
  (ii) $I(f + g) = I(f) + I(g)$.
  (iii) If $f \leq g$, then $I(f) \leq I(g)$.

Proof This is left as an exercise. $\qquad \square$

Definition 4.4 (Integral) If $f : \Omega \to [0, \infty]$ is measurable, then we define the **integral** of f with respect to μ by

$$\int f \, d\mu := \sup \{ I(g) : g \in E^+, \ g \leq f \}.$$

Remark 4.5 By Lemma 4.3(iii), we have $I(f) = \int f \, d\mu$ for any $f \in \mathbb{E}^+$. Hence the integral is an extension of the map I from $\mathbb{E}^+$ to the set of nonnegative measurable functions. ◊

If $f, g : \Omega \to \overline{\mathbb{R}}$ with $f(\omega) \leq g(\omega)$ for any $\omega \in \Omega$, then we write $f \leq g$. Analogously, we write $f \geq 0$ and so on. On the other hand, we write "$f \leq g$ almost everywhere" if the weaker condition holds that there exists a μ-null set N such that $f(\omega) \leq g(\omega)$ for any $\omega \in N^c$.

Lemma 4.6 *Let $f, g, f_1, f_2, \ldots$ be measurable maps $\Omega \to [0, \infty]$. Then:*

(i) *(Monotonicity) If $f \leq g$, then $\int f \, d\mu \leq \int g \, d\mu$.*
(ii) *(Monotone convergence) If $f_n \uparrow f$, then the integrals also converge: $\int f_n \, d\mu \uparrow \int f \, d\mu$.*
(iii) *(Linearity) If $\alpha, \beta \in [0, \infty]$, then*

$$\int (\alpha f + \beta g) \, d\mu = \alpha \int f \, d\mu + \beta \int g \, d\mu,$$

where we use the convention $\infty \cdot 0 := 0$.

Proof

(i) This is immediate from the definition of the integral.
(ii) By (i), we have

$$\lim_{n \to \infty} \int f_n \, d\mu = \sup_{n \in \mathbb{N}} \int f_n \, d\mu \leq \int f \, d\mu.$$

Hence we only have to show $\int f \, d\mu \leq \sup_{n \in \mathbb{N}} \int f_n \, d\mu$.

Let $g \in \mathbb{E}^+$ with $g \leq f$. It is enough to show that

$$\sup_{n \in \mathbb{N}} \int f_n \, d\mu \geq \int g \, d\mu. \tag{4.2}$$

Assume that the simple function g has the normal representation $g = \sum_{i=1}^{N} \alpha_i \mathbb{1}_{A_i}$ for some $\alpha_1, \ldots, \alpha_N \in (0, \infty)$ and mutually disjoint sets $A_1, \ldots, A_N \in \mathcal{A}$. For any $\varepsilon > 0$ and $n \in \mathbb{N}$, define the set

$$B_n^\varepsilon = \{ f_n \geq (1 - \varepsilon) \, g \}.$$

Since $f_n \uparrow f \geq g$, we have $B_n^\varepsilon \uparrow \Omega$ for any $\varepsilon > 0$. Hence, by (i), for any $\varepsilon > 0$,

$$\int f_n \, d\mu \geq \int \left((1 - \varepsilon) \, g \, \mathbb{1}_{B_n^\varepsilon} \right) d\mu$$

$$= \sum_{i=1}^{N} (1 - \varepsilon) \, \alpha_i \, \mu(A_i \cap B_n^\varepsilon)$$

$$\overset{n \to \infty}{\longrightarrow} \sum_{i=1}^{N} (1 - \varepsilon) \, \alpha_i \, \mu(A_i) = (1 - \varepsilon) \int g \, d\mu.$$

Letting $\varepsilon \downarrow 0$ implies (4.2) and hence the claim (ii).

(iii) By Theorem 1.96, any nonnegative measurable map is a monotone limit of simple functions. Hence there are sequences $(f_n)_{n \in \mathbb{N}}$ and $(g_n)_{n \in \mathbb{N}}$ in $\mathbb{E}^+$ such that $f_n \uparrow f$ and $g_n \uparrow g$. Thus also $(\alpha f_n + \beta g_n) \uparrow \alpha f + \beta g$. By (ii) and Lemma 4.3, this implies

$$\int (\alpha f + \beta g) \, d\mu = \lim_{n \to \infty} \int (\alpha f_n + \beta g_n) \, d\mu$$

$$= \alpha \lim_{n \to \infty} \int f_n \, d\mu + \beta \lim_{n \to \infty} \int g_n \, d\mu = \alpha \int f \, d\mu + \beta \int g \, d\mu. \quad \square$$

For any measurable map $f : \Omega \to \overline{\mathbb{R}}$, we have $f^+ \leq |f|$ and $f^- \leq |f|$, which implies $\int f^\pm \, d\mu \leq \int |f| \, d\mu$. In particular, if $\int |f| \, d\mu < \infty$, then also $\int f^- \, d\mu < \infty$ and $\int f^+ \, d\mu < \infty$. Thus we can make the following definition that is the final definition for the integral of measurable functions.

Definition 4.7 (Integral of measurable functions) *A measurable function* $f : \Omega \to \overline{\mathbb{R}}$ *is called* μ-***integrable*** *if* $\int |f| \, d\mu < \infty$. *We write*

$$\mathcal{L}^1(\mu) := \mathcal{L}^1(\Omega, \mathcal{A}, \mu) := \left\{ f : \Omega \to \overline{\mathbb{R}} : f \text{ is measurable and } \int |f| \, d\mu < \infty \right\}.$$

For $f \in \mathcal{L}^1(\mu)$, *we define the integral of* f *with respect to* μ *by*

$$\int f(\omega) \, \mu(d\omega) := \int f \, d\mu := \int f^+ \, d\mu - \int f^- \, d\mu. \tag{4.3}$$

If we only have $\int f^- \, d\mu < \infty$ *or* $\int f^+ \, d\mu < \infty$, *then we also define* $\int f \, d\mu$ *by* (4.3). *Here the values* $+\infty$ *and* $-\infty$, *respectively, are possible.*

For $A \in \mathcal{A}$, *we define* $\displaystyle\int_A f \, d\mu := \int (f \, \mathbb{1}_A) \, d\mu.$

Theorem 4.8 *Let $f : \Omega \to [0, \infty]$ be a measurable map.*

(i) We have $f = 0$ almost everywhere if and only if $\int f \, d\mu = 0$.
(ii) If $\int f \, d\mu < \infty$, then $f < \infty$ almost everywhere.

Proof

(i) " $\Longrightarrow$ " Assume $f = 0$ almost everywhere. Let $N = \{\omega : f(\omega) > 0\}$. Then $f \leq \infty \cdot \mathbb{1}_N$ and $n\mathbb{1}_N \uparrow \infty \cdot \mathbb{1}_N$. From Lemma 4.6(i) and (ii), we infer

$$0 \leq \int f \, d\mu \leq \int (\infty \cdot \mathbb{1}_N) \, d\mu = \lim_{n \to \infty} \int n\mathbb{1}_N \, d\mu = 0.$$

" $\Longleftarrow$ " Let $N_n = \{f \geq \frac{1}{n}\}, n \in \mathbb{N}$. Then $N_n \uparrow N$ and

$$0 = \int f \, d\mu \geq \int \frac{1}{n} \mathbb{1}_{N_n} \, d\mu = \frac{\mu(N_n)}{n}.$$

Hence $\mu(N_n) = 0$ for any $n \in \mathbb{N}$ and thus $\mu(N) = 0$.

(ii) Let $A = \{\omega : f(\omega) = \infty\}$. For $n \in \mathbb{N}$, we have $\frac{1}{n} f \, \mathbb{1}_{\{f \geq n\}} \geq \mathbb{1}_{\{f \geq n\}}$. Hence Lemma 4.6(i) implies

$$\mu(A) = \int \mathbb{1}_A \, d\mu \leq \int \mathbb{1}_{\{f \geq n\}} \, d\mu \leq \frac{1}{n} \int f \mathbb{1}_{\{f \geq n\}} \, d\mu \leq \frac{1}{n} \int f \, d\mu \overset{n \to \infty}{\longrightarrow} 0. \quad \square$$

Theorem 4.9 (Properties of the integral) *Let $f, g \in \mathcal{L}^1(\mu)$.*

(i) (Monotonicity) If $f \leq g$ almost everywhere, then $\int f \, d\mu \leq \int g \, d\mu$.
In particular, if $f = g$ almost everywhere, then $\int f \, d\mu = \int g \, d\mu$.
(ii) (Triangle inequality) $\left| \int f \, d\mu \right| \leq \int |f| \, d\mu$.
(iii) (Linearity) If $\alpha, \beta \in \mathbb{R}$, then $\alpha f + \beta g \in \mathcal{L}^1(\mu)$ and

$$\int (\alpha f + \beta g) \, d\mu = \alpha \int f \, d\mu + \beta \int g \, d\mu.$$

This equation also holds if at most one of the integrals $\int f \, d\mu$ and $\int g \, d\mu$ is infinite.

Proof

(i) Clearly, $f^+ \leq g^+$ a.e., hence $(f^+ - g^+)^+ = 0$ a.e. By Theorem 4.8, we get $\int (f^+ - g^+)^+ \, d\mu = 0$. Since $f^+ \leq g^+ + (f^+ - g^+)^+$ (not only a.e.), we infer from Lemma 4.6(i) and (iii)

$$\int f^+ \, d\mu \leq \int \left(g^+ + (f^+ - g^+)^+\right) d\mu = \int g^+ \, d\mu.$$

Similarly, we use $f^- \geq g^-$ a.e. to obtain

$$\int f^- \, d\mu \geq \int g^- \, d\mu.$$

This implies

$$\int f \, d\mu = \int f^+ \, d\mu - \int f^- \, d\mu \leq \int g^+ \, d\mu - \int g^- \, d\mu = \int g \, d\mu.$$

(ii) Since $f^+ + f^- = |f|$, Lemma 4.6(iii) yields

$$\left| \int f \, d\mu \right| = \left| \int f^+ \, d\mu - \int f^- \, d\mu \right| \leq \int f^+ \, d\mu + \int f^- \, d\mu$$

$$= \int \left(f^+ + f^- \right) d\mu = \int |f| \, d\mu.$$

(iii) Since $|\alpha f + \beta g| \leq |\alpha| \cdot |f| + |\beta| \cdot |g|$, Lemma 4.6(i) and (iii) yield that $\alpha f + \beta g \in \mathcal{L}^1(\mu)$. In order to show linearity, it is enough to check the following three properties.

(a) $\int (f + g) \, d\mu = \int f \, d\mu + \int g \, d\mu$.
(b) $\int \alpha f \, d\mu = \alpha \int f \, d\mu$ for $\alpha \geq 0$.
(c) $\int (-f) \, d\mu = - \int f \, d\mu$.

(a) We have $(f + g)^+ - (f + g)^- = f + g = f^+ - f^- + g^+ - g^-$; hence $(f + g)^+ + f^- + g^- = (f + g)^- + f^+ + g^+$. By Lemma 4.6(iii), we infer

$$\int (f+g)^+ \, d\mu + \int f^- \, d\mu + \int g^- \, d\mu = \int (f+g)^- \, d\mu + \int f^+ \, d\mu + \int g^+ \, d\mu.$$

Hence

$$\int (f + g) \, d\mu = \int (f + g)^+ \, d\mu - \int (f + g)^- \, d\mu$$

$$= \int f^+ \, d\mu - \int f^- \, d\mu + \int g^+ \, d\mu - \int g^- \, d\mu$$

$$= \int f \, d\mu + \int g \, d\mu.$$

(b) For $\alpha \geq 0$, we have

$$\int \alpha f \, d\mu = \int \alpha f^+ \, d\mu - \int \alpha f^- \, d\mu = \alpha \int f^+ \, d\mu - \alpha \int f^- \, d\mu = \alpha \int f \, d\mu.$$

(c) We have

$$\int (-f)\,d\mu = \int (-f)^+\,d\mu - \int (-f)^-\,d\mu$$

$$= \int f^-\,d\mu - \int f^+\,d\mu = -\int f\,d\mu.$$

The supplementary statement is simple and is left as an exercise. □

Theorem 4.10 (Image measure) *Let $(\Omega, \mathcal{A})$ and $(\Omega', \mathcal{A}')$ be measurable spaces, let μ be a measure on $(\Omega, \mathcal{A})$ and let $X : \Omega \to \Omega'$ be measurable. Let $\mu' = \mu \circ X^{-1}$ be the image measure of μ under the map X. Assume that $f : \Omega' \to \overline{\mathbb{R}}$ is μ'-integrable. Then $f \circ X \in \mathcal{L}^1(\mu)$ and*

$$\int (f \circ X)\,d\mu = \int f\,d(\mu \circ X^{-1}).$$

In particular, if X is a random variable on $(\Omega, \mathcal{A}, \mathbf{P})$, then

$$\int f(x)\,\mathbf{P}[X \in dx] := \int f(x)\,\mathbf{P}_X[dx] = \int f\,d\mathbf{P}_X = \int f(X(\omega))\,\mathbf{P}[d\omega].$$

Proof This is left as an exercise. □

Example 4.11 (Discrete measure space) Let $(\Omega, \mathcal{A})$ be a discrete measurable space and let $\mu = \sum_{\omega \in \Omega} \alpha_\omega \delta_\omega$ for certain numbers $\alpha_\omega \geq 0$, $\omega \in \Omega$. A map $f : \Omega \to \mathbb{R}$ is integrable if and only if $\sum_{\omega \in \Omega} |f(\omega)|\,\alpha_\omega < \infty$. In this case,

$$\int f\,d\mu = \sum_{\omega \in \Omega} f(\omega)\,\alpha_\omega. \quad \Diamond$$

Definition 4.12 (Lebesgue integral) *Let λ be the Lebesgue measure on $\mathbb{R}^n$ and let $f : \mathbb{R}^n \to \mathbb{R}$ be measurable with respect to $\mathcal{B}^*(\mathbb{R}^n) - \mathcal{B}(\mathbb{R})$ (here $\mathcal{B}^*(\mathbb{R}^n)$ is the Lebesgue σ-algebra; see Example 1.71) and λ-integrable. Then we call*

$$\int f\,d\lambda$$

the **Lebesgue integral** *of f. If $A \in \mathcal{B}(\mathbb{R}^n)$ and $f : \mathbb{R}^n \to \mathbb{R}$ is measurable (or $f : A \to \mathbb{R}$ is $\mathcal{B}^*(\mathbb{R}^n)\big|_A - \mathcal{B}(\mathbb{R})$-measurable and hence $f\,\mathbb{1}_A$ is $\mathcal{B}^*(\mathbb{R}^n) - \mathcal{B}(\mathbb{R})$-measurable), then we write*

$$\int_A f\,d\lambda := \int f\,\mathbb{1}_A\,d\lambda.$$

Definition 4.13 *Let μ be a measure on $(\Omega, \mathcal{A})$ and let $f : \Omega \to [0, \infty)$ be a measurable map. Define the measure ν by*

$$\nu(A) := \int (\mathbb{1}_A \, f) \, d\mu \quad \text{for } A \in \mathcal{A}.$$

*We say that $f\mu := \nu$ has **density** f with respect to μ.*

Remark 4.14 We still have to show that ν is a measure. To this end, we check the conditions of Theorem 1.36. Clearly, $\nu(\emptyset) = 0$. Finite additivity follows from additivity of the integral (Lemma 4.6(iii)). Lower semicontinuity follows from the monotone convergence theorem (Theorem 4.20). ◊

Theorem 4.15 *We have $g \in \mathcal{L}^1(f\mu)$ if and only if $(gf) \in \mathcal{L}^1(\mu)$. In this case,*

$$\int g \, d(f\mu) = \int (gf) \, d\mu.$$

Proof First note that the statement holds for indicator functions. Then, with the usual arguments, extend it step by step first to simple functions, then to nonnegative measurable functions and finally to signed measurable functions.   □

Definition 4.16 *For measurable $f : \Omega \to \overline{\mathbb{R}}$, define*

$$\|f\|_p := \left(\int |f|^p \, d\mu \right)^{1/p}, \quad \text{if } p \in [1, \infty),$$

and

$$\|f\|_\infty := \inf \left\{ K \geq 0 : \mu(\{|f| > K\}) = 0 \right\}.$$

Further, for any $p \in [1, \infty]$, define the vector space

$$\mathcal{L}^p(\mu) := \left\{ f : \Omega \to \overline{\mathbb{R}} \text{ is measurable and } \|f\|_p < \infty \right\}.$$

Theorem 4.17 *The map $\| \cdot \|_1$ is a seminorm on $\mathcal{L}^1(\mu)$; that is, for all $f, g \in \mathcal{L}^1(\mu)$ and $\alpha \in \mathbb{R}$,*

$$\|\alpha f\|_1 = |\alpha| \cdot \|f\|_1,$$

$$\|f + g\|_1 \leq \|f\|_1 + \|g\|_1, \tag{4.4}$$

$$\|f\|_1 \geq 0 \text{ for all } f \quad \text{and} \quad \|f\|_1 = 0 \text{ if } f = 0 \text{ a.e.}$$

Proof The first and the third statements follow from Theorem 4.9(iii) and Theorem 4.8(i). The second statement follows from Theorem 4.9(i) since $|f + g| \leq$

$|f| + |g|$; hence

$$\|f + g\|_1 = \int |f + g|\, d\mu \le \int |f|\, d\mu + \int |g|\, d\mu = \|f\|_1 + \|g\|_1. \qquad \square$$

Remark 4.18 In fact, $\|\cdot\|_p$ is a seminorm on $\mathcal{L}^p(\mu)$ for all $p \in [1, \infty]$. Linearity and positivity are obvious, and the triangle inequality is a consequence of Minkowski's inequality, which we will show in Theorem 7.17. $\Diamond$

Theorem 4.19 *Let $\mu(\Omega) < \infty$ and $1 \le p' \le p \le \infty$. Then $\mathcal{L}^p(\mu) \subset \mathcal{L}^{p'}(\mu)$ and the canonical inclusion $i : \mathcal{L}^p(\mu) \hookrightarrow \mathcal{L}^{p'}(\mu)$, $f \mapsto f$ is continuous.*

Proof Let $f \in \mathcal{L}^\infty(\mu)$ and $p' \in [1, \infty)$. Then $|f|^{p'} \le \|f\|_\infty^{p'}$ almost everywhere; hence

$$\int |f|^{p'}\, d\mu \le \int \|f\|_\infty^{p'}\, d\mu = \|f\|_\infty^{p'} \cdot \mu(\Omega) < \infty.$$

Thus $\|f - g\|_{p'} \le \mu(\Omega)^{1/p'} \|f - g\|_\infty$ for $f, g \in \mathcal{L}^\infty(\mu)$ and hence i is continuous.

Now let $p, p' \in [1, \infty)$ with $p' < p$ and let $f \in \mathcal{L}^p(\mu)$. Then $|f|^{p'} \le 1 + |f|^p$; hence

$$\int |f|^{p'}\, d\mu \le \mu(\Omega) + \int |f|^p\, d\mu < \infty.$$

Finally, let $f, g \in \mathcal{L}^p(\mu)$. For any $c > 0$, we have

$$|f - g|^{p'} = |f - g|^{p'} \mathbb{1}_{\{|f-g| \le c\}} + |f - g|^{p'} \mathbb{1}_{\{|f-g| > c\}} \le c^{p'} + c^{p'-p}|f - g|^p.$$

In particular, letting $c = \|f - g\|_p$ we obtain

$$\|f - g\|_{p'} \le \left(c^{p'} \mu(\Omega) + c^{p'-p} \|f - g\|_p^p \right)^{1/p'} = (1 + \mu(\Omega))^{1/p'} \|f - g\|_p.$$

Hence, also in this case, i is continuous. $\qquad \square$

Takeaways The integral was defined first for functions which take only finitely many values. For more general measurable functions, the integral was then defined as the limit of integrals of approximating elementary functions. The full procedure is rather technical and does not allow for a smooth intuitive description. From an abstract point of view, the integral is monotone and linear and fulfills the triangle inequality, which allows to use it to define normed vector spaces of functions.

Exercise 4.1.1 Let $f : \mathbb{R} \to \mathbb{R}$ be defined by $f(x) = e^{-x}\mathbb{1}_{[0,\infty)}(x)$, and let λ the Lebesgue measure on $\mathbb{R}$.

(i) Find a sequence (f_n) of elementary functions such that $f_n \uparrow f$.
(ii) Compute $\int f_n \, d\lambda$ and determine $\int f \, d\lambda$ as a limit of integrals. ♣

Exercise 4.1.2 (Sequence spaces) Now we do not assume $\mu(\Omega) < \infty$. Assume there exists an $a > 0$ such that for any $A \in \mathcal{A}$ either $\mu(A) = 0$ or $\mu(A) \geq a$. Show that the reverse inclusion to Theorem 4.19 holds,

$$\mathcal{L}^{p'}(\mu) \subset \mathcal{L}^p(\mu) \quad \text{if } 1 \leq p' \leq p \leq \infty. \tag{4.5}$$

♣

Exercise 4.1.3 Let $1 \leq p' < p \leq \infty$ and let μ be σ-finite but not finite. Show that $\mathcal{L}^p(\mu) \setminus \mathcal{L}^{p'}(\mu) \neq \emptyset$. ♣

4.2   Monotone Convergence and Fatou's Lemma

What are the conditions that allow the interchange of limit and integral? In this section, we derive two simple criteria that prepare us for important applications such as the law of large numbers (Chap. 5). More general criteria will be presented in Chap. 6.

Theorem 4.20 (Monotone convergence, Beppo Levi theorem) *Let $f_1, f_2, \ldots \in \mathcal{L}^1(\mu)$ and let $f : \Omega \to \overline{\mathbb{R}}$ be measurable. Assume $f_n \uparrow f$ a.e. for $n \to \infty$. Then*

$$\lim_{n\to\infty} \int f_n \, d\mu = \int f \, d\mu,$$

where both sides can equal $+\infty$.

Proof Let $N \subset \Omega$ be a null set such that $f_n(\omega) \uparrow f(\omega)$ for all $\omega \in N^c$. The functions $f_n' := (f_n - f_1)\mathbb{1}_{N^c}$ and $f' := (f - f_1)\mathbb{1}_{N^c}$ are nonnegative and fulfill $f_n' \uparrow f'$. By Lemma 4.6(ii), we have $\int f_n' \, d\mu \overset{n\to\infty}{\longrightarrow} \int f' \, d\mu$. Since $f_n = f_n' + f_1$ a.e. and $f = f' + f_1$ a.e., Theorem 4.9(iii) implies

$$\int f_n \, d\mu = \int f_1 \, d\mu + \int f_n' \, d\mu \overset{n\to\infty}{\longrightarrow} \int f_1 \, d\mu + \int f' \, d\mu = \int f \, d\mu. \qquad \square$$

Theorem 4.21 (Fatou's lemma) *Let $f \in \mathcal{L}^1(\mu)$ and let $f_1, f_2, \ldots$ be measurable with $f_n \geq f$ a.e. for all $n \in \mathbb{N}$. Then*

$$\int \left(\liminf_{n\to\infty} f_n \right) d\mu \leq \liminf_{n\to\infty} \int f_n \, d\mu.$$

Proof By considering $(f_n - f)_{n \in \mathbb{N}}$, we may assume $f_n \geq 0$ a.e. for all $n \in \mathbb{N}$. Define

$$g_n := \inf_{m \geq n} f_m.$$

Then $g_n \uparrow \liminf_{m \to \infty} f_m$ as $n \to \infty$, and hence by the monotone convergence theorem (Lemma 4.6(ii)) and by monotonicity, $g_n \leq f_n$ (thus $\int g_n \, d\mu \leq \int f_n \, d\mu$),

$$\int \liminf_{n \to \infty} f_n \, d\mu = \lim_{n \to \infty} \int g_n \, d\mu \leq \liminf_{n \to \infty} \int f_n \, d\mu. \qquad \square$$

Example 4.22 (Petersburg game) By a concrete example, we show that in Fatou's lemma the assumption of an integrable minorant is essential. Consider a gamble in a casino where in each round the player's bet either gets doubled or lost. For example, roulette is such a game. If the player bets on "red", she gets the stake back doubled if the ball lands in a red pocket. Otherwise the bet is lost (for the player, not for the casino). There are 37 pockets (in European roulettes), 18 of which are red, 18 are black and one is green (the zero). Hence, by symmetry, the chance of winning should be $p = 18/37 < \frac{1}{2}$. Now assume the gamble is played again and again. We can model this on a probability space $(\Omega, \mathcal{A}, \mathbf{P})$ where $\Omega = \{-1, 1\}^{\mathbb{N}}$, $\mathcal{A} = (2^{\{-1,1\}})^{\otimes \mathbb{N}}$ is the σ-algebra generated by the cylinder sets $[\omega_1, \ldots, \omega_n]$ and $\mathbf{P} = ((1 - p)\delta_{-1} + p\delta_1)^{\otimes \mathbb{N}}$ is the product measure. Denote by $D_n : \Omega \to \{-1, 1\}$, $\omega \mapsto \omega_n$ the result of the nth game (for $n \in \mathbb{N}$). If in the ith game the player makes a (random) stake of H_i euros, then the cumulative profit after the nth game is

$$S_n = \sum_{i=1}^{n} H_i D_i.$$

Now assume the gambler adopts the following doubling strategy. In the first round, the stake is $H_1 = 1$. If she wins, then she does not bet any money in the subsequent games; that is, $H_n = 0$ for all $n \geq 2$ if $D_1 = 1$. On the other hand, if she loses, then in the second game she doubles the stake; that is, $H_2 = 2$ if $D_1 = -1$. If she wins the second game, she leaves the casino and otherwise doubles the stake again and so on. Hence we can describe the strategy by the formula

$$H_n = \begin{cases} 0, & \text{if there is an } i \in \{1, \ldots, n-1\} \text{ with } D_i = 1, \\ 2^{n-1}, & \text{else.} \end{cases}$$

Note that H_n depends on $D_1, \ldots, D_{n-1}$ only. That is, it is measurable with respect to $\sigma(D_1, \ldots, D_{n-1})$. Clearly, it is a crucial requirement for any strategy that the decision for the next stake depend only on the information available at that time and not depend on the future results of the gamble.

The probability of no win until the nth game is $(1 - p)^n$; hence $\mathbf{P}[S_n = 1 - 2^n] = (1 - p)^n$ and $\mathbf{P}[S_n = 1] = 1 - (1 - p)^n$. Hence we expect an average gain of

$$\int S_n \, d\mathbf{P} = (1 - p)^n (1 - 2^n) + (1 - (1 - p)^n) = 1 - (2(1 - p))^n \leq 0$$

since $p \leq \frac{1}{2}$ (in the profitable casinos). We define

$$S = \begin{cases} -\infty, & \text{if } -1 = D_1 = D_2 = \ldots, \\ 1, & \text{else.} \end{cases}$$

Then $S_n \overset{n \to \infty}{\longrightarrow} S$ a.s. but $\lim_{n \to \infty} \int S_n \, d\mathbf{P} < \int S \, d\mathbf{P} = 1$ since $S = 1$ a.s. By Fatou's lemma, this is possible only if there is no integrable minorant for the sequence $(S_n)_{n \in \mathbb{N}}$. If we define $\tilde{S} := \inf\{S_n : n \in \mathbb{N}\}$, then indeed

$$\mathbf{P}[\tilde{S} = 1 - 2^{n-1}] = \mathbf{P}[D_1 = \ldots = D_{n-1} = -1 \text{ and } D_n = 1] = p(1 - p)^{n-1}.$$

Hence $\int \tilde{S} \, d\mathbf{P} = \sum_{n=1}^{\infty} (1 - 2^{n-1}) \, p(1 - p)^{n-1} = -\infty$ since $p \leq \frac{1}{2}$. ◊

Takeaways Assume we are given a pointwise convergent sequence of non-negative functions. Then the limit (inferior) of the integrals is at least as large as the integral of the limit (Fatou's lemma). In the case of monotone convergence we have equality. As an example where inequality holds, instead of the standard example from calculus textbooks ($f_n = n \cdot \mathbb{1}_{(0,1/n)}$, $f = 0$), we studied a game of hazard that we will encounter in a different context later.

Exercise 4.2.1 Let $(\Omega, \mathcal{A}, \mu)$ be a measure space and let $f \in \mathcal{L}^1(\mu)$. Show that for any $\varepsilon > 0$, there is an $A \in \mathcal{A}$ with $\mu(A) < \infty$ and $\left| \int_A f \, d\mu - \int f \, d\mu \right| < \varepsilon$. ♣

Exercise 4.2.2 Let $f_1, f_2, \ldots \in \mathcal{L}^1(\mu)$ be nonnegative and such that $\lim_{n \to \infty} \int f_n \, d\mu$ exists. Assume there exists a measurable f with $f_n \overset{n \to \infty}{\longrightarrow} f$ μ-almost everywhere. Show that $f \in \mathcal{L}^1(\mu)$ and

$$\lim_{n \to \infty} \int |f_n - f| \, d\mu = \lim_{n \to \infty} \int f_n \, d\mu - \int f \, d\mu. \quad ♣$$

Exercise 4.2.3 Let $f \in \mathcal{L}^1([0, \infty), \lambda)$ be a Lebesgue integrable function on $[0, \infty)$. Show that for λ-almost all $t \in [0, \infty)$ the series $\sum_{n=1}^{\infty} f(nt)$ converges absolutely. ♣

Exercise 4.2.4 Let λ be the Lebesgue measure on $\mathbb{R}$ and let A be a Borel set with $\lambda(A) < \infty$. Show that for any $\varepsilon > 0$, there is a compact set $C \subset A$, a closed set

$D \subset \mathbb{R} \setminus A$ and a continuous map $\varphi : \mathbb{R} \to [0, 1]$ with $\mathbb{1}_C \leq \varphi \leq \mathbb{1}_{\mathbb{R} \setminus D}$ and such that $\| \mathbb{1}_A - \varphi \|_1 < \varepsilon$.
Hint: Use the regularity of Lebesgue measure (Remark 1.67). ♣

Exercise 4.2.5 Let λ be the Lebesgue measure on $\mathbb{R}$, $p \in [1, \infty)$ and let $f \in \mathcal{L}^p(\lambda)$. Show that for any $\varepsilon > 0$, there is a continuous function $h : \mathbb{R} \to \mathbb{R}$ such that $\| f - h \|_p < \varepsilon$.
Hint: Use Exercise 4.2.4 to show the assertion first for indicator functions, then for simple functions and finally for general $f \in \mathcal{L}^p(\lambda)$. ♣

Exercise 4.2.6 Let λ be the Lebesgue measure on $\mathbb{R}$, $p \in [1, \infty)$ and let $f \in \mathcal{L}^p(\lambda)$. A map $h : \mathbb{R} \to \mathbb{R}$ is called a **step function** if there exist $n \in \mathbb{N}$ and numbers $t_0 < t_1 < \ldots < t_n$ and $\alpha_1, \ldots, \alpha_n$ such that $h = \sum_{k=1}^{n} \alpha_k \mathbb{1}_{(t_{k-1}, t_k]}$.
 Show that for any $\varepsilon > 0$, there exists a step function h such that $\| f - h \|_p < \varepsilon$.
Hint: Use the approximation theorem for measures (Theorem 1.65) with the semiring of left open intervals to show the assertion first for measurable indicator functions. Then use the approximation arguments as in Exercise 4.2.5. ♣

4.3 Lebesgue Integral Versus Riemann Integral

We show that for Riemann integrable functions the Lebesgue integral and the Riemann integral coincide.
 Let $I = [a, b] \subset \mathbb{R}$ be an interval and let λ be the Lebesgue measure on I. Further, consider sequences $t = (t^n)_{n \in \mathbb{N}}$ of partitions $t^n = (t_i^n)_{i=0,\ldots,n}$ of I (i.e., $a = t_0^n < t_1^n < \ldots < t_n^n = b$) that get finer and finer. That is,

$$|t^n| := \max\{t_i^n - t_{i-1}^n : i = 1, \ldots, n\} \overset{n \to \infty}{\longrightarrow} 0.$$

Assume that for any $n \in \mathbb{N}$, the partition t^{n+1} is a *refinement* of t^n; that is, $\{t_0^n, \ldots, t_n^n\} \subset \{t_0^{n+1}, \ldots, t_{n+1}^{n+1}\}$.
 For any function $f : I \to \mathbb{R}$ and any $n \in \mathbb{N}$, define the nth lower sum and upper sum, respectively, by

$$L_n^t(f) := \sum_{i=1}^{n} (t_i^n - t_{i-1}^n) \inf f\big([t_{i-1}^n, t_i^n)\big),$$

$$U_n^t(f) := \sum_{i=1}^{n} (t_i^n - t_{i-1}^n) \sup f\big([t_{i-1}^n, t_i^n)\big).$$

A function $f : I \to \mathbb{R}$ is called Riemann integrable if there exists a t such that the limits of the lower sums and upper sums are finite and coincide. In this case, the value of the limit does not depend on the choice of t, and the Riemann integral of f

is defined as (see, e.g., [149])

$$\int_a^b f(x)\, dx := \lim_{n\to\infty} L_n^t(f) = \lim_{n\to\infty} U_n^t(f). \tag{4.6}$$

Theorem 4.23 (Riemann integral and Lebesgue integral) *Let $f : I \to \mathbb{R}$ be Riemann integrable on $I = [a, b]$. Then f is Lebesgue integrable on I with integral*

$$\int_I f\, d\lambda = \int_a^b f(x)\, dx.$$

Proof Choose t such that (4.6) holds. By assumption, there is an $n \in \mathbb{N}$ with $|L_n^t(f)| < \infty$ and $|U_n^t(f)| < \infty$. Hence f is bounded. We can thus replace f by $f + \|f\|_\infty$ and hence assume that $f \geq 0$. Define

$$g_n := f(b)\, \mathbb{1}_{\{b\}} + \sum_{i=1}^n (\inf f([t_{i-1}^n, t_i^n)))\, \mathbb{1}_{[t_{i-1}^n, t_i^n)},$$

$$h_n := f(b)\, \mathbb{1}_{\{b\}} + \sum_{i=1}^n (\sup f([t_{i-1}^n, t_i^n)))\, \mathbb{1}_{[t_{i-1}^n, t_i^n)}.$$

As t^{n+1} is a refinement of t^n, we have $g_n \leq g_{n+1} \leq h_{n+1} \leq h_n$. Hence there exist g and h with $g_n \uparrow g$ and $h_n \downarrow h$. By construction, we have $g \leq h$ and

$$\int_I g\, d\lambda = \lim_{n\to\infty} \int_I g_n\, d\lambda = \lim_{n\to\infty} L_n^t(f)$$

$$= \lim_{n\to\infty} U_n^t(f) = \lim_{n\to\infty} \int_I h_n\, d\lambda = \int_I h\, d\lambda.$$

Hence $h = g$ λ-a.e. By construction, $g \leq f \leq h$, and as limits of simple functions, g and h are $\mathcal{B}(I) - \mathcal{B}(\mathbb{R})$-measurable. This implies that, for any $\alpha \in \mathbb{R}$, the set

$$\{f \leq \alpha\} = (\{g \leq \alpha\} \cap \{g = h\}) \uplus (\{f \leq \alpha\} \cap \{g \neq h\})$$

is the union of a $\mathcal{B}(I)$-set with a subset of a null set and is hence in $\mathcal{B}(I)^*$ (the Lebesgue completion of $\mathcal{B}(I)$). Hence f is $\mathcal{B}(I)^*$-measurable. By the monotone convergence theorem (Theorem 4.20), we conclude

$$\int_I f\, d\lambda = \lim_{n\to\infty} \int_I g_n\, d\lambda = \int_a^b f(x)\, dx. \qquad \square$$

Example 4.24 Let $f : [0, 1] \to \mathbb{R}, x \mapsto \mathbb{1}_{\mathbb{Q}}$. Then clearly f is not Riemann integrable since $L_n(f) = 0$ and $U_n(f) = 1$ for all $n \in \mathbb{N}$. On the other hand, f is Lebesgue integrable with integral $\int_{[0,1]} f \, d\lambda = 0$ because $\mathbb{Q} \cap [0, 1]$ is a null set. $\Diamond$

Remark 4.25 An improperly Riemann integrable function f on a one-sided open interval $I = (a, b]$ or $I = [0, \infty)$ is not necessarily Lebesgue integrable. Indeed, the improper integral $\int_0^\infty f(x) \, dx := \lim_{n \to \infty} \int_0^n f(x) \, dx$ is defined by a limit procedure that respects the *geometry* of $\mathbb{R}$. The Lebesgue integral does not do that. For example, the function $f : [0, \infty) \to \mathbb{R}, x \mapsto \frac{1}{1+x} \sin(x)$ is improperly Riemann integrable but is not Lebesgue integrable since $\int_{[0,\infty)} |f| \, d\lambda = \infty$. $\Diamond$

Reflection Consider the function $f(x) = 1/x, x \in [-1, 1] \setminus \{0\}, f(0) = 0$. Cauchy's principal value of the integral $\int_{-1}^1 f(x) dx$ is defined as

$$
\lim_{n \to \infty} \left(\int_{-1}^{-1/n} \frac{1}{x} \, dx + \int_{1/n}^1 \frac{1}{x} \, dx \right) = 0.
$$

Why is this kind of limit incompatible with the concept of the Lebesgue integral? ♠

On the one hand, improperly Riemann integrable functions need not be Lebesgue integrable. On the other hand, there are Lebesgue integrable functions that are not Riemann integrable (such as $\mathbb{1}_{\mathbb{Q}}$). The geometric interpretation is that the Riemann integral respects the geometry of the integration *domain* by being defined via slimmer and slimmer vertical rectangles (Fig. 4.1). On the other hand, the Lebesgue integral respects the geometry of the *range* by being defined via slimmer and slimmer horizontal strips. In particular, the Lebesgue integral does not make any assumption on the geometry of the domain and is thus more universal than the Riemann integral. In order to underline this, we present the following theorem that will also be useful later.

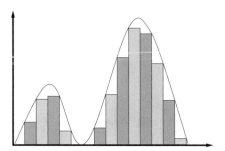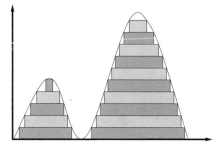

Fig. 4.1 For the Riemann integral, the area under the curve is approximated by rectangles of a fixed breadth (left hand side). The Lebesgue integral approximates the area by the measure of the levels sets (right hand side).

Theorem 4.26 *Let* $f : \Omega \to \mathbb{R}$ *be measurable and* $f \geq 0$ *almost everywhere.*
Then

$$\sum_{n=1}^{\infty} \mu(\{f \geq n\}) \leq \int f \, d\mu \leq \sum_{n=0}^{\infty} \mu(\{f > n\}) \tag{4.7}$$

and

$$\int f \, d\mu = \int_0^{\infty} \mu(\{f \geq t\}) \, dt. \tag{4.8}$$

Proof Define $f' = \lfloor f \rfloor$ and $f'' = \lceil f \rceil$. Then $f' \leq f \leq f''$ and hence $\int f' \, d\mu \leq \int f \, d\mu \leq \int f'' \, d\mu$. Now the first inequality of (4.7) follows from

$$\int f' \, d\mu = \sum_{k=1}^{\infty} \mu(\{f' = k\}) \cdot k = \sum_{k=1}^{\infty} \sum_{n=1}^{k} \mu(\{f' = k\})$$

$$= \sum_{n=1}^{\infty} \sum_{k=n}^{\infty} \mu(\{f' = k\})$$

$$= \sum_{n=1}^{\infty} \mu(\{f' \geq n\}) = \sum_{n=1}^{\infty} \mu(\{f \geq n\}).$$

Similarly, we infer the second inequality in (4.7) from

$$\int f'' \, d\mu = \sum_{n=1}^{\infty} \mu(\{f'' \geq n\}) = \sum_{n=1}^{\infty} \mu(\{f > n - 1\}).$$

If $g(t) := \mu(\{f \geq t\}) = \infty$ for some $t > 0$, then both sides in (4.8) equal ∞.
Hence, in the following, assume $g(t) < \infty$ for all $t > 0$.

For $\varepsilon > 0$ and $k \in \mathbb{N}$, define $g^{\varepsilon} := g \wedge g(\varepsilon)$, $f^{\varepsilon} := f \, \mathbb{1}_{\{f \geq \varepsilon\}}$ and $f_k^{\varepsilon} = 2^k f^{\varepsilon}$ as
well as

$$\alpha_k^{\varepsilon} := 2^{-k} \sum_{n=1}^{\infty} \mu(\{f^{\varepsilon} \geq n 2^{-k}\}).$$

Then $\alpha_k^{\varepsilon} \overset{k \to \infty}{\longrightarrow} \int_0^{\infty} g^{\varepsilon}(t) \, dt$. Furthermore, by (4.7) (with f replaced by f_k^{ε}), we
have

$$\alpha_k^{\varepsilon} = 2^{-k} \sum_{n=1}^{\infty} \mu(\{f_k^{\varepsilon} \geq n\}) \leq \int f^{\varepsilon} \, d\mu$$

$$\leq 2^{-k} \sum_{n=0}^{\infty} \mu(\{f_k^{\varepsilon} > n\}) = 2^{-k} \sum_{n=0}^{\infty} \mu(\{f^{\varepsilon} > n 2^{-k}\}) \leq \alpha_k^{\varepsilon} + 2^{-k} g(\varepsilon).$$

Since $2^{-k} g(\varepsilon) \overset{k\to\infty}{\longrightarrow} 0$, we get $\int_0^\infty g^\varepsilon(t)\,dt = \int f^\varepsilon\,d\mu$. Since $f^\varepsilon \uparrow f$ and $g^\varepsilon \uparrow g$ for $\varepsilon \downarrow 0$, the monotone convergence theorem implies (4.8). $\square$

> **Takeaways** A Riemann integrable function on a compact interval is Lebesgue integrable and the integrals coincide. For nonnegative functions, the Lebesgue integral can be computed via a kind of partial integration formula (Theorem 4.26).

Exercise 4.3.1 Use Theorem 4.26 to compute $\int_0^1 \log(x)\,dx$ and $\int_0^\pi \sin(x)\,dx$. ♣

Exercise 4.3.2 Let $f : [0, 1] \to \mathbb{R}$ be bounded. Show that f is (properly) Riemann integrable if and only if f is λ-a.e. continuous. ♣

Exercise 4.3.3 If $f : [0, 1] \to \mathbb{R}$ is Riemann integrable, then f is Lebesgue measurable. Give an example that shows that f need not be Borel measurable. (*Hint:* Without proof, use the existence of a subset of $[0, 1]$ that is not Borel measurable. Based on this, construct a set that is not Borel and whose closure is a null set.) ♣

Exercise 4.3.4 Let $f : [0, 1] \to (0, \infty)$ be Riemann integrable. Without using the equivalence of the Lebesgue integral and the Riemann integral, show that $\int_0^1 f(x)\,dx > 0$. ♣

Chapter 5
Moments and Laws of Large Numbers

The most important characteristic quantities of random variables are the median, expectation and variance. For large n, the expectation describes the typical approximate value of the arithmetic mean $(X_1 + \ldots + X_n)/n$ of i.i.d. random variables (law of large numbers). In Chap. 15, we will see how the variance determines the size of the typical deviations of the arithmetic mean from the expectation.

5.1 Moments

In the following, let $(\Omega, \mathcal{A}, \mathbf{P})$ be a probability space.

Definition 5.1 *Let X be a real-valued random variable.*

(i) *If $X \in \mathcal{L}^1(\mathbf{P})$, then X is called **integrable** and we call*

$$\mathbf{E}[X] := \int X \, d\mathbf{P}$$

*the **expectation** or **mean** of X. If $\mathbf{E}[X] = 0$, then X is called **centered**. More generally, we also write $\mathbf{E}[X] = \int X \, d\mathbf{P}$ if only X^- or X^+ is integrable.*

(ii) *If $n \in \mathbb{N}$ and $X \in \mathcal{L}^n(\mathbf{P})$, then the quantities*

$$m_k := \mathbf{E}\left[X^k\right], \quad M_k := \mathbf{E}\left[|X|^k\right] \quad \text{for any } k = 1, \ldots, n,$$

*are called the kth **moments** and kth **absolute moments**, respectively, of X.*

(iii) *If $X \in \mathcal{L}^2(\mathbf{P})$, then X is called **square integrable** and*

$$\mathbf{Var}[X] := \mathbf{E}\left[X^2\right] - \mathbf{E}[X]^2$$

© The Editor(s) (if applicable) and The Author(s), under exclusive license
to Springer Nature Switzerland AG 2020
A. Klenke, *Probability Theory*, Universitext,
https://doi.org/10.1007/978-3-030-56402-5_5

*is the **variance** of X. The number* $\sigma := \sqrt{\text{Var}[X]}$ *is called the **standard deviation** of X. Formally, we sometimes write* $\text{Var}[X] = \infty$ *if* $\text{E}[X^2] = \infty$.

(iv) If $X, Y \in \mathcal{L}^2(\mathbf{P})$, *then we define the **covariance** of X and Y by*

$$\text{Cov}[X, Y] := \text{E}\big[(X - \text{E}[X])(Y - \text{E}[Y])\big].$$

*X and Y are called **uncorrelated** if* $\text{Cov}[X, Y] = 0$ *and **correlated** otherwise.*

Remark 5.2

  (i) The definition in (ii) is sensible since, by virtue of Theorem 4.19, $X \in \mathcal{L}^n(\mathbf{P})$ implies that $M_k < \infty$ for all $k = 1, \ldots, n$.
 (ii) If $X, Y \in \mathcal{L}^2(\mathbf{P})$, then $XY \in \mathcal{L}^1(\mathbf{P})$ since $|XY| \leq X^2 + Y^2$. Hence the definition in (iv) makes sense and we have

$$\text{Cov}[X, Y] = \text{E}[XY] - \text{E}[X]\,\text{E}[Y].$$

In particular, $\text{Var}[X] = \text{Cov}[X, X]$. ◊

We collect the most important rules of expectations in a theorem. All of these properties are direct consequences of the corresponding properties of the integral.

Theorem 5.3 (Rules for expectations) *Let* $X, Y, X_n, Z_n, n \in \mathbb{N}$, *be real integrable random variables on* $(\Omega, \mathcal{A}, \mathbf{P})$.

  (i) *If* $\mathbf{P}_X = \mathbf{P}_Y$, *then* $\text{E}[X] = \text{E}[Y]$.
 (ii) *(Linearity) Let* $c \in \mathbb{R}$. *Then* $cX \in \mathcal{L}^1(\mathbf{P})$ *and* $X + Y \in \mathcal{L}^1(\mathbf{P})$ *as well as*

$$\text{E}[cX] = c\,\text{E}[X] \quad \text{and} \quad \text{E}[X + Y] = \text{E}[X] + \text{E}[Y].$$

(iii) *If* $X \geq 0$ *almost surely, then*

$$\text{E}[X] = 0 \quad \Longleftrightarrow \quad X = 0 \quad \text{almost surely.}$$

 (iv) *(Monotonicity) If* $X \leq Y$ *almost surely, then* $\text{E}[X] \leq \text{E}[Y]$ *with equality if and only if* $X = Y$ *almost surely.*
  (v) *(Triangle inequality)* $\big|\text{E}[X]\big| \leq \text{E}\big[|X|\big]$.
 (vi) *If* $X_n \geq 0$ *almost surely for all* $n \in \mathbb{N}$, *then* $\text{E}\Big[\sum_{n=1}^{\infty} X_n\Big] = \sum_{n=1}^{\infty} \text{E}[X_n]$.
(vii) *If* $Z_n \uparrow Z$ *for some Z, then* $\text{E}[Z] = \lim_{n\to\infty} \text{E}[Z_n] \in (-\infty, \infty]$.

Again probability theory comes into play when independence enters the stage; that is, when we exit the realm of linear integration theory.

Theorem 5.4 (Independent $\mathcal{L}(\mathbf{P})$**-random variables are uncorrelated)** *Let* $X, Y \in \mathcal{L}^1(\mathbf{P})$ *be independent. Then* $(X\,Y) \in \mathcal{L}^1(\mathbf{P})$ *and* $\text{E}[XY] = \text{E}[X]\,\text{E}[Y]$. *In particular, independent square integrable random variables are uncorrelated.*

Proof Assume first that X and Y take only finitely many values. Then XY also takes only finitely many values and thus $XY \in \mathcal{L}^1(\mathbf{P})$. It follows that

$$\mathbf{E}[XY] = \sum_{z \in \mathbb{R} \setminus \{0\}} z \, \mathbf{P}[XY = z]$$

$$= \sum_{z \in \mathbb{R} \setminus \{0\}} \sum_{x \in \mathbb{R} \setminus \{0\}} x \frac{z}{x} \, \mathbf{P}[X = x, \, Y = z/x]$$

$$= \sum_{y \in \mathbb{R} \setminus \{0\}} \sum_{x \in \mathbb{R} \setminus \{0\}} xy \, \mathbf{P}[X = x] \, \mathbf{P}[Y = y]$$

$$= \left(\sum_{x \in \mathbb{R}} x \, \mathbf{P}[X = x] \right) \left(\sum_{y \in \mathbb{R}} y \, \mathbf{P}[Y = y] \right)$$

$$= \mathbf{E}[X] \, \mathbf{E}[Y].$$

For $N \in \mathbb{N}$, the random variables $X_N := \left(2^{-N} \lfloor 2^N |X| \rfloor \right) \wedge N$ and $Y_N := \left(2^{-N} \lfloor 2^N |Y| \rfloor \right) \wedge N$ take only finitely many values and are independent as well. Furthermore, $X_N \uparrow |X|$ and $Y_N \uparrow |Y|$. By the monotone convergence theorem (Theorem 4.20), we infer

$$\mathbf{E}[|XY|] = \lim_{N \to \infty} \mathbf{E}[X_N Y_N] = \lim_{N \to \infty} \mathbf{E}[X_N] \, \mathbf{E}[Y_N]$$

$$= \left(\lim_{N \to \infty} \mathbf{E}[X_N] \right) \left(\lim_{N \to \infty} \mathbf{E}[Y_N] \right) = \mathbf{E}[|X|] \, \mathbf{E}[|Y|] < \infty.$$

Hence $XY \in \mathcal{L}^1(\mathbf{P})$. Furthermore, we have shown the claim in the case where X and Y are nonnegative. Hence (and since each of the families $\{X^+, Y^+\}$, $\{X^-, Y^+\}$, $\{X^+, Y^-\}$ and $\{X^-, Y^-\}$ is independent) we obtain

$$\mathbf{E}[XY] = \mathbf{E}[(X^+ - X^-)(Y^+ - Y^-)]$$

$$= \mathbf{E}[X^+ Y^+] - \mathbf{E}[X^- Y^+] - \mathbf{E}[X^+ Y^-] + \mathbf{E}[X^- Y^-]$$

$$= \mathbf{E}[X^+] \, \mathbf{E}[Y^+] - \mathbf{E}[X^-] \, \mathbf{E}[Y^+] - \mathbf{E}[X^+] \, \mathbf{E}[Y^-] + \mathbf{E}[X^-] \, \mathbf{E}[Y^-]$$

$$= \mathbf{E}[X^+ - X^-] \, \mathbf{E}[Y^+ - Y^-] = \mathbf{E}[X] \, \mathbf{E}[Y]. \qquad \square$$

Theorem 5.5 (Wald's identity) *Let* $T, X_1, X_2, \ldots$ *be independent real random variables in* $\mathcal{L}^1(\mathbf{P})$. *Let* $\mathbf{P}[T \in \mathbb{N}_0] = 1$ *and assume that* $X_1, X_2, \ldots$ *are identically distributed. Define*

$$S_T := \sum_{i=1}^{T} X_i.$$

Then $S_T \in \mathcal{L}^1(\mathbf{P})$ *and* $\mathbf{E}[S_T] = \mathbf{E}[T] \, \mathbf{E}[X_1]$.

Proof Define $S_n = \sum_{i=1}^{n} X_i$ for $n \in \mathbb{N}_0$. Then $S_T = \sum_{n=1}^{\infty} S_n \mathbb{1}_{\{T=n\}}$. By Remark 2.15, the random variables S_n and $\mathbb{1}_{\{T=n\}}$ are independent for any $n \in \mathbb{N}$ and thus uncorrelated. This implies (using the triangle inequality; see Theorem 5.3(v))

$$\mathbf{E}\big[|S_T|\big] = \sum_{n=1}^{\infty} \mathbf{E}\big[|S_n| \, \mathbb{1}_{\{T=n\}}\big] = \sum_{n=1}^{\infty} \mathbf{E}\big[|S_n|\big]\, \mathbf{E}\big[\mathbb{1}_{\{T=n\}}\big]$$

$$\leq \sum_{n=1}^{\infty} \mathbf{E}\big[|X_1|\big]\, n\, \mathbf{P}[T=n] = \mathbf{E}[|X_1|]\,\mathbf{E}[T].$$

The same computation without absolute values yields the remaining part of the claim.   □

We collect some basic properties of the variance.

Theorem 5.6 *Let* $X \in \mathcal{L}^2(\mathbf{P})$. *Then:*

(i) $\mathbf{Var}[X] = \mathbf{E}\big[(X - \mathbf{E}[X])^2\big] \geq 0$.
(ii) $\mathbf{Var}[X] = 0 \iff X = \mathbf{E}[X]$ *almost surely.*
(iii) *The map* $f : \mathbb{R} \to \mathbb{R}$, $x \mapsto \mathbf{E}\big[(X - x)^2\big]$ *is minimal at* $x_0 = \mathbf{E}[X]$ *with* $f(\mathbf{E}[X]) = \mathbf{Var}[X]$.

Proof

(i) This is a direct consequence of Remark 5.2(ii).
(ii) By Theorem 5.3(iii), we have $\mathbf{E}\big[(X - \mathbf{E}[X])^2\big] = 0 \iff (X - \mathbf{E}[X])^2 = 0$ a.s.
(iii) Clearly, $f(x) = \mathbf{E}[X^2] - 2x\,\mathbf{E}[X] + x^2 = \mathbf{Var}[X] + (x - \mathbf{E}[X])^2$.   □

Theorem 5.7 *The map* $\mathbf{Cov} : \mathcal{L}^2(\mathbf{P}) \times \mathcal{L}^2(\mathbf{P}) \to \mathbb{R}$ *is a positive semidefinite symmetric bilinear form and* $\mathbf{Cov}[X, Y] = 0$ *if* Y *is almost surely constant. The detailed version of this concise statement is: Let* $X_1, \ldots, X_m, Y_1, \ldots, Y_n \in \mathcal{L}^2(\mathbf{P})$ *and* $\alpha_1, \ldots, \alpha_m, \beta_1, \ldots, \beta_n \in \mathbb{R}$ *as well as* $d, e \in \mathbb{R}$. *Then*

$$\mathbf{Cov}\left[d + \sum_{i=1}^{m} \alpha_i X_i, e + \sum_{j=1}^{n} \beta_j Y_j\right] = \sum_{i,j} \alpha_i \beta_j\, \mathbf{Cov}[X_i, Y_j]. \qquad (5.1)$$

In particular, $\mathbf{Var}[\alpha X] = \alpha^2 \mathbf{Var}[X]$ *and the* **Bienaymé formula** *holds,*

$$\mathbf{Var}\left[\sum_{i=1}^{m} X_i\right] = \sum_{i=1}^{m} \mathbf{Var}[X_i] + \sum_{\substack{i,j=1 \\ i \neq j}}^{m} \mathbf{Cov}[X_i, X_j]. \qquad (5.2)$$

For uncorrelated $X_1, \ldots, X_m$, *we have* $\mathbf{Var}\left[\sum_{i=1}^{m} X_i\right] = \sum_{i=1}^{m} \mathbf{Var}[X_i]$.

Proof

$$\mathbf{Cov}\left[d + \sum_{i=1}^{m} \alpha_i X_i, \; e + \sum_{j=1}^{n} \beta_j Y_j\right]$$

$$= \mathbf{E}\left[\left(\sum_{i=1}^{m} \alpha_i (X_i - \mathbf{E}[X_i])\right)\left(\sum_{j=1}^{n} \beta_j (Y_j - \mathbf{E}[Y_j])\right)\right]$$

$$= \sum_{i=1}^{m}\sum_{j=1}^{n} \alpha_i \beta_j \, \mathbf{E}\left[(X_i - \mathbf{E}[X_i])(Y_j - \mathbf{E}[Y_j])\right]$$

$$= \sum_{i=1}^{m}\sum_{j=1}^{n} \alpha_i \beta_j \, \mathbf{Cov}[X_i, Y_j]. \qquad \square$$

Theorem 5.8 (Cauchy–Schwarz inequality) *If $X, Y \in \mathcal{L}^2(\mathbf{P})$, then*

$$\left(\mathbf{Cov}[X, Y]\right)^2 \leq \mathbf{Var}[X]\,\mathbf{Var}[Y].$$

Equality holds if and only if there are $a, b, c \in \mathbb{R}$ with $|a| + |b| + |c| > 0$ and such that $aX + bY + c = 0$ a.s.

Proof The Cauchy–Schwarz inequality holds for any positive semidefinite bilinear form and hence in particular for the covariance map. Using the notation of variance and covariance, a simple proof looks like this:

Case 1: $\mathbf{Var}[Y] = 0$. Here the statement is trivial (choose $a = 0$, $b = 1$ and $c = -\mathbf{E}[Y]$).

Case 2: $\mathbf{Var}[Y] > 0$. Let $\theta := -\frac{\mathbf{Cov}[X,Y]}{\mathbf{Var}[Y]}$. Then, by Theorem 5.6(i),

$$0 \leq \mathbf{Var}[X + \theta Y]\,\mathbf{Var}[Y] = \left(\mathbf{Var}[X] + 2\theta\,\mathbf{Cov}[X, Y] + \theta^2\,\mathbf{Var}[Y]\right)\mathbf{Var}[Y]$$

$$= \mathbf{Var}[X]\,\mathbf{Var}[Y] - \mathbf{Cov}[X, Y]^2$$

with equality if and only if $X + \theta Y$ is a.s. constant. Now let $a = 1$, $b = \theta$ and $c = -\mathbf{E}[X] - b\,\mathbf{E}[Y]$. $\qquad \square$

Example 5.9

(i) Let $p \in [0, 1]$ and $X \sim \mathrm{Ber}_p$. Then

$$\mathbf{E}[X^2] = \mathbf{E}[X] = \mathbf{P}[X = 1] = p$$

and thus $\mathbf{Var}[X] = p(1 - p)$.

(ii) Let $n \in \mathbb{N}$ and $p \in [0, 1]$. Let X be binomially distributed, $X \sim b_{n,p}$. Then

$$\mathbf{E}[X] = \sum_{k=0}^{n} k \mathbf{P}[X = k] = \sum_{k=0}^{n} k \binom{n}{k} p^k (1 - p)^{n-k}$$

$$= np \cdot \sum_{k=1}^{n} \binom{n-1}{k-1} p^{k-1} (1 - p)^{(n-1)-(k-1)} = np.$$

Furthermore,

$$\mathbf{E}[X(X - 1)] = \sum_{k=0}^{n} k(k - 1) \mathbf{P}[X = k]$$

$$= \sum_{k=0}^{n} k(k - 1) \binom{n}{k} p^k (1 - p)^{n-k}$$

$$= np \cdot \sum_{k=1}^{n} (k - 1) \binom{n-1}{k-1} p^{k-1} (1 - p)^{(n-1)-(k-1)}$$

$$= n(n - 1)p^2 \cdot \sum_{k=2}^{n} \binom{n-2}{k-2} p^{k-2} (1 - p)^{(n-2)-(k-2)}$$

$$= n(n - 1)p^2.$$

Hence $\mathbf{E}[X^2] = \mathbf{E}[X(X - 1)] + \mathbf{E}[X] = n^2 p^2 + np(1 - p)$ and thus $\mathbf{Var}[X] = np(1 - p)$.

The statement can be derived more simply than by direct computation if we make use of the fact that $b_{n,p} = b_{1,p}^{*n}$ (see Example 3.4(ii)). That is (see Theorem 2.31), $\mathbf{P}_X = \mathbf{P}_{Y_1 + \ldots + Y_n}$, where $Y_1, \ldots, Y_n$ are independent and $Y_i \sim \text{Ber}_p$ for any $i = 1, \ldots, n$. Hence

$$\mathbf{E}[X] = n\mathbf{E}[Y_1] = np,$$

$$\mathbf{Var}[X] = n\mathbf{Var}[Y_1] = np(1 - p). \tag{5.3}$$

(iii) Let $\mu \in \mathbb{R}$ and $\sigma^2 > 0$, and let X be normally distributed, $X \sim \mathcal{N}_{\mu,\sigma^2}$. Then

$$\mathbf{E}[X] = \frac{1}{\sqrt{2\pi\sigma^2}} \int_{-\infty}^{\infty} x\, e^{-(x-\mu)^2/(2\sigma^2)}\, dx$$

$$= \frac{1}{\sqrt{2\pi\sigma^2}} \int_{-\infty}^{\infty} (x+\mu)\, e^{-x^2/(2\sigma^2)}\, dx \qquad (5.4)$$

$$= \mu + \frac{1}{\sqrt{2\pi\sigma^2}} \int_{-\infty}^{\infty} x\, e^{-x^2/(2\sigma^2)}\, dx = \mu.$$

Similarly, we get $\mathbf{Var}[X] = \mathbf{E}[X^2] - \mu^2 = \ldots = \sigma^2$.

(iv) Let $\theta > 0$ and let X be exponentially distributed, $X \sim \exp_\theta$. Then

$$\mathbf{E}[X] = \theta \int_0^{\infty} x\, e^{-\theta x}\, dx = \frac{1}{\theta},$$

$$\mathbf{Var}[X] = -\theta^{-2} + \theta \int_0^{\infty} x^2\, e^{-\theta x}\, dx = \theta^{-2}\left(-1 + \int_0^{\infty} x^2 e^{-x}\, dx\right) = \theta^{-2}. \quad \Diamond$$

$$(5.5)$$

Theorem 5.10 (Blackwell–Girshick) *Let $T, X_1, X_2, \ldots$ be independent real random variables in $\mathcal{L}^2(\mathbf{P})$. Let $\mathbf{P}[T \in \mathbb{N}_0] = 1$ and let $X_1, X_2, \ldots$ be identically distributed. Define*

$$S_T := \sum_{i=1}^{T} X_i.$$

Then $S_T \in \mathcal{L}^2(\mathbf{P})$ and

$$\mathbf{Var}[S_T] = \mathbf{E}[X_1]^2\, \mathbf{Var}[T] + \mathbf{E}[T]\, \mathbf{Var}[X_1].$$

Proof Define $S_n = \sum_{i=1}^n X_i$ for $n \in \mathbb{N}$. Then (as in the proof of Wald's identity) S_n and $\mathbb{1}_{\{T=n\}}$ are independent; hence S_n^2 and $\mathbb{1}_{\{T=n\}}$ are uncorrelated and thus

$$\mathbf{E}\left[S_T^2\right] = \sum_{n=0}^{\infty} \mathbf{E}\left[\mathbb{1}_{\{T=n\}} S_n^2\right]$$

$$= \sum_{n=0}^{\infty} \mathbf{E}[\mathbb{1}_{\{T=n\}}]\, \mathbf{E}\left[S_n^2\right]$$

$$= \sum_{n=0}^{\infty} \mathbf{P}[T = n]\left(\mathbf{Var}[S_n] + \mathbf{E}[S_n]^2\right)$$

$$= \sum_{n=0}^{\infty} \mathbf{P}[T = n]\left(n\, \mathbf{Var}[X_1] + n^2\, \mathbf{E}[X_1]^2\right)$$

$$= \mathbf{E}[T]\, \mathbf{Var}[X_1] + \mathbf{E}[T^2]\, \mathbf{E}[X_1]^2.$$

By Wald's identity (Theorem 5.5), we have $\mathbf{E}[S_T] = \mathbf{E}[T]\,\mathbf{E}[X_1]$; hence

$$\mathbf{Var}[S_T] = \mathbf{E}\left[S_T^2\right] - \mathbf{E}[S_T]^2 = \mathbf{E}[T]\,\mathbf{Var}[X_1] + \left(\mathbf{E}[T^2] - \mathbf{E}[T]^2\right)\mathbf{E}[X_1]^2,$$

as claimed.   □

Reflection In the proof of Theorem 5.10, where did we use the independence of T? Check that if $\mathbf{E}[X_i] = 0$ for all $i \in \mathbb{N}$, then instead of independence of T, it is enough to postulate: $\{T \leq n\}$ is independent of $X_{n+1}, X_{n+2}, \ldots$ for all n. ♠♠

Takeaways Moments are important characteristics of probability distributions. We have seen a formula for the first and second moment of a sum of random variables, even if the number of summands is random itself. Independent random variables are uncorrelated. In this case, the formulas for the second moments of sums are particularly simple.

Exercise 5.1.1 Let X be a nonnegative random variable with finite second moment. Use the Cauchy-Schwarz inequality for X and $\mathbb{1}_{\{X>0\}}$ in order to show the Paley-Zygmund inequality

$$\mathbf{P}[X > 0] \geq \frac{\mathbf{E}[X]^2}{\mathbf{E}[X^2]}. \quad ♣$$

Exercise 5.1.2 Let X be an integrable real random variable whose distribution $\mathbf{P}_X$ has a density f (with respect to the Lebesgue measure λ). Show (using Theorem 4.15) that

$$\mathbf{E}[X] = \int_{\mathbb{R}} x f(x)\, \lambda(dx). \quad ♣$$

Exercise 5.1.3 Let $X \sim \beta_{r,s}$ be a Beta-distributed random variable with parameters $r, s > 0$ (see Example 1.107(ii)). Show that

$$\mathbf{E}[X^n] = \prod_{k=0}^{n-1} \frac{r+k}{r+s+k} \qquad \text{for any } n \in \mathbb{N}. \quad ♣$$

Exercise 5.1.4 Let $X_1, X_2, \ldots$ be i.i.d. nonnegative random variables. By virtue of the Borel–Cantelli lemma, show that

$$\limsup_{n\to\infty} \frac{1}{n} X_n = \begin{cases} 0 \text{ a.s.,} & \text{if } \mathbf{E}[X_1] < \infty, \\ \infty \text{ a.s.,} & \text{if } \mathbf{E}[X_1] = \infty. \end{cases} \quad \clubsuit$$

Exercise 5.1.5 Let $X_1, X_2, \ldots$ be i.i.d. nonnegative random variables. By virtue of the Borel–Cantelli lemma, show that for any $c \in (0, 1)$

$$\sum_{n=1}^{\infty} e^{X_n} c^n \begin{cases} < \infty \text{ a.s.,} & \text{if } \mathbf{E}[X_1] < \infty, \\ = \infty \text{ a.s.,} & \text{if } \mathbf{E}[X_1] = \infty. \end{cases} \quad \clubsuit$$

5.2 Weak Law of Large Numbers

Theorem 5.11 (Markov inequality, Chebyshev inequality) *Let X be a real random variable and let $f : [0, \infty) \to [0, \infty)$ be monotone increasing. Then for any $\varepsilon > 0$ with $f(\varepsilon) > 0$, the **Markov inequality** holds,*

$$\mathbf{P}\big[|X| \geq \varepsilon\big] \leq \frac{\mathbf{E}[f(|X|)]}{f(\varepsilon)}.$$

*In the special case $f(x) = x^2$, we get $\mathbf{P}\big[|X| \geq \varepsilon\big] \leq \varepsilon^{-2}\,\mathbf{E}\big[X^2\big]$. In particular, if $X \in \mathcal{L}^2(\mathbf{P})$, the **Chebyshev inequality** holds:*

$$\mathbf{P}\big[|X - \mathbf{E}[X]| \geq \varepsilon\big] \leq \varepsilon^{-2}\,\mathrm{Var}[X].$$

Proof We have

$$\mathbf{E}[f(|X|)] \geq \mathbf{E}\big[f(|X|)\,\mathbb{1}_{\{f(|X|)\geq f(\varepsilon)\}}\big]$$

$$\geq \mathbf{E}\big[f(\varepsilon)\,\mathbb{1}_{\{f(|X|)\geq f(\varepsilon)\}}\big]$$

$$\geq f(\varepsilon)\,\mathbf{P}\big[|X| \geq \varepsilon\big]. \qquad \square$$

Definition 5.12 *Let $(X_n)_{n\in\mathbb{N}}$ be a sequence of real random variables in $\mathcal{L}^1(\mathbf{P})$ and let $\tilde{S}_n = \sum_{i=1}^{n}(X_i - \mathbf{E}[X_i])$.*

*(i) We say that $(X_n)_{n\in\mathbb{N}}$ fulfills the **weak law of large numbers** if*

$$\lim_{n\to\infty} \mathbf{P}\left[\left|\frac{1}{n}\tilde{S}_n\right| > \varepsilon\right] = 0 \quad \text{for any } \varepsilon > 0.$$

*(ii) We say that $(X_n)_{n\in\mathbb{N}}$ fulfills the **strong law of large numbers** if*

$$\mathbf{P}\left[\limsup_{n\to\infty}\left|\frac{1}{n}\tilde{S}_n\right| = 0\right] = 1.$$

Remark 5.13 The strong law of large numbers implies the weak law. Indeed, if $A_n^\varepsilon := \left\{\left|\frac{1}{n}\tilde{S}_n\right| > \varepsilon\right\}$ and $A = \left\{\limsup_{n\to\infty}\left|\frac{1}{n}\tilde{S}_n\right| > 0\right\}$, then clearly

$$A = \bigcup_{m\in\mathbb{N}}\limsup_{n\to\infty} A_n^{1/m};$$

hence $\mathbf{P}\left[\limsup_{n\to\infty} A_n^\varepsilon\right] = 0$ for $\varepsilon > 0$. By Fatou's lemma (Theorem 4.21), we obtain

$$\limsup_{n\to\infty}\mathbf{P}\left[A_n^\varepsilon\right] = 1 - \liminf_{n\to\infty}\mathbf{E}\left[\mathbb{1}_{(A_n^\varepsilon)^c}\right]$$

$$\leq 1 - \mathbf{E}\left[\liminf_{n\to\infty}\mathbb{1}_{(A_n^\varepsilon)^c}\right] = \mathbf{E}\left[\limsup_{n\to\infty}\mathbb{1}_{A_n^\varepsilon}\right] = 0. \quad \lozenge$$

Theorem 5.14 *Let $X_1, X_2, \ldots$ be uncorrelated random variables in $\mathcal{L}^2(\mathbf{P})$ with $V := \sup_{n\in\mathbb{N}}\mathbf{Var}[X_n] < \infty$. Then $(X_n)_{n\in\mathbb{N}}$ fulfills the weak law of large numbers. More precisely, for any $\varepsilon > 0$, we have*

$$\mathbf{P}\left[\left|\frac{1}{n}\tilde{S}_n\right| \geq \varepsilon\right] \leq \frac{V}{\varepsilon^2 n} \quad \text{for all } n \in \mathbb{N}. \tag{5.6}$$

Reflection Let S_n be the sum of the numbers shown when rolling a die n-times. Then S_n/n should be close to 3.5 for large n - but close in which sense? The weak law of large numbers states that the distribution of S_n/n is concentrated in the vicinity of 3.5 (Fig. 5.1). On the other, the strong law of large numbers claims that for fixed ω, we have $S_n(\omega)/n \xrightarrow{n\to\infty} 3.5$ (Fig. 5.2).♠

Proof Without loss of generality, assume $\mathbf{E}[X_i] = 0$ for all $i \in \mathbb{N}$ and thus $\tilde{S}_n = X_1 + \cdots + X_n$. By Bienaymé's formula (Theorem 5.7), we obtain

$$\mathbf{Var}\left[\frac{1}{n}\tilde{S}_n\right] = n^{-2}\sum_{i=1}^{n}\mathbf{Var}[X_i] \leq \frac{V}{n}.$$

By Chebyshev's inequality (Theorem 5.11), for any $\varepsilon > 0$,

$$\mathbf{P}\left[|\tilde{S}_n/n| \geq \varepsilon\right] \leq \frac{V}{\varepsilon^2 n} \xrightarrow{n\to\infty} 0. \qquad \square$$

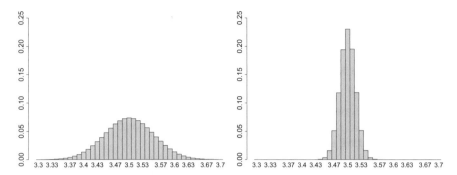

Fig. 5.1 Rolling a die n times: Probabilities for S_n/n. Left hand side: $n = 1000$, Right hand side $n = 10\,000$. The bars show the probabilities for values in $[x, x + 0.01)$, $3.3 \le x \le 3.7$.

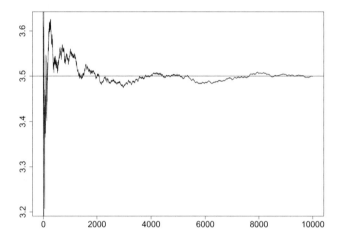

Fig. 5.2 Rolling a die n times: For a fixed realisation, the values of S_n/n converge to 3.5. We have n on the horizontal axis.

Example 5.15 (Weierstraß's approximation theorem) Let $f : [0, 1] \to \mathbb{R}$ be a continuous map. By Weierstraß's approximation theorem, there exist polynomials f_n of degree at most n such that

$$\| f_n - f \|_\infty \overset{n \to \infty}{\longrightarrow} 0,$$

where $\| f \|_\infty := \sup\{|f(x)| : x \in [0, 1]\}$ denotes the supremum norm of $f \in C([0, 1])$ (the space of continuous functions $[0, 1] \to \mathbb{R}$).

We present a probabilistic proof of this theorem. For $n \in \mathbb{N}$, define the polynomial f_n by

$$f_n(x) := \sum_{k=0}^{n} f(k/n) \binom{n}{k} x^k (1-x)^{n-k} \quad \text{for } x \in [0, 1].$$

f_n is called the **Bernstein polynomial** of order n.

Fix $\varepsilon > 0$. As f is continuous on the compact interval $[0, 1]$, f is uniformly continuous. Hence there exists a $\delta > 0$ such that

$$|f(x) - f(y)| < \varepsilon \quad \text{for all } x, y \in [0, 1] \text{ with } |x - y| < \delta.$$

Now fix $p \in [0, 1]$ and let $X_1, X_2, \ldots$ be independent random variables with $X_i \sim \text{Ber}_p$, $i \in \mathbb{N}$. Then $S_n := X_1 + \ldots + X_n \sim b_{n,p}$ and thus

$$\mathbf{E}[f(S_n/n)] = \sum_{k=0}^{n} f(k/n) \mathbf{P}[S_n = k] = f_n(p).$$

We get

$$|f(S_n/n) - f(p)| \leq \varepsilon + 2\|f\|_\infty \mathbb{1}_{\{|(S_n/n)-p|\geq\delta\}}$$

and thus (by Theorem 5.14 with $V = p(1-p) \leq \frac{1}{4}$)

$$|f_n(p) - f(p)| \leq \mathbf{E}[|f(S_n/n) - f(p)|]$$
$$\leq \varepsilon + 2\|f\|_\infty \mathbf{P}\left[\left|\frac{S_n}{n} - p\right| \geq \delta\right]$$
$$\leq \varepsilon + \frac{\|f\|_\infty}{2\delta^2 n}$$

for any $p \in [0, 1]$. Hence $\|f_n - f\|_\infty \xrightarrow{n \to \infty} 0$. $\Diamond$

Takeaways Laws of large numbers show that sums of very many random variables approach their expected value. One can use second moments and Chebyshev's inequality to establish a weak law of large numbers.

Exercise 5.2.1 (Bernstein–Chernov bound) Let $n \in \mathbb{N}$ and $p_1, \ldots, p_n \in [0, 1]$. Let $X_1, \ldots, X_n$ be independent random variables with $X_i = \text{Ber}_{p_i}$ for any $i = 1, \ldots, n$. Define $S_n = X_1 + \ldots + X_n$ and $m := \mathbf{E}[S_n]$. Show that, for any $\delta > 0$, the following two estimates hold:

$$\mathbf{P}[S_n \geq (1+\delta)m] \leq \left(\frac{e^\delta}{(1+\delta)^{1+\delta}}\right)^m$$

and

$$\mathbf{P}\big[S_n \leq (1-\delta)m\big] \leq \exp\left(-\frac{\delta^2 m}{2}\right).$$

Hint: For S_n, use Markov's inequality with $f(x) = e^{\lambda x}$ for some $\lambda > 0$ and then find the λ that optimizes the bound. ♣

5.3 Strong Law of Large Numbers

We show Etemadi's version [47] of the strong law of large numbers for identically distributed, pairwise independent random variables. There is a zoo of strong laws of large numbers, each of which varies in the exact assumptions it makes on the underlying sequence of random variables. For example, the assumption that the random variables be identically distributed can be waived if other assumptions are introduced such as bounded variances. We do not strive for completeness but show only a few of the statements.

In order to illustrate the method of the proof of Etemadi's theorem, we first present (and prove) a strong law of large numbers under stronger assumptions.

Theorem 5.16 *Let $X_1, X_2, \ldots \in \mathcal{L}^2(\mathbf{P})$ be pairwise independent (that is, X_i and X_j are independent for all $i, j \in \mathbb{N}$ with $i \neq j$) and identically distributed. Then $(X_n)_{n\in\mathbb{N}}$ fulfills the strong law of large numbers.*

Proof The random variables $(X_n^+)_{n\in\mathbb{N}}$ and $(X_n^-)_{n\in\mathbb{N}}$ again form pairwise independent families of square integrable random variables (compare Remark 2.15(ii)). Hence, it is enough to consider $(X_n^+)_{n\in\mathbb{N}}$. Thus we henceforth assume $X_n \geq 0$ almost surely for all $n \in \mathbb{N}$.

Let $S_n = X_1 + \ldots + X_n$ for $n \in \mathbb{N}$. Fix $\varepsilon > 0$. For any $n \in \mathbb{N}$, define $k_n = \lfloor (1+\varepsilon)^n \rfloor \geq \frac{1}{2}(1+\varepsilon)^n$. Then, by Chebyshev's inequality (Theorem 5.11),

$$\sum_{n=1}^{\infty} \mathbf{P}\left[\left|\frac{S_{k_n}}{k_n} - \mathbf{E}[X_1]\right| \geq (1+\varepsilon)^{-n/4}\right] \leq \sum_{n=1}^{\infty} (1+\varepsilon)^{n/2} \operatorname{Var}\left[k_n^{-1} S_{k_n}\right]$$

$$= \sum_{n=1}^{\infty} (1+\varepsilon)^{n/2} k_n^{-1} \operatorname{Var}[X_1]$$

$$\leq 2\operatorname{Var}[X_1] \sum_{n=1}^{\infty} (1+\varepsilon)^{-n/2} < \infty. \tag{5.7}$$

Thus, by the Borel–Cantelli lemma, for **P**-a.a. ω, there is an $n_0 = n_0(\omega)$ such that

$$\left| \frac{S_{k_n}}{k_n} - \mathbf{E}[X_1] \right| < (1 + \varepsilon)^{-n/4} \quad \text{for all } n \geq n_0,$$

whence

$$\limsup_{n \to \infty} \left| k_n^{-1} S_{k_n} - \mathbf{E}[X_1] \right| = 0 \quad \text{almost surely.}$$

Note that $k_{n+1} \leq (1 + 2\varepsilon)k_n$ for sufficiently large $n \in \mathbb{N}$. For $l \in \{k_n, \ldots, k_{n+1}\}$, we get

$$\frac{1}{1 + 2\varepsilon} k_n^{-1} S_{k_n} \leq k_{n+1}^{-1} S_{k_n} \leq l^{-1} S_l \leq k_n^{-1} S_{k_{n+1}} \leq (1 + 2\varepsilon) k_{n+1}^{-1} S_{k_{n+1}}.$$

Now $1 - (1 + 2\varepsilon)^{-1} \leq 2\varepsilon$ implies

$$\limsup_{l \to \infty} \left| l^{-1} S_l - \mathbf{E}[X_1] \right| \leq \limsup_{n \to \infty} \left| k_n^{-1} S_{k_n} - \mathbf{E}[X_1] \right| + 2\varepsilon \limsup_{n \to \infty} k_n^{-1} S_{k_n}$$

$$\leq 2\varepsilon \, \mathbf{E}[X_1] \quad \text{almost surely.}$$

Hence the strong law of large numbers is in force. $\qquad \square$

Reflection The proof of the previous theorem made use of the fact that $(X_n^+)_{n \in \mathbb{N}}$ and $(X_n^-)_{n \in \mathbb{N}}$ are uncorrelated families. Why is it not enough to assume in the theorem that $(X_n)_{n \in \mathbb{N}}$ be uncorrelated (instead of pairwise independent)? ♠

The similarity of the variance estimates in the weak law of large numbers and in (5.7) suggests that in the preceding theorem the condition that the random variables $X_1, X_2, \ldots$ be identically distributed could be replaced by the condition that the variances be bounded (see Exercise 5.3.1).

We can weaken the condition in Theorem 5.16 in a different direction by requiring integrability only instead of square integrability of the random variables.

Theorem 5.17 (Etemadi's strong law of large numbers (1981)) *Let* $X_1, X_2, \ldots$ $\in \mathcal{L}^1(\mathbf{P})$ *be pairwise independent and identically distributed. Then* $(X_n)_{n \in \mathbb{N}}$ *fulfills the strong law of large numbers.*

We follow the proof in [39, Section 2.4]. Define $\mu = \mathbf{E}[X_1]$ and $S_n = X_1 + \ldots + X_n$. We start with some preparatory lemmas. (For the "a.s." notation see Definition 1.68.)

Lemma 5.18 *For* $n \in \mathbb{N}$, *define* $Y_n := X_n \mathbb{1}_{\{|X_n| \leq n\}}$ *and* $T_n = Y_1 + \cdots + Y_n$. *The sequence* $(X_n)_{n \in \mathbb{N}}$ *fulfills the strong law of large numbers if* $T_n/n \xrightarrow{n \to \infty} \mu$ *a.s.*

Proof By Theorem 4.26, we have $\sum\limits_{n=1}^{\infty} \mathbf{P}\big[|X_n| > n\big] \le \mathbf{E}\big[|X_1|\big] < \infty$. Thus, by the Borel–Cantelli lemma,

$$\mathbf{P}\big[X_n \neq Y_n \text{ for infinitely many } n\big] = 0.$$

Hence there is an $n_0 = n_0(\omega)$ with $X_n = Y_n$ for all $n \ge n_0$, whence for $n \ge n_0$

$$\frac{T_n - S_n}{n} = \frac{T_{n_0} - S_{n_0}}{n} \xrightarrow{n \to \infty} 0. \qquad \square$$

Lemma 5.19 $2x \sum\limits_{n>x} n^{-2} \le 4$ *for all* $x \ge 0$.

Proof For $m \in \mathbb{N}$, by comparison with the corresponding integral, we get

$$\sum_{n=m}^{\infty} n^{-2} \le m^{-2} + \int_m^{\infty} t^{-2}\,dt = m^{-2} + m^{-1} \le \frac{2}{m}. \qquad \square$$

Lemma 5.20 $\sum\limits_{n=1}^{\infty} \dfrac{\mathbf{E}\big[Y_n^2\big]}{n^2} \le 4\,\mathbf{E}[|X_1|]$.

Proof By Theorem 4.26, $\mathbf{E}\big[Y_n^2\big] = \int_0^{\infty} \mathbf{P}\big[Y_n^2 > t\big]\,dt$. Substituting $x = \sqrt{t}$, we obtain

$$\mathbf{E}\Big[Y_n^2\Big] = \int_0^{\infty} 2x\,\mathbf{P}[|Y_n| > x]\,dx \le \int_0^{n} 2x\,\mathbf{P}[|X_1| > x]\,dx.$$

By Lemma 5.19, for $m \to \infty$,

$$f_m(x) = \left(\sum_{n=1}^{m} n^{-2}\,\mathbb{1}_{\{x<n\}}\right) 2x\,\mathbf{P}[|X_1| > x] \uparrow f(x) \le 4\,\mathbf{P}[|X_1| > x].$$

Hence, by the monotone limit theorem, we can interchange the summation and the integral and obtain

$$\sum_{n=1}^{\infty} \frac{\mathbf{E}\big[Y_n^2\big]}{n^2} \le \sum_{n=1}^{\infty} n^{-2} \int_0^{\infty} \mathbb{1}_{\{x<n\}}\, 2x\,\mathbf{P}[|X_1| > x]\,dx$$

$$= \int_0^{\infty} \left(\sum_{n=1}^{\infty} n^{-2}\,\mathbb{1}_{\{x<n\}}\right) 2x\,\mathbf{P}[|X_1| > x]\,dx$$

$$\le 4 \int_0^{\infty} \mathbf{P}[|X_1| > x]\,dx = 4\,\mathbf{E}[|X_1|]. \qquad \square$$

Proof of Theorem 5.17 As in the proof of Theorem 5.16, it is enough to consider the case $X_n \geq 0$. Fix $\varepsilon > 0$ and let $\alpha = 1 + \varepsilon$. For $n \in \mathbb{N}$, define $k_n = \lfloor \alpha^n \rfloor$. Note that $k_n \geq \alpha^n / 2$. Hence, for all $m \in \mathbb{N}$ (with $n_0 = \lceil \log m / \log \alpha \rceil$),

$$\sum_{n: k_n \geq m} k_n^{-2} \leq 4 \sum_{n=n_0}^{\infty} \alpha^{-2n} = 4\alpha^{-2n_0}(1 - \alpha^{-2})^{-1} \leq 4(1 - \alpha^{-2})^{-1} m^{-2}. \tag{5.8}$$

The aim is to employ Lemma 5.20 to refine the estimate (5.7) for $(Y_n)_{n \in \mathbb{N}}$ and $(T_n)_{n \in \mathbb{N}}$. For $\delta > 0$, Chebyshev's inequality yields (together with (5.8))

$$\sum_{n=1}^{\infty} \mathbf{P}\big[\big|T_{k_n} - \mathbf{E}[T_{k_n}]\big| > \delta\, k_n\big] \leq \delta^{-2} \sum_{n=1}^{\infty} \frac{\mathbf{Var}\big[T_{k_n}\big]}{k_n^2}$$

$$= \delta^{-2} \sum_{n=1}^{\infty} k_n^{-2} \sum_{m=1}^{k_n} \mathbf{Var}[Y_m] = \delta^{-2} \sum_{m=1}^{\infty} \mathbf{Var}[Y_m] \sum_{n: k_n \geq m} k_n^{-2}$$

$$\leq 4(1 - \alpha^{-2})^{-1} \delta^{-2} \sum_{m=1}^{\infty} m^{-2}\, \mathbf{E}\big[Y_m^2\big] < \infty \quad \text{by Lemma 5.20.}$$

(In the third step, we could change the order of summation since all summands are nonnegative.) Letting $\delta \downarrow 0$, we infer by the Borel–Cantelli lemma

$$\lim_{n \to \infty} \frac{T_{k_n} - \mathbf{E}\big[T_{k_n}\big]}{k_n} = 0 \quad \text{almost surely.} \tag{5.9}$$

By the monotone convergence theorem (Theorem 4.20), we have

$$\mathbf{E}[Y_n] = \mathbf{E}\big[X_1 \mathbb{1}_{\{X_1 \leq n\}}\big] \overset{n \to \infty}{\longrightarrow} \mathbf{E}[X_1].$$

Hence $\mathbf{E}[T_{k_n}]/k_n \overset{n \to \infty}{\longrightarrow} \mathbf{E}[X_1]$. By (5.9), we also have $T_{k_n}/k_n \overset{n \to \infty}{\longrightarrow} \mathbf{E}[X_1]$ a.s. As in the proof of Theorem 5.16, we also get (since $Y_n \geq 0$)

$$\lim_{l \to \infty} \frac{T_l}{l} = \mathbf{E}[X_1] \quad \text{almost surely.}$$

By Lemma 5.18, this implies the claim of Theorem 5.17. $\qquad\qquad\qquad\qquad\qquad\square$

Reflection In Etemadi's theorem, we assumed that the random variables $X_1, X_2, \ldots$ are identically distributed. Come up with an example that shows that this condition cannot simply be dropped. ♠♠

Example 5.21 (Monte Carlo integration) Let $f : [0, 1] \to \mathbb{R}$ be a function and assume we want to determine the value of its integral $I := \int_0^1 f(x)\, dx$ numerically.

Assume that the computer generates numbers $X_1, X_2, \ldots$ that can be considered as independent random numbers, uniformly distributed on $[0, 1]$. For $n \in \mathbb{N}$, define the estimated value

$$\widehat{I}_n := \frac{1}{n} \sum_{i=1}^{n} f(X_i).$$

Assuming $f \in \mathcal{L}^1([0, 1])$, the strong law of large numbers yields $\widehat{I}_n \overset{n \to \infty}{\longrightarrow} I$ a.s.

Note that the last theorem made no statement on the *speed of convergence*. That is, we do not have control on the quantity $\mathbf{P}[|\widehat{I}_n - I| > \varepsilon]$. In order to get more precise estimates for the integral, we need additional information; for example, the value $V_1 := \int f^2(x)\,dx - I^2$ if $f \in \mathcal{L}^2([0, 1])$. (For bounded f, V_1 can easily be bounded.) Indeed, in this case, $\mathbf{Var}[\widehat{I}_n] = V_1/n$; hence, by Chebyshev's inequality,

$$\mathbf{P}\left[|\widehat{I}_n - I| > \varepsilon\,n^{-1/2}\right] \leq V_1/\varepsilon^2.$$

Hence the error is at most of order $n^{-1/2}$. The central limit theorem will show that the error is indeed exactly of this order.

If f is smooth in some sense, then the usual numerical procedures yield better orders of convergence. Hence **Monte Carlo simulation** should be applied only if all other methods fail. This is the case in particular if $[0, 1]$ is replaced by $G \subset \mathbb{R}^d$ for very large d. $\Diamond$

Definition 5.22 (Empirical distribution function) *Let $X_1, X_2, \ldots$ be real random variables. The map $F_n : \mathbb{R} \to [0, 1]$, $x \mapsto \frac{1}{n} \sum_{i=1}^{n} \mathbb{1}_{(-\infty,x]}(X_i)$ is called the **empirical distribution function** of $X_1, \ldots, X_n$.*

Theorem 5.23 (Glivenko–Cantelli) *Let $X_1, X_2, \ldots$ be i.i.d. real random variables with distribution function F, and let F_n, $n \in \mathbb{N}$, be the empirical distribution functions. Then*

$$\limsup_{n \to \infty} \sup_{x \in \mathbb{R}} |F_n(x) - F(x)| = 0 \quad \text{almost surely.}$$

Proof Fix $x \in \mathbb{R}$ and let $Y_n(x) = \mathbb{1}_{(-\infty,x]}(X_n)$ and $Z_n(x) = \mathbb{1}_{(-\infty,x)}(X_n)$ for $n \in \mathbb{N}$. Additionally, define the left-sided limits $F(x-) = \lim_{y \uparrow x} F(y)$ and similarly for F_n. Then each of the families $(Y_n(x))_{n \in \mathbb{N}}$ and $(Z_n(x))_{n \in \mathbb{N}}$ is independent. Furthermore, $\mathbf{E}[Y_n(x)] = \mathbf{P}[X_n \leq x] = F(x)$ and $\mathbf{E}[Z_n(x)] = \mathbf{P}[X_n < x] = F(x-)$. By the strong law of large numbers, we thus have

$$F_n(x) = \frac{1}{n} \sum_{i=1}^{n} Y_i(x) \overset{n \to \infty}{\longrightarrow} F(x) \quad \text{almost surely}$$

and

$$F_n(x-) = \frac{1}{n} \sum_{i=1}^{n} Z_i(x) \xrightarrow{n\to\infty} F(x-) \quad \text{almost surely.}$$

Formally, define $F(-\infty) = 0$ and $F(\infty) = 1$. Fix some $N \in \mathbb{N}$ and define

$$x_j := \inf\{x \in \overline{\mathbb{R}} : F(x) \geq j/N\}, \quad j = 0, \ldots, N,$$

and

$$R_n := \max_{j=1,\ldots,N-1} \left(\left|F_n(x_j) - F(x_j)\right| + \left|F_n(x_j-) - F(x_j-)\right|\right).$$

As shown above, $R_n \xrightarrow{n\to\infty} 0$ almost surely. For $x \in (x_{j-1}, x_j)$, we have (by definition of x_j)

$$F_n(x) \leq F_n(x_j-) \leq F(x_j-) + R_n \leq F(x) + R_n + \frac{1}{N}$$

and

$$F_n(x) \geq F_n(x_{j-1}) \geq F(x_{j-1}) - R_n \geq F(x) - R_n - \frac{1}{N}.$$

Hence

$$\limsup_{n\to\infty} \sup_{x\in\mathbb{R}} |F_n(x) - F(x)| \leq \frac{1}{N} + \limsup_{n\to\infty} R_n = \frac{1}{N}.$$

Letting $N \to \infty$, the claim follows.   □

Example 5.24 (Shannon's theorem) Consider a source of information that sends a sequence of independent random symbols $X_1, X_2, \ldots$ drawn from a finite alphabet E (that is, from an arbitrary finite set E). Let p_e be the probability of the symbol $e \in E$. Formally, the $X_1, X_2, \ldots$ are i.i.d. E-valued random variables with $\mathbf{P}[X_i = e] = p_e$ for $e \in E$.

For any $\omega \in \Omega$ and $n \in \mathbb{N}$, let

$$\pi_n(\omega) := \prod_{i=1}^{n} p_{X_i(\omega)}$$

be the probability that the observed sequence $X_1(\omega), \ldots, X_n(\omega)$ occurs. Define $Y_n(\omega) := -\log(p_{X_n(\omega)})$. Then $(Y_n)_{n\in\mathbb{N}}$ is i.i.d. and $\mathbf{E}[Y_n] = H(p)$, where

$$H(p) := -\sum_{e\in E} p_e \log(p_e)$$

is the **entropy** of the distribution $p = (p_e)_{e \in E}$ (compare Definition 5.25). By the strong law of large numbers, we infer Shannon's theorem:

$$-\frac{1}{n}\log \pi_n = \frac{1}{n}\sum_{i=1}^{n} Y_i \xrightarrow{n \to \infty} H(p) \quad \text{almost surely.} \quad \Diamond$$

*Entropy and Source Coding Theorem**

We briefly discuss the importance of π_n and the entropy. How can we quantify the *information* inherent in a message $X_1(\omega), \ldots, X_n(\omega)$? This information can be measured by the length of the shortest sequence of zeros and ones by which the message can be encoded. Of course, you do not want to invent a new code for every message but rather use one code that allows for the shortest average coding of the messages for the particular information source. To this end, associate with each symbol $e \in E$ a sequence of zeros and ones that when concatenated yield the message. The length $l(e)$ of the sequence that codes for e may depend on e. Hence, for efficiency, those symbols that appear more often get a shorter code than the more rare symbols. The Morse alphabet is constructed similarly (the letters "e" and "t", which are the most frequent letters in English, have the shortest codes ("dot" and "dash"), and the rare letter "q" has the code "dash-dash-dot-dash"). However, the Morse code also consists of gaps of different lengths that signal ends of letters and words. As we want to use only zeros and ones (and no gap-like symbols), we have to arrange the code in such a way that no code is the beginning of the code of a different symbol. For example, we could not encode one symbol with 0110 and a different one with 011011. A code that fulfills this condition is called a **binary prefix code**. Denote by $c(e) \in \{0, 1\}^{l(e)}$ the code of e, where $l(e)$ is its length. We can represent the codes of all letters in a tree.

Let us construct a code $C = (c(e), \ e \in E)$ that is efficient in the sense that it minimizes the expected length of the code (of a random symbol)

$$L_p(C) := \sum_{e \in E} p_e \, l(e).$$

We first define a specific code and then show that it is almost optimal. As a first step, we enumerate $E = \{e_1, \ldots, e_N\}$ such that $p_{e_1} \geq p_{e_2} \geq \ldots \geq p_{e_N}$. Define $\ell(e) \in \mathbb{N}$ for any $e \in E$ by

$$2^{-\ell(e)} \leq p_e < 2^{-\ell(e)+1}.$$

Let $\tilde{p}_e = 2^{-\ell(e)}$ for any $e \in E$ and let $\tilde{q}_k = \sum_{l < k} \tilde{p}_{e_l}$ for $k = 1, \ldots, N$.

By construction, $\ell(e_l) \leq \ell(e_k)$ for all $l \leq k$; hence the binary representation of $\tilde{q}_k$ has at most $\ell(e_k)$ digits:

$$\tilde{q}_k = \sum_{i=1}^{\ell(e_k)} c_i(e_k) \, 2^{-i}.$$

Here the numbers $c_1(e_k), \ldots, c_{\ell(e_k)}(e_k) \in \{0, 1\}$ are uniquely determined.

Clearly, $\tilde{q}_l \geq \tilde{q}_k + 2^{-\ell(e_k)}$ for any $l > k$; hence

$$\big(c_1(e_k), \ldots, c_{\ell(e_k)}(e_k)\big) \neq \big(c_1(e_l), \ldots, c_{\ell(e_k)}(e_l)\big) \quad \text{for all } l > k.$$

Thus $C = (c(e), \ e \in E)$ is a prefix code.

For any $b > 0$ and $x > 0$, denote by $\log_b(x) := \frac{\log(x)}{\log(b)}$ the logarithm of x to base b. By construction, $-\log_2(p_e) \leq l(e) \leq 1 - \log_2(p_e)$. Hence the expected length is

$$-\sum_{e \in E} p_e \, \log_2(p_e) \;\leq\; L_p(C) \;\leq\; 1 - \sum_{e \in E} p_e \, \log_2(p_e).$$

The length of this code for the first n symbols of our random information source is thus approximately $-\sum_{k=1}^{n} \log_2(p_{X_k}(\omega)) = -\log_2 \pi_n(\omega)$. Here we have the connection to Shannon's theorem. That theorem thus makes a statement about the length of a binary prefix code needed to transmit a long message.

Now, is the code constructed above optimal, or are there codes with smaller mean length? The answer is given by the source coding theorem for which we prepare with a definition and a lemma.

Definition 5.25 (Entropy) *Let $p = (p_e)_{e \in E}$ be a probability distribution on the countable set E. For $b > 0$, define*

$$H_b(p) := -\sum_{e \in E} p_e \log_b(p_e)$$

*with the convention $0 \log_b(0) := 0$. We call $H(p) := H_e(p)$ ($e = 2.71 \ldots$ Euler's number) the **entropy** and $H_2(p)$ the **binary entropy** of p.*

Note that, for infinite E, the entropy need not be finite.

Lemma 5.26 (Entropy inequality) *Let b and p be as above. Further, let q be a sub-probability distribution; that is, $q_e \geq 0$ for all $e \in E$ and $\sum_{e \in E} q_e \leq 1$. Then*

$$H_b(p) \leq -\sum_{e \in E} p_e \log_b(q_e) \tag{5.10}$$

with equality if and only if $H_b(p) = \infty$ or $q = p$.

Proof Without loss of generality, we can do the computation with $b = e$; that is, with the natural logarithm. Note that $\log(1 + x) \leq x$ for $x > -1$ with equality if and only if $x = 0$. If in (5.10) the left-hand side is finite, then we can subtract the right-hand side from the left-hand side and obtain

$$H(p) + \sum_{e \in E} p_e \log(q_e) = \sum_{e:\, p_e > 0} p_e \log(q_e/p_e)$$

$$= \sum_{e:\, p_e > 0} p_e \log \left(1 + \frac{q_e - p_e}{p_e}\right)$$

$$\leq \sum_{e:\, p_e > 0} p_e \frac{q_e - p_e}{p_e} = \sum_{e \in E} (q_e - p_e) \leq 0.$$

If $q \neq p$, then there is an $e \in E$ with $p_e > 0$ and $q_e \neq p_e$. If this is the case, then strict inequality holds if $H(p) < \infty$.   □

Theorem 5.27 (Source coding theorem) *Let* $p = (p_e)_{e \in E}$ *be a probability distribution on the finite alphabet* E. *For any binary prefix code* $C = (c(e),\, e \in E)$, *we have* $L_p(C) \geq H_2(p)$. *Furthermore, there is a binary prefix code* C *with* $L_p(C) \leq H_2(p) + 1$.

Proof The second part of the theorem was shown in the above construction. Now assume that a prefix code is given. Let $L = \max_{e \in E} l(e)$. For $e \in E$, let

$$C_L(e) = \left\{c \in \{0, 1\}^L : c_k = c_k(e) \text{ for } k \leq l(e)\right\}$$

the set of all dyadic sequences of length L that start like $c(e)$. Since we have a prefix code, the sets $C_L(e)$, $e \in E$, are pairwise disjoint and $\bigcup_{e \in E} C_L(e) \subset \{0, 1\}^L$. Hence, if we define $q_e := 2^{-l(e)}$, then (note that $\#C_L(e) = 2^{L-l(e)}$)

$$\sum_{e \in E} q_e = 2^{-L} \sum_{e \in E} \#C_L(e) \leq 1.$$

By Lemma 5.26, we have $L_p(C) = \sum_{e \in E} p_e l(e) = -\sum_{e \in E} p_e \log_2(q_e) \geq H_2(p)$.   □

Takeaways For random variables with second moments, a strong law of large numbers can be shown using the Borel-Cantelli lemma and Chebyshev's inequality first on an subsequence and then on the full sequence. For pairwise independent random variables with first moment, we could establish a strong law of large number via an involved truncation procedure which allows to use second moments estimates.

Exercise 5.3.1 Show the following improvement of Theorem 5.16: If $X_1, X_2, \ldots \in \mathcal{L}^2(\mathbf{P})$ are pairwise independent with bounded variances, then $(X_n)_{n \in \mathbb{N}}$ fulfills the strong law of large numbers. ♣

Exercise 5.3.2 Let $(X_n)_{n\in\mathbb{N}}$ be a sequence of independent identically distributed random variables with $\frac{1}{n}(X_1 + \ldots + X_n) \xrightarrow{n\to\infty} Y$ almost surely for some random variable Y. Show that $X_1 \in \mathcal{L}^1(\mathbf{P})$ and $Y = \mathbf{E}[X_1]$ almost surely.
Hint: First show that

$$\mathbf{P}\big[|X_n| > n \text{ for infinitely many } n\big] = 0 \quad \Longleftrightarrow \quad X_1 \in \mathcal{L}^1(\mathbf{P}). \quad \clubsuit$$

Exercise 5.3.3 Let E be a finite set and let p be a probability vector on E. Show that the entropy $H(p)$ is minimal (in fact, zero) if $p = \delta_e$ for some $e \in E$. It is maximal (in fact, $\log(\#E)$) if p is the uniform distribution on E. $\clubsuit$

Exercise 5.3.4 (Subadditivity of Entropy) For $i = 1, 2$, let E^i be a finite set and p^i a probability vector on E^i. Let p be a probability vector on $E^1 \times E^2$ with marginals p^1 and p^2. That is,

$$\sum_{e^2 \in E^2} p_{(e^1, e^2)} = p^1_{e^1} \quad \text{and} \quad \sum_{f^1 \in E^1} p_{(f^1, f^2)} = p^2_{f^2} \quad \text{for all } e^1 \in E^1, \ f^2 \in E^2.$$

Show that $H(p) \leq H(p^1) + H(p^2)$. $\clubsuit$

Exercise 5.3.5 Let $b \in \{2, 3, 4, \ldots\}$. A b-adic prefix code is defined in a similar way as a binary prefix code; however, instead of 0 and 1, now all numbers $0, 1, \ldots, b-1$ are admissible. Show that the statement of the source coding theorem holds for b-adic prefix codes with $H_2(p)$ replaced by $H_b(p)$. $\clubsuit$

Exercise 5.3.6 We want to check the efficiency of the Morse alphabet. To this end we need a table of the Morse code as well as the frequencies of the letters in a typical text. The following frequencies for letters in German texts are taken from [11, p. 10]. The frequencies for other languages can be found easily, e.g., at Wikipedia.

Letter	Morse code	Frequency	Letter	Morse code	Frequency
A	. -	0.0651	N	- .	0.0978
B	- . . .	0.0189	O	- -	0.0251
C	- . - .	0.0306	P	. - - .	0.0079
D	- . .	0.0508	Q	- . -	0.0002
E	.	0.1740	R	. - .	0.07
F	. . - .	0.0166	S	. . .	0.0727
G	- .	0.0301	T	-	0.0615
H	. . . .	0.0476	U	. . -	0.0435
I	. .	0.0755	V	. . . -	0.0067
J	. - -	0.0027	W	. -	0.0189
K	- . -	0.0121	X	- . . -	0.0003
L	. - . .	0.0344	Y	- . -	0.0004
M	-	0.0253	Z	- . .	0.0113

Here '.' denotes a short signal while '-' denotes a long signal. Each letter is finished by a pause sign. Thus the Morse code can be interpreted as a ternary prefix code.

Determine the average code length of a letter and compare it with the entropy H_3 in order to check the efficiency of the Morse code. ♣

Exercise 5.3.7 Let $m \in (0, \infty)$ and let

$$W_m = \left\{ p = (p_k)_{k \in \mathbb{N}_0} \text{ is a probability measure on } \mathbb{N}_0 \text{ and } \sum_{k=0}^{\infty} k p_k = m \right\}$$

be the set of probability measures on $\mathbb{N}_0$ with expectation m.

(i) Show that there exists a $p_{\max} \in W_m$ that maximises the entropy; that is, $H(p_{\max}) = \sup_{p \in W_m} H(p)$.
(ii) Compute $p_{\max}$ explicitly. ♣

5.4 Speed of Convergence in the Strong LLN

In the weak law of large numbers, we had a statement on the speed of convergence (Theorem 5.14). In the strong law of large numbers, however, we did not. As we required only first moments, in general, we cannot expect to get any useful statements. However, if we assume the existence of higher moments, we get reasonable estimates on the rate of convergence.

The core of the weak law of large numbers is Chebyshev's inequality. Here we present a stronger inequality that claims the same bound but now for the maximum over all partial sums until a fixed time.

Theorem 5.28 (Kolmogorov's inequality) *Let $n \in \mathbb{N}$ and let $X_1, X_2, \ldots, X_n$ be independent random variables with $\mathbf{E}[X_i] = 0$ and $\mathbf{Var}[X_i] < \infty$ for $i = 1, \ldots, n$. Further, let $S_k = X_1 + \ldots + X_k$ for $k = 1, \ldots, n$. Then, for any $t > 0$,*

$$\mathbf{P}\big[\max\{S_k : k = 1, \ldots, n\} \geq t \big] \leq \frac{\mathbf{Var}[S_n]}{t^2 + \mathbf{Var}[S_n]}. \tag{5.11}$$

Furthermore, Kolmogorov's inequality holds:

$$\mathbf{P}\big[\max\{|S_k| : k = 1, \ldots, n\} \geq t \big] \leq t^{-2}\, \mathbf{Var}[S_n]. \tag{5.12}$$

a generalization of Kolmogorov's inequality.

Proof We decompose the probability space according to the first time τ at which the partial sums exceed the value t. Hence, let

$$\tau := \min\left\{k \in \{1, \ldots, n\} : S_k \geq t\right\}$$

and $A_k = \{\tau = k\}$ for $k = 1, \ldots, n$. Further, let

$$A = \biguplus_{k=1}^{n} A_k = \left\{\max\{S_k : k = 1, \ldots, n\} \geq t\right\}.$$

Let $c \geq 0$. The random variable $(S_k + c)\,\mathbb{1}_{A_k}$ is $\sigma(X_1, \ldots, X_k)$-measurable and $S_n - S_k$ is $\sigma(X_{k+1}, \ldots, X_n)$-measurable. By Theorem 2.26, the two random variables are independent, and

$$\mathbf{E}\left[(S_k + c)\,\mathbb{1}_{A_k}(S_n - S_k)\right] = \mathbf{E}\left[(S_k + c)\,\mathbb{1}_{A_k}\right]\mathbf{E}\left[S_n - S_k\right] = 0.$$

Clearly, the events $A_1, \ldots, A_n$ are pairwise disjoint; hence $\sum_{k=1}^{n}\mathbb{1}_{A_k} = \mathbb{1}_A \leq 1$. We thus obtain

$$\mathbf{Var}[S_n] + c^2 = \mathbf{E}\left[(S_n + c)^2\right]$$

$$\geq \mathbf{E}\left[\sum_{k=1}^{n}(S_n + c)^2\,\mathbb{1}_{A_k}\right] = \sum_{k=1}^{n}\mathbf{E}\left[(S_n + c)^2\,\mathbb{1}_{A_k}\right]$$

$$= \sum_{k=1}^{n}\mathbf{E}\left[\left((S_k + c)^2 + 2(S_k + c)(S_n - S_k) + (S_n - S_k)^2\right)\mathbb{1}_{A_k}\right] \qquad (5.13)$$

$$= \sum_{k=1}^{n}\mathbf{E}\left[(S_k + c)^2\,\mathbb{1}_{A_k}\right] + \sum_{k=1}^{n}\mathbf{E}\left[(S_n - S_k)^2\,\mathbb{1}_{A_k}\right]$$

$$\geq \sum_{k=1}^{n}\mathbf{E}\left[(S_k + c)^2\,\mathbb{1}_{A_k}\right].$$

Since $c \geq 0$, we have $(S_k + c)^2\,\mathbb{1}_{A_k} \geq (t + c)^2\,\mathbb{1}_{A_k}$. Hence we can continue (5.13) to get

$$\mathbf{Var}[S_n] + c^2 \geq \sum_{k=1}^{n}\mathbf{E}\left[(t + c)^2\,\mathbb{1}_{A_k}\right] = (t + c)^2\,\mathbf{P}[A].$$

For $c = \mathbf{Var}[S_n]/t \geq 0$, we obtain

$$\mathbf{P}[A] \leq \frac{\mathbf{Var}[S_n] + c^2}{(t + c)^2} = \frac{c(t + c)}{(t + c)^2} = \frac{tc}{t^2 + tc} = \frac{\mathbf{Var}[S_n]}{t^2 + \mathbf{Var}[S_n]}.$$

This shows (5.11). In order to show (5.12), choose

$$\bar\tau := \min\big\{k \in \{1,\dots,n\} : |S_k| \ge t\big\}.$$

Let $\bar A_k = \{\bar\tau = k\}$ and $\bar A = \{\bar\tau \le n\}$. We cannot now continue (5.13) as above with $c > 0$. However, if we choose $c = 0$, then $S_k^2 \mathbb{1}_{\bar A_k} \ge t^2 \mathbb{1}_{\bar A_k}$. The same calculation as in (5.13) does then yield $\mathbf{P}[\bar A] \le t^{-2}\,\mathbf{Var}[S_n]$. □

From Kolmogorov's inequality, we derive the following sharpening of the strong law of large numbers.

Theorem 5.29 *Let $X_1, X_2, \dots$ be independent random variables with $\mathbf{E}[X_n] = 0$ for any $n \in \mathbb{N}$ and $V := \sup\{\mathbf{Var}[X_n] : n \in \mathbb{N}\} < \infty$. Then, for any $\varepsilon > 0$,*

$$\limsup_{n\to\infty} \frac{|S_n|}{n^{1/2}(\log(n))^{(1/2)+\varepsilon}} = 0 \quad \text{almost surely.}$$

Proof Let $k_n = 2^n$ and $l(n) = n^{1/2}(\log(n))^{(1/2)+\varepsilon}$ for $n \in \mathbb{N}$. Then we have $l(k_{n+1})/l(k_n) \overset{n\to\infty}{\longrightarrow} \sqrt 2$. Hence, for $n \in \mathbb{N}$ sufficiently large and $k \in \mathbb{N}$ with $k_{n-1} \le k \le k_n$, we have $|S_k|/l(k) \le 2|S_k|/l(k_n)$. Hence, it is enough to show for every $\delta > 0$ that

$$\limsup_{n\to\infty} l(k_n)^{-1} \max\{|S_k| : k \le k_n\} \le \delta \quad \text{almost surely.} \tag{5.14}$$

For $\delta > 0$ and $n \in \mathbb{N}$, define $A_n^\delta := \big\{ \max\{|S_k| : k \le k_n\} > \delta\, l(k_n)\big\}$. Kolmogorov's inequality yields

$$\sum_{n=1}^{\infty} \mathbf{P}\big[A_n^\delta\big] \le \sum_{n=1}^{\infty} \delta^{-2}(l(k_n))^{-2}\, V\, k_n = \frac{V}{\delta^2(\log 2)^{1+2\varepsilon}} \sum_{n=1}^{\infty} n^{-1-2\varepsilon} < \infty.$$

The Borel–Cantelli lemma then gives $\mathbf{P}\big[\limsup_{n\to\infty} A_n^\delta\big] = 0$ and hence (5.14). □

In Chap. 22, we will see that for independent identically distributed, square integrable, centered random variables $X_1, X_2, \dots$, the following strengthening holds,

$$\limsup_{n\to\infty} \frac{|S_n|}{\sqrt{2n\,\mathbf{Var}[X_1]\,\log(\log(n))}} = 1 \quad \text{almost surely.}$$

Hence, in this case, the speed of convergence is known precisely. If the $X_1, X_2, \dots$ are not independent but only pairwise independent, then the rate of convergence deteriorates, although not drastically. Here we cite without proof a theorem that was found independently by Rademacher [141] and Menshov [113].

Theorem 5.30 (Rademacher–Menshov) *Let $X_1, X_2, \ldots$ be uncorrelated, square integrable, centered random variables and let $(a_n)_{n \in \mathbb{N}}$ be an increasing sequence of nonnegative numbers such that*

$$\sum_{n=1}^{\infty} (\log n)^2 a_n^{-2} \, \mathbf{Var}[X_n] < \infty. \tag{5.15}$$

Then $\limsup\limits_{n \to \infty} \left| a_n^{-1} \sum\limits_{k=1}^{n} X_k \right| = 0$ *almost surely.*

Proof See, for example, [128].   □

Remark 5.31 Condition (5.15) is sharp in the sense that for any increasing sequence $(a_n)_{n \in \mathbb{N}}$ with $\sum_{n=1}^{\infty} a_n^{-2} (\log n)^2 = \infty$, there exists a sequence of pairwise independent, square integrable, centered random variables $X_1, X_2, \ldots$ with $\mathbf{Var}[X_n] = 1$ for all $n \in \mathbb{N}$ such that

$$\limsup_{n \to \infty} \left| a_n^{-1} \sum_{k=1}^{n} X_k \right| = \infty \quad \text{almost surely.}$$

See [22]. There an example of [164] (see also [165, 166]) for orthogonal series is developed further. See also [117]. ◊

For random variables with infinite variance, the statements about the rate of convergence naturally get weaker. For example (see [8]), see the following theorem.

Theorem 5.32 (Baum and Katz [8]) *Let $\gamma > 1$ and let $X_1, X_2, \ldots$ be i.i.d. Define $S_n = X_1 + \ldots + X_n$ for $n \in \mathbb{N}$. Then*

$$\sum_{n=1}^{\infty} n^{\gamma-2} \, \mathbf{P}[|S_n|/n > \varepsilon] < \infty \text{ for any } \varepsilon > 0 \iff \mathbf{E}[|X_1|^{\gamma}] < \infty \text{ and } \mathbf{E}[X_1] = 0.$$

> **Takeaways** Kolmogorov's inequality gives bounds for the maximum of partial sums of random variables similar to Chebyshev's inequality for one random variable. We have used this inequality in order to give an (almost sharp) upper bound on the speed of convergence in the strong law of large numbers (Theorem 5.29). Later, with a lot of additional effort, we will achieve a sharp bound in Theorem 22.11.

Exercise 5.4.1 Let $X_1, \ldots, X_n$ be independent real random variables and let $S_k = X_1 + \ldots + X_k$ for $k = 1, \ldots, n$. Show that for $t > 0$ **Etemadi's inequality** holds:

$$\mathbf{P}\left[\max_{k=1,\ldots,n} |S_k| \geq t \right] \leq 3 \max_{k=1,\ldots,n} \mathbf{P}[|S_k| \geq t/3]. \quad ♣$$

5.5 The Poisson Process

We develop a model for the number of clicks of a Geiger counter in the (time) interval $I = (a, b]$. The number of clicks should obey the following rules. It should

- be random and independent for disjoint intervals,
- be homogeneous in time in the sense that the number of clicks in $I = (a, b]$ has the same distribution as the number of clicks in $c + I = (a + c, b + c]$,
- have finite expectation, and
- have no double points: At any point of time, the counter makes at most one click.

We formalize these requirements by introducing the following notation:

$$\mathcal{I} := \big\{ (a, b] : a, b \in [0, \infty), \ a \leq b \big\},$$

$$\ell((a, b]) := b - a \qquad \text{(the length of the interval } I = (a, b]).$$

For $I \in \mathcal{I}$, let N_I be the number of clicks after time a but no later than b. In particular, we define $N_t := N_{(0,t]}$ as the total number of clicks until time t. The above requirements translate to: $(N_I, \ I \in \mathcal{I})$ being a family of random variables with values in $\mathbb{N}_0$ and with the following properties:

(P1) $N_{I \cup J} = N_I + N_J$ if $I \cap J = \emptyset$ and $I \cup J \in \mathcal{I}$.
(P2) The distribution of N_I depends only on the length of I: $\mathbf{P}_{N_I} = \mathbf{P}_{N_J}$ for all $I, J \in \mathcal{I}$ with $\ell(I) = \ell(J)$.
(P3) If $\mathcal{J} \subset \mathcal{I}$ with $I \cap J = \emptyset$ for all $I, J \in \mathcal{J}$ with $I \neq J$, then $(N_J, \ J \in \mathcal{J})$ is an independent family.
(P4) For any $I \in \mathcal{I}$, we have $\mathbf{E}[N_I] < \infty$.
(P5) $\limsup_{\varepsilon \downarrow 0} \varepsilon^{-1} \mathbf{P}[N_\varepsilon \geq 2] = 0$.

The meaning of (P5) is explained by the following calculation. Define

$$\lambda := \limsup_{\varepsilon \downarrow 0} \varepsilon^{-1} \mathbf{P}[N_\varepsilon \geq 2].$$

For any $n \in \mathbb{N}$ and $\varepsilon > 0$, we have

$$\mathbf{P}[N_{2^{-n}} \geq 2] \geq \lfloor 2^{-n}/\varepsilon \rfloor \, \mathbf{P}[N_\varepsilon \geq 2] - \lfloor 2^{-n}/\varepsilon \rfloor^2 \, \mathbf{P}[N_\varepsilon \geq 2]^2.$$

Hence

$$2^n \, \mathbf{P}[N_{2^{-n}} \geq 2] \ \geq \ \lambda - 2^{-n}\lambda^2 \ \overset{n \to \infty}{\longrightarrow} \ \lambda.$$

Then (because $(1 - a_k/k)^k \xrightarrow{k\to\infty} e^{-a}$ if $a_k \xrightarrow{k\to\infty} a$)

$$\mathbf{P}\big[\text{there is a double click in } (0, 1]\big] = \lim_{n\to\infty} \mathbf{P}\left[\bigcup_{k=0}^{2^n-1} \{N_{(k\,2^{-n},(k+1)2^{-n}]} \geq 2\}\right]$$

$$= 1 - \lim_{n\to\infty} \mathbf{P}\left[\bigcap_{k=0}^{2^n-1} \{N_{(k\,2^{-n},(k+1)2^{-n}]} \leq 1\}\right]$$

$$= 1 - \lim_{n\to\infty} \prod_{k=0}^{2^n-1} \mathbf{P}\big[N_{(k\,2^{-n},(k+1)2^{-n}]} \leq 1\big]$$

$$= 1 - \lim_{n\to\infty} \big(1 - \mathbf{P}[N_{2^{-n}} \geq 2]\big)^{2^n}$$

$$= 1 - e^{-\lambda}.$$

Hence we have to postulate $\lambda = 0$. This, however, is exactly (P5).

The following theorem shows that properties (P1)–(P5) characterize the random variables $(N_I, \ I \in \mathcal{I})$ uniquely and that they form a Poisson process.

Definition 5.33 (Poisson process) *A family $(N_t, \ t \geq 0)$ of $\mathbb{N}_0$-valued random variables is called a **Poisson process** with intensity $\alpha \geq 0$ if $N_0 = 0$ and if:*

(i) *For any $n \in \mathbb{N}$ and any choice of $n + 1$ numbers $0 = t_0 < t_1 < \ldots < t_n$, the family $(N_{t_i} - N_{t_{i-1}}, \ i = 1, \ldots, n)$ is independent.*

(ii) *For $t > s \geq 0$, the difference $N_t - N_s$ is Poisson-distributed with parameter $\alpha(t - s)$; that is,*

$$\mathbf{P}[N_t - N_s = k] = e^{-\alpha(t-s)} \frac{(\alpha(t - s))^k}{k!} \quad \text{for all } k \in \mathbb{N}_0.$$

See Fig. 5.3 for a computer simulation of a Poisson process.

The *existence* of the Poisson process has not yet been shown. We come back to this point in Theorem 5.36.

Fig. 5.3 Simulation of a Poisson process with rate $\alpha = 0.5$.

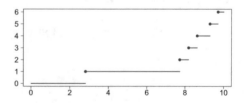

Theorem 5.34 *If $(N_I,\ I \in \mathcal{I})$ has properties (P1)–(P5), then $(N_{(0,t]},\ t \geq 0)$ is a Poisson process with intensity $\alpha := \mathbf{E}[N_{(0,1]}]$. If, on the other hand, $(N_t,\ t \geq 0)$ is a Poisson process, then $(N_t - N_s,\ (s,t] \in \mathcal{I})$ has properties (P1)–(P5).*

Proof First assume that $(N_t,\ t \geq 0)$ is a Poisson process with intensity $\alpha \geq 0$. Then, for $I = (a, b]$, clearly $\mathbf{P}_{N_I} = \mathrm{Poi}_{\alpha(b-a)} = \mathrm{Poi}_{\alpha \ell(I)}$. Hence (P2) holds. By (i), we have (P3). Clearly, $\mathbf{E}[N_I] = \alpha\,\ell(I) < \infty$; thus we have (P4). Finally, $\mathbf{P}[N_\varepsilon \geq 2] = 1 - e^{-\alpha\varepsilon} - \alpha\,\varepsilon\,e^{-\alpha\varepsilon} = f(0) - f(\alpha\varepsilon)$, where $f(x) := e^{-x} + xe^{-x}$. The derivative is $f'(x) = -xe^{-x}$, whence

$$\lim_{\varepsilon \downarrow 0} \varepsilon^{-1}\, \mathbf{P}[N_\varepsilon \geq 2] \;=\; -\alpha f'(0) \;=\; 0.$$

This implies (P5).

Now assume that $(N_I,\ I \in \mathcal{I})$ fulfills (P1)–(P5). Define $\alpha(t) := \mathbf{E}[N_t]$. Then (owing to (P2))

$$\alpha(s+t) = \mathbf{E}\big[N_{(0,s]} + N_{(s,s+t]}\big] = \mathbf{E}\big[N_{(0,s]}\big] + \mathbf{E}\big[N_{(0,t]}\big] = \alpha(s) + \alpha(t).$$

As $t \mapsto \alpha(t)$ is monotone increasing, this implies linearity: $\alpha(t) = t\,\alpha(1)$ for any $t \geq 0$. Letting $\alpha := \alpha(1)$, we obtain $\mathbf{E}[N_I] = \alpha\,\ell(I)$. It remains to show that $\mathbf{P}_{N_t} = \mathrm{Poi}_{\alpha t}$. In order to apply the Poisson approximation theorem (Theorem 3.7), for fixed $n \in \mathbb{N}$, we decompose the interval $(0, t]$ into 2^n disjoint intervals of equal length,

$$I^n(k) := \big((k-1)2^{-n}t,\, k2^{-n}t\big], \qquad k = 1, \ldots, 2^n.$$

Now define $X^n(k) := N_{I^n(k)}$ and

$$\overline{X}^n(k) := \begin{cases} 1, & \text{if } X^n(k) \geq 1, \\ 0, & \text{else.} \end{cases}$$

By properties (P2) and (P3), the random variables $(X^n(k),\ k = 1, \ldots, 2^n)$ are independent and identically distributed. Hence also $(\overline{X}^n(k),\ k = 1, \ldots, 2^n)$ are i.i.d., namely $\overline{X}^n(k) \sim \mathrm{Ber}_{p_n}$, where $p_n = \mathbf{P}[N_{2^{-n}t} \geq 1]$.

Finally, let $N_t^n := \sum_{k=1}^{2^n} \overline{X}^n(k)$. Then $N_t^n \sim b_{2^n, p_n}$. Clearly, $N_t^{n+1} - N_t^n \geq 0$. Now, by (P5),

$$\mathbf{P}\big[N_t \neq N_t^n\big] \;\leq\; \sum_{k=1}^{2^n} \mathbf{P}\big[X^n(k) \geq 2\big] \;=\; 2^n\,\mathbf{P}[N_{2^{-n}t} \geq 2] \;\overset{n\to\infty}{\longrightarrow}\; 0. \qquad (5.16)$$

Hence $\mathbf{P}\Big[N_t = \lim\limits_{n\to\infty} N_t^n\Big] = 1$. By the monotone convergence theorem, we get

$$\alpha t \;=\; \mathbf{E}[N_t] \;=\; \lim_{n\to\infty} \mathbf{E}\big[N_t^n\big] \;=\; \lim_{n\to\infty} p_n\, 2^n.$$

Using the Poisson approximation theorem (Theorem 3.7), we infer that, for any $l \in \mathbb{N}_0$,

$$\mathbf{P}[N_t = l] \;=\; \lim_{n \to \infty} \mathbf{P}\left[N_t^n = l\right] \;=\; \mathrm{Poi}_{\alpha t}(\{l\}).$$

Hence $\mathbf{P}_{N_t} = \mathrm{Poi}_{\alpha t}$.   □

Reflection Why do we need condition (iii) in the definition of the Poisson process? Consider a Poisson process $(N_t)_{t \geq 0}$ and choose an independent exponentially distributed random variable T (it would suffice for T to have a density). Now define $\tilde{N}_t = N_t$ for $t \neq T$ and let $\tilde{N}_T = 0$. The process $\tilde{N}$ fulfills (i) and (ii), but not (iii). In fact, $t \mapsto \tilde{N}_t$ is not monotone. A second possibility to spoil (iii) is to define $\overline{N}_t := \sup_{r < t} N_r$ für $t > 0$ und $\overline{N}_0 = 0$. In this case, $t \mapsto \overline{N}_t$ is monotone but it is not right continuous, although (i) and (ii) hold.

   In either case, where does the proof of Theorem 5.34 fail? ♠♠

At this point, we still have to show that there are Poisson processes at all. We present a general two-step construction principle that will be used in a similar form later in Chap. 24 in a more general setting. In the first step, we determine the (random) number of jumps in $(0, 1]$. In the second step, we distribute these jumps uniformly and independently on $(0, 1]$. Strictly speaking, this gives the Poisson process only on the time interval $(0, 1]$, but it is clear how to move on: We perform the same procedure independently for each of the intervals $(1, 2]$, $(2, 3]$ and so on and then collect the jumps (see also Exercise 5.5.1).

   Let $\alpha > 0$ and let L be a Poi_α random variable. Further, let $X_1, X_2, \ldots$ be independent random variables, that are uniformly distributed on $(0, 1]$, i.e., $X_k \sim \mathcal{U}_{(0,1]}$ for each k. We assume that $\{L, X_1, X_2, \ldots\}$ is an independent family of random variables. We now define $N = (N_t)_{t \in [0,1]}$ by

$$N_t := \sum_{l=1}^{L} \mathbb{1}_{(0,t]}(X_l) \quad \text{for } t \in [0, 1]. \tag{5.17}$$

Theorem 5.35 *The family N of random variables defined in (5.17) is a Poisson process with intensity α (and time set $[0, 1]$).*

Proof We have to show that the increments of N in finitely many pairwise disjoint intervals are independent and Poisson distributed. Hence let $m \in \mathbb{N}$ and $0 = t_0 < t_1 < \ldots < t_m = 1$. We use the abbreviations $p_i := t_i - t_{i-1}$ and $\lambda_i = \alpha \cdot (t_i - t_{i-1})$ and show that

$$(N_{t_i} - N_{t_{i-1}})_{i=1,\ldots,m} \quad \text{is independent} \tag{5.18}$$

and

$$N_{t_i} - N_{t_{i-1}} \sim \mathrm{Poi}_{\lambda_i} \quad \text{for all } i = 1, \ldots, m. \tag{5.19}$$

This is equivalent to showing that for each choice of $k_1, \ldots, k_m \in \mathbb{N}_0$, we have

$$\mathbf{P}[N_{t_i} - N_{t_{i-1}} = k_i \text{ for any } i = 1, \ldots, m] = \prod_{i=1}^{m}\left(e^{-\lambda_i}\frac{\lambda_i^{k_i}}{k_i!}\right). \tag{5.20}$$

Write

$$M_{n,i} := \#\{l \le n : t_{i-1} < X_l \le t_i\} = \sum_{l=1}^{n}\mathbb{1}_{(t_{i-1},t_i]}(X_l).$$

By Exercise 2.2.3, the vector $(M_{n,1}, \ldots, M_{n,m})$ is multinomially distributed with parameters n and $p = (p_1, \ldots, p_m)$. That is, if we assume $n := k_1 + \ldots + k_m$, then

$$\mathbf{P}[M_{n,1} = k_1, \ldots, M_{n,m} = k_m] = \frac{n!}{k_1! \cdots k_m!}\, p_1^{k_1} \cdots p_m^{k_m}.$$

In order to show (5.20), note that the event in (5.20) implies $L = n$ and that L and $(M_{n,1}, \ldots, M_{n,m})$ are independent. Hence we have

$$\mathbf{P}[N_{t_i} - N_{t_{i-1}} = k_i \text{ for } i = 1, \ldots, m]$$
$$= \mathbf{P}[\{N_{t_i} - N_{t_{i-1}} = k_i \text{ for } i = 1, \ldots, m\} \cap \{L = n\}]$$
$$= \mathbf{P}[\{M_{n,1} = k_1, \ldots, M_{n,m} = k_m\} \cap \{L = n\}]$$
$$= \mathbf{P}[M_{n,1} = k_1, \ldots, M_{n,m} = k_m] \cdot \mathbf{P}[L = n]$$
$$= \frac{n!}{k_1! \cdots k_m!}\, p_1^{k_1} \cdots p_m^{k_m}\, e^{-\alpha}\frac{\alpha^n}{n!}$$
$$= \prod_{i=1}^{m}\left(e^{-\lambda_i}\frac{\lambda_i^{k_i}}{k_i!}\right). \qquad \square$$

We close this section by presenting a further, rather elementary and instructive construction of the Poisson process based on specifying the waiting times between the clicks of the Geiger counter, or, more formally, between the points of discontinuity of the map $t \mapsto N_t(\omega)$. At time s, what is the probability that we have to wait another t time units (or longer) for the next click? Since we modeled the clicks as a Poisson process with intensity α, this probability can easily be computed:

$$\mathbf{P}[N_{(s,s+t]} = 0] = e^{-\alpha t}.$$

Hence the waiting time for the next click is exponentially distributed with parameter α. Furthermore, the waiting times should be independent. We now take the waiting times as the starting point and, based on them, construct the Poisson process.

Let $W_1, W_2, \ldots$ be an independent family of exponentially distributed random variables with parameter $\alpha > 0$; hence $\mathbf{P}[W_n > x] = e^{-\alpha x}$. We define

$$T_n := \sum_{k=1}^{n} W_k$$

and interpret W_n as the waiting time between the $(n-1)$th click and the nth click. T_n is the time of the nth click. Appealing to this intuition we define the number of clicks until time t by

$$N_t := \#\{n \in \mathbb{N}_0 : T_n \le t\}.$$

Hence

$$\{N_t = k\} = \{T_k \le t < T_{k+1}\}.$$

In particular, N_t is a random variable; that is, measurable.

Theorem 5.36 *The family $(N_t,\ t \ge 0)$ is a Poisson process with intensity α.*

Proof (We follow the proof in [59, Theorem 3.34]) We must show that for any $n \in \mathbb{N}$ and any sequence $0 = t_0 < t_1 < \ldots < t_n$, we have that $(N_{t_i} - N_{t_{i-1}}, i = 1, \ldots, n)$ is independent and $N_{t_i} - N_{t_{i-1}} \sim \mathrm{Poi}_{\alpha(t_i - t_{i-1})}$. We are well aware that it is not enough to show this for the case $n = 2$ only. However, the notational complications become overwhelming for $n \ge 3$, and the idea for general $n \in \mathbb{N}$ becomes clear in the case $n = 2$. Hence we restrict ourselves to the case $n = 2$.

Hence we show for $0 < s < t$ and $l, k \in \mathbb{N}_0$ that

$$\mathbf{P}[N_s = k,\ N_t - N_s = l] = \left(e^{-\alpha s} \frac{(\alpha s)^k}{k!} \right) \left(e^{-\alpha(t-s)} \frac{(\alpha(t-s))^l}{l!} \right). \qquad (5.21)$$

This implies that N_s and $(N_t - N_s)$ are independent. Furthermore, by summing over $k \in \mathbb{N}_0$, this yields $N_t - N_s \sim \mathrm{Poi}_{\alpha(t-s)}$.

By Corollary 2.22, the distribution $\mathbf{P}_{(W_1, \ldots, W_{k+l+1})}$ has the density

$$x \mapsto \alpha^{k+l+1} e^{-\alpha S_{k+l+1}(x)},$$

where $S_n(x) := x_1 + \ldots + x_n$. It is sufficient to consider $l \ge 1$ since we get the $l = 0$ term from the fact that the probability measure has total mass one. Hence, let $l \ge 1$. We compute

$$\mathbf{P}[N_s = k,\ N_t - N_s = l] = \mathbf{P}[T_k \le s < T_{k+1},\ T_{k+l} \le t < T_{k+l+1}]$$

$$= \int_0^\infty \cdots \int_0^\infty dx_1 \cdots dx_{k+l+1}$$

$$\alpha^{k+l+1} e^{-\alpha S_{k+l+1}(x)} \mathbb{1}_{\{S_k(x) \le s < S_{k+1}(x)\}} \mathbb{1}_{\{S_{k+l}(x) \le t < S_{k+l+1}(x)\}}.$$

Starting with x_{k+l+1}, we integrate successively. In the first step, substitute $z = S_{k+l+1}(x)$ to obtain

$$\int_0^\infty dx_{k+l+1}\, \alpha\, e^{-\alpha S_{k+l+1}(x)}\, \mathbb{1}_{\{S_{k+l+1}(x)>t\}} = \int_t^\infty dz\, \alpha\, e^{-\alpha z} = e^{-\alpha t}.$$

Now keep $x_1, \ldots, x_k$ fixed and substitute for the remaining variables by letting $y_1 = S_{k+1}(x) - s$, $y_2 = x_{k+2}, \ldots, y_l = x_{k+l}$ to obtain

$$\int_0^\infty \cdots \int_0^\infty dx_{k+1} \cdots dx_{k+l}\, \mathbb{1}_{\{s < S_{k+1}(x) \leq S_{k+l} \leq t\}}$$

$$= \int_0^\infty \cdots \int_0^\infty dy_1 \cdots dy_l\, \mathbb{1}_{\{y_1+\ldots+y_l \leq t-s\}} = \frac{(t-s)^l}{l!}.$$

(The last identity can be obtained, for example, by induction on l.) Now integrate the remaining variables $x_1, \ldots, x_k$ to get

$$\int_0^\infty \cdots \int_0^\infty dx_1 \cdots dx_k\, \mathbb{1}_{\{S_k(x) \leq s\}} = \frac{s^k}{k!}.$$

In total, we have

$$\mathbf{P}[N_s = k,\ N_t - N_s = l] = e^{-\alpha t}\, \alpha^{k+l}\, \frac{s^k}{k!}\, \frac{(t-s)^l}{l!};$$

hence (5.21) holds.   □

> **Takeaways** Assume that events, e.g. radioactive decays, happen at probability $\approx \alpha \cdot (b-a)$ (for some $\alpha > 0$) in a small time interval $(a, b]$. Also assume that the events in disjoints intervals come independently. Then the waiting times between events are exponentially distributed with parameter α and the total number of events up to time t is described by a Poisson process.

Exercise 5.5.1 Let $L_n, X_k^n, k, n \in \mathbb{N}$ be independent random variables with $L_n \sim \mathrm{Poi}_\alpha$ and $X_k^n \sim \mathcal{U}_{(n-1,n]}$ (the uniform distribution on $(n-1, n]$) for all $k, n \in \mathbb{N}$. Define

$$N_t := \#\{(k, n) \in \mathbb{N}^2 : k \leq L_n \text{ and } X_k^n \leq t\}.$$

Show that $(N_t)_{t \geq 0}$ is a Poisson process with intensity α. ♣

Exercise 5.5.2 Let $T > 0$ and let $X_1, X_2, \ldots$ be i.i.d. random variables that are uniformly distributed on $[0, 1]$. Let

$$N := \max \left\{ n \in \mathbb{N}_0 : X_1 + \ldots + X_n \leq T \right\}$$

and compute $\mathbf{E}[N]$. ♣

Chapter 6
Convergence Theorems

In the strong and the weak laws of large numbers, we implicitly introduced the notions of almost sure convergence and convergence in probability of random variables. We saw that almost sure convergence implies convergence in measure/probability. This chapter is devoted to a systematic treatment of almost sure convergence, convergence in measure and convergence of integrals. The key role for connecting convergence in measure and convergence of integrals is played by the concept of uniform integrability.

6.1 Almost Sure and Measure Convergence

In the following, $(\Omega, \mathcal{A}, \mu)$ will be a σ-finite measure space. We first define in metric spaces almost sure convergence and convergence in measure and then compare both concepts. To this end, we need two lemmas that ensure that the distance function associated with two measurable maps is again measurable. In the following, let (E, d) be a separable metric space with Borel σ-algebra $\mathcal{B}(E)$. "Separable" means that there exists a countable dense subset. For $x \in E$ and $r > 0$, denote by $B_r(x) = \{y \in E : d(x, y) < r\}$ the ball with radius r centered at x.

Lemma 6.1 *Let $f, g : \Omega \to E$ be measurable with respect to $\mathcal{A} - \mathcal{B}(E)$. Then the map $H : \Omega \to [0, \infty), \omega \mapsto d(f(\omega), g(\omega))$ is $\mathcal{A} - \mathcal{B}([0, \infty))$-measurable.*

Proof Let $F \subset E$ be countable and dense. By the triangle inequality, $d(x, z) + d(z, y) \geq d(x, y)$ for all $x, y \in E$ and $z \in F$. Let $(z_n)_{n \in \mathbb{N}}$ be a sequence in F with $z_n \overset{n \to \infty}{\longrightarrow} x$. Since d is continuous, we have $d(x, z_n) + d(z_n, y) \overset{n \to \infty}{\longrightarrow} d(x, y)$. Putting things together, we infer $\inf_{z \in F}(d(x, z) + d(z, y)) = d(x, y)$. Since $x \mapsto d(x, z)$ is continuous and hence measurable, the maps $f_z, g_z : \Omega \to [0, \infty)$ with

© The Editor(s) (if applicable) and The Author(s), under exclusive license
to Springer Nature Switzerland AG 2020
A. Klenke, *Probability Theory*, Universitext,
https://doi.org/10.1007/978-3-030-56402-5_6

$f_z(\omega) = d(f(\omega), z)$ and $g_z(\omega) = d(g(\omega), z)$ are also measurable. Thus $f_z + g_z$ and $H = \inf_{z \in F}(f_z + g_z)$ are measurable.

(A somewhat more systematic proof is based on the fact that (f, g) is $\mathcal{A} - \mathcal{B}(E \times E)$-measurable (this will follow from Theorem 14.8) and that $d : E \times E \to [0, \infty)$ is continuous and hence $\mathcal{B}(E \times E) - \mathcal{B}([0, \infty))$-measurable. As a composition of measurable maps, $\omega \mapsto d(f(\omega), g(\omega))$ is measurable.)   □

Let $f, f_1, f_2, \dots : \Omega \to E$ be measurable with respect to $\mathcal{A} - \mathcal{B}(E)$.

Definition 6.2 We say that $(f_n)_{n \in \mathbb{N}}$ converges to f

(i) in μ-**measure** (or, briefly, in measure), symbolically $f_n \xrightarrow{\text{meas}} f$, if

$$\mu(\{d(f, f_n) > \varepsilon\} \cap A) \xrightarrow{n \to \infty} 0$$

for all $\varepsilon > 0$ and all $A \in \mathcal{A}$ with $\mu(A) < \infty$, and

(ii) μ-**almost everywhere** (a.e.), symbolically $f_n \xrightarrow{\text{a.e.}} f$, if there exists a μ-null set $N \in \mathcal{A}$ such that

$$d(f(\omega), f_n(\omega)) \xrightarrow{n \to \infty} 0 \quad \text{for any } \omega \in \Omega \setminus N.$$

If μ is a probability measure, then convergence in μ-measure is also called **convergence in probability**. If $(f_n)_{n \in \mathbb{N}}$ converges a.e., then we also say that $(f_n)_{n \in \mathbb{N}}$ converges **almost surely** (a.s.) and write $f_n \xrightarrow{\text{a.s.}} f$. Sometimes we will drop the qualifications "almost everywhere" and "almost surely".

Remark 6.3 Let $A_1, A_2, \dots \in \mathcal{A}$ with $A_n \uparrow \Omega$ and $\mu(A_n) < \infty$ for any $n \in \mathbb{N}$. Then a.e. convergence is equivalent to a.e. convergence on each A_n. ◊

Remark 6.4 Almost everywhere convergence implies convergence in measure: For $\varepsilon > 0$, define

$$D_n(\varepsilon) = \{d(f, f_m) > \varepsilon \text{ for some } m \geq n\}.$$

Then $D(\varepsilon) := \bigcap_{n=1}^{\infty} D_n(\varepsilon) \subset N$, where N is the null set from the definition of almost everywhere convergence. Upper semicontinuity of μ implies

$$\mu(D_n(\varepsilon) \cap A) \xrightarrow{n \to \infty} \mu(D(\varepsilon) \cap A) = 0$$

for any $A \in \mathcal{A}$ with $\mu(A) < \infty$. ◊

Remark 6.5 Almost everywhere convergence and convergence in measure determine the limit up to equality almost everywhere. Indeed, let $f_n \xrightarrow{\text{meas}} f$ and $f_n \xrightarrow{\text{meas}} g$. Let $A_1, A_2, \dots \in \mathcal{A}$ with $A_n \uparrow \Omega$ and $\mu(A_n) < \infty$ for any $n \in \mathbb{N}$.

Then (since $d(f, g) \leq d(f, f_n) + d(g, f_n)$), for any $m \in \mathbb{N}$ and $\varepsilon > 0$,

$$\mu\big(A_m \cap \{d(f, g) > \varepsilon\}\big)$$

$$\leq \mu\big(A_m \cap \{d(f, f_n) > \varepsilon/2\}\big) + \mu\big(A_m \cap \{d(g, f_n) > \varepsilon/2\}\big) \overset{n\to\infty}{\longrightarrow} 0.$$

Hence $\mu\big(\{d(f, g) > 0\}\big) = 0.$ $\Diamond$

Reflection Maybe it comes as a surprise that in the definition of stochastic convergence, the set A of finite measure pops up. It is used to localise the convergence. Consider the case $\Omega = \mathbb{R}$, μ the Lebesgue measure and $f_n := \mathbb{1}_{[n,n+1]}$, $f \equiv 0$. With our definition of stochastic convergence we have $f_n \overset{\text{meas}}{\longrightarrow} f$. However, if we do not intersect with the set A, then stochastic convergence would fail, although we still had $f_n \overset{\text{a.e.}}{\longrightarrow} f$. Hence convergence almost everywhere would not imply stochastic convergence. Now this causes trouble in many places and so we chose a definition where this implication holds. ♠

Remark 6.6 In general, convergence in measure does not imply almost everywhere convergence. Indeed, let $(X_n)_{n\in\mathbb{N}}$ be an independent family of random variables with $X_n \sim \mathrm{Ber}_{1/n}$. Then $X_n \overset{n\to\infty}{\longrightarrow} 0$ in probability but the Borel–Cantelli lemma implies $\limsup_{n\to\infty} X_n = 1$ almost surely. $\Diamond$

Theorem 6.7 *Let $A_1, A_2, \ldots \in \mathcal{A}$ with $A_N \uparrow \Omega$ and $\mu(A_N) < \infty$ for all $N \in \mathbb{N}$. For measurable $f, g : \Omega \to E$, let*

$$\tilde{d}(f, g) := \sum_{N=1}^{\infty} \frac{2^{-N}}{1 + \mu(A_N)} \int_{A_N} \big(1 \wedge d(f(\omega), g(\omega))\big)\, \mu(d\omega). \tag{6.1}$$

Then $\tilde{d}$ is a metric that induces convergence in measure: If $f, f_1, f_2, \ldots$ are measurable, then

$$f_n \overset{\text{meas}}{\longrightarrow} f \quad \Longleftrightarrow \quad \tilde{d}(f, f_n) \overset{n\to\infty}{\longrightarrow} 0.$$

Proof For $N \in \mathbb{N}$, define

$$\tilde{d}_N(f, g) := \int_{A_N} \big(1 \wedge d(f(\omega), g(\omega))\big)\, \mu(d\omega).$$

Then $\tilde{d}(f, f_n) \overset{n\to\infty}{\longrightarrow} 0$ if and only if $\tilde{d}_N(f, f_n) \overset{n\to\infty}{\longrightarrow} 0$ for all $N \in \mathbb{N}$.
"$\Longrightarrow$" Assume $f_n \overset{\text{meas}}{\longrightarrow} f$. Then, for any $\varepsilon \in (0, 1)$,

$$\tilde{d}_N(f, f_n) \leq \mu\big(A_N \cap \{d(f, f_n) > \varepsilon\}\big) + \varepsilon\,\mu(A_N) \overset{n\to\infty}{\longrightarrow} \varepsilon\,\mu(A_N).$$

Letting $\varepsilon \downarrow 0$ yields $\tilde{d}_N(f, f_n) \overset{n\to\infty}{\longrightarrow} 0$.

"$\Longleftarrow$" Assume $\tilde{d}(f, f_n) \overset{n\to\infty}{\longrightarrow} 0$. Let $B \in \mathcal{A}$ with $\mu(B) < \infty$. Fix $\delta > 0$ and choose $N \in \mathbb{N}$ large enough that $\mu(B \setminus A_N) < \delta$. Then, for $\varepsilon \in (0, 1)$,

$$\mu\big(B \cap \{d(f, f_n) > \varepsilon\}\big) \leq \delta + \mu\big(A_N \cap \{d(f, f_n) > \varepsilon\}\big)$$

$$\leq \delta + \varepsilon^{-1} \tilde{d}_N(f, f_n) \overset{n\to\infty}{\longrightarrow} \delta.$$

Letting $\delta \downarrow 0$ yields $\mu\big(B \cap \{d(f, f_n) > \varepsilon\}\big) \overset{n\to\infty}{\longrightarrow} 0$; hence $f_n \overset{\mathrm{meas}}{\longrightarrow} f$. □

Consider the most prominent case $E = \mathbb{R}$ equipped with the Euclidean metric. Here the integral is the basis for another concept of convergence.

Definition 6.8 (Mean convergence) Let $f, f_1, f_2, \ldots \in \mathcal{L}^1(\mu)$. We say that the sequence $(f_n)_{n \in \mathbb{N}}$ converges **in mean** to f, symbolically

$$f_n \overset{L^1}{\longrightarrow} f,$$

if $\| f_n - f \|_1 \overset{n\to\infty}{\longrightarrow} 0$.

Remark 6.9 If $f_n \overset{L^1}{\longrightarrow} f$, then in particular $\int f_n \, d\mu \overset{n\to\infty}{\longrightarrow} \int f \, d\mu$. ◊

Remark 6.10 If $f_n \overset{L^1}{\longrightarrow} f$ and $f_n \overset{L^1}{\longrightarrow} g$, then $f = g$ almost everywhere. Indeed, by the triangle inequality, $\| f - g \|_1 \leq \| f_n - f \|_1 + \| f_n - g \|_1 \overset{n\to\infty}{\longrightarrow} 0$. ◊

Remark 6.11 Both L^1-convergence and almost everywhere convergence imply convergence in measure. All other implications are incorrect in general. ◊

Theorem 6.12 (Fast convergence) *Let (E, d) be a separable metric space. In order for the sequence $(f_n)_{n \in \mathbb{N}}$ of measurable maps $\Omega \to E$ to converge almost everywhere, it is sufficient that one of the following conditions holds.*

(i) $E = \mathbb{R}$ and there is a $p \in [1, \infty)$ with $f_n \in \mathcal{L}^p(\mu)$ for all $n \in \mathbb{N}$ and there is an $f \in \mathcal{L}^p(\mu)$ with $\displaystyle\sum_{n=1}^{\infty} \| f_n - f \|_p^p < \infty$.

(ii) There is a measurable f with $\displaystyle\sum_{n=1}^{\infty} \mu(A \cap \{d(f, f_n) > \varepsilon\}) < \infty$ for all $\varepsilon > 0$ and for all $A \in \mathcal{A}$ with $\mu(A) < \infty$.
  In both cases, we have $f_n \overset{n\to\infty}{\longrightarrow} f$ almost everywhere.
(iii) E is complete and there is a summable sequence $(\varepsilon_n)_{n \in \mathbb{N}}$ such that

$$\sum_{n=1}^{\infty} \mu(A \cap \{d(f_n, f_{n+1}) > \varepsilon_n\}) < \infty \quad \text{for all } A \in \mathcal{A} \text{ with } \mu(A) < \infty.$$

Proof Clearly, condition (i) implies (ii) since Markov's inequality yields that
$\mu(\{|f - f_n| > \varepsilon\}) \leq \varepsilon^{-p} \|f - f_n\|_p^p$.

By Remark 6.3, it is enough to consider the case $\mu(\Omega) < \infty$.

Assume (ii). Let $B_n(\varepsilon) = \{d(f, f_n) > \varepsilon\}$ and $B(\varepsilon) = \limsup_{n\to\infty} B_n(\varepsilon)$. By the Borel–Cantelli lemma, $\mu(B(\varepsilon)) = 0$. Let $N = \bigcup_{n=1}^{\infty} B(1/n)$. Then $\mu(N) = 0$ and $f_n(\omega) \overset{n\to\infty}{\longrightarrow} f(\omega)$ for any $\omega \in \Omega \setminus N$.

Assume (iii). Let $B_n = \{d(f_n, f_{n+1}) > \varepsilon_n\}$ and $B = \limsup_{n\to\infty} B_n$. Then $\mu(B) = 0$ and $(f_n(\omega))_{n\in\mathbb{N}}$ is a Cauchy sequence in E for any $\omega \in \Omega \setminus B$. Since E is complete, the limit $f(\omega) := \lim_{n\to\infty} f_n(\omega)$ exists. For $\omega \in B$, define $f(\omega) = 0$. $\qquad\square$

Corollary 6.13 *Let (E, d) be a separable metric space. Let $f, f_1, f_2, \ldots$ be measurable maps $\Omega \to E$. Then the following statements are equivalent.*

(i) *$f_n \overset{n\to\infty}{\longrightarrow} f$ in measure.*
(ii) *For any subsequence of $(f_n)_{n\in\mathbb{N}}$, there exists a sub-subsequence that converges to f almost everywhere.*

Proof **"(ii) $\Longrightarrow$ (i)"** Assume that (i) does not hold. Let $\tilde{d}$ be a metric that induces convergence in measure (see Theorem 6.7). Then there exists an $\varepsilon > 0$ and a subsequence $(f_{n_k})_{k\in\mathbb{N}}$ with $\tilde{d}(f_{n_k}, f) > \varepsilon$ for all $k \in \mathbb{N}$. Clearly, no subsequence of $(f_{n_k})_{k\in\mathbb{N}}$ converges to f in measure; hence neither converges almost everywhere.

"(i) $\Longrightarrow$ (ii)" Now assume (i). Let $A_1, A_2, \ldots \in \mathcal{A}$ with $A_N \uparrow \Omega$ and $\mu(A_N) < \infty$ for any $N \in \mathbb{N}$. Since $f_{n_k} \overset{\text{meas}}{\longrightarrow} f$ for $k \to \infty$, we can choose a subsequence $(f_{n_{k_l}})_{l\in\mathbb{N}}$ such that $\mu\big(A_l \cap \big(d(f, f_{n_{k_l}}) > 1/l\big)\big) < 2^{-l}$ for any $l \in \mathbb{N}$. Hence, for each $N \in \mathbb{N}$, we have

$$\sum_{l=1}^{\infty} \mu\Big(A_N \cap \Big(d\big(f, f_{n_{k_l}}\big) > \frac{1}{l}\Big)\Big) \leq N\,\mu(A_N) + \sum_{l=N+1}^{\infty} 2^{-l} < \infty.$$

By Theorem 6.12(ii), $(f_{n_{k_l}})_{l\in\mathbb{N}}$ converges to f almost everywhere on A_N. By Remark 6.3, $(f_{n_{k_l}})_{l\in\mathbb{N}}$ converges to f almost everywhere. $\qquad\square$

Corollary 6.14 *Let $(\Omega, \mathcal{A}, \mu)$ be a measure space in which almost everywhere convergence and convergence in measure do not coincide. Then there does not exist a topology on the set of measurable maps $\Omega \to E$ that induces almost everywhere convergence.*

Proof Assume that there does exist a topology that induces almost everywhere convergence. Let $f, f_1, f_2, \ldots$ be measurable maps with the property that $f_n \overset{\text{meas}}{\longrightarrow} f$, but not $f_n \overset{n\to\infty}{\longrightarrow} f$ almost everywhere. Now let U be an open set that contains f, but with $f_n \notin U$ for infinitely many $n \in \mathbb{N}$. Hence, let $(f_{n_k})_{k\in\mathbb{N}}$ be a subsequence with $f_{n_k} \notin U$ for all $k \in \mathbb{N}$. Since $f_{n_k} \overset{k\to\infty}{\longrightarrow} f$ in measure, by Corollary 6.13, there exists a further subsequence $(f_{n_{k_l}})_{l\in\mathbb{N}}$ of $(f_{n_k})_{k\in\mathbb{N}}$ with $f_{n_{k_l}} \overset{l\to\infty}{\longrightarrow} f$ almost everywhere. However, then $f_{n_{k_l}} \in U$ for all but finitely many l, which yields a contradiction! $\square$

Corollary 6.15 *Let (E, d) be a separable complete metric space. Let $(f_n)_{n \in \mathbb{N}}$ be a Cauchy sequence in measure in E; that is, for any $A \in \mathcal{A}$ with $\mu(A) < \infty$ and any $\varepsilon > 0$, we have*

$$\mu\big(A \cap \{d(f_m, f_n) > \varepsilon\}\big) \longrightarrow 0 \quad \text{for } m, n \to \infty.$$

Then $(f_n)_{n \in \mathbb{N}}$ converges in measure.

Proof Without loss of generality, we may assume $\mu(\Omega) < \infty$. Choose a subsequence $(f_{n_k})_{k \in \mathbb{N}}$ such that

$$\mu\big(\{d(f_n, f_{n_k}) > 2^{-k}\}\big) < 2^{-k} \quad \text{for all } n \geq n_k.$$

By Theorem 6.12(iii), there is an f with $f_{n_k} \overset{k \to \infty}{\longrightarrow} f$ almost everywhere; hence, in particular, $\mu(\{d(f_{n_k}, f) > \varepsilon/2\}) \overset{k \to \infty}{\longrightarrow} 0$ for all $\varepsilon > 0$. Now

$$\mu(\{d(f_n, f) > \varepsilon\}) \leq \mu(\{d(f_{n_k}, f_n) > \varepsilon/2\}) + \mu(\{d(f_{n_k}, f) > \varepsilon/2\}).$$

If k is large enough that $2^{-k} < \varepsilon/2$ and if $n \geq n_k$, then the first summand is smaller than 2^{-k}. Hence we have $\mu(\{d(f_n, f) > \varepsilon\}) \overset{n \to \infty}{\longrightarrow} 0$; that is, $f_n \overset{\text{meas}}{\longrightarrow} f$. $\square$

> **Takeaways** Almost everywhere (almost sure) convergence implies stochastic convergence. Also L^1-convergence implies stochastic convergence. The opposite implications hold only under an additional condition of summability (see Theorem 6.12).

Exercise 6.1.1 Let Ω be countable. Show that convergence in probability implies almost everywhere convergence. ♣

Exercise 6.1.2 Give an example of a sequence that

(i) converges in L^1 but not almost everywhere,
(ii) converges almost everywhere but not in L^1. ♣

Exercise 6.1.3 (Egorov's theorem (1911)) Let $(\Omega, \mathcal{A}, \mu)$ be a finite measure space and let $f_1, f_2, \ldots$ be measurable functions that converge to some f almost everywhere. Show that, for every $\varepsilon > 0$, there is a set $A \in \mathcal{A}$ with $\mu(\Omega \setminus A) < \varepsilon$ and $\sup_{\omega \in A} |f_n(\omega) - f(\omega)| \overset{n \to \infty}{\longrightarrow} 0$. ♣

Exercise 6.1.4 Let $(X_i)_{i \in \mathbb{N}}$ be independent, square integrable random variables with $\mathbf{E}[X_i] = 0$ for all $i \in \mathbb{N}$.

(i) Show that $\sum_{i=1}^{\infty} \mathbf{Var}[X_i] < \infty$ implies that there exists a real random variable X with $\sum_{i=1}^{n} X_i \overset{n \to \infty}{\longrightarrow} X$ almost surely.
(ii) Does the converse implication hold in (i)? ♣

6.2 Uniform Integrability

From the preceding section, we can conclude that convergence in measure plus existence of L^1 limit points implies L^1-convergence. Hence convergence in measure plus relative sequential compactness in L^1 yields convergence in L^1. In this section, we study a criterion for relative sequential compactness in L^1, the so-called uniform integrability.

Definition 6.16 A family $\mathcal{F} \subset \mathcal{L}^1(\mu)$ is called **uniformly integrable** if

$$\inf_{0 \le g \in \mathcal{L}^1(\mu)} \sup_{f \in \mathcal{F}} \int (|f| - g)^+ \, d\mu = 0. \tag{6.2}$$

Theorem 6.17 *The family $\mathcal{F} \subset \mathcal{L}^1(\mu)$ is uniformly integrable if and only if*

$$\inf_{0 \le \tilde{g} \in \mathcal{L}^1(\mu)} \sup_{f \in \mathcal{F}} \int_{\{|f| > \tilde{g}\}} |f| \, d\mu = 0. \tag{6.3}$$

If $\mu(\Omega) < \infty$, then uniform integrability is equivalent to either of the following two conditions:

(i) $\displaystyle \inf_{a \in [0, \infty)} \sup_{f \in \mathcal{F}} \int (|f| - a)^+ \, d\mu = 0,$

(ii) $\displaystyle \inf_{a \in [0, \infty)} \sup_{f \in \mathcal{F}} \int_{\{|f| > a\}} |f| \, d\mu = 0.$

Proof Clearly, $(|f| - g)^+ \le |f| \cdot \mathbb{1}_{\{|f| > g\}}$; hence (6.3) implies uniform integrability. Now assume (6.2). For $\varepsilon > 0$, choose $g_\varepsilon \in \mathcal{L}^1(\mu)$ such that

$$\sup_{f \in \mathcal{F}} \int (|f| - g_\varepsilon)^+ \, d\mu \le \varepsilon. \tag{6.4}$$

Define $\tilde{g}_\varepsilon = 2 g_{\varepsilon/2}$. Then, for $f \in \mathcal{F}$,

$$\int_{\{|f| > \tilde{g}_\varepsilon\}} |f| \, d\mu \le \int_{\{|f| > \tilde{g}_\varepsilon\}} (|f| - g_{\varepsilon/2})^+ \, d\mu + \int_{\{|f| > \tilde{g}_\varepsilon\}} g_{\varepsilon/2} \, d\mu.$$

By construction, $\int_{\{|f|>\widetilde{g}_\varepsilon\}}(|f|-g_{\varepsilon/2})^+\,d\mu \le \varepsilon/2$ and

$$g_{\varepsilon/2}\,\mathbb{1}_{\{|f|>\widetilde{g}_\varepsilon\}} \le (|f|-g_{\varepsilon/2})^+\,\mathbb{1}_{\{|f|>\widetilde{g}_\varepsilon\}};$$

hence also

$$\int_{\{|f|>\widetilde{g}_\varepsilon\}} g_{\varepsilon/2}\,d\mu \le \int_{\{|f|>\widetilde{g}_\varepsilon\}}(|f|-g_{\varepsilon/2})^+\,d\mu \le \frac{\varepsilon}{2}.$$

Summing up, we have

$$\sup_{f\in\mathcal{F}} \int_{\{|f|>\widetilde{g}_\varepsilon\}} |f|\,d\mu \le \varepsilon. \tag{6.5}$$

Clearly, (ii) implies (i). If $\mu(\Omega) < \infty$, then (i) implies uniform integrability of $\mathcal{F}$ since the infimum is taken over the smaller set of constant functions. We still have to show that uniform integrability implies (ii). Accordingly, assume $\mathcal{F}$ is uniformly integrable (but not necessarily $\mu(\Omega) < \infty$). For any $\varepsilon > 0$ (and g_ε and $\widetilde{g}_\varepsilon$ as above), choose a_ε such that $\int_{\{\widetilde{g}_{\varepsilon/2}>a_\varepsilon\}}\widetilde{g}_{\varepsilon/2}\,d\mu < \frac{\varepsilon}{2}$. Then

$$\int_{\{|f|>a_\varepsilon\}} |f|\,d\mu \le \int_{\{|f|>\widetilde{g}_{\varepsilon/2}\}} |f|\,d\mu + \int_{\{\widetilde{g}_{\varepsilon/2}>a_\varepsilon\}} \widetilde{g}_{\varepsilon/2}\,d\mu \; < \; \varepsilon. \qquad \square$$

Theorem 6.18

(i) If $\mathcal{F} \subset \mathcal{L}^1(\mu)$ is a finite set, then $\mathcal{F}$ is uniformly integrable.

(ii) If $\mathcal{F}, \mathcal{G} \subset \mathcal{L}^1(\mu)$ are uniformly integrable, then $(f + g : f \in \mathcal{F}, g \in \mathcal{G})$, $(f - g : f \in \mathcal{F}, g \in \mathcal{G})$ and $\{|f| : f \in \mathcal{F}\}$ are also uniformly integrable.

(iii) If $\mathcal{F}$ is uniformly integrable and if, for any $g \in \mathcal{G}$, there exists an $f \in \mathcal{F}$ with $|g| \le |f|$, then $\mathcal{G}$ is also uniformly integrable.

Proof The proof is simple and is left as an exercise. $\square$

The following theorem describes a very useful criterion for uniform integrability. We will use it in many places.

Theorem 6.19 *For finite μ, $\mathcal{F} \subset \mathcal{L}^1(\mu)$ is uniformly integrable if and only if there is a measurable function $H : [0,\infty) \to [0,\infty)$ with $\lim_{x\to\infty} H(x)/x = \infty$ and*

$$\sup_{f\in\mathcal{F}} \int H(|f|)\,d\mu < \infty.$$

H can be chosen to be monotone increasing and convex.

Proof " $\Longleftarrow$ "   Assume there is an H with the advertised properties. Then $K_a :=$ $\inf_{x \geq a} \frac{H(x)}{x} \uparrow \infty$ if $a \uparrow \infty$. Hence, for $a > 0$,

$$\sup_{f \in \mathcal{F}} \int_{\{|f| \geq a\}} |f| \, d\mu \leq \frac{1}{K_a} \sup_{f \in \mathcal{F}} \int_{\{|f| \geq a\}} H(|f|) \, d\mu$$

$$\leq \frac{1}{K_a} \sup_{f \in \mathcal{F}} \int H(|f|) \, d\mu \overset{a \to \infty}{\longrightarrow} 0.$$

" $\Longrightarrow$ "   Assume $\mathcal{F}$ is uniformly integrable. As we have $\mu(\Omega) < \infty$, by Theorem 6.17, there exists a sequence $a_n \uparrow \infty$ with

$$\sup_{f \in \mathcal{F}} \int (|f| - a_n)^+ \, d\mu < 2^{-n}.$$

Define

$$H(x) = \sum_{n=1}^{\infty} (x - a_n)^+ \qquad \text{for any } x \geq 0.$$

As a sum of convex functions, H is convex. Further, for any $n \in \mathbb{N}$ and $x \geq 2a_n$, $H(x)/x \geq \sum_{k=1}^{n}(1 - a_k/x)^+ \geq n/2$; hence we have $H(x)/x \uparrow \infty$. Finally, by monotone convergence, for any $f \in \mathcal{F}$,

$$\int H(|f(\omega)|) \, \mu(d\omega) = \sum_{n=1}^{\infty} \int (|f| - a_n)^+ \, d\mu \leq \sum_{n=1}^{\infty} 2^{-n} = 1. \qquad \square$$

Reflection   In the above theorem, why did we need that μ is finite? Can you come up with a counterexample for the case $\mu(\Omega) = \infty$? Which part of the theorem would still hold? ♠

Recall the notation $\| \cdot \|_p$ from Definition 4.16.

Definition 6.20   Let $p \in [1, \infty]$. A family $\mathcal{F} \subset \mathcal{L}^p(\mu)$ is called **bounded in $\mathcal{L}^p(\mu)$** if $\sup\{\|f\|_p : f \in \mathcal{F}\} < \infty$.

Corollary 6.21   *Let $\mu(\Omega) < \infty$ and $p > 1$. If $\mathcal{F}$ is bounded in $\mathcal{L}^p(\mu)$, then $\mathcal{F}$ is uniformly integrable.*

Proof   Apply Theorem 6.19 with the convex map $H(x) = x^p$. $\qquad \square$

Corollary 6.22   *If $(X_i)_{i \in I}$ is a family of square integrable random variables with*

$$\sup\{|\mathbf{E}[X_i]| : i \in I\} < \infty \qquad and \qquad \sup\{\mathbf{Var}[X_i] : i \in I\} < \infty,$$

then $(X_i)_{i \in I}$ is uniformly integrable.

Proof Since $\mathbf{E}[X_i^2] = \mathbf{E}[X_i]^2 + \mathbf{Var}[X_i]$, $i \in I$, is bounded, this follows from Corollary 6.21 with $p = 2$.   □

Lemma 6.23 *There is a map $h \in \mathcal{L}^1(\mu)$ with $h > 0$ almost everywhere.*

Proof Let $A_1, A_2, \ldots, \in \mathcal{A}$ with $A_n \uparrow \Omega$ and $\mu(A_n) < \infty$ for all $n \in \mathbb{N}$. Define

$$h = \sum_{n=1}^{\infty} 2^{-n} (1 + \mu(A_n))^{-1} \mathbb{1}_{A_n}.$$

Then $h > 0$ almost everywhere and $\int h \, d\mu = \sum_{n=1}^{\infty} 2^{-n} \frac{\mu(A_n)}{1+\mu(A_n)} \leq 1$.   □

Theorem 6.24 *A family $\mathcal{F} \subset \mathcal{L}^1(\mu)$ is uniformly integrable if and only if the following two conditions are fulfilled.*

(i) $C := \sup\limits_{f \in \mathcal{F}} \int |f| \, d\mu < \infty$.

(ii) There is a function $0 \leq h \in \mathcal{L}^1(\mu)$ such that for any $\varepsilon > 0$, there is a $\delta(\varepsilon) > 0$ with

$$\sup_{f \in \mathcal{F}} \int_A |f| \, d\mu \leq \varepsilon \quad \text{for all } A \in \mathcal{A} \text{ such that } \int_A h \, d\mu < \delta(\varepsilon).$$

If $\mu(\Omega) < \infty$, then (ii) is equivalent to (iii):

(iii) For all $\varepsilon > 0$, there is a $\delta(\varepsilon) > 0$ such that

$$\sup_{f \in \mathcal{F}} \int_A |f| \, d\mu \leq \varepsilon \quad \text{for all } A \in \mathcal{A} \text{ with } \mu(A) < \delta(\varepsilon).$$

Proof "$\Longrightarrow$" Let $\mathcal{F}$ be uniformly integrable. Let $h \in \mathcal{L}^1(\mu)$ with $h > 0$ a.e. Let $\varepsilon > 0$ and let $\widetilde{g}_{\varepsilon/3}$ be an $\varepsilon/3$-bound for $\mathcal{F}$ (as in (6.5)). Since $\{\widetilde{g}_{\varepsilon/3} \geq \alpha h\} \downarrow \emptyset$ for $\alpha \to \infty$, for sufficiently large $\alpha = \alpha(\varepsilon)$, we have

$$\int_{\{\widetilde{g}_{\varepsilon/3} \geq \alpha h\}} \widetilde{g}_{\varepsilon/3} \, d\mu < \frac{\varepsilon}{3}.$$

Letting $\delta(\varepsilon) := \frac{\varepsilon}{3\alpha(\varepsilon)}$, we get for any $A \in \mathcal{A}$ with $\int_A h \, d\mu < \delta(\varepsilon)$ and any $f \in \mathcal{F}$,

$$\int_A |f| \, d\mu \leq \int_{\{|f| > \widetilde{g}_{\varepsilon/3}\}} |f| \, d\mu + \int_A \widetilde{g}_{\varepsilon/3} \, d\mu$$

$$\leq \frac{\varepsilon}{3} + \alpha \int_A h \, d\mu + \int_{\{\widetilde{g}_{\varepsilon/3} \geq \alpha h\}} \widetilde{g}_{\varepsilon/3} \, d\mu \leq \varepsilon.$$

Hence we have shown (ii). In the above computation, let $A = \Omega$ to obtain

$$\int |f|\,d\mu \leq \frac{2\varepsilon}{3} + \alpha \int h\,d\mu < \infty.$$

Hence we have also shown (i).

"$\Longleftarrow$" Assume (i) and (ii). Let $\varepsilon > 0$. Choose h and $\delta(\varepsilon) > 0$ as in (ii) and C as in (i). Define $\tilde{h} = \frac{C}{\delta(\varepsilon)}h$. Then

$$\int_{\{|f|>\tilde{h}\}} h\,d\mu = \frac{\delta(\varepsilon)}{C} \int_{\{|f|>\tilde{h}\}} \tilde{h}\,d\mu \leq \frac{\delta(\varepsilon)}{C} \int |f|\,d\mu \leq \delta(\varepsilon);$$

hence, by assumption, $\int_{\{|f|>\tilde{h}\}} |f|\,d\mu < \varepsilon$.

"(ii) $\Longrightarrow$ (iii)" Assume (ii). Let $\varepsilon > 0$ and choose $\delta = \delta(\varepsilon)$ as in (ii). Choose $K < \infty$ large enough that $\int_{\{h\geq K\}} h\,d\mu < \delta/2$. For all $A \in \mathcal{A}$ with $\mu(A) < \delta/(2K)$, we obtain

$$\int_A h\,d\mu \leq K\mu(A) + \int_{\{h\geq K\}} h\,d\mu < \delta;$$

hence $\int_A |f|\,d\mu \leq \varepsilon$ for all $f \in \mathcal{F}$.

"(iii) $\Longrightarrow$ (ii)" Assume (iii) and $\mu(\Omega) < \infty$. Then $h \equiv 1$ serves the purpose. $\qquad\square$

We come to the main theorem of this section.

Theorem 6.25 *Let* $\{f_n : n \in \mathbb{N}\} \subset \mathcal{L}^1(\mu)$. *The following statements are equivalent.*

(i) *There is an* $f \in \mathcal{L}^1(\mu)$ *with* $f_n \overset{n\to\infty}{\longrightarrow} f$ *in* L^1.

(ii) *$(f_n)_{n\in\mathbb{N}}$ is an $\mathcal{L}^1(\mu)$-Cauchy sequence; that is,* $\|f_n - f_m\|_1 \longrightarrow 0$ *for* $m, n \to \infty$.

(iii) *$(f_n)_{n\in\mathbb{N}}$ is uniformly integrable and there is a measurable map f such that* $f_n \overset{\text{meas}}{\longrightarrow} f$.

The limits in (i) and (iii) coincide.

Proof "(i) $\Longrightarrow$ (ii)" This is evident.

"(ii) $\Longrightarrow$ (iii)" For any $\varepsilon > 0$, there is an $n_\varepsilon \in \mathbb{N}$ such that $\|f_n - f_{n_\varepsilon}\|_1 < \varepsilon$ for all $n \geq n_\varepsilon$. Hence $\|(|f_n| - |f_{n_\varepsilon}|)^+\|_1 < \varepsilon$ for all $n \geq n_\varepsilon$. Thus $g_\varepsilon = \max\{|f_1|, \dots, |f_{n_\varepsilon}|\}$ is an ε-bound for $(f_n)_{n\in\mathbb{N}}$ (as in (6.4)). For $\varepsilon > 0$, let

$$\mu(\{|f_m - f_n| > \varepsilon\}) \leq \varepsilon^{-1}\|f_m - f_n\|_1 \longrightarrow 0 \qquad \text{for } m, n \to \infty.$$

Thus $(f_n)_{n\in\mathbb{N}}$ is also a Cauchy sequence in measure; hence it converges in measure by Corollary 6.15.

"(iii) $\Longrightarrow$ (i)" Let f be the limit in measure of the sequence $(f_n)_{n\in\mathbb{N}}$. Assume that $(f_n)_{n\in\mathbb{N}}$ does not converge to f in L^1. Then there is an $\varepsilon > 0$ and a subsequence $(f_{n_k})_{k\in\mathbb{N}}$ with

$$\|f - f_{n_k}\|_1 > 2\varepsilon \quad \text{for all } k \in \mathbb{N}. \tag{6.6}$$

Here we define $\|f - f_{n_k}\|_1 = \infty$ if $f \notin L^1(\mu)$. By Corollary 6.13, there is a subsequence $(f_{n_k'})_{k\in\mathbb{N}}$ of $(f_{n_k})_{k\in\mathbb{N}}$ with $f_{n_k'} \overset{k\to\infty}{\longrightarrow} f$ almost everywhere. By Fatou's lemma (Theorem 4.21) with 0 as a minorant, we thus get

$$\int |f|\, d\mu \leq \liminf_{k\to\infty} \int |f_{n_k'}|\, d\mu < \infty.$$

Hence $f \in L^1(\mu)$. By Theorem 6.18(ii) (with $\mathcal{G} = \{f\}$), we obtain that the family $(f - f_{n_k'})_{k\in\mathbb{N}}$ is uniformly integrable; hence there is a $0 \leq g \in L^1(\mu)$ such that $\int (|f - f_{n_k'}| - g)^+\, d\mu < \varepsilon$. Define

$$g_k = |f_{n_k'} - f| \wedge g \quad \text{for } k \in \mathbb{N}.$$

Then $g_k \overset{k\to\infty}{\longrightarrow} 0$ almost everywhere and $g - g_k \geq 0$. By Fatou's lemma,

$$\limsup_{k\to\infty} \int g_k\, d\mu = \int g\, d\mu - \liminf_{k\to\infty} \int (g - g_k)\, d\mu$$

$$\leq \int g\, d\mu - \int \left(\lim_{k\to\infty} (g - g_k) \right) d\mu = 0.$$

Since $|f - f_{n_k'}| = (|f - f_{n_k'}| - g)^+ + g_k$, this implies that

$$\limsup_{k\to\infty} \|f - f_{n_k'}\|_1 \leq \limsup_{k\to\infty} \int \left(|f - f_{n_k'}| - g \right)^+ d\mu + \limsup_{k\to\infty} \int g_k\, d\mu \leq \varepsilon,$$

contradicting (6.6). $\square$

Corollary 6.26 (Lebesgue's convergence theorem, dominated convergence) *Let f be measurable and let $(f_n)_{n\in\mathbb{N}}$ be a sequence in $L^1(\mu)$ with $f_n \overset{n\to\infty}{\longrightarrow} f$ in measure. Assume that there is an integrable dominating function $0 \leq g \in L^1(\mu)$ with $|f_n| \leq g$ almost everywhere for all $n \in \mathbb{N}$. Then $f \in L^1(\mu)$ and $f_n \overset{n\to\infty}{\longrightarrow} f$ in L^1; hence in particular $\int f_n\, d\mu \overset{n\to\infty}{\longrightarrow} \int f\, d\mu$.*

Proof This is a consequence of Theorem 6.25, as the dominating function ensures uniform integrability of the sequence $(f_n)_{n\in\mathbb{N}}$. $\square$

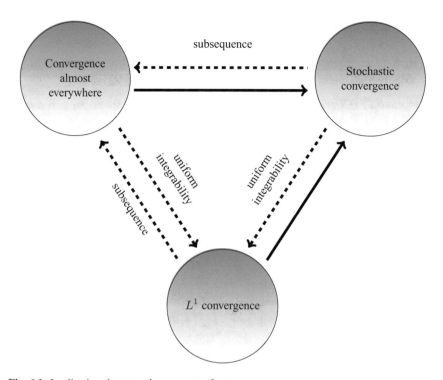

Fig. 6.1 Implications between the concepts of convergence.

Takeaways Loosely speaking, a family of functions is uniformly integrable if the main contributions to the integrals of those functions do not come from extremely large values of the functions. Together with stochastic convergence, uniform integrability is equivalent to L^1-convergence (see Fig. 6.1). As a consequence we get Lebesgue's dominated convergence theorem.

Exercise 6.2.1 Let $H \in \mathcal{L}^1(\mu)$ with $H > 0$ μ-a.e. (see Lemma 6.23) and let (E, d) be a separable metric space. For measurable $f, g : \Omega \to E$, define

$$d_H(f, g) := \int \left(1 \wedge d(f(\omega), g(\omega))\right) H(\omega)\, \mu(d\omega).$$

(i) Show that d_H is a metric that induces convergence in measure.
(ii) Show that d_H is complete if (E, d) is complete. ♣

6.3   Exchanging Integral and Differentiation

We study how properties such as continuity and differentiability of functions of two variables behave under integration with respect to one of the variables.

Theorem 6.27 (Continuity lemma) *Let (E, d) be a metric space, $x_0 \in E$ and let $f : \Omega \times E \to \mathbb{R}$ be a map with the following properties.*

(i) *For any $x \in E$, the map $\omega \mapsto f(\omega, x)$ is in $\mathcal{L}^1(\mu)$.*
(ii) *For almost all $\omega \in \Omega$, the map $x \mapsto f(\omega, x)$ is continuous at the point x_0.*
(iii) *There is a map $h \in \mathcal{L}^1(\mu)$, $h \geq 0$, such that $|f(\cdot, x)| \leq h$ μ-a.e. for all $x \in E$.*

Then the map $F : E \to \mathbb{R}$, $x \mapsto \int f(\omega, x)\, \mu(d\omega)$ is continuous at x_0.

Proof Let $(x_n)_{n \in \mathbb{N}}$ be a sequence in E with $\lim\limits_{n \to \infty} x_n = x_0$. Define $f_n = f(\cdot, x_n)$. By assumption, $|f_n| \leq h$ and $f_n \overset{n \to \infty}{\longrightarrow} f(\cdot, x_0)$ almost everywhere. By the dominated convergence theorem (Corollary 6.26), we get

$$F(x_n) = \int f_n \, d\mu \overset{n \to \infty}{\longrightarrow} \int f(\cdot, x_0)\, d\mu = F(x_0).$$

Hence F is continuous at x_0. $\square$

Reflection Find an example that shows that condition (iii) in Theorem 6.27 cannot simply be dropped. ♠

Theorem 6.28 (Differentiation lemma) *Let $I \subset \mathbb{R}$ be a nontrivial open interval and let $f : \Omega \times I \to \mathbb{R}$ be a map with the following properties.*

(i) *For any $x \in E$, the map $\omega \mapsto f(\omega, x)$ is in $\mathcal{L}^1(\mu)$.*
(ii) *For almost all $\omega \in \Omega$, the map $I \to \mathbb{R}$, $x \mapsto f(\omega, x)$ is differentiable with derivative f'.*
(iii) *There is a map $h \in \mathcal{L}^1(\mu)$, $h \geq 0$, such that $|f'(\cdot, x)| \leq h$ μ-a.e. for all $x \in I$.*

Then, for any $x \in I$, $f'(\cdot, x) \in \mathcal{L}^1(\mu)$ and the function $F : x \mapsto \int f(\omega, x)\, \mu(d\omega)$ is differentiable with derivative

$$F'(x) = \int f'(\omega, x)\, \mu(d\omega).$$

Proof Let $x_0 \in I$ and let $(x_n)_{n \in \mathbb{N}}$ be a sequence in I with $x_n \neq x_0$ for all $n \in \mathbb{N}$ and such that $\lim\limits_{n \to \infty} x_n = x_0$. We show that, along the sequence $(x_n)_{n \in \mathbb{N}}$, the difference quotients converge. Define

$$g_n(\omega) = \frac{f(\omega, x_n) - f(\omega, x_0)}{x_n - x_0} \qquad \text{for all } \omega \in \Omega.$$

By assumption (ii), we have

$$g_n \overset{n \to \infty}{\longrightarrow} f'(\cdot, x_0) \qquad \mu\text{-almost everywhere.}$$

By the mean value theorem of calculus, for all $n \in \mathbb{N}$ and for almost all $\omega \in \Omega$, there exists a $y_n(\omega) \in I$ with $g_n(\omega) = f'(\omega, y_n(\omega))$. In particular, $|g_n| \leq h$ almost everywhere for all $n \in \mathbb{N}$. By the dominated convergence theorem (Corollary 6.26), the limiting function $f'(\cdot, x_0)$ is in $\mathcal{L}^1(\mu)$ and

$$\lim_{n \to \infty} \frac{F(x_n) - F(x_0)}{x_n - x_0} = \lim_{n \to \infty} \int g_n(\omega) \, \mu(d\omega) = \int f'(\omega, x_0) \, \mu(d\omega). \qquad \square$$

Example 6.29 (Laplace transform) Let X be a nonnegative random variable on $(\Omega, \mathcal{A}, \mathbf{P})$. Using the notation of Theorem 6.28, let $I = [0, \infty)$ and $f(x, \lambda) = e^{-\lambda x}$ for $\lambda \in I$. Then

$$F(\lambda) = \mathbf{E}\left[e^{-\lambda X}\right]$$

is infinitely often differentiable in $(0, \infty)$. The first two derivatives of F are $F'(\lambda) = -\mathbf{E}[X e^{-\lambda X}]$ and $F''(\lambda) = \mathbf{E}[(X^2)e^{-\lambda X}]$. Successively, we get the nth derivative $F^{(n)}(\lambda) = \mathbf{E}[(-X)^n e^{-\lambda X}]$. By monotone convergence, we get

$$\mathbf{E}[X] = -\lim_{\lambda \downarrow 0} F'(\lambda) \tag{6.7}$$

and

$$\mathbf{E}[X^n] = (-1)^n \lim_{\lambda \downarrow 0} F^{(n)}(\lambda) \qquad \text{for all } n \in \mathbb{N}. \tag{6.8}$$

Indeed, for $\varepsilon > 0$ and $I = (\varepsilon, \infty)$, we have

$$\sup_{x \geq 0, \, \lambda \in I} \left| \frac{d}{d\lambda} f(x, \lambda) \right| = \sup_{x \geq 0, \, \lambda \in I} x \, e^{-\lambda x} = \varepsilon^{-1} e^{-1} < \infty.$$

Thus F fulfills the assumptions of Theorem 6.28. Inductively, we get the statement for $F^{(n)}$ since

$$\left| \frac{d^n}{d\lambda^n} f(x, \lambda) \right| \leq (n/\varepsilon)^n e^{-n} < \infty \qquad \text{for } x \geq 0 \text{ and } \lambda \geq \varepsilon. \quad \Diamond$$

Takeaways Consider a function of two variables that is continuous or differentiable with respect to one variable. Take the integral with respect to the other variable. We have used the convergence theorems from the last section to show that the integral is continuous or differentiable, respectively, if a regularity assumption is fulfilled. In this case, integral and derivative commute.

Exercise 6.3.1 Let X be a random variable on $(\Omega, \mathcal{A}, \mathbf{P})$ and let

$$\Lambda(t) := \log\left(\mathbf{E}\left[e^{tX}\right]\right) \quad \text{for all } t \in \mathbb{R}.$$

Show that $D := \{t \in \mathbb{R} : \Lambda(t) < \infty\}$ is a nonempty interval and that Λ is infinitely often differentiable in the interior of D. ♣

Chapter 7
L^p-Spaces and the Radon–Nikodym Theorem

In this chapter, we study the spaces of functions whose pth power is integrable. In Sect. 7.2, we first derive some of the important inequalities (Hölder, Minkowski, Jensen) and then in Sect. 7.3 investigate the case $p = 2$ in more detail. Apart from the inequalities, the important results for probability theory are Lebesgue's decomposition theorem and the Radon–Nikodym theorem in Sect. 7.4. At first reading, some readers might wish to skip some of the more analytic parts of this chapter.

7.1 Definitions

We always assume that $(\Omega, \mathcal{A}, \mu)$ is a σ-finite measure space. In Definition 4.16, for measurable $f : \Omega \to \overline{\mathbb{R}}$, we defined

$$\|f\|_p := \left(\int |f|^p \, d\mu \right)^{1/p} \qquad \text{for } p \in [1, \infty)$$

and

$$\|f\|_\infty := \inf \big\{ K \geq 0 : \mu(|f| > K) = 0 \big\}.$$

Further, we defined the spaces of functions where these expressions are finite:

$$\mathcal{L}^p(\Omega, \mathcal{A}, \mu) = \mathcal{L}^p(\mathcal{A}, \mu) = \mathcal{L}^p(\mu) = \{f : \Omega \to \overline{\mathbb{R}} \text{ measurable and } \|f\|_p < \infty\}.$$

We saw that $\| \cdot \|_1$ is a seminorm on $\mathcal{L}^1(\mu)$. Here our first goal is to change $\| \cdot \|_p$ into a proper norm for all $p \in [1, \infty]$. Apart from the fact that we still have to show

A. Klenke, *Probability Theory*, Universitext,
https://doi.org/10.1007/978-3-030-56402-5_7

the triangle inequality, to this end, we have to change the space a little bit since we only have

$$\|f - g\|_p = 0 \quad \Longleftrightarrow \quad f = g \quad \mu\text{-a.e.}$$

For a proper norm (that is, not only a seminorm), the left-hand side has to imply equality (not only a.e.) of f and g. Hence we now consider f and g as equivalent if $f = g$ almost everywhere. Thus let

$$\mathcal{N} = \{f \text{ is measurable and } f = 0 \quad \mu\text{-a.e.}\}.$$

For any $p \in [1, \infty]$, $\mathcal{N}$ is a subvector space of $\mathcal{L}^p(\mu)$. Thus formally we can build the factor space. This is the standard procedure in order to change a seminorm into a proper norm.

Definition 7.1 (Factor space)   For any $p \in [1, \infty]$, define

$$L^p(\Omega, \mathcal{A}, \mu) := \mathcal{L}^p(\Omega, \mathcal{A}, \mu)/\mathcal{N} = \{\bar{f} := f + \mathcal{N} : f \in \mathcal{L}^p(\mu)\}.$$

For $\bar{f} \in L^p(\mu)$, define $\|\bar{f}\|_p = \|f\|_p$ for any $f \in \bar{f}$. Also let $\int \bar{f} \, d\mu = \int f \, d\mu$ if this expression is defined for f.

Note that $\|\bar{f}\|_p$ and $\int \bar{f} \, d\mu$ do not depend on the choice of the representative $f \in \bar{f}$. Recall from Theorem 4.19 that $\int \bar{f} \, d\mu$ is well-defined if $f \in \mathcal{L}^p(\mu)$ and if μ is finite but it need not be if μ is infinite.

We first investigate convergence with respect to $\|\cdot\|_p$. To this end, we extend the corresponding theorem (Theorem 6.25) on convergence with respect to $\|\cdot\|_1$.

Definition 7.2   Let $p \in [1, \infty]$ and $f, f_1, f_2, \ldots \in L^p(\mu)$. If $\|f_n - f\|_p \overset{n\to\infty}{\longrightarrow} 0$, then we say that $(f_n)_{n\in\mathbb{N}}$ converges to f in $L^p(\mu)$ and we write $f_n \overset{L^p}{\longrightarrow} f$.

Theorem 7.3   *Let $p \in [1, \infty]$ and $f_1, f_2, \ldots \in L^p(\mu)$. Then the following statements are equivalent:*

(i)   There is an $f \in L^p(\mu)$ with $f_n \overset{L^p}{\longrightarrow} f$.
(ii) $(f_n)_{n\in\mathbb{N}}$ is a Cauchy sequence in $L^p(\mu)$.

If $p < \infty$, then, in addition, (i) and (ii) are equivalent to:

(iii) $(|f_n|^p)_{n\in\mathbb{N}}$ is uniformly integrable and there exists a measurable f with $f_n \overset{\text{meas}}{\longrightarrow} f$.

The limits in (i) and (iii) coincide.

Proof   For $p = \infty$, the equivalence of (i) and (ii) is a simple consequence of the triangle inequality.

Now let $p \in [1, \infty)$. The proof is similar to the proof of Theorem 6.25.

"(i) $\Longrightarrow$ (ii)" Note that $|x + y|^p \leq 2^p (|x|^p + |y|^p)$ for all $x, y \in \mathbb{R}$. Hence

$$\| f_m - f_n \|_p^p \leq 2^p \left(\| f_m - f \|_p^p + \| f_n - f \|_p^p \right) \stackrel{n \to \infty}{\longrightarrow} 0 \quad \text{for } m, n \to \infty.$$

"(ii) $\Longrightarrow$ (iii)" This works as in the proof of Theorem 6.25.

"(iii) $\Longrightarrow$ (i)" Since $|f_n|^p \stackrel{n \to \infty}{\longrightarrow} |f|^p$ in measure, by Theorem 6.25, we have $|f|^p \in \mathcal{L}^1(\mu)$ and hence $f \in \mathcal{L}^p(\mu)$. For $n \in \mathbb{N}$, define $g_n = |f_n - f|^p$. Then $g_n \stackrel{n \to \infty}{\longrightarrow} 0$ in measure, and $(g_n)_{n \in \mathbb{N}}$ is uniformly integrable since $g_n \leq 2^p (|f_n|^p + |f|^p)$. Hence we get (by Theorem 6.25) $\| f_n - f \|_p^p = \| g_n \|_1 \stackrel{n \to \infty}{\longrightarrow} 0$. □

Takeaways We have adapted the convergence theorems of the last chapter to L^p convergence. This is the starting point for the investigation of topological properties of L^p spaces in the subsequent sections.

Exercise 7.1.1 Let $f : \Omega \to \mathbb{R}$ be measurable. Show that the following hold.

(i) If $\int |f|^p \, d\mu < \infty$ for some $p \in (0, \infty)$, then $\| f \|_p \stackrel{p \to \infty}{\longrightarrow} \| f \|_\infty$.
(ii) The integrability condition in (i) cannot be waived. ♣

Exercise 7.1.2 Let $p \in (1, \infty)$, $f \in \mathcal{L}^p(\lambda)$, where λ is the Lebesgue measure on $\mathbb{R}$. Let $T : \mathbb{R} \to \mathbb{R}, x \mapsto x + 1$. Show that

$$\frac{1}{n} \sum_{k=0}^{n-1} f \circ T^k \stackrel{n \to \infty}{\longrightarrow} 0 \quad \text{in } L^p(\lambda). ♣$$

7.2 Inequalities and the Fischer–Riesz Theorem

We present one of the most important inequalities of probability theory, Jensen's inequality for convex functions, and indicate how to derive from it Hölder's inequality and Minkowski's inequality. They in turn yield the triangle inequality for $\| \cdot \|_p$ and help in determining the dual space of $L^p(\mu)$. However, for the formal proofs of the latter inequalities, we will follow a different route.

Before stating Jensen's inequality, we give a primer on the basics of convexity of sets and functions.

Definition 7.4 A subset G of a vector space (or of an affine linear space) is called **convex** if, for any two points $x, y \in G$ and any $\lambda \in [0, 1]$, we have $\lambda x + (1 - \lambda) y \in G$.

Example 7.5

(i) The convex subsets of $\mathbb{R}$ are the intervals.
(ii) A linear subspace of a vector space is convex.
(iii) The set of all probability measures on a measurable space is a convex set. ◊

Definition 7.6 Let G be a convex set. A map $\varphi : G \to \mathbb{R}$ is called **convex** if for any two points $x, y \in G$ and any $\lambda \in [0, 1]$, we have

$$\varphi\big(\lambda x + (1 - \lambda)y\big) \leq \lambda \varphi(x) + (1 - \lambda)\varphi(y).$$

φ is called **concave** if $(-\varphi)$ is convex.

Let $I \subset \mathbb{R}$ be an interval. Let $\varphi : I \to \mathbb{R}$ be continuous and in the interior I° twice continuously differentiable with second derivative φ''. Then φ is convex if and only if $\varphi''(x) \geq 0$ for all $x \in I^\circ$. To put it differently, the first derivative φ' of a convex function is a monotone increasing function. In the next theorem, we will see that this is still true even if φ is not twice continuously differentiable when we pass to the right-sided derivative $D^+\varphi$ (or to the left-sided derivative), which we show always exists.

Theorem 7.7 *Let $I \subset \mathbb{R}$ be an interval with interior I° and let $\varphi : I \to \mathbb{R}$ be a convex map. Then:*

(i) φ is continuous on I° and hence measurable with respect to $\mathcal{B}(I)$.
(ii) For $x \in I^\circ$, define the function of difference quotients

$$g_x(y) := \frac{\varphi(y) - \varphi(x)}{y - x} \quad \text{for } y \in I \setminus \{x\}.$$

Then g_x is monotone increasing and there exist the left-sided and right-sided derivatives

$$D^-\varphi(x) := \lim_{y \uparrow x} g_x(y) = \sup\{g_x(y) : y < x\}$$

and

$$D^+\varphi(x) := \lim_{y \downarrow x} g_x(y) = \inf\{g_x(y) : y > x\}.$$

(iii) For $x \in I^\circ$, we have $D^-\varphi(x) \leq D^+\varphi(x)$ and

$$\varphi(x) + (y - x)t \leq \varphi(y) \text{ for any } y \in I \quad \Longleftrightarrow \quad t \in [D^-\varphi(x), D^+\varphi(x)].$$

Hence $D^-\varphi(x)$ and $D^+\varphi(x)$ are the minimal and maximal slopes of a tangent at x.

(iv) *The maps $x \mapsto D^-\varphi(x)$ and $x \mapsto D^+\varphi(x)$ are monotone increasing. $x \mapsto D^-\varphi(x)$ is left continuous and $x \mapsto D^+\varphi(x)$ is right continuous. We have $D^-\varphi(x) = D^+\varphi(x)$ at all points of continuity of $D^-\varphi$ and $D^+\varphi$.*

(v) *φ is differentiable at x if and only if $D^-\varphi(x) = D^+\varphi(x)$. In this case, the derivative is $\varphi'(x) = D^+\varphi(x)$.*

(vi) *φ is almost everywhere differentiable and $\varphi(b) - \varphi(a) = \int_a^b D^+\varphi(x)\,dx$ for $a, b \in I^\circ$.*

Proof

(i) Let $x \in I^\circ$. Assume that $\liminf_{h\downarrow 0} \varphi(x-h) \le \varphi(x) - \varepsilon$ for some $\varepsilon > 0$. Since φ is convex, for $y \in I^\circ$ such that $y > x$, we have

$$\varphi(y) \ge \varphi(x) + \frac{y-x}{h}\left(\varphi(x) - \varphi(x-h)\right) \quad \text{for all } h > 0 \text{ with } x - h \in I^\circ.$$

Combining this with the assumption, we get $\varphi(y) = \infty$ for all $y > x$. Hence the assumption was false. A similar argument for the right-hand side yields continuity of φ at x.

(ii) Monotonicity is implied by convexity. The other claims are evident.

(iii) By monotonicity of g_x, we have $D^-\varphi(x) \le D^+\varphi(x)$. By construction, $\varphi(x) + (y-x)t \le \varphi(y)$ for all $y < x$ if and only if $t \ge D^-\varphi(x)$. The inequality holds for all $y > x$ if and only if $t \le D^+\varphi(x)$.

(iv) For $\varepsilon > 0$, by the convexity, the map $x \mapsto g_x(x + \varepsilon)$ is monotone increasing and is continuous by (i). Being an infimum of monotone increasing and continuous functions the map $x \mapsto D^+\varphi(x)$ is monotone increasing and right continuous. The statement for $D^-\varphi$ follows similarly. As $x \mapsto g_x(y)$ is monotone, we get $D^+\varphi(x') \ge D^-\varphi(x') \ge D^+\varphi(x)$ for $x' > x$. If $D^+\varphi$ is continuous at x, then $D^-\varphi(x) = D^+\varphi(x)$.

(v) This is obvious since $D^-\varphi$ and $D^+\varphi$ are the limits of the sequences of slopes of the left-sided and right-sided secant lines, respectively.

(vi) For $\varepsilon > 0$, let $A_\varepsilon = \{x \in I : D^+\varphi(x) \ge \varepsilon + \lim_{y\uparrow x} D^+\varphi(y)\}$ be the set of points of discontinuity of size at least ε. For any two points $a, b \in I$ with $a < b$, we have $\#(A_\varepsilon \cap (a, b)) \le \varepsilon^{-1}(D^+\varphi(b) - D^+\varphi(a))$; hence $A_\varepsilon \cap (a, b)$ is a finite set. Thus A_ε is countable. Hence also $A = \bigcup_{n=1}^\infty A_{1/n}$ is countable and thus a null set. By (iv) and (v), φ is differentiable in $I^\circ \setminus A$ with derivative $D^+\varphi$. $\qquad\square$

If I is an interval, then a map $g : I \to \mathbb{R}$ is called *affine linear* if there are numbers $a, b \in \mathbb{R}$ such that $g(x) = ax + b$ for all $x \in I$. If $\varphi : I \to \mathbb{R}$ is a map, then we write

$$L(\varphi) := \{g : I \to \mathbb{R} \text{ is affine linear and } g \le \varphi\}.$$

As a shorthand, we write $\sup L(\varphi)$ for the map $x \mapsto \sup\{f(x) : f \in L(\varphi)\}$.

Corollary 7.8 *Let $I \subset \mathbb{R}$ be an open interval and let $\varphi : I \to \mathbb{R}$ be a map. Then the following are equivalent.*

  (i) φ is convex.
  (ii) For any $x_0 \in I$, there exists a $g \in L(\varphi)$ with $g(x_0) = \varphi(x_0)$.
  (iii) $L(\varphi)$ is nonempty and $\varphi = \sup L(\varphi)$.
  (iv) There is a sequence $(g_n)_{n\in\mathbb{N}}$ in $L(\varphi)$ with $\varphi = \lim_{n\to\infty} \max\{g_1, \ldots, g_n\}$.

Furthermore, if $I \subset \mathbb{R}$ is an interval (not necessarily open) and $\varphi : I \to \mathbb{R}$ is convex, then we still have $L(\varphi) \neq \emptyset$.

Proof "(ii) $\Longrightarrow$ (iii) $\Longleftrightarrow$ (iv)"   This is obvious.

"(iii) $\Longrightarrow$ (i)"   The supremum of convex functions is convex and any affine linear map is convex. Hence $\sup L(\varphi)$ is convex if $L(\varphi) \neq \emptyset$.

"(i) $\Longrightarrow$ (ii)"   By Theorem 7.7(iii), for any $x_0 \in I$, the map

$$x \mapsto \varphi(x_0) + (x - x_0)\, D^+\varphi(x_0)$$

is in $L(\varphi)$.

Since for the implication "(i) $\Longrightarrow$ (ii)" we did not need that I is open, we get the supplementary statement. $\square$

Theorem 7.9 (Jensen's inequality) *Let $I \subset \mathbb{R}$ be an interval and let X be an I-valued random variable with $\mathbf{E}[|X|] < \infty$. If φ is convex, then $\mathbf{E}[\varphi(X)^-] < \infty$ and*

$$\mathbf{E}[\varphi(X)] \geq \varphi(\mathbf{E}[X]).$$

Proof As $L(\varphi) \neq \emptyset$ by Corollary 7.8, we can choose numbers $a, b \in \mathbb{R}$ such that $ax + b \leq \varphi(x)$ for all $x \in I$. Hence

$$\mathbf{E}[\varphi(X)^-] \leq \mathbf{E}[(aX + b)^-] \leq |b| + |a| \cdot \mathbf{E}[|X|] < \infty.$$

We distinguish the cases where $\mathbf{E}[X]$ is in the interior I° or at the boundary ∂I.
Case 1. If $\mathbf{E}[X] \in I^\circ$, then let $t^+ := D^+\varphi(\mathbf{E}[X])$ be the maximal slope of a tangent of φ at $\mathbf{E}[X]$. Then $\varphi(x) \geq t^+ \cdot (x - \mathbf{E}[X]) + \varphi(\mathbf{E}[X])$ for all $x \in I$; hence

$$\mathbf{E}[\varphi(X)] \geq t^+\, \mathbf{E}[X - \mathbf{E}[X]] + \mathbf{E}[\varphi(\mathbf{E}[X])] = \varphi(\mathbf{E}[X]).$$

Case 2. If $\mathbf{E}[X] \in \partial I$, then $X = \mathbf{E}[X]$ a.s.; hence $\mathbf{E}[\varphi(X)] = \mathbf{E}[\varphi(\mathbf{E}[X])] = \varphi(\mathbf{E}[X])$. $\square$

Jensen's inequality can be extended to $\mathbb{R}^n$. To this end, we need a representation of convex functions of many variables as a supremum of affine linear functions. Recall that a function $g : \mathbb{R}^n \to \mathbb{R}$ is called affine linear if there is an $a \in \mathbb{R}^n$ and a $b \in \mathbb{R}$ such that $g(x) = \langle a, x \rangle + b$ for all x. Here $\langle \cdot, \cdot \rangle$ denotes the usual inner product on $\mathbb{R}^n$.

Theorem 7.10 *Let $G \subset \mathbb{R}^n$ be open and convex and let $\varphi : G \to \mathbb{R}$ be a map. Then Corollary 7.8 holds with I replaced by G. If φ is convex, then φ is continuous and hence measurable. If φ is twice continuously differentiable, then φ is convex if and only if the Hessian matrix is positive semidefinite.*

Proof As we need these statements only in the proof of the multidimensional Jensen inequality, which will not play a central role in the following, we only give references for the proofs. In Rockafellar's book [146], continuity follows from Theorem 10.1, and the statements of Corollary 7.8 follow from Theorem 12.1 and Theorem 18.8. The claim about the Hessian matrix can be found in Theorem 4.5.

$\qquad\square$

Theorem 7.11 (Jensen's inequality in $\mathbb{R}^n$) *Let $G \subset \mathbb{R}^n$ be a convex set and let $X_1, \ldots, X_n$ be integrable real random variables with $\mathbf{P}[(X_1, \ldots, X_n) \in G] = 1$. Further, let $\varphi : G \to \mathbb{R}$ be convex. Then $\mathbf{E}[\varphi(X_1, \ldots, X_n)^-] < \infty$ and*

$$\mathbf{E}\big[\varphi(X_1, \ldots, X_n)\big] \geq \varphi(\mathbf{E}[X_1], \ldots, \mathbf{E}[X_n]).$$

Proof First consider the case where G is open. Here, the argument is similar to the proof of Theorem 7.9. Let $g \in L(\varphi)$ with

$$g\big(\mathbf{E}[X_1], \ldots, \mathbf{E}[X_n]\big) = \varphi\big(\mathbf{E}[X_1], \ldots, \mathbf{E}[X_n]\big).$$

As $g \leq \varphi$ is linear, we get

$$\mathbf{E}\big[\varphi(X_1, \ldots, X_n)\big] \geq \mathbf{E}[g(X_1, \ldots, X_n)] = g\big(\mathbf{E}[X_1], \ldots, \mathbf{E}[X_n]\big).$$

Integrability of $\varphi(X_1, \ldots, X_n)^-$ can be derived in a similar way to the one-dimensional case.

Now consider the general case where G is not necessarily open. Here the problem that arises when $(\mathbf{E}[X_1], \ldots, \mathbf{E}[X_n]) \in \partial G$ is a bit more tricky than in the one-dimensional case since ∂G can have flat pieces that in turn, however, are convex. Hence one cannot infer that $(X_1, \ldots, X_n)$ equals its expectation almost surely. We only sketch the argument. First infer that $(X_1, \ldots, X_n)$ is almost surely in one of those flat pieces. This piece is necessarily of dimension smaller than n. Now restrict φ to that flat piece and inductively reduce its dimension until reaching a point, the case that has already been treated above. Details can be found, e.g., in [37, Theorem 10.2.6].

$\qquad\square$

Example 7.12 Let X be a real random variable with $\mathbf{E}[X^2] < \infty$, $I = \mathbb{R}$ and $\varphi(x) = x^2$. By Jensen's inequality, we get

$$\mathbf{Var}[X] = \mathbf{E}[X^2] - (\mathbf{E}[X])^2 \geq 0. \quad \Diamond$$

Example 7.13 Let $G = [0, \infty) \times [0, \infty)$, $\alpha \in (0, 1)$ and $\varphi(x, y) = x^\alpha y^{1-\alpha}$. Then φ is concave (exercise!); hence, for nonnegative random variables X and Y with finite expectation (by Theorem 7.11),

$$\mathbf{E}\left[X^\alpha Y^{1-\alpha}\right] \leq (\mathbf{E}[X])^\alpha \, (\mathbf{E}[Y])^{1-\alpha}. \quad \Diamond$$

Example 7.14 Let G, X and Y be as in Example 7.13. Let $p \in (1, \infty)$. Then $\psi(x, y) = \left(x^{1/p} + y^{1/p}\right)^p$ is concave. Hence (by Theorem 7.11)

$$\left(\mathbf{E}[X]^{1/p} + \mathbf{E}[Y]^{1/p}\right)^p \geq \mathbf{E}\left[\left(X^{1/p} + Y^{1/p}\right)^p\right]. \quad \Diamond$$

Before we present Hölder's inequality and Minkowski's inequality, we need a preparatory lemma.

Lemma 7.15 (Young's inequality) *For* $p, q \in (1, \infty)$ *with* $\frac{1}{p} + \frac{1}{q} = 1$ *and for* $x, y \in [0, \infty)$,

$$xy \leq \frac{x^p}{p} + \frac{y^q}{q}. \tag{7.1}$$

Proof Fix $y \in [0, \infty)$ and define $f(x) := \dfrac{x^p}{p} + \dfrac{y^q}{q} - xy$ for $x \in [0, \infty)$. f is twice continuously differentiable in $(0, \infty)$ with derivatives $f'(x) = x^{p-1} - y$ and $f''(x) = (p - 1)x^{p-2}$. In particular, f is strictly convex and hence assumes its (unique) minimum at $x_0 = y^{1/(p-1)}$. By assumption, $q = \frac{p}{p-1}$; hence $x_0^p = y^q$ and thus

$$f(x_0) = \left(\frac{1}{p} + \frac{1}{q}\right) y^q - y^{1/(p-1)}y = 0. \qquad \square$$

Theorem 7.16 (Hölder's inequality) *Let* $p, q \in [1, \infty]$ *with* $\frac{1}{p} + \frac{1}{q} = 1$ *and* $f \in \mathcal{L}^p(\mu)$, $g \in \mathcal{L}^q(\mu)$. *Then* $(fg) \in \mathcal{L}^1(\mu)$ *and*

$$\|fg\|_1 \leq \|f\|_p \cdot \|g\|_q.$$

Proof The cases $p = 1$ and $p = \infty$ are trivial. Hence, let $p \in (1, \infty)$. Let $f \in \mathcal{L}^p(\mu)$ and $g \in \mathcal{L}^q(\mu)$ be nontrivial. By passing to $f/\|f\|_p$ and $g/\|g\|_q$, we may assume that $\|f\|_p = \|g\|_q = 1$. By Lemma 7.15, we have

$$\|fg\|_1 = \int |f| \cdot |g| \, d\mu \leq \frac{1}{p} \int |f|^p \, d\mu + \frac{1}{q} \int |g|^q \, d\mu$$

$$= \frac{1}{p} + \frac{1}{q} = 1 = \|f\|_p \cdot \|g\|_q. \qquad \square$$

Theorem 7.17 (Minkowski's inequality) *For $p \in [1, \infty]$ and $f, g \in \mathcal{L}^p(\mu)$,*

$$\|f + g\|_p \leq \|f\|_p + \|g\|_p. \tag{7.2}$$

Proof The case $p = \infty$ is trivial. Hence, let $p \in [1, \infty)$. The left-hand side in (7.2) does not decrease if we replace f and g by $|f|$ and $|g|$. Hence we may assume $f \geq 0$ and $g \geq 0$ and (to avoid trivialities) $\|f + g\|_p > 0$.

Now $(f + g)^p \leq 2^p (f^p \vee g^p) \leq 2^p (f^p + g^p)$; hence $f + g \in \mathcal{L}^p(\mu)$. Apply Hölder's inequality to $f \cdot (f + g)^{p-1}$ and to $g \cdot (f + g)^{p-1}$ to get

$$
\begin{aligned}
\|f + g\|_p^p &= \int (f + g)^p \, d\mu = \int f(f+g)^{p-1} \, d\mu + \int g(f+g)^{p-1} \, d\mu \\
&\leq \|f\|_p \cdot \|(f+g)^{p-1}\|_q + \|g\|_p \cdot \|(f+g)^{p-1}\|_q \\
&= (\|f\|_p + \|g\|_p) \cdot \|f + g\|_p^{p-1}.
\end{aligned}
$$

Note that in the last step, we used the fact that $p - p/q = 1$. Dividing both sides by $\|f + g\|_p^{p-1}$ yields (7.2). $\qquad\square$

In Theorem 7.17, we verified the triangle inequality and hence that $\| \cdot \|_p$ is a norm. Theorem 7.3 says that this norm is complete (i.e., every Cauchy sequence converges). A complete normed vector space is called a **Banach space**. Summing up, we have shown the following theorem.

Theorem 7.18 (Fischer–Riesz) $(L^p(\mu), \| \cdot \|_p)$ *is a Banach space for every* $p \in [1, \infty]$.

Takeaways In this section, we have encountered the most prominent integral inequalities: Jensen's inequality for convex functions, Hölder's inequality and Minkowski's inequality. Together with the topological considerations in Sect. 7.1, these inequalities enabled us to show the celebrated Fischer-Riesz theorem.

Exercise 7.2.1 Show Hölder's inequality by applying Jensen's inequality to the function of Example 7.13. ♣

Exercise 7.2.2 Show Minkowski's inequality by applying Jensen's inequality to the function of Example 7.14. ♣

Exercise 7.2.3 Let X be a real random variable and let $p, q \in (1, \infty)$ with $\frac{1}{p} + \frac{1}{q} = 1$. Show that X is in $\mathcal{L}^p(\mathbf{P})$ if and only if there exists a $C < \infty$ such that $|\mathbf{E}[XY]| \leq C \|Y\|_q$ for any bounded random variable Y. ♣

7.3   Hilbert Spaces

In this section, we study the case $p = 2$ in more detail. The main goal is the representation theorem for continuous linear functionals on Hilbert spaces due to Riesz and Fréchet. This theorem is a cornerstone for a functional analytic proof of the Radon–Nikodym theorem in Sect. 7.4.

Definition 7.19 Let V be a real vector space. A map $\langle \cdot, \cdot \rangle : V \times V \to \mathbb{R}$ is called an **inner product** if:

(i) (Linearity) $\langle x, \alpha y + z \rangle = \alpha \langle x, y \rangle + \langle x, z \rangle$ for all $x, y, z \in V$ and $\alpha \in \mathbb{R}$.
(ii) (Symmetry) $\langle x, y \rangle = \langle y, x \rangle$ for all $x, y \in V$.
(iii) (Positive definiteness) $\langle x, x \rangle > 0$ for all $x \in V \setminus \{0\}$.

If only (i) and (ii) hold and $\langle x, x \rangle \geq 0$ for all x, then $\langle \cdot, \cdot \rangle$ is called a positive semidefinite symmetric bilinear form, or a **semi-inner product**.

   If $\langle \cdot, \cdot \rangle$ is an inner product, then $(V, \langle \cdot, \cdot \rangle)$ is called a (real) **Hilbert space** if the norm defined by $\|x\| := \langle x, x \rangle^{1/2}$ is complete; that is, if $(V, \| \cdot \|)$ is a Banach space.

Definition 7.20 For $f, g \in \mathcal{L}^2(\mu)$, define

$$\langle f, g \rangle := \int fg \, d\mu.$$

For $\bar{f}, \bar{g} \in L^2(\mu)$, define $\langle \bar{f}, \bar{g} \rangle := \langle f, g \rangle$, where $f \in \bar{f}$ and $g \in \bar{g}$.

Note that this definition is independent of the particular choices of the representatives of f and g.

Theorem 7.21 $\langle \cdot, \cdot \rangle$ *is an inner product on* $L^2(\mu)$ *and a semi-inner product on* $\mathcal{L}^2(\mu)$. *In addition,* $\|f\|_2 = \langle f, f \rangle^{1/2}$.

Proof This is left as an exercise.   □

As a corollary to Theorem 7.18, we get the following.

Corollary 7.22 $(L^2(\mu), \langle \cdot, \cdot \rangle)$ *is a real Hilbert space.*

Lemma 7.23 *If* $\langle \cdot, \cdot \rangle$ *is a semi-inner product on the real vector space* V, *then* $\langle \cdot, \cdot \rangle : V \times V \to \mathbb{R}$ *is continuous (with respect to the product topology of the topology on* V *that is generated by the pseudo-metric* $d(x, y) = \langle x - y, x - y \rangle^{1/2}$).

Proof This is obvious.   □

Definition 7.24 (Orthogonal complement) Let V be a real vector space with inner product $\langle \cdot, \cdot \rangle$. If $W \subset V$, then the orthogonal complement of W is the following linear subspace of V:

$$W^{\perp} := \{v \in V : \langle v, w \rangle = 0 \text{ for all } w \in W\}.$$

Theorem 7.25 (Orthogonal decomposition) *Let* $(V, \langle \cdot, \cdot \rangle)$ *be a Hilbert space and let* $W \subset V$ *be a closed linear subspace. For any* $x \in V$, *there is a unique representation* $x = y + z$ *where* $y \in W$ *and* $z \in W^\perp$.

Proof Let $x \in V$ and $c := \inf\{\|x - w\| : w \in W\}$. Further, let $(w_n)_{n \in \mathbb{N}}$ be a sequence in W with $\|x - w_n\| \overset{n \to \infty}{\longrightarrow} c$. The parallelogram law yields

$$\|w_m - w_n\|^2 = 2\|w_m - x\|^2 + 2\|w_n - x\|^2 - 4\left\|\frac{1}{2}(w_m + w_n) - x\right\|^2.$$

As W is linear, we have $(w_m + w_n)/2 \in W$; hence $\|\frac{1}{2}(w_m + w_n) - x\| \geq c$. Thus $(w_n)_{n \in \mathbb{N}}$ is a Cauchy sequence: $\|w_m - w_n\| \longrightarrow 0$ if $m, n \to \infty$.

Since V is complete and W is closed, W is also complete; hence there is a $y \in W$ with $w_n \overset{n \to \infty}{\longrightarrow} y$. Now let $z := x - y$. Then $\|z\| = \lim_{n \to \infty} \|w_n - x\| = c$ by continuity of the norm (Lemma 7.23).

Consider an arbitrary $w \in W \setminus \{0\}$. We define $\varrho := \langle z, w \rangle / \|w\|^2$ and get $y + \varrho w \in W$; hence

$$c^2 \leq \|x - (y + \varrho w)\|^2 = \|z\|^2 + \varrho^2 \|w\|^2 + 2\varrho \langle z, w \rangle = c^2 - \varrho^2 \|w\|^2.$$

Concluding, we have $\langle z, w \rangle = 0$ for all $w \in W$ and thus $z \in W^\perp$.

Uniqueness of the decomposition is easy: If $x = y' + z'$ is an orthogonal decomposition, then $y - y' \in W$ and $z - z' \in W^\perp$ as well as $y - y' + z - z' = 0$; hence

$$0 = \|y - y' + z - z'\|^2 = \|y - y'\|^2 + \|z - z'\|^2 + 2\langle y - y', z - z' \rangle$$
$$= \|y - y'\|^2 + \|z - z'\|^2,$$

whence $y = y'$ and $z = z'$. $\qquad\square$

Theorem 7.26 (Riesz–Fréchet representation theorem) *Let* $(V, \langle \cdot, \cdot \rangle)$ *be a Hilbert space and let* $F : V \to \mathbb{R}$ *be a map. Then the following are equivalent.*

(i) F is continuous and linear.
(ii) There is an $f \in V$ with $F(x) = \langle x, f \rangle$ for all $x \in V$.

The element $f \in V$ in (ii) is uniquely determined.

Proof "(ii) $\Longrightarrow$ (i)" For any $f \in V$, by definition of the inner product, the map $x \mapsto \langle x, f \rangle$ is linear. By Lemma 7.23, this map is also continuous.

"(i) $\Longrightarrow$ (ii)" If $F \equiv 0$, then choose $f = 0$. Now assume F is not identically zero. As F is continuous, the kernel $W := F^{-1}(\{0\})$ is a closed (proper) linear subspace of V. Let $v \in V \setminus W$ and let $v = y + z$ for $y \in W$ and $z \in W^\perp$ be the orthogonal decomposition of v. Then $z \neq 0$ and $F(z) = F(v) - F(y) = F(v) \neq 0$. Hence we can define $u := z/F(z) \in W^\perp$. Clearly, $F(u) = 1$ and for any $x \in V$, we have $F(x - F(x)u) = F(x) - F(x)F(u) = 0$; hence $x - F(x)u \in W$ and thus

$\langle x - F(x)u, u \rangle = 0$. Consequently, $F(x) = \langle x, u \rangle / \|u\|^2$. Now define $f := u / \|u\|^2$. Then $F(x) = \langle x, f \rangle$ for all $x \in V$.

"Uniqueness"   Let $\langle x, f \rangle = \langle x, g \rangle$ for all $x \in V$. Letting $x = f - g$, we get $0 = \langle f - g, f - g \rangle$; hence $f = g$. $\qquad\qquad\qquad\qquad\qquad\qquad\square$

Reflection Find an example of a discontinuous linear map $F : V \to \mathbb{R}$. ♠

In the following section, we will need the representation theorem for the space $\mathcal{L}^2(\mu)$, which, unlike $L^2(\mu)$, is not a Hilbert space. However, with a little bit of *abstract nonsense*, one can apply the preceding theorem to $\mathcal{L}^2(\mu)$. Recall that $\mathcal{N} = \{ f \in \mathcal{L}^2(\mu) : \langle f, f \rangle = 0 \}$ is the subspace of functions that equal zero almost everywhere. Let $L^2(\mu) = \mathcal{L}^2(\mu)/\mathcal{N}$ be the factor space. This is a special case of the situation where $(V, \langle \cdot, \cdot \rangle)$ is a linear space with complete semi-inner product. In this case, $\mathcal{N} := \{ v \in V : \langle v, v \rangle = 0 \}$ and $V_0 = V/\mathcal{N} := \{ f + \mathcal{N} : f \in V \}$. Denote $\langle v + \mathcal{N}, w + \mathcal{N} \rangle_0 := \langle v, w \rangle$ to obtain a Hilbert space $(V_0, \langle \cdot, \cdot \rangle_0)$.

Corollary 7.27 *Let $(V, \langle \cdot, \cdot \rangle)$ be a linear vector space with complete semi-inner product. The map $F : V \to \mathbb{R}$ is continuous and linear if and only if there is an $f \in V$ with $F(x) = \langle x, f \rangle$ for all $x \in V$.*

Proof One implication is trivial. Hence, let F be continuous and linear. Then $F(0) = 0$ since F is linear. Note that $F(v) = F(0) = 0$ for all $v \in \mathcal{N}$ since F is continuous. Indeed, v lies in *every* open neighborhood of 0; hence F assumes at v the same value as at 0. Thus F induces a continuous linear map $F_0 : V_0 \to \mathbb{R}$ by $F_0(x + \mathcal{N}) = F(x)$. By Theorem 7.26, there is an $f + \mathcal{N} \in V_0$ with $F_0(x + \mathcal{N}) = \langle x + \mathcal{N}, f + \mathcal{N} \rangle_0$ for all $x + \mathcal{N} \in V_0$. However, $F(x) = \langle x, f \rangle$ for all $x \in V$ by the definition of F_0 and $\langle \cdot, \cdot \rangle_0$. $\qquad\square$

Corollary 7.28 *The map $F : \mathcal{L}^2(\mu) \to \mathbb{R}$ is continuous and linear if and only if there is an $f \in \mathcal{L}^2(\mu)$ with $F(g) = \int gf \, d\mu$ for all $g \in \mathcal{L}^2(\mu)$.*

Proof The space $\mathcal{L}^2(\mu)$ fulfills the conditions of Corollary 7.27. $\qquad\qquad\square$

> **Takeaways** We recognised the function spaces L^2 as Hilbert spaces. In Hilbert spaces, continuous linear maps can be represented as a scalar product with some fixed vector (Riesz-Fréchet theorem).

Exercise 7.3.1 (Fourier series) For $n \in \mathbb{N}_0$, define $S_n, C_n : [0, 1] \to [0, 1]$ by $S_n(x) = \sin(2\pi n x)$, $C_n(x) = \cos(2\pi n x)$. For two square summable sequences $(a_n)_{n \in \mathbb{N}}$ and $(b_n)_{n \in \mathbb{N}_0}$, let $h_{a,b} := b_0 + \sum_{n=1}^{\infty} (a_n S_n + b_n C_n)$. Further, let W be the vector space of such $h_{a,b}$.

Show the following:

(i) The functions $C_0, S_n, C_n, n \in \mathbb{N}$ form an orthogonal system in $L^2([0, 1], \lambda)$.
(ii) The series defining $h_{a,b}$ converges in $L^2([0, 1], \lambda)$.
(iii) W is a closed linear subspace of $L^2([0, 1], \lambda)$.

(iv) $W = L^2([0, 1], \lambda)$. More precisely, for any $f \in L^2([0, 1], \lambda)$, there exist uniquely defined square summable sequences $(a_n)_{n \in \mathbb{N}}$ and $(b_n)_{n \in \mathbb{N}_0}$ such that $f = h_{a,b}$. Furthermore, $\|f\|_2^2 = b_0^2 + \sum_{n=1}^{\infty}(a_n^2 + b_n^2)$.

Hint: Show (iv) first for step functions (see Exercise 4.2.6). ♣

7.4 Lebesgue's Decomposition Theorem

In this section, we employ the properties of Hilbert spaces that we derived in the last section in order to decompose a measure into a singular part and a part that is absolutely continuous, both with respect to a second given measure. Furthermore, we show that the absolutely continuous part has a density. Let μ and ν be measures on $(\Omega, \mathcal{A})$. By Definition 4.13, a measurable function $f : \Omega \to [0, \infty)$ is called a **density** of ν with respect to μ if

$$\nu(A) := \int f \, \mathbb{1}_A \, d\mu \quad \text{for all } A \in \mathcal{A}. \tag{7.3}$$

On the other hand, for any measurable $f : \Omega \to [0, \infty)$, equation (7.3) defines a measure ν on $(\Omega, \mathcal{A})$. In this case, we also write

$$\nu = f\mu \quad \text{and} \quad f = \frac{d\nu}{d\mu}. \tag{7.4}$$

For example, the normal distribution $\nu = \mathcal{N}_{0,1}$ has the density $f(x) = \frac{1}{\sqrt{2\pi}} e^{-x^2/2}$ with respect to the Lebesgue measure $\mu = \lambda$ on $\mathbb{R}$.

If $g : \Omega \to [0, \infty]$ is measurable, then (by Theorem 4.15)

$$\int g \, d\nu = \int gf \, d\mu. \tag{7.5}$$

Hence $g \in \mathcal{L}^1(\nu)$ if and only if $gf \in \mathcal{L}^1(\mu)$, and in this case (7.5) holds.

If $\nu = f\mu$, then $\nu(A) = 0$ for all $A \in \mathcal{A}$ with $\mu(A) = 0$. The situation is quite the opposite for, e.g., the Poisson distribution $\mu = \mathrm{Poi}_\varrho$ with parameter $\varrho > 0$ and $\nu = \mathcal{N}_{0,1}$. Here $\mathbb{N}_0 \subset \mathbb{R}$ is a ν-null set with $\mu(\mathbb{R} \setminus \mathbb{N}_0) = 0$. We say that ν is **singular** to μ.

The main goal of this chapter is to show that an arbitrary σ-finite measure ν on a measurable space $(\Omega, \mathcal{A})$ can be decomposed into a part that is singular to the σ-finite measure μ and a part that has a density with respect to μ (Lebesgue's decomposition theorem, Theorem 7.33).

Theorem 7.29 (Uniqueness of the density) *Let v be σ-finite. If f_1 and f_2 are densities of v with respect to μ, then $f_1 = f_2$ μ-almost everywhere. In particular, the density $\frac{dv}{d\mu}$ is unique up to equality μ-almost everywhere.*

Proof Let $E_n \uparrow \Omega$ with $v(E_n) < \infty$, $n \in \mathbb{N}$. Let $A_n = E_n \cap \{f_1 > f_2\}$ for $n \in \mathbb{N}$. Then $v(A_n) < \infty$; hence

$$0 = v(A_n) - v(A_n) = \int_{A_n} (f_1 - f_2)\, d\mu.$$

By Theorem 4.8(i), $f_2\, \mathbb{1}_{A_n} = f_1\, \mathbb{1}_{A_n}$ μ-a.e. As $f_1 > f_2$ on A_n, we infer $\mu(A_n) = 0$ and

$$\mu(\{f_1 > f_2\}) = \mu\left(\bigcup_{n\in\mathbb{N}} A_n\right) = 0.$$

Similarly, we get $\mu(\{f_1 < f_2\}) = 0$; hence $f_1 = f_2$ μ-a.e. $\square$

Definition 7.30 Let μ and v be two measures on $(\Omega, \mathcal{A})$.

(i) v is called **absolutely continuous** with respect to μ (symbolically $v \ll \mu$) if

$$v(A) = 0 \quad \text{for all } A \in \mathcal{A} \text{ with } \mu(A) = 0. \tag{7.6}$$

   The measures μ and v are called **equivalent** (symbolically $\mu \approx v$) if $v \ll \mu$ and $\mu \ll v$.

(ii) μ is called **singular** to v (symbolically $\mu \perp v$) if there exists an $A \in \mathcal{A}$ such that $\mu(A) = 0$ and $v(\Omega \setminus A) = 0$.

Remark 7.31 Clearly, $\mu \perp v \iff v \perp \mu$. $\Diamond$

Example 7.32

(i) Let μ be a measure on $(\mathbb{R}, \mathcal{B}(\mathbb{R}))$ with density f with respect to the Lebesgue measure λ. Then $\mu(A) = \int_A f\, d\lambda = 0$ for every $A \in \mathcal{A}$ with $\lambda(A) = 0$; hence $\mu \ll \lambda$. If λ-almost everywhere $f > 0$, then $\mu(A) = \int_A f\, d\lambda > 0$ if $\lambda(A) > 0$; hence $\mu \approx \lambda$. If $\lambda(\{f = 0\}) > 0$, then (since $\mu(\{f = 0\}) = 0$) $\lambda \not\ll \mu$.

(ii) Consider the Bernoulli distributions Ber_p and Ber_q for $p, q \in [0, 1]$. If $p \in (0, 1)$, then $\mathrm{Ber}_q \ll \mathrm{Ber}_p$. If $p \in \{0, 1\}$, then $\mathrm{Ber}_q \ll \mathrm{Ber}_p$ if and only if $p = q$, and $\mathrm{Ber}_q \perp \mathrm{Ber}_p$ if and only if $q = 1 - p$.

(iii) Consider the Poisson distributions Poi_α and Poi_β for $\alpha, \beta \geq 0$. We have $\mathrm{Poi}_\alpha \ll \mathrm{Poi}_\beta$ if and only if $\beta > 0$ or $\alpha = 0$.

(iv) Consider the infinite product measures (see Theorem 1.64) $(\mathrm{Ber}_p)^{\otimes\mathbb{N}}$ and $(\mathrm{Ber}_q)^{\otimes\mathbb{N}}$ on $\Omega = \{0, 1\}^\mathbb{N}$. Then $(\mathrm{Ber}_p)^{\otimes\mathbb{N}} \perp (\mathrm{Ber}_q)^{\otimes\mathbb{N}}$ if $p \neq q$. Indeed, for $n \in \mathbb{N}$, let $X_n((\omega_1, \omega_2, \ldots)) = \omega_n$ be the projection of Ω to the nth coordinate. Then under $(\mathrm{Ber}_r)^{\otimes\mathbb{N}}$ the family $(X_n)_{n\in\mathbb{N}}$ is independent and Bernoulli-distributed with parameter r (see Example 2.18). By the strong law of large numbers, for any $r \in \{p, q\}$, there exists a measurable set $A_r \subset \Omega$ with $(\mathrm{Ber}_r)^{\otimes \mathbb{N}}(\Omega \setminus A_r) = 0$ and

$$\lim_{n \to \infty} \frac{1}{n} \sum_{i=1}^{n} X_i(\omega) = r \quad \text{for all } \omega \in A_r.$$

In particular, $A_p \cap A_q = \emptyset$ if $p \neq q$ and thus $(\mathrm{Ber}_p)^{\otimes \mathbb{N}} \perp (\mathrm{Ber}_q)^{\otimes \mathbb{N}}$. $\Diamond$

Theorem 7.33 (Lebesgue's decomposition theorem) *Let μ and ν be σ-finite measures on $(\Omega, \mathcal{A})$. Then ν can be uniquely decomposed into an absolutely continuous part ν_a and a singular part ν_s (with respect to μ):*

$$\nu = \nu_a + \nu_s, \quad \text{where } \nu_a \ll \mu \text{ and } \nu_s \perp \mu.$$

ν_a has a density with respect to μ, and $\dfrac{d\nu_a}{d\mu}$ is $\mathcal{A}$-measurable and finite μ-a.e.

Corollary 7.34 (Radon–Nikodym theorem) *Let μ and ν be σ-finite measures on $(\Omega, \mathcal{A})$. Then*

$$\nu \text{ has a density w.r.t. } \mu \quad \Longleftrightarrow \quad \nu \ll \mu.$$

*In this case, $\frac{d\nu}{d\mu}$ is $\mathcal{A}$-measurable and finite μ-a.e. $\frac{d\nu}{d\mu}$ is called the **Radon–Nikodym derivative** of ν with respect to μ.*

Proof One direction is trivial. Hence, let $\nu \ll \mu$. By Theorem 7.33, we get that $\nu = \nu_a$ has a density with respect to μ. $\qquad \square$

Proof (of Theorem 7.33) The idea goes back to von Neumann. We follow the exposition in [37].

By the usual exhaustion arguments, we can restrict ourselves to the case where μ and ν are finite. By Theorem 4.19, the canonical inclusion $i : \mathcal{L}^2(\Omega, \mathcal{A}, \mu + \nu) \hookrightarrow \mathcal{L}^1(\Omega, \mathcal{A}, \mu + \nu)$ is continuous. Since $\nu \leq \mu + \nu$, the linear functional $\mathcal{L}^2(\Omega, \mathcal{A}, \mu + \nu) \to \mathbb{R}, h \mapsto \int h \, d\nu$ is continuous. By the Riesz–Fréchet theorem (here Corollary 7.28), there exists a $g \in \mathcal{L}^2(\Omega, \mathcal{A}, \mu + \nu)$ such that

$$\int h \, d\nu = \int hg \, d(\mu + \nu) \quad \text{for all } h \in \mathcal{L}^2(\Omega, \mathcal{A}, \mu + \nu) \tag{7.7}$$

or equivalently

$$\int f(1 - g) \, d(\mu + \nu) = \int f \, d\mu \quad \text{for all } f \in \mathcal{L}^2(\Omega, \mathcal{A}, \mu + \nu). \tag{7.8}$$

If in (7.7) we choose $h = \mathbb{1}_{\{g<0\}}$, then we get that $(\mu+\nu)$-almost everywhere $g \geq 0$. Similarly, with $f = \mathbb{1}_{\{g>1\}}$ in (7.8), we obtain that $(\mu+\nu)$-almost everywhere $g \leq 1$. Hence $0 \leq g \leq 1$.

Now let $f \geq 0$ be measurable and let $(f_n)_{n\in\mathbb{N}}$ be a sequence of nonnegative functions in $\mathcal{L}^2(\Omega, \mathcal{A}, \mu + \nu)$ with $f_n \uparrow f$. By the monotone convergence theorem (applied to the measure $(1 - g)(\mu+\nu)$; that is, the measure with density $(1 - g)$ with respect to $\mu + \nu$), we obtain that (7.8) holds for all measurable $f \geq 0$. Similarly, we get (7.7) for all measurable $h \geq 0$.

Let $E := g^{-1}(\{1\})$. If we let $f = \mathbb{1}_E$ in (7.8), then we get $\mu(E) = 0$. Define the measures ν_a and ν_s for $A \in \mathcal{A}$ by

$$\nu_a(A) := \nu(A \setminus E) \quad \text{and} \quad \nu_s(A) := \nu(A \cap E).$$

Clearly, $\nu = \nu_a + \nu_s$ and $\nu_s(\Omega \setminus E) = 0$; hence $\nu_s \perp \mu$. If now $A \cap E = \emptyset$ and $\mu(A) = 0$, then $\int \mathbb{1}_A\, d\mu = 0$. Hence, by (7.8), also $\int_A (1 - g)\, d(\mu + \nu) = 0$. On the other hand, we have $1 - g > 0$ on A; hence $\mu(A) + \nu(A) = 0$ and thus $\nu_a(A) = \nu(A) = 0$. If, more generally, B is measurable with $\mu(B) = 0$, then $\mu(B \setminus E) = 0$; hence, as shown above, $\nu_a(B) = \nu_a(B \setminus E) = 0$. Consequently, $\nu_a \ll \mu$ and $\nu = \nu_a + \nu_s$ is the decomposition we wanted to construct.

In order to obtain the density of ν_a with respect to μ, we define $f := \dfrac{g}{1 - g} \mathbb{1}_{\Omega \setminus E}$. For any $A \in \mathcal{A}$, by (7.8) and (7.7) with $h = \mathbb{1}_{A\setminus E}$,

$$\int_A f\, d\mu = \int_{A \cap E^c} g\, d(\mu + \nu) = \nu(A \setminus E) = \nu_a(A).$$

Hence $f = \dfrac{d\nu_a}{d\mu}$. $\square$

Reflection Why do we assume σ-finiteness of the measures in the Radon-Nikodym theorem? Find an example that shows that this assumption cannot be dropped. ♠

Takeaways Loosely speaking, a σ-finite measure μ has a density with respect to a σ-finite measure ν if locally μ is a multiple of ν. A necessary and sufficient condition for this to be true is that μ vanishes on the sets where ν vanishes. The actual density could be gained using Hilbert space theory, in particular the Riesz-Fréchet representation theorem.

Exercise 7.4.1 For every $x \in (0, 1]$, let $x = (0, x_1 x_2 x_3 \ldots) := \sum_{n=1}^{\infty} x_n 2^{-n}$ be the dyadic expansion (with $\limsup_{n\to\infty} x_n = 1$ for definiteness). Define a map $F : (0, 1] \to (0, 1]$ by

$$F(x) = (0, x_1 x_1 x_2 x_2 x_3 x_3 \ldots) = \sum_{n=1}^{\infty} 3\, x_n\, 4^{-n}.$$

Let U be a random variable that is uniformly distributed on $(0, 1]$ and denote by $\mu := \mathbf{P}_{U \circ F^{-1}}$ the distribution of $F(U)$.

Show that the probability measure μ has a continuous distribution function and that μ is singular to the Lebesgue measure $\lambda|_{(0,1]}$. ♣

Exercise 7.4.2 Let $n \in \mathbb{N}$ and $p, q \in [0, 1]$. For which values of p and q do we have $b_{n,p} \ll b_{n,q}$? Compute the Radon–Nikodym derivative $\frac{db_{n,p}}{db_{n,q}}$. ♣

7.5 Supplement: Signed Measures

In this section, we show the decomposition theorems for signed measures (Hahn, Jordan) and deliver an alternative proof for Lebesgue's decomposition theorem. We owe some of the proofs to [89].

Definition 7.35 Let μ and ν be two measures on $(\Omega, \mathcal{A})$. ν is called **totally continuous** with respect to μ if, for any $\varepsilon > 0$, there exists a $\delta > 0$ such that for all $A \in \mathcal{A}$

$$\mu(A) < \delta \quad \text{implies} \quad \nu(A) < \varepsilon. \tag{7.9}$$

Remark 7.36 The definition of total continuity is similar to that of uniform integrability (see Theorem 6.24(iii)), at least for finite μ. We will come back to this connection in the framework of the martingale convergence theorem that will provide an alternative proof of the Radon–Nikodym theorem (Corollary 7.34). ◇

Theorem 7.37 *Let μ and ν be measures on $(\Omega, \mathcal{A})$. If ν is totally continuous with respect to μ, then $\nu \ll \mu$. If $\nu(\Omega) < \infty$, then the converse also holds.*

Proof "$\Longrightarrow$" Let ν be totally continuous with respect to μ. Let $A \in \mathcal{A}$ with $\mu(A) = 0$. For all $\varepsilon > 0$, by assumption, $\nu(A) < \varepsilon$; hence $\nu(A) = 0$ and thus $\nu \ll \mu$.

"$\Longleftarrow$" Let ν be finite but not totally continuous with respect to μ. Then there exist an $\varepsilon > 0$ and sets $A_n \in \mathcal{A}$ with $\mu(A_n) < 2^{-n}$ but $\nu(A_n) \geq \varepsilon$ for all $n \in \mathbb{N}$.

Define $A := \limsup_{n \to \infty} A_n = \bigcap_{n=1}^{\infty} \bigcup_{k=n}^{\infty} A_k$. Then

$$\mu(A) = \lim_{n \to \infty} \mu\left(\bigcup_{k=n}^{\infty} A_k\right) \leq \lim_{n \to \infty} \sum_{k=n}^{\infty} \mu(A_k) \leq \lim_{n \to \infty} \sum_{k=n}^{\infty} 2^{-k} = 0.$$

Since ν is finite and upper semicontinuous (Theorem 1.36), we have

$$\nu(A) = \lim_{n\to\infty} \nu\left(\bigcup_{k=n}^{\infty} A_k\right) \geq \inf_{n\in\mathbb{N}} \nu(A_n) \geq \varepsilon > 0.$$

Thus $\nu \not\ll \mu$.   □

Example 7.38 In the converse implication of the theorem, the assumption of finiteness is essential. For example, let $\mu = \mathcal{N}_{0,1}$ be the standard normal distribution on $\mathbb{R}$ and let ν be the Lebesgue measure on $\mathbb{R}$. Then ν has the density $f(x) = \sqrt{2\pi}\, e^{x^2/2}$ with respect to μ. In particular, we have $\nu \ll \mu$. On the other hand, $\mu([n,\infty)) \overset{n\to\infty}{\longrightarrow} 0$ and $\nu([n,\infty)) = \infty$ for any $n \in \mathbb{N}$. Hence ν is not totally continuous with respect to μ. ◇

Example 7.39 Let $(\Omega, \mathcal{A})$ be a measurable space and let μ and ν be finite measures on $(\Omega, \mathcal{A})$. Denote by $\mathcal{Z}$ the set of finite partitions of Ω into pairwise disjoint measurable sets. That is, $Z \in \mathcal{Z}$ is a finite subset of $\mathcal{A}$ such that the sets $C \in Z$ are pairwise disjoint and $\bigcup_{C\in Z} C = \Omega$ for all Z. For $Z \in \mathcal{Z}$, define a function $f_Z : \Omega \to \mathbb{R}$ by

$$f_Z(\omega) = \sum_{C\in Z:\, \mu(C)>0} \frac{\nu(C)}{\mu(C)} \mathbb{1}_C(\omega).$$

We show that the following three statements are equivalent.

(i) The family $(f_Z : Z \in \mathcal{Z})$ is uniformly integrable in $\mathcal{L}^1(\mu)$ and $\int f_Z\, d\mu = \nu(\Omega)$ for any $Z \in \mathcal{Z}$.

(ii) $\nu \ll \mu$.

(iii) ν is totally continuous with respect to μ.

The equivalence of (ii) and (iii) was established in the preceding theorem. If (ii) holds, then, for all $Z \in \mathcal{Z}$,

$$\int f_Z\, d\mu = \sum_{C\in Z:\, \mu(C)>0} \nu(C) = \nu(\Omega)$$

since $\nu(C) = 0$ for those C that do not appear in the sum. Now fix $\varepsilon > 0$. Since (ii) implies (iii), there is a $\delta' > 0$ such that $\nu(A) < \varepsilon/2$ for all $A \in \mathcal{A}$ with $\mu(A) \leq \delta'$. Let $K := \nu(\Omega)/\delta'$ and $\delta < \varepsilon/(2K)$. Then

$$\mu\left(\bigcup_{C\in Z:\, K\mu(C)\leq\nu(C)} C\right) = \sum_{C\in Z:\, K\mu(C)\leq\nu(C)} \mu(C) \leq \frac{1}{K}\nu(\Omega) = \delta';$$

hence

$$\sum_{C \in Z: K\mu(C) \le v(C)} v(C) = v\left(\bigcup_{C \in Z: K\mu(C) \le v(C)} C\right) < \frac{\varepsilon}{2}.$$

We conclude that for all $A \in \mathcal{A}$ with $\mu(A) < \delta$,

$$\int_A f_Z \, d\mu = \sum_{C \in Z: \mu(C) > 0} \mu(A \cap C) \frac{v(C)}{\mu(C)}$$

$$= \sum_{0 < K\mu(C) \le v(C)} \mu(A \cap C) \frac{v(C)}{\mu(C)} + \sum_{K\mu(C) > v(C)} \mu(A \cap C) \frac{v(C)}{\mu(C)}$$

$$\le \frac{\varepsilon}{2} + \sum_{K\mu(C) > v(C)} K \mu(A \cap C) \le \frac{\varepsilon}{2} + K \mu(A) < \varepsilon.$$

Hence $(f_Z, \ Z \in \mathcal{Z})$ is uniformly integrable by Theorem 6.24(iii).

Now assume (i). If $\mu = 0$, then $\int f \, d\mu = 0$ for all f; hence $v(\Omega) = 0$ and thus $v \ll \mu$. Hence, let $\mu \neq 0$. Let $A \in \mathcal{A}$ with $\mu(A) = 0$. Then $Z = \{A, A^c\} \in \mathcal{Z}$ and $f_Z = \mathbb{1}_{A^c} v(A^c)/\mu(A^c)$. By assumption, $v(\Omega) = \int f_Z \, d\mu = v(A^c)$; hence $v(A) = 0$ and thus $v \ll \mu$. $\Diamond$

Definition 7.40 (Signed measure) A set function $\varphi : \mathcal{A} \to \mathbb{R}$ is called a **signed measure** on $(\Omega, \mathcal{A})$ if it is σ-additive; that is, if for any sequence of pairwise disjoint sets $A_1, A_2, \ldots \in \mathcal{A}$,

$$\varphi\left(\biguplus_{n=1}^{\infty} A_n\right) = \sum_{n=1}^{\infty} \varphi(A_n). \tag{7.10}$$

The set of all signed measures will be denoted by $\mathcal{M}^{\pm} = \mathcal{M}^{\pm}(\Omega, \mathcal{A})$.

Remark 7.41

(i) If φ is a signed measure, then in (7.10) we automatically have absolute convergence. Indeed, the value of the left-hand side does not change if we change the order of the sets $A_1, A_2, \ldots$. In order for this to hold for the right-hand side, by Weierstraß's theorem on rearrangements of series, the series has to converge absolutely. In particular, for any sequence $(A_n)_{n \in \mathbb{N}}$ of pairwise disjoint sets, we have $\lim_{n \to \infty} \sum_{k=n}^{\infty} |\varphi(A_k)| = 0$.

(ii) If $\varphi \in \mathcal{M}^{\pm}$, then $\varphi(\emptyset) = 0$ since $\mathbb{R} \ni v(\emptyset) = \sum_{n \in \mathbb{N}} v(\emptyset)$.

(iii) In general, $\varphi \in \mathcal{M}^{\pm}$ is not σ-subadditive. $\Diamond$

Example 7.42 If μ^+, μ^- are finite measures, then $\varphi := \mu^+ - \mu^- \in \mathcal{M}^{\pm}$. We will see that every signed measure has such a representation. $\Diamond$

Theorem 7.43 (Hahn's decomposition theorem) *Let φ be a signed measure. Then there is a set $\Omega^+ \in \mathcal{A}$ with $\varphi(A) \geq 0$ for all $A \in \mathcal{A}$, $A \subset \Omega^+$ and $\varphi(A) \leq 0$ for all $A \in \mathcal{A}$, $A \subset \Omega^- := \Omega \setminus \Omega^+$. Such a decomposition $\Omega = \Omega^- \uplus \Omega^+$ is called a Hahn decomposition of Ω (with respect to φ).*

Proof Let $\alpha := \sup \{\varphi(A) : A \in \mathcal{A}\}$. We have to show that φ attains the maximum α; that is, there exists an $\Omega^+ \in \mathcal{A}$ with $\varphi(\Omega^+) = \alpha$. If this is the case, then $\alpha \in \mathbb{R}$ and for $A \subset \Omega^+$, $A \in \mathcal{A}$, we would have

$$\alpha \geq \varphi(\Omega^+ \setminus A) = \varphi(\Omega^+) - \varphi(A) = \alpha - \varphi(A);$$

hence $\varphi(A) \geq 0$. For $A \subset \Omega^-$, $A \in \mathcal{A}$, we would have $\varphi(A) \leq 0$ since

$$\alpha \geq \varphi(\Omega^+ \cup A) = \varphi(\Omega^+) + \varphi(A) = \alpha + \varphi(A).$$

We now construct Ω^+ with $\varphi(\Omega^+) = \alpha$. Let $(A_n)_{n \in \mathbb{N}}$ be a sequence in $\mathcal{A}$ with $\alpha = \lim_{n \to \infty} \varphi(A_n)$. Let $A := \bigcup_{n=1}^{\infty} A_n$. As each A_n could still contain "portions with negative mass", we cannot simply choose $\Omega^+ = A$. Rather, we have to peel off the negative portions layer by layer.

Define $A_n^0 := A_n$, $A_n^1 := A \setminus A_n$, and let

$$\mathcal{P}_n := \left\{ \bigcap_{i=1}^{n} A_i^{s(i)} : s \in \{0, 1\}^n \right\}$$

be the partition of A that is generated by $A_1, \ldots, A_n$. Clearly, for any $B, C \in \mathcal{P}_n$, either $B = C$ or $B \cap C = \emptyset$ holds. In addition, we have $A_n = \underset{\substack{B \in \mathcal{P}_n \\ B \subset A_n}}{\biguplus} B$. Define

$$\mathcal{P}_n^- := \{B \in \mathcal{P}_n : \varphi(B) < 0\}, \qquad \mathcal{P}_n^+ := \mathcal{P}_n \setminus \mathcal{P}_n^-$$

and

$$C_n := \bigcup_{B \in \mathcal{P}_n^+} B.$$

Due to the finite additivity of φ, we have

$$\varphi(A_n) = \sum_{\substack{B \in \mathcal{P}_n \\ B \subset A_n}} \varphi(B) \leq \sum_{\substack{B \in \mathcal{P}_n^+ \\ B \subset A_n}} \varphi(B) \leq \sum_{B \in \mathcal{P}_n^+} \varphi(B) = \varphi(C_n).$$

For $m \leq n$, let $E_m^n = C_m \cup \ldots \cup C_n$. Hence, for $m < n$, we have $E_m^n \setminus E_m^{n-1} \subset C_n$ and thus

$$E_m^n \setminus E_m^{n-1} = \biguplus_{\substack{B \in \mathcal{P}_n^+ \\ B \subset E_m^n \setminus E_m^{n-1}}} B.$$

In particular, this implies $\varphi(E_m^n \setminus E_m^{n-1}) \geq 0$. For $E_m := \bigcup_{n \geq m} C_n$, we also have $E_m^n \uparrow E_m \; (n \to \infty)$ and

$$\varphi(A_m) \leq \varphi(C_m) = \varphi(E_m^m) \leq \varphi(E_m^m) + \sum_{n=m+1}^{\infty} \varphi(E_m^n \setminus E_m^{n-1})$$

$$= \varphi\left(E_m^m \cup \bigcup_{n=m+1}^{\infty}(E_m^n \setminus E_m^{n-1})\right) = \varphi\left(\bigcup_{n=m}^{\infty} E_m^n\right) = \varphi(E_m).$$

Now define $\Omega^+ = \bigcap_{m=1}^{\infty} E_m$; hence $E_m \downarrow \Omega^+$. Then

$$\varphi(E_m) = \varphi\left(\Omega^+ \uplus \biguplus_{n \geq m}(E_n \setminus E_{n+1})\right)$$

$$= \varphi(\Omega^+) + \sum_{n=m}^{\infty} \varphi(E_n \setminus E_{n+1}) \stackrel{m \to \infty}{\longrightarrow} \varphi(\Omega^+).$$

In the last step, we used Remark 7.41(i). Summing up, we have

$$\alpha = \lim_{m \to \infty} \varphi(A_m) \leq \lim_{m \to \infty} \varphi(E_m) = \varphi(\Omega^+).$$

However, by definition, $\alpha \geq \varphi(\Omega^+)$; hence $\alpha = \varphi(\Omega^+)$. This finishes the proof. □

Corollary 7.44 (Jordan's decomposition theorem) *Assume $\varphi \in \mathcal{M}^{\pm}(\Omega, \mathcal{A})$ is a signed measure. Then there exist uniquely determined finite measures φ^+, φ^- with $\varphi = \varphi^+ - \varphi^-$ and $\varphi^+ \perp \varphi^-$.*

Proof Let $\Omega = \Omega^+ \uplus \Omega^-$ be a Hahn decomposition. Define $\varphi^+(A) := \varphi(A \cap \Omega^+)$ and $\varphi^-(A) := -\varphi(A \cap \Omega^-)$.

The uniqueness of the decomposition is trivial. □

Corollary 7.45 Let $\varphi \in \mathcal{M}^\pm(\Omega, \mathcal{A})$ and let $\varphi = \varphi^+ - \varphi^-$ be the Jordan decomposition of φ. Let $\Omega = \Omega^+ \uplus \Omega^-$ be a Hahn decomposition of Ω. Then

$$\|\varphi\|_{TV} := \sup\left\{\varphi(A) - \varphi(\Omega \setminus A) : A \in \mathcal{A}\right\}$$
$$= \varphi(\Omega^+) - \varphi(\Omega^-)$$
$$= \varphi^+(\Omega) + \varphi^-(\Omega)$$

defines a norm on $\mathcal{M}^\pm(\Omega, \mathcal{A})$, the so-called **total variation norm**.

Proof We only have to show the triangle inequality. Let $\varphi_1, \varphi_2 \in \mathcal{M}^\pm$. Let $\Omega = \Omega^+ \uplus \Omega^-$ be a Hahn decomposition with respect to $\varphi := \varphi_1 + \varphi_2$ and let $\Omega = \Omega_i^+ \uplus \Omega_i^-$ be a Hahn decomposition with respect to φ_i, $i = 1, 2$. Then

$$\|\varphi_1 + \varphi_2\|_{TV} = \varphi_1(\Omega^+) - \varphi_1(\Omega^-) + \varphi_2(\Omega^+) - \varphi_2(\Omega^-)$$
$$\leq \varphi_1(\Omega_1^+) - \varphi_1(\Omega_1^-) + \varphi_2(\Omega_2^+) - \varphi_2(\Omega_2^-)$$
$$= \|\varphi_1\|_{TV} + \|\varphi_2\|_{TV}. \qquad\qquad \square$$

Reflection An important object in stochastic analysis is a random set function φ on $\mathcal{B}([0, 1])$ with the property that $\varphi(A) \sim \mathcal{N}_{0,\lambda(A)}$ is normally distributed and $\varphi(A)$ and $\varphi(B)$ are independent on disjoint sets A and B. With a little effort it is possible to construct φ as a (random) additive set function. However, φ cannot be a (random) signed measure. Why? ♠♠♠

With a lemma, we prepare for an alternative proof of Lebesgue's decomposition theorem (Theorem 7.33).

Lemma 7.46 Let μ, ν be finite measures on $(\Omega, \mathcal{A})$ that are not mutually singular; in short, $\mu \not\perp \nu$. Then there is an $A \in \mathcal{A}$ with $\mu(A) > 0$ and an $\varepsilon > 0$ with

$$\varepsilon\mu(E) \leq \nu(E) \quad \text{for all } E \in \mathcal{A} \text{ with } E \subset A.$$

Proof For $n \in \mathbb{N}$, let $\Omega = \Omega_n^+ \uplus \Omega_n^-$ be a Hahn decomposition for $(\nu - \frac{1}{n}\mu) \in \mathcal{M}^\pm$. Define $M := \bigcap_{n \in \mathbb{N}} \Omega_n^-$. Clearly, $(\nu - \frac{1}{n}\mu)(M) \leq 0$; hence $\nu(M) \leq \frac{1}{n}\mu(M)$ for all $n \in \mathbb{N}$ and thus $\nu(M) = 0$. Since $\mu \not\perp \nu$, we get $\mu(\Omega \setminus M) = \mu(\bigcup_{n \in \mathbb{N}} \Omega_n^+) > 0$. Thus $\mu(\Omega_{n_0}^+) > 0$ for some $n_0 \in \mathbb{N}$. Define $A := \Omega_{n_0}^+$ and $\varepsilon := \frac{1}{n_0}$. Then $\mu(A) > 0$ and $(\nu - \varepsilon\mu)(E) \geq 0$ for all $E \subset A$, $E \in \mathcal{A}$. $\qquad \square$

Alternative proof of Theorem 7.33. We show only the existence of a decomposition. By choosing a suitable sequence $\Omega_n \uparrow \Omega$, we can assume that ν is finite. Consider the set of functions

$$\mathcal{G} := \left\{ g : \Omega \to [0, \infty] : g \text{ is measurable and } \int_A g \, d\mu \leq \nu(A) \text{ for all } A \in \mathcal{A} \right\},$$

and define

$$\gamma := \sup \left\{ \int g \, d\mu : g \in \mathcal{G} \right\}.$$

Our aim is to find a maximal element f in $\mathcal{G}$ (i.e., an f for which $\int f \, d\mu = \gamma$). This f will be the density of ν_a.

Clearly, $0 \in \mathcal{G}$; hence $\mathcal{G} \neq \emptyset$. Furthermore,

$$f, g \in \mathcal{G} \qquad \text{implies} \qquad f \vee g \in \mathcal{G}. \tag{7.11}$$

Indeed, letting $E := \{f \geq g\}$, for all $A \in \mathcal{A}$, we have

$$\int_A (f \vee g) \, d\mu = \int_{A \cap E} f \, d\mu + \int_{A \setminus E} g \, d\mu \leq \nu(A \cap E) + \nu(A \setminus E) = \nu(A).$$

Choose a sequence $(g_n)_{n \in \mathbb{N}}$ in $\mathcal{G}$ such that $\int g_n \, d\mu \xrightarrow{n \to \infty} \gamma$, and define the function $f_n = g_1 \vee \ldots \vee g_n$. Now (7.11) implies $f_n \in \mathcal{G}$. Letting $f := \sup\{f_n : n \in \mathbb{N}\}$, the monotone convergence theorem yields

$$\int_A f \, d\mu = \sup_{n \in \mathbb{N}} \int_A f_n \, d\mu \leq \nu(A) \qquad \text{for all } A \in \mathcal{A}$$

(that is, $f \in \mathcal{G}$), and

$$\int f \, d\mu = \sup_{n \in \mathbb{N}} \int f_n \, d\mu \geq \sup_{n \in \mathbb{N}} \int g_n \, d\mu = \gamma.$$

Hence $\int f \, d\mu = \gamma \leq \nu(\Omega)$. Now define, for any $A \in \mathcal{A}$,

$$\nu_a(A) := \int_A f \, d\mu \quad \text{and} \quad \nu_s(A) := \nu(A) - \nu_a(A).$$

By construction, $\nu_a \ll \mu$ is a finite measure with density f with respect to μ. Since $f \in \mathcal{G}$, we have $\nu_s(A) = \nu(A) - \int_A f \, d\mu \geq 0$ for all $A \in \mathcal{A}$, and thus also ν_s is a finite measure. It remains to show $\nu_s \perp \mu$.

At this point we use Lemma 7.46. Assume that we had $\nu_s \not\perp \mu$. Then there would be an $\varepsilon > 0$ and an $A \in \mathcal{A}$ with $\mu(A) > 0$ such that $\varepsilon \mu(E) \leq \nu_s(E)$ for all $E \subset A$, $E \in \mathcal{A}$. Then, for $B \in \mathcal{A}$, we would have

$$\int_B (f + \varepsilon \mathbb{1}_A) \, d\mu = \int_B f \, d\mu + \varepsilon \mu(A \cap B)$$

$$\leq \nu_a(B) + \nu_s(A \cap B) \leq \nu_a(B) + \nu_s(B) = \nu(B).$$

In other words, $(f + \varepsilon \mathbb{1}_A) \in \mathcal{G}$ and thus $\int (f + \varepsilon \mathbb{1}_A) \, d\mu = \gamma + \varepsilon \mu(A) > \gamma$, contradicting the definition of γ. Hence in fact $\nu_s \perp \mu$.

Takeaways A signed measure is a finite measure that can also assume negative values. A signed measure can be written as the difference of two finite mutually singular measures (Jordan decomposition). With the help of signed measures we could present a different approach to proving the Radon–Nikodym theorem.

Exercise 7.5.1 Let μ be a σ-finite measure on $(\Omega, \mathcal{A})$ and let φ be a signed measure on $(\Omega, \mathcal{A})$. Show that, analogously to the Radon–Nikodym theorem, the following two statements are equivalent:

(i) $\varphi(A) = 0$ for all $A \in \mathcal{A}$ with $\mu(A) = 0$.
(ii) There is an $f \in \mathcal{L}^1(\mu)$ with $\varphi = f\mu$; hence $\int_A f \, d\mu = \varphi(A)$ for all $A \in \mathcal{A}$.
♣

Exercise 7.5.2 Let μ, ν, α be finite measures on $(\Omega, \mathcal{A})$ with $\nu \ll \mu \ll \alpha$.

(i) Show the chain rule for the Radon–Nikodym derivative:

$$\frac{d\nu}{d\alpha} = \frac{d\nu}{d\mu} \frac{d\mu}{d\alpha} \qquad \alpha\text{-a.e.}$$

(ii) Show that $f := \frac{d\nu}{d(\mu+\nu)}$ exists and that $\frac{d\nu}{d\mu} = \frac{f}{1-f}$ holds μ-a.e. ♣

7.6   Supplement: Dual Spaces

By the Riesz–Fréchet theorem (Theorem 7.26), every continuous linear functional $F : L^2(\mu) \to \mathbb{R}$ has a representation $F(g) = \langle f, g \rangle$ for some $f \in L^2(\mu)$. On the other hand, for any $f \in L^2(\mu)$, the map $L^2(\mu) \to \mathbb{R}$, $g \mapsto \langle f, g \rangle$ is continuous and linear. Hence $L^2(\mu)$ is canonically isomorphic to its topological dual space $(L^2(\mu))'$. This dual space is defined as follows.

Definition 7.47 (Dual space) Let $(V, \| \cdot \|)$ be a Banach space. The **dual space** V' of V is defined by

$$V' := \{F : V \to \mathbb{R} \text{ is continuous and linear}\}.$$

For $F \in V'$, we define $\|F\|' := \sup\{|F(f)| : \|f\| = 1\}$.

Remark 7.48 As F is continuous, for any $\delta > 0$, there exists an $\varepsilon > 0$ such that $|F(f)| < \delta$ for all $f \in V$ with $\|f\| < \varepsilon$. Hence $\|F\|' \le \delta/\varepsilon < \infty$. ◊

We are interested in the case $V = L^p(\mu)$ for $p \in [1, \infty]$ and write $\|F\|'_p$ for the norm of $F \in V'$. In the particular case $V = L^2(\mu)$, by the Cauchy–Schwarz inequality, we have $\|F\|'_2 = \|f\|_2$. This can be generalized:

Lemma 7.49 *Let $p, q \in [1, \infty]$ with $\frac{1}{p} + \frac{1}{q} = 1$. The canonical map*

$$\kappa : L^q(\mu) \to (L^p(\mu))'$$

$$\kappa(f)(g) = \int fg \, d\mu \quad \text{for } f \in L^q(\mu), g \in L^p(\mu)$$

is an isometry; that is, $\|\kappa(f)\|'_p = \|f\|_q$.

Proof We show equality by showing the two inequalities separately.

"$\leq$" This follows from Hölder's inequality.

"$\geq$" For any admissible pair p, q and all $f \in L^q(\mu)$, $g \in L^p(\mu)$, by the definition of the operator norm, $\|\kappa(f)\|'_p \|g\|_p \geq |\int fg \, d\mu|$. Define the sign function $\text{sign}(x) = \mathbb{1}_{(0,\infty)}(x) - \mathbb{1}_{(-\infty,0)}(x)$. Replacing g by $\tilde{g} := |g| \, \text{sign}(f)$ (note that $\|\tilde{g}\|_p = \|g\|_p$), we obtain

$$\|\kappa(f)\|'_p \|g\|_p \geq \left| \int f\tilde{g} \, d\mu \right| = \|fg\|_1. \tag{7.12}$$

First consider the case $q = 1$ and $f \in \mathcal{L}^1(\mu)$. Applying (7.12) with $g \equiv 1 \in \mathcal{L}^\infty(\mu)$ yields $\|\kappa(f)\|'_\infty \geq \|f\|_1$.

Now let $q \in (1, \infty)$. Let $g := |f|^{q-1}$. Since $\frac{q-1}{q} = \frac{1}{p}$, we have

$$\|\kappa(f)\|'_p \cdot \|g\|_p \geq \|fg\|_1 = \||f|^q\|_1 = \|f\|_q^q = \|f\|_q \cdot \|f\|_q^{q-1} = \|f\|_q \cdot \|g\|_p.$$

Finally, let $q = \infty$. Without loss of generality, assume $\|f\|_\infty \in (0, \infty)$. Let $\varepsilon > 0$. Then there exists an $A_\varepsilon \in \mathcal{A}$ with $0 < \mu(A_\varepsilon) < \infty$ such that

$$A_\varepsilon \subset \{|f| > (1 - \varepsilon)\|f\|_\infty\}.$$

If we let $g = \frac{1}{\mu(A_\varepsilon)} \mathbb{1}_{A_\varepsilon}$, then $\|g\|_1 = 1$ and $\|\kappa(f)\|'_1 \geq \|fg\|_1 \geq (1 - \varepsilon)\|f\|_\infty$. $\square$

Theorem 7.50 *Let $p \in [1, \infty)$ and assume $\frac{1}{p} + \frac{1}{q} = 1$. Then $L^q(\mu)$ is isomorphic to its dual space $(L^p(\mu))'$ by virtue of the isometry κ.*

Proof The proof makes use of the Radon–Nikodym theorem (Corollary 7.34). However, here we only sketch the proof since we do not want to go into the details of signed measures and signed contents. A signed content ν is an additive set function that is the difference $\nu = \nu^+ - \nu^-$ of two finite contents. This definition is parallel to that of a signed measure that is the difference of two finite measures.

As κ is an isometry, κ in particular is injective. Hence we only have to show that κ is surjective. Let $F \in (L^p(\mu))'$. Then $\nu(A) = F(\mathbb{1}_A)$ is a signed content on $\mathcal{A}$ and we have

$$|\nu(A)| \leq \|F\|'_p \, (\mu(A))^{1/p}.$$

Since μ is $\emptyset$-continuous, ν is also $\emptyset$-continuous and is thus a signed measure on $\mathcal{A}$. We even have $\nu \ll \mu$. By the Radon–Nikodym theorem (Corollary 7.34) (applied to the measures ν^- and ν^+; see Exercise 7.5.1), ν admits a density with respect to μ; that is, a measurable function f with $\nu = f\mu$.

Let

$$\mathbb{E}_f := \big\{ g : g \text{ is a simple function with } \mu(g \neq 0) < \infty \big\}$$

and let

$$\mathbb{E}_f^+ := \big\{ g \in \mathbb{E}_f : g \geq 0 \big\}.$$

Then, for $g \in \mathbb{E}_f$,

$$F(g) = \int gf \, d\mu. \tag{7.13}$$

In order to show that (7.13) holds for all $g \in L^p(\mu)$, we first show $f \in \mathcal{L}^q(\mu)$. To this end, we distinguish two cases.

Case 1: $p = 1$.   For every $\alpha > 0$,

$$\mu(\{|f| > \alpha\}) \leq \frac{1}{\alpha} \nu(\{|f| > \alpha\})$$

$$= \frac{1}{\alpha} F(\mathbb{1}_{\{|f|>\alpha\}}) \leq \frac{1}{\alpha} \|F\|'_1 \cdot \|\mathbb{1}_{\{|f|>\alpha\}}\|_1 = \frac{1}{\alpha} \|F\|'_1 \cdot \mu(\{|f| > \alpha\}).$$

This implies $\mu(\{|f| > \alpha\}) = 0$ if $\alpha > \|F\|'_1$; hence $\|f\|_\infty \leq \|F\|'_1 < \infty$.

Case 2: $p \in (1, \infty)$.   By Theorem 1.96, there are $g_1, g_2, \ldots \in \mathbb{E}_f^+$ such that $g_n \uparrow |f|$ μ-a.e. Define $h_n = \text{sign}(f)(g_n)^{q-1} \in \mathbb{E}_f$; hence

$$\|g_n\|_q^q \leq \int h_n f \, d\mu = F(h_n)$$

$$\leq \|F\|'_p \cdot \|h_n\|_p = \|F\|'_p \cdot (\|g_n\|_q)^{q-1}.$$

Thus we have $\|g_n\|_q \leq \|F\|'_p$. Monotone convergence (Theorem 4.20) now yields $\|f\|_q \leq \|F\|'_p < \infty$; hence $f \in \mathcal{L}^q(\mu)$.

Concluding, the map $\widetilde{F} : g \mapsto \int gf \, d\mu$ is in $(L^p(\mu))'$, and $\widetilde{F}(g) = F(g)$ for every $g \in \mathbb{E}_f$. Since $\widetilde{F}$ is continuous and $\mathbb{E}_f \subset L^p(\mu)$ is dense, we get $\widetilde{F} = F$. $\square$

Remark 7.51 For $p = \infty$, the statement of Theorem 7.50 is false in general. (For finite $\mathcal{A}$, the claim is trivially true even for $p = \infty$.) For example, let $\Omega = \mathbb{N}$, $\mathcal{A} = 2^{\Omega}$ and let μ be the counting measure. Thus we consider sequence spaces $\ell^p = L^p(\mathbb{N}, 2^{\mathbb{N}}, \mu)$. For the subspace $\ell^K \subset \ell^{\infty}$ of convergent sequences, $F : \ell^K \to \mathbb{R}$, $(a_n)_{n \in \mathbb{N}} \mapsto \lim_{n \to \infty} a_n$ is a continuous linear functional. By the Hahn–Banach theorem of functional analysis (see, e.g., [87] or [174]), F can be extended to a continuous linear functional on ℓ^{∞}. However, clearly there is no sequence $(b_n)_{n \in \mathbb{N}} \in \ell^1$ with

$$F((a_n)_{n \in \mathbb{N}}) = \sum_{m=1}^{\infty} a_m b_m. \ \Diamond$$

Takeaways L^q is the dual space to L^p, if $\frac{1}{p} + \frac{1}{q} = 1$ and $p \in [1, \infty)$. This beautiful theorem could be shown using the tools that we developed in the previous sections for other purposes.

Exercise 7.6.1 Show that $\mathbb{E}_f \subset L^p(\mu)$ is dense if $p \in [1, \infty)$. ♣

Chapter 8
Conditional Expectations

If there is partial information on the outcome of a random experiment, the probabilities for the possible events may change. The concept of conditional probabilities and conditional expectations formalizes the corresponding calculus.

8.1 Elementary Conditional Probabilities

Example 8.1 We throw a die and consider the events

$$A := \{\text{the face shows an odd number}\},$$

$$B := \{\text{the face shows three or smaller}\}.$$

Clearly, $\mathbf{P}[A] = \frac{1}{2}$ and $\mathbf{P}[B] = \frac{1}{2}$. However, what is the probability that A occurs if we already know that B occurs?

We model the experiment on the probability space $(\Omega, \mathcal{A}, \mathbf{P})$, where $\Omega = \{1, \ldots, 6\}$, $\mathcal{A} = 2^{\Omega}$ and $\mathbf{P}$ is the uniform distribution on Ω. Then

$$A = \{1, 3, 5\} \quad \text{and} \quad B = \{1, 2, 3\}.$$

If we know that B has occurred, it is plausible to assume the uniform distribution on the remaining possible outcomes; that is, on $\{1, 2, 3\}$. Thus we define a new probability measure $\mathbf{P}_B$ on $(B, 2^B)$ by

$$\mathbf{P}_B[C] = \frac{\#C}{\#B} \quad \text{for } C \subset B.$$

© The Editor(s) (if applicable) and The Author(s), under exclusive license
to Springer Nature Switzerland AG 2020
A. Klenke, *Probability Theory*, Universitext,
https://doi.org/10.1007/978-3-030-56402-5_8

By assigning the points in $\Omega \setminus B$ probability zero (since they are impossible if B has occurred), we can extend $\mathbf{P}_B$ to a measure on Ω:

$$\mathbf{P}[C \,|\, B] \;:=\; \mathbf{P}_B[C \cap B] \;=\; \frac{\#(C \cap B)}{\#B} \quad \text{for } C \subset \Omega.$$

In this way, we get $\mathbf{P}[A \,|\, B] = \dfrac{\#\{1, 3\}}{\#\{1, 2, 3\}} = \dfrac{2}{3}$. $\Diamond$

Motivated by this example, we make the following definition.

Definition 8.2 (Conditional probability)   Let $(\Omega, \mathcal{A}, \mathbf{P})$ be a probability space and $B \in \mathcal{A}$. We define the **conditional probability given** B for any $A \in \mathcal{A}$ by

$$\mathbf{P}[A \,|\, B] = \begin{cases} \dfrac{\mathbf{P}[A \cap B]}{\mathbf{P}[B]}, & \text{if } \mathbf{P}[B] > 0, \\[2mm] \quad 0, & \text{otherwise.} \end{cases} \tag{8.1}$$

Remark 8.3   The specification in (8.1) for the case $\mathbf{P}[B] = 0$ is arbitrary and is of no importance. $\Diamond$

Theorem 8.4   *If* $\mathbf{P}[B] > 0$*, then* $\mathbf{P}[\,\cdot\,|\, B]$ *is a probability measure on* $(\Omega, \mathcal{A})$.

Proof   This is obvious. $\square$

Theorem 8.5   *Let* $A, B \in \mathcal{A}$ *with* $\mathbf{P}[A],\ \mathbf{P}[B] > 0$. *Then*

$$A, B \ \text{are independent} \quad \Longleftrightarrow \quad \mathbf{P}[A \,|\, B] = \mathbf{P}[A] \quad \Longleftrightarrow \quad \mathbf{P}[B \,|\, A] = \mathbf{P}[B].$$

Proof   This is trivial! $\square$

Theorem 8.6 (Summation formula)   *Let* I *be a countable set and let* $(B_i)_{i \in I}$ *be pairwise disjoint sets with* $\mathbf{P}\left[\biguplus_{i \in I} B_i \right] = 1$. *Then, for any* $A \in \mathcal{A}$,

$$\mathbf{P}[A] = \sum_{i \in I} \mathbf{P}[A \,|\, B_i] \, \mathbf{P}[B_i]. \tag{8.2}$$

Proof   Due to the σ-additivity of $\mathbf{P}$, we have

$$\mathbf{P}[A] = \mathbf{P}\left[\biguplus_{i \in I} (A \cap B_i) \right] = \sum_{i \in I} \mathbf{P}[A \cap B_i] = \sum_{i \in I} \mathbf{P}[A \,|\, B_i] \mathbf{P}[B_i]. \qquad \square$$

Theorem 8.7 (Bayes' formula)   *Let* I *be a countable set and let* $(B_i)_{i \in I}$ *be pairwise disjoint sets with* $\mathbf{P}\left[\biguplus_{i \in I} B_i \right] = 1$. *Then, for any* $A \in \mathcal{A}$ *with* $\mathbf{P}[A] > 0$ *and any* $k \in I$,

$$\mathbf{P}[B_k \,|\, A] = \frac{\mathbf{P}[A \,|\, B_k] \, \mathbf{P}[B_k]}{\sum_{i \in I} \mathbf{P}[A \,|\, B_i] \, \mathbf{P}[B_i]}. \tag{8.3}$$

Proof We have

$$\mathbf{P}[B_k\,|\,A] = \frac{\mathbf{P}[B_k \cap A]}{\mathbf{P}[A]} = \frac{\mathbf{P}[A\,|\,B_k]\,\mathbf{P}[B_k]}{\mathbf{P}[A]}.$$

Now use the expression in (8.2) for $\mathbf{P}[A]$. □

Example 8.8 In the production of certain electronic devices, a fraction of 2% of the production is defective. A quick test detects a defective device with probability 95%; however, with probability 10% it gives a false alarm for an intact device.

If the test gives an alarm, what is the probability that the device just tested is indeed defective?

We formalize the description given above. Let

$$A := \{\text{device is declared as defective}\},$$
$$B := \{\text{device is defective}\},$$

and

$$\mathbf{P}[B] = 0.02, \qquad\qquad \mathbf{P}[B^c] = 0.98,$$
$$\mathbf{P}[A\,|\,B] = 0.95, \qquad \mathbf{P}[A\,|\,B^c] = 0.1.$$

Bayes' formula yields

$$\mathbf{P}[B\,|\,A] = \frac{\mathbf{P}[A\,|\,B]\,\mathbf{P}[B]}{\mathbf{P}[A\,|\,B]\,\mathbf{P}[B] + \mathbf{P}[A\,|\,B^c]\,\mathbf{P}[B^c]}$$

$$= \frac{0.95 \cdot 0.02}{0.95 \cdot 0.02 + 0.1 \cdot 0.98} = \frac{19}{117} \approx 0.162.$$

On the other hand, the probability that a device that was not classified as defective is in fact defective is

$$\mathbf{P}[B\,|\,A^c] = \frac{0.05 \cdot 0.02}{0.05 \cdot 0.02 + 0.9 \cdot 0.98} = \frac{1}{883} \approx 0.00113. \quad \Diamond$$

Now let $X \in \mathcal{L}^1(\mathbf{P})$. If $A \in \mathcal{A}$, then clearly also $\mathbb{1}_A X \in \mathcal{L}^1(\mathbf{P})$. We define

$$\mathbf{E}[X;\,A] := \mathbf{E}[\mathbb{1}_A\,X]. \tag{8.4}$$

If $\mathbf{P}[A] > 0$, then $\mathbf{P}[\,\cdot\,|\,A]$ is a probability measure. Since $\mathbb{1}_A X \in \mathcal{L}^1(\mathbf{P})$, we have $X \in \mathcal{L}^1(\mathbf{P}[\,\cdot\,|\,A])$. Hence we can define the expectation of X with respect to $\mathbf{P}[\,\cdot\,|\,A]$.

Definition 8.9 Let $X \in \mathcal{L}^1(\mathbf{P})$ and $A \in \mathcal{A}$. Then we define

$$\mathbf{E}[X|A] := \int X(\omega)\,\mathbf{P}[d\omega|A] = \begin{cases} \dfrac{\mathbf{E}[\mathbb{1}_A X]}{\mathbf{P}[A]}, & \text{if } \mathbf{P}[A] > 0, \\[2mm] 0, & \text{else.} \end{cases} \tag{8.5}$$

Clearly, $\mathbf{P}[B|A] = \mathbf{E}[\mathbb{1}_B|A]$ for all $B \in \mathcal{A}$.

Consider now the situation that we studied with the summation formula for conditional probabilities. Hence, let I be a countable set and let $(B_i)_{i\in I}$ be pairwise disjoint events with $\biguplus_{i\in I} B_i = \Omega$. We define $\mathcal{F} := \sigma(B_i, \, i \in I)$. For $X \in \mathcal{L}^1(\mathbf{P})$, we define a map $\mathbf{E}[X|\mathcal{F}] : \Omega \to \mathbb{R}$ by

$$\mathbf{E}[X|\mathcal{F}](\omega) = \mathbf{E}[X|B_i] \quad\Longleftrightarrow\quad B_i \ni \omega. \tag{8.6}$$

Lemma 8.10 *The map* $\mathbf{E}[X|\mathcal{F}]$ *has the following properties.*

(i) $\mathbf{E}[X|\mathcal{F}]$ *is* $\mathcal{F}$-*measurable.*

(ii) $\mathbf{E}[X|\mathcal{F}] \in \mathcal{L}^1(\mathbf{P})$, *and for any* $A \in \mathcal{F}$, *we have* $\int_A \mathbf{E}[X|\mathcal{F}]\,d\mathbf{P} = \int_A X\,d\mathbf{P}$.

Proof

(i) Let f be the map $f : \Omega \to I$ with

$$f(\omega) = i \quad\Longleftrightarrow\quad B_i \ni \omega.$$

Further, let $g : I \to \mathbb{R}$, $i \mapsto \mathbf{E}[X|B_i]$. Since I is discrete, g is measurable. Since f is $\mathcal{F}$-measurable, $\mathbf{E}[X|\mathcal{F}] = g \circ f$ is also $\mathcal{F}$-measurable.

(ii) Let $A \in \mathcal{F}$ and $J \subset I$ with $A = \biguplus_{j\in J} B_j$. Let $J' := \{i \in J : \mathbf{P}[B_i] > 0\}$. Hence

$$\int_A \mathbf{E}[X|\mathcal{F}]\,d\mathbf{P} = \sum_{i\in J'} \mathbf{P}[B_i]\,\mathbf{E}[X|B_i] = \sum_{i\in J'} \mathbf{E}[\mathbb{1}_{B_i} X] = \int_A X\,d\mathbf{P}. \qquad \square$$

Takeaways We have developed the notion of the (elementary) conditional probability and have established two simple but important formulas: the summation formula and Bayes' formula. We have also reformulated the elementary conditional expectation and highlighted those of its properties that allow for a generalisation to conditional expectations given σ-algebras.

Exercise 8.1.1 (Lack of memory of the exponential distribution) Let $X > 0$ be a strictly positive random variable and let $\theta > 0$. Show that X is exponentially distributed if and only if

$$\mathbf{P}[X > t + s \,|\, X > s] = \mathbf{P}[X > t] \quad \text{for all } s, t \geq 0.$$

In particular, $X \sim \exp_\theta$ if and only if $\mathbf{P}[X > t + s \,|\, X > s] = e^{-\theta t}$ for all $s, t \geq 0$.
♣

Exercise 8.1.2 Consider a theatre with n seats that is fully booked for this evening. Each of the n people entering the theatre (one by one) has a seat reservation. However, the first person is absent-minded and takes a seat at random. Any subsequent person takes his or her reserved seat if it is free and otherwise picks a free seat at random.

(i) What is the probability that the last person gets his or her reserved seat?
(ii) What is the probability that the kth person gets his or her reserved seat? ♣

8.2 Conditional Expectations

Let X be a random variable that is uniformly distributed on $[0, 1]$. Assume that if we know the value $X = x$, the random variables $Y_1, \ldots, Y_n$ are independent and Ber_x-distributed. So far, with our machinery we can only deal with conditional probabilities of the type $\mathbf{P}[\,\cdot\,|\,X \in [a, b]]$, $a < b$ (since $X \in [a, b]$ has positive probability). How about $\mathbf{P}[Y_1 = \ldots = Y_n = 1 \,|\, X = x]$? Intuitively, this should be x^n. We thus need a notion of conditional probabilities that allows us to deal with conditioning on events with probability zero and that is consistent with our intuition. In the next section, we will see that in the current example this can be done using transition kernels. First, however, we have to consider a more general situation.

In the following, $\mathcal{F} \subset \mathcal{A}$ will be a sub-σ-algebra and $X \in \mathcal{L}^1(\Omega, \mathcal{A}, \mathbf{P})$. In analogy with Lemma 8.10, we make the following definition.

Definition 8.11 (Conditional expectation) A random variable Y is called a **conditional expectation** of X given $\mathcal{F}$, symbolically $\mathbf{E}[X \,|\, \mathcal{F}] := Y$, if:

(i) Y is $\mathcal{F}$-measurable.
(ii) For any $A \in \mathcal{F}$, we have $\mathbf{E}[X \mathbb{1}_A] = \mathbf{E}[Y \mathbb{1}_A]$.

For $B \in \mathcal{A}$, $\mathbf{P}[B \,|\, \mathcal{F}] := \mathbf{E}[\mathbb{1}_B \,|\, \mathcal{F}]$ is called a **conditional probability** of B given the σ-algebra $\mathcal{F}$.

Theorem 8.12 $\mathbf{E}[X \,|\, \mathcal{F}]$ *exists and is unique (up to equality almost surely).*

Since conditional expectations are defined only up to equality a.s., all equalities with conditional expectations are understood as equalities a.s., even if we do not say so explicitly.

Proof

Uniqueness Let Y and Y' be random variables that fulfill (i) and (ii). Let $A = \{Y > Y'\} \in \mathcal{F}$. Then, by (ii),

$$0 = \mathbf{E}[Y \mathbb{1}_A] - \mathbf{E}[Y' \mathbb{1}_A] = \mathbf{E}[(Y - Y') \mathbb{1}_A].$$

Since $(Y - Y') \mathbb{1}_A \geq 0$, we have $\mathbf{P}[A] = 0$; hence $Y \leq Y'$ almost surely. Similarly, we get $Y \geq Y'$ almost surely.

Existence Let $X^+ = X \vee 0$ and $X^- = X^+ - X$. By

$$Q^\pm(A) := \mathbf{E}[X^\pm \mathbb{1}_A] \quad \text{for all } A \in \mathcal{F},$$

we define two finite measures on $(\Omega, \mathcal{F})$. Clearly, $Q^\pm \ll \mathbf{P}$; hence the Radon–Nikodym theorem (Corollary 7.34) yields the existence of $\mathcal{F}$-measurable densities $Y^\pm$ such that

$$Q^\pm(A) = \int_A Y^\pm \, d\mathbf{P} = \mathbf{E}[Y^\pm \mathbb{1}_A].$$

Now define $Y = Y^+ - Y^-$.

$\square$

Definition 8.13 If Y is a random variable and $X \in \mathcal{L}^1(\mathbf{P})$, then we define $\mathbf{E}[X \,|\, Y] := \mathbf{E}[X \,|\, \sigma(Y)]$.

Theorem 8.14 (Properties of the conditional expectation) *Let $(\Omega, \mathcal{A}, \mathbf{P})$ and let X be as above. Let $\mathcal{G} \subset \mathcal{F} \subset \mathcal{A}$ be σ-algebras and let $Y \in \mathcal{L}^1(\Omega, \mathcal{A}, \mathbf{P})$. Then:*

(i) *(**Linearity**)* $\mathbf{E}[\lambda X + Y \,|\, \mathcal{F}] = \lambda \mathbf{E}[X \,|\, \mathcal{F}] + \mathbf{E}[Y \,|\, \mathcal{F}]$.

(ii) *(**Monotonicity**)* If $X \geq Y$ a.s., then $\mathbf{E}[X \,|\, \mathcal{F}] \geq \mathbf{E}[Y \,|\, \mathcal{F}]$.

(iii) *If $\mathbf{E}[|XY|] < \infty$ and Y is measurable with respect to $\mathcal{F}$, then*

$$\mathbf{E}[XY \,|\, \mathcal{F}] = Y \, \mathbf{E}[X \,|\, \mathcal{F}] \qquad and \qquad \mathbf{E}[Y \,|\, \mathcal{F}] = \mathbf{E}[Y \,|\, Y] = Y.$$

(iv) *(**Tower property**)* $\mathbf{E}[\mathbf{E}[X \,|\, \mathcal{F}] \,|\, \mathcal{G}] = \mathbf{E}[\mathbf{E}[X \,|\, \mathcal{G}] \,|\, \mathcal{F}] = \mathbf{E}[X \,|\, \mathcal{G}]$.

(v) *(**Triangle inequality**)* $\mathbf{E}[|X| \,|\, \mathcal{F}] \geq |\mathbf{E}[X \,|\, \mathcal{F}]|$.

(vi) *(**Independence**)* If $\sigma(X)$ and $\mathcal{F}$ are independent, then $\mathbf{E}[X \,|\, \mathcal{F}] = \mathbf{E}[X]$.

(vii) *If $\mathbf{P}[A] \in \{0, 1\}$ for any $A \in \mathcal{F}$, then $\mathbf{E}[X \,|\, \mathcal{F}] = \mathbf{E}[X]$.*

(viii) *(**Dominated convergence**)* Assume $Y \in \mathcal{L}^1(\mathbf{P})$, $Y \geq 0$ and $(X_n)_{n \in \mathbb{N}}$ is a sequence of random variables with $|X_n| \leq Y$ for $n \in \mathbb{N}$ and such that $X_n \overset{n \to \infty}{\longrightarrow} X$ a.s. Then

$$\lim_{n \to \infty} \mathbf{E}[X_n \,|\, \mathcal{F}] = \mathbf{E}[X \,|\, \mathcal{F}] \quad a.s. \text{ and in } L^1(\mathbf{P}). \tag{8.7}$$

Proof

(i) The right-hand side is $\mathcal{F}$-measurable; hence, for $A \in \mathcal{F}$,

$$
\begin{aligned}
\mathbf{E}\big[\mathbb{1}_A \left(\lambda \mathbf{E}[X \mid \mathcal{F}] + \mathbf{E}[Y \mid \mathcal{F}]\right)\big] &= \lambda \mathbf{E}\big[\mathbb{1}_A \, \mathbf{E}[X \mid \mathcal{F}]\big] + \mathbf{E}\big[\mathbb{1}_A \, \mathbf{E}[Y \mid \mathcal{F}]\big] \\
&= \lambda \mathbf{E}[\mathbb{1}_A \, X] + \mathbf{E}[\mathbb{1}_A \, Y] \\
&= \mathbf{E}\big[\mathbb{1}_A \left(\lambda X + Y\right)\big].
\end{aligned}
$$

(ii) Let $A = \{\mathbf{E}[X \mid \mathcal{F}] < \mathbf{E}[Y \mid \mathcal{F}]\} \in \mathcal{F}$. Since we have $X \geq Y$, we get $\mathbf{E}[\mathbb{1}_A (X - Y)] \geq 0$ and thus $\mathbf{P}[A] = 0$.

(iii) First assume $X \geq 0$ and $Y \geq 0$. For $n \in \mathbb{N}$, define $Y_n = 2^{-n} \lfloor 2^n Y \rfloor$. Then $Y_n \uparrow Y$ and $Y_n \mathbf{E}[X \mid \mathcal{F}] \uparrow Y \mathbf{E}[X \mid \mathcal{F}]$ (since $\mathbf{E}[X \mid \mathcal{F}] \geq 0$ by (ii)). By the monotone convergence theorem (Lemma 4.6(ii)),

$$
\mathbf{E}\big[\mathbb{1}_A \, Y_n \, \mathbf{E}[X \mid \mathcal{F}]\big] \overset{n \to \infty}{\longrightarrow} \mathbf{E}\big[\mathbb{1}_A \, Y \, \mathbf{E}[X \mid \mathcal{F}]\big].
$$

On the other hand,

$$
\begin{aligned}
\mathbf{E}\big[\mathbb{1}_A \, Y_n \, \mathbf{E}[X \mid \mathcal{F}]\big] &= \sum_{k=1}^{\infty} \mathbf{E}\big[\mathbb{1}_A \, \mathbb{1}_{\{Y_n = k \, 2^{-n}\}} \, k \, 2^{-n} \, \mathbf{E}[X \mid \mathcal{F}]\big] \\
&= \sum_{k=1}^{\infty} \mathbf{E}\big[\mathbb{1}_A \, \mathbb{1}_{\{Y_n = k \, 2^{-n}\}} \, k \, 2^{-n} \, X\big] \\
&= \mathbf{E}\big[\mathbb{1}_A \, Y_n \, X\big] \overset{n \to \infty}{\longrightarrow} \mathbf{E}[\mathbb{1}_A \, Y X].
\end{aligned}
$$

Hence $\mathbf{E}[\mathbb{1}_A \, Y \, \mathbf{E}[X \mid \mathcal{F}]] = \mathbf{E}[\mathbb{1}_A \, Y X]$. In the general case, write $X = X^+ - X^-$ and $Y = Y^+ - Y^-$ and exploit the linearity of the conditional expectation.

(iv) The second equality follows from (iii) with $Y = \mathbf{E}[X \mid \mathcal{G}]$ and $X = 1$. Now let $A \in \mathcal{G}$. Then, in particular, $A \in \mathcal{F}$; hence

$$
\mathbf{E}\big[\mathbb{1}_A \mathbf{E}[\mathbf{E}[X \mid \mathcal{F}] \mid \mathcal{G}]\big] = \mathbf{E}\big[\mathbb{1}_A \mathbf{E}[X \mid \mathcal{F}]\big] = \mathbf{E}[\mathbb{1}_A \, X] = \mathbf{E}\big[\mathbb{1}_A \, \mathbf{E}[X \mid \mathcal{G}]\big].
$$

(v) This follows from (i) and (ii) with $X = X^+ - X^-$.

(vi) Trivially, $\mathbf{E}[X]$ is measurable with respect to $\mathcal{F}$. Let $A \in \mathcal{F}$. Then X and $\mathbb{1}_A$ are independent; hence $\mathbf{E}[\mathbf{E}[X \mid \mathcal{F}] \mathbb{1}_A] = \mathbf{E}[X \mathbb{1}_A] = \mathbf{E}[X] \mathbf{E}[\mathbb{1}_A]$.

(vii) For any $A \in \mathcal{F}$ and $B \in \mathcal{A}$, we have $\mathbf{P}[A \cap B] = 0$ if $\mathbf{P}[A] = 0$, and $\mathbf{P}[A \cap B] = \mathbf{P}[B]$ if $\mathbf{P}[A] = 1$. Hence $\mathcal{F}$ and $\mathcal{A}$ are independent and thus $\mathcal{F}$ is independent of any sub-σ-algebra of $\mathcal{A}$. In particular, $\mathcal{F}$ and $\sigma(X)$ are independent. Hence the claim follows from (vi).

(viii) Let $|X_n| \leq Y$ for any $n \in \mathbb{N}$ and $X_n \overset{n \to \infty}{\longrightarrow} X$ almost surely. Define $Z_n := \sup_{k \geq n} |X_k - X|$. Then $0 \leq Z_n \leq 2Y$ and $Z_n \overset{\text{a.s.}}{\longrightarrow} 0$. By Corollary 6.26 (dominated convergence), we have $\mathbf{E}[Z_n] \overset{n \to \infty}{\longrightarrow} 0$; hence, by the triangle inequality,

$$\mathbf{E}\big[\big|\mathbf{E}[X_n\,|\,\mathcal{F}]-\mathbf{E}[X\,|\,\mathcal{F}]\big|\big] \le \mathbf{E}[\mathbf{E}[|X_n-X|\,\big|\,\mathcal{F}]] = \mathbf{E}[|X_n-X|] \le \mathbf{E}[Z_n] \overset{n\to\infty}{\longrightarrow} 0.$$

However, this is the $L^1(\mathbf{P})$-convergence in (8.7). As $(Z_n)_{n\in\mathbb{N}}$ is decreasing, by (ii) also $(\mathbf{E}[Z_n\,\big|\,\mathcal{F}])_{n\in\mathbb{N}}$ decreases to some limit, say, Z. By Fatou's lemma,

$$\mathbf{E}[Z] \le \lim_{n\to\infty} \mathbf{E}[\mathbf{E}[Z_n\,|\,\mathcal{F}]] = \lim_{n\to\infty}\mathbf{E}[Z_n] = 0.$$

Hence $Z = 0$ and thus $\mathbf{E}[Z_n\,\big|\,\mathcal{F}] \overset{n\to\infty}{\longrightarrow} 0$ almost surely. However, by (v),

$$\big|\mathbf{E}[X_n\,\big|\,\mathcal{F}] - \mathbf{E}[X\,\big|\,\mathcal{F}]\big| \le \mathbf{E}[Z_n\,\big|\,\mathcal{F}]. \qquad\qquad \square$$

Reflection Can we relax the condition in (viii) that the X_n be dominated to uniform integrability of $(X_n)_{n\in\mathbb{N}}$? ♠♠♠

Remark 8.15 Intuitively, $\mathbf{E}[X\,|\,\mathcal{F}]$ is the best prediction we can make for the value of X if we only have the information of the σ-algebra $\mathcal{F}$. For example, if $\sigma(X) \subset \mathcal{F}$ (that is, if we know X already), then $\mathbf{E}[X\,|\,\mathcal{F}] = X$, as shown in (iii). At the other end of the spectrum is the case where X and $\mathcal{F}$ are independent; that is, where knowledge of $\mathcal{F}$ does not give any information on X. Here the best prediction for X is its mean; hence $\mathbf{E}[X] = \mathbf{E}[X\,|\,\mathcal{F}]$, as shown in (vi).

What exactly do we mean by "best prediction"? For square integrable random variables X, by the best prediction for X we will understand the $\mathcal{F}$-measurable random variable that minimizes the L^2-distance from X. The next corollary shows that the conditional expectation is in fact this minimizer. ◊

Remark 8.16 Let $X : \Omega \to \mathbb{R}$ be a random variable such that $X^- \in \mathcal{L}^1(\mathbf{P})$. We can define the conditional expectation as the monotone limit

$$\mathbf{E}[X\,|\,\mathcal{F}] := \lim_{n\to\infty} \mathbf{E}[X_n\,|\,\mathcal{F}],$$

where $-X^- \le X_1$ and $X_n \uparrow X$. Due to the monotonicity of the conditional expectation (Theorem 8.14(ii)) it is easy to show that the limit does not depend on the choice of the sequence (X_n) and that it fulfills the conditions of Definition 8.11. Analogously, we can define the conditional expectation $X^+ \in \mathcal{L}^1(\mathbf{P})$. For this generalization of the conditional expectation, we still have $\mathbf{E}[X\,|\,\mathcal{F}] \le \mathbf{E}[Y\,|\,\mathcal{F}]$ a.s. if $Y \ge X$ a.s. (see Exercise 8.2.1). ◊

Corollary 8.17 (Conditional expectation as projection) *Let $\mathcal{F} \subset \mathcal{A}$ be a σ-algebra and let X be a random variable with $\mathbf{E}[X^2] < \infty$. Then $\mathbf{E}[X\,|\,\mathcal{F}]$ is the orthogonal projection of X on $\mathcal{L}^2(\Omega, \mathcal{F}, \mathbf{P})$. That is, for any $\mathcal{F}$-measurable Y with $\mathbf{E}[Y^2] < \infty$,*

$$\mathbf{E}\big[(X - Y)^2\big] \ge \mathbf{E}\big[(X - \mathbf{E}[X\,|\,\mathcal{F}])^2\big]$$

with equality if and only if $Y = \mathbf{E}[X\,|\,\mathcal{F}]$.

Proof First assume that $\mathbf{E}[\mathbf{E}[X\,|\,\mathcal{F}]^2] < \infty$. (In Theorem 8.20, we will see that we have $\mathbf{E}[\mathbf{E}[X\,|\,\mathcal{F}]^2] \leq \mathbf{E}[X^2]$, but here we want to keep the proof self-contained.) Let Y be $\mathcal{F}$-measurable and assume $\mathbf{E}[Y^2] < \infty$. Then, by the Cauchy–Schwarz inequality, we have $\mathbf{E}[|XY|] < \infty$. Thus, using the tower property, we infer $\mathbf{E}[XY] = \mathbf{E}[\mathbf{E}[X\,|\,\mathcal{F}]Y]$ and $\mathbf{E}\big[X\mathbf{E}[X\,|\,\mathcal{F}]\big] = \mathbf{E}\big[\mathbf{E}[X\mathbf{E}[X\,|\,\mathcal{F}]\,|\,\mathcal{F}]\big] = \mathbf{E}\big[\mathbf{E}[X\,|\,\mathcal{F}]^2\big]$. Summing up, we have

$$
\mathbf{E}\big[(X - Y)^2\big] - \mathbf{E}\Big[\big(X - \mathbf{E}[X\,|\,\mathcal{F}]\big)^2\Big]
$$

$$
= \mathbf{E}\Big[X^2 - 2XY + Y^2 - X^2 + 2X\mathbf{E}[X\,|\,\mathcal{F}] - \mathbf{E}[X\,|\,\mathcal{F}]^2\Big]
$$

$$
= \mathbf{E}\Big[Y^2 - 2Y\,\mathbf{E}[X\,|\,\mathcal{F}] + \mathbf{E}[X\,|\,\mathcal{F}]^2\Big]
$$

$$
= \mathbf{E}\Big[\big(Y - \mathbf{E}[X\,|\,\mathcal{F}]\big)^2\Big] \geq 0.
$$

For the case $\mathbf{E}[\mathbf{E}[X\,|\,\mathcal{F}]^2] < \infty$, we are done. Hence, it suffices to show that this condition follows from the assumption $\mathbf{E}[X^2] < \infty$. For $N \in \mathbb{N}$, define the truncated random variables $|X| \wedge N$. Clearly, we have $\mathbf{E}[\mathbf{E}[|X| \wedge N\,|\,\mathcal{F}]^2] \leq N^2$. By what we have shown already (with X replaced by $|X| \wedge N$ and with $Y = 0 \in \mathcal{L}^2(\Omega, \mathcal{F}, \mathbf{P})$), and using the elementary inequality $a^2 \leq 2(a - b)^2 + 2b^2$, $a, b \in \mathbb{R}$, we infer

$$
\mathbf{E}\Big[\mathbf{E}[|X| \wedge N\,|\,\mathcal{F}]^2\Big] \leq 2\mathbf{E}\Big[\big((|X| \wedge N) - \mathbf{E}[|X| \wedge N\,|\,\mathcal{F}]\big)^2\Big] + 2\mathbf{E}\big[(|X| \wedge N)^2\big]
$$

$$
\leq 4\mathbf{E}\big[(|X| \wedge N)^2\big] \leq 4\mathbf{E}[X^2].
$$

By Theorem 8.14(ii) and (viii), we get $\mathbf{E}[|X| \wedge N\,|\,\mathcal{F}] \uparrow \mathbf{E}[|X|\,|\,\mathcal{F}]$ for $N \to \infty$. By the triangle inequality (Theorem 8.14(v)) and the monotone convergence theorem (Theorem 4.20), we conclude

$$
\mathbf{E}\big[\mathbf{E}[X\,|\,\mathcal{F}]^2\big] \leq \mathbf{E}\big[\mathbf{E}[|X|\,|\,\mathcal{F}]^2\big] = \lim_{N \to \infty} \mathbf{E}\big[\mathbf{E}[|X| \wedge N\,|\,\mathcal{F}]^2\big] \leq 4\mathbf{E}[X^2] < \infty.
$$

This completes the proof.   □

Example 8.18 Let $X, Y \in \mathcal{L}^1(\mathbf{P})$ be independent. Then

$$
\mathbf{E}[X + Y\,|\,Y] = \mathbf{E}[X\,|\,Y] + \mathbf{E}[Y\,|\,Y] = \mathbf{E}[X] + Y. \quad \Diamond
$$

Example 8.19 Let $X_1, \ldots, X_N$ be independent with $E[X_i] = 0$, $i = 1, \ldots, N$. For $n = 1, \ldots, N$, define $\mathcal{F}_n := \sigma(X_1, \ldots, X_n)$ and $S_n := X_1 + \ldots + X_n$. Then, for $n \geq m$,

$$
\begin{aligned}
E[S_n \mid \mathcal{F}_m] &= E[X_1 \mid \mathcal{F}_m] + \ldots + E[X_n \mid \mathcal{F}_m] \\
&= X_1 + \ldots + X_m + E[X_{m+1}] + \ldots + E[X_n] \\
&= S_m.
\end{aligned}
$$

By Theorem 8.14(iv), since $\sigma(S_m) \subset \mathcal{F}_m$, we have

$$
E[S_n \mid S_m] = E\big[E[S_n \mid \mathcal{F}_m] \mid S_m\big] = E[S_m \mid S_m] = S_m. \quad \Diamond
$$

Next we show Jensen's inequality for conditional expectations.

Theorem 8.20 (Jensen's inequality) *Let $I \subset \mathbb{R}$ be an interval, let $\varphi : I \to \mathbb{R}$ be convex and let X be an I-valued random variable on $(\Omega, \mathcal{A}, \mathbf{P})$. Further, let $E[|X|] < \infty$ and let $\mathcal{F} \subset \mathcal{A}$ be a σ-algebra. Then*

$$
\infty \geq E[\varphi(X) \mid \mathcal{F}] \geq \varphi(E[X \mid \mathcal{F}]).
$$

Proof For the existence of $E[\varphi(X) \mid \mathcal{F}]$ with values in $(-\infty, \infty]$ note that $\varphi(X)^- \in \mathcal{L}^1(\mathbf{P})$ and see Remark 8.16. By Exercise 8.2.2, we have $E[X \mid \mathcal{F}] \in I$ a.s., hence $\varphi(E[X \mid \mathcal{F}])$ is well-defined.

(Recall from Definition 1.68 the jargon words "almost surely on A".) Note that $X = E[X \mid \mathcal{F}]$ on the event $\{E[X \mid \mathcal{F}]$ is a boundary point of $I\}$; hence here the claim is trivial. Indeed, without loss of generality, assume 0 is the left boundary of I and $A := \{E[X \mid \mathcal{F}] = 0\}$. As X assumes values in $I \subset [0, \infty)$, we have $0 \leq E[X \mathbb{1}_A] = E[E[X \mid \mathcal{F}] \mathbb{1}_A] = 0$; hence $X \mathbb{1}_A = 0$. The case of a right boundary point is similar.

Hence now consider the event $B := \{E[X \mid \mathcal{F}]$ is an interior point of $I\}$. For every interior point $x \in I$, let $D^+\varphi(x)$ be the maximal slope of a tangent of φ at x; i.e., the maximal number t with $\varphi(y) \geq (y - x)t + \varphi(x)$ for all $y \in I$ (see Theorem 7.7).

For each $x \in I^\circ$, there exists a $\mathbf{P}$-null set N_x such that, for every $\omega \in B \setminus N_x$, we have

$$
\begin{aligned}
E[\varphi(X) \mid \mathcal{F}](\omega) &\geq \varphi(x) + E[D^+\varphi(x)(X - x) \mid \mathcal{F}](\omega) \\
&= \varphi(x) + D^+\varphi(x)(E[X \mid \mathcal{F}](\omega) - x) =: \psi_\omega(x).
\end{aligned} \tag{8.8}
$$

Let $V := \mathbb{Q} \cap I^\circ$. Then $N := \bigcup_{x \in V} N_x$ is a $\mathbf{P}$-null set and (8.8) holds for every $\omega \in B \setminus N$ and every $x \in V$.

The map $x \mapsto D^+\varphi(x)$ is right continuous (by Theorem 7.7(iv)). Therefore $x \mapsto \psi_\omega(x)$ is also right continuous. Hence, for every $\omega \in B \setminus N$, we have

$$\varphi\big(\mathbf{E}[X|\mathcal{F}](\omega)\big) = \psi_\omega\big(\mathbf{E}[X|\mathcal{F}](\omega)\big)$$

$$\leq \sup_{x \in I^\circ} \psi_\omega(x) = \sup_{x \in V} \psi_\omega(x) \leq \mathbf{E}\big[\varphi(X)|\mathcal{F}\big](\omega). \qquad (8.9)$$

$\square$

Corollary 8.21 *Let $p \in [1, \infty]$ and let $\mathcal{F} \subset \mathcal{A}$ be a sub-σ-algebra. Then the map*

$$\mathcal{L}^p(\Omega, \mathcal{A}, \mathbf{P}) \to \mathcal{L}^p(\Omega, \mathcal{F}, \mathbf{P}), \qquad X \mapsto \mathbf{E}[X|\mathcal{F}],$$

is a contraction (that is, $\|\mathbf{E}[X|\mathcal{F}]\|_p \leq \|X\|_p$) and thus continuous. Hence, for $X, X_1, X_2, \ldots \in \mathcal{L}^p(\Omega, \mathcal{A}, \mathbf{P})$ with $\|X_n - X\|_p \xrightarrow{n \to \infty} 0$,

$$\big\|\mathbf{E}[X_n|\mathcal{F}] - \mathbf{E}[X|\mathcal{F}]\big\|_p \xrightarrow{n \to \infty} 0.$$

Proof For $p \in [1, \infty)$, use Jensen's inequality with $\varphi(x) = |x|^p$. For $p = \infty$, note that $|\mathbf{E}[X|\mathcal{F}]| \leq \mathbf{E}[|X||\mathcal{F}] \leq \mathbf{E}[\|X\|_\infty|\mathcal{F}] = \|X\|_\infty$. $\square$

Corollary 8.22 *Let $(X_i, i \in I)$ be uniformly integrable and let $(\mathcal{F}_j, j \in J)$ be a family of sub-σ-algebras of $\mathcal{A}$. Define $X_{i,j} := \mathbf{E}[X_i|\mathcal{F}_j]$. Then $(X_{i,j}, (i, j) \in I \times J)$ is uniformly integrable. In particular, for $X \in \mathcal{L}^1(\mathbf{P})$, the family $(\mathbf{E}[X|\mathcal{F}_j], j \in J)$ is uniformly integrable.*

Proof By Theorem 6.19, there exists a monotone increasing convex function f with the property that $f(x)/x \to \infty$, $x \to \infty$ and $L := \sup_{i \in I} \mathbf{E}[f(|X_i|)] < \infty$. Then $x \mapsto f(|x|)$ is convex; hence, by Jensen's inequality,

$$\mathbf{E}\big[f(|X_{i,j}|)\big] = \mathbf{E}\big[f\big(|\mathbf{E}[X_i|\mathcal{F}_j]|\big)\big] \leq L < \infty.$$

Thus $(X_{i,j}, (i, j) \in I \times J)$ is uniformly integrable by Theorem 6.19. $\square$

Example 8.23 Let μ and ν be finite measures with $\nu \ll \mu$. Let $f = d\nu/d\mu$ be the Radon–Nikodym derivative and let $I = \{\mathcal{F} \subset \mathcal{A} : \mathcal{F} \text{ is a } \sigma\text{-algebra}\}$. Consider the measures $\mu\big|_\mathcal{F}$ and $\nu\big|_\mathcal{F}$ that are restricted to $\mathcal{F}$. Then $\nu\big|_\mathbb{C}\mathcal{F} \ll \mu\big|_\mathbb{C}\mathcal{F}$ (since in $\mathcal{F}$ there are fewer μ-null sets); hence the Radon–Nikodym derivative $f_\mathcal{F} := d\nu\big|_\mathbb{C}\mathcal{F}/d\mu\big|_\mathbb{C}\mathcal{F}$ exists. Then $(f_\mathcal{F} : \mathcal{F} \in I)$ is uniformly integrable (with respect to μ). (For finite σ-algebras $\mathcal{F}$, this was shown in Example 7.39.) Indeed, let $\mathbf{P} = \mu/\mu(\Omega)$ and $\mathbf{Q} = \nu/\mu(\Omega)$. Then $f_\mathcal{F} = d\mathbf{Q}\big|_\mathbb{C}\mathcal{F}/d\mathbf{P}\big|_\mathbb{C}\mathcal{F}$. For any $F \in \mathcal{F}$, we thus have $\mathbf{E}[f_\mathcal{F} \mathbb{1}_F] = \int_F f_\mathcal{F} \, d\mathbf{P} = \mathbf{Q}(F) = \int_F f \, d\mathbf{P} = \mathbf{E}[f \mathbb{1}_F]$; hence $f_\mathcal{F} = \mathbf{E}[f|\mathcal{F}]$. By the preceding corollary, $(f_\mathcal{F} : \mathcal{F} \in I)$ is uniformly integrable with respect to $\mathbf{P}$ and thus also with respect to μ. $\Diamond$

Takeaways The conditional expectation of a random variable X given a σ-algebra $\mathcal{F}$ is the best prediction on X that can be made given the information coded in $\mathcal{F}$ (at least if X has a second moment). On the technical side, the conditional expectation is constructed via the Radon-Nikodym theorem. It shares the main properties of ordinary expectations (linearity, triangle inequality, monotone and dominated convergence, Jensen's inequality) and in addition has the so-called tower property.

Exercise 8.2.1 Show the assertions of Remark 8.16. ♣

Exercise 8.2.2 Let $I \subset \mathbb{R}$ be an arbitrary interval and let $X \in \mathcal{L}^1(\Omega, \mathcal{A}, \mathbf{P})$ be a random variable such that $X \in I$ a.s. For $\mathcal{F} \subset \mathcal{A}$, show that $\mathbf{E}[X|\mathcal{F}] \in I$ a.s. Is this statement still true if we require only $X^- \in \mathcal{L}^1(\Omega, \mathcal{A}, \mathbf{P})$ instead of $X \in \mathcal{L}^1(\Omega, \mathcal{A}, \mathbf{P})$? ♣

Exercise 8.2.3 (Bayes' formula) Let $A \in \mathcal{A}$ and $B \in \mathcal{F} \subset \mathcal{A}$. Show that

$$\mathbf{P}[B|A] = \frac{\int_B \mathbf{P}[A|\mathcal{F}] \, d\mathbf{P}}{\int \mathbf{P}[A|\mathcal{F}] \, d\mathbf{P}}.$$

If $\mathcal{F}$ is generated by pairwise disjoint sets $B_1, B_2, \ldots$, then this is exactly Bayes' formula of Theorem 8.7. ♣

Exercise 8.2.4 Give an example for $\mathbf{E}[\mathbf{E}[X|\mathcal{F}]|\mathcal{G}] \neq \mathbf{E}[\mathbf{E}[X|\mathcal{G}]|\mathcal{F}]$. ♣

Exercise 8.2.5 Show the conditional Markov inequality: For monotone increasing $f : [0, \infty) \to [0, \infty)$ and $\varepsilon > 0$ with $f(\varepsilon) > 0$,

$$\mathbf{P}\big[|X| \geq \varepsilon |\mathcal{F}\big] \leq \frac{\mathbf{E}\big[f(|X|)\big|\mathcal{F}\big]}{f(\varepsilon)}. \quad ♣$$

Exercise 8.2.6 Show the conditional Cauchy–Schwarz inequality: For square integrable random variables X, Y,

$$\mathbf{E}[XY|\mathcal{F}]^2 \leq \mathbf{E}[X^2|\mathcal{F}] \, \mathbf{E}[Y^2|\mathcal{F}]. \quad ♣$$

Exercise 8.2.7 Let $X_1, \ldots, X_n$ be integrable i.i.d. random variables. Let $S_n = X_1 + \ldots + X_n$. Show that

$$\mathbf{E}[X_i | S_n] = \frac{1}{n} S_n \quad \text{for every } i = 1, \ldots, n. \quad ♣$$

Exercise 8.2.8 Let X_1 and X_2 be independent and exponentially distributed with parameter $\theta > 0$. Compute $\mathbf{E}[X_1 \wedge X_2|X_1]$. ♣

Exercise 8.2.9 Let X and Y be real random variables with joint density f and let $h : \mathbb{R} \to \mathbb{R}$ be measurable with $\mathbf{E}[|h(X)|] < \infty$. Denote by λ the Lebesgue measure on $\mathbb{R}$.

(i) Show that almost surely

$$\mathbf{E}[h(X)|Y] = \frac{\int h(x) f(x, Y) \, \lambda(dx)}{\int f(x, Y) \, \lambda(dx)}.$$

(ii) Let X and Y be independent and $\exp_\theta$-distributed for some $\theta > 0$. Compute $\mathbf{E}[X \mid X + Y]$ and $\mathbf{P}[X \le x \mid X + Y]$ for $x \ge 0$. ♣

8.3 Regular Conditional Distribution

Let X be a random variable with values in a measurable space $(E, \mathcal{E})$. With our machinery, so far we can define the conditional probability $\mathbf{P}[A \mid X]$ for *fixed* $A \in \mathcal{A}$ only. However, we would like to define *for every* $x \in E$ a probability measure $\mathbf{P}[\,\cdot\, | X = x]$ such that for any $A \in \mathcal{A}$, we have $\mathbf{P}[A|X] = \mathbf{P}[A|X = x]$ on $\{X = x\}$. In this section, we show how to do this.

For example, we are interested in a two-stage random experiment. At the first stage, we manipulate a coin *at random* such that the probability of a success (i.e., "head") is X. At the second stage, we toss the coin n times independently with outcomes $Y_1, \ldots, Y_n$. Hence the "conditional distribution of $(Y_1, \ldots, Y_n)$ given $\{X = x\}$" should be $(\mathrm{Ber}_x)^{\otimes n}$.

Let X be as above and let Z be a $\sigma(X)$-measurable real random variable. By the factorization lemma (Corollary 1.97 with $f = X$ and $g = Z$), there is a map $\varphi : E \to \mathbb{R}$ such that

$$\varphi \text{ is } \mathcal{E} - \mathcal{B}(\mathbb{R})\text{-measurable} \quad \text{and} \quad \varphi(X) = Z. \tag{8.10}$$

If X is surjective, then φ is determined uniquely. In this case, we denote $Z \circ X^{-1} := \varphi$ (even if the inverse map X^{-1} itself does not exist).

Definition 8.24 Let $Y \in \mathcal{L}^1(\mathbf{P})$ and $X : (\Omega, \mathcal{A}) \to (E, \mathcal{E})$. We define the conditional expectation of Y given $X = x$ by $\mathbf{E}[Y \mid X = x] := \varphi(x)$, where φ is the function from (8.10) with $Z = \mathbf{E}[Y \mid X]$.

Analogously, define $\mathbf{P}[A \mid X = x] = \mathbf{E}[\mathbb{1}_A \mid X = x]$ for $A \in \mathcal{A}$.

For a fixed set $B \in \mathcal{A}$ with $\mathbf{P}[B] > 0$, the conditional probability $\mathbf{P}[\,\cdot\, | B]$ is a probability measure. Is this true also for $\mathbf{P}[\,\cdot\, | X = x]$? The question is a bit tricky since for every given $A \in \mathcal{A}$, the expression $\mathbf{P}[A \mid X = x]$ is defined for almost all x only; that is, up to x in a null set that may, however, depend on A. Since there are uncountably many $A \in \mathcal{A}$ in general, we could not simply unite all the exceptional sets for any A. However, if the σ-algebra $\mathcal{A}$ can be approximated by countably many A sufficiently well, then there is hope.

Our first task is to give precise definitions. Then we present the theorem that justifies our hope.

Definition 8.25 (Transition kernel, Markov kernel) Let $(\Omega_1, \mathcal{A}_1)$, $(\Omega_2, \mathcal{A}_2)$ be measurable spaces. A map $\kappa : \Omega_1 \times \mathcal{A}_2 \to [0, \infty]$ is called a $(\sigma\text{-})$finite **transition kernel** (from Ω_1 to Ω_2) if:

(i) $\omega_1 \mapsto \kappa(\omega_1, A_2)$ is $\mathcal{A}_1$-measurable for any $A_2 \in \mathcal{A}_2$.
(ii) $A_2 \mapsto \kappa(\omega_1, A_2)$ is a $(\sigma\text{-})$finite measure on $(\Omega_2, \mathcal{A}_2)$ for any $\omega_1 \in \Omega_1$.

If in (ii) the measure is a probability measure for all $\omega_1 \in \Omega_1$, then κ is called a **stochastic kernel** or a **Markov kernel**. If in (ii) we also have $\kappa(\omega_1, \Omega_2) \leq 1$ for any $\omega_1 \in \Omega_1$, then κ is called sub-Markov or substochastic.

Remark 8.26 It is sufficient to check property (i) in Definition 8.25 for sets A_2 from a π-system $\mathcal{E}$ that generates $\mathcal{A}_2$ and that either contains Ω_2 or a sequence $E_n \uparrow \Omega_2$. Indeed, in this case,

$$\mathcal{D} := \big\{ A_2 \in \mathcal{A}_2 : \omega_1 \mapsto \kappa(\omega_1, A_2) \text{ is } \mathcal{A}_1\text{-measurable} \big\}$$

is a λ-system (exercise!). Since $\mathcal{E} \subset \mathcal{D}$, by the π–λ theorem (Theorem 1.19), $\mathcal{D} = \sigma(\mathcal{E}) = \mathcal{A}_2$. $\Diamond$

Example 8.27

(i)   Let $(\Omega_1, \mathcal{A}_1)$ and $(\Omega_2, \mathcal{A}_2)$ be discrete measurable spaces and let $(K_{ij})_{\substack{i \in \Omega_1 \\ j \in \Omega_2}}$
      be a matrix with nonnegative entries and finite row sums

$$K_i := \sum_{j \in \Omega_2} K_{ij} < \infty \quad \text{for } i \in \Omega_1.$$

Then we can define a finite transition kernel from Ω_1 to Ω_2 by $\kappa(i, A) = \sum_{j \in A} K_{ij}$. κ is stochastic if $K_i = 1$ for all $i \in \Omega_1$. It is substochastic if $K_i \leq 1$ for all $i \in \Omega_1$.

(ii) If μ_2 is a finite measure on Ω_2, then $\kappa(\omega_1, \cdot) \equiv \mu_2$ is a finite transition kernel.

(iii) $\kappa(x, \cdot) = \mathrm{Poi}_x$ is a stochastic kernel from $[0, \infty)$ to $\mathbb{N}_0$ (note that $x \mapsto \mathrm{Poi}_x(A)$ is continuous and hence measurable for all $A \subset \mathbb{N}_0$).

(iv) Let μ be a distribution on $\mathbb{R}^n$ and let X be a random variable with $\mathbf{P}_X = \mu$. Then $\kappa(x, \cdot) = \mathbf{P}[X + x \in \cdot] = \delta_x * \mu$ defines a stochastic kernel from $\mathbb{R}^n$ to $\mathbb{R}^n$. Indeed, the sets $(-\infty, y]$, $y \in \mathbb{R}^n$ form an $\cap$-stable generator of $\mathcal{B}(\mathbb{R}^n)$ and $x \mapsto \kappa(x, (-\infty, y]) = \mu((-\infty, y - x])$ is left continuous and hence measurable. Hence, by Remark 8.26, $x \mapsto \kappa(x, A)$ is measurable for all $A \in \mathcal{B}(\mathbb{R}^n)$. $\Diamond$

Definition 8.28 Let Y be a random variable with values in a measurable space $(E, \mathcal{E})$ and let $\mathcal{F} \subset \mathcal{A}$ be a sub-σ-algebra. A stochastic kernel $\kappa_{Y,\mathcal{F}}$ from $(\Omega, \mathcal{F})$ to $(E, \mathcal{E})$ is called a **regular conditional distribution** of Y given $\mathcal{F}$ if

$$\kappa_{Y,\mathcal{F}}(\omega, B) = \mathbf{P}[\{Y \in B\}|\mathcal{F}](\omega)$$

for **P**-almost all $\omega \in \Omega$ and for all $B \in \mathcal{E}$; that is, if

$$\int \mathbb{1}_B(Y)\,\mathbb{1}_A\,d\mathbf{P} = \int \kappa_{Y,\mathcal{F}}(\,\cdot\,, B)\,\mathbb{1}_A\,d\mathbf{P} \quad \text{for all } A \in \mathcal{F}, \ B \in \mathcal{E}. \tag{8.11}$$

Consider the special case where $\mathcal{F} = \sigma(X)$ for a random variable X (with values in an arbitrary measurable space $(E', \mathcal{E}')$). Then the stochastic kernel

$$(x, A) \mapsto \kappa_{Y,X}(x, A) = \mathbf{P}[\{Y \in A\}|X = x] = \kappa_{Y,\sigma(X)}\big(X^{-1}(x), A\big)$$

(the function from the factorization lemma with an arbitrary value for $x \notin X(\Omega)$) is called a regular conditional distribution of Y given X.

Theorem 8.29 (Regular conditional distributions in $\mathbb{R}$) *Let $Y : (\Omega, \mathcal{A}) \to \big(\mathbb{R}, \mathcal{B}(\mathbb{R})\big)$ be real-valued. Then there exists a regular conditional distribution $\kappa_{Y,\mathcal{F}}$ of Y given $\mathcal{F}$.*

Proof The strategy of the proof consists in constructing a measurable version of the distribution function of the conditional distribution of Y by first defining it for rational values (up to a null set) and then extending it to the real numbers.

For $r \in \mathbb{Q}$, let $F(r, \cdot)$ be a version of the conditional probability $\mathbf{P}[Y \in (-\infty, r]\,|\,\mathcal{F}]$. For $r \leq s$, clearly $\mathbb{1}_{\{Y \in (-\infty, r]\}} \leq \mathbb{1}_{\{Y \in (-\infty, s]\}}$. Hence, by Theorem 8.14(ii) (monotonicity of the conditional expectation), there is a null set $A_{r,s} \in \mathcal{F}$ with

$$F(r, \omega) \leq F(s, \omega) \quad \text{for all } \omega \in \Omega \setminus A_{r,s}. \tag{8.12}$$

By Theorem 8.14(viii) (dominated convergence), there are null sets $B_r \in \mathcal{F}, r \in \mathbb{Q}$, and $C \in \mathcal{F}$ such that

$$\lim_{n \to \infty} F\left(r + \frac{1}{n}, \omega\right) = F(r, \omega) \quad \text{for all } \omega \in \Omega \setminus B_r \tag{8.13}$$

as well as

$$\inf_{n \in \mathbb{N}} F(-n, \omega) = 0 \quad \text{and} \quad \sup_{n \in \mathbb{N}} F(n, \omega) = 1 \quad \text{for all } \omega \in \Omega \setminus C. \tag{8.14}$$

Let $N := \big(\bigcup_{r,s \in \mathbb{Q}} A_{r,s}\big) \cup \big(\bigcup_{r \in \mathbb{Q}} B_r\big) \cup C$. For $\omega \in \Omega \setminus N$, define

$$\tilde{F}(z, \omega) := \inf\big\{F(r, \omega) : r \in \mathbb{Q}, \ r > z\big\} \quad \text{for all } z \in \mathbb{R}.$$

By construction, $\tilde{F}(\,\cdot\,, \omega)$ is monotone increasing and right continuous. By (8.12) and (8.13), we have

$$\tilde{F}(z, \omega) = F(z, \omega) \quad \text{for all } z \in \mathbb{Q} \text{ and } \omega \in \Omega \setminus N. \tag{8.15}$$

Therefore, by (8.14), $\tilde{F}(\,\cdot\,, \omega)$ is a distribution function for any $\omega \in \Omega \setminus N$. For $\omega \in N$, define $\tilde{F}(\,\cdot\,, \omega) = F_0$, where F_0 is an arbitrary but fixed distribution function.

For any $\omega \in \Omega$, let $\kappa(\omega, \cdot)$ be the probability measure on $(\Omega, \mathcal{A})$ with distribution function $\tilde{F}(\,\cdot\,, \omega)$. Then, for $r \in \mathbb{Q}$ and $B = (-\infty, r]$,

$$\omega \mapsto \kappa(\omega, B) = F(r, \omega)\, \mathbb{1}_{N^c}(\omega) + F_0(r)\, \mathbb{1}_N(\omega) \tag{8.16}$$

is $\mathcal{F}$-measurable. Now $\{(-\infty, r], \ r \in \mathbb{Q}\}$ is a π-system that generates $\mathcal{B}(\mathbb{R})$. By Remark 8.26, measurability holds for all $B \in \mathcal{B}(\mathbb{R})$ and hence κ is identified as a stochastic kernel.

We still have to show that κ is a version of the conditional distribution. For $A \in \mathcal{F}$, $r \in \mathbb{Q}$ and $B = (-\infty, r]$, by (8.16),

$$\int_A \kappa(\omega, B)\, \mathbf{P}[d\omega] \; = \; \int_A \mathbf{P}\big[Y \in B \,\big|\, \mathcal{F}\big]\, d\mathbf{P} \; = \; \mathbf{P}\big[A \cap \{Y \in B\}\big].$$

As functions of B, both sides are finite measures on $\mathcal{B}(\mathbb{R})$ that coincide on the $\cap$-stable generator $\{(-\infty, r], \ r \in \mathbb{Q}\}$. By the uniqueness theorem (Lemma 1.42), we thus have equality for all $B \in \mathcal{B}(\mathbb{R})$. Hence $\mathbf{P}$-a.s. $\kappa(\,\cdot\,, B) = \mathbf{P}[Y \in B \,|\, \mathcal{F}]$ and thus $\kappa = \kappa_{Y, \mathcal{F}}$. $\qquad\qquad\qquad\qquad\qquad\qquad\qquad\qquad\qquad\qquad\qquad\qquad\qquad\square$

Example 8.30 Let Z_1, Z_2 be independent Poisson random variables with parameters $\lambda_1, \lambda_2 \geq 0$. One can show (exercise!) that (with $Y = Z_1$ and $X = Z_1 + Z_2$)

$$\mathbf{P}[Z_1 = k \,|\, Z_1 + Z_2 = n] = b_{n,p}(k) \quad \text{for } k = 0, \ldots, n,$$

where $p = \frac{\lambda_1}{\lambda_1 + \lambda_2}$. $\Diamond$

This example could still be treated by elementary means. The full strength of the result is displayed in the following examples.

Example 8.31 Let X and Y be real random variables with joint density f (with respect to Lebesgue measure λ^2 on $\mathbb{R}^2$). For $x \in \mathbb{R}$, define

$$f_X(x) = \int_{\mathbb{R}} f(x, y)\, \lambda(dy).$$

Clearly, $f_X(x) > 0$ for $\mathbf{P}_X$-a.a. $x \in \mathbb{R}$ and f_X^{-1} is the density of the absolutely continuous part of the Lebesgue measure λ with respect to $\mathbf{P}_X$. The regular conditional distribution of Y given X has density

$$\frac{\mathbf{P}[Y \in dy \,|\, X = x]}{dy} = f_{Y|X}(x, y) := \frac{f(x, y)}{f_X(x)} \quad \text{for } \mathbf{P}_X[dx]\text{-a.a. } x \in \mathbb{R}.$$
(8.17)

Indeed, by Fubini's theorem (Theorem 14.19), the map $x \mapsto \int_B f_{Y|X}(x, y)\, \lambda(dy)$ is measurable for all $B \in \mathcal{B}(\mathbb{R})$ and for $A, B \in \mathcal{B}(\mathbb{R})$, we have

$$\int_A \mathbf{P}[X \in dx] \int_B f_{Y|X}(x, y)\, \lambda(dy)$$
$$= \int_A \mathbf{P}[X \in dx]\, f_X(x)^{-1} \int_B f(x, y)\, \lambda(dy)$$
$$= \int_A \lambda(dx) \int_B f(x, y)\, \lambda(dy)$$
$$= \int_{A \times B} f\, d\lambda^2 = \mathbf{P}[X \in A,\, Y \in B]. \quad \Diamond$$

Example 8.32 Let $\mu_1, \mu_2 \in \mathbb{R}$, $\sigma_1, \sigma_2 > 0$ and let Z_1, Z_2 be independent and $\mathcal{N}_{\mu_i, \sigma_i^2}$-distributed ($i = 1, 2$). Then there exists a regular conditional distribution

$$\mathbf{P}[Z_1 \in \cdot \,|\, Z_1 + Z_2 = x] \quad \text{for } x \in \mathbb{R}.$$

If we define $X = Z_1 + Z_2$ and $Y = Z_1$, then $(X, Y) \sim \mathcal{N}_{\mu, \Sigma}$ is bivariate normally distributed with covariance matrix $\Sigma := \begin{pmatrix} \sigma_1^2 + \sigma_2^2 & \sigma_1^2 \\ \sigma_1^2 & \sigma_1^2 \end{pmatrix}$ and with $\mu := \begin{pmatrix} \mu_1 + \mu_2 \\ \mu_1 \end{pmatrix}$. Note that

$$\Sigma^{-1} = (\sigma_1^2 \sigma_2^2)^{-1} \begin{pmatrix} \sigma_1^2 & -\sigma_1^2 \\ -\sigma_1^2 & \sigma_1^2 + \sigma_2^2 \end{pmatrix} = (\sigma_1^2 \sigma_2^2)^{-1} B^T B,$$

where $B = \begin{pmatrix} \sigma_1 & -\sigma_1 \\ 0 & \sigma_2 \end{pmatrix}$. Hence (X, Y) has the density (see Example 1.105(ix))

$$
\begin{aligned}
f(x, y) &= \det(2\pi\ \Sigma)^{-1/2} \exp\left(-\frac{1}{2\sigma_1^2\sigma_2^2} \left\| B \begin{pmatrix} x - (\mu_1 + \mu_2) \\ y - \mu_1 \end{pmatrix} \right\|^2 \right) \\
&= \left(4\pi^2\sigma_1^2\sigma_2^2\right)^{-1/2} \exp\left(-\frac{\sigma_1^2(y - (x - \mu_2))^2 + \sigma_2^2(y - \mu_1)^2}{2\sigma_1^2\sigma_2^2}\right) \\
&= C_x \exp\left(-(y - \mu_x)^2/2\sigma_x^2\right).
\end{aligned}
$$

Here C_x is a normalising constant and

$$
\mu_x = \mu_1 + \frac{\sigma_1^2}{\sigma_1^2 + \sigma_2^2}(x - \mu_1 - \mu_2) \quad \text{and} \quad \sigma_x^2 = \frac{\sigma_1^2\sigma_2^2}{\sigma_1^2 + \sigma_2^2}.
$$

By (8.17), $\mathbf{P}[Z_1 \in \cdot\,|\,Z_1 + Z_2 = x]$ has the density

$$
y \mapsto f_{Y|X}(x, y) = \frac{C_x}{f_X(x)} \exp\left(-\frac{(y - \mu_x)^2}{2\sigma_x^2}\right),
$$

hence

$$
\mathbf{P}[Z_1 \in \cdot\,|\,Z_1 + Z_2 = x] = \mathcal{N}_{\mu_x, \sigma_x^2} \quad \text{for almost all } x \in \mathbb{R}. \quad \Diamond
$$

Example 8.33 If X and Y are independent real random variables, then for $\mathbf{P}_X$-almost all $x \in \mathbb{R}$

$$
\mathbf{P}[X + Y \in \cdot\,|\,X = x] = \delta_x * \mathbf{P}_Y. \quad \Diamond
$$

The situation is not completely satisfying as we have made the very restrictive assumption that Y is real-valued. Originally we were also interested in the situation where Y takes values in $\mathbb{R}^n$ or in even more general spaces. We now extend the result to a larger class of ranges for Y.

Definition 8.34 Two measurable spaces $(E, \mathcal{E})$ and $(E', \mathcal{E}')$ are called **isomorphic** if there exists a bijective map $\varphi : E \to E'$ such that φ is $\mathcal{E} - \mathcal{E}'$-measurable and the inverse map φ^{-1} is $\mathcal{E}' - \mathcal{E}$-measurable. Then we say that φ is an isomorphism of measurable spaces. If in addition μ and μ' are measures on $(E, \mathcal{E})$ and $(E', \mathcal{E}')$ and if $\mu' = \mu \circ \varphi^{-1}$, then φ is an isomorphism of measure spaces, and the measure spaces $(E, \mathcal{E}, \mu)$ and $(E', \mathcal{E}', \mu')$ are called isomorphic.

Definition 8.35 A measurable space $(E, \mathcal{E})$ is called a **Borel space** if there exists a Borel set $B \in \mathcal{B}(\mathbb{R})$ such that $(E, \mathcal{E})$ and $(B, \mathcal{B}(B))$ are isomorphic.

A separable topological space whose topology is induced by a complete metric is called a **Polish space**. In particular, $\mathbb{R}^d$, $\mathbb{Z}^d$, $\mathbb{R}^{\mathbb{N}}$, $(C([0, 1]), \| \cdot \|_\infty)$ and so forth are Polish. Closed subsets of Polish spaces are again Polish. We come back to Polish spaces in the context of convergence of measures in Chap. 13. Without proof, we present the following topological result (see, e.g., [37, Theorem 13.1.1]).

Theorem 8.36 *Let E be a Polish space with Borel σ-algebra $\mathcal{E}$. Then $(E, \mathcal{E})$ is a Borel space.*

Theorem 8.37 (Regular conditional distribution) *Let $\mathcal{F} \subset \mathcal{A}$ be a sub-σ-algebra. Let Y be a random variable with values in a Borel space $(E, \mathcal{E})$ (hence, for example, E Polish, $E = \mathbb{R}^d$, $E = \mathbb{R}^\infty$, $E = C([0, 1])$, etc.). Then there exists a regular conditional distribution $\kappa_{Y, \mathcal{F}}$ of Y given $\mathcal{F}$.*

Proof Let $B \in \mathcal{B}(\mathbb{R})$ and let $\varphi : E \to B$ be an isomorphism of measurable spaces. By Theorem 8.29, we obtain the regular conditional distribution $\kappa_{Y', \mathcal{F}}$ of the real random variable $Y' = \varphi \circ Y$. Now define $\kappa_{Y, \mathcal{F}}(\omega, A) = \kappa_{Y', \mathcal{F}}(\omega, \varphi(A))$ for $A \in \mathcal{E}$.
□

To conclude, we pick up again the example with which we started. Now we can drop the quotation marks from the statement and write it down formally. Hence, let X be uniformly distributed on $[0, 1]$. Given $X = x$, let $(Y_1, \dots, Y_n)$ be independent and Ber_x-distributed. Define $Y = (Y_1, \dots, Y_n)$. By Theorem 8.37 (with $E = \{0, 1\}^n \subset \mathbb{R}^n$), a regular conditional distribution exists:

$$\kappa_{Y, X}(x, \cdot) = \mathbf{P}[Y \in \cdot \,|\, X = x] \quad \text{for } x \in [0, 1].$$

Indeed, for almost all $x \in [0, 1]$,

$$\mathbf{P}[Y \in \cdot \,|\, X = x] = (\mathrm{Ber}_x)^{\otimes n}.$$

Theorem 8.38 *Let X be a random variable on $(\Omega, \mathcal{A}, \mathbf{P})$ with values in a Borel space $(E, \mathcal{E})$. Let $\mathcal{F} \subset \mathcal{A}$ be a σ-algebra and let $\kappa_{X, \mathcal{F}}$ be a regular conditional distribution of X given $\mathcal{F}$. Further, let $f : E \to \mathbb{R}$ be measurable and $\mathbf{E}[|f(X)|] < \infty$. Then*

$$\mathbf{E}[f(X)|\mathcal{F}](\omega) = \int f(x)\, \kappa_{X, \mathcal{F}}(\omega, dx) \quad \text{for } \mathbf{P}\text{-almost all } \omega. \tag{8.18}$$

Proof We check that the right-hand side in (8.18) has the properties of the conditional expectation.

It is enough to consider the case $f \geq 0$. By approximating f by simple functions, we see that the right-hand side in (8.18) is $\mathcal{F}$-measurable (see Lemma 14.23 for a formal argument). Hence, by Theorem 1.96, there exist sets $A_1, A_2, \dots \in \mathcal{E}$ and numbers $\alpha_1, \alpha_2, \dots \geq 0$ such that

$$g_n := \sum_{i=1}^{n} \alpha_i\, \mathbb{1}_{A_i} \xrightarrow{n \to \infty} f.$$

Now, for any $n \in \mathbb{N}$ and $B \in \mathcal{F}$,

$$
\begin{aligned}
\mathbf{E}[g_n(X) \mathbb{1}_B] &= \sum_{i=1}^{n} \alpha_i \, \mathbf{P}[\{X \in A_i\} \cap B] \\
&= \sum_{i=1}^{n} \alpha_i \int_B \mathbf{P}[\{X \in A_i\} \,|\, \mathcal{F}] \, \mathbf{P}[d\omega] \\
&= \sum_{i=1}^{n} \alpha_i \int_B \kappa_{X,\mathcal{F}}(\omega, A_i) \, \mathbf{P}[d\omega] \\
&= \int_B \sum_{i=1}^{n} \alpha_i \, \kappa_{X,\mathcal{F}}(\omega, A_i) \, \mathbf{P}[d\omega] \\
&= \int_B \left(\int g_n(x) \, \kappa_{X,\mathcal{F}}(\omega, dx) \right) \mathbf{P}[d\omega].
\end{aligned}
$$

By the monotone convergence theorem, for almost all ω, the inner integral converges to $\int f(x)\kappa_{X,\mathcal{F}}(\omega, dx)$. Applying the monotone convergence theorem once more, we get

$$
\mathbf{E}[f(X)\mathbb{1}_B] \;=\; \lim_{n\to\infty} \mathbf{E}[g_n(X)\mathbb{1}_B] \;=\; \int_B \int f(x)\,\kappa_{X,\mathcal{F}}(\omega, dx)\,\mathbf{P}[d\omega]. \qquad \square
$$

Takeaways Consider the conditional probability of some event B given a σ-algebra. If it is chosen such that as a function of B it is a probability measure (almost surely), then it is called a regular version of the conditional probabilities. The existence is nontrivial as there can be uncountably many events B and the conditional probability is defined only up to null sets. So it is an important theorem that a regular version of the conditional probabilities exists at least on Polish spaces (like $\mathbb{R}^d$).

Exercise 8.3.1 Let $(E, \mathcal{E})$ be a Borel space and let μ be an atom-free measure (that is, $\mu(\{x\}) = 0$ for any $x \in E$). Show that for any $A \in \mathcal{E}$ and any $n \in \mathbb{N}$, there exist pairwise disjoint sets $A_1, \dots, A_n \in \mathcal{E}$ with $\biguplus_{k=1}^{n} A_k = A$ and $\mu(A_k) = \mu(A)/n$ for any $k = 1, \dots, n$. ♣

Exercise 8.3.2 Let $p, q \in (1, \infty)$ with $\frac{1}{p} + \frac{1}{q} = 1$ and let $X \in \mathcal{L}^p(\mathbf{P})$ and $Y \in \mathcal{L}^q(\mu)$. Let $\mathcal{F} \subset \mathcal{A}$ be a σ-algebra. Use the preceding theorem to show the conditional version of Hölder's inequality:

$$
\mathbf{E}\big[|XY| \,\big|\, \mathcal{F}\big] \;\leq\; \mathbf{E}\big[|X|^p \,\big|\, \mathcal{F}\big]^{1/p} \, \mathbf{E}\big[|Y|^q \,\big|\, \mathcal{F}\big]^{1/q} \quad \text{almost surely.} \quad ♣
$$

Exercise 8.3.3 Assume the random variable (X, Y) is uniformly distributed on the disc $B := \{(x, y) \in \mathbb{R}^2 : x^2 + y^2 \leq 1\}$ and on $[-1, 1]^2$, respectively.

(i) In both cases, determine the conditional distribution of Y given $X = x$.
(ii) Let $R := \sqrt{X^2 + Y^2}$ and $\Theta = \arctan(Y/X)$. In both cases, determine the conditional distribution of Θ given $R = r$. ♣

Exercise 8.3.4 Let $A \subset \mathbb{R}^n$ be a Borel measurable set of finite Lebesgue measure $\lambda(A) \in (0, \infty)$ and let X be uniformly distributed on A (see Example 1.75). Let $B \subset A$ be measurable with $\lambda(B) > 0$. Show that the conditional distribution of X given $\{X \in B\}$ is the uniform distribution on B. ♣

Exercise 8.3.5 (Borel's paradox) Consider the Earth as a ball (as widely accepted nowadays). Let X be a random point that is uniformly distributed on the surface. Let Θ be the longitude and let Φ be the latitude of X. A little differently from the usual convention, assume that Θ takes values in $[0, \pi)$ and Φ in $[-\pi, \pi)$. Hence, for fixed Θ, a complete great circle is described when Φ runs through its domain. Now, given Θ, is Φ uniformly distributed on $[-\pi, \pi)$? One could conjecture that any point on the great circle is equally likely. However, this is not the case! If we thicken the great circle slightly such that its longitudes range from Θ to $\Theta + \varepsilon$ (for a small ε), on the equator it is thicker (measured in meters) than at the poles. If we let $\varepsilon \to 0$, intuitively we should get the conditional probabilities as proportional to the thickness (in metres).

(i) Show that $\mathbf{P}[\{\Phi \in \cdot\}|\Theta = \theta]$ for almost all θ has the density $\frac{1}{4}|\cos(\phi)|$ for $\phi \in [-\pi, \pi)$.
(ii) Show that $\mathbf{P}[\{\Theta \in \cdot\}|\Phi = \phi] = \mathcal{U}_{[0,\pi)}$ for almost all ϕ.

Hint: Show that Θ and Φ are independent, and compute the distributions of Θ and Φ. ♣

Exercise 8.3.6 (Rejection sampling for generating random variables) Let E be a countable set and let P and Q be probability measures on E. Assume there is a $c > 0$ with

$$f(e) := \frac{Q(\{e\})}{P(\{e\})} \leq c \quad \text{for all } e \in E \text{ with } P(\{e\}) > 0.$$

Let $X_1, X_2, \ldots$ be independent random variables with distribution P. Let $U_1, U_2, \ldots$ be i.i.d. random variables that are independent of $X_1, X_2, \ldots$ and that are uniformly distributed on $[0, 1]$. Let N be the smallest (random) nonnegative integer n such that $U_n \leq f(X_n)/c$ and define $Y := X_N$.
   Show that Y has distribution Q.

Remark This method for generating random variables with a given distribution Q is called *rejection sampling*, as it can also be described as follows. The random variable X_1 is a proposal for the value of Y. This proposal is accepted with probability $f(X_1)/c$ and is rejected otherwise. If the first proposal is rejected, the game starts afresh with proposal X_2 and so on. ♣

Exercise 8.3.7 Let E be a Polish space and let $P, Q \in \mathcal{M}_1(\mathbb{R})$. Let $c > 0$ with $f := \frac{dQ}{dP} \leq c$ P-almost surely. Show the statement analogous to Exercise 8.3.6. ♣

Exercise 8.3.8 Show that $(\mathbb{R}, \mathcal{B}(\mathbb{R}))$ and $(\mathbb{R}^n, \mathcal{B}(\mathbb{R}^n))$ are isomorphic. Conclude that every Borel set $B \in \mathcal{B}(\mathbb{R}^n)$ is a Borel space. ♣

Chapter 9
Martingales

One of the most important concepts of modern probability theory is the martingale, which formalizes the notion of a fair game. In this chapter, we first lay the foundations for the treatment of general stochastic processes. We then introduce martingales and the discrete stochastic integral. We close with an application to a model from mathematical finance.

9.1 Processes, Filtrations, Stopping Times

We introduce the fundamental technical terms for the investigation of stochastic processes (including martingales). In order to be able to recycle the terms later in a more general context, we go for greater generality than is necessary for the treatment of martingales only.

In the following, let (E, τ) be a Polish space with Borel σ-algebra $\mathcal{E}$. Further, let $(\Omega, \mathcal{F}, \mathbf{P})$ be a probability space and let $I \subset \mathbb{R}$ be arbitrary. We are mostly interested in the cases $I = \mathbb{N}_0$, $I = \mathbb{Z}$, $I = [0, \infty)$ and I an interval.

Definition 9.1 (Stochastic process) Let $I \subset \mathbb{R}$. A family of random variables $X = (X_t,\ t \in I)$ (on $(\Omega, \mathcal{F}, \mathbf{P})$) with values in $(E, \mathcal{E})$ is called a **stochastic process** with index set (or time set) I and range E.

Remark 9.2 Sometimes families of random variables with more general index sets are called stochastic processes. We come back to this with the Poisson point process in Chap. 24. $\Diamond$

Remark 9.3 Following a certain tradition, we will often denote a stochastic process by $X = (X_t)_{t \in I}$ if we want to emphasize the "time evolution" aspect rather than the formal notion of a family of random variables. Formally, both objects are of course the same. $\Diamond$

© The Editor(s) (if applicable) and The Author(s), under exclusive license to Springer Nature Switzerland AG 2020
A. Klenke, *Probability Theory*, Universitext,
https://doi.org/10.1007/978-3-030-56402-5_9

Example 9.4 Let $I = \mathbb{N}_0$ and let $(Y_n, \ n \in \mathbb{N})$ be a family of i.i.d. $\text{Rad}_{1/2}$-random variables on a probability space $(\Omega, \mathcal{F}, \mathbf{P})$; that is, random variables with

$$\mathbf{P}[Y_n = 1] = \mathbf{P}[Y_n = -1] = \frac{1}{2}.$$

Let $E = \mathbb{Z}$ (with the discrete topology) and let

$$X_t = \sum_{n=1}^{t} Y_n \qquad \text{for all } t \in \mathbb{N}_0.$$

$(X_t, \ t \in \mathbb{N}_0)$ is called a **symmetric simple random walk** on $\mathbb{Z}$. ◊

Example 9.5 The Poisson process $X = (X_t)_{t \geq 0}$ with intensity $\alpha > 0$ (see Sect. 5.5) is a stochastic process with range $\mathbb{N}_0$. ◊

We introduce some further terms.

Definition 9.6 If X is a random variable (or a stochastic process), we write $\mathcal{L}[X] = \mathbf{P}_X$ for the distribution of X. If $\mathcal{G} \subset \mathcal{F}$ is a σ-algebra, then we write $\mathcal{L}[X|\mathcal{G}]$ for the regular conditional distribution of X given $\mathcal{G}$.

Definition 9.7 An E-valued stochastic process $X = (X_t)_{t \in I}$ is called

(i) **real-valued** if $E = \mathbb{R}$,

(ii) a process with **independent increments** if X is real-valued and for all $n \in \mathbb{N}$ and all $t_0, \ldots, t_n \in I$ with $t_0 < t_1 < \ldots < t_n$, we have that

$$(X_{t_i} - X_{t_{i-1}})_{i=1,\ldots,n} \text{ is independent,}$$

(iii) a **Gaussian process** if X is real-valued and for all $n \in \mathbb{N}$ and $t_1, \ldots, t_n \in I$,

$$(X_{t_1}, \ldots, X_{t_n}) \text{ is } n\text{-dimensional normally distributed, and}$$

(iv) **integrable** (respectively **square integrable**) if X is real-valued and $\mathbf{E}[|X_t|] < \infty$ (respectively $\mathbf{E}[(X_t)^2] < \infty$) for all $t \in I$.

Now assume that $I \subset \mathbb{R}$ is closed under addition. Then X is called

(v) **stationary** if $\mathcal{L}[(X_{s+t})_{t \in I}] = \mathcal{L}[(X_t)_{t \in I}]$ for all $s \in I$, and

(vi) a process with **stationary increments** if X is real-valued and

$$\mathcal{L}[X_{s+t+r} - X_{t+r}] = \mathcal{L}[X_{s+r} - X_r] \qquad \text{for all } r, s, t \in I.$$

(If $0 \in I$, then it is enough to consider $r = 0$.)

Example 9.8

(i) The Poisson process with intensity θ and the random walk on $\mathbb{Z}$ are processes with stationary independent increments.
(ii) If X_t, $t \in I$, are i.i.d. random variables, then $(X_t)_{t \in I}$ is stationary.
(iii) Let $(X_n)_{n \in \mathbb{Z}}$ be real-valued and stationary and let $k \in \mathbb{N}$ and $c_0, \ldots, c_k \in \mathbb{R}$. Define

$$Y_n := \sum_{i=0}^{k} c_i X_{n-i}.$$

Then $Y = (Y_n)_{n \in \mathbb{Z}}$ is a stationary process. If $c_0, \ldots, c_k \geq 0$ and $c_0 + \ldots + c_k = 1$, then Y is called the **moving average** of X (with weights $c_0, \ldots, c_k$). $\Diamond$

The following two definitions make sense also for more general index sets I that are partially ordered. However, we restrict ourselves to the case $I \subset \mathbb{R}$.

Definition 9.9 (Filtration) Let $\mathbb{F} = (\mathcal{F}_t, t \in I)$ be a family of σ-algebras with $\mathcal{F}_t \subset \mathcal{F}$ for all $t \in I$. $\mathbb{F}$ is called a **filtration** if $\mathcal{F}_s \subset \mathcal{F}_t$ for all $s, t \in I$ with $s \leq t$.

Definition 9.10 A stochastic process $X = (X_t, t \in I)$ is called **adapted** to the filtration $\mathbb{F}$ if X_t is $\mathcal{F}_t$-measurable for all $t \in I$. If $\mathcal{F}_t = \sigma(X_s, s \leq t)$ for all $t \in I$, then we denote by $\mathbb{F} = \sigma(X)$ the filtration that is generated by X.

Remark 9.11 Clearly, a stochastic process is always adapted to the filtration it generates. Hence the generated filtration is the smallest filtration to which the process is adapted. $\Diamond$

Definition 9.12 (Predictable) Let $I = \mathbb{N}_0$ or $I = \mathbb{N}$. A stochastic process $X = (X_n, n \in I)$ is called **predictable** (or **previsible**) with respect to the filtration $\mathbb{F} = (\mathcal{F}_n, n \in \mathbb{N}_0)$ if X_0 is constant (if $I = \mathbb{N}_0$) and if for every $n \in \mathbb{N}$,

$$X_n \text{ is } \mathcal{F}_{n-1}\text{-measurable.}$$

Example 9.13 Let $I = \mathbb{N}_0$ and let $Y_1, Y_2, \ldots$ be real random variables. For $n \in \mathbb{N}_0$, define $X_n := \sum_{m=1}^{n} Y_m$. Let

$$\mathcal{F}_0 = \{\emptyset, \Omega\} \quad \text{and} \quad \mathcal{F}_n = \sigma(Y_1, \ldots, Y_n) \quad \text{for } n \in \mathbb{N}.$$

Then $\mathbb{F} = (\mathcal{F}_n, n \in \mathbb{N}_0) = \sigma(Y)$ is the filtration generated by $Y = (Y_n)_{n \in \mathbb{N}}$ and X is adapted to $\mathbb{F}$; hence $\sigma(X) \subset \mathbb{F}$. Clearly, $(Y_1, \ldots, Y_n)$ is measurable with respect to $\sigma(X_1, \ldots, X_n)$; hence $\sigma(Y) \subset \sigma(X)$, and thus also $\mathbb{F} = \sigma(X)$.

Now let $\tilde{X}_n := \sum_{m=1}^{n} \mathbb{1}_{[0,\infty)}(Y_m)$. Then $\tilde{X}$ is also adapted to $\mathbb{F}$; however, in general, $\mathbb{F} \supsetneq \sigma(\tilde{X})$. $\Diamond$

Example 9.14 Let $I = \mathbb{N}_0$ and let $D_1, D_2, \ldots$ be i.i.d. $\mathrm{Rad}_{1/2}$-distributed random variables (that is, $\mathbf{P}[D_i = -1] = \mathbf{P}[D_i = 1] = \frac{1}{2}$ for all $i \in \mathbb{N}$). Let $D = (D_i)_{i \in \mathbb{N}}$ and $\mathbb{F} = \sigma(D)$. We interpret D_i as the result of a bet that gives a gain or loss of one euro for every euro we put at stake. Just before each gamble we decide how much money we bet. Let H_n be the number of euros to bet in the nth gamble. Clearly, H_n may only depend on the results of the gambles that happened earlier, but not on D_m for any $m \geq n$. To put it differently, there must be a function $F_n : \{-1, 1\}^{n-1} \rightarrow \mathbb{N}$ such that $H_n = F_n(D_1, \ldots, D_{n-1})$. (For example, for the Petersburg game (Example 4.22) we had $F_n(x_1, \ldots, x_{n-1}) = 2^{n-1} \mathbb{1}_{\{x_1 = x_2 = \ldots = x_{n-1} = 0\}}$.) Hence H is predictable. On the other hand, any predictable H has the form $H_n = F_n(D_1, \ldots, D_{n-1})$, $n \in \mathbb{N}$, for certain functions $F_n : \{-1, 1\}^{n-1} \rightarrow \mathbb{N}$. Hence any predictable H is an admissible gambling strategy. $\Diamond$

Definition 9.15 (Stopping time) A random variable τ with values in $I \cup \{\infty\}$ is called a **stopping time** (with respect to $\mathbb{F}$) if for any $t \in I$

$$\{\tau \leq t\} \in \mathcal{F}_t.$$

The idea is that $\mathcal{F}_t$ reflects the knowledge of an observer at time t. Whether or not $\{\tau \leq t\}$ is true can thus be determined on the basis of the information available at time t.

Theorem 9.16 *Let I be countable. τ is a stopping time if and only if $\{\tau = t\} \in \mathcal{F}_t$ for all $t \in I$.*

Proof This is left as an exercise! $\square$

Example 9.17 Let $I = \mathbb{N}_0$ (or, more generally, let $I \subset [0, \infty)$ be right-discrete; that is, $t < \inf I \cap (t, \infty)$ for all $t \geq 0$, and hence I in particular is countable) and let $K \subset \mathbb{R}$ be measurable. Let X be an adapted real-valued stochastic process. Consider the first time that X is in K:

$$\tau_K := \inf\{t \in I : X_t \in K\}.$$

It is intuitively clear that τ_K should be a stopping time since we can determine by observation up to time t whether $\{\tau_K \leq t\}$ occurs. Formally, we argue that $\{X_s \in K\} \in \mathcal{F}_s \subset \mathcal{F}_t$ for all $s \leq t$. Hence also the countable union of these sets is in $\mathcal{F}_t$:

$$\{\tau_K \leq t\} = \bigcup_{s \in I \cap [0,t]} \{X_s \in K\} \in \mathcal{F}_t.$$

Consider now the random time $\widetilde{\tau} := \sup\{t \in I : X_t \in K\}$ of the last visit of X to K. For a fixed time t, on the basis of previous observations, we cannot determine whether X is already in K for the last time. For this we would have to rely on prophecy. Hence, in general, $\widetilde{\tau}$ is not a stopping time. $\Diamond$

Lemma 9.18 *Let $I \subset [0, \infty)$ be closed under addition and let σ and τ be stopping times. Then:*

(i) $\sigma \vee \tau$ and $\sigma \wedge \tau$ are stopping times.
(ii) If $\sigma, \tau \geq 0$, then $\sigma + \tau$ is also a stopping time.
(iii) If $s \geq 0$, then $\tau + s$ is a stopping time. However, in general, $\tau - s$ is not.

Before we present the (simple) formal proof, we state that in particular (i) and (iii) are properties we would expect of stopping times. With (i), the interpretation is clear. For (iii), note that $\tau - s$ peeks into the future by s time units (in fact, $\{\tau - s \leq t\} \in \mathcal{F}_{t+s}$), while $\tau + s$ looks back s time units. For stopping times, however, only retrospection is allowed.

Proof

(i) For $t \in I$, we have $\{\sigma \vee \tau \leq t\} = \{\sigma \leq t\} \cap \{\tau \leq t\} \in \mathcal{F}_t$ and $\{\sigma \wedge \tau \leq t\} = \{\sigma \leq t\} \cup \{\tau \leq t\} \in \mathcal{F}_t$.

(ii) Let $t \in I$. By (9.18), $\tau \wedge t$ and $\sigma \wedge t$ are stopping times for any $t \in I$. In particular, $\{\tau \wedge t \leq s\} \in \mathcal{F}_s \subset \mathcal{F}_t$ for any $s \leq t$. On the other hand, we have $\tau \wedge t \leq s$ for $s > t$. Hence $\tau' := (\tau \wedge t) + \mathbb{1}_{\{\tau > t\}}$ and $\sigma' := (\sigma \wedge t) + \mathbb{1}_{\{\sigma > t\}}$ (and thus $\tau' + \sigma'$) are $\mathcal{F}_t$-measurable. We conclude $\{\tau + \sigma \leq t\} = \{\tau' + \sigma' \leq t\} \in \mathcal{F}_t$.

(iii) For $\tau + s$, this is a consequence of (9.18) (with the stopping time $\sigma \equiv s$). For $\tau - s$, since τ is a stopping time, we have $\{\tau - s \leq t\} = \{\tau \leq t + s\} \in \mathcal{F}_{t+s}$. However, in general, $\mathcal{F}_{t+s}$ is a strict superset of $\mathcal{F}_t$; hence $\tau - s$ is not a stopping time.

$\square$

Definition 9.19 Let τ be a stopping time. Then

$$\mathcal{F}_\tau := \left\{A \in \mathcal{F} : A \cap \{\tau \leq t\} \in \mathcal{F}_t \text{ for any } t \in I\right\}$$

is called the σ-**algebra of** τ-**past**.

Example 9.20 Let $I = \mathbb{N}_0$ (or let $I \subset [0, \infty)$ be right-discrete; compare Example 9.17) and let X be an adapted real-valued stochastic process. Let $K \in \mathbb{R}$ and let $\tau = \inf\{t : X_t \geq K\}$ be the stopping time of first entrance in $[K, \infty)$. Consider the events $A = \{\sup\{X_t : t \in I\} > K - 5\}$ and $B = \{\sup\{X_t : t \in I\} > K + 5\}$.

Clearly, $\{\tau \leq t\} \subset A$ for all $t \in I$; hence $A \cap \{\tau \leq t\} = \{\tau \leq t\} \in \mathcal{F}_t$. Thus $A \in \mathcal{F}_\tau$. However, in general, $B \notin \mathcal{F}_\tau$ since up to time τ, we cannot decide whether X will ever exceed $K + 5$. $\Diamond$

Lemma 9.21 *If σ and τ are stopping times with $\sigma \leq \tau$, then $\mathcal{F}_\sigma \subset \mathcal{F}_\tau$.*

Proof Let $A \in \mathcal{F}_\sigma$ and $t \in I$. Then $A \cap \{\sigma \leq t\} \in \mathcal{F}_t$. Now $\{\tau \leq t\} \in \mathcal{F}_t$ since τ is a stopping time. Since $\sigma \leq \tau$, we thus get

$$A \cap \{\tau \leq t\} = \left(A \cap \{\sigma \leq t\}\right) \cap \{\tau \leq t\} \in \mathcal{F}_t.$$

$\square$

Definition 9.22 If $\tau < \infty$ is a stopping time, then we define $X_\tau(\omega) := X_{\tau(\omega)}(\omega)$.

Lemma 9.23 *Let I be countable, let X be adapted and let $\tau < \infty$ be a stopping time. Then X_τ is measurable with respect to $\mathcal{F}_\tau$.*

Proof Let A be measurable and $t \in I$. Hence $\{\tau = s\} \cap X_s^{-1}(A) \in \mathcal{F}_s \subset \mathcal{F}_t$ for all $s \leq t$. Thus

$$X_\tau^{-1}(A) \cap \{\tau \leq t\} = \bigcup_{\substack{s \in I \\ s \leq t}} \left(\{\tau = s\} \cap X_s^{-1}(A) \right) \in \mathcal{F}_t.$$

$\square$

For uncountable I and for fixed ω, in general, the map $I \to E$, $t \mapsto X_t(\omega)$ is not measurable; hence neither is the composition X_τ always measurable. Here one needs assumptions on the regularity of the *paths* $t \mapsto X_t(\omega)$; for example, right continuity. We come back to this point in Chap. 21 and leave this as a warning for the time being.

Takeaways We have got acquainted to the notions stochastic process, filtration, adapted, stopping time, and σ-algebra of τ-past. These concepts form the basic vocabulary for the description of stochastic processes, in particular martingales, in the subsequent sections.

9.2   Martingales

Everyone who does not own a casino would agree without hesitation that the successive payment of gains $Y_1, Y_2, \ldots$, such that $Y_1, Y_2, \ldots$ are i.i.d. with $\mathbf{E}[Y_1] = 0$, could be considered a fair game consisting of consecutive rounds. In this case, the process X of partial sums $X_n = Y_1 + \ldots + Y_n$ is integrable and $\mathbf{E}[X_n \,|\, \mathcal{F}_m] = X_m$ if $m < n$ (where $\mathbb{F} = \sigma(X)$). We want to use this equation for the conditional expectations as the defining equation for a fair game that in the following will be called a martingale. Note that, in particular, this definition does not require that the individual payments be independent or identically distributed. This makes the notion quite a bit more flexible. The momentousness of the following concept will become manifest only gradually.

Definition 9.24 Let $(\Omega, \mathcal{F}, \mathbf{P})$ be a probability space, $I \subset \mathbb{R}$, and let $\mathbb{F}$ be a filtration. Let $X = (X_t)_{t \in I}$ be a real-valued, adapted stochastic process with $\mathbf{E}[|X_t|] < \infty$ for all $t \in I$. X is called (with respect to $\mathbb{F}$) a

martingale	if $\mathbf{E}[X_t \,|\, \mathcal{F}_s] = X_s$ for all $s, t \in I$ with $t > s$,
submartingale	if $\mathbf{E}[X_t \,|\, \mathcal{F}_s] \geq X_s$ for all $s, t \in I$ with $t > s$,
supermartingale	if $\mathbf{E}[X_t \,|\, \mathcal{F}_s] \leq X_s$ for all $s, t \in I$ with $t > s$.

Remark 9.25 Clearly, for a martingale, the map $t \mapsto \mathbf{E}[X_t]$ is constant, for submartingales it is monotone increasing and for supermartingales it is monotone decreasing. $\Diamond$

Remark 9.26 The etymology of the term *martingale* has not been resolved completely. The French *la martingale* (originally Provençal *martegalo*, named after the town *Martiques*) in equitation means "a piece of rein used in jumping and cross country riding". Sometimes the ramified shape, in particular of the *running martingale* (French *la martingale à anneaux*), is considered as emblematic for the doubling strategy in the Petersburg game.

This doubling strategy itself is the second meaning of *la martingale*. Starting here, a shift in the meaning towards the mathematical notion seems plausible. A different derivation, in contrast to the appearance, is based on the function of the rein, which is to "check the upward movement of the horse's head". Thus the notion of a martingale might first have been used for general gambling strategies (checking the movements of chance) and later for the doubling strategy in particular. $\Diamond$

Remark 9.27 If $I = \mathbb{N}$, $I = \mathbb{N}_0$ or $I = \mathbb{Z}$, then it is enough to consider at each instant s only $t = s + 1$. In fact, by the tower property of the conditional expectation (Theorem 8.14(iv)), we get

$$\mathbf{E}[X_{s+2} \,|\, \mathcal{F}_s] = \mathbf{E}\big[\mathbf{E}[X_{s+2} \,|\, \mathcal{F}_{s+1}] \,|\, \mathcal{F}_s\big].$$

Thus, if the defining equality (or inequality) holds for any time step of size one, by induction it holds for all times. $\Diamond$

Remark 9.28 If we do not explicitly mention the filtration $\mathbb{F}$, we tacitly assume that $\mathbb{F}$ is generated by X; that is, $\mathcal{F}_t = \sigma(X_s, \ s \le t)$. $\Diamond$

Remark 9.29 Let $\mathbb{F}$ and $\mathbb{F}'$ be filtrations with $\mathcal{F}_t \subset \mathcal{F}'_t$ for all t, and let X be an $\mathbb{F}'$-(sub-, super-) martingale that is adapted to $\mathbb{F}$. Then X is also a (sub-, super-) martingale with respect to the smaller filtration $\mathbb{F}$. Indeed, for $s < t$ and for the case of a submartingale,

$$\mathbf{E}[X_t \,|\, \mathcal{F}_s] = \mathbf{E}[\mathbf{E}[X_t \,|\, \mathcal{F}'_s] \,|\, \mathcal{F}_s] \ge \mathbf{E}[X_s \,|\, \mathcal{F}_s] = X_s.$$

In particular, an $\mathbb{F}$-(sub-, super-) martingale X is always a (sub-, super-) martingale with respect to its own filtration $\sigma(X)$. $\Diamond$

Reflection If X is a martingale with respect to some filtration $\mathbb{F}$, then X is adapted to any larger filtration $\mathbb{F}' \supset \mathbb{F}$ but it is not necessarily an $\mathbb{F}'$-martingale. Why? ♠

Example 9.30 Let $Y_1, \ldots, Y_N$ be independent random variables with $\mathbf{E}[Y_t] = 0$ for all $t = 1, \ldots, N$. Let $\mathcal{F}_t := \sigma(Y_1, \ldots, Y_t)$ and $X_t := \sum_{s=1}^{t} Y_s$. Then X is adapted and integrable, and $E[Y_r | \mathcal{F}_s] = 0$ for $r > s$. Hence, for $t > s$,

$$E[X_t | \mathcal{F}_s] = E[X_s | \mathcal{F}_s] + E[X_t - X_s | \mathcal{F}_s] = X_s + \sum_{r=s+1}^{t} E[Y_r | \mathcal{F}_s] = X_s.$$

Thus, X is an $\mathbb{F}$-martingale.

Similarly, X is a submartingale if $E[Y_t] \geq 0$ for all t, and a supermartingale if $E[Y_t] \leq 0$ for all t. $\lozenge$

Reflection In the previous example one might be tempted to assume that the Y_i are uncorrelated instead of independent. Why is this not enough? ♠

Example 9.31 Consider the situation of the preceding example; however, now with $E[Y_t] = 1$ and $X_t = \prod_{s=1}^{t} Y_s$ for $t \in \mathbb{N}_0$. By Theorem 5.4, $Y_1 \cdot Y_2$ is integrable. Inductively, we get $E[|X_t|] < \infty$ for all $t \in \mathbb{N}_0$. Evidently, X is adapted to $\mathbb{F}$ and for all $s \in \mathbb{N}_0$, we have

$$E[X_{s+1} | \mathcal{F}_s] = E[X_s Y_{s+1} | \mathcal{F}_s] = X_s E[Y_{s+1} | \mathcal{F}_s] = X_s.$$

Hence X is an $\mathbb{F}$-martingale. $\lozenge$

Theorem 9.32

   (i) *X is a supermartingale if and only if $(-X)$ is a submartingale.*
   (ii) *Let X and Y be martingales and let $a, b \in \mathbb{R}$. Then $(aX + bY)$ is a martingale.*
   (iii) *Let X and Y be supermartingales and $a, b \geq 0$. Then $(aX + bY)$ is a supermartingale.*
   (iv) *Let X and Y be supermartingales. Then $Z := X \wedge Y = (\min(X_t, Y_t))_{t \in I}$ is a supermartingale.*
   (v) *If $(X_t)_{t \in \mathbb{N}_0}$ is a supermartingale and $E[X_T] \geq E[X_0]$ for some $T \in \mathbb{N}_0$, then $(X_t)_{t \in \{0,\ldots,T\}}$ is a martingale. If there exists a sequence $T_N \to \infty$ with $E[X_{T_N}] \geq E[X_0]$, then X is a martingale.*

Proof **(i), (ii) and (iii)** These are evident.

   (iv) Since $|Z_t| \leq |X_t| + |Y_t|$, we have $E[|Z_t|] < \infty$ for all $t \in I$. expectation (Theorem 8.14(ii)), for $t > s$, we have $E[Z_t | \mathcal{F}_s] \leq E[Y_t | \mathcal{F}_s] \leq Y_s$ and $E[Z_t | \mathcal{F}_s] \leq E[X_t | \mathcal{F}_s] \leq X_s$; hence $E[Z_t | \mathcal{F}_s] \leq X_s \wedge Y_s = Z_s$.

   (v) For $t \leq T$, let $Y_t := E[X_T | \mathcal{F}_t]$. Then Y is a martingale and $Y_t \leq X_t$. Hence

$$E[X_0] \leq E[X_T] = E[Y_T] = E[Y_t] \leq E[X_t] \leq E[X_0].$$

(The first inequality holds by assumption.) We infer that $Y_t = X_t$ almost surely for all t and thus $(X_t)_{t \in \{0,\ldots,T\}}$ is a martingale.

Let $T_N \to \infty$ with $E[X_{T_N}] \geq E[X_0]$ for all $N \in \mathbb{N}$. Then, for any $t > s \geq 0$, there is an $N \in \mathbb{N}$ with $T_N > t$. Hence, $E[X_t | \mathcal{F}_s] = X_s$ and X is a martingale. $\square$

Remark 9.33 Many statements about supermartingales hold *mutatis mutandis* for submartingales. For example, in the preceding theorem, claim (i) holds with the words "submartingale" and "supermartingale" interchanged, claim (iv) holds for submartingales if the minimum is replaced by a maximum, and so on. We often do not give the statements both for submartingales and for supermartingales. Instead, we choose representatively one case. Note, however, that those statements that we make explicitly about martingales usually cannot be adapted easily to sub- or supermartingales (such as (ii) in the preceding theorem). ◊

Corollary 9.34 *Let X be a submartingale and $a \in \mathbb{R}$. Then $(X - a)^+$ is a submartingale.*

Proof Clearly, 0 and $Y = X - a$ are submartingales. By (iv), $(X - a)^+ = Y \vee 0$ is also a submartingale. □

Theorem 9.35 *Let X be a martingale and let $\varphi : \mathbb{R} \to \mathbb{R}$ be a convex function.*

(i) If

$$\mathbf{E}[\varphi(X_t)^+] < \infty \quad \text{for all } t \in I, \tag{9.1}$$

then $(\varphi(X_t))_{t \in I}$ is a submartingale.
(ii) If $t^* := \sup(I) \in I$, then $\mathbf{E}[\varphi(X_{t^*})^+] < \infty$ implies (9.1).
(iii) In particular, if $p \geq 1$ and $\mathbf{E}[|X_t|^p] < \infty$ for all $t \in I$, then $(|X_t|^p)_{t \in I}$ is a submartingale.

Proof

(i) We always have $\mathbf{E}[\varphi(X_t)^-] < \infty$ (Theorem 7.9); hence, by assumption, $\mathbf{E}[|\varphi(X_t)|] < \infty$ for all $t \in I$. Jensen's inequality (Theorem 8.20) then yields, for $t > s$,

$$\mathbf{E}[\varphi(X_t) \,|\, \mathcal{F}_s] \geq \varphi(\mathbf{E}[X_t \,|\, \mathcal{F}_s]) = \varphi(X_s).$$

(ii) Since φ is convex, so is $x \mapsto \varphi(x)^+$. Furthermore, by assumption, we have $\mathbf{E}[\varphi(X_{t^*})^+] < \infty$; hence Jensen's inequality implies that, for all $t \in I$,

$$\mathbf{E}[\varphi(X_t)^+] = \mathbf{E}\big[\varphi\big(\mathbf{E}[X_{t^*} \,|\, \mathcal{F}_t]\big)^+\big] \leq \mathbf{E}\big[\mathbf{E}[\varphi(X_{t^*})^+ \,|\, \mathcal{F}_t]\big] = \mathbf{E}\big[\varphi(X_{t^*})^+\big] < \infty.$$

(iii) This is evident since $x \mapsto |x|^p$ is convex.

□

Example 9.36 (See Example 9.4) Symmetric simple random walk X on $\mathbb{Z}$ is a square integrable martingale. Hence $(X_n^2)_{n \in \mathbb{N}_0}$ is a submartingale. ◊

Takeaways A martingale is a mathematical model for a fair game of many rounds. Partial sums of independent centred random variables are an important example. Submartingales are favourable games (the mean future is better than the present) and supermartingales are unfavourable games (the mean future is not as good as the present). Convex functions of martingales are submartingales.

Exercise 9.2.1 Let Y be a random variable with $\mathbf{E}[|Y|] < \infty$ and let $\mathbb{F}$ be a filtration as well as

$$X_t := \mathbf{E}[Y \,|\, \mathcal{F}_t] \quad \text{for all } t \in I.$$

Show that X is an $\mathbb{F}$-martingale. ♣

Exercise 9.2.2 Let $(X_n)_{n \in \mathbb{N}_0}$ be a predictable $\mathbb{F}$-martingale. Show that $X_n = X_0$ almost surely for all $n \in \mathbb{N}_0$. ♣

Exercise 9.2.3 Show that the claim of Theorem 9.35 continues to hold if X is only a submartingale but if φ is in addition assumed to be monotone increasing. Give an example that shows that the monotonicity of φ is essential. (Compare Corollary 9.34.) ♣

Exercise 9.2.4 (Azuma's inequality) Show the following.

(i) If X is a random variable with $|X| \leq 1$ a.s., then there is a random variable Y with values in $\{-1, +1\}$ and with $\mathbf{E}[Y \,|\, X] = X$.

(ii) For X as in (i) with $\mathbf{E}[X] = 0$, infer that (using Jensen's inequality)

$$\mathbf{E}[e^{\lambda X}] \leq \cosh(\lambda) \leq e^{\lambda^2/2} \quad \text{for all } \lambda \in \mathbb{R}.$$

(iii) If $(M_n)_{n \in \mathbb{N}_0}$ is a martingale with $M_0 = 0$ and if there is a sequence $(c_k)_{k \in \mathbb{N}}$ of nonnegative numbers with $|M_n - M_{n-1}| \leq c_n$ a.s. for all $n \in \mathbb{N}$, then

$$\mathbf{E}[e^{\lambda M_n}] \leq \exp\left(\frac{1}{2}\lambda^2 \sum_{k=1}^{n} c_k^2\right).$$

(iv) Under the assumptions of (iii), **Azuma's inequality** holds:

$$\mathbf{P}[|M_n| \geq \lambda] \leq 2 \exp\left(-\frac{\lambda^2}{2\sum_{k=1}^{n} c_k^2}\right) \quad \text{for all } \lambda \geq 0.$$

Hint: Use Markov's inequality for $f(x) = e^{\gamma x}$ and choose the optimal γ. ♣

9.3 Discrete Stochastic Integral

So far we have encountered a martingale as the process of partial sums of gains of a fair game. This game can also be the price of a stock that is traded at discrete times on a stock exchange. With this interpretation, it is particularly evident that it is natural to construct new stochastic processes by considering investment strategies for the stock. The value of the portfolio, which is the new stochastic process, changes as the stock price changes. It is the price multiplied by the number of stocks in the portfolio. In order to describe such processes formally, we introduce the following notion.

Definition 9.37 (Discrete stochastic integral)   Let $(X_n)_{n \in \mathbb{N}_0}$ be an $\mathbb{F}$-adapted real process and let $(H_n)_{n \in \mathbb{N}}$ be a real-valued and $\mathbb{F}$-predictable process. The discrete stochastic integral of H with respect to X is the stochastic process $H \cdot X$ defined by

$$(H \cdot X)_n := \sum_{m=1}^{n} H_m (X_m - X_{m-1}) \quad \text{for } n \in \mathbb{N}_0. \tag{9.2}$$

If X is a martingale, then $H \cdot X$ is also called the **martingale transform** of X.

Remark 9.38 Clearly, $H \cdot X$ is adapted to $\mathbb{F}$. ◊

Let X be a (possibly unfair) game where $X_n - X_{n-1}$ is the gain per euro in the nth round. We interpret H_n as the number of euros we bet in the nth game. H is then a **gambling strategy**. Clearly, the value of H_n has to be decided at time $n - 1$; that is, *before* the result of X_n is known. In other words, H must be predictable.

Now assume that X is a fair game (that is, a martingale) and H is **locally bounded** (that is, each H_n is bounded). Then (since $\mathbf{E}[X_{n+1} - X_n \,|\, \mathcal{F}_n] = 0$)

$$\mathbf{E}[(H \cdot X)_{n+1} \,|\, \mathcal{F}_n] = \mathbf{E}[(H \cdot X)_n + H_{n+1}(X_{n+1} - X_n) \,|\, \mathcal{F}_n]$$
$$= (H \cdot X)_n + H_{n+1}\, \mathbf{E}[X_{n+1} - X_n \,|\, \mathcal{F}_n]$$
$$= (H \cdot X)_n.$$

Thus $H \cdot X$ is a martingale. The following theorem says that the converse also holds; that is, X is a martingale if, for sufficiently many predictable processes, the stochastic integral is a martingale.

Theorem 9.39 (Stability theorem)   *Let $(X_n)_{n \in \mathbb{N}_0}$ be an adapted, real-valued stochastic process with $\mathbf{E}[|X_0|] < \infty$.*

  (i) *X is a martingale if and only if, for any locally bounded predictable process H, the stochastic integral $H \cdot X$ is a martingale.*
 (ii) *X is a submartingale (supermartingale) if and only if $H \cdot X$ is a submartingale (supermartingale) for any locally bounded predictable $H \geq 0$.*

Proof

(i) " $\Longrightarrow$ "   This has been shown in the discussion above.
    " $\Longleftarrow$ "   Fix an $n_0 \in \mathbb{N}$, and let $H_n = \mathbb{1}_{\{n=n_0\}}$. Then $(H \cdot X)_{n_0-1} = 0$; hence

$$0 = \mathbf{E}\big[(H \cdot X)_{n_0} \,\big|\, \mathcal{F}_{n_0-1}\big] = \mathbf{E}\big[X_{n_0} \,\big|\, \mathcal{F}_{n_0-1}\big] - X_{n_0-1}.$$

(ii) This is similar to (i).

$\square$

The preceding theorem says, in particular, that we cannot find any locally bounded gambling strategy that transforms a martingale (or, if we are bound to nonnegative gambling strategies, as we are in real life, a supermartingale) into a submartingale. Quite the contrary is suggested by the many invitations to play all kinds of "sure winning systems" in lotteries.

Example 9.40 (Petersburg game) We continue Example 9.14 (see also Example 4.22). Define $X_n := D_1 + \ldots + D_n$ for $n \in \mathbb{N}_0$. Then X is a martingale. The gambling strategy $H_n := 2^{n-1} \mathbb{1}_{\{D_1=D_2=\ldots=D_{n-1}=-1\}}$ for $n \in \mathbb{N}$ and $H_0 = 1$ is predictable and locally bounded. Let $S_n := \sum_{i=1}^{n} H_i D_i = (H \cdot X)_n$ be the gain after n rounds. Then S is a martingale by the preceding theorem. In particular, we get (as shown already in Example 4.22) that $\mathbf{E}[S_n] = 0$ for all $n \in \mathbb{N}$. We will come back later to the point that this superficially contrasts with $S_n \xrightarrow{n \to \infty} 1$ a.s. (see Example 11.6).

For the moment, note that the martingale $S' = (1 - S_n)_{n \in \mathbb{N}_0}$, just like the one in Example 9.31, has the structure of a product of independent random variables with expectation 1. In fact, $S'_n = \prod_{i=1}^{n}(1 - D_i)$. $\Diamond$

> **Takeaways** Assume that the price of a risky asset at discrete trading times $n = 0, 1, 2, \ldots$ is a martingale. Also assume that we follow a bounded trading strategy that cannot use future information. Then the value of our portfolio is described by a discrete stochastic integral which is again a martingale.

9.4 Discrete Martingale Representation Theorem and the CRR Model

By virtue of the stochastic integral, we have transformed a martingale X via a gambling strategy H into a new martingale $H \cdot X$. Let us change the perspective and ask: For fixed X, which are the martingales Y (with $Y_0 = 0$) that can be obtained as discrete stochastic integrals of X with a suitable gambling strategy $H = H(Y)$? Possibly all martingales Y? This is not the case, in general, as the example below indicates. However, we will see that all martingales can be represented as stochastic integrals if the increments $X_{n+1} - X_n$ can take only two values (given $X_1, \ldots, X_n$). In this case, we give a representation theorem and use it to discuss the fair price for a European call option in the stock market model of Cox–Ross–Rubinstein. This model is rather simple and describes an idealized market (no transaction costs, fractional numbers of stocks tradeable and so on). For extensive literature on stochastic aspects of mathematical finance, we refer to the textbooks [9, 42, 48, 57, 86, 102, 121] or [160].

Example 9.41   Consider the very simple martingale $X = (X_n)_{n=0,1}$ with only two time points. Let $X_0 = 0$ almost surely and $\mathbf{P}[X_1 = -1] = \mathbf{P}[X_1 = 0] = \mathbf{P}[X_1 = 1] = \frac{1}{3}$. Let $Y_0 = 0$. Further, let $Y_1 = 2$ if $X_1 = 1$ and $Y_1 = -1$ otherwise. Then Y is manifestly a $\sigma(X)$-martingale. However, there is no number H_1 such that $H_1 X_1 = Y_1$. $\Diamond$

Let $T \in \mathbb{N}$ be a fixed time. If $(Y_n)_{n=0,1,\ldots,T}$ is an $\mathbb{F}$-martingale, then $Y_n = \mathbf{E}[Y_T \mid \mathcal{F}_n]$ for all $n \leq T$. An $\mathbb{F}$-martingale Y is thus determined uniquely by the terminal values Y_T (and vice versa). Let X be a martingale. As $(H \cdot X)$ is a martingale, the representation problem for martingales is thus reduced to the problem of representing an integrable random variable $V := Y_T$ as $v_0 + (H \cdot X)_T$, where $v_0 = \mathbf{E}[Y_T]$.

We saw that, in general, this is not possible if the differences $X_{n+1} - X_n$ take three (or more) different values. Hence we now consider the case where only two values are possible. Here, at each time step, a system of two linear equations with two unknowns has to be solved. In the case where $X_{n+1} - X_n$ takes three values, the system has three equations and is thus overdetermined.

Definition 9.42 (Binary model)   A stochastic process $X_0, \ldots, X_T$ is called **binary splitting** or a **binary model** if there exist random variables $D_1, \ldots, D_T$ with values in $\{-1, +1\}$ and functions $f_n : \mathbb{R}^{n-1} \times \{-1, +1\} \to \mathbb{R}$ for $n = 1, \ldots, T$, as well as $x_0 \in \mathbb{R}$ such that $X_0 = x_0$ and

$$X_n = f_n(X_1, \ldots, X_{n-1}, D_n) \quad \text{for any } n = 1, \ldots, T.$$

By $\mathbb{F} = \sigma(X)$, we denote the filtration generated by X.

Note that X_n depends only on $X_1, \ldots, X_{n-1}$ and D_n but not on the full information inherent in the values $D_1, \ldots, D_n$. If the latter were the case, a ramification into more than two values in one time step would be possible.

Theorem 9.43 (Representation theorem)   *Let X be a binary model and let V_T be an $\mathcal{F}_T$-measurable random variable. Then there exists a bounded predictable process H and a $v_0 \in \mathbb{R}$ with $V_T = v_0 + (H \cdot X)_T$.*

Note that $\mathbb{F}$ is the filtration generated by X, not the, possibly larger, filtration generated by $D_1, \ldots, D_T$. For the latter case, the claim of the theorem would be incorrect since, loosely speaking, with H we can bet on X but not on the D_i.

Proof We show that there exist $\mathcal{F}_{T-1}$-measurable random variables V_{T-1} and H_T such that $V_T = V_{T-1} + H_T(X_T - X_{T-1})$. By a backward induction, this yields the claim.

Since V_T is $\mathcal{F}_T$-measurable, by the factorization lemma (Corollary 1.97) there exists a function $g_T : \mathbb{R}^T \to \mathbb{R}$ with $V_T = g_T(X_1, \ldots, X_T)$. Define

$$X_T^{\pm} = f_T(X_1, \ldots, X_{T-1}, \pm 1) \quad \text{and} \quad V_T^{\pm} = g_T(X_1, \ldots, X_{T-1}, X_T^{\pm}).$$

Each of these four random variables is manifestly $\mathcal{F}_{T-1}$-measurable. Hence we are looking for solutions V_{T-1} and H_T of the following system of linear equations:

$$\begin{aligned} V_{T-1} + H_T(X_T^- - X_{T-1}) = V_T^-, \\ V_{T-1} + H_T(X_T^+ - X_{T-1}) = V_T^+. \end{aligned} \tag{9.3}$$

By construction, $X_T^+ - X_T^- \neq 0$ if $V_T^+ - V_T^- \neq 0$. Hence we can solve (9.3) and get

$$H_T := \begin{cases} \frac{V_T^+ - V_T^-}{X_T^+ - X_T^-}, & \text{if } X_T^+ \neq X_T^-, \\ 0, & \text{else,} \end{cases}$$

and $V_{T-1} = V_T^+ - H_T(X_T^+ - X_{T-1}) = V_T^- - H_T(X_T^- - X_{T-1})$. $\quad\square$

We now want to interpret X as the market price of a stock and V_T as the payment of a financial derivative on X, a so-called **contingent claim** or, briefly, claim. For example, V_T could be a (European) **call option** with *maturity T* and *strike price $K \geq 0$*. In this case, we have $V_T = (X_T - K)^+$. Economically speaking, the European call gives the buyer the right (but not the obligation) to buy one stock at time T at price K (from the issuer of the option). As typically the option is exercised only if the market price at time T is larger than K (and then gives a profit of $X_T - K$ as the stock could be sold at price X_T on the market), the value of the option is indeed $V_T = (X_T - K)^+$.

At the stock exchanges, not only are stocks traded but also derivatives on stocks. Hence, what is the fair price $\pi(V_T)$ for which a trader would offer (and buy) the contingent claim V_T? If there exists a strategy H and a v_0 such that $V_T = v_0 + (H \cdot X)_T$, then the trader can sell the claim for v_0 (at time 0) and replicate the claim by building a portfolio that follows the trading strategy H. In this case, the claim V_T is called **replicable** and the strategy H is called a **hedging strategy**, or briefly a hedge. A market in which every claim can be replicated is called complete. In this sense, the binary model is a complete market.

If there was a second strategy H' and a second v_0' with $v_0' + (H' \cdot X)_T = V_T$, then, in particular, $v_0 - v_0' = ((H' - H) \cdot X)_T$. If we had $v_0 > v_0'$, then the trader could follow the strategy $H' - H$ (which gives a final payment of $V_T - V_T = 0$) and make a sure profit of $v_0 - v_0'$. In the opposite case, $v_0 < v_0'$, the strategy $H - H'$

ensures a risk-free profit. Such a risk-free profit (or **free lunch** in economic jargon) is called an **arbitrage**. It is reasonable to assume that a market gives no opportunity for an arbitrage. Hence the fair price $\pi(V_T)$ is determined uniquely once there is *one* trading strategy H and a v_0 such that $V_T = v_0 + (H \cdot X)_T$.

Until now, we have not assumed that X is a martingale. However, if X is a martingale, then $(H \cdot X)$ is a martingale with $(H \cdot X)_0 = 0$; hence clearly $\mathbf{E}[(H \cdot X)_T] = 0$. Thus

$$\pi(V_T) = v_0 = \mathbf{E}[V_T]. \tag{9.4}$$

Since, in this case, v_0 does not depend on the trading strategy and is hence unique, the market is automatically arbitrage-free. A finite market is thus arbitrage-free if and only if there exists an equivalent martingale (to be defined below). In this case, uniqueness of this martingale is equivalent to completeness of the market ("the fundamental theorem of asset pricing" by Harrison–Pliska [68]). In larger markets, equivalence holds only with a somewhat more flexible notion of arbitrage (see [30]).

Now if X is not a martingale, then in some cases, we can replace X by a different process X' that *is* a martingale and such that the distributions $\mathbf{P}_X$ and $\mathbf{P}_{X'}$ are equivalent; that is, have the same null sets. Such a process is called an *equivalent martingale*, and $\mathbf{P}_{X'}$ is called an *equivalent martingale measure*. A trading strategy that replicates V_T with respect to X also replicates V_T with respect to X'. In particular, the fair price does not change if we pass to the equivalent martingale X'. Thus we can compute $\pi(V_T)$ by applying (9.4) to the equivalent martingale.

While here this is only of interest in that it simplifies the computation of fair prices, it has an economic interpretation as a measure for the market prices that we would see if all traders were risk-neutral; that is, for traders who price a future payment by its mean value. Typically, however, traders are risk-averse and thus real market prices include a discount due to the inherent risk.

Now we consider one model in greater detail.

Definition 9.44 Let $T \in \mathbb{N}$, $a \in (-1, 0)$ and $b > 0$ as well as $p \in (0, 1)$. Further, let $D_1, \ldots, D_T$ be i.i.d. Rad_p random variables (that is, $\mathbf{P}[D_1 = 1] = 1 - \mathbf{P}[D_1 = -1] = p$). We let $X_0 = x_0 > 0$ and for $n = 1, \ldots, T$, define

$$X_n = \begin{cases} (1 + b)\, X_{n-1}, & \text{if } D_n = +1, \\[2mm] (1 + a)\, X_{n-1}, & \text{if } D_n = -1. \end{cases}$$

X is called the **multi-period binomial model** or the **Cox–Ross–Rubinstein model** (without interest returns).

As we have shown already, the CRR model is complete. Further, with the choice $p = p^* := \frac{a}{a-b}$, we can change X into a martingale. Hence the model is also arbitrage-free (for all $p \in (0, 1)$). Now we want to compute the price of a European call option $V_T := (X_T - K)^+$ explicitly. To this end, we can assume $p = p^*$.

Letting $A := \min\{i \in \mathbb{N}_0 : (1+b)^i(1+a)^{T-i}x_0 > K\}$, we get

$$\pi(V_T) = \mathbf{E}_{p^*}[V_T] = \sum_{i=0}^{T} b_{T,p^*}(\{i\}) \left[(1+b)^i(1+a)^{T-i}x_0 - K\right]^+$$

$$= x_0 \sum_{i=A}^{T} \binom{T}{i}(p^*)^i(1-p^*)^{T-i}\left[(1+b)^i(1+a)^{T-i}\right] - K \sum_{i=A}^{T} b_{T,p^*}(\{i\}).$$

If we define $p' = (1+b)p^*$, then $p' \in (0,1)$ and $1 - p' = (1-p^*)(1+a)$. We thus obtain the Cox–Ross–Rubinstein formula

$$\pi(V_T) = x_0\, b_{T,p'}(\{A,\ldots,T\}) - K\, b_{T,p^*}(\{A,\ldots,T\}). \tag{9.5}$$

This is the discrete analogue of the celebrated Black–Scholes formula for option pricing in certain time-continuous markets.

Takeaways Consider a stochastic process that can take only two values at time $t+1$ given the full history up to time t. Such a process is said to be binary splitting. The main theorem says that any function of the history up to a given time t can be represented as a discrete stochastic integral with respect to this binary splitting process. Rephrased to the language of financial markets this means: If the price of a risky asset is given by a binary splitting process than there is a hedging strategy for any contingent claim. As a by-product we get a discrete version of the celebrated Black-Scholes formula.

Chapter 10
Optional Sampling Theorems

In Chap. 9 we saw that martingales are transformed into martingales if we apply certain admissible gambling strategies. In this chapter, we establish a similar stability property for martingales that are stopped at a random time. In order also to obtain these results for submartingales and supermartingales, in the first section, we start with a decomposition theorem for adapted processes. We show the optional sampling and optional stopping theorems in the second section. The chapter finishes with the investigation of random stopping times with an infinite time horizon.

10.1 Doob Decomposition and Square Variation

Let $X = (X_n)_{n \in \mathbb{N}_0}$ be an adapted process with $\mathbf{E}[|X_n|] < \infty$ for all $n \in \mathbb{N}_0$. We will decompose X into a sum consisting of a martingale and a predictable process. To this end, for $n \in \mathbb{N}_0$, define

$$M_n := X_0 + \sum_{k=1}^{n} \left(X_k - \mathbf{E}[X_k \,|\, \mathcal{F}_{k-1}] \right) \tag{10.1}$$

and

$$A_n := \sum_{k=1}^{n} \left(\mathbf{E}[X_k \,|\, \mathcal{F}_{k-1}] - X_{k-1} \right).$$

Evidently, $X_n = M_n + A_n$. By construction, A is predictable with $A_0 = 0$, and M is a martingale since

$$\mathbf{E}[M_n - M_{n-1} \,|\, \mathcal{F}_{n-1}] = \mathbf{E}\left[X_n - \mathbf{E}[X_n \,|\, \mathcal{F}_{n-1}] \,\big|\, \mathcal{F}_{n-1}\right] = 0.$$

© The Editor(s) (if applicable) and The Author(s), under exclusive license
to Springer Nature Switzerland AG 2020
A. Klenke, *Probability Theory*, Universitext,
https://doi.org/10.1007/978-3-030-56402-5_10

Theorem 10.1 (Doob decomposition) *Let* $X = (X_n)_{n \in \mathbb{N}_0}$ *be an adapted inte-grable process. Then there exists a unique decomposition* $X = M + A$, *where* A *is predictable with* $A_0 = 0$ *and* M *is a martingale. This representation of* X *is called the* **Doob decomposition**. X *is a submartingale if and only if* A *is monotone increasing.*

Proof We only have to show uniqueness of the decomposition. Hence, let $X = M + A = M' + A'$ be two such decompositions. Then $M - M' = A' - A$ is a predictable martingale; hence (see Exercise 9.2.2) $M_n - M'_n = M_0 - M'_0 = 0$ for all $n \in \mathbb{N}_0$. $\qquad \qquad \square$

Example 10.2 Let $I = \mathbb{N}_0$ or $I = \{0, \ldots, N\}$. Let $(X_n)_{n \in I}$ be a square integrable $\mathbb{F}$-martingale (that is, $\mathbf{E}[X_n^2] < \infty$ for all $n \in I$). By Theorem 9.35, $Y := (X_n^2)_{n \in I}$ is a submartingale. Let $Y = M + A$ be the Doob decomposition of Y. Then $(X_n^2 - A_n)_{n \in I}$ is a martingale. Furthermore, $\mathbf{E}[X_{i-1} X_i \,|\, \mathcal{F}_{i-1}] = X_{i-1} \mathbf{E}[X_i \,|\, \mathcal{F}_{i-1}] = X_{i-1}^2$; hence (as in (10.1))

$$
\begin{aligned}
A_n &= \sum_{i=1}^n \left(\mathbf{E}[X_i^2 \,|\, \mathcal{F}_{i-1}] - X_{i-1}^2 \right) \\
&= \sum_{i=1}^n \left(\mathbf{E}[(X_i - X_{i-1})^2 \,|\, \mathcal{F}_{i-1}] - 2X_{i-1}^2 + 2\,\mathbf{E}[X_{i-1} X_i \,|\, \mathcal{F}_{i-1}] \right) \\
&= \sum_{i=1}^n \mathbf{E}\left[(X_i - X_{i-1})^2 \,|\, \mathcal{F}_{i-1} \right]. \quad \Diamond
\end{aligned}
\tag{10.2}
$$

Definition 10.3 Let $(X_n)_{n \in I}$ be a square integrable $\mathbb{F}$-martingale. The unique predictable process A for which $(X_n^2 - A_n)_{n \in I}$ becomes a martingale is called the **square variation process** of X and is denoted by $(\langle X \rangle_n)_{n \in I} := A$.

By the preceding example, we conclude the following theorem.

Theorem 10.4 *Let* X *be as in Definition 10.3. Then, for* $n \in \mathbb{N}_0$,

$$
\langle X \rangle_n = \sum_{i=1}^n \mathbf{E}\left[(X_i - X_{i-1})^2 \,|\, \mathcal{F}_{i-1} \right]
\tag{10.3}
$$

and

$$
\mathbf{E}[\langle X \rangle_n] = \mathbf{Var}[X_n - X_0].
\tag{10.4}
$$

Remark 10.5 If Y and A are as in Example 10.2, then A is monotone increasing since $(X_n^2)_{n \in I}$ is a submartingale (see Theorem 10.1). Therefore, A is sometimes called the **increasing process** of Y. $\Diamond$

Example 10.6 Let $Y_1, Y_2, \ldots$ be independent, square integrable, centered random variables. Then $X_n := Y_1 + \ldots + Y_n$ defines a square integrable martingale with $\langle X \rangle_n = \sum_{i=1}^{n} \mathbf{E}[Y_i^2]$. In fact, $A_n = \sum_{i=1}^{n} \mathbf{E}[Y_i^2 \mid Y_1, \ldots, Y_{i-1}] = \sum_{i=1}^{n} \mathbf{E}[Y_i^2]$ (as in Example 10.2).

Note that in order for $\langle X \rangle$ to have the simple form as in Example 10.6, it is not enough for the random variables $Y_1, Y_2, \ldots$ to be uncorrelated. ◊

Example 10.7 Let $Y_1, Y_2, \ldots$ be independent, square integrable random variables with $\mathbf{E}[Y_n] = 1$ for all $n \in \mathbb{N}$. Let $X_n := \prod_{i=1}^{n} Y_i$ for $n \in \mathbb{N}_0$. Then $X = (X_n)_{n \in \mathbb{N}_0}$ is a square integrable martingale with respect to $\mathbb{F} = \sigma(X)$ (why?) and

$$\mathbf{E}\big[(X_n - X_{n-1})^2 \mid \mathcal{F}_{n-1}\big] = \mathbf{E}\big[(Y_n - 1)^2 X_{n-1}^2 \mid \mathcal{F}_{n-1}\big] = \mathbf{Var}[Y_n]\, X_{n-1}^2.$$

Hence $\langle X \rangle_n = \sum_{i=1}^{n} \mathbf{Var}[Y_i]\, X_{i-1}^2$. We see that the square variation process can indeed be a truly random process. ◊

Example 10.8 Let $(X_n)_{n \in \mathbb{N}_0}$ be the one-dimensional symmetric simple random walk

$$X_n = \sum_{i=1}^{n} R_i \quad \text{for all } n \in \mathbb{N}_0,$$

where $R_1, R_2, R_3, \ldots$ are i.i.d. and $\sim \mathrm{Rad}_{1/2}$; that is,

$$\mathbf{P}[R_i = 1] = 1 - \mathbf{P}[R_i = -1] = \frac{1}{2}.$$

Clearly, X is a martingale and hence $|X|$ is a submartingale. Let $|X| = M + A$ be Doob's decomposition of $|X|$. Then

$$A_n = \sum_{i=1}^{n} \big(\mathbf{E}[|X_i| \mid \mathcal{F}_{i-1}] - |X_{i-1}|\big).$$

Now

$$|X_i| = \begin{cases} |X_{i-1}| + R_i, & \text{if } X_{i-1} > 0, \\ |X_{i-1}| - R_i, & \text{if } X_{i-1} < 0, \\ \qquad\quad 1, & \text{if } X_{i-1} = 0. \end{cases}$$

Therefore,

$$\mathbf{E}[|X_i| \mid \mathcal{F}_{i-1}] = \begin{cases} |X_{i-1}|, & \text{if } |X_{i-1}| \neq 0, \\ \quad 1, & \text{if } |X_{i-1}| = 0. \end{cases}$$

The process

$$A_n = \#\{i \le n - 1 : |X_i| = 0\}$$

is the so-called **local time** of X at 0. We conclude that (since $\mathbf{P}[X_{2j} = 0] = \binom{2j}{j}4^{-j}$ and $\mathbf{P}[X_{2j+1} = 0] = 0$)

$$\mathbf{E}[|X_n|] = \mathbf{E}\big[\#\{i \le n - 1 : X_i = 0\}\big]$$

$$= \sum_{i=0}^{n-1} \mathbf{P}[X_i = 0] = \sum_{j=0}^{\lfloor(n-1)/2\rfloor} \binom{2j}{j}4^{-j}. \quad \Diamond \qquad (10.5)$$

Example 10.9 We want to generalize the preceding example further. Evidently, we did not use (except in the last formula) the fact that X is a random walk. Rather, we just used the fact that the differences $(\Delta X)_n := X_n - X_{n-1}$ take only the values -1 and $+1$. Hence, now let X be a martingale with $|X_n - X_{n-1}| = 1$ almost surely for all $n \in \mathbb{N}$ and with $X_0 = x_0 \in \mathbb{Z}$ almost surely. Let $f : \mathbb{Z} \to \mathbb{R}$ be an arbitrary map. Then $Y := (f(X_n))_{n\in\mathbb{N}_0}$ is an integrable adapted process (since $|f(X_n)| \le \max_{x\in\{x_0-n,\dots,x_0+n\}} |f(x)|$). In order to compute Doob's decomposition of Y, define the first and second discrete derivatives of f:

$$f'(x) := \frac{f(x+1) - f(x-1)}{2}$$

and

$$f''(x) := f(x-1) + f(x+1) - 2f(x).$$

Further, let $F'_n := f'(X_{n-1})$ and $F''_n := f''(X_{n-1})$. By computing the cases $X_n = X_{n-1} - 1$ and $X_n = X_{n-1} + 1$ separately, we see that for all $n \in \mathbb{N}$

$$f(X_n) - f(X_{n-1}) = \frac{f(X_{n-1}+1) - f(X_{n-1}-1)}{2}(X_n - X_{n-1})$$

$$+ \frac{1}{2}f(X_{n-1}-1) + \frac{1}{2}f(X_{n-1}+1) - f(X_{n-1})$$

$$= f'(X_{n-1})(X_n - X_{n-1}) + \frac{1}{2}f''(X_{n-1})$$

$$= F'_n \cdot (X_n - X_{n-1}) + \frac{1}{2}F''_n.$$

Summing up, we get the **discrete Itô formula**:

$$f(X_n) = f(x_0) + \sum_{i=1}^{n} f'(X_{i-1})(X_i - X_{i-1}) + \sum_{i=1}^{n} \frac{1}{2} f''(X_{i-1})$$

$$= f(x_0) + (F' \cdot X)_n + \sum_{i=1}^{n} \frac{1}{2} F_i''.$$

(10.6)

Here $F' \cdot X$ is the discrete stochastic integral (see Definition 9.37). Now $M :=$ $f(x_0) + F' \cdot X$ is a martingale by Theorem 9.39 since F' is predictable (and since $|F_n'| \leq \max_{x \in \{x_0 - n, \dots, x_0 + n\}} |F'(x)|$), and $A := \left(\sum_{i=1}^{n} \frac{1}{2} F_i'' \right)_{n \in \mathbb{N}_0}$ is predictable. Hence $f(X) := (f(X_n))_{n \in \mathbb{N}_0} = M + A$ is the Doob decomposition of $f(X)$. In particular, $f(X)$ is a submartingale if $f''(x) \geq 0$ for all $x \in \mathbb{Z}$; that is, if f is convex. We knew this already from Theorem 9.35; however, here we could also quantify how much $f(X)$ differs from a martingale.

In the special cases $f(x) = x^2$ and $f(x) = |x|$, the second derivative is $f''(x) = 2$ and $f''(x) = 2 \cdot \mathbb{1}_{\{0\}}(x)$, respectively. Thus, from (10.6), we recover the statements of Theorem 10.4 and Example 10.8.

Later we will derive a formula similar to (10.6) for stochastic processes in continuous time (see Sect. 25.3). $\Diamond$

> **Takeaways** For a submartingale, the mean future is better than the present. Subtracting the differences for each time step, we decompose a submartingale into a sum of a martingale and a monotone increasing predictable process. This is the so-called Doob decomposition. If X is an L^2-martingale, then X^2 is a submartingale and the corresponding increasing process is called the variance process or square variation process.

10.2 Optional Sampling and Optional Stopping

Lemma 10.10 *Let $I \subset \mathbb{R}$ be countable, let $(X_t)_{t \in I}$ be a martingale, let $T \in I$ and let τ be a stopping time with $\tau \leq T$. Then $X_\tau = \mathbf{E}[X_T \,|\, \mathcal{F}_\tau]$ and, in particular, $\mathbf{E}[X_\tau] = \mathbf{E}[X_0]$.*

Proof It is enough to show that $\mathbf{E}[X_T \, \mathbb{1}_A] = \mathbf{E}[X_\tau \, \mathbb{1}_A]$ for all $A \in \mathcal{F}_\tau$. By the definition of $\mathcal{F}_\tau$, we have $\{\tau = t\} \cap A \in \mathcal{F}_t$ for all $t \in I$. Hence

$$\mathbf{E}[X_\tau \, \mathbb{1}_A] = \sum_{t \leq T} \mathbf{E}[X_t \, \mathbb{1}_{\{\tau=t\} \cap A}] = \sum_{t \leq T} \mathbf{E}\big[\mathbf{E}[X_T \,|\, \mathcal{F}_t] \, \mathbb{1}_{\{\tau=t\} \cap A}\big]$$

$$= \sum_{t \leq T} \mathbf{E}[X_T \, \mathbb{1}_A \, \mathbb{1}_{\{\tau=t\}}] = \mathbf{E}[X_T \, \mathbb{1}_A].$$ $\square$

Theorem 10.11 (Optional sampling theorem) *Let* $X = (X_n)_{n \in \mathbb{N}_0}$ *be a supermartingale and let* $\sigma \leq \tau$ *be stopping times.*

(i) *Assume there exists a* $T \in \mathbb{N}$ *with* $\tau \leq T$. *Then*

$$X_\sigma \geq \mathbf{E}[X_\tau \,|\, \mathcal{F}_\sigma],$$

and, in particular, $\mathbf{E}[X_\sigma] \geq \mathbf{E}[X_\tau]$. *If* X *is a martingale, then equality holds in each case.*

(ii) *If* X *is nonnegative and if* $\tau < \infty$ *a.s., then we have* $\mathbf{E}[X_\tau] \leq \mathbf{E}[X_0] < \infty$, $\mathbf{E}[X_\sigma] \leq \mathbf{E}[X_0] < \infty$ *and* $X_\sigma \geq \mathbf{E}[X_\tau \,|\, \mathcal{F}_\sigma]$.

(iii) *Assume that, more generally,* X *is only adapted and integrable. Then* X *is a martingale if and only if* $\mathbf{E}[X_\tau] = \mathbf{E}[X_0]$ *for any bounded stopping time* τ.

Proof

(i) Let $X = M + A$ be Doob's decomposition of X. Hence A is predictable and monotone decreasing, $A_0 = 0$, and M is a martingale. Applying Lemma 10.10 to M yields

$$X_\sigma = A_\sigma + M_\sigma = \mathbf{E}[A_\sigma + M_T \,|\, \mathcal{F}_\sigma]$$

$$\geq \mathbf{E}[A_\tau + M_T \,|\, \mathcal{F}_\sigma] = \mathbf{E}[A_\tau + \mathbf{E}[M_T \,|\, \mathcal{F}_\tau] \,|\, \mathcal{F}_\sigma]$$

$$= \mathbf{E}[A_\tau + M_\tau \,|\, \mathcal{F}_\sigma] = \mathbf{E}[X_\tau \,|\, \mathcal{F}_\sigma].$$

Here we used $\mathcal{F}_\tau \supset \mathcal{F}_\sigma$, the tower property and the monotonicity of the conditional expectation (see Theorem 8.14).

(ii) We have $X_{\tau \wedge n} \xrightarrow{n \to \infty} X_\tau$ almost surely. By (i), we get $\mathbf{E}[X_{\tau \wedge n}] \leq \mathbf{E}[X_0]$ for any $n \in \mathbb{N}$. Using Fatou's lemma, we infer

$$\mathbf{E}[X_\tau] \leq \liminf_{n \to \infty} \mathbf{E}[X_{\tau \wedge n}] \leq \mathbf{E}[X_0] < \infty.$$

Similarly, we can show that $\mathbf{E}[X_\sigma] \leq \mathbf{E}[X_0]$.

Now, let $m, n \in \mathbb{N}$ with $m \geq n$. Part (i) applied to the bounded stopping times $\tau \wedge m \geq \sigma \wedge n$ yields

$$X_{\sigma \wedge n} \geq \mathbf{E}[X_{\tau \wedge m} \,|\, \mathcal{F}_{\sigma \wedge n}].$$

Now $\{\sigma < n\} \cap A \in \mathcal{F}_{\sigma \wedge n}$ for $A \in \mathcal{F}_\sigma$. Hence

$$\mathbf{E}\big[X_\sigma \, \mathbb{1}_{\{\sigma < n\} \cap A}\big] = \mathbf{E}\big[X_{\sigma \wedge n} \, \mathbb{1}_{\{\sigma < n\} \cap A}\big] \geq \mathbf{E}\big[X_{\tau \wedge m} \, \mathbb{1}_{\{\sigma < n\} \cap A}\big].$$

Using Fatou's lemma, we get

$$\mathbf{E}[X_\tau \, \mathbb{1}_{\{\sigma < n\} \cap A}] \leq \liminf_{m \to \infty} \mathbf{E}\big[X_{\tau \wedge m} \, \mathbb{1}_{\{\sigma < n\} \cap A}\big] \leq \mathbf{E}\big[X_\sigma \, \mathbb{1}_{\{\sigma < n\} \cap A}\big].$$

Monotone convergence (for $n \to \infty$) thus yields $\mathbf{E}[X_\tau \, \mathbb{1}_A] \leq \mathbf{E}[X_\sigma \, \mathbb{1}_A]$.

(iii) If X is a martingale, then the claim follows from Lemma 10.10. Now assume that $\mathbf{E}[X_\tau] = \mathbf{E}[X_0]$ for any bounded stopping time τ. Let $t > s$ and $A \in \mathcal{F}_s$. It is enough to show that $\mathbf{E}[X_t \, \mathbb{1}_A] = \mathbf{E}[X_s \, \mathbb{1}_A]$. Define $\tau = s \, \mathbb{1}_A + t \mathbb{1}_{A^c}$. Then τ is a bounded stopping time. However, by assumption,

$$\mathbf{E}[X_t \, \mathbb{1}_A] = \mathbf{E}[X_t] - \mathbf{E}[X_t \, \mathbb{1}_{A^c}] = \mathbf{E}[X_0] - \mathbf{E}[X_\tau] + \mathbf{E}[X_s \, \mathbb{1}_A] = \mathbf{E}[X_s \, \mathbb{1}_A]. \qquad \square$$

Corollary 10.12 *Let X be a martingale (respectively a submartingale), and assume $(\tau_N)_{N \in \mathbb{N}}$ is a monotone increasing sequence of bounded stopping times (hence $\tau_N \leq T_N$, $N \in \mathbb{N}$ for some $T_N \in \mathbb{N}$). Then $(X_{\tau_N})_{N \in \mathbb{N}}$ is a martingale (respectively a submartingale) with respect to the filtration $(\mathcal{F}_{\tau_N})_{N \in \mathbb{N}}$.*

Definition 10.13 (Stopped process) Let $I \subset \mathbb{R}$ be countable, let $(X_t)_{t \in I}$ be adapted and let τ be a stopping time. We define the **stopped process** X^τ by

$$X_t^\tau = X_{\tau \wedge t} \quad \text{for any } t \in I.$$

Further, let $\mathbb{F}^\tau$ be the filtration $\mathbb{F}^\tau = (\mathcal{F}_t^\tau)_{t \in I} = (\mathcal{F}_{\tau \wedge t})_{t \in I}$.

Remark 10.14 X^τ is adapted both to $\mathbb{F}$ and to $\mathbb{F}^\tau$. $\Diamond$

Theorem 10.15 (Optional stopping) *Let $(X_n)_{n \in \mathbb{N}_0}$ be a (sub-, super-) martingale with respect to $\mathbb{F}$ and let τ be a stopping time. Then X^τ is a (sub-, super-) martingale both with respect to $\mathbb{F}$ and with respect to $\mathbb{F}^\tau$.*

Proof We give the proof only for the case where X is a submartingale. The other cases are similar since there $(-X)$ is a submartingale.

For each $n \in \mathbb{N}_0$, we have

$$\mathbf{E}\big[|X_n^\tau|\big] \leq \mathbf{E}\big[\max\{|X_m| : m \leq n\}\big] \leq \mathbf{E}\big[|X_0|\big] + \ldots + \mathbf{E}[|X_n|] < \infty.$$

Hence X^τ is integrable.

Let X be a submartingale. Since $\{\tau > n - 1\} \in \mathcal{F}_{n-1}$, we have

$$\begin{aligned}
\mathbf{E}[X_n^\tau - X_{n-1}^\tau \,|\, \mathcal{F}_{n-1}] &= \mathbf{E}[X_{\tau \wedge n} - X_{\tau \wedge (n-1)} \,|\, \mathcal{F}_{n-1}] \\
&= \mathbf{E}[(X_n - X_{n-1})\,\mathbb{1}_{\{\tau > n-1\}} \,|\, \mathcal{F}_{n-1}] \\
&= \mathbb{1}_{\{\tau > n-1\}}\, \mathbf{E}[X_n - X_{n-1} \,|\, \mathcal{F}_{n-1}] \\
&\geq 0, \quad \text{since } X \text{ is an } \mathbb{F}\text{-submartingale.}
\end{aligned}$$

Therefore, X^τ is an $\mathbb{F}$-submartingale. As X^τ is adapted to $\mathbb{F}^\tau$ and since $\mathbb{F}^\tau$ is the smaller filtration, X^τ is also an $\mathbb{F}^\tau$-submartingale (see Remark 9.29).   □

Example 10.16 Let X be a symmetric simple random walk on $\mathbb{Z}$ (see Example 10.8). Let $a, b \in \mathbb{Z}$ with $a < 0$, $b > 0$ and let

$$\tau_a = \inf\{t \geq 0 : X_t = a\}, \quad \tau_b = \inf\{t \geq 0 : X_t = b\} \quad \text{and} \quad \tau_{a,b} = \tau_a \wedge \tau_b.$$

$\tau_{a,b}$ is a stopping time by Lemma 9.18. Let $A = \{\tau_{a,b} = \tau_a\}$ be the event where X hits a before hitting b. We want to compute $\mathbf{P}[A]$. By Exercise 2.3.1, almost surely $\limsup_{n \to \infty} X_n = \infty$ and $\liminf_{n \to \infty} X_n = -\infty$. Therefore, almost surely $\tau_a < \infty$ and $\tau_b < \infty$. By the optional stopping theorem, $X^{\tau_{a,b}}$ is a martingale. Since $\tau_{a,b} \wedge n \xrightarrow{n \to \infty} \tau_{a,b}$ almost surely, we get $X_n^{\tau_{a,b}} \xrightarrow{n \to \infty} X_{\tau_{a,b}}$ almost surely. As $|X_n^{\tau_{a,b}}|$ is bounded by $b - a$, we can infer that $X_n^{\tau_{a,b}} \xrightarrow{n \to \infty} X_{\tau_{a,b}}$ also in L^1. Thus

$$\begin{aligned}
0 = \lim_{n \to \infty} \mathbf{E}\left[X_n^{\tau_{a,b}}\right] = \mathbf{E}\left[X_{\tau_{a,b}}\right] &= a \cdot \mathbf{P}\left[\tau_{a,b} = \tau_a\right] + b \cdot \mathbf{P}\left[\tau_{a,b} = \tau_b\right] \\
&= b + (a - b)\, \mathbf{P}[\tau_{a,b} = \tau_a].
\end{aligned}$$

We conclude that $\mathbf{P}\left[\tau_{a,b} = \tau_a\right] = \dfrac{b}{b - a}$. ◇

Example 10.17 Finally, we use our machinery in order to compute $\mathbf{E}[\tau_{a,b}]$ and $\mathbf{E}[\tau_a]$. The square variation process $\langle X \rangle$ (compare Definition 10.3) is given by

$$\langle X \rangle_n = \sum_{i=1}^{n} \mathbf{E}\left[(X_i - X_{i-1})^2 \,|\, \mathcal{F}_{i-1}\right] = n;$$

hence $\left(X_n^2 - n\right)_{n \in \mathbb{N}_0}$ is a martingale. By the optional stopping theorem,

$$0 = \mathbf{E}\left[X_{\tau_{a,b} \wedge n}^2 - (\tau_{a,b} \wedge n)\right] \quad \text{for all } n \in \mathbb{N}_0.$$

Monotone convergence yields

$$\mathbf{E}\left[\tau_{a,b}\right] = \mathbf{E}[X_{\tau_{a,b}}^2] = a^2\, \mathbf{P}\left[\tau_{a,b} = \tau_a\right] + b^2\, \mathbf{P}\left[\tau_{a,b} = \tau_b\right] = |a| \cdot b.$$

In order to compute $\mathbf{E}[\tau_a]$, note that $\tau_{a,b} \uparrow \tau_a$ almost surely if $b \to \infty$. The monotone convergence theorem thus yields $\mathbf{E}[\tau_a] = \lim_{b \to \infty} \mathbf{E}[\tau_{a,b}] = \infty$. $\Diamond$

Remark 10.18 Evidently, $X_{\tau_b} = b > 0$. Hence $X_0 < \mathbf{E}\left[X_{\tau_b} \big| \mathcal{F}_0\right] = b$. The claim of the optional sampling theorem may thus fail, in general, if the stopping time is unbounded. $\Diamond$

Example 10.19 (Gambler's ruin problem) Consider a game of two persons, A and B. In each round, a coin is tossed. Depending on the outcome, either A gets a euro from B or vice versa. The game endures until one of the players is ruined. For simplicity, we assume that in the beginning A has $k_A \in \mathbb{N}$ euros while B has $k_B = N - k_A$ euros, where $N \in \mathbb{N}$, $N \geq k_A$. We want to know the probability of B's ruin. In Example 10.16 we saw that for a fair coin this probability is k_A/N. Now we allow the coin to be unfair.

Hence, let $Y_1, Y_2, \ldots$ be i.i.d. and $\sim \mathrm{Rad}_p$ (that is, $\mathbf{P}[Y_i = 1] = 1 - \mathbf{P}[Y_i = -1] = p$) for some $p \in (0, 1) \setminus \{\frac{1}{2}\}$. Denote by $X_n := k_B + \sum_{i=1}^{n} Y_i$ the running total for B after n rounds, where formally we assume that the game continues even after one player is ruined. We define as above τ_0, τ_N and $\tau_{0,N}$ as the times of first entrance of X into $\{0\}$, $\{N\}$ and $\{0, N\}$, respectively. The ruin probability of B thus is $p_B^N := \mathbf{P}[\tau_{0,N} = \tau_0]$. Since X is not a martingale (except for the case $p = \frac{1}{2}$ that was excluded), we use a trick to construct a martingale. Define a new process Z by $Z_n := r^{X_n} = r^{k_B} \prod_{i=1}^{n} r^{Y_i}$, where $r > 0$ has to be chosen so that Z becomes a martingale. By Example 9.31, this is the case if and only if

$$\mathbf{E}[r^{Y_1}] = pr + (1 - p)r^{-1} = 1;$$

hence, if $r = 1$ or $r = \frac{1-p}{p}$. Evidently, the choice $r = 1$ is useless (as Z does not yield any information on X); hence we assume $r = \frac{1-p}{p}$. We thus get

$$\tau_0 = \inf \left\{n \in \mathbb{N}_0 : Z_n = 1\right\} \quad \text{and} \quad \tau_N = \inf \left\{n \in \mathbb{N}_0 : Z_n = r^N\right\}.$$

(Note that here we cannot argue as above in order to show that $\tau_0 < \infty$ and $\tau_N < \infty$ almost surely. In fact, for $p \neq \frac{1}{2}$, *only one* of the statements holds. However, using, for example, the strong law of large numbers, we obtain that $\liminf_{n \to \infty} X_n = \infty$ (and thus $\tau_N < \infty$) almost surely if $p > \frac{1}{2}$. Similarly, $\tau_0 < \infty$ almost surely if $p < \frac{1}{2}$.) As in Example 10.16, the optional stopping theorem yields $r^{k_B} = Z_0 = \mathbf{E}[Z_{\tau_{0,N}}] = p_B^N + (1 - p_B^N)r^N$. Therefore, the probability of B's ruin is

$$p_B^N = \frac{r^{k_B} - r^N}{1 - r^N}. \tag{10.7}$$

If the game is advantageous for B (that is, $p > \frac{1}{2}$), then $r < 1$. In this case, in the limit $N \to \infty$ (with constant k_B),

$$p_B^\infty := \lim_{N \to \infty} p_B^N = r^{k_B}. \tag{10.8}$$

◇

Takeaways For a martingale, by definition, the conditional expected value at a later time given the information available at the present time is just the value at the present time. Here we have shown that this remains true for random times as long as they are bounded stopping times (optional sampling theorem). For general stopping times, we have shown that the stopped martingale is again a martingale (optional stopping theorem). The corresponding statements also hold for submartingales and supermartingales.

Exercise 10.2.1 Let X be a square integrable martingale with square variation process $\langle X \rangle$. Let τ be a finite stopping time. Show the following:

(i) If $\mathbf{E}[\langle X \rangle_\tau] < \infty$, then

$$\mathbf{E}\big[(X_\tau - X_0)^2\big] = \mathbf{E}\big[\langle X \rangle_\tau\big] \quad \text{and} \quad \mathbf{E}\big[X_\tau\big] = \mathbf{E}\big[X_0\big]. \tag{10.9}$$

(ii) If $\mathbf{E}[\langle X \rangle_\tau] = \infty$, then both equalities in (10.9) may fail. ♣

Exercise 10.2.2 We consider a situation that is more general than the one in the preceding example by assuming only that $Y_1, Y_2, \ldots$ are i.i.d. integrable random variables that are not almost surely constant (and $X_n = Y_1 + \ldots + Y_n$). We further assume that there is a $\delta > 0$ such that $\mathbf{E}[\exp(\theta Y_1)] < \infty$ for all $\theta \in (-\delta, \delta)$. Define a map $\psi : (-\delta, \delta) \to \mathbb{R}$ by $\theta \mapsto \log\big(\mathbf{E}[\exp(\theta Y_1)]\big)$ and the process Z^θ by $Z_n^\theta := \exp(\theta X_n - n\psi(\theta))$ for $n \in \mathbb{N}_0$. Show the following:

(i) Z^θ is a martingale for all $\theta \in (-\delta, \delta)$.
(ii) ψ is strictly convex.
(iii) $\mathbf{E}\big[\sqrt{Z_n^\theta}\big] \overset{n \to \infty}{\longrightarrow} 0$ for $\theta \neq 0$.
(iv) $Z_n^\theta \overset{n \to \infty}{\longrightarrow} 0$ almost surely.

We may interpret Y_n as the difference between the premiums and the payments of an insurance company at time n. If the initial capital of the company is $k_0 > 0$, then $k_0 + X_n$ is the account balance at time n. We are interested in the ruin probability

$$p(k_0) = \mathbf{P}\big[\inf\{X_n + k_0 : n \in \mathbb{N}_0\} < 0\big]$$

depending on the initial capital.

It can be assumed that the premiums are calculated such that $\mathbf{E}[Y_1] > 0$. Show that if the equation $\psi(\theta) = 0$ has a solution $\theta^* \neq 0$, then $\theta^* < 0$. Show further that in this case, the **Cramér–Lundberg inequality** holds:

$$p(k_0) \leq \exp(\theta^* k_0). \tag{10.10}$$

Equality holds if $k_0 \in \mathbb{N}$ and if Y_i assumes only the values -1 and 1. In this case, we get (10.8) with $r = \exp(\theta^*)$. ♣

10.3 Uniform Integrability and Optional Sampling

We extend the optional sampling theorem to unbounded stopping times. We will see that this is possible if the underlying martingale is uniformly integrable (compare Definition 6.16).

Lemma 10.20 *Let* $(X_n)_{n \in \mathbb{N}_0}$ *be a uniformly integrable martingale. Then the family* $(X_\tau : \tau$ *is a finite stopping time) is uniformly integrable.*

Proof By Theorem 6.19, there exists a monotone increasing, convex function $f : [0, \infty) \to [0, \infty)$ with $\liminf_{x \to \infty} f(x)/x = \infty$ and $L := \sup_{n \in \mathbb{N}_0} \mathbf{E}[f(|X_n|)] < \infty$. If $\tau < \infty$ is a finite stopping time, then by the optional sampling theorem for bounded stopping times (Theorem 10.11 with $\tau = n$ and $\sigma = \tau \wedge n$), $\mathbf{E}[X_n \mid \mathcal{F}_{\tau \wedge n}] = X_{\tau \wedge n}$. Since $\{\tau \leq n\} \in \mathcal{F}_{\tau \wedge n}$, Jensen's inequality yields

$$
\begin{aligned}
\mathbf{E}\big[f(|X_\tau|)\, \mathbb{1}_{\{\tau \leq n\}}\big] &= \mathbf{E}\big[f(|X_{\tau \wedge n}|)\, \mathbb{1}_{\{\tau \leq n\}}\big] \\
&\leq \mathbf{E}\big[\mathbf{E}\big[f(|X_n|) \mid \mathcal{F}_{\tau \wedge n}\big] \mathbb{1}_{\{\tau \leq n\}}\big] \\
&= \mathbf{E}\big[f(|X_n|)\, \mathbb{1}_{\{\tau \leq n\}}\big] \leq L.
\end{aligned}
$$

Hence $\mathbf{E}[f(|X_\tau|)] \leq L$. By Theorem 6.19, the family

$$\{X_\tau, \ \tau \text{ is a finite stopping time}\}$$

is uniformly integrable.   □

Theorem 10.21 (Optional sampling and uniform integrability) *Let* $(X_n, n \in \mathbb{N}_0)$ *be a uniformly integrable martingale (respectively supermartingale) and let* $\sigma \leq \tau$ *be finite stopping times. Then* $\mathbf{E}[|X_\tau|] < \infty$ *and* $X_\sigma = \mathbf{E}[X_\tau \mid \mathcal{F}_\sigma]$ *(respectively* $X_\sigma \geq \mathbf{E}[X_\tau \mid \mathcal{F}_\sigma]$).

Proof First let X be a martingale. We have $\{\sigma \leq n\} \cap F \in \mathcal{F}_{\sigma \wedge n}$ for all $F \in \mathcal{F}_\sigma$. Hence, by the optional sampling theorem (Theorem 10.11),

$$\mathbf{E}\big[X_{\tau \wedge n}\, \mathbb{1}_{\{\sigma \leq n\} \cap F}\big] = \mathbf{E}\big[X_{\sigma \wedge n}\, \mathbb{1}_{\{\sigma \leq n\} \cap F}\big].$$

By Lemma 10.20, $(X_{\sigma \wedge n}, n \in \mathbb{N}_0)$ and thus $(X_{\sigma \wedge n} \mathbb{1}_{\{\sigma \leq n\} \cap F}, n \in \mathbb{N}_0)$ are uniformly integrable. Similarly, this holds for X_τ. Therefore, by Theorem 6.25,

$$\mathbf{E}[X_\tau \, \mathbb{1}_F] = \lim_{n \to \infty} \mathbf{E}\big[X_{\tau \wedge n} \, \mathbb{1}_{\{\sigma \leq n\} \cap F}\big] = \lim_{n \to \infty} \mathbf{E}\big[X_{\sigma \wedge n} \, \mathbb{1}_{\{\sigma \leq n\} \cap F}\big] = \mathbf{E}[X_\sigma \, \mathbb{1}_F].$$

We conclude that $\mathbf{E}[X_\tau \,|\, \mathcal{F}_\sigma] = X_\sigma$.

Now let X be a supermartingale and let $X = M + A$ be its Doob decomposition; that is, M is a martingale and $A \leq 0$ is predictable and decreasing. Since

$$\mathbf{E}[|A_n|] = \mathbf{E}[-A_n] \leq \mathbf{E}[|X_n - X_0|] \leq \mathbf{E}[|X_0|] + \sup_{m \in \mathbb{N}_0} \mathbf{E}[|X_m|] < \infty,$$

we have $A_n \downarrow A_\infty$ for some $A_\infty \leq 0$ with $\mathbf{E}[-A_\infty] < \infty$ (by the monotone convergence theorem). Hence A and thus $M = X - A$ are uniformly integrable (Theorem 6.18(ii)). Therefore,

$$\mathbf{E}[|X_\tau|] \leq \mathbf{E}[-A_\tau] + \mathbf{E}[|M_\tau|] \leq \mathbf{E}[-A_\infty] + \mathbf{E}[|M_\tau|] < \infty.$$

Furthermore,

$$\begin{aligned}
\mathbf{E}[X_\tau \,|\, \mathcal{F}_\sigma] &= \mathbf{E}[M_\tau \,|\, \mathcal{F}_\sigma] + \mathbf{E}[A_\tau \,|\, \mathcal{F}_\sigma] \\
&= M_\sigma + A_\sigma + \mathbf{E}[(A_\tau - A_\sigma) \,|\, \mathcal{F}_\sigma] \\
&\leq M_\sigma + A_\sigma = X_\sigma. \qquad \qquad \square
\end{aligned}$$

Corollary 10.22 *Let X be a uniformly integrable martingale (respectively supermartingale) and let $\tau_1 \leq \tau_2 \leq \ldots$ be finite stopping times. Then $(X_{\tau_n})_{n \in \mathbb{N}}$ is a martingale (respectively supermartingale).*

Reflection Find an example that shows that uniform integrability is essential in Theorem 10.21. ♠

Takeaways The defining property for (sub-, super-) martingales can be extended to unbounded (but finite) stopping times if and only if the processes are uniformly integrable.

Chapter 11
Martingale Convergence Theorems and Their Applications

We became familiar with martingales $X = (X_n)_{n \in \mathbb{N}_0}$ as fair games and found that under certain transformations (optional stopping, discrete stochastic integral) martingales turn into martingales. In this chapter, we will see that under weak conditions (non-negativity or uniform integrability) martingales converge almost surely. Furthermore, the martingale structure implies L^p-convergence under assumptions that are (formally) weaker than those of Chap. 7. The basic ideas of this chapter are Doob's inequality (Theorem 11.2) and the upcrossing inequality (Lemma 11.3).

11.1 Doob's Inequality

With Kolmogorov's inequality (Theorem 5.28), we became acquainted with an inequality that bounds the probability of large values of the maximum of a square integrable process with independent centered increments. Here we want to improve this inequality in two directions. On the one hand, we replace the independent increments by the assumption that the process of partial sums is a martingale. On the other hand, we can manage with less than second moments; alternatively, we can get better bounds if we have higher moments.

Let $I \subset \mathbb{N}_0$ and let $X = (X_n)_{n \in I}$ be a stochastic process. For $n \in \mathbb{N}$, we denote

$$X_n^* = \sup\{X_k : k \leq n\} \quad \text{and} \quad |X|_n^* = \sup\{|X_k| : k \leq n\}.$$

Lemma 11.1 *If X is a submartingale, then, for all $\lambda > 0$,*

$$\lambda \, \mathbf{P}\big[X_n^* \geq \lambda\big] \; \leq \; \mathbf{E}\big[X_n \, \mathbb{1}_{\{X_n^* \geq \lambda\}}\big] \; \leq \; \mathbf{E}\big[|X_n| \, \mathbb{1}_{\{X_n^* \geq \lambda\}}\big].$$

Proof The second inequality is trivial. For the first one, let

$$\tau := \inf\{k \in I : X_k \geq \lambda\} \wedge n.$$

By Theorem 10.11 (optional sampling theorem),

$$\mathbf{E}[X_n] \geq \mathbf{E}[X_\tau] = \mathbf{E}\big[X_\tau \, \mathbb{1}_{\{X_n^* \geq \lambda\}}\big] + \mathbf{E}\big[X_\tau \, \mathbb{1}_{\{X_n^* < \lambda\}}\big]$$

$$\geq \lambda \, \mathbf{P}\big[X_n^* \geq \lambda\big] + \mathbf{E}\big[X_n \, \mathbb{1}_{\{X_n^* < \lambda\}}\big].$$

(Note that $\tau = n$ if $X_n^* < \lambda$.) Now subtract $\mathbf{E}\big[X_n \, \mathbb{1}_{\{X_n^* < \lambda\}}\big]$.   □

Theorem 11.2 (Doob's L^p-inequality) *Let X be a martingale or a positive submartingale.*

(i) *For any $p \geq 1$ and $\lambda > 0$,*

$$\lambda^p \, \mathbf{P}\big[|X|_n^* \geq \lambda\big] \leq \mathbf{E}\big[|X_n|^p\big].$$

(ii) *For any $p > 1$,*

$$\mathbf{E}\big[|X_n|^p\big] \leq \mathbf{E}\big[(|X|_n^*)^p\big] \leq \left(\frac{p}{p-1}\right)^p \mathbf{E}\big[|X_n|^p\big].$$

Proof We follow the proof in [145]. As all the statements in (i) and (ii) are trivially true if $\mathbf{E}[|X|_n^p] = \infty$, we may and will assume that $\mathbf{E}[|X_n|^p] < \infty$.

(i) By Theorem 9.35, $(|X_n|^p)_{n \in I}$ is a submartingale, and the claim follows by Lemma 11.1.

(ii) The first inequality is trivial. By Lemma 11.1, for the second inequality, we have

$$\lambda \mathbf{P}\big[|X|_n^* \geq \lambda\big] \leq \mathbf{E}\big[|X_n| \, \mathbb{1}_{\{|X|_n^* \geq \lambda\}}\big].$$

Hence, for any $K > 0$,

$$\mathbf{E}\big[(|X|_n^* \wedge K)^p\big] = \mathbf{E}\left[\int_0^{|X|_n^* \wedge K} p \, \lambda^{p-1} \, d\lambda\right]$$

$$= \mathbf{E}\left[\int_0^K p \, \lambda^{p-1} \, \mathbb{1}_{\{|X|_n^* \geq \lambda\}} \, d\lambda\right]$$

$$= \int_0^K p \, \lambda^{p-1} \, \mathbf{P}[|X|_n^* \geq \lambda] \, d\lambda$$

$$\leq \int_0^K p\,\lambda^{p-2}\,\mathbf{E}\big[|X_n|\,\mathbb{1}_{\{|X|_n^* \geq \lambda\}}\big]\,d\lambda$$

$$= p\,\mathbf{E}\left[|X_n|\int_0^{|X|_n^* \wedge K} \lambda^{p-2}\,d\lambda\right]$$

$$= \frac{p}{p-1}\,\mathbf{E}\big[|X_n|\cdot(|X|_n^* \wedge K)^{p-1}\big].$$

Hölder's inequality then yields

$$\mathbf{E}\big[(|X|_n^* \wedge K)^p\big] \;\leq\; \frac{p}{p-1}\,\mathbf{E}\big[(|X|_n^* \wedge K)^p\big]^{(p-1)/p}\cdot\mathbf{E}\big[|X_n|^p\big]^{1/p}.$$

We raise both sides to the pth power and divide by $\mathbf{E}\big[(|X|_n^* \wedge K)^p\big]^{p-1}$ (here we need the truncation at K to make sure we divide by a finite number) to obtain

$$\mathbf{E}\big[(|X|_n^* \wedge K)^p\big] \;\leq\; \left(\frac{p}{p-1}\right)^p \mathbf{E}\big[|X_n|^p\big].$$

Finally, let $K \to \infty$.

$\square$

Reflection Check that Kolmogorov's inequality (5.12) is a special case of (i) for $p = 2$. ♠

> **Takeaways** Martingales, respectively sub- and supermartingales, have so much structure that the maximal value over a finite trajectory can be estimated by the final value only; at least in expectation or as a pth moment.

Exercise 11.1.1 Let $(X_n)_{n\in\mathbb{N}_0}$ be a submartingale or a supermartingale. Use Theorem 11.2 and Doob's decomposition to show that, for all $n \in \mathbb{N}$ and $\lambda > 0$,

$$\lambda\,\mathbf{P}\big[|X|_n^* \geq \lambda\big] \leq 6\,\mathbf{E}[|X_0|] + 4\,\mathbf{E}[|X_n|]. \quad \clubsuit$$

11.2 Martingale Convergence Theorems

In this section, we present the usual martingale convergence theorems and give a few small examples. We start with the core of the martingale convergence theorems, the so-called upcrossing inequality.

Let $\mathbb{F} = (\mathcal{F}_n)_{n\in\mathbb{N}_0}$ be a filtration and $\mathcal{F}_\infty = \sigma\big(\bigcup_{n\in\mathbb{N}_0}\mathcal{F}_n\big)$. Let $(X_n)_{n\in\mathbb{N}_0}$ be real-valued and adapted to $\mathbb{F}$. Let $a, b \in \mathbb{R}$ with $a < b$. If we think of X as a stock

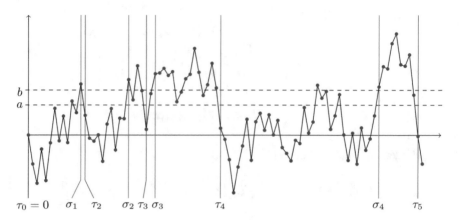

Fig. 11.1 Four upcrossings over $[a, b]$ are completed. One more has just started.

price, it would be a sensible trading strategy to buy the stock when its price has fallen below a and to sell it when it exceeds b at least if we knew for sure that the price would always rise above the level b again. Each time the price makes such an *upcrossing* from a to b, we make a profit of at least $b - a$. If we get a bound on the maximal profit we can make, dividing it by $b - a$ gives a bound on the maximal number of such upcrossings. If this number is finite for all $a < b$, then the price has to converge as $n \to \infty$.

Let us get into the details. Define stopping times $\sigma_0 \equiv 0$ and (see Fig. 11.1)

$$\tau_k := \inf\{n \geq \sigma_{k-1} : \ X_n \leq a\} \quad \text{for } k \in \mathbb{N},$$

$$\sigma_k := \inf\{n \geq \tau_k : \ X_n \geq b\} \quad \text{for } k \in \mathbb{N}.$$

Note that $\tau_k = \infty$ if $\sigma_{k-1} = \infty$, and $\sigma_k = \infty$ if $\tau_k = \infty$. We say that X has its kth **upcrossing** over $[a, b]$ between τ_k and σ_k if $\sigma_k < \infty$. For $n \in \mathbb{N}$, define

$$U_n^{a,b} := \sup\{k \in \mathbb{N}_0 : \ \sigma_k \leq n\}$$

as the number of upcrossings over $[a, b]$ until time n.

Lemma 11.3 (Upcrossing inequality) *Let* $(X_n)_{n\in\mathbb{N}_0}$ *be a submartingale. Then*

$$\mathbf{E}\left[U_n^{a,b}\right] \leq \frac{\mathbf{E}[(X_n - a)^+] - \mathbf{E}[(X_0 - a)^+]}{b - a}.$$

Proof Recall the discrete stochastic integral $H \cdot X$ from Definition 9.37. Formally, the intimated trading strategy H is described for $m \in \mathbb{N}_0$ by

$$H_m := \begin{cases} 1, & \text{if } m \in \{\tau_k + 1, \ldots, \sigma_k\} \text{ for some } k \in \mathbb{N}, \\ 0, & \text{else.} \end{cases}$$

H is nonnegative and predictable since, for all $m \in \mathbb{N}$,

$$\{H_m = 1\} = \bigcup_{k=1}^{\infty} \left(\{\tau_k \leq m - 1\} \cap \{\sigma_k > m - 1\} \right),$$

and each of the events is in $\mathcal{F}_{m-1}$. Define $Y = \max(X, a)$. If $k \in \mathbb{N}$ and $\sigma_k < \infty$, then clearly $Y_{\sigma_i} - Y_{\tau_i} = Y_{\sigma_i} - a \geq b - a$ for all $i \leq k$; hence

$$(H \cdot Y)_{\sigma_k} = \sum_{i=1}^{k} \sum_{j=\tau_i+1}^{\sigma_i} (Y_j - Y_{j-1}) = \sum_{i=1}^{k} (Y_{\sigma_i} - Y_{\tau_i}) \geq k(b - a).$$

For $j \in \{\sigma_k, \ldots, \tau_{k+1}\}$, we have $(H \cdot Y)_j = (H \cdot Y)_{\sigma_k}$. On the other hand, for $j \in \{\tau_k + 1, \ldots, \sigma_k\}$, we have $(H \cdot Y)_j \geq (H \cdot Y)_{\tau_k} = (H \cdot Y)_{\sigma_{k-1}}$. Hence $(H \cdot Y)_n \geq (b - a) U_n^{a,b}$ for all $n \in \mathbb{N}$.

By Corollary 9.34, Y is a submartingale, and (by Theorem 9.39) so are $H \cdot Y$ and $(1 - H) \cdot Y$. Now $Y_n - Y_0 = (1 \cdot Y)_n = (H \cdot Y)_n + ((1 - H) \cdot Y)_n$; hence

$$\mathbf{E}[Y_n - Y_0] \geq \mathbf{E}\left[(H \cdot Y)_n \right] \geq (b - a) \mathbf{E}\left[U_n^{a,b} \right]. \qquad \square$$

Theorem 11.4 (Martingale convergence theorem)   *Let $(X_n)_{n \in \mathbb{N}_0}$ be a submartingale with $\sup\{\mathbf{E}[X_n^+] : n \geq 0\} < \infty$. Then there exists an $\mathcal{F}_{\infty}$-measurable random variable X_{∞} with $\mathbf{E}[|X_{\infty}|] < \infty$ and $X_n \overset{n \to \infty}{\longrightarrow} X_{\infty}$ almost surely.*

Proof   Let $a < b$. Since $\mathbf{E}[(X_n - a)^+] \leq |a| + \mathbf{E}[X_n^+]$, by Lemma 11.3,

$$\mathbf{E}[U_n^{a,b}] \leq \frac{|a| + \mathbf{E}[X_n^+]}{b - a}.$$

Manifestly, the monotone limit $U^{a,b} := \lim_{n \to \infty} U_n^{a,b}$ exists. By assumption, we have $\mathbf{E}\left[U^{a,b} \right] = \lim_{n \to \infty} \mathbf{E}[U_n^{a,b}] < \infty$. In particular, $\mathbf{P}\left[U^{a,b} < \infty \right] = 1$. Define the $\mathcal{F}_{\infty}$-measurable events

$$C^{a,b} = \left\{ \liminf_{n \to \infty} X_n < a \right\} \cap \left\{ \limsup_{n \to \infty} X_n > b \right\} \subset \left\{ U^{a,b} = \infty \right\}$$

and

$$C = \bigcup_{\substack{a,b \in \mathbb{Q} \\ a < b}} C^{a,b}.$$

Then $\mathbf{P}\left[C^{a,b} \right] = 0$ and thus also $\mathbf{P}[C] = 0$. However, by construction, $(X_n)_{n \in \mathbb{N}}$ is convergent on C^c. Hence there exists the almost sure limit $X_{\infty} = \lim_{n \to \infty} X_n$. Each X_n is $\mathcal{F}_{\infty}$-measurable; hence X_{∞} also is $\mathcal{F}_{\infty}$-measurable.

By Fatou's lemma,

$$\mathbf{E}[X_\infty^+] \leq \sup\left\{\mathbf{E}[X_n^+] : n \geq 0\right\} < \infty.$$

On the other hand (since X is a submartingale), again by Fatou's lemma,

$$\mathbf{E}[X_\infty^-] \leq \liminf_{n\to\infty}\mathbf{E}[X_n^-] = \liminf_{n\to\infty}\left(\mathbf{E}[X_n^+] - \mathbf{E}[X_n]\right)$$

$$\leq \sup\left\{\mathbf{E}[X_n^+] : n \in \mathbb{N}_0\right\} - \mathbf{E}[X_0] < \infty. \qquad \square$$

Corollary 11.5 *If X is a nonnegative supermartingale, then there is an $\mathcal{F}_\infty$-measurable random variable $X_\infty \geq 0$ with $\mathbf{E}[X_\infty] \leq \mathbf{E}[X_0]$ and $X_n \xrightarrow{n\to\infty} X_\infty$ a.s.*

Proof The preceding theorem with $(-X)$ establishes X_∞ as the almost sure limit. Fatou's lemma yields

$$\mathbf{E}[X_\infty] \leq \liminf_{n\to\infty}\mathbf{E}[X_n] \leq \mathbf{E}[X_0]. \qquad \square$$

Reflection Why do we need nonnegativity of X in Corollary 11.5? Find an example of a supermartingale that does not converge. ♠

Example 11.6 Let S_n be the account balance in the Petersburg game after the nth round (see Example 9.40). Then S is a martingale and $S_n \leq 1$ almost surely for any n. Hence the assumptions of Theorem 11.4 are fulfilled and $(S_n)_{n\in\mathbb{N}_0}$ converges to a finite random variable almost surely for $n \to \infty$. Since the account changes as long as stakes are put up (that is, as long as $S_n < 1$), we get $\lim_{n\to\infty} S_n = 1$ almost surely.

Since $\mathbf{E}[S_n] = 0$ for all $n \in \mathbb{N}_0$, this convergence cannot hold in L^1. This observation tallies with the fact that S is not uniformly integrable. ◇

For uniformly integrable martingales, a stronger convergence theorem holds.

Theorem 11.7 (Convergence theorem for uniformly integrable martingales)
Let $(X_n)_{n\in\mathbb{N}_0}$ be a uniformly integrable $\mathbb{F}$- (sub-, super-) martingale. Then there exists an $\mathcal{F}_\infty$-measurable integrable random variable X_∞ with $X_n \xrightarrow{n\to\infty} X_\infty$ a.s. and in L^1. Furthermore:

- $X_n = \mathbf{E}[X_\infty \,|\, \mathcal{F}_n]$ *for all $n \in \mathbb{N}$ if X is a martingale.*
- $X_n \leq \mathbf{E}[X_\infty \,|\, \mathcal{F}_n]$ *for all $n \in \mathbb{N}$ if X is a submartingale.*
- $X_n \geq \mathbf{E}[X_\infty \,|\, \mathcal{F}_n]$ *for all $n \in \mathbb{N}$ if X is a supermartingale.*

Remark 11.8 The statement of Theorem 11.7 can be reformulated as: The process $(X_n)_{n\in\mathbb{N}_0\cup\{\infty\}}$ is a (sub-, super-) martingale with respect to $(\mathcal{F}_n)_{n\in\mathbb{N}_0\cup\{\infty\}}$. ◇

Proof We give the proof for the case where X is a submartingale. Uniform integrability implies $\sup\{\mathbf{E}[X_n^+] : n \geq 0\} < \infty$. By Theorem 11.4, the almost sure limit X_∞ exists. Hence $\mathbf{E}[|X_n - X_\infty|] \xrightarrow{n\to\infty} 0$ by Theorem 6.25. By Corollary 8.21, the L^1-convergence of (X_n) implies the L^1-convergence of the conditional expectations: $\mathbf{E}\big[\big|\mathbf{E}[X_n\,|\,\mathcal{F}_m] - \mathbf{E}[X_\infty\,|\,\mathcal{F}_m]\big|\big] \overset{n\to\infty}{\longrightarrow} 0$. Thus, by the triangle inequality,

$$\left|\mathbf{E}\big[\big(\mathbf{E}[X_\infty\,|\,\mathcal{F}_m] - X_m\big)^-\big] - \mathbf{E}\big[\big(\mathbf{E}[X_n\,|\,\mathcal{F}_m] - X_m\big)^-\big]\right|$$

$$\leq \mathbf{E}\Big[\big|\mathbf{E}[X_\infty\,|\,\mathcal{F}_m] - \mathbf{E}[X_n\,|\,\mathcal{F}_m]\big|\Big] \overset{n\to\infty}{\longrightarrow} 0.$$

As X is a submartingale, we have $\big(\mathbf{E}[X_n\,|\,\mathcal{F}_m] - X_m\big)^- = 0$ for $n \geq m$. Therefore, $\mathbf{E}\big[\big(\mathbf{E}[X_\infty\,|\,\mathcal{F}_m] - X_m\big)^-\big] = 0$ and thus $\mathbf{E}[X_\infty\,|\,\mathcal{F}_m] - X_m \geq 0$ almost surely.   □

Corollary 11.9 *Let $X \geq 0$ be a martingale and let $X_\infty = \lim\limits_{n\to\infty} X_n$. Then $\mathbf{E}[X_\infty] = \mathbf{E}[X_0]$ if and only if X is uniformly integrable.*

Proof This is a direct consequence of Theorem 6.25.   □

Let $p \in [1, \infty)$. A real-valued stochastic process $(X_i)_{i\in I}$ is called L^p-bounded if $\sup_{i\in I} \mathbf{E}[|X_i|^p] < \infty$ (Definition 6.20). In general, for $(|X_i|^p)_{i\in I}$ to be uniformly integrable it is not enough that $(X_i)_{i\in I}$ be L^p-bounded. However, if X is a martingale and if $p > 1$, then Doob's inequality implies that the statements are equivalent. In particular, in this case, almost sure convergence implies convergence in L^p.

Theorem 11.10 (L^p**-convergence theorem for martingales**) *Let $p > 1$ and let $(X_n)_{n\in\mathbb{N}_0}$ be an L^p-bounded martingale. Then there exists an $\mathcal{F}_\infty$-measurable random variable X_∞ with $\mathbf{E}[|X_\infty|^p] < \infty$ and $X_n \overset{n\to\infty}{\longrightarrow} X_\infty$ almost surely and in L^p. In particular, $(|X_n|^p)_{n\in\mathbb{N}_0}$ is uniformly integrable.*

Proof By Corollary 6.21, X is uniformly integrable. Hence the almost sure limit X_∞ exists. By Doob's inequality (Theorem 11.2), for all $n \in \mathbb{N}$,

$$\mathbf{E}\big[\sup\{|X_k|^p : k \leq n\}\big] \leq \left(\frac{p}{p-1}\right)^p \mathbf{E}[|X_n|^p].$$

Therefore,

$$\mathbf{E}\big[\sup\{|X_k|^p : k \in \mathbb{N}_0\}\big] \leq \left(\frac{p}{p-1}\right)^p \sup\big\{\mathbf{E}[|X_n|^p] : n \in \mathbb{N}_0\big\} < \infty.$$

Hence, in particular, $(|X_n|^p)_{n\in\mathbb{N}_0}$ is uniformly integrable.

Since $|X_n - X_\infty|^p \leq 2^p \sup\{|X_m|^p : m \in \mathbb{N}_0\}$, dominated convergence yields

$$\mathbf{E}\big[|X_\infty|^p\big] < \infty \quad\text{and}\quad \mathbf{E}\big[|X_n - X_\infty|^p\big] \overset{n\to\infty}{\longrightarrow} 0.$$ □

Reflection Let (X_n) be an L^p-bounded sequence of random variables that is almost surely convergent. In general, this does not imply that (X_n) also converges in L^p.

However, for martingales, it does as we have seen in the above theorem. Exactly where in the proof did we use the martingale property? ♠

For the case of square integrable martingales, there is a convenient criterion for L^2-boundedness that we record as a corollary (see Definition 10.3).

Corollary 11.11 *Let X be a square integrable martingale with square variation process $\langle X \rangle$. Then the following four statements are equivalent:*

(i) $\sup_{n \in \mathbb{N}} \mathbf{E}[X_n^2] < \infty$.
(ii) $\lim_{n \to \infty} \mathbf{E}[\langle X \rangle_n] < \infty$.
(iii) X *converges in* L^2.
(iv) X *converges almost surely and in* L^2.

Proof "((i)) $\iff$ ((ii))" Since $\mathbf{Var}[X_n - X_0] = \mathbf{E}[\langle X \rangle_n]$ (see Theorem 10.4), X is bounded in L^2 if and only if ((ii)) holds.
"((iv)) $\implies$ ((iii)) $\implies$ ((i))" This is trivial.
"((i)) $\implies$ ((iv))" This is the statement of Theorem 11.10.   □

Remark 11.12 In general, the statement of Theorem 11.10 fails for $p = 1$. See Exercise 11.2.1. ◊

Lemma 11.13 *Let X be a square integrable martingale with square variation process $\langle X \rangle$, and let τ be a stopping time. Then the stopped process X^τ has square variation process $\langle X^\tau \rangle = \langle X \rangle^\tau := (\langle X \rangle_{\tau \wedge n})_{n \in \mathbb{N}_0}$.*

Proof This is left as an exercise.   □

If in Corollary 11.11 we do not assume that the *expectations* of the square variation are bounded but only that the square variation is *almost surely* bounded, then we still get that X converges almost surely (albeit not in L^2).

Theorem 11.14 *If X is a square integrable martingale with $\sup_{n \in \mathbb{N}} \langle X \rangle_n < \infty$ almost surely, then X converges almost surely.*

Proof Without loss of generality, we can assume that $X_0 = 0$, otherwise consider the martingale $(X_n - X_0)_{n \in \mathbb{N}_0}$, which has the same square variation process. For $K > 0$, let

$$\tau_K := \inf\{n \in \mathbb{N} : \langle X \rangle_{n+1} \geq K\}.$$

This is a stopping time since $\langle X \rangle$ is predictable. Evidently, $\sup_{n \in \mathbb{N}} \langle X \rangle_{\tau_K \wedge n} \leq K$ almost surely. By Corollary 11.11, the stopped process X^{τ_K} converges almost surely (and in L^2) to a random variable that we denote by $X_\infty^{\tau_K}$. By assumption, $\mathbf{P}[\tau_K = \infty] \to 1$ for $K \to \infty$; hence X converges almost surely.   □

Example 11.15 Let X be a symmetric simple random walk on $\mathbb{Z}$. That is, $X_n = \sum_{k=1}^{n} R_k$, where $R_1, R_2, \ldots$ are i.i.d. and $\sim \mathrm{Rad}_{1/2}$:

$$\mathbf{P}[R_1 = 1] = \mathbf{P}[R_1 = -1] = \frac{1}{2}.$$

Then X is a martingale; however, $\limsup_{n\to\infty} X_n = \infty$ and $\liminf_{n\to\infty} X_n = -\infty$. Therefore, X does not even converge improperly. By the martingale convergence theorem, this is consonant with the fact that X is not uniformly integrable. ◊

Example 11.16 (Voter model, due to [28, 75]) Consider a simple model that describes the behavior of opportunistic voters who are capable of only one out of two opinions, say 0 and 1. Let $\Lambda \subset \mathbb{Z}^d$ be a set that we interpret as the sites at each of which there is one voter. For simplicity, assume that $\Lambda = \{0, \ldots, L - 1\}^d$ for some $L \in \mathbb{N}$. Let $x(i) \in \{0, 1\}$ be the opinion of the voter at site $i \in \Lambda$ and denote by $x \in \{0, 1\}^\Lambda$ a generic state of the whole population. We now assume that the individual opinions may change at discrete time steps. At any time $n \in \mathbb{N}_0$, one site I_n out of Λ is chosen at random and the individual at that site reconsiders his or her opinion. To this end, the voter chooses a neighbor $I_n + N_n \in \Lambda$ (with *periodic boundary conditions*; that is, with addition modulo L in each coordinate) at random and adopts his or her opinion. We thus get a random sequence $(X_n)_{n\in\mathbb{N}_0}$ of states in $\{0, 1\}^\Lambda$ that represents the random evolution of the opinions of the whole colony. See Fig. 11.2 for a simulation of the voter model.

For a formal description of this model, let $(I_n)_{n\in\mathbb{N}}$ and $(N_n)_{n\in\mathbb{N}}$ be independent random variables. For any $n \in \mathbb{N}$, I_n is uniformly distributed on Λ and N_n is uniformly distributed on the set $\mathcal{N} := \{i \in \mathbb{Z}^d : \|i\|_2 = 1\}$ of the $2d$ nearest neighbors of the origin. Furthermore, $x = X_0 \in \{0, 1\}^\Lambda$ is the initial state. The states at later times are defined inductively by

$$X_n(i) = \begin{cases} X_{n-1}(i), & \text{if } I_n \neq i, \\ X_{n-1}(I_n + N_n), & \text{if } I_n = i. \end{cases}$$

Of course, the behavior over small periods of time is determined by the perils of randomness. However, in the long run, we might see certain patterns. To be more specific, the question is: In the long run, will there be a consensus of all individuals or will competing opinions persist?

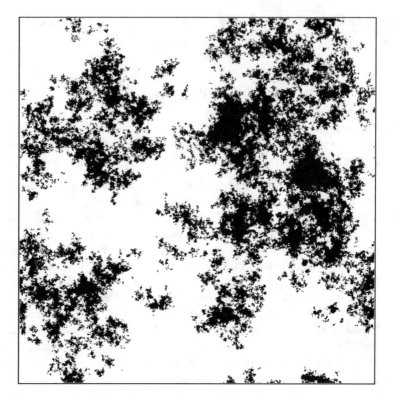

Fig. 11.2 Snapshot of a voter model on an 800×800 torus. The black dots are the Ones.

Let $M_n := \sum_{i \in \Lambda} X_n(i)$ be the total number of individuals of opinion 1 at time n. Let $\mathbb{F}$ be the filtration $\mathbb{F} = (\mathcal{F}_n)_{n \in \mathbb{N}_0}$, where $\mathcal{F}_n = \sigma(I_k, N_k : k \leq n)$ for all $n \in \mathbb{N}_0$. Then M is adapted to $\mathbb{F}$ and

$$\mathbf{E}[M_n \,|\, \mathcal{F}_{n-1}] = M_{n-1} - \mathbf{E}[X_{n-1}(I_n) \,|\, \mathcal{F}_{n-1}] + \mathbf{E}[X_{n-1}(I_n + N_n) \,|\, \mathcal{F}_{n-1}]$$

$$= M_{n-1} - \sum_{i \in \Lambda} \mathbf{P}[I_n = i]\, X_{n-1}(i) + \sum_{i \in \Lambda} \mathbf{P}[I_n + N_n = i]\, X_{n-1}(i)$$

$$= M_{n-1}$$

since $\mathbf{P}[I_n = i] = \mathbf{P}[I_n + N_n = i] = L^{-d}$ for all $i \in \Lambda$. Hence M is a bounded $\mathbb{F}$-martingale and thus converges almost surely and in L^1 to a random variable M_∞. Since M takes only integer values, there is a (random) n_0 such that $M_n = M_{n_0}$ for all $n \geq n_0$. However, then also $X_n = X_{n_0}$ for all $n \geq n_0$. Manifestly, no state x with $x \not\equiv 0$ and $x \not\equiv 1$ is stable. In fact, if x is not constant and if $i, j \in \Lambda$ are neighbors with $x(i) \neq x(j)$, then

$$\mathbf{P}[X_n \neq X_{n-1} \,|\, X_{n-1} = x] \geq \mathbf{P}[I_{n-1} = i, N_{n-1} = j - i] = L^{-d}(2d)^{-1}.$$

This implies $M_\infty \in \{0, L^d\}$. Now $\mathbf{E}[M_\infty] = M_0$; hence we have

$$\mathbf{P}\big[M_\infty = L^d\big] = \frac{M_0}{L^d} \quad \text{and} \quad \mathbf{P}\big[M_\infty = 0\big] = 1 - \frac{M_0}{L^d}.$$

Thus, eventually there will be a consensus of all individuals, and the probability that the surviving opinion is $e \in \{0, 1\}$ is the initial frequency of opinion e.

We could argue more formally to show that only the constant states are stable: Let $\langle M \rangle$ be the square variation process of M. Then

$$\langle M \rangle_n = \sum_{k=1}^{n} \mathbf{E}\big[\mathbb{1}_{\{M_k \neq M_{k-1}\}}\,\big|\,\mathcal{F}_{k-1}\big] = \sum_{k=1}^{n} \mathbf{P}\big[X_{k-1}(I_k) \neq X_{k-1}(I_k + N_k)\,\big|\,\mathcal{F}_{k-1}\big].$$

Hence

$$L^{2d} \geq \mathbf{Var}[M_n] = \mathbf{E}[\langle M \rangle_n]$$

$$= \sum_{k=1}^{n} \mathbf{P}[X_{k-1}(I_k) \neq X_{k-1}(I_k + N_k)]$$

$$\geq (2d)^{-1} L^{-d} \sum_{k=1}^{n} \mathbf{P}[M_{k-1} \notin \{0, L^d\}].$$

Therefore, $\sum_{k=1}^{\infty} \mathbf{P}[M_{k-1} \notin \{0, L^d\}] \leq 2d L^{3d} < \infty$, and so, by the Borel–Cantelli lemma, $M_\infty \in \{0, L^d\}$. $\Diamond$

Example 11.17 (Radon–Nikodym theorem) With the aid of the martingale convergence theorem, we give an alternative proof of the Radon–Nikodym theorem (Corollary 7.34).

Let $(\Omega, \mathcal{F}, \mathbf{P})$ be a probability space and let Q be another probability measure on $(\Omega, \mathcal{A})$. We assume that $\mathcal{F}$ is countably generated; that is, there exist countably many sets $A_1, A_2, \ldots \in \mathcal{F}$ such that $\mathcal{F} = \sigma(\{A_1, A_2, \ldots\})$. For example, this is the case if $\mathcal{F}$ is the Borel σ-algebra on a Polish space. For the case $\Omega = \mathbb{R}^d$, one could take the open balls with rational radii, centered at points with rational coordinates (compare Remark 1.24).

We construct a filtration $\mathbb{F} = (\mathcal{F}_n)_{n \in \mathbb{N}}$ by letting $\mathcal{F}_n := \sigma(\{A_1, \ldots, A_n\})$. Evidently, $\#\mathcal{F}_n < \infty$ for all $n \in \mathbb{N}$. More precisely, there exists a unique finite subset $Z_n \subset \mathcal{F}_n \setminus \{\emptyset\}$ such that $B = \biguplus_{\substack{C \in Z_n \\ C \subset B}} C$ for any $B \in \mathcal{F}_n$. Z_n decomposes $\mathcal{F}_n$ into its "atoms". Finally, define a stochastic process $(X_n)_{n \in \mathbb{N}}$ by

$$X_n := \sum_{C \in Z_n : \mathbf{P}[C] > 0} \frac{Q(C)}{\mathbf{P}[C]} \mathbb{1}_C.$$

Clearly, X is adapted to $\mathbb{F}$. Let $B \in \mathcal{F}_n$ and $m \geq n$. For any $C \in Z_m$, either $C \cap B = \emptyset$ or $C \subset B$. Hence

$$\mathbf{E}[X_m \mathbb{1}_B] = \sum_{C \in Z_m: \mathbf{P}[C] > 0} \frac{Q(C)}{\mathbf{P}[C]} \mathbf{P}[C \cap B] = \sum_{C \in Z_m: C \subset B} Q(C) = Q(B). \qquad (11.1)$$

In particular, X is an $\mathbb{F}$-martingale.

Now assume that Q is absolutely continuous with respect to P. By Example 7.39, this implies that X is uniformly integrable. By the martingale convergence theorem, X converges $\mathbf{P}$-almost surely and in $L^1(\mathbf{P})$ to a random variable X_∞. By (11.1), we have $\mathbf{E}[X_\infty \mathbb{1}_B] = Q(B)$ for all $B \in \bigcup_{n \in \mathbb{N}} \mathcal{F}_n$ and thus also for all $B \in \mathcal{F}$. Therefore, X_∞ is the Radon–Nikodym density of Q with respect to $\mathbf{P}$.

Note that for this proof we did not presume the existence of conditional expectations (rather we constructed them explicitly for finite σ-algebras); that is, we did not resort to the Radon–Nikodym theorem in a hidden way.

It could be objected that this argument works only for probability measures. However, this flaw can easily be remedied. Let μ and ν be arbitrary (but nonzero) σ-finite measures. Then there exist measurable functions $g, h : \Omega \to (0, \infty)$ with $\int g \, d\mu = 1$ and $\int h \, d\nu = 1$. Define $\mathbf{P} = g\mu$ and $Q = h\nu$. Clearly, $Q \ll \mathbf{P}$ if $\nu \ll \mu$. In this case, $\frac{g}{h} X_\infty$ is a version of the Radon–Nikodym derivative $\frac{d\nu}{d\mu}$.

The restriction that $\mathcal{F}$ is countably generated can also be dropped. Using the approximation theorems for measures, it can be shown that there is always a countably generated σ-algebra $\mathcal{G} \subset \mathcal{F}$ such that for any $A \in \mathcal{F}$, there is a $B \in \mathcal{G}$ with $\mathbf{P}[A \triangle B] = 0$. This can be employed to prove the general case. We do not give the details but refer to [170, Chapter 14.13]. $\Diamond$

> **Takeaways** For (sub-) martingales, Doob's upcrossing lemma bounds the number of upcrossings over a given interval. As a consequence, we get almost sure convergence of nonnegative (super-) martingales. Uniformly integrable (sub-, super-) martingales converge almost surely and in L^1. For $p > 1$, L^p-bounded martingales also converge in L^p.

Exercise 11.2.1 For $p = 1$, the statement of Theorem 11.10 may fail. Give an example of a nonnegative martingale X with $\mathbf{E}[X_n] = 1$ for all $n \in \mathbb{N}$ but such that $X_n \overset{n \to \infty}{\longrightarrow} 0$ almost surely. ♣

Exercise 11.2.2 Let $X_1, X_2, \ldots$ be independent, square integrable random variables with $\sum_{n=1}^{\infty} \frac{1}{n^2} \mathbf{Var}[X_n] < \infty$. Use the martingale convergence theorem to show the strong law of large numbers for $(X_n)_{n \in \mathbb{N}}$. ♣

Exercise 11.2.3 Give an example of a square integrable martingale that converges almost surely but not in L^2. ♣

Exercise 11.2.4 Show that in Theorem 11.14 the converse implication may fail. That is, there exists a square integrable martingale X that converges almost surely but without $\lim_{n \to \infty} \langle X \rangle_n < \infty$ almost surely. ♣

Exercise 11.2.5 Show the following converse of Theorem 11.14. Let $L > 0$ and let $(X_n)_{n \in \mathbb{N}}$ be a martingale with the property

$$|X_{n+1} - X_n| \leq L \quad \text{a.s.} \tag{11.2}$$

Define the events

$$C := \left\{ (X_n)_{n \in \mathbb{N}} \text{ converges as } n \to \infty \right\},$$

$$A^+ := \left\{ \limsup_{n \to \infty} X_n < \infty \right\},$$

$$A^- := \left\{ \liminf_{n \to \infty} X_n > -\infty \right\},$$

$$F := \left\{ \sup_{n \in \mathbb{N}} \langle X \rangle_n < \infty \right\}.$$

Show that

$$C = A^+ = A^- = F \pmod{\mathbf{P}}.$$

Here equality of events (mod $\mathbf{P}$) means that the events differ at most by a $\mathbf{P}$-null set (see Definition 1.68(iii)).
Hint: Use the stopping times $\sigma_K = \inf\{n \in \mathbb{N} : |X_n| \geq K\}$; $\sigma_K^{\pm} = \inf\{n \in \mathbb{N} : \pm X_n \geq K\}$ and τ_K as in the proof of Theorem 11.14. ♣

Exercise 11.2.6 Let the notation be as in Exercise 11.2.5. However, instead of (11.2) we make the weaker assumption

$$\mathbf{E}\left[\sup_{n \in \mathbb{N}} |X_{n+1} - X_n| \right] < \infty. \tag{11.3}$$

Show that

$$C = A^+ = A^- \pmod{\mathbf{P}}.$$

Hint: Use suitable stopping times ϱ_K and apply the martingale convergence theorem (Theorem 11.4) to the stopped process X^{ϱ_K}. ♣

Exercise 11.2.7 (Conditional Borel–Cantelli lemma) Let $(\mathcal{F}_n)_{n \in \mathbb{N}_0}$ be a filtration and let $(A_n)_{n \in \mathbb{N}}$ be events with $A_n \in \mathcal{F}_n$ for all $n \in \mathbb{N}$. Define $A_\infty = \left\{ \sum_{n=1}^{\infty} \mathbf{P}[A_n | \mathcal{F}_{n-1}] = \infty \right\}$ and $A^* = \limsup_{n \to \infty} A_n$. Show the conditional Borel–Cantelli lemma: $\mathbf{P}[A_\infty \triangle A^*] = 0$.

Hint: Apply Exercise 11.2.5 to $X_n = \sum_{n=1}^{\infty} (\mathbb{1}_{A_n} - \mathbf{P}[A_n \,|\, \mathcal{F}_{n-1}])$. ♣

Exercise 11.2.8 Let $p \in [0, 1]$ and let $X = (X_n)_{n \in \mathbb{N}_0}$ be a stochastic process with values in $[0, 1]$. Assume that for all $n \in \mathbb{N}_0$, given $X_0, \ldots, X_n$, we have

$$X_{n+1} = \begin{cases} 1 - p + pX_n & \text{with probability } X_n, \\ pX_n & \text{with probability } 1 - X_n. \end{cases}$$

Show that X is a martingale that converges almost surely. Compute the distribution of the almost sure limit $\lim_{n \to \infty} X_n$. ♣

Exercise 11.2.9 Let $f \in \mathcal{L}^1(\lambda)$, where λ is the restriction of the Lebesgue measure to $[0, 1]$. Let $I_{n,k} = [k\, 2^{-n}, (k + 1)\, 2^{-n})$ for $n \in \mathbb{N}$ and $k = 0, \ldots, 2^n - 1$. Define $f_n : [0, 1] \to \mathbb{R}$ by

$$f_n(x) = 2^n \int_{I_{k,n}} f \, d\lambda, \quad \text{if } k \text{ is chosen such that } x \in I_{k,n}.$$

Show that $f_n(x) \overset{n \to \infty}{\longrightarrow} f(x)$ for λ-almost all $x \in [0, 1]$. ♣

Exercise 11.2.10 Assume that $\mathbb{F} = (\mathcal{F}_n)_{n \in \mathbb{N}}$ is a filtration on the probability space $(\Omega, \mathcal{A}, \mathbf{P})$. Let $\mathcal{F}_\infty := \sigma(\mathcal{F}_n : n \in \mathbb{N})$, and let $\mathcal{M}$ be the vector space of uniformly integrable $\mathbb{F}$-martingales. Show that the map $\Phi : \mathcal{L}^1(\mathcal{F}_\infty) \to \mathcal{M}$, $X_\infty \mapsto (\mathbf{E}[X_\infty \,|\, \mathcal{F}_n])_{n \in \mathbb{N}}$ is an isomorphism of vector spaces. ♣

11.3   Example: Branching Process

Let $p = (p_k)_{k \in \mathbb{N}_0}$ be a probability vector on $\mathbb{N}_0$ and let $(Z_n)_{n \in \mathbb{N}_0}$ be the Galton–Watson process with one ancestor and offspring distribution p (see Definition 3.9). For convenience, we recall the construction of Z. Let $(X_{n,i})_{n \in \mathbb{N}_0, \, i \in \mathbb{N}}$ be i.i.d. random variables with $\mathbf{P}[X_{1,1} = k] = p_k$ for $k \in \mathbb{N}_0$. Let $Z_0 = 1$ and inductively define

$$Z_{n+1} = \sum_{i=1}^{Z_n} X_{n,i} \quad \text{for } n \in \mathbb{N}_0.$$

We interpret Z_n as the size of a population at time n and $X_{n,i}$ as the number of offspring of the ith individual of the nth generation.

Let $m := \mathbf{E}[X_{1,1}] < \infty$ be the expected number of offspring of an individual and let $\sigma^2 := \mathbf{Var}[X_{1,1}] \in (0, \infty)$ be its variance. Let $\mathcal{F}_n := \sigma(X_{k,i} : k < n, \, i \in \mathbb{N})$. Then Z is adapted to $\mathbb{F}$. Define $W_n = m^{-n} Z_n$.

Lemma 11.18 *W is a martingale. In particular, $\mathbf{E}[Z_n] = m^n$ for all $n \in \mathbb{N}$.*

Proof We compute the conditional expectation for $n \in \mathbb{N}_0$:

$$\mathbf{E}[W_{n+1} \mid \mathcal{F}_n] = m^{-(n+1)} \mathbf{E}[Z_{n+1} \mid \mathcal{F}_n]$$

$$= m^{-(n+1)} \mathbf{E}\left[\sum_{i=1}^{Z_n} X_{n,i} \mid \mathcal{F}_n\right]$$

$$= m^{-(n+1)} \sum_{k=1}^{\infty} \mathbf{E}\left[\mathbb{1}_{\{Z_n=k\}} k \cdot X_{n,i} \mid \mathcal{F}_n\right]$$

$$= m^{-n} \sum_{k=1}^{\infty} \mathbf{E}\left[k \cdot \mathbb{1}_{\{Z_n=k\}} \mid \mathcal{F}_n\right]$$

$$= m^{-n} Z_n = W_n.$$

$\square$

Theorem 11.19 *Let* $\mathbf{Var}[X_{1,1}] \in (0, \infty)$. *The a.s. limit* $W_\infty = \lim\limits_{n\to\infty} W_n$ *exists and*

$$m > 1 \quad \Longleftrightarrow \quad \mathbf{E}[W_\infty] = 1 \quad \Longleftrightarrow \quad \mathbf{E}[W_\infty] > 0.$$

Proof W_∞ exists since $W \geq 0$ is a martingale. If $m \leq 1$, then $(Z_n)_{n\in\mathbb{N}}$ converges a.s. to some random variable Z_∞. Note that Z_∞ is the only choice since $\sigma^2 > 0$.

Now let $m > 1$. Since $\mathbf{E}[Z_{n-1}] = m^{n-1}$ (Lemma 11.18), by the Blackwell–Girshick formula (Theorem 5.10),

$$\mathbf{Var}[W_n] = m^{-2n}\left(\sigma^2 \mathbf{E}[Z_{n-1}] + m^2 \mathbf{Var}[Z_{n-1}]\right)$$

$$= \sigma^2 m^{-(n+1)} + \mathbf{Var}[W_{n-1}].$$

Inductively, we get $\mathbf{Var}[W_n] = \sigma^2 \sum\limits_{k=2}^{n+1} m^{-k} \leq \dfrac{\sigma^2 m}{m-1} < \infty$. Hence W is bounded in L^2 and Theorem 11.10 yields $W_n \to W_\infty$ in L^2 and thus in L^1. In particular, $\mathbf{E}[W_\infty] = \mathbf{E}[W_0] = 1$. $\square$

Reflection Typically, the distribution of W cannot be computed explicitly. However, here we sketch a particular situation where this is possible. Assume that the offspring distribution is given by $\mathbf{P}[X_{1,1} = k] = \frac{1}{3}(2/3)^k$, $k \in \mathbb{N}_0$. In Lemma 3.10, we have seen how to iterate the generating function $\psi(s) = \mathbf{E}[s^{X_{1,1}}] = 1/(3 - 2s)$

so as to get $\psi_n(s) = \mathbf{E}[s^{Z_n}] = \frac{(2-2^n)s+2^n-1}{(2-2^{n+1})s+2^{n+1}-1}$. Compare also Lemma 21.44. By letting $s = e^{-\lambda/2^n}$, we obtain

$$\lim_{n\to\infty} \mathbf{E}[e^{\lambda Z_n/2^n}] = \frac{1+\lambda}{1+2\lambda} = \frac{1}{2} + \frac{1}{2}\frac{1}{1+2\lambda} = \mathbf{E}[e^{-\lambda W}],$$

where in the last step we assumed that $\mathbf{P}_W = \frac{1}{2}\delta_0 + \frac{1}{2}\exp_{1/2}$. ♠♠♠

The proof of Theorem 11.19 was simple due to the assumption of finite variance of the offspring distribution. However, there is a much stronger statement that here we can only quote (see [95], and see [111] for a modern proof).

Theorem 11.20 (Kesten–Stigum [95]) *Let $m > 1$. Then*

$$\mathbf{E}[W_\infty] = 1 \quad \Longleftrightarrow \quad \mathbf{E}[W_\infty] > 0 \quad \Longleftrightarrow \quad \mathbf{E}[X_{1,1}\log(X_{1,1})^+] < \infty.$$

Takeaways A Galton-Watson branching process (Z_n) with mean offspring number $m > 1$ has a positive chance to survive and in this case grows indefinitely. If the offspring number has a second moment, then Z_n grows of order m^n and Z_n/m^n is uniformly integrable. Hence Z_n/m^n is an almost surely and L^1-convergent martingale. In general, the growth to infinity of Z_n can also be slower than m^n.

Chapter 12
Backwards Martingales and Exchangeability

With many data acquisitions, such as telephone surveys, the order in which the data come does not matter. Mathematically, we say that a family of random variables is *exchangeable* if the joint distribution does not change under finite permutations. De Finetti's structural theorem says that an infinite family of E-valued exchangeable random variables can be described by a two-stage experiment. At the first stage, a probability distribution $\varXi$ on E is drawn at random. At the second stage, i.i.d. random variables with distribution $\varXi$ are implemented.

We first define the notion of exchangeability. Then we consider backwards martingales and prove the convergence theorem for them. This is the cornerstone for the proof of de Finetti's theorem.

12.1 Exchangeable Families of Random Variables

Definition 12.1 Let I be an arbitrary index set and let E be a Polish space. A family $(X_i)_{i \in I}$ of random variables with values in E is called **exchangeable** if

$$\mathcal{L}\big[(X_{\varrho(i)})_{i \in I}\big] = \mathcal{L}\big[(X_i)_{i \in I}\big]$$

for any finite permutation $\varrho : I \to I$.

Recall that a finite permutation is a bijection $\varrho : I \to I$ that leaves all but finitely many points unchanged.

A. Klenke, *Probability Theory*, Universitext, https://doi.org/10.1007/978-3-030-56402-5_12

Strictly speaking, the definition of exchangeability makes sense only if we identify $Y = (X_i)_{i \in I}$ and $Y' = (X_{\varrho(i)})_{i \in I}$ as random variables on the product space E^I. The investigation of product spaces and their σ-algebras is, however, postponed to Chap. 14.

Remark 12.2 Clearly, the following are equivalent.

(i) $(X_i)_{i \in I}$ is exchangeable.
(ii) Let $n \in \mathbb{N}$ and assume $i_1, \ldots, i_n \in I$ are pairwise distinct and $j_1, \ldots, j_n \in I$ are pairwise distinct. Then we have $\mathcal{L}[(X_{i_1}, \ldots, X_{i_n})] = \mathcal{L}[(X_{j_1}, \ldots, X_{j_n})]$.

In particular ($n = 1$), exchangeable random variables are identically distributed. ◊

Example 12.3

(i) If $(X_i)_{i \in I}$ is i.i.d., then $(X_i)_{i \in I}$ is exchangeable.
(ii) Consider an urn with N balls, M of which are black. Successively draw without replacement all of the balls and define

$$X_n := \begin{cases} 1, & \text{if the } n\text{th ball is black,} \\ 0, & \text{else.} \end{cases}$$

Then $(X_n)_{n=1,\ldots,N}$ is exchangeable. Indeed, this follows by elementary combinatorics since for any choice $x_1, \ldots, x_N \in \{0, 1\}$ with $x_1 + \ldots + x_N = M$, we have

$$\mathbf{P}[X_1 = x_1, \ldots, X_N = x_N] = \frac{1}{\binom{N}{M}}.$$

This formula can be derived formally via a small computation with conditional probabilities. As we will need a similar computation for Pólya's urn model in Example 12.29, we give the details here. Let $s_k = x_1 + \ldots + x_k$ for $k = 0, \ldots, N$ and let

$$g_k(x) = \begin{cases} M - s_k, & \text{if } x = 1, \\ N - M + s_k - k, & \text{if } x = 0. \end{cases}$$

Then $\mathbf{P}[X_1 = x_1] = g_0(x_1)/N$ and

$$\mathbf{P}[X_{k+1} = x_{k+1} \mid X_1 = x_1, \ldots, X_k = x_k] = \frac{g_k(x_{k+1})}{N - k} \qquad \text{for } k = 1, \ldots, N - 1.$$

Clearly, $g_k(0) = N - M - l$, where $l = \#\{i \leq k : x_i = 0\}$. Therefore,

$$\mathbf{P}[X_1 = x_1, \ldots, X_N = x_N]$$

$$= \mathbf{P}[X_1 = x_1] \prod_{k=1}^{N-1} \mathbf{P}[X_{k+1} = x_{k+1} \mid X_1 = x_1, \ldots, X_k = x_k]$$

$$= \frac{1}{N!} \prod_{k=0}^{N-1} g_k(x_{k+1}) = \frac{1}{N!} \prod_{k: x_k=1} g_k(1) \prod_{k: x_k=0} g_k(0)$$

$$= \frac{1}{N!} \prod_{l=0}^{M-1} (M - l) \prod_{l=0}^{N-1} (N - M - l) = \frac{M!\,(N - M)!}{N!}.$$

(iii) Let Y be a random variable with values in $[0, 1]$. Assume that, given Y, the random variables $(X_i)_{i \in I}$ are independent and Ber_Y-distributed. That is, for any finite $J \subset I$,

$$\mathbf{P}[X_j = 1 \ \text{for all} \ \ j \in J \mid Y] = Y^{\#J}.$$

Then $(X_i)_{i \in I}$ is exchangeable. ◊

Reflection Find an example of an exchangeable family $(X_n)_{n \in \mathbb{N}}$ of $\{0, 1\}$-valued random variables that is not independent.♠

Let $X = (X_n)_{n \in \mathbb{N}}$ be a stochastic process with values in a Polish space E. Let $S(n)$ be the set of permutations $\varrho : \{1, \ldots, n\} \to \{1, \ldots, n\}$. We consider ϱ also as a map $\mathbb{N} \to \mathbb{N}$ by defining $\varrho(k) = k$ for $k > n$. For $\varrho \in S(n)$ and $x = (x_1, \ldots, x_n) \in E^n$, denote $x^\varrho = (x_{\varrho(1)}, \ldots, x_{\varrho(n)})$. Similarly, for $x \in E^{\mathbb{N}}$, denote $x^\varrho = (x_{\varrho(1)}, x_{\varrho(2)}, \ldots) \in E^{\mathbb{N}}$. Let E' be another Polish space. For measurable maps $f : E^n \to E'$ and $F : E^{\mathbb{N}} \to E'$, define the maps f^ϱ and F^ϱ by $f^\varrho(x) = f(x^\varrho)$ and $F^\varrho(x) = F(x^\varrho)$. Further, we write $f(x) = f(x_1, \ldots, x_n)$ for $x \in E^n$ and for $x \in E^{\mathbb{N}}$.

Definition 12.4

(i) A map $f : E^n \to E'$ is called **symmetric** if $f^\varrho = f$ for all $\varrho \in S(n)$.
(ii) A map $F : E^{\mathbb{N}} \to E'$ is called n-symmetric if $F^\varrho = F$ for all $\varrho \in S(n)$. F is called symmetric if F is n-symmetric for all $n \in \mathbb{N}$.

Example 12.5

(i) For $x \in \mathbb{R}^{\mathbb{N}}$, define the nth arithmetic mean by $a_n(x) = \frac{1}{n} \sum_{i=1}^{n} x_i$. Clearly, a_n is an n-symmetric map (but not m-symmetric for any $m > n$). Furthermore, $\bar{a}(x) := \limsup_{n \to \infty} a_n(x)$ defines a symmetric map $\mathbb{R}^{\mathbb{N}} \to \mathbb{R} \cup \{-\infty, +\infty\}$.
(ii) The map $s : \mathbb{R}^{\mathbb{N}} \to [0, \infty]$, $x \mapsto \sum_{i=1}^{\infty} |x_i|$ is symmetric. Unlike $\bar{a}$, the value of s depends on every coordinate if it is finite.

(iii) For $x \in E^{\mathbb{N}}$, define the nth empirical distribution by $\xi_n(x) = \frac{1}{n} \sum_{i=1}^{n} \delta_{x_i}$ (recall that δx_i is the Dirac measure at the point x_i). Clearly, ξ_n is an n-symmetric map.

(iv) Let $k \in \mathbb{N}$ and let $\varphi : E^k \to \mathbb{R}$ be a map. The nth symmetrized average

$$A_n(\varphi) : E^{\mathbb{N}} \to \mathbb{R}, \quad x \mapsto \frac{1}{n!} \sum_{\varrho \in S(n)} \varphi(x^{\varrho}) \tag{12.1}$$

is an n-symmetric map. $\Diamond$

Definition 12.6 Let $X = (X_n)_{n \in \mathbb{N}}$ be a stochastic process with values in E. For $n \in \mathbb{N}$, define

$$\mathcal{E}'_n := \sigma(F : \ F : E^{\mathbb{N}} \to \mathbb{R} \ \text{is measurable and n-symmetric})$$

and let $\mathcal{E}_n := X^{-1}(\mathcal{E}'_n)$ be the σ-algebra of events that are invariant under all permutations $\varrho \in S(n)$. Further, let

$$\mathcal{E}' := \bigcap_{n=1}^{\infty} \mathcal{E}'_n = \sigma\left(F : \ F : E^{\mathbb{N}} \to \mathbb{R} \ \text{is measurable and symmetric}\right)$$

and let $\mathcal{E}_n := \bigcap_{n=1}^{\infty} \mathcal{E}_n = X^{-1}(\mathcal{E}')$ be the σ-algebra of exchangeable events for X, or briefly the **exchangeable σ-algebra**.

Remark 12.7 If $A \in \sigma(X_n, \ n \in \mathbb{N})$ is an event, then there is a measurable $B \subset E^{\mathbb{N}}$ with $A = \{X \in B\}$. If we denote $A^{\varrho} = \{X^{\varrho} \in B\}$ for $\varrho \in S(n)$, then $\mathcal{E}_n = \{A : A^{\varrho} = A \ \text{for all} \ \varrho \in S(n)\}$. This justifies the name "exchangeable event". $\Diamond$

Remark 12.8 If we write $\Xi_n(\omega) := \xi_n(X(\omega)) = \frac{1}{n} \sum_{i=1}^{n} \delta_{X_i(\omega)}$ for the nth empirical distribution, then, by Exercise 12.1.1, we have $\mathcal{E}_n \supset \sigma(\Xi_n)$ and $\mathcal{E}_n = \sigma(\Xi_n, X_{n+1}, X_{n+2}, \ldots)$.. $\Diamond$

Remark 12.9 Denote by $\mathcal{T} = \bigcap_{n \in \mathbb{N}} \sigma(X_{n+1}, X_{n+2}, \ldots)$ the tail σ-algebra. Then $\mathcal{T} \subset \mathcal{E}$, and strict inclusion is possible.

Indeed, evidently $\sigma(X_{n+1}, X_{n+2}, \ldots) \subset \mathcal{E}_n$ for $n \in \mathbb{N}$; hence $\mathcal{T} \subset \mathcal{E}$. Now let $E = \{0, 1\}$ and let $X_1, X_2, \ldots$ be independent random variables with $\mathbf{P}[X_n = 1] \in (0, 1)$ for all $n \in \mathbb{N}$. The random variable $S := \sum_{n=1}^{\infty} X_n$ is measurable with respect to $\mathcal{E}$ but not with respect to $\mathcal{T}$. $\Diamond$

Theorem 12.10 *Let $X = (X_n)_{n \in \mathbb{N}}$ be exchangeable. If $\varphi : E^{\mathbb{N}} \to \mathbb{R}$ is measurable and if $\mathbf{E}[|\varphi(X)|] < \infty$, then for all $n \in \mathbb{N}$ and all $\varrho \in S(n)$,*

$$\mathbf{E}[\varphi(X) | \mathcal{E}_n] = \mathbf{E}[\varphi(X^{\varrho}) | \mathcal{E}_n]. \tag{12.2}$$

In particular,

$$\mathbf{E}[\varphi(X)\,|\,\mathcal{E}_n] = A_n(\varphi) := \frac{1}{n!} \sum_{\varrho \in S(n)} \varphi(X^\varrho). \tag{12.3}$$

Proof Let $A \in \mathcal{E}_n$. Then there exists a $B \in \mathcal{E}'_n$ such that $A = X^{-1}(B)$. Let $F = \mathbb{1}_B$. Then $F \circ X = \mathbb{1}_A$. By the definition of $\mathcal{E}_n$, the map $F : E^{\mathbb{N}} \to \mathbb{R}$ is measurable, n-symmetric and bounded. Therefore,

$$\mathbf{E}\big[\varphi(X)F(X)\big] = \mathbf{E}\big[\varphi(X^\varrho)F(X^\varrho)\big] = \mathbf{E}\big[\varphi(X^\varrho)F(X)\big].$$

Here we used the exchangeability of X in the first equality and the symmetry of F in the second equality. From this (12.2) follows. However, $A_n(\varphi)$ is $\mathcal{E}_n$-measurable and hence

$$\mathbf{E}\big[\varphi(X)\,|\,\mathcal{E}_n\big] = \mathbf{E}\left[\frac{1}{n!} \sum_{\varrho \in S(n)} \varphi(X^\varrho)\,\Big|\,\mathcal{E}_n\right] = \frac{1}{n!} \sum_{\varrho \in S(n)} \varphi(X^\varrho). \qquad \square$$

Heuristic for the Structure of Exchangeable Families

Consider a finite exchangeable family $X_1, \ldots, X_N$ of E-valued random variables. For $n \leq N$, what is the conditional distribution of $(X_1, \ldots, X_n)$ given Ξ_N? For any measurable $A \subset E$, $\{X_i \in A\}$ occurs for exactly $N\Xi_N(A)$ of the $i \in \{1, \ldots, N\}$, where the order does not change the probability. Hence we are in the situation of drawing colored balls *without* replacement. More precisely, let the pairwise distinct points $e_1, \ldots, e_k \in E$ be the atoms of Ξ_N and let $N_1, \ldots, N_k$ be the corresponding absolute frequencies. Hence $\Xi_N = \sum_{i=1}^{k}(N_i/N)\delta_{e_i}$. We thus deal with balls of k different colors and with N_i balls of the ith color. We draw n of these balls without replacement but respecting the order. Up to the order, the resulting distribution is thus the generalized hypergeometric distribution (see (1.19) on page 48). Hence, for pairwise disjoint, measurable sets $A_1, \ldots, A_k$ with $\biguplus_{l=1}^{k} A_l = E$, for $i_1, \ldots, i_n \in \{1, \ldots, k\}$, pairwise distinct $j_1, \ldots, j_n \in \{1, \ldots, N\}$ and with the convention $m_l := \#\{r \in \{1, \ldots, n\} : i_r = l\}$ for $l \in \{1, \ldots, k\}$, we have

$$\mathbf{P}\big[X_{j_r} \in A_{i_r} \text{ for all } r = 1, \ldots, n\,\big|\,\Xi_N\big] = \frac{1}{(N)_n} \prod_{l=1}^{k} \big(N\Xi_N(A_l)\big)_{m_l}. \tag{12.4}$$

Here we defined $(n)_l := n(n-1)\cdots(n-l+1)$.

What happens if we let $N \to \infty$? For simplicity, assume that for all $l = 1, \ldots, k$, the limit $\varXi_\infty(A_l) = \lim_{N \to \infty} \varXi_N(A_l)$ exists in a suitable sense. Then (12.4) formally becomes

$$\mathbf{P}\big[X_{j_r} \in A_{i_r} \text{ for all } r = 1, \ldots, n \,\big|\, \varXi_\infty\big] = \prod_{l=1}^{k} \varXi_\infty(A_l)^{m_l}. \tag{12.5}$$

Drawing without replacements thus asymptotically turns into drawing *with* replacements. Hence the random variables $X_1, X_2, \ldots$ are independent with distribution $\varXi_\infty$ given $\varXi_\infty$.

For a formal proof along the lines of this heuristic, see Sect. 13.4.

In order to formulate (and prove) this statement (de Finetti's theorem) rigorously in Sect. 12.3, we need some more technical tools (e.g., the notion of conditional independence). A further tool will be the convergence theorem for backwards martingales that will be formulated in Sect. 12.2. For further reading on exchangeable random variables, we refer to [4, 33, 98, 105].

Takeaways An event that is described by a sequence $X_1, X_2, \ldots$ of random variables is called symmetric if finitely many of the X_i can be permuted without changing the event. The event is called n-symmetric if we allow only permutations of $X_1, \ldots, X_n$ for some n. The exchangeable events form the so-called exchangeable σ-algebra $\mathcal{E}$. Every terminal event is symmetric. The family $X = (X_n)_{n \in \mathbb{N}}$ is called exchangeable if finitely many of the X_i can be permuted without changing the distribution of the family.

Exercise 12.1.1 Let $n \in \mathbb{N}$. Show that every symmetric function $f : E^n \to \mathbb{R}$ can be written in the form $f(x) = g\big(\frac{1}{n} \sum_{i=1}^{n} \delta_{x_i}\big)$, where g has to be chosen appropriately (depending on f). ♣

Exercise 12.1.2 Derive equation (12.4) formally. ♣

Exercise 12.1.3 Let $X_1, \ldots, X_n$ be exchangeable, square integrable random variables. Show that

$$\mathbf{Cov}[X_1, X_2] \geq -\frac{1}{n-1} \mathbf{Var}[X_1]. \tag{12.6}$$

For $n \geq 2$, give a nontrivial example for equality in (12.6). ♣

Exercise 12.1.4 Let $X_1, X_2, X_3 \ldots$ be exchangeable, square integrable random variables. Show that $\mathbf{Cov}[X_1, X_2] \geq 0$. ♣

Exercise 12.1.5 Show that for all $n \in \mathbb{N} \setminus \{1\}$, there is an exchangeable family of random variables $X_1, \ldots, X_n$ that cannot be extended to an infinite exchangeable family $X_1, X_2, \ldots$. ♣

12.2 Backwards Martingales

The concepts of filtration and martingale do not require the index set I (interpreted as time) to be a subset of $[0, \infty)$. Hence we can consider the case $I = -\mathbb{N}_0$.

Definition 12.11 (Backwards martingale) Let $\mathbb{F} = (\mathcal{F}_n)_{n \in -\mathbb{N}_0}$ be a filtration. A stochastic process $X = (X_n)_{n \in \mathbb{N}_0}$ is called a **backwards martingale** with respect to $\mathbb{F}$ if $X = (X_{-n})_{n \in -\mathbb{N}_0}$ is an $\mathbb{F}$-martingale.

Remark 12.12 A backwards martingale is always uniformly integrable. This follows from Corollary 8.22 and the fact that $X_{-n} = \mathbf{E}[X_0 \mid \mathcal{F}_{-n}]$ for any $n \in \mathbb{N}_0$. ◇

Example 12.13 Let $X_1, X_2, \ldots$ be exchangeable real random variables. For $n \in \mathbb{N}$, let $\mathcal{F}_{-n} = \mathcal{E}_n$ and

$$Y_n = \frac{1}{n} \sum_{i=1}^{n} X_i.$$

We show that $(Y_{-n})_{n \in \mathbb{N}}$ is an $\mathbb{F}$-backwards martingale. Clearly, Y is adapted. Furthermore, by Theorem 12.10 (with $k = n$ and $\varphi(X_1, \ldots, X_n) = \frac{1}{n-1}(X_1 + \ldots + X_{n-1})$),

$$\mathbf{E}\left[Y_{n-1} \mid \mathcal{F}_{-n}\right] = \frac{1}{n!} \sum_{\varrho \in S(n)} \frac{1}{n-1} \left(X_{\varrho(1)} + \ldots + X_{\varrho(n-1)}\right) = Y_n.$$

Now replace $\mathbb{F}$ by the smaller filtration $\mathbb{G} = (\mathcal{G}_n)_{n \in -\mathbb{N}}$ that is defined by $\mathcal{G}_{-n} = \sigma(Y_{-n}, X_{n+1}, X_{n+2}, \ldots) = \sigma(Y_{-n}, Y_{-n-1}, Y_{-n-2}, \ldots)$ for $n \in \mathbb{N}$. This is the filtration generated by Y; thus Y is also a $\mathbb{G}$-backwards martingale (see Remark 9.29). ◇

Let $a < b$ and $n \in \mathbb{N}$. Let $U_{-n}^{a,b}$ be the number of upcrossings of X over $[a, b]$ between times $-n$ and 0. Further, let $U^{a,b} = \lim_{n \to \infty} U_{-n}^{a,b}$. By the upcrossing inequality (Lemma 11.3), we have $\mathbf{E}[U_{-n}^{a,b}] \leq \frac{1}{b-a}\mathbf{E}[(X_0 - a)^+]$; hence $\mathbf{P}[U^{a,b} < \infty] = 1$. As in the proof of the martingale convergence theorem (Theorem 11.4), we infer the following.

Theorem 12.14 (Convergence theorem for backwards martingales) *Let $(X_n)_{n \in -\mathbb{N}_0}$ be a martingale with respect to $\mathbb{F} = (\mathcal{F}_n)_{n \in -\mathbb{N}_0}$. Then there exists $X_{-\infty} = \lim\limits_{n \to \infty} X_{-n}$ almost surely and in L^1. Furthermore, $X_{-\infty} = \mathbf{E}[X_0 \mid \mathcal{F}_{-\infty}]$, where $\mathcal{F}_{-\infty} = \bigcap\limits_{n=1}^{\infty} \mathcal{F}_{-n}$.*

Example 12.15 Let $X_1, X_2, \ldots$ be exchangeable, integrable random variables. Further, let $\mathcal{T} = \bigcap_{n=1}^{\infty} \sigma(X_m, \, m \geq n)$ be the tail σ-algebra of $X_1, X_2, \ldots$ and let $\mathcal{E}$ be the exchangeable σ-algebra. Then $\mathbf{E}[X_1 \mid \mathcal{T}] = \mathbf{E}[X_1 \mid \mathcal{E}]$ a.s. and

$$\frac{1}{n} \sum_{i=1}^{n} X_i \overset{n \to \infty}{\longrightarrow} \mathbf{E}[X_1 \mid \mathcal{E}] \quad \text{a.s. and in } L^1.$$

Indeed, if we let $Y_n := \frac{1}{n} \sum\limits_{i=1}^{n} X_i$, then (by Example 12.13) $(Y_n)_{n \in \mathbb{N}}$ is a backwards martingale with respect to $(\mathcal{F}_n)_{n \in -\mathbb{N}} = (\mathcal{E}_{-n})_{n \in -\mathbb{N}}$ and thus

$$Y_n \overset{n \to \infty}{\longrightarrow} Y_\infty = \mathbf{E}[X_1 \mid \mathcal{E}] \quad \text{a.s. and in } L^1.$$

Now, by Example 2.36(ii), Y_∞ is $\mathcal{T}$-measurable; hence (since $\mathcal{T} \subset \mathcal{E}$ and by virtue of the tower property of conditional expectation) $Y_\infty = \mathbf{E}[X_1 \mid \mathcal{T}]$. ◊

Example 12.16 (Strong law of large numbers) If $Z_1, Z_2, \ldots$ are real and i.i.d. with $\mathbf{E}[|Z_1|] < \infty$, then

$$\frac{1}{n} \sum_{i=1}^{n} Z_i \overset{n \to \infty}{\longrightarrow} \mathbf{E}[Z_1] \quad \text{almost surely.}$$

By Kolmogorov's 0-1 law (Theorem 2.37), the tail σ-algebra $\mathcal{T}$ is trivial; hence we have

$$\mathbf{E}[Z_1 \mid \mathcal{T}] = \mathbf{E}[Z_1] \quad \text{almost surely.}$$

In Corollary 12.19, we will see that in the case of independent random variables, $\mathcal{E}$ is also **P**-trivial. This implies $\mathbf{E}[Z_1 \mid \mathcal{E}] = \mathbf{E}[Z_1]$. ◊

We close this section with a generalization of Example 12.15 to mean values of functions of $k \in \mathbb{N}$ variables. This conclusion from the convergence theorem for backwards martingales will be used in an essential way in the next section.

Theorem 12.17 *Let $X = (X_n)_{n \in \mathbb{N}}$ be an exchangeable family of random variables with values in E. Assume that $k \in \mathbb{N}$ and let $\varphi : E^k \to \mathbb{R}$ be measurable with $\mathbf{E}[|\varphi(X_1, \ldots, X_k)|] < \infty$. Denote $\varphi(X) = \varphi(X_1, \ldots, X_k)$ and let $A_n(\varphi) := \frac{1}{n!} \sum_{\varrho \in S(n)} \varphi(X^\varrho)$. Then*

$$\mathbf{E}[\varphi(X)\,|\,\mathcal{E}] = \mathbf{E}[\varphi(X)\,|\,\mathcal{T}] = \lim_{n\to\infty} A_n(\varphi) \quad \textit{a.s. and in } L^1. \tag{12.7}$$

Proof By Theorem 12.10, $A_n(\varphi) = \mathbf{E}[\varphi(X)\,|\,\mathcal{E}_n]$. Hence $(A_n(\varphi))_{n\ge k}$ is a backwards martingale with respect to $(\mathcal{E}_{-n})_{n\in -\mathbb{N}}$. Hence, by Theorem 12.14,

$$A_n(\varphi) \overset{n\to\infty}{\longrightarrow} \mathbf{E}\big[\varphi(X)\,\big|\,\mathcal{E}\big] \quad \text{a.s. and in } L^1. \tag{12.8}$$

As for the arithmetic mean (Example 12.16), we can argue that $\lim_{n\to\infty} A_n(\varphi)$ is $\mathcal{T}$-measurable. Indeed,

$$\limsup_{n\to\infty} \frac{\#\big\{\varrho \in S(n) : \varrho^{-1}(i) \le l \text{ for some } i \in \{1,\ldots,k\}\big\}}{n!} = 0 \quad \text{for all } l \in \mathbb{N}.$$

Thus, for large n, the dependence of $A_n(\varphi)$ on the first l coordinates is negligible. Together with (12.8), we get (12.7). $\qquad\square$

Corollary 12.18 *Let* $X = (X_n)_{n\in\mathbb{N}}$ *be exchangeable. Then, for any* $A \in \mathcal{E}$ *there exists a* $B \in \mathcal{T}$ *with* $\mathbf{P}[A \triangle B] = 0$.

Note that $\mathcal{T} \subset \mathcal{E}$; hence the statement is trivially true if the roles of $\mathcal{E}$ and $\mathcal{T}$ are interchanged.

Proof Since $\mathcal{E} \subset \sigma(X_1, X_2, \ldots)$, by the approximation theorem for measures, there exists a sequence of measurable sets $(A_k)_{k\in\mathbb{N}}$ with $A_k \in \sigma(X_1, \ldots, X_k)$ and such that $\mathbf{P}[A \triangle A_k] \overset{k\to\infty}{\longrightarrow} 0$. Choose $(k_l)_{l\in\mathbb{N}}$ such that $\sum_{l=1}^{\infty} \mathbf{P}[A \triangle A_{k_l}] < \infty$, hence $\mathbb{1}_{A_{k_l}} \overset{l\to\infty}{\longrightarrow} \mathbb{1}_A$ almost surely. Let $C_k \subset E^k$ be measurable with

$$A_k = \{(X_1, \ldots, X_k) \in C_k\}$$

for all $k \in \mathbb{N}$. Letting $\varphi_k := \mathbb{1}_{C_k}$, Theorem 12.17 implies that

$$\mathbb{1}_A = \mathbf{E}[\mathbb{1}_A\,|\,\mathcal{E}] = \mathbf{E}\Big[\lim_{l\to\infty} \varphi_{k_l}(X)\,\Big|\,\mathcal{E}\Big] = \lim_{l\to\infty} \mathbf{E}[\varphi_{k_l}(X)\,|\,\mathcal{E}]$$

$$= \lim_{l\to\infty} \mathbf{E}[\varphi_{k_l}(X)\,|\,\mathcal{T}] =: \psi \quad \text{almost surely.}$$

Hence there is a $\mathcal{T}$-measurable function ψ with $\psi = \mathbb{1}_A$ almost surely. We can assume that $\psi = \mathbb{1}_B$ for some $B \in \mathcal{T}$. $\qquad\square$

As a further application, we get the 0-1 law of Hewitt and Savage [72].

Corollary 12.19 (0-1 law of Hewitt–Savage) *Let* $X_1, X_2, \ldots$ *be i.i.d. random variables. Then the exchangeable* σ-*algebra is* $\mathbf{P}$-*trivial; that is,* $\mathbf{P}[A] \in \{0, 1\}$ *for all* $A \in \mathcal{E}$.

Proof By Kolmogorov's 0-1 law (Theorem 2.37), $\mathcal{T}$ is trivial. Hence the claim follows immediately from Corollary 12.18. $\qquad\square$

Takeaways A backwards martingale $(Y_n)_{n\in\mathbb{N}_0}$ is a stochastic process such that $(Y_{-m})_{m\in-\mathbb{N}_0}$ is a martingale. Backwards martingales are uniformly integrable and converge almost surely and in L^1. For functions of exchangeable random variables $X_1, X_2, \ldots$, averaging over all permutations of $X_1, \ldots, X_n$ defines a backwards martingale. As a consequence, the terminal σ-algebra and the exchangeable σ-algebra coincide (mod $\mathbf{P}$). In particular, for i.i.d. random variables, the exchangeable σ-algebra is $\mathbf{P}$-trivial.

12.3   De Finetti's Theorem

In this section, we show the structural theorem for countably infinite exchangeable families that was heuristically motivated at the end of Sect. 12.1. Hence we shall show that a countably infinite exchangeable family of random variables is an i.i.d. family given the exchangeable σ-algebra $\mathcal{E}$. Furthermore, we compute the conditional distribution of the individual random variables. As a first step, we define conditional independence formally (see [25, Chapter 7.3]).

Definition 12.20 (Conditional independence) Let $(\Omega, \mathcal{F}, \mathbf{P})$ be a probability space, let $\mathcal{A} \subset \mathcal{F}$ be a sub-σ-algebra and let $(\mathcal{A}_i)_{i\in I}$ be an arbitrary family of sub-σ-algebras of $\mathcal{F}$. Assume that for any finite $J \subset I$, any choice of $A_j \in \mathcal{A}_j$ and for all $j \in J$,

$$\mathbf{P}\left[\bigcap_{j\in J} A_j \,\middle|\, \mathcal{A}\right] = \prod_{j\in J} \mathbf{P}\left[A_j \,\middle|\, \mathcal{A}\right] \quad \text{almost surely.} \tag{12.9}$$

Then the family $(\mathcal{A}_i)_{i\in I}$ is called **independent given** $\mathcal{A}$.

A family $(X_i)_{i\in I}$ of random variables on $(\Omega, \mathcal{F}, \mathbf{P})$ is called independent (and identically distributed) given $\mathcal{A}$ if the generated σ-algebras $(\sigma(X_i))_{i\in I}$ are independent given $\mathcal{A}$ (and the conditional distributions $\mathbf{P}[X_i \in \cdot \,|\, \mathcal{A}]$ are equal).

Example 12.21 Any family $(\mathcal{A}_i)_{i\in I}$ of sub-σ-algebras of $\mathcal{F}$ is independent given $\mathcal{F}$. Indeed, letting $A = \bigcap_{j\in J} A_j$,

$$\mathbf{P}[A \,|\, \mathcal{F}] = \mathbb{1}_A = \prod_{j\in J} \mathbb{1}_{A_j} = \prod_{j\in J} \mathbf{P}\left[A_j \,|\, \mathcal{F}\right] \quad \text{almost surely.} \qquad \Diamond$$

Example 12.22 If $(\mathcal{A}_i)_{i\in I}$ is an independent family of σ-algebras and if $\mathcal{A}$ is trivial, then $(\mathcal{A}_i)_{i\in I}$ is independent given $\mathcal{A}$. $\Diamond$

Example 12.23 There is no "monotonicity" for conditional independence in the following sense: If $\mathcal{F}_1$, $\mathcal{F}_2$ and $\mathcal{F}_3$ are σ-algebras with $\mathcal{F}_1 \subset \mathcal{F}_2 \subset \mathcal{F}_3$ and such that $(\mathcal{A}_i)_{i \in I}$ is independent given $\mathcal{F}_1$ as well as given $\mathcal{F}_3$, then this does not imply independence given $\mathcal{F}_2$.

In order to illustrate this, assume that X and Y are nontrivial independent real random variables. Let $\mathcal{F}_1 = \{\emptyset, \Omega\}$, $\mathcal{F}_2 = \sigma(X + Y)$ and $\mathcal{F}_3 = \sigma(X, Y)$. Then $\sigma(X)$ and $\sigma(Y)$ are independent given $\mathcal{F}_1$ as well as given $\mathcal{F}_3$ but not given $\mathcal{F}_2$. ◇

Let $X = (X_n)_{n \in \mathbb{N}}$ be a stochastic process on a probability space $(\Omega, \mathcal{F}, \mathbf{P})$ with values in a Polish space E. Let $\mathcal{E}$ be the exchangeable σ-algebra and let $\mathcal{T}$ be the tail σ-algebra.

Theorem 12.24 (de Finetti) *The family $X = (X_n)_{n \in \mathbb{N}}$ is exchangeable if and only if there exists a σ-algebra $\mathcal{A} \subset \mathcal{F}$ such that $(X_n)_{n \in \mathbb{N}}$ is i.i.d. given $\mathcal{A}$. In this case, $\mathcal{A}$ can be chosen to equal the exchangeable σ-algebra $\mathcal{E}$ or the tail-σ-algebra $\mathcal{T}$.*

Proof "$\Longrightarrow$"   Let X be exchangeable and let $\mathcal{A} = \mathcal{E}$ or $\mathcal{A} = \mathcal{T}$. For any $n \in \mathbb{N}$, let $f_n : E \to \mathbb{R}$ be a bounded measurable map. Let

$$\varphi_k(x_1, \dots, x_k) = \prod_{i=1}^{k} f_i(x_i) \quad \text{for any } k \in \mathbb{N}.$$

Let $A_n(\varphi)$ be the symmetrized average from Theorem 12.17. Then

$$A_n(\varphi_{k-1}) A_n(f_k) = \frac{1}{n!} \sum_{\varrho \in S(n)} \varphi_{k-1}(X^\varrho) \frac{1}{n} \sum_{i=1}^{n} f_k(X_i)$$

$$= \frac{1}{n!} \sum_{\varrho \in S(n)} \varphi_k(X^\varrho) + R_{n,k} = A_n(\varphi_k) + R_{n,k},$$

where

$$|R_{n,k}| \leq 2 \|\varphi_{k-1}\|_\infty \cdot \|f_k\|_\infty \cdot \frac{1}{n!} \frac{1}{n} \sum_{\varrho \in S(n)} \sum_{i=1}^{n} \mathbb{1}_{\{i \in \{\varrho(1), \dots, \varrho(k-1)\}\}}$$

$$= 2 \|\varphi_{k-1}\|_\infty \cdot \|f_k\|_\infty \cdot \frac{k-1}{n} \overset{n \to \infty}{\longrightarrow} 0.$$

Together with Theorem 12.17, we conclude that

$$A_n(\varphi_{k-1}) A_n(f_k) \overset{n \to \infty}{\longrightarrow} \mathbf{E}\big[\varphi_k(X_1, \dots, X_k)\big|\mathcal{A}\big] \quad \text{a.s. and in } L^1.$$

On the other hand, again by Theorem 12.17,

$$A_n(\varphi_{k-1}) \overset{n \to \infty}{\longrightarrow} \mathbf{E}\big[\varphi_{k-1}(X_1, \dots, X_{k-1})\big|\mathcal{A}\big]$$

and

$$A_n(f_k) \xrightarrow{n \to \infty} \mathbf{E}\big[f_k(X_1)\,\big|\,\mathcal{A}\big].$$

Hence

$$\mathbf{E}\big[\varphi_k(X_1,\ldots,X_k)\,\big|\,\mathcal{A}\big] = \mathbf{E}\big[\varphi_{k-1}(X_1,\ldots,X_{k-1})\,\big|\,\mathcal{A}\big]\,\mathbf{E}\big[f_k(X_1)\,\big|\,\mathcal{A}\big].$$

Thus we get inductively

$$\mathbf{E}\left[\prod_{i=1}^{k} f_i(X_i)\,\bigg|\,\mathcal{A}\right] = \prod_{i=1}^{k} \mathbf{E}\big[f_i(X_1)\,\big|\,\mathcal{A}\big].$$

Therefore, X is i.i.d. given $\mathcal{A}$.

" $\Longleftarrow$ " Now let X be i.i.d. given $\mathcal{A}$ for a suitable σ-algebra $\mathcal{A} \subset \mathcal{F}$. For any bounded measurable function $\varphi : E^n \to \mathbb{R}$ and for any $\varrho \in S(n)$, we have $\mathbf{E}[\varphi(X)\,|\,\mathcal{A}] = \mathbf{E}[\varphi(X^\varrho)\,|\,\mathcal{A}]$. Hence

$$\mathbf{E}[\varphi(X)] = \mathbf{E}\big[\mathbf{E}[\varphi(X)\,|\,\mathcal{A}]\big] = \mathbf{E}\big[\mathbf{E}[\varphi(X^\varrho)\,|\,\mathcal{A}]\big] = \mathbf{E}[\varphi(X^\varrho)],$$

whence X is exchangeable. $\square$

Denote by $\mathcal{M}_1(E)$ the set of probability measures on E equipped with the topology of weak convergence (see Definition 13.12 and Remark 13.14). That is, a sequence $(\mu_n)_{n \in \mathbb{N}}$ in $\mathcal{M}_1(E)$ converges weakly to a $\mu \in \mathcal{M}_1(E)$ if and only if $\int f \, d\mu_n \xrightarrow{n \to \infty} \int f \, d\mu$ for any bounded continuous function $f : E \to \mathbb{R}$. We will study weak convergence in Chap. 13 in greater detail. At this point, we use the topology only to make $\mathcal{M}_1(E)$ a measurable space, namely with the Borel σ-algebra $\mathcal{B}(\mathcal{M}_1(E))$. Now we can study random variables with values in $\mathcal{M}_1(E)$, so-called random measures (compare also Sect. 24.1). For $x \in E^\mathbb{N}$, let $\xi_n(x) = \frac{1}{n} \sum_{i=1}^{n} \delta_{x_i} \in \mathcal{M}_1(E)$.

Definition 12.25 The random measure

$$\Xi_n := \xi_n(X) := \frac{1}{n} \sum_{i=1}^{n} \delta_{X_i}$$

is called the **empirical distribution** of $X_1, \ldots, X_n$.

Assume the conditions of Theorem 12.24 are in force.

Theorem 12.26 (de Finetti representation theorem) *The family $X = (X_n)_{n \in \mathbb{N}}$ is exchangeable if and only if there is a σ-algebra $\mathcal{A} \subset \mathcal{F}$ and an $\mathcal{A}$-measurable random variable $\Xi_\infty : \Omega \to \mathcal{M}_1(E)$ with the property that given Ξ_∞, $(X_n)_{n \in \mathbb{N}}$ is i.i.d. with $\mathcal{L}[X_1\,|\,\Xi_\infty] = \Xi_\infty$. In this case, we can choose $\mathcal{A} = \mathcal{E}$ or $\mathcal{A} = \mathcal{T}$.*

Proof "$\Longleftarrow$"   This follows as in the proof of Theorem 12.24.
"$\Longrightarrow$"   Let X be exchangeable. Then, by Theorem 12.24, there exists a σ-algebra $\mathcal{A} \subset \mathcal{F}$ such that $(X_n)_{n \in \mathbb{N}}$ is i.i.d. given $\mathcal{A}$. As E is Polish, there exists a regular conditional distribution (see Theorem 8.37) $\Xi_\infty := \mathcal{L}[X_1 | \mathcal{A}]$. For measurable $A_1, \ldots, A_n \subset E$, we have $\mathbf{P}[X_i \in A_i | \mathcal{A}] = \Xi_\infty(A_i)$ for all $i = 1, \ldots, n$; hence

$$
\mathbf{P}\left[\bigcap_{i=1}^n \{X_i \in A_i\} \,\bigg|\, \Xi_\infty \right] = \mathbf{E}\left[\mathbf{P}\left[\bigcap_{i=1}^n \{X_i \in A_i\} \,\bigg|\, \mathcal{A} \right] \,\bigg|\, \Xi_\infty \right]
$$

$$
= \mathbf{E}\left[\prod_{i=1}^n \Xi_\infty(A_i) \,\bigg|\, \Xi_\infty \right] = \prod_{i=1}^n \Xi_\infty(A_i).
$$

Therefore, $\mathcal{L}[X | \Xi_\infty] = \Xi_\infty^{\otimes \mathbb{N}}$. $\square$

Remark 12.27

(i) In the case considered in the previous theorem, by the strong law of large numbers, for any bounded continuous function $f : E \to \mathbb{R}$,

$$
\int f \, d\Xi_n \overset{n \to \infty}{\longrightarrow} \int f \, d\Xi_\infty \quad \text{almost surely.}
$$

If in addition E is locally compact (e.g., $E = \mathbb{R}^d$), then one can even show that

$$
\Xi_n \overset{n \to \infty}{\longrightarrow} \Xi_\infty \quad \text{almost surely.}
$$

(ii) For finite families of random variables there is no perfect analog of de Finetti's theorem. See [33] for a detailed treatment of finite exchangeable families. $\Diamond$

Example 12.28 Let $(X_n)_{n \in \mathbb{N}}$ be exchangeable and assume $X_n \in \{0, 1\}$. Then there exists a random variable $Y : \Omega \to [0, 1]$ such that, for all finite $J \subset \mathbb{N}$,

$$
\mathbf{P}\big[X_j = 1 \ \text{for all} \ j \in J \,\big|\, Y \big] = Y^{\#J}.
$$

In other words, $(X_n)_{n \in \mathbb{N}}$ is independent given Y and Ber_Y-distributed. Compare Example 12.3(iii). $\Diamond$

Example 12.29 (Pólya's urn model) (See Example 14.41, compare also [17, 135] and [58].) Consider an urn with a total of N balls among which M are black and $M - N$ are white. At each step, a ball is drawn and is returned to the urn together with an *additional* ball of the same color. Let

$$
X_n := \begin{cases} 1, & \text{if the } n\text{th ball is black,} \\ 0, & \text{else,} \end{cases}
$$

and let $S_n = \sum_{i=1}^{n} X_i$. Then

$$\mathbf{P}\big[X_n = 1 \,\big|\, X_1, X_2, \ldots, X_{n-1}\big] = \frac{S_{n-1} + M}{N + n - 1}.$$

Inductively, for $x_1, \ldots, x_n \in \{0, 1\}$ and $s_k = \sum_{i=1}^{k} x_i$, we get

$$\mathbf{P}\big[X_i = x_i \ \text{for any} \ i = 1, \ldots, n\big]$$

$$= \prod_{i \leq n:\, x_i = 1} \frac{M + s_{i-1}}{N + i - 1} \prod_{i \leq n:\, x_i = 0} \frac{N + i - 1 - M - s_{i-1}}{N + i - 1}$$

$$= \frac{(N-1)!}{(N-1+n)!} \cdot \frac{(M + s_n - 1)!}{(M-1)!} \frac{\big(N - M - 1 + (n - s_n)\big)!}{(N - M - 1)!}.$$

The right-hand side depends on s_n only and not on the order of the $x_1, \ldots, x_n$. Hence $(X_n)_{n \in \mathbb{N}}$ is exchangeable. Let $Z = \lim_{n \to \infty} \frac{1}{n} S_n$. Then $(X_n)_{n \in \mathbb{N}}$ is i.i.d. Ber$_Z$-distributed given Z. Hence (see Example 12.28)

$$\mathbf{E}\big[Z^n\big] = \mathbf{E}\big[\mathbf{P}\big[X_1 = \cdots = X_n = 1 \,\big|\, Z\big]\big]$$

$$= \mathbf{P}[S_n = n]$$

$$= \frac{(N-1)!}{(M-1)!} \frac{(M+n-1)!}{(N+n-1)!} \quad \text{for all} \ n \in \mathbb{N}.$$

By Exercise 5.1.3, these are the moments of the Beta distribution $\beta_{M, N-M}$ on $[0, 1]$ with parameters $(M, N - M)$ (see Example 1.107(ii)). A distribution on $[0, 1]$ is uniquely characterized by its moments (see Theorem 15.4). Hence $Z \sim \beta_{M, N-M}$. $\Diamond$

Takeaways Consider a two-step random experiment where in the first step we choose a probability measure Ξ_∞ on some space E. In the second step, we draw i.i.d. random variables $X_1, X_2, \ldots$ with distribution Ξ_∞. Then the sequence $X_1, X_2, \ldots$ is exchangeable. Now, de Finetti's theorem states that any infinite and exchangeable family can be constructed in this way and that Ξ_∞ is measurable with respect to the exchangeable σ-algebra. For finite families, this is not true.

Exercise 12.3.1 Let $(X_n)_{n \in \mathbb{Z}}$ be an exchangeable family of $\{0, 1\}$-valued random variables.

(i) Show that the distribution of $(X_n)_{n \in \mathbb{Z}}$ is uniquely determined by the values

$$m_n := \mathbf{E}[X_1 \cdot X_2 \cdots X_n], \qquad n \in \mathbb{N}.$$

(ii) Conclude that for any random variable Y on $[0, 1]$, the distribution is uniquely determined by its moments $m_n := \mathbf{E}[Y^n]$, $n \in \mathbb{N}$. ♣

Chapter 13
Convergence of Measures

One focus of probability theory is distributions that are the result of an interplay of a large number of random impacts. Often a useful approximation can be obtained by taking a limit of such distributions, for example, a limit where the number of impacts goes to infinity. With the Poisson distribution, we have encountered such a limit distribution that occurs as the number of very rare events when the number of possibilities goes to infinity (see Theorem 3.7). In many cases, it is necessary to rescale the original distributions in order to capture the behavior of the essential fluctuations, e.g., in the central limit theorem. While these theorems work with real random variables, we will also see limit theorems where the random variables take values in more general spaces such as the space of continuous functions when we model the path of the random motion of a particle.

In this chapter, we provide the abstract framework for the investigation of convergence of measures. We introduce the notion of weak convergence of probability measures on general (mostly Polish) spaces and derive the fundamental properties. The reader will profit from a solid knowledge of point set topology. Thus we start with a short overview of some topological definitions and theorems.

We do not strive for the greatest generality but rather content ourselves with the key theorems for probability theory. For further reading, we recommend [14] and [82].

At first reading, the reader might wish to skip this rather analytically flavored chapter. In this case, for the time being it suffices to get acquainted with the definitions of weak convergence and tightness (Definitions 13.12 and 13.26), as well as with the statements of the Portemanteau theorem (Theorem 13.16) and Prohorov's theorem (Theorem 13.29).

A. Klenke, *Probability Theory*, Universitext, https://doi.org/10.1007/978-3-030-56402-5_13

13.1 A Topology Primer

Excursively, we present some definitions and facts from point set topology. For details, see, e.g., [90].

In the following, let (E, τ) be a topological space with the Borel σ-algebra $\mathcal{E} = \mathcal{B}(E)$ (compare Definitions 1.20 and 1.21). We will also assume that (E, τ) is a **Hausdorff space**; that is, for any two points $x, y \in E$ with $x \neq y$, there exist disjoint open sets U, V such that $x \in U$ and $y \in V$.

For $A \subset E$, we denote by $\overline{A}$ the **closure** of A, by A° the **interior** and by ∂A the **boundary** of A. A set $A \subset E$ is called **dense** if $\overline{A} = E$.

(E, τ) is called **metrizable** if there exists a metric d on E such that τ is induced by the open balls $B_\varepsilon(x) := \{y \in E : d(x, y) < \varepsilon\}$. A metric d on E is called **complete** if any Cauchy sequence with respect to d converges in E. (E, τ) is called **completely metrizable** if there exists a complete metric on E that induces τ. If (E, d) is a metric space and $A, B \subset E$, then we write $d(A, B) = \inf\{d(x, y) : x \in A, \ y \in B\}$ and $d(x, B) := d(\{x\}, B)$ for $x \in E$.

A metrizable space (E, τ) is called **separable** if there exists a countable dense subset of E. Separability in metrizable spaces is equivalent to the existence of a **countable base of the topology**; that is, a countable set $\mathcal{U} \subset \tau$ with $A = \bigcup_{U \in \mathcal{U}: U \subset A} U$ for all $A \in \tau$. (For example, choose the ε-balls centered at the points of a countable subset and let ε run through the positive rational numbers.) A compact metric space is always separable (simply choose for each $n \in \mathbb{N}$ a finite cover $\mathcal{U}_n \subset \tau$ comprising balls of radius $\frac{1}{n}$ and then take $\mathcal{U} := \bigcup_{n \in \mathbb{N}} \mathcal{U}_n$).

A set $A \subset E$ is called **compact** if each open cover $\mathcal{U} \subset \tau$ of A (that is, $A \subset \bigcup_{U \in \mathcal{U}} U$) has a finite subcover; that is, a finite $\mathcal{U}' \subset \mathcal{U}$ with $A \subset \bigcup_{U \in \mathcal{U}'} U$. Compact sets are closed. By the Heine–Borel theorem, a subset of $\mathbb{R}^d$ is compact if and only if it is bounded and closed. $A \subset E$ is called **relatively compact** if $\overline{A}$ is compact. On the other hand, A is called **sequentially compact** (respectively **relatively sequentially compact**) if any sequence $(x_n)_{n \in \mathbb{N}}$ with values in A has a subsequence $(x_{n_k})_{k \in \mathbb{N}}$ that converges to some $x \in A$ (respectively $x \in \overline{A}$). In metrizable spaces, the notions *compact* and *sequentially compact* coincide. A set $A \subset E$ is called σ-**compact** if A is a countable union of compact sets. E is called **locally compact** if any point $x \in E$ has an open neighborhood whose closure is compact. A locally compact, separable metric space is manifestly σ-compact and there even exists a countable basis $\mathcal{U}'$ of the topology consisting of relatively compact open sets. (In fact: Choose an arbitrary countable base $\mathcal{U}$ of the topology. For any $x \in E$, there exists a relatively compact neighborhood $B_x \ni x$. Each B_x is a union of elements of $\mathcal{U}$ each of which is then also relatively compact. Hence, there exists a relatively compact $U_x \in \mathcal{U}$ with $x \in U_x$. Now let $\mathcal{U}' := \{U_x : x \in E\} \subset \mathcal{U}$.)

If E is a locally compact metric space and if $U \subset E$ is open and $K \subset U$ is compact, then there exists a compact set L with $K \subset L^\circ \subset L \subset U$. (For example, for any $x \in K$, take an open ball $B_{\varepsilon_x}(x)$ of radius $\varepsilon_x > 0$ that is contained in U and that is relatively compact. By making ε_x smaller (if necessary), one can assume that the closure of this ball is contained in U. As K is compact, there are finitely many points $x_1, \ldots, x_n \in K$ with $K \subset V := \bigcup_{i=1}^{n} B_{\varepsilon_{x_i}}(x_i)$. By construction, $L = \overline{V} \subset U$ is compact.) Specializing on the case $U = E$, we get that for any compact set K, there exists a relatively compact open set $L^\circ \supset K$.

We present one type of topological space that is of particular importance in probability theory in a separate definition.

Definition 13.1 A topological space (E, τ) is called a **Polish space** if it is separable and if there exists a complete metric that induces the topology τ.

Examples of Polish spaces are countable discrete spaces (however, not $\mathbb{Q}$ with the usual topology), the Euclidean spaces $\mathbb{R}^n$, and the space $C([0, 1])$ of continuous functions $[0, 1] \to \mathbb{R}$, equipped with the supremum norm $\| \cdot \|_\infty$. In practice, all spaces that are of importance in probability theory are Polish spaces.

Let (E, d) be a metric space. A set $A \subset E$ is called **totally bounded** if, for any $\varepsilon > 0$, there exist finitely many points $x_1, \ldots, x_n \in A$ such that $A \subset \bigcup_{i=1}^{n} B_\varepsilon(x_i)$. Evidently, compact sets are totally bounded. In Polish spaces, a partial converse is true.

Lemma 13.2 *Let (E, τ) be a Polish space with complete metric d. A subset $A \subset E$ is totally bounded with respect to d if and only if A is relatively compact.*

Proof This is left as an exercise.   □

In the following, let (E, τ) be a topological space with Borel σ-algebra $\mathcal{E} = \mathcal{B}(E) := \sigma(\tau)$ and with complete metric d. For measures on $(E, \mathcal{E})$, we introduce the following notions of regularity.

Definition 13.3 A σ-finite measure μ on $(E, \mathcal{E})$ is called

(i) **locally finite** or a **Borel measure** if, for any point $x \in E$, there exists an open neighborhood $U \ni x$ such that $\mu(U) < \infty$,

(ii) **inner regular** if

$$\mu(A) = \sup \{ \mu(K) : K \subset A \text{ is compact} \} \quad \text{for all } A \in \mathcal{E}, \tag{13.1}$$

(iii) **outer regular** if

$$\mu(A) = \inf \{ \mu(U) : U \supset A \text{ is open} \} \quad \text{for all } A \in \mathcal{E}, \tag{13.2}$$

(iv) **regular** if μ is inner and outer regular, and

(v) a **Radon measure** if μ is an inner regular Borel measure.

Definition 13.4 We introduce the following spaces of measures on E:

$$\mathcal{M}(E) := \{\text{Radon measures on } (E, \mathcal{E})\},$$

$$\mathcal{M}_f(E) := \{\text{finite measures on } (E, \mathcal{E})\},$$

$$\mathcal{M}_1(E) := \{\mu \in \mathcal{M}_f(E) : \mu(E) = 1\},$$

$$\mathcal{M}_{\leq 1}(E) := \{\mu \in \mathcal{M}_f(E) : \mu(E) \leq 1\}.$$

The elements of $\mathcal{M}_{\leq 1}(E)$ are called **sub-probability measures** on E.

Further, we agree on the following notation for spaces of continuous functions:

$$C(E) := \{f : E \to \mathbb{R} \text{ is continuous}\},$$

$$C_b(E) := \{f \in C(E) \text{ is bounded}\},$$

$$C_c(E) := \{f \in C(E) \text{ has compact support}\} \subset C_b(E).$$

Recall that the support of a real function f is $\overline{f^{-1}(\mathbb{R} \setminus \{0\})}$.

Unless otherwise stated, the vector spaces $C(E)$, $C_b(E)$ and $C_c(E)$ are equipped with the supremum norm.

Lemma 13.5 *If E is Polish and $\mu \in \mathcal{M}_f(E)$, then for any $\varepsilon > 0$, there is a compact set $K \subset E$ with $\mu(E \setminus K) < \varepsilon$.*

Proof Let $\varepsilon > 0$. For each $n \in \mathbb{N}$, there exists a sequence $x_1^n, x_2^n, \ldots \in E$ with $E = \bigcup_{i=1}^{\infty} B_{1/n}(x_i^n)$. Fix $N_n \in \mathbb{N}$ such that $\mu\left(E \setminus \bigcup_{i=1}^{N_n} B_{1/n}(x_i^n)\right) < \dfrac{\varepsilon}{2^n}$. Define

$$A := \bigcap_{n=1}^{\infty} \bigcup_{i=1}^{N_n} B_{1/n}(x_i^n).$$

By construction, A is totally bounded. Since E is Polish, $\overline{A}$ is compact. Furthermore, it follows that $\mu\left(E \setminus \overline{A}\right) \leq \mu\left(E \setminus A\right) < \sum_{n=1}^{\infty} \varepsilon 2^{-n} = \varepsilon$. $\qquad\square$

Theorem 13.6 *If E is Polish and if $\mu \in \mathcal{M}_f(E)$, then μ is regular. In particular, in this case, $\mathcal{M}_f(E) \subset \mathcal{M}(E)$.*

Proof (**Outer regularity**)

Step 1. Let $B \subset E$ be closed and let $\varepsilon > 0$. Let d be a complete metric on E. For $\delta > 0$, let

$$B_\delta := \{x \in E : d(x, B) < \delta\}$$

be the open δ-neighborhood of B. As B is closed, we have $\bigcap_{\delta>0} B_\delta = B$. Since μ is upper semicontinuous (Theorem 1.36), there is a $\delta > 0$ such that $\mu(B_\delta) \leq \mu(B) + \varepsilon$.
Step 2. Let $B \in \mathcal{E}$ and $\varepsilon > 0$. Consider the class of sets

$$\mathcal{A} := \{V \cap C : V \subset E \text{ open}, \ C \subset E \text{ closed}\}.$$

Clearly, we have $\mathcal{E} = \sigma(\mathcal{A})$. It is easy to check that $\mathcal{A}$ is a semiring. Hence by the approximation theorem for measures (Theorem 1.65), there are mutually disjoint sets $A_n = V_n \cap C_n \in \mathcal{A}$, $n \in \mathbb{N}$, such that $B \subset A := \bigcup_{n=1}^\infty A_n$ and $\mu(A) \leq \mu(B) + \varepsilon/2$. As shown in the first step, for any $n \in \mathbb{N}$, there is an open set $W_n \supset C_n$ such that $\mu(W_n) \leq \mu(C_n) + \varepsilon\, 2^{-n-1}$. Hence also $U_n := V_n \cap W_n$ is open. Let $B \subset U := \bigcup_{n=1}^\infty U_n$. We conclude that $\mu(U) \leq \mu(A) + \sum_{n=1}^\infty \varepsilon\, 2^{-n-1} \leq \mu(B) + \varepsilon$.

Inner regularity Replacing B by B^c, the outer regularity yields the existence of a closed set $D \subset B$ with $\mu(B \setminus D) < \varepsilon/2$. By Lemma 13.5, there exists a compact set K with $\mu(K^c) < \varepsilon/2$. Define $C = D \cap K$. Then $C \subset B$ is compact and $\mu(B \setminus C) < \varepsilon$. Hence μ is also inner regular.

$\square$

Corollary 13.7 *The Lebesgue measure λ on $\mathbb{R}^d$ is a regular Radon measure. However, not all σ-finite measures on $\mathbb{R}^d$ are regular.*

Proof Clearly, $\mathbb{R}^d$ is Polish and λ is locally finite. Let $A \in \mathcal{B}(\mathbb{R}^d)$ and $\varepsilon > 0$. There is an increasing sequence $(K_n)_{n\in\mathbb{N}}$ of compact sets with $K_n \uparrow \mathbb{R}^d$. Since any K_n is bounded, we have $\lambda(K_n) < \infty$. Hence, by the preceding theorem, for any $n \in \mathbb{N}$, there exists an open set $U_n \supset A \cap K_n$ with $\lambda(U_n \setminus A) < \varepsilon/2^n$. Thus $\lambda(U \setminus A) < \varepsilon$ for the open set $U := \bigcup_{n\in\mathbb{N}} U_n$.

If $\lambda(A) < \infty$, then there exists an $n \in \mathbb{N}$ with $\lambda(A \setminus K_n) < \varepsilon/2$. By the preceding theorem, there exists a compact set $C \subset A \cap K_n$ with $\lambda((A \cap K_n) \setminus C) < \varepsilon/2$. Therefore, $\lambda(A \setminus C) < \varepsilon$.

If, on the other hand, $\lambda(A) = \infty$, then for any $L > 0$, we have to find a compact set $C \subset A$ with $\lambda(C) > L$. However, $\lambda(A \cap K_n) \xrightarrow{n\to\infty} \infty$; hence there exists an $n \in \mathbb{N}$ with $\lambda(A \cap K_n) > L + 1$. By what we have shown already, there exists a compact set $C \subset A \cap K_n$ with $\lambda((A \cap K_n) \setminus C) < 1$; hence $\lambda(C) > L$.

Finally, consider the measure $\mu = \sum_{q\in\mathbb{Q}} \delta_q$. Clearly, this measure is σ-finite; however, it is neither locally finite nor outer regular. $\square$

Definition 13.8 Let (E, d_E) and (F, d_F) be metric spaces. A function $f : E \to F$ is called **Lipschitz continuous** if there exists a constant $K < \infty$, the so-called Lipschitz constant, with $d_F(f(x), f(y)) \leq K \cdot d_E(x, y)$ for all $x, y \in E$. Denote by $\mathrm{Lip}_K(E; F)$ the space of Lipschitz continuous functions with constant K and by $\mathrm{Lip}(E; F) = \bigcup_{K>0} \mathrm{Lip}_K(E; F)$ the space of Lipschitz continuous functions on E.

We abbreviate $\mathrm{Lip}_K(E) := \mathrm{Lip}_K(E; \mathbb{R})$ and $\mathrm{Lip}(E) := \mathrm{Lip}(E; \mathbb{R})$.

Definition 13.9 Let $\mathcal{F} \subset \mathcal{M}(E)$ be a family of Radon measures. A family $\mathcal{C}$ of measurable maps $E \to \mathbb{R}$ is called a **separating family** for $\mathcal{F}$ if, for any two measures $\mu, \nu \in \mathcal{F}$, the following holds:

$$\left(\int f \, d\mu = \int f \, d\nu \quad \text{for all } f \in C \cap \mathcal{L}^1(\mu) \cap \mathcal{L}^1(\nu) \right) \quad \Longrightarrow \quad \mu = \nu.$$

Lemma 13.10 *Let (E, d) be a metric space. For any closed set $A \subset E$ and any $\varepsilon > 0$, there is a Lipschitz continuous map $\rho_{A,\varepsilon} : E \to [0, 1]$ with Lipschitz constant $1/\varepsilon$ and*

$$\rho_{A,\varepsilon}(x) = \begin{cases} 1, & \text{if } x \in A, \\ 0, & \text{if } d(x, A) \geq \varepsilon. \end{cases}$$

Proof Let $\varphi : \mathbb{R} \to [0, 1]$, $t \mapsto (t \vee 0) \wedge 1$. For $x \in E$, define $\rho_{A,\varepsilon}(x) = 1 - \varphi(\varepsilon^{-1} d(x, A))$. $\qquad\square$

Theorem 13.11 *Let (E, d) be a metric space.*

(i). $\mathrm{Lip}_1(E; [0, 1])$ is separating for $\mathcal{M}_f(E)$ and for $\mathcal{M}(E)$.
(ii). If, in addition, E is locally compact, then $C_c(E) \cap \mathrm{Lip}_1(E; [0, 1])$ is separating for $\mathcal{M}(E)$.

Proof (i)   **Case $\mathcal{M}_f(E)$.** Let $\mu_1, \mu_2 \in \mathcal{M}_f(E)$ be such that $\int f \, d\mu_1 = \int f \, d\mu_2$ for all $f \in \mathrm{Lip}_1(E; [0, 1])$. It is enough to show that $\mu_1(C) = \mu_2(C)$ for all closed sets $C \subset E$ as the closed sets form a $\cap$-stable generator of the Borel σ-algebra that contains E. Since μ_1 and μ_2 are finite measures, for the function $\rho_{C,\varepsilon}$ from Lemma 13.10, we have

$$0 \leq \rho_{C,\varepsilon} \leq 1 \in \mathcal{L}^1(\mu_i), \quad i = 1, 2, \quad \text{for all } \varepsilon > 0.$$

Note that $\rho_{C,\varepsilon} \xrightarrow{\varepsilon \to 0} \mathbb{1}_C$. Hence by the dominated convergence theorem (Corollary 6.26), we have $\mu_i(C) = \lim_{\varepsilon \to 0} \int \rho_{C,\varepsilon} \, d\mu_i$. For any $\varepsilon \in (0, 1]$, we have $\varepsilon \rho_{C,\varepsilon} \in \mathrm{Lip}_1(E; [0, 1])$. We conclude that

$$\int \rho_{C,\varepsilon} \, d\mu_1 = \varepsilon^{-1} \int (\varepsilon \rho_{C,\varepsilon}) \, d\mu_1 = \varepsilon^{-1} \int (\varepsilon \rho_{C,\varepsilon}) \, d\mu_2 = \int \rho_{C,\varepsilon} \, d\mu_2.$$

This implies $\mu_1(C) = \mu_2(C)$; hence $\mu_1 = \mu_2$.

(i) **Case $\mathcal{M}(E)$.**   In contrast to the case $\mathcal{M}_f(E)$, the function 1 is not integrable. Hence, we modify the approach by showing that it is enough to consider compact sets K instead of C and by showing that there exists an open set $U \supset K$ and a $\delta > 0$ such that

$$\rho_{K,\varepsilon} \leq \mathbb{1}_U \in \mathcal{L}^1(\mu_i), \quad i = 1, 2 \quad \text{for all } \varepsilon \in (0, \delta). \tag{13.3}$$

Arguing as above, this will show that $\mu_1(K) = \mu_2(K)$ and will hence conclude the proof.

Here come the details. Assume $\mu_1, \mu_2 \in \mathcal{M}(E)$ are measures with $\int f\, d\mu_1 = \int f\, d\mu_2$ for all $f \in \mathrm{Lip}_1(E; [0, 1])$. If $A \in \mathcal{E}$, then

$$\mu_i(A) = \sup\{\mu_i(K) : K \subset A \text{ is compact}\}$$

since the Radon measure μ_i is inner regular ($i = 1, 2$). Hence, it is enough to show that $\mu_1(K) = \mu_2(K)$ for any compact set K.

Now let $K \subset E$ be compact. Since μ_1 and μ_2 are locally finite, for every $x \in K$, there exists an open set $U_x \ni x$ with $\mu_1(U_x) < \infty$ and $\mu_2(U_x) < \infty$. Since K is compact, we can find finitely many points $x_1, \ldots, x_n \in K$ such that $K \subset U := \bigcup_{j=1}^{n} U_{x_j}$. By construction, $\mu_i(U) < \infty$; hence $\mathbb{1}_U \in L^1(\mu_i)$ for $i = 1, 2$. Since U^c is closed and since $U^c \cap K = \emptyset$, we get $\delta := d(U^c, K) > 0$. Let $\rho_{K,\varepsilon}$ be the map from Lemma 13.10. By construction, equation (13.3) holds and the proof is complete. (ii)

If E is locally compact, then in ((i)) we can choose the neighborhoods U_x to be relatively compact. Hence U is relatively compact; thus $\rho_{K,\varepsilon}$ has compact support and is thus in $C_c(E)$ for all $\varepsilon \in (0, \delta)$. $\qquad\square$

Takeaways A Polish space is a separable topological space that allows for a complete metric, e.g., the euclidian space $\mathbb{R}^d$. Polish spaces are standard spaces of measure theory and probability theory. Radon measures are inner regular Borel measures (locally finite measures). Finite measures on Polish spaces are Radon measures. The class of Lipschitz-continuous functions is a separating family for finite measures and for Radon measures. On locally compact spaces, also Lipschitz continuous functions with compact support form a separating class.

Exercise 13.1.1

(i) Show that $C([0, 1])$ has a separable dense subset.

(ii) Show that the space $(C_b([0, \infty)), \|\cdot\|_\infty)$ of bounded continuous functions, equipped with the supremum norm, is not separable.

(iii) Show that the space $C_c([0, \infty))$ of continuous functions with compact support, equipped with the supremum norm, is separable. ♣

Exercise 13.1.2 Let μ be a locally finite measure. Show that $\mu(K) < \infty$ for any compact set K. ♣

Exercise 13.1.3 (Lusin's theorem) Let Ω be a Polish space, let μ be a finite measure on $(\Omega, \mathcal{B}(\Omega))$ and let $f : \Omega \to \mathbb{R}$ be a map. Show that the following two statements are equivalent:

(i) There is a Borel measurable map $g : \Omega \to \mathbb{R}$ with $f = g$ μ-almost everywhere.

(ii) For any $\varepsilon > 0$, there is a compact set K_ε with $\mu(\Omega \setminus K_\varepsilon) < \varepsilon$ such that the restricted function $f|_{K_\varepsilon}$ is continuous. ♣

Exercise 13.1.4 Let $\mathcal{U}$ be a family of intervals in $\mathbb{R}$ such that $W := \bigcup_{U \in \mathcal{U}} U$ has finite Lebesgue measure $\lambda(W)$. Show that for any $\varepsilon > 0$, there exist finitely many pairwise disjoint sets $U_1, \ldots, U_n \in \mathcal{U}$ with

$$\sum_{i=1}^{n} \lambda(U_i) > \frac{1-\varepsilon}{3} \lambda(W).$$

Hint: Choose a finite family $\mathcal{U}' \subset \mathcal{U}$ such that $\bigcup_{U \in \mathcal{U}'} U$ has Lebesgue measure at least $(1 - \varepsilon)\lambda(W)$. Choose a maximal sequence $\mathcal{U}''$ (sorted by decreasing lengths) of *disjoint* intervals and show that each $U \in \mathcal{U}'$ is in $(x - 3a, x + 3a)$ for some $(x - a, x + a) \in \mathcal{U}''$. ♣

Exercise 13.1.5 Let $C \subset \mathbb{R}^d$ be an open, bounded and convex set and assume that $\mathcal{U} \subset \{x + rC : x \in \mathbb{R}^d, r > 0\}$ is such that $W := \bigcup_{U \in \mathcal{U}} U$ has finite Lebesgue measure $\lambda^d(W)$. Show that for any $\varepsilon > 0$, there exist finitely many pairwise disjoint sets $U_1, \ldots, U_n \in \mathcal{U}$ such that

$$\sum_{i=1}^{n} \lambda^d(U_i) > \frac{1-\varepsilon}{3^d} \lambda(W).$$

Show by a counterexample that the condition of similarity of the open sets in $\mathcal{U}$ is essential. ♣

Exercise 13.1.6 Let μ be a Radon measure on $\mathbb{R}^d$ and let $A \in \mathcal{B}(\mathbb{R}^d)$ be a μ-null set. Let $C \subset \mathbb{R}^d$ be bounded, convex and open with $0 \in C$. Use Exercise 13.1.5 to show that

$$\lim_{r \downarrow 0} \frac{\mu(x + rC)}{r^d} = 0 \quad \text{for } \lambda^d\text{-almost all } x \in A.$$

Conclude that if F is the distribution function of a Stieltjes measure μ on $\mathbb{R}$ and if $A \in \mathcal{B}(\mathbb{R})$ is a μ-null set, then $\frac{d}{dx} F(x) = 0$ for λ-almost all $x \in A$. ♣

Exercise 13.1.7 (Fundamental theorem of calculus) (Compare [37].) Let $f \in \mathcal{L}^1(\mathbb{R}^d)$, $\mu = f \lambda^d$ and let $C \subset \mathbb{R}^d$ be open, convex and bounded with $0 \in C$. Show that

$$\lim_{r \downarrow 0} \frac{\mu(x + rC)}{r^d \lambda^d(C)} = f(x) \quad \text{for } \lambda^d\text{-almost all } x \in \mathbb{R}^d.$$

For the case $d = 1$, conclude the fundamental theorem of calculus:

$$\frac{d}{dx} \int_{[0,x]} f \, d\lambda = f(x) \quad \text{for } \lambda\text{-almost all } x \in \mathbb{R}.$$

Hint: Use Exercise 13.1.6 with $\mu_q(dx) = (f(x) - q)^+ \lambda^d(dx)$ for $q \in \mathbb{Q}$, as well as the inequality

$$\frac{\mu(x + rC)}{r^d \lambda^d(C)} \leq q + \frac{\mu_q(x + rC)}{r^d \lambda^d(C)}. \quad \clubsuit$$

Exercise 13.1.8 Similarly as in Corollary 13.7, show the following: Let E be a σ-compact polish space and let μ be a measure on E. Then μ is a Radon measure if and only if $\mu(K) < \infty$ for any compact $K \subset E$. $\clubsuit$

Exercise 13.1.9 Show that the set of rationals $\mathbb{Q}$ (with the standard topology) is not a Polish space. To this end, fill in the details in the following sketch.

We assume that d is a metric on $\mathbb{Q}$ that induces the standard topology and such that $(\mathbb{Q}, d)$ is complete. We aim at a contradiction.

Let $q_1, q_2, \ldots$ be an arbitrary enumeration of $\mathbb{Q}$. Choose $r_1 \in \mathbb{Q}$ and $\varepsilon_1 > 0$ such that $q_1 \notin B_{\varepsilon_1}(r_1)$. Now choose successively $r_{i+1} \in B_{\varepsilon_i/2}(r_i)$ and $\varepsilon_{i+1} \in (0, \varepsilon_i/2)$ such that $q_{i+1} \notin B_{\varepsilon_{i+1}}(r_{i+1})$. As d is complete, there is an $x \in \mathbb{Q}$ with $\{x\} = \bigcap_{i=1}^{\infty} B_{\varepsilon_i}(r_i)$. On the other hand, by construction, we have $q_k \notin \bigcap_{i=1}^{\infty} B_{\varepsilon_i}(r_i)$ for all $k \in \mathbb{N}$. $\clubsuit$

13.2 Weak and Vague Convergence

In Theorem 13.11, we saw that integrals of bounded continuous functions f determine a Radon measure on a metric space (E, d). If E is locally compact, it is enough to consider f with compact support. This suggests that we can use $C_b(E)$ and $C_c(E)$ as classes of test functions in order to define the convergence of measures.

Definition 13.12 (Weak and vague convergence) Let E be a metric space.

(i) Let $\mu, \mu_1, \mu_2, \ldots \in \mathcal{M}_f(E)$. We say that $(\mu_n)_{n \in \mathbb{N}}$ **converges weakly** to μ, formally $\mu_n \overset{n \to \infty}{\longrightarrow} \mu$ (weakly) or $\mu = \underset{n \to \infty}{\text{w-lim}}\, \mu_n$, if

$$\int f \, d\mu_n \overset{n \to \infty}{\longrightarrow} \int f \, d\mu \quad \text{for all } f \in C_b(E).$$

(ii) Let $\mu, \mu_1, \mu_2, \ldots \in \mathcal{M}(E)$. We say that $(\mu_n)_{n \in \mathbb{N}}$ **converges vaguely** to μ, formally $\mu_n \overset{n \to \infty}{\longrightarrow} \mu$ (vaguely) or $\mu = \underset{n \to \infty}{\text{v-lim}}\, \mu_n$, if

$$\int f \, d\mu_n \overset{n \to \infty}{\longrightarrow} \int f \, d\mu \quad \text{for any } f \in C_c(E).$$

Remark 13.13 By Theorem 13.11, the weak limit is unique. By Theorems 13.6 and 13.11, the same holds for the vague limit if E is Polish and locally compact. ◊

Remark 13.14

(i) In functional analysis the notion of weak convergence is somewhat different. Starting from a normed vector space X (here the space of finite signed measures with the total variation norm), consider the space X' of continuous linear functionals $X \to \mathbb{R}$. The sequence (μ_n) in X converges *weakly* to $\mu \in X$, if $\Phi(\mu_n) \stackrel{n\to\infty}{\longrightarrow} \Phi(\mu)$ for every $\Phi \in X'$. In the case of finite signed measures this is equivalent to: (μ_n) is bounded and $\mu_n(A) \stackrel{n\to\infty}{\longrightarrow} \mu(A)$ for any measurable A (see [38, Theorem IV.9.5]). Comparing this to Theorem 13.16(vi), we see that the functional analysis notion of weak convergence is stronger than ours in Definition 13.12.

(ii) Weak convergence (as introduced in Definition 13.12) induces on $\mathcal{M}_f(E)$ the **weak topology** τ_w. This is the coarsest topology such that for all $f \in C_b(E)$, the map $\mathcal{M}_f(E) \to \mathbb{R}$, $\mu \mapsto \int f\,d\mu$ is continuous. In functional analysis, τ_w corresponds to the so-called weak*-topology. Starting from a normed vector space X (here $X = C_b(E)$ with the norm $\| \cdot \|_\infty$), we define the weak*-topology on the dual space X' by writing $\mu_n \stackrel{n\to\infty}{\longrightarrow} \mu$ if and only if $\mu_n(x) \stackrel{n\to\infty}{\longrightarrow} \mu(x)$ for all $x \in X$. Clearly, each μ defines a continuous linear form on $C_b(E)$ by $f \mapsto \mu(f) := \int f\,d\mu$. Hence $\mathcal{M}_f(E) \subset C_b(E)'$. This implies that τ_w is the trace of the weak*-topology on $\mathcal{M}_f(E)$.

(iii) If E is separable, then it can be shown that $(\mathcal{M}_f(E), \tau_w)$ is metrizable; for example, by virtue of the so-called **Prohorov metric**. This is defined by

$$d_P(\mu, \nu) := \max\{d'_P(\mu, \nu), d'_P(\nu, \mu)\}, \tag{13.4}$$

where

$$d'_P(\mu, \nu) := \inf\{\varepsilon > 0 : \mu(B) \le \nu(B^\varepsilon) + \varepsilon \text{ for any } B \in \mathcal{B}(E)\}, \tag{13.5}$$

and where $B^\varepsilon = \{x : d(x, B) < \varepsilon\}$; see, e.g., [14, Appendix III, Theorem 5]. (It can be shown that $d'_P(\mu, \nu) = d'_P(\nu, \mu)$ if $\mu, \nu \in \mathcal{M}_1(E)$.) If E is locally compact and Polish, then $(\mathcal{M}_f(E), \tau_w)$ is again Polish (see [136, page 167]).

(iv) Similarly, the **vague topology** τ_v on $\mathcal{M}(E)$ is the coarsest topology such that for all $f \in C_c(E)$, the map $\mathcal{M}(E) \to \mathbb{R}$, $\mu \mapsto \int f\,d\mu$ is continuous. If E is locally compact, then $(\mathcal{M}(E), \tau_v)$ is a Hausdorff space. If, in addition, E is Polish, then $(\mathcal{M}(E), \tau_v)$ is again Polish (see, e.g., [82, Section 15.7]). ◊

While weak convergence implies convergence of the total masses (since $1 \in C_b(E)$), with vague convergence a mass defect (but not a mass gain) can be experienced in the limit.

Lemma 13.15 *Let E be a locally compact Polish space and let $\mu, \mu_1, \mu_2, \ldots \in \mathcal{M}(E)$ be measures such that $\mu_n \stackrel{n\to\infty}{\longrightarrow} \mu$ vaguely. Then*

$$\mu(E) \leq \liminf_{n\to\infty} \mu_n(E).$$

Proof Let $(f_N)_{N\in\mathbb{N}}$ be a sequence in $C_c(E; [0, 1])$ with $f_N \uparrow 1$. Then

$$
\begin{aligned}
\mu(E) &= \sup_{N\in\mathbb{N}} \int f_N \, d\mu \\
&= \sup_{N\in\mathbb{N}} \lim_{n\to\infty} \int f_N \, d\mu_n \\
&\leq \liminf_{n\to\infty} \sup_{N\in\mathbb{N}} \int f_N \, d\mu_n \\
&= \liminf_{n\to\infty} \mu_n(E). \qquad \square
\end{aligned}
$$

Clearly, the sequence $(\delta_{1/n})_{n\in\mathbb{N}}$ of probability measures on $\mathbb{R}$ converges weakly to δ_0; however, not in total variation norm. Indeed, for the closed set $(-\infty, 0]$, we have $\lim_{n\to\infty} \delta_{1/n}((-\infty, 0]) = 0 < 1 = \delta_0((-\infty, 0])$. Loosely speaking, at the boundaries of closed sets, mass can immigrate but not emigrate. The opposite is true for open sets: $\lim_{n\to\infty} \delta_{1/n}((0, \infty)) = 1 > 0 = \delta_0((0, \infty))$. Here mass can emigrate but not immigrate. In fact, weak convergence can be characterized by this property. In the following theorem, a whole bunch of such statements will be hung on a coat hanger (French: *portemanteau*).

For measurable $g : \Omega \to \mathbb{R}$, let U_g be the set of points of discontinuity of g. Recall from Exercise 1.1.3 that U_g is Borel measurable.

Theorem 13.16 (Portemanteau) *Let E be a metric space and let $\mu, \mu_1, \mu_2, \ldots \in \mathcal{M}_{\leq 1}(E)$. The following are equivalent.*

(i) $\mu = \text{w-}\lim_{n\to\infty} \mu_n$.

(ii) $\int f \, d\mu_n \stackrel{n\to\infty}{\longrightarrow} \int f \, d\mu$ *for all bounded Lipschitz continuous f.*

(iii) $\int f \, d\mu_n \stackrel{n\to\infty}{\longrightarrow} \int f \, d\mu$ *for all bounded measurable f with $\mu(U_f) = 0$.*

(iv) $\liminf_{n\to\infty} \mu_n(E) \geq \mu(E)$ *and* $\limsup_{n\to\infty} \mu_n(F) \leq \mu(F)$ *for all closed $F \subset E$.*

(v) $\limsup_{n\to\infty} \mu_n(E) \leq \mu(E)$ *and* $\liminf_{n\to\infty} \mu_n(G) \geq \mu(G)$ *for all open $G \subset E$.*

(vi) $\lim_{n\to\infty} \mu_n(A) = \mu(A)$ *for all measurable A with $\mu(\partial A) = 0$.*

If E is locally compact and Polish, then in addition each of the following is equivalent to the previous statements.

(vii) $\mu = \text{v-}\lim_{n\to\infty} \mu_n$ *and* $\mu(E) = \lim_{n\to\infty} \mu_n(E)$.

(viii) $\mu = \text{v-}\lim_{n\to\infty} \mu_n$ *and* $\mu(E) \geq \limsup_{n\to\infty} \mu_n(E)$.

Proof "**(iv)** $\Longleftrightarrow$ **(v)** $\Longrightarrow$ **(vi)**"   This is trivial.
"**(iii)** $\Longrightarrow$ **(i)** $\Longrightarrow$ **(ii)**"   This is trivial.
"**(ii)** $\Longrightarrow$ **(iv)**"   Convergence of the total masses follows by using the test function $1 \in \mathrm{Lip}(E; [0, 1])$. Let F be closed and let $\rho_{F,\varepsilon}$ be as in Lemma 13.10. Then

$$\limsup_{n\to\infty} \mu_n(F) \leq \inf_{\varepsilon>0} \lim_{n\to\infty} \int \rho_{F,\varepsilon}\, d\mu_n = \inf_{\varepsilon>0} \int \rho_{F,\varepsilon}\, d\mu = \mu(F)$$

since $\rho_{F,\varepsilon}(x) \overset{\varepsilon\to0}{\longrightarrow} \mathbb{1}_F(x)$ for all $x \in E$.
"**(viii)** $\Longrightarrow$ **(vii)**"   This is obvious by Lemma 13.15.
"**(i)** $\Longrightarrow$ **(vii)**"   This is clear since $C_c(E) \subset C_b(E)$ and $1 \in C_b(E)$.
"**(vii)** $\Longrightarrow$ **(v)**"   Let G be open and $\varepsilon > 0$. Since μ is inner regular (Theorem 13.6), there is a compact set $K \subset G$ with $\mu(G) - \mu(K) < \varepsilon$. As E is locally compact, there is a compact set L with $K \subset L^\circ \subset L \subset G$. Let $\delta := d(K, L^c) > 0$ and let $\rho_{K,\delta}$ be as in Lemma 13.10. Then $\mathbb{1}_K \leq \rho_{K,\delta} \leq \mathbb{1}_L$; hence $\rho_{K,\delta} \in C_c(E)$ and thus

$$\liminf_{n\to\infty} \mu_n(G) \geq \liminf_{n\to\infty} \int \rho_{K,\delta}\, d\mu_n = \int \rho_{K,\delta}\, d\mu \geq \mu(K) \geq \mu(G) - \varepsilon.$$

Letting $\varepsilon \to 0$, we get (v).
"**(vi)** $\Longrightarrow$ **(iii)**"   Let $f : E \to \mathbb{R}$ be bounded and measurable with $\mu(U_f) = 0$. We make the elementary observation that for all $D \subset \mathbb{R}$,

$$\partial f^{-1}(D) \subset f^{-1}(\partial D) \cup U_f. \tag{13.6}$$

Indeed, if f is continuous at $x \in E$, then for any $\delta > 0$, there is an $\varepsilon(\delta) > 0$ with $f(B_{\varepsilon(\delta)}(x)) \subset B_\delta(f(x))$. If $x \in \partial f^{-1}(D)$, then there are $y \in f^{-1}(D) \cap B_{\varepsilon(\delta)}(x)$ and $z \in f^{-1}(D^c) \cap B_{\varepsilon(\delta)}(x)$. Therefore, $f(y) \in B_\delta(f(x)) \cap D \neq \emptyset$ and $f(z) \in B_\delta(f(x)) \cap D^c \neq \emptyset$; hence $f(x) \in \partial D$.
   Let $\varepsilon > 0$. Evidently, the set $A := \{y \in \mathbb{R} : \mu\left(f^{-1}(\{y\})\right) > 0\}$ of atoms of the finite measure $\mu \circ f^{-1}$ is at most countable. Hence, there exist $N \in \mathbb{N}$ and $y_0 \leq -\|f\|_\infty < y_1 < \ldots < y_{N-1} < \|f\|_\infty < y_N$ such that

$$y_i \in \mathbb{R} \setminus A \quad \text{and} \quad |y_{i+1} - y_i| < \varepsilon \quad \text{for all } i.$$

Let $E_i = f^{-1}([y_{i-1}, y_i))$ for $i = 1, \ldots, N$. Then $E = \biguplus_{i=1}^N E_i$ and by (13.6),

$$\mu(\partial E_i) \leq \mu\left(f^{-1}(\{y_{i-1}\})\right) + \mu\left(f^{-1}(\{y_i\})\right) + \mu(U_f) = 0.$$

Therefore,

$$\limsup_{n\to\infty} \int f\, d\mu_n \leq \limsup_{n\to\infty} \sum_{i=1}^N \mu_n(E_i) \cdot y_i = \sum_{i=1}^N \mu(E_i) \cdot y_i \leq \varepsilon + \int f\, d\mu.$$

We let $\varepsilon \to 0$ and obtain $\limsup\limits_{n\to\infty} \int f \, d\mu_n \le \int f \, d\mu$. Finally, consider $(-f)$ to obtain the reverse inequality $\liminf\limits_{n\to\infty} \int f \, d\mu_n \ge \int f \, d\mu$.   □

Definition 13.17   Let $X, X_1, X_2, \dots$ be random variables with values in E. We say that $(X_n)_{n\in\mathbb{N}}$ **converges in distribution** to X, formally $X_n \overset{\mathcal{D}}{\longrightarrow} X$ or $X_n \overset{n\to\infty}{\Longrightarrow} X$, if the distributions converge weakly and hence if $\mathbf{P}_X = \operatorname*{w-lim}\limits_{n\to\infty} \mathbf{P}_{X_n}$. Sometimes we write $X_n \overset{\mathcal{D}}{\longrightarrow} \mathbf{P}_X$ or $X_n \overset{n\to\infty}{\Longrightarrow} \mathbf{P}_X$ if we want to specify only the distribution $\mathbf{P}_X$ but not the random variable X.

Theorem 13.18 (Slutzky's theorem)   *Let $X, X_1, X_2, \dots$ and $Y_1, Y_2, \dots$ be random variables with values in E. Assume $X_n \overset{\mathcal{D}}{\longrightarrow} X$ and $d(X_n, Y_n) \overset{n\to\infty}{\longrightarrow} 0$ in probability. Then $Y_n \overset{\mathcal{D}}{\longrightarrow} X$.*

Proof   Let $f : E \to \mathbb{R}$ be bounded and Lipschitz continuous with constant K. Then

$$\left| f(x) - f(y) \right| \le K \, d(x, y) \wedge 2 \, \|f\|_\infty \quad \text{for all } x, y \in E.$$

Dominated convergence yields $\limsup\limits_{n\to\infty} \mathbf{E}\left[\left| f(X_n) - f(Y_n) \right| \right] = 0$. Hence we have

$$\limsup_{n\to\infty} \left| \mathbf{E}[f(Y_n)] - \mathbf{E}[f(X)] \right|$$

$$\le \limsup_{n\to\infty} \left| \mathbf{E}[f(X)] - \mathbf{E}[f(X_n)] \right| + \limsup_{n\to\infty} \left| \mathbf{E}[f(X_n)] - f(Y_n)] \right| = 0. \quad □$$

Corollary 13.19   *If $X_n \overset{n\to\infty}{\longrightarrow} X$ in probability, then $X_n \overset{\mathcal{D}}{\longrightarrow} X$, $n \to \infty$. The converse is false in general.*

Example 13.20   If $X, X_1, X_2, \dots$ are i.i.d. (with nontrivial distribution), then trivially $X_n \overset{\mathcal{D}}{\longrightarrow} X$ but not $X_n \overset{n\to\infty}{\longrightarrow} X$ in probability.   ◊

Recall the definition of a distribution function of a probability measure from Definition 1.59.

Definition 13.21   Let $F, F_1, F_2, \dots$ be distribution functions of probability measures on $\mathbb{R}$. We say that $(F_n)_{n\in\mathbb{N}}$ **converges weakly** to F, formally $F_n \overset{n\to\infty}{\Longrightarrow} F$, $F_n \overset{\mathcal{D}}{\longrightarrow} F$ or $F = \operatorname*{w-lim}\limits_{n\to\infty} F_n$, if

$$F(x) = \lim_{n\to\infty} F_n(x) \quad \text{for all points of continuity } x \text{ of } F. \tag{13.7}$$

If $F, F_1, F_2, \dots$ are distribution functions of sub-probability measures, then we define $F(\infty) := \lim_{x\to\infty} F(x)$ and for weak convergence require in addition $F(\infty) \ge \limsup_{n\to\infty} F_n(\infty)$.

Note that (13.7) implies $F(\infty) \leq \liminf_{n\to\infty} F_n(\infty)$. Hence, if $F_n \xrightarrow{D} F$, then $F(\infty) = \lim_{n\to\infty} F_n(\infty)$.

Example 13.22 If F is the distribution function of a probability measure on $\mathbb{R}$ and $F_n(x) := F(x+n)$ for $x \in \mathbb{R}$, then $(F_n)_{n\in\mathbb{N}}$ converges pointwise to 1. However, this is not a distribution function, as 1 does not converge to 0 for $x \to -\infty$. On the other hand, if $G_n(x) = F(x-n)$, then $(G_n)_{n\in\mathbb{N}}$ converges pointwise to $G \equiv 0$. However, $G(\infty) = 0 < \limsup_{n\to\infty} G_n(\infty) = 1$; hence we do not have weak convergence here either. Indeed, in each case, there is a mass defect in the limit (in the case of the F_n on the left and in the case of the G_n on the right). However, the definition of weak convergence of distribution functions is constructed so that no mass defect occurs in the limit. $\Diamond$

Theorem 13.23 *Let* $\mu, \mu_1, \mu_2, \ldots \in \mathcal{M}_{\leq 1}(\mathbb{R})$ *with corresponding distribution functions* $F, F_1, F_2, \ldots$. *The following are equivalent.*

(i) $\mu = \operatorname*{w-lim}_{n\to\infty} \mu_n$.

(ii) $F_n \xrightarrow{D} F$.

Proof "(i) $\Longrightarrow$ (ii)"   Let F be continuous at x. Then $\mu\big(\partial(-\infty, x]\big) = \mu(\{x\}) = 0$. By Theorem 13.16, $F_n(x) = \mu_n\left((-\infty, x]\right) \xrightarrow{n\to\infty} \mu((-\infty, x]) = F(x)$.

"(ii) $\Longrightarrow$ (i)"   Let $f \in \mathrm{Lip}_1(\mathbb{R}; [0, 1])$. By Theorem 13.16, it is enough to show that

$$\int f \, d\mu_n \xrightarrow{n\to\infty} \int f \, d\mu. \tag{13.8}$$

Let $\varepsilon > 0$. Fix $N \in \mathbb{N}$ and choose $N + 1$ points of continuity $y_0 < y_1 < \ldots < y_N$ of F such that $F(y_0) < \varepsilon$, $F(y_N) > F(\infty) - \varepsilon$ and $y_i - y_{i-1} < \varepsilon$ for all i. Then

$$\int f \, d\mu_n \leq \big(F_n(y_0) + F_n(\infty) - F_n(y_N)\big) + \sum_{i=1}^{N}(f(y_i) + \varepsilon)(F_n(y_i) - F_n(y_{i-1})).$$

By assumption, $\lim_{n\to\infty} F_n(\infty) = F(\infty)$ and $F_n(y_i) \xrightarrow{n\to\infty} F(y_i)$ for every $i = 0, \ldots, N$; hence

$$\limsup_{n\to\infty} \int f \, d\mu_n \leq 3\varepsilon + \sum_{i=1}^{N} f(y_i)\big(F(y_i) - F(y_{i-1})\big) \leq 4\varepsilon + \int f \, d\mu.$$

Therefore,

$$\limsup_{n\to\infty} \int f \, d\mu_n \leq \int f \, d\mu.$$

Replacing f by $(1 - f)$, we get (13.8). $\square$

Corollary 13.24 *Let* $X, X_1, X_2, \ldots$ *be real random variables with distribution functions* $F, F_1, F_2, \ldots$ *Then the following are equivalent.*

(i) $X_n \xrightarrow{D} X$.

(ii) $\mathbf{E}[f(X_n)] \xrightarrow{n\to\infty} \mathbf{E}[f(X)]$ *for all* $f \in C_b(\mathbb{R})$.

(iii) $F_n \xrightarrow{D} F$.

How stable is weak convergence if we pass to image measures under some map φ? Clearly, we need a certain continuity of φ at least at those points where the limit measure puts mass. The following theorem formalizes this idea and will come in handy in many applications.

Theorem 13.25 (Continuous mapping theorem) *Let* (E_1, d_1) *and* (E_2, d_2) *be metric spaces and let* $\varphi : E_1 \to E_2$ *be measurable. Denote by* U_φ *the set of points of discontinuity of* φ.

(i) *If* $\mu, \mu_1, \mu_2, \ldots \in \mathcal{M}_{\leq 1}(E_1)$ *with* $\mu(U_\varphi) = 0$ *and* $\mu_n \xrightarrow{n\to\infty} \mu$ *weakly, then* $\mu_n \circ \varphi^{-1} \xrightarrow{n\to\infty} \mu \circ \varphi^{-1}$ *weakly.*

(ii) *If* $X, X_1, X_2, \ldots$ *are* E_1-*valued random variables with* $\mathbf{P}[X \in U_\varphi] = 0$ *and* $X_n \xrightarrow{D} X$, *then* $\varphi(X_n) \xrightarrow{D} \varphi(X)$.

Proof First note that $U_\varphi \subset E_1$ is Borel measurable by Exercise 1.1.3. Hence the conditions make sense.

(i) Let $f \in C_b(E_2)$. Then $f \circ \varphi$ is bounded and measurable and $U_{f\circ\varphi} \subset U_\varphi$; hence $\mu(U_{f\circ\varphi}) = 0$. By Theorem 13.16,

$$\lim_{n\to\infty} \int f \, d(\mu_n \circ \varphi^{-1}) = \lim_{n\to\infty} \int (f \circ \varphi) \, d\mu_n$$

$$= \int (f \circ \varphi) \, d\mu = \int f \, d(\mu \circ \varphi^{-1}).$$

(ii) This is obvious since $\mathbf{P}_{\varphi(X)} = \mathbf{P}_X \circ \varphi^{-1}$. $\qquad\qquad\qquad\qquad\square$

Takeaways Weak convergence of measures is defined via convergence of integrals of bounded continuous test functions. If the test functions are also assumed to have compact support, we get vague convergence of measures. Roughly speaking, the difference is that vague convergence does not imply convergence of total masses. In fact, vague convergence is a sensible notion even for infinite measures. The most important properties of weak and vague convergence are summarised in the portemanteau theorem and the continuous mapping theorem. For probability measures on $\mathbb{R}$, weak convergence is tantamount to convergence of distribution functions at all points of continuity of the limiting function.

Exercise 13.2.1 Recall d'_P from (13.5). Show that $d_P(\mu, v) = d'_P(\mu, v) = d'_P(v, \mu)$ for all $\mu, v \in \mathcal{M}_1(E)$. ♣

Exercise 13.2.2 Show that the topology of weak convergence on $\mathcal{M}_f(E)$ is coarser than the topology induced on $\mathcal{M}_f(E)$ by the total variation norm (see Corollary 7.45). That is, $\|\mu_n - \mu\|_{TV} \overset{n\to\infty}{\longrightarrow} 0$ implies $\mu_n \overset{n\to\infty}{\longrightarrow} \mu$ weakly. ♣

Exercise 13.2.3 Let $E = \mathbb{R}$ and $\mu_n = \frac{1}{n}\sum_{k=0}^{n}\delta_{k/n}$. Let $\mu = \lambda\big|_{[0,1]}$ be the Lebesgue measure restricted to $[0, 1]$. Show that $\mu = \text{w-lim}_{n\to\infty}\mu_n$. ♣

Exercise 13.2.4 Let $E = \mathbb{R}$ and λ be the Lebesgue measure on $\mathbb{R}$. For $n \in \mathbb{N}$, let $\mu_n = \lambda\big|_{[-n,n]}$. Show that $\lambda = \text{v-lim}_{n\to\infty}\mu_n$ but that $(\mu_n)_{n\in\mathbb{N}}$ does not converge weakly. ♣

Exercise 13.2.5 Let $E = \mathbb{R}$ and $\mu_n = \delta_n$ for $n \in \mathbb{N}$. Show that $\text{v-lim}_{n\to\infty}\mu_n = 0$ but that $(\mu_n)_{n\in\mathbb{N}}$ does not converge weakly. ♣

Exercise 13.2.6 (Lévy metric)   For two probability distribution functions F and G on $\mathbb{R}$, define the Lévy distance by

$$d(F, G) = \inf\big\{\varepsilon \geq 0 : G(x - \varepsilon) - \varepsilon \leq F(x) \leq G(x + \varepsilon) + \varepsilon \text{ for all } x \in \mathbb{R}\big\}.$$

Show the following:

  (i) d is a metric on the set of distribution functions.
 (ii) $F_n \overset{n\to\infty}{\Longrightarrow} F$ if and only if $d(F_n, F) \overset{n\to\infty}{\longrightarrow} 0$.
(iii) For every $P \in \mathcal{M}_1(\mathbb{R})$, there is a sequence $(P_n)_{n\in\mathbb{N}}$ in $\mathcal{M}_1(\mathbb{R})$ such that each P_n has finite support and such that $P_n \overset{n\to\infty}{\Longrightarrow} P$. ♣

Exercise 13.2.7 We can extend the notions of *weak convergence* and *vague convergence* to signed measures; that is, to differences $\varphi := \mu^+ - \mu^-$ of measures from $\mathcal{M}_f(E)$ and $\mathcal{M}(E)$, respectively, by repeating the words of Definition 13.12 for these classes. Show that the topology of weak convergence is not metrizable in general.
Hint: Consider $E = [0, 1]$.

  (i) For $n \in \mathbb{N}$, define $\varphi_n = \delta_{1/n} - \delta_{2/n}$. Show that, for any $C > 0$, $(C\varphi_n)_{n\in\mathbb{N}}$ converges weakly to the zero measure.
 (ii) Assume there is a metric that induces weak convergence. Show that then there would be a sequence $(C_n)_{n\in\mathbb{N}}$ with $C_n \uparrow \infty$ and $0 = \text{w-lim}_{n\to\infty}(C_n\varphi_n)$.
(iii) Choose an $f \in C([0, 1])$ with $f(2^{-n}) = (-1)^n C_n^{-1/2}$ for any $n \in \mathbb{N}$, and show that $\big(\int f\, d(C_n\varphi_n)\big)_{n\in\mathbb{N}}$ does not converge to zero.
(iv) Use this construction to contradict the assumption of metrizability. ♣

Exercise 13.2.8 Show that (13.4) defines a metric on $\mathcal{M}_1(E)$ and that this metric induces the topology of weak convergence. ♣

Exercise 13.2.9 Show the implication "(vi) $\Longrightarrow$ (iv)" of Theorem 13.16 directly. ♣

Exercise 13.2.10 Let $X, X_1, X_2, \ldots$ and $Y_1, Y_2, \ldots$ be real random variables. Assume $\mathbf{P}_{Y_n} = \mathcal{N}_{0,1/n}$ for all $n \in \mathbb{N}$. Show that $X_n \overset{\mathcal{D}}{\longrightarrow} X$ if and only if $X_n + Y_n \overset{\mathcal{D}}{\longrightarrow} X$. ♣

Exercise 13.2.11 For each $n \in \mathbb{N}$, let X_n be a geometrically distributed random variable with parameter $p_n \in (0, 1)$. How must we choose the sequence $(p_n)_{n\in\mathbb{N}}$ in order that $\mathbf{P}_{X_n/n}$ converges weakly to the exponential distribution with parameter $\alpha > 0$? ♣

Exercise 13.2.12 Let $X, X_1, X_2, \ldots$ be real random variables with $X_n \overset{n\to\infty}{\Longrightarrow} X$. Show the following.

(i) $\mathbf{E}[|X|] \leq \liminf_{n\to\infty} \mathbf{E}[|X_n|]$.
(ii) Let $r > p > 0$. If $\sup_{n\in\mathbb{N}} \mathbf{E}[|X_n|^r] < \infty$, then $\mathbf{E}[|X|^p] = \lim_{n\to\infty} \mathbf{E}[|X_n|^p]$. ♣

Exercise 13.2.13 Let $F, F_1, F_2, \ldots$ be probability distribution functions on $\mathbb{R}$, and assume $F_n \overset{n\to\infty}{\Longrightarrow} F$. Let $F^{-1}(u) = \inf\{x \in \mathbb{R} : F(x) \geq u\}$, $u \in (0, 1)$, be the left continuous inverse of F (see the proof of Theorem 1.104). Show that

$$F_n^{-1}(u) \overset{n\to\infty}{\longrightarrow} F^{-1}(u) \text{ at every point of continuity } u \text{ of } F^{-1}.$$

Conclude that $F^{-1}(u) \overset{n\to\infty}{\longrightarrow} F^{-1}(u)$ for Lebesgue almost all $u \in (0, 1)$. ♣

Exercise 13.2.14 Let $\mu, \mu_1, \mu_2, \ldots \in \mathcal{M}_1(\mathbb{R})$ with $\mu_n \overset{n\to\infty}{\longrightarrow} \mu$ weakly. Show that there exists a probability space $(\Omega, \mathcal{A}, \mathbf{P})$ and real random variables $X, X_1, X_2, \ldots$ on $(\Omega, \mathcal{A}, \mathbf{P})$ with distributions $\mathbf{P}_X = \mu$ and $\mathbf{P}_{X_n} = \mu_n$, $n \in \mathbb{N}$, such that

$$X_n \overset{n\to\infty}{\longrightarrow} X \quad \mathbf{P}\text{-a.s.}$$

Hint: Use Exercise 13.2.13. ♣

Exercise 13.2.15 Let (E, d) be a metric space and let $\mu, \mu_1, \mu_2, \ldots$ be probability measures on E. A measurable map $f : E \to \mathbb{R}$ is called uniformly integrable with respect to $(\mu_n)_{n\in\mathbb{N}}$, if

$$\inf_{a>0} \sup_{n\in\mathbb{N}} \int_{\{|f|>a\}} |f| \, d\mu_n = 0.$$

Let f be continuous and uniformly integrable with respect to $(\mu_n)_{n\in\mathbb{N}}$ and assume that $\mu_n \xrightarrow{n\to\infty} \mu$ weakly. Show that $\int |f|\,d\mu < \infty$ and that

$$\int f\,d\mu_n \xrightarrow{n\to\infty} \int f\,d\mu.$$

Hint: Apply Exercise 13.2.14 to the image measures $\mu_n \circ f^{-1}$. ♣

13.3   Prohorov's Theorem

In the following, let E be a Polish space with Borel σ-algebra $\mathcal{E}$. A fundamental question is: When does a sequence $(\mu_n)_{n\in\mathbb{N}}$ of measures on $(E, \mathcal{E})$ converge weakly or does at least have a weak limit point? Evidently, a necessary condition is that $(\mu_n(E))_{n\in\mathbb{N}}$ is bounded. Hence, without loss of generality, we will consider only sequences in $\mathcal{M}_{\leq 1}(E)$. However, this condition is not sufficient for the existence of weak limit points, as for example the sequence $(\delta_n)_{n\in\mathbb{N}}$ of probability measures on $\mathbb{R}$ does not have a weak limit point (although it converges vaguely to the zero measure). This example suggests that we also have to make sure that no mass "vanishes at infinity". The idea will be made precise by the notion of *tightness*.

We start this section by presenting as the main result Prohorov's theorem [136]. We give the proof first for the special case $E = \mathbb{R}$ and then come to a couple of applications. The full proof of the general case is deferred to the end of the section.

Definition 13.26 (Tightness) A family $\mathcal{F} \subset \mathcal{M}_f(E)$ is called **tight** if, for any $\varepsilon > 0$, there exists a compact set $K \subset E$ such that

$$\sup\{\mu(E \setminus K) : \mu \in \mathcal{F}\} < \varepsilon.$$

Remark 13.27 If E is Polish, then by Lemma 13.5, every singleton $\{\mu\} \subset \mathcal{M}_f(E)$ is tight and thus so is every finite family. ◊

Example 13.28

(i) If E is compact, then $\mathcal{M}_1(E)$ and $\mathcal{M}_{\leq 1}(E)$ are tight.
(ii) If $(X_i)_{i\in I}$ is an arbitrary family of random variables with

$$C := \sup\{\mathbf{E}[|X_i|] : i \in I\} < \infty,$$

   then $\{\mathbf{P}_{X_i} : i \in I\}$ is tight. Indeed, for $\varepsilon > 0$ and $K = [-C/\varepsilon, C/\varepsilon]$, by Markov's inequality, $\mathbf{P}_{X_i}(\mathbb{R} \setminus K) = \mathbf{P}[|X_i| > C/\varepsilon] \leq \varepsilon$.
(iii) The family $(\delta_n)_{n\in\mathbb{N}}$ of probability measures on $\mathbb{R}$ is not tight.
(iv) The family $(\mathcal{U}_{[-n,n]})_{n\in\mathbb{N}}$ of uniform distributions on the intervals $[-n, n]$, regarded as measures on $\mathbb{R}$, is not tight. ◊

Recall that a family $\mathcal{F}$ of measures is called weakly relatively sequentially compact if every sequence in $\mathcal{F}$ has a weak limit point (in the closure of $\mathcal{F}$).

Theorem 13.29 (Prohorov's theorem (1956)) *Let (E, d) be a metric space and $\mathcal{F} \subset \mathcal{M}_{\leq 1}(E)$. Then:*

(i) $\mathcal{F}$ is tight $\implies$ $\mathcal{F}$ is weakly relatively sequentially compact.
(ii) If E is Polish, then also the converse holds:

$$\mathcal{F} \text{ is tight} \impliedby \mathcal{F} \text{ is weakly relatively sequentially compact.}$$

Corollary 13.30 *Let E be a compact metric space. Then the sets $\mathcal{M}_{\leq 1}(E)$ and $\mathcal{M}_1(E)$ are weakly sequentially compact.*

Corollary 13.31 *If E is a locally compact separable metric space, then $\mathcal{M}_{\leq 1}(E)$ is vaguely sequentially compact.*

Proof Let $(\mu_n)_{n \in \mathbb{N}}$ be a sequence in $\mathcal{M}_{\leq 1}(E)$. We have to show that there exists a vaguely convergent subsequence.

As E is locally compact and separable, E is σ-compact. Let $U_1, U_2, \ldots \subset E$ be relatively compact open sets covering E. Inductively, we define relatively compact open sets $W_n \uparrow E$ with $\overline{W_n} \subset W_{n+1}$ for all $n \in \mathbb{N}$. Let $W_1 := U_1$. Having defined W_n, we choose a relatively open set $L_n \supset \overline{W_n}$ and define $W_{n+1} := L_n \cup U_{n+1}$. Note that for any compact $C \subset E$, there exists an $N(C) \in \mathbb{N}$ such that $C \subset W_n$ for all $n \geq N(C)$.

Applying Prohorov's theorem (i.e., Corollary 13.30) to the measures $(\mu_k \mathbb{1}_{\overline{W_n}})_{k \in \mathbb{N}}$, for each $n \in \mathbb{N}$, we can choose a sequence $(k_l^n)_{l \in \mathbb{N}}$ and a measure $\tilde{\mu}_n := \underset{l \to \infty}{\text{w-lim}}\, \mu_{k_l^n} \mathbb{1}_{\overline{W_n}}$ whose support lies in $\overline{W_n}$. We may assume that the sequences $(k_l^n)_{l \in \mathbb{N}}$ were chosen successively such that (k_l^{n+1}) is a subsequence of (k_l^n).

Note that we have $\tilde{\mu}_n(\overline{W_n}) \leq \tilde{\mu}_{n+1}(\overline{W_n})$, but equality does not hold in general.

For $f \in C_c(E)$, there exists an $n_0 \in \mathbb{N}$ such that the support of f is contained in W_{n_0}. Hence, for $m \geq n \geq n_0$, we have

$$\int f \, d\tilde{\mu}_n = \lim_{l \to \infty} \int f \mathbb{1}_{\overline{W_n}} \, d\mu_{k_l^n}$$

$$= \lim_{l \to \infty} \int f \mathbb{1}_{\overline{W_n}} \, d\mu_{k_l^m}$$

$$= \lim_{l \to \infty} \int f \mathbb{1}_{\overline{W_m}} \, d\mu_{k_l^m} = \int f \, d\tilde{\mu}_m$$

and thus

$$\int f \, d\tilde{\mu}_n = \lim_{m \to \infty} \int f \, d\mu_{k_m^m}.$$

This implies that for any measurable relatively compact set $A \subset E$, we have

$$\tilde{\mu}_m(A) = \tilde{\mu}_{N(\overline{A})}(A) \quad \text{for any } m \geq N(\overline{A}).$$

For any measurable set $A \subset E$, define

$$\mu(A) := \sup_{n \in \mathbb{N}} \sup_{m > n} \tilde{\mu}_m(A \cap W_n) = \sup_{n \in \mathbb{N}} \tilde{\mu}_{n+1}(A \cap W_n).$$

It is easy to check that μ is a lower semicontinuous content and is hence a measure (see Theorem 1.36). By construction, for any $f \in C_c(E)$, we infer

$$\int f \, d\mu = \lim_{n \to \infty} \int f \, d\mu_{k_n^n}.$$

Concluding, we have $\mu = \underset{n \to \infty}{\text{v-lim}}\, \mu_{k_n^n}$.   □

Remark 13.32 The implication (ii) in Theorem 13.29 is less useful but a lot simpler to prove. Here we need that E is Polish since clearly every singleton is weakly compact but is tight only under additional assumptions; for example, if E is Polish (see Lemma 13.5). ◊

Proof (of Theorem 13.29(ii)) We start as in the proof of Lemma 13.5. Let $\{x_1, x_2, \ldots\} \subset E$ be dense. For $n \in \mathbb{N}$, define $A_{n,N} := \bigcup_{i=1}^{N} B_{1/n}(x_i)$. Then $A_{n,N} \uparrow E$ for $N \to \infty$ for all $n \in \mathbb{N}$. Let

$$\delta := \sup_{n \in \mathbb{N}} \inf_{N \in \mathbb{N}} \sup_{\mu \in \mathcal{F}} \mu(A_{n,N}^c).$$

Then there is an $n \in \mathbb{N}$ such that for any $N \in \mathbb{N}$, there is a $\mu_N \in \mathcal{F}$ with $\mu_N(A_{n,N}^c) \geq \delta/2$. As $\mathcal{F}$ is weakly relatively sequentially compact, $(\mu_N)_{N \in \mathbb{N}}$ has a weakly convergent subsequence $(\mu_{N_k})_{k \in \mathbb{N}}$ whose weak limit will be denoted by $\mu \in \mathcal{M}_{\leq 1}(E)$. By the Portemanteau theorem (Theorem 13.16(iv)), for any $N \in \mathbb{N}$,

$$\mu(A_{n,N}^c) \geq \liminf_{k \to \infty} \mu_{N_k}(A_{n,N}^c) \geq \liminf_{k \to \infty} \mu_{N_k}(A_{n,N_k}^c) \geq \delta/2.$$

On the other hand, $A_{n,N}^c \downarrow \emptyset$ for $N \to \infty$; hence $\mu(A_{n,N}^c) \overset{N \to \infty}{\longrightarrow} 0$. Thus $\delta = 0$.

Now fix $\varepsilon > 0$. By the above, for any $n \in \mathbb{N}$, we can choose an $N_n' \in \mathbb{N}$ such that $\mu(A_{n,N_n'}^c) < \varepsilon/2^n$ for all $\mu \in \mathcal{F}$. By construction, the set $A := \bigcap_{n=1}^{\infty} A_{n,N_n'}$ is totally bounded and hence relatively compact. Further, for every $\mu \in \mathcal{F}$,

$$\mu\big((\overline{A})^c\big) \leq \mu(A^c) \leq \sum_{n=1}^{\infty} \mu(A_{n,N_n'}^c) \leq \varepsilon.$$

Hence $\mathcal{F}$ is tight.   □

The other implication in Prohorov's theorem is more difficult to prove, especially in the case of a general metric space. For this reason, we first give a proof only for the case $E = \mathbb{R}$ and come to applications before proving the difficult implication in the general situation.

The problem consists in finding a candidate for a weak limit point. For distributions on $\mathbb{R}$, the problem is equivalent to finding a weak limit point for a sequence of distribution functions. Here Helly's theorem is the tool. It is based on a diagonal sequence argument that will be recycled later in the proof of Prohorov's theorem in the general case.

Let

$$V = \{F : \mathbb{R} \to \mathbb{R} \text{ is right continuous, monotone increasing and bounded}\}$$

be the set of distribution functions of finite measures on $\mathbb{R}$.

Theorem 13.33 (Helly's theorem) *Let $(F_n)_{n\in\mathbb{N}}$ be a uniformly bounded sequence in V. Then there exists an $F \in V$ and a subsequence $(F_{n_k})_{k\in\mathbb{N}}$ with*

$$F_{n_k}(x) \overset{k\to\infty}{\longrightarrow} F(x) \text{ at all points of continuity of } F.$$

Proof We use a diagonal sequence argument. Choose an enumeration of the rational numbers $\mathbb{Q} = \{q_1, q_2, q_3, \dots\}$. By the Bolzano–Weierstraß theorem, the sequence $(F_n(q_1))_{n\in\mathbb{N}}$ has a convergent subsequence $(F_{n_k^1}(q_1))_{k\in\mathbb{N}}$. Analogously, we find a subsequence $(n_k^2)_{k\in\mathbb{N}}$ of $(n_k^1)_{k\in\mathbb{N}}$ such that $(F_{n_k^2}(q_2))_{k\in\mathbb{N}}$ converges. Inductively, we obtain subsequences $(n_k^1) \supset (n_k^2) \supset (n_k^3) \supset \dots$ such that $(F_{n_k^l}(q_l))_{k\in\mathbb{N}}$ converges for all $l \in \mathbb{N}$. Now define $n_k := n_k^k$. Then $(F_{n_k}(q))_{k\in\mathbb{N}}$ converges for all $q \in \mathbb{Q}$. Define $\widetilde{F}(q) = \lim_{k\to\infty} F_{n_k}(q)$ and

$$F(x) = \inf\{\widetilde{F}(q) : q \in \mathbb{Q} \text{ with } q > x\}.$$

As $\widetilde{F}$ is monotone increasing, F is right continuous and monotone increasing.

If F is continuous at x, then for every $\varepsilon > 0$, there exist numbers $q^-, q^+ \in \mathbb{Q}$, $q^- < x < q^+$ with $\widetilde{F}(q^-) \geq F(x)-\varepsilon$ and $\widetilde{F}(q^+) \leq F(x)+\varepsilon$. By construction,

$$\limsup_{k\to\infty} F_{n_k}(x) \leq \lim_{k\to\infty} F_{n_k}(q^+) = \widetilde{F}(q^+) \leq F(x) + \varepsilon.$$

Hence $\limsup_{k\to\infty} F_{n_k}(x) \leq F(x)$. A similar argument for q^- yields $\liminf_{k\to\infty} F_{n_k}(x) \geq F(x)$. $\square$

Reflection Check that in the above proof, in general, we do not have $F(q) = \widetilde{F}(q)$ for all $q \in \mathbb{Q}$.◆

Proof (of Theorem 13.29(i) for the case $E = \mathbb{R}$) Assume $\mathcal{F}$ is tight and $(\mu_n)_{n\in\mathbb{N}}$ is a sequence in $\mathcal{F}$ with distribution functions $F_n : x \mapsto \mu_N((-\infty, x])$. By Helly's theorem, there is a monotone right continuous function $F : \mathbb{R} \to [0, 1]$ and a subsequence $(F_{n_k})_{k\in\mathbb{N}}$ of $(F_n)_{n\in\mathbb{N}}$ with $F_{n_k}(x) \overset{k\to\infty}{\longrightarrow} F(x)$ at all points of continuity x of F. We will show

(i) F is the distribution function of a (sub-) probability measure. That is, we have
$F(-\infty) = 0$.
(ii) We have $F(\infty) \geq \limsup_{k\to\infty} F_{n_k}(\infty)$.

By Theorem 13.23, this is enough to conclude the proof.

As $\mathcal{F}$ is tight, for every $\varepsilon > 0$, there is a $K < \infty$ with $F_n(x) - F_n(-\infty) < \varepsilon$ for all $n \in \mathbb{N}$ and $x < -K$. If $x > K$ is a point of continuity of F, then

$$0 = \liminf_{k\to\infty} F_{n_k}(-\infty) \geq \liminf_{k\to\infty} F_{n_k}(x) - \varepsilon$$

$$= F(x) - \varepsilon \geq F(-\infty) - \varepsilon \geq -\varepsilon.$$

This shows (i).

As $\mathcal{F}$ is tight, for every $\varepsilon > 0$, there is a $K < \infty$ with $F_n(\infty) - F_n(x) < \varepsilon$ for all $n \in \mathbb{N}$ and $x > K$. If $x > K$ is a point of continuity of F, then $\limsup_{k\to\infty} F_{n_k}(\infty) \leq \limsup_{k\to\infty} F_{n_k}(x) + \varepsilon = F(x) + \varepsilon \leq F(\infty) + \varepsilon$. This shows (ii). $\qquad\square$

Reflection Find an example that shows that without the tightness assumption, we need not have $F(-\infty) = 0$ nor $F(\infty) = 1$. ♠

We come to a first application of Prohorov's theorem. The full strength of that theorem will become manifest when suitable separating classes of functions are at our disposal. We come back to this point in more detail in Chap. 15.

Theorem 13.34 Let E be Polish and let $\mu, \mu_1, \mu_2, \ldots \in \mathcal{M}_{\leq 1}(E)$. Then the following are equivalent.

(i) $\mu = \text{w-}\lim\limits_{n\to\infty} \mu_n$.
(ii) $(\mu_n)_{n\in\mathbb{N}}$ is tight, and there is a separating family $\mathcal{C} \subset C_b(E)$ such that

$$\int f\, d\mu = \lim_{n\to\infty} \int f\, d\mu_n \quad \text{for all } f \in \mathcal{C}. \tag{13.9}$$

Proof "(i) $\Longrightarrow$ (ii)" By the simple implication in Prohorov's theorem (Theorem 13.29(ii)), weak convergence implies tightness.

"(ii) $\Longrightarrow$ (i)" Let $(\mu_n)_{n\in\mathbb{N}}$ be tight and let $\mathcal{C} \subset C_b(E)$ be a separating class with (13.9). Assume that $(\mu_n)_{n\in\mathbb{N}}$ does not converge weakly to μ. Then there are $\varepsilon > 0$, $f \in C_b(E)$ and $(n_k)_{k\in\mathbb{N}}$ with $n_k \uparrow \infty$ and such that

$$\left| \int f\, d\mu_{n_k} - \int f\, d\mu \right| > \varepsilon \quad \text{for all } k \in \mathbb{N}. \tag{13.10}$$

By Prohorov's theorem, there exists a $\nu \in \mathcal{M}_{\le 1}(E)$ and a subsequence $(n'_k)_{k\in\mathbb{N}}$ of $(n_k)_{k\in\mathbb{N}}$ with $\mu_{n'_k} \to \nu$ weakly. Due to (13.10), we have $\left| \int f \, d\mu - \int f \, d\nu \right| \ge \varepsilon$; hence $\mu \ne \nu$. On the other hand,

$$\int h \, d\mu = \lim_{k\to\infty} \int h \, d\mu_{n'_k} = \int h \, d\nu \quad \text{for all } h \in \mathcal{C};$$

hence $\mu = \nu$. This contradicts the assumption and thus (i) holds. $\qquad\square$

We want to shed some more light on the connection between weak and vague convergence.

Theorem 13.35 *Let E be a locally compact Polish space and let $\mu, \mu_1, \mu_2, \ldots \in \mathcal{M}_f(E)$. Then the following are equivalent.*

(i) $\mu = \text{w-}\lim\limits_{n\to\infty} \mu_n$.

(ii) $\mu = \text{v-}\lim\limits_{n\to\infty} \mu_n$ *and* $\mu(E) = \lim\limits_{n\to\infty} \mu_n(E)$.

(iii) $\mu = \text{v-}\lim\limits_{n\to\infty} \mu_n$ *and* $\mu(E) \ge \limsup\limits_{n\to\infty} \mu_n(E)$.

(iv) $\mu = \text{v-}\lim\limits_{n\to\infty} \mu_n$ *and* $\{\mu_n, \ n \in \mathbb{N}\}$ *is tight.*

Proof **"(i) $\iff$ (ii) $\iff$ (iii)"**   This follows by the Portemanteau theorem.
"(ii) $\implies$ (iv)"   It is enough to show that for any $\varepsilon > 0$, there is a compact set $K \subset E$ with $\limsup_{n\to\infty} \mu_n(E \setminus K) \le \varepsilon$. As μ is regular (Theorem 13.6), there is a compact set $L \subset E$ with $\mu(E \setminus L) < \varepsilon$. Since E is locally compact, there exists a compact set $K \subset E$ with $K^\circ \supset L$ and a $\rho_{L,K} \in C_c(E)$ with $\mathbb{1}_L \le \rho_{L,K}(x) \le \mathbb{1}_K$. Therefore,

$$\limsup_{n\to\infty} \mu_n(E \setminus K) \le \limsup_{n\to\infty} \left(\mu_n(E) - \int \rho_{L,K} \, d\mu_n \right)$$

$$= \mu(E) - \int \rho_{L,K} \, d\mu \le \mu(E \setminus L) < \varepsilon.$$

"(iv) $\implies$ (i)"   Let $L \subset E$ be compact with $\mu_n(E \setminus L) \le 1$ for all $n \in \mathbb{N}$. Let $\rho \in C_c(E)$ with $\rho \ge \mathbb{1}_L$. Since $\int \rho \, d\mu_n$ converges by assumption, we thus have

$$\sup_{n\in\mathbb{N}} \mu_n(E) \le 1 + \sup_{n\in\mathbb{N}} \mu_n(L) \le 1 + \sup_{n\in\mathbb{N}} \int \rho \, d\mu_n < \infty.$$

Hence also

$$C := \max(\mu(E), \sup\{\mu_n(E) : n \in \mathbb{N}\}) < \infty,$$

and we can pass to μ/C and μ_n/C. Thus, without loss of generality assume that all measures are in $\mathcal{M}_{\le 1}(E)$. As $C_c(E)$ is a separating class for $\mathcal{M}_{\le 1}(E)$ (see Theorem 13.11), (i) follows by Theorem 13.34. $\qquad\square$

Proof of Prohorov's theorem, Part (i), general case There are two main routes for proving Prohorov's theorem in the general situation. One possibility is to show the claim first for measures on $\mathbb{R}^d$. (We have done this already for $d = 1$, see Exercise 13.3.4 for $d \geq 2$.) In a second step, the statement is lifted to sequence spaces $\mathbb{R}^{\mathbb{N}}$. Finally, in the third step, an embedding of E into $\mathbb{R}^{\mathbb{N}}$ is constructed. For a detailed description, see [12] or [83].

Here we follow the alternative route as described in [13] (and later [14]) or [44]. The main point of this proof consists in finding a candidate for a weak limit point for the family $\mathcal{F}$. This candidate will be constructed first as a content on a countable class of sets. From this an outer measure will be derived. Finally, we show that closed sets are measurable with respect to this outer measure. As you see, the argument follows a pattern similar to the proof of Carathéodory's theorem.

Let (E, d) be a metric space and let $\mathcal{F} \subset \mathcal{M}_{\leq 1}(E)$ be tight. Then there exists an increasing sequence $K_1 \subset K_2 \subset K_3 \subset \dots$ of compact sets in E such that $\mu(K_n^c) < \frac{1}{n}$ for all $\mu \in \mathcal{F}$ and all $n \in \mathbb{N}$. Define $E' := \bigcup_{n=1}^{\infty} K_n$. Then E' is a σ-compact metric space and therefore in particular, separable. By construction, $\mu(E \setminus E') = 0$ for all $\mu \in \mathcal{F}$. Thus, any μ can be regarded as a measure on E'. Without loss of generality, we may hence assume that E is σ-compact and thus separable. Hence there exists a countable base $\mathcal{U}$ of the topology $\tau\big|_E$ on E; that is, a countable set $\mathcal{U}$ of open sets such that $A = \bigcup_{U \in \mathcal{U},\, U \subset A} U$ for any open $A \subset E$. Define

$$\mathcal{C}' := \{ \overline{U} \cap K_n : U \in \mathcal{U},\ n \in \mathbb{N} \}$$

and

$$\mathcal{C} := \left\{ \bigcup_{n=1}^{N} C_n : N \in \mathbb{N} \text{ and } C_1, \dots, C_N \in \mathcal{C}' \right\}.$$

Clearly, $\mathcal{C}$ is a countable set of compact sets in E, and $\mathcal{C}$ is stable under formation of unions. Any K_n possesses a finite covering with sets from $\mathcal{U}$; hence $K_n \in \mathcal{C}$.

Now let $(\mu_n)_{n \in \mathbb{N}}$ be a sequence in $\mathcal{F}$. By virtue of the diagonal sequence argument (see the proof of Helly's theorem, Theorem 13.33), we can find a subsequence $(\mu_{n_k})_{k \in \mathbb{N}}$ such that for all $C \in \mathcal{C}$, there exists the limit

$$\alpha(C) := \lim_{k \to \infty} \mu_{n_k}(C). \tag{13.11}$$

Assume that we can show that there is a measure μ on the Borel σ-algebra $\mathcal{E}$ of E such that

$$\mu(A) = \sup \{ \alpha(C) : C \in \mathcal{C} \text{ with } C \subset A \} \quad \text{for all } A \subset E \text{ open.} \tag{13.12}$$

Then

$$\mu(E) \geq \sup_{n\in\mathbb{N}} \alpha(K_n) = \sup_{n\in\mathbb{N}} \lim_{k\to\infty} \mu_{n_k}(K_n)$$

$$\geq \sup_{n\in\mathbb{N}} \limsup_{k\to\infty} \left(\mu_{n_k}(E) - \frac{1}{n} \right)$$

$$= \limsup_{k\to\infty} \mu_{n_k}(E).$$

Furthermore, for open A and for $C \in \mathcal{C}$ with $C \subset A$,

$$\alpha(C) = \lim_{k\to\infty} \mu_{n_k}(C) \leq \liminf_{k\to\infty} \mu_{n_k}(A),$$

hence $\mu(A) \leq \liminf_{k\to\infty} \mu_{n_k}(A)$. By the Portemanteau theorem (Theorem 13.16), $\mu = \text{w-}\lim_{k\to\infty} \mu_{n_k}$; hence $\mathcal{F}$ is recognized as weakly relatively sequentially compact. It remains to show that there exists a measure μ on $(E, \mathcal{E})$ that satisfies (13.12).

Clearly, the set function α on $\mathcal{C}$ is monotone, additive and subadditive:

$$\begin{aligned}
\alpha(C_1) &\leq \alpha(C_2), & &\text{if } C_1 \subset C_2, \\
\alpha(C_1 \cup C_2) &= \alpha(C_1) + \alpha(C_2), & &\text{if } C_1 \cap C_2 = \emptyset, \\
\alpha(C_1 \cup C_2) &\leq \alpha(C_1) + \alpha(C_2).
\end{aligned} \qquad (13.13)$$

We define

$$\beta(A) := \sup \{\alpha(C) : C \in \mathcal{C} \text{ with } C \subset A\} \quad \text{for } A \subset E \text{ open}$$

and

$$\mu^*(G) := \inf \{\beta(A) : A \supset G \text{ is open}\} \quad \text{for } G \in 2^E.$$

Manifestly, $\beta(A) = \mu^*(A)$ for any open A. It is enough to show (Steps 1–3 below) that μ^* is an outer measure (see Definition 1.46) and that (Step 4) the σ-algebra of μ^*-measurable sets (see Definition 1.48 and Lemma 1.52) contains the closed sets and thus $\mathcal{E}$. Indeed, Lemma 1.52 would then imply that μ^* is a measure on the σ-algebra of μ^*-measurable sets and the restricted measure $\mu := \mu^*|_{\mathcal{E}}$ fulfills $\mu(A) = \mu^*(A) = \beta(A)$ for all open A. Hence equation (13.12) holds.

Evidently, $\mu^*(\emptyset) = 0$ and μ^* is monotone. In order to show that μ^* is an outer measure, it only remains to check that μ^* is σ-subadditive.

Step 1 (Finite subadditivity of β). Let $A_1, A_2 \subset E$ be open and let $C \in \mathcal{C}$ with $C \subset A_1 \cup A_2$. Let $n \in N$ with $C \subset K_n$. Define two sets

$$B_1 := \{x \in C : d(x, A_1^c) \geq d(x, A_2^c)\},$$
$$B_2 := \{x \in C : d(x, A_1^c) \leq d(x, A_2^c)\}.$$

Evidently, $B_1 \subset A_1$ and $B_2 \subset A_2$. As $x \mapsto d(x, A_i^c)$ is continuous for $i = 1, 2$, the closed subsets B_1 and B_2 of C are compact. Hence $d(B_1, A_1^c) > 0$. Thus there exists an open set D_1 with $B_1 \subset D_1 \subset \overline{D}_1 \subset A_1$. (One could choose D_1 as the union of the sets of a finite covering of B_1 with balls of radius $d(B_1, A_1^c)/2$. These balls, as well as their closures, are subsets of A_1.) Let $\mathcal{U}_{D_1} := \{U \in \mathcal{U} : U \subset D_1\}$. Then $B_1 \subset D_1 = \bigcup_{U \in \mathcal{U}_{D_1}} U$. Now choose a finite subcovering $\{U_1, \ldots, U_N\} \subset \mathcal{U}_{D_1}$ of B_1 and define $C_1 := \bigcup_{i=1}^{N} \overline{U}_i \cap K_n$. Then $B_1 \subset C_1 \subset A_1$ and $C_1 \in \mathcal{C}$. Similarly, choose $C_2 \in \mathcal{C}$ with $B_2 \subset C_2 \subset A_2$. Thus

$$\alpha(C) \leq \alpha(C_1 \cup C_2) \leq \alpha(C_1) + \alpha(C_2) \leq \beta(A_1) + \beta(A_2).$$

Hence also

$$\beta(A_1 \cup A_2) = \sup\{\alpha(C) : C \in \mathcal{C} \text{ with } C \subset A_1 \cup A_2\} \leq \beta(A_1) + \beta(A_2).$$

Step 2 (σ-subadditivity of β). Let $A_1, A_2, \ldots$ be open sets and let $C \in \mathcal{C}$ with $C \subset \bigcup_{i=1}^{\infty} A_i$. As C is compact, there exists an $n \in \mathbb{N}$ with $C \subset \bigcup_{i=1}^{n} A_i$. As shown above, β is subadditive; thus

$$\alpha(C) \leq \beta\left(\bigcup_{i=1}^{n} A_i\right) \leq \sum_{i=1}^{\infty} \beta(A_i).$$

Taking the supremum over such C yields

$$\beta\left(\bigcup_{i=1}^{\infty} A_i\right) = \sup\left\{\alpha(C) : C \in \mathcal{C} \text{ with } C \subset \bigcup_{i=1}^{\infty} A_i\right\} \leq \sum_{i=1}^{\infty} \beta(A_i).$$

Step 3 (σ-subadditivity of μ^*). Let $G_1, G_2, \ldots \in 2^E$. Let $\varepsilon > 0$. For any $n \in \mathbb{N}$ choose an open set $A_n \supset G_n$ with $\beta(A_n) < \mu^*(G_n) + \varepsilon/2^n$. By the σ-subadditivity of β,

$$\mu^*\left(\bigcup_{n=1}^{\infty} G_n\right) \leq \beta\left(\bigcup_{n=1}^{\infty} A_n\right) \leq \sum_{n=1}^{\infty} \beta(A_n) \leq \varepsilon + \sum_{n=1}^{\infty} \mu^*(G_n).$$

Letting $\varepsilon \downarrow 0$ yields $\mu^*\left(\bigcup_{n=1}^{\infty} G_n\right) \leq \sum_{n=1}^{\infty} \mu^*(G_n)$. Hence μ^* is an outer measure.

Step 4 (Closed sets are μ^*-measurable). By Lemma 1.49, a set $B \subset E$ is μ^*-measurable if and only if

$$\mu^*(B \cap G) + \mu^*(B^c \cap G) \leq \mu^*(G) \quad \text{for all } G \in 2^E.$$

Taking the infimum over all open sets $A \supset G$, it is enough to show that for every open B and every open $A \subset E$,

$$\mu^*(B \cap A) + \mu^*(B^c \cap A) \leq \beta(A). \tag{13.14}$$

Let $\varepsilon > 0$. Choose $C_1 \in \mathcal{C}$ with $C_1 \subset A \cap B^c$ and $\alpha(C_1) > \beta(A \cap B^c) - \varepsilon$. Further, let $C_2 \in \mathcal{C}$ with $C_2 \subset A \cap C_1^c$ and $\alpha(C_2) > \beta(A \cap C_1^c) - \varepsilon$. Since $C_1 \cap C_2 = \emptyset$ and $C_1 \cup C_2 \subset A$, we get

$$\beta(A) \geq \alpha(C_1 \cup C_2) = \alpha(C_1) + \alpha(C_2) \geq \beta(A \cap B^c) + \beta(A \cap C_1^c) - 2\varepsilon$$
$$\geq \mu^*(A \cap B^c) + \mu^*(A \cap B) - 2\varepsilon.$$

Letting $\varepsilon \to 0$, we get (13.14). This completes the proof of Prohorov's theorem.

$\square$

Takeaways A family of measures is called tight if for larger and larger compacts, there is arbitrarily little mass outside the compact. By Prohorov's theorem, in Polish spaces, tightness is equivalent to relative sequential compactness. Since usually tightness is easier to check, we have a powerful tool for showing the existence of accumulation points.

Exercise 13.3.1 Show that a family $\mathcal{F} \subset \mathcal{M}_f(\mathbb{R})$ is tight if and only if there exists a measurable map $f : \mathbb{R} \to [0, \infty)$ such that $f(x) \to \infty$ for $|x| \to \infty$ and $\sup_{\mu \in \mathcal{F}} \int f \, d\mu < \infty$. ♣

Exercise 13.3.2 Let $L \subset \mathbb{R} \times (0, \infty)$ and let $\mathcal{F} = \{\mathcal{N}_{\mu, \sigma^2} : (\mu, \sigma^2) \in L\}$ be a family of normal distributions with parameters in L. Show that $\mathcal{F}$ is tight if and only if L is bounded. ♣

Exercise 13.3.3 If P is a probability measure on $[0, \infty)$ with $m_P := \int x \, P(dx) \in (0, \infty)$, then we define the **size-biased distribution** $\widehat{P}$ on $[0, \infty)$ by

$$\widehat{P}(A) = m_P^{-1} \int_A x \, P(dx). \tag{13.15}$$

Now let $(X_i)_{i \in I}$ be a family of random variables on $[0, \infty)$ with $\mathbf{E}[X_i] = 1$. Show that $(\widehat{\mathbf{P}_{X_i}})_{i \in I}$ is tight if and only if $(X_i)_{i \in I}$ is uniformly integrable. ♣

Exercise 13.3.4 (Helly's theorem in $\mathbb{R}^d$)   Let $x = (x^1, \ldots, x^d) \in \mathbb{R}^d$ and $y = (y^1, \ldots, y^d) \in \mathbb{R}^d$. Recall the notation $x \leq y$ if $x^i \leq y^i$ for all $i = 1, \ldots, d$. A map $F : \mathbb{R}^d \to \mathbb{R}$ is called monotone increasing if $F(x) \leq F(y)$ whenever $x \leq y$. F is called right continuous if $F(x) = \lim_{n \to \infty} F(x_n)$ for all $x \in \mathbb{R}^d$ and every sequence $(x_n)_{n \in \mathbb{N}}$ in $\mathbb{R}^d$ with $x_1 \geq x_2 \geq x_3 \geq \ldots$ and $x = \lim_{n \to \infty} x_n$. By V_d denote the set of monotone increasing, bounded right continuous functions on $\mathbb{R}^d$.

(i)   Show the validity of Helly's theorem with V replaced by V_d.
(ii) Conclude that Prohorov's theorem holds for $E = \mathbb{R}^d$. ♣

13.4   Application: A Fresh Look at de Finetti's Theorem

(After an idea of Götz Kersting.) Let E be a Polish space and let $X_1, X_2, \ldots$ be an exchangeable sequence of random variables with values in E. As an alternative to the backwards martingale argument of Sect. 12.3, here we give a different proof of de Finetti's theorem (Theorem 12.26). Recall that de Finetti's theorem states that there exists a random probability measure Ξ on E such that, given Ξ, the random variables $X_1, X_2, \ldots$ are independent and Ξ-distributed. For $x = (x_1, x_2, \ldots) \in E^{\mathbb{N}}$, let $\xi_n(x) := \frac{1}{n} \sum_{l=1}^{n} \delta_{x_l}$ be the empirical distribution of $x_1, \ldots, x_n$. Let

$$\mu_{n,k}(x) := \xi_n(x)^{\otimes k} = n^{-k} \sum_{i_1, \ldots, i_k = 1}^{n} \delta_{(x_{i_1}, \ldots, x_{i_k})}$$

be the distribution on E^k that describes k-fold independent sampling *with replacement* (respecting the order) from $(x_1, \ldots, x_n)$. Let

$$\nu_{n,k}(x) := \frac{(n-k)!}{n!} \sum_{\substack{i_1, \ldots, i_k = 1 \\ \#\{i_1, \ldots, i_k\} = k}}^{n} \delta_{(x_{i_1}, \ldots, x_{i_k})}$$

be the distribution on E^k that describes k-fold independent sampling *without replacement* (respecting the order) from $(x_1, \ldots, x_n)$. For all $x \in E^{\mathbb{N}}$,

$$\|\mu_{n,k}(x) - \nu_{n,k}(x)\|_{TV} \leq R_{n,k} := \frac{k(k-1)}{n}.$$

Indeed, the probability $p_{n,k}$ that we do not see any ball twice when drawing k balls (with replacement) from n different balls is

$$p_{n,k} = \prod_{l=1}^{k-1}(1 - l/n)$$

and thus $R_{n,k} \geq 2(1 - p_{n,k})$. We therefore obtain the rather intuitive statement that as $n \to \infty$ the distributions of k-samples with replacement and without replacement, respectively, become the same:

$$\lim_{n \to \infty} \sup_{x \in E^{\mathbb{N}}} \|\mu_{n,k}(x) - \nu_{n,k}(x)\|_{TV} = 0.$$

Now let $f_1, \ldots, f_k \in C_b(E)$ and $F(x_1, \ldots, x_k) := f_1(x_1) \cdots f_k(x_k)$. As the sequence $X_1, X_2, \ldots$ is exchangeable, for any choice of pairwise distinct numbers $1 \leq i_1, \ldots, i_k \leq n$,

$$\mathbf{E}[F(X_1, \ldots, X_k)] = \mathbf{E}[F(X_{i_1}, \ldots, X_{i_k})].$$

Averaging over all choices $i_1, \ldots, i_k$, we get

$$\mathbf{E}\big[f_1(X_1) \cdots f_k(X_k)\big] = \mathbf{E}\big[F(X_1, \ldots, X_k)\big] = \mathbf{E}\Big[\int F \, d\nu_{n,k}(X)\Big].$$

Hence

$$\Big|\mathbf{E}\big[f_1(X_1) \cdots f_k(X_k)\big] - \mathbf{E}\Big[\int f_1 \, d\xi_n(X) \cdots \int f_k \, d\xi_n(X)\Big]\Big|$$

$$= \Big|\mathbf{E}\Big[\int F \, d\nu_{n,k}(X)\Big] - \mathbf{E}\Big[\int F \, d\mu_{n,k}(X)\Big]\Big|$$

$$\leq \|F\|_\infty R_{n,k} \xrightarrow{n \to \infty} 0.$$

We will exploit the following criterion for tightness of subsets of $\mathcal{M}_1(\mathcal{M}_1(E))$.

Exercise 13.4.1 Show that a subset $\mathcal{K} \subset \mathcal{M}_1(\mathcal{M}_1(E))$ is tight if and only if, for any $\varepsilon > 0$, there exists a compact set $K \subset E$ with the property

$$\tilde{\mu}\big(\{\mu \in \mathcal{M}_1(E) : \mu(K^c) > \varepsilon\}\big) < \varepsilon \quad \text{for all } \tilde{\mu} \in \mathcal{K}. \quad \clubsuit$$

Since E is Polish, $\mathbf{P}_{X_1}$ is tight. Hence, for any $\varepsilon > 0$, there exists a compact set $K \subset E$ with $\mathbf{P}[X_1 \in K^c] < \varepsilon^2$. Therefore,

$$\mathbf{P}[\xi_n(X)(K^c) > \varepsilon] \;\leq\; \varepsilon^{-1}\,\mathbf{E}[\xi_n(X)(K^c)] \;=\; \varepsilon^{-1}\,\mathbf{P}[X_1 \in K^c] \;\leq\; \varepsilon.$$

Hence the family $(\mathbf{P}_{\xi_n(X)})_{n \in \mathbb{N}}$ is tight. Let Ξ_∞ be a random variable (with values in $\mathcal{M}_1(E)$) such that $\mathbf{P}_{\Xi_\infty} = \underset{l \to \infty}{\text{w-lim}}\, \mathbf{P}_{\xi_{n_l}(X)}$ for a suitable subsequence $(n_l)_{l \in \mathbb{N}}$. The map $\xi \mapsto \int F\, d\xi = \int f_1\, d\xi \cdots \int f_k\, d\xi$ is bounded and (as a product of continuous maps) is continuous with respect to the topology of weak convergence on $\mathcal{M}_1(E)$; hence it is in $C_b(\mathcal{M}_1(E))$. Thus

$$\mathbf{E}\!\left[\int F\, d\Xi_\infty^{\otimes k}\right] \;=\; \lim_{l \to \infty} \mathbf{E}\!\left[\int f_1\, d\xi_{n_l}(X) \cdots \int f_k\, d\xi_{n_l}(X)\right]$$

$$\;=\; \mathbf{E}\big[f_1(X_1) \cdots f_k(X_k)\big].$$

Note that the limit does not depend on the choice of the subsequence and is thus unique. Summarising, we have

$$\mathbf{E}\big[f_1(X_1) \cdots f_k(X_k)\big] = \mathbf{E}\!\left[\int f_1\, d\Xi_\infty \cdots \int f_k\, d\Xi_\infty\right].$$

Since the distribution of $(X_1, \dots, X_k)$ is uniquely determined by integrals of the above type, we conclude that $\mathbf{P}_{(X_1,\dots,X_k)} = \mathbf{P}_{\Xi_\infty^{\otimes k}}$. In other words, $(X_1, \dots, X_k) \overset{\mathcal{D}}{=} (Y_1, \dots, Y_k)$, where, given Ξ_∞, the random variables $Y_1, \dots, Y_k$ are independent with distribution Ξ_∞.

Takeaways Consider an exchangeable family $X_1, X_2, \dots$ of random variables. The empirical distributions of the first n random variables yield a tight family which, by Prohorov's theorem, has a limit point. This approach allows for an independent proof of de Finetti's theorem.

Exercise 13.4.2 Show that a family $(X_n)_{n \in \mathbb{N}}$ of random variables is exchangeable if and only if, for every choice of natural numbers $1 \leq n_1 < n_2 < n_3 \dots$, we have

$$(X_1, X_2, \dots) \overset{\mathcal{D}}{=} (X_{n_1}, X_{n_2}, \dots).$$

Warning: One of the implications is rather difficult to show. ♣

Chapter 14
Probability Measures on Product Spaces

As a motivation, consider the following example. Let X be a random variable that is uniformly distributed on $[0, 1]$. As soon as we know the value of X, we toss n times a coin that has probability X for a success. Denote the results by $Y_1, \ldots, Y_n$.

How can we construct a probability space on which all these random variables are defined? One possibility is to construct $n + 1$ independent random variables $Z_0, \ldots, Z_n$ that are uniformly distributed on $[0, 1]$ (see, e.g., Corollary 2.23 for the construction). Then define $X = Z_0$ and

$$Y_k = \begin{cases} 1, & \text{if } Z_k < X, \\ 0, & \text{if } Z_k \geq X. \end{cases}$$

Intuitively, this fits well with our idea that the $Y_1, \ldots, Y_n$ are independent as soon as we know X and record a success with probability X.

In the above description, we have constructed by hand a **two-stage experiment**. At the first stage, we determine the value of X. At the second stage, depending on the value of X, the values of $Y = (Y_1, \ldots, Y_n)$ are determined. Clearly, this construction makes use of the specific structure of the problem. However, we now want to develop a *systematic* framework for the description and construction of multi-stage experiments. In contrast to Chap. 2, here the random variables need not be independent. In addition, we also want to construct *systematically* infinite families of random variables with given (joint) distributions.

In the first section, we start with products of measurable spaces. Then we come to finite products of measure spaces and product measures with transition kernels. Finally, we consider infinite products of probability spaces. The main result is Kolmogorov's extension theorem.

A. Klenke, *Probability Theory*, Universitext, https://doi.org/10.1007/978-3-030-56402-5_14

14.1　Product Spaces

Definition 14.1 (Product space) Let $(\Omega_i,\ i \in I)$ be an arbitrary family of sets. Denote by $\Omega = \underset{i\in I}{\times}\, \Omega_i$ the set of maps $\omega : I \to \underset{i\in I}{\bigcup}\, \Omega_i$ such that $\omega(i) \in \Omega_i$ for all $i \in I$. Ω is called the **product** of the spaces $(\Omega_i,\ i \in I)$, or briefly the **product space**. If, in particular, all the Ω_i are equal, say $\Omega_i = \Omega_0$, then we write $\Omega = \underset{i\in I}{\times}\, \Omega_i = \Omega_0^I$.

Example 14.2

(i) If $\Omega_1 = \{1, \ldots, 6\}$ and $\Omega_2 = \{1, 2, 3\}$, then

$$\Omega_1 \times \Omega_2 = \big\{\omega = (\omega_1, \omega_2) : \omega_1 \in \{1, \ldots, 6\},\ \omega_2 \in \{1, 2, 3\}\big\}.$$

(ii) If $\Omega_0 = \mathbb{R}$ and $I = \{1, 2, 3\}$, then $\mathbb{R}^{\{1,2,3\}}$ is isomorphic to the customary $\mathbb{R}^3$.
(iii) If $\Omega_0 = \mathbb{R}$ and $I = \mathbb{N}$, then $\mathbb{R}^\mathbb{N}$ is the space of sequences $(\omega(n),\ n \in \mathbb{N})$ in $\mathbb{R}$.
(iv) If $I = \mathbb{R}$ and $\Omega_0 = \mathbb{R}$, then $\mathbb{R}^\mathbb{R}$ is the set of maps $\mathbb{R} \to \mathbb{R}$. ◊

Definition 14.3 (Coordinate maps) If $i \in I$, then $X_i : \Omega \to \Omega_i,\ \omega \mapsto \omega(i)$ denotes the ith **coordinate map**. More generally, for $J \subset J' \subset I$, the restricted map

$$X_J^{J'} : \underset{j\in J'}{\times}\, \Omega_j \longrightarrow \underset{j\in J}{\times}\, \Omega_j, \qquad \omega' \mapsto \omega'\big|_J \tag{14.1}$$

is called the canonical projection. In particular, we write $X_J := X_J^I$.

Definition 14.4 (Product-σ-algebra) Let $(\Omega_i, \mathcal{A}_i),\ i \in I$, be measurable spaces. The **product-σ-algebra**

$$\mathcal{A} = \bigotimes_{i\in I} \mathcal{A}_i$$

is the smallest σ-algebra on Ω such that for every $i \in I$, the coordinate map X_i is measurable with respect to $\mathcal{A} - \mathcal{A}_i$; that is,

$$\mathcal{A} = \sigma\big(X_i,\ i \in I\big) := \sigma\big(X_i^{-1}(\mathcal{A}_i),\ i \in I\big).$$

If $(\Omega_i, \mathcal{A}_i) = (\Omega_0, \mathcal{A}_0)$ for all $i \in I$, then we also write $\mathcal{A} = \mathcal{A}_0^{\otimes I}$.
For $J \subset I$, let $\Omega_J := \underset{j\in J}{\times}\, \Omega_j$ and $\mathcal{A}_J = \underset{j\in J}{\bigotimes}\, \mathcal{A}_j$.

Remark 14.5 The concept of the product-σ-algebra is similar to that of the **product topology**: If $((\Omega_i, \tau_i),\ i \in I)$ are topological spaces, then the product topology τ on $\Omega = \underset{i\in I}{\times}\, \Omega_i$ is the coarsest topology with respect to which all coordinate maps $X_i : \Omega \longrightarrow \Omega_i$ are continuous. ◊

Definition 14.6 Let $I \neq \emptyset$ be an arbitrary index set, let $(E, \mathcal{E})$ be a measurable space, let $(\Omega, \mathcal{A}) = (E^I, \mathcal{E}^{\otimes I})$ and let $X_t : \Omega \to E$ be the coordinate map for every $t \in I$. Then the family $(X_t)_{t \in I}$ is called the **canonical process** on $(\Omega, \mathcal{A})$.

Lemma 14.7 *Let $\emptyset \neq J \subset I$. Then X_J^I is measurable with respect to $\mathcal{A}_I - \mathcal{A}_J$.*

Proof For any $j \in J$, $X_j = X_j^J \circ X_J^I$ is measurable with respect to $\mathcal{A} - \mathcal{A}_j$. Thus, by Corollary 1.82, X_J^I is measurable.   ☐

Theorem 14.8 *Let I be countable, and for every $i \in I$, let (Ω_i, τ_i) be Polish with Borel σ-algebra $\mathcal{B}_i = \sigma(\tau_i)$. Let τ be the product topology on $\Omega = \underset{i \in I}{\times} \Omega_i$ and $\mathcal{B} = \sigma(\tau)$.*
   Then (Ω, τ) is Polish and $\mathcal{B} = \underset{i \in I}{\bigotimes} \mathcal{B}_i$. In particular, $\mathcal{B}(\mathbb{R}^d) = \mathcal{B}(\mathbb{R})^{\otimes d}$ for $d \in \mathbb{N}$.

Proof Without loss of generality, assume $I = \mathbb{N}$. For $i \in \mathbb{N}$, let d_i be a complete metric that induces τ_i. It is easy to check that

$$d(\omega, \omega') := \sum_{i=1}^{\infty} 2^{-i} \, \frac{d_i(\omega(i), \omega'(i))}{1 + d_i(\omega(i), \omega'(i))} \tag{14.2}$$

is a complete metric on Ω that induces τ.
   Now for any $i \in \mathbb{N}$, let $D_i \subset \Omega_i$ be a countable dense subset and let $y_i \in D_i$ be an arbitrary point. It is easy to see that the set

$$D = \left\{ x \in \underset{i \in \mathbb{N}}{\times} D_i : x_i \neq y_i \text{ only finitely often} \right\}$$

is a countable dense subset of Ω. Hence Ω is separable and thus Polish.
   Now, for any $i \in I$, let $\beta_i = \{ B_\varepsilon(x_i) : x_i \in D_i, \, \varepsilon \in \mathbb{Q}^+ \}$ be a countable base of the topology of Ω_i consisting of ε-balls. Define

$$\beta := \bigcup_{N=1}^{\infty} \left\{ \bigcap_{i=1}^{N} X_i^{-1}(B_i) : B_1 \in \beta_1, \dots, B_N \in \beta_N \right\}.$$

Then β is a countable base of the topology τ; hence any open set $A \subset \Omega$ is a (countable) union of sets in $\beta \subset \bigotimes_{i \in \mathbb{N}} \mathcal{B}_i$. Hence $\tau \subset \bigotimes_{i \in \mathbb{N}} \mathcal{B}_i$ and thus $\mathcal{B} \subset \bigotimes_{i \in \mathbb{N}} \mathcal{B}_i$.
   On the other hand, each X_i is continuous and thus measurable with respect to $\mathcal{B} - \mathcal{B}_i$. Therefore, $\mathcal{B} \supset \bigotimes_{i \in \mathbb{N}} \mathcal{B}_i$.   ☐

Definition 14.9 (Cylinder sets) For any $i \in I$, let $\mathcal{E}_i \subset \mathcal{A}_i$ be a subclass of the class of measurable sets.
   For any $A \in \mathcal{A}_J$, $X_J^{-1}(A) \subset \Omega$ is called a **cylinder set** with base J. The set of such cylinder sets is denoted by $\mathcal{Z}_J$. In particular, if $A = \times_{j \in J} A_j$ for certain

$A_j \in \mathcal{A}_j$, then $X_J^{-1}(A)$ is called a **rectangular cylinder** with base J. The set of such rectangular cylinders will be denoted by $\mathcal{Z}_J^R$. The set of such rectangular cylinders for which in addition $A_j \in \mathcal{E}_j$ for all $j \in J$ holds will be denoted by $\mathcal{Z}_J^{\mathcal{E},R}$.

Write

$$\mathcal{Z} = \bigcup_{J \subset I \text{ finite}} \mathcal{Z}_J, \tag{14.3}$$

and similarly define $\mathcal{Z}^R$ and $\mathcal{Z}^{\mathcal{E},R}$. Further, define

$$\mathcal{Z}_*^R = \bigcup_{N=1}^{\infty} \left\{ \bigcup_{n=1}^{N} A_n : A_1, \dots, A_N \in \mathcal{Z}^R \right\}$$

and similarly $\mathcal{Z}_*^{\mathcal{E},R}$.

Remark 14.10 Every $\mathcal{Z}_J$ is a σ-algebra, and $\mathcal{Z}$ and $\mathcal{Z}_*^R$ are algebras. Furthermore, $\bigotimes_{i \in I} \mathcal{A}_i = \sigma(\mathcal{Z})$. ◊

Lemma 14.11 *If every $\mathcal{E}_i$ is a π-system, then $\mathcal{Z}^{\mathcal{E},R}$ is a π-system.*

Proof This is left as an exercise.   □

Theorem 14.12 *For any $i \in I$, let $\mathcal{E}_i \subset \mathcal{A}_i$ be a generator of $\mathcal{A}_i$.*

(i) $\displaystyle\bigotimes_{j \in J} \mathcal{A}_j = \sigma\left(\underset{j \in J}{\times} E_j : E_j \in \mathcal{E}_j \cup \{\Omega_j\} \right)$ *for every countable $J \subset I$.*

(ii) $\displaystyle\bigotimes_{i \in I} \mathcal{A}_i = \sigma(\mathcal{Z}^R) = \sigma(\mathcal{Z}^{\mathcal{E},R})$.

(iii) *Let μ be a σ-finite measure on $\mathcal{A}$, and assume every $\mathcal{E}_i$ is also a π-system. Furthermore, assume there is a sequence $(E_n)_{n \in \mathbb{N}}$ in $\mathcal{Z}^{\mathcal{E},R}$ with $E_n \uparrow \Omega$ and $\mu(E_n) < \infty$ for all $n \in \mathbb{N}$ (this condition is satisfied, for example, if μ is finite and $\Omega_i \in \mathcal{E}_i$ for all $i \in I$). Then μ is uniquely determined by the values $\mu(A)$ for all $A \in \mathcal{Z}^{\mathcal{E},R}$.*

Proof

(i) Let $\mathcal{A}_J' = \sigma\left(\underset{j \in J}{\times} E_j : E_j \in \mathcal{E}_j \cup \{\Omega_j\} \text{ for every } j \in J \right)$. Note that

$$\underset{j \in J}{\times} E_j = \bigcap_{j \in J} (X_j^J)^{-1}(E_j) \in \mathcal{A}_J,$$

hence $\mathcal{A}_J' \subset \mathcal{A}_J$. On the other hand, $(X_j^J)^{-1}(E_j) \in \mathcal{A}_J'$ for all $j \in J$ and $E_j \in \mathcal{E}_j$. Since $\mathcal{E}_i$ is a generator of $\mathcal{A}_i$, we have $(X_j^J)^{-1}(A_j) \in \mathcal{A}_J'$ for all $A_j \in \mathcal{A}_j$, and hence $\mathcal{A}_J \subset \mathcal{A}_J'$.

(ii) Evidently, $\mathcal{Z}^{\mathcal{E},R} \subset \mathcal{Z}^R \subset \mathcal{A}$; hence also $\sigma(\mathcal{Z}^{\mathcal{E},R}) \subset \sigma(\mathcal{Z}^R) \subset \mathcal{A}$. By Theorem 1.81, we have $\sigma\left(\mathcal{Z}^{\mathcal{E},R}_{\{i\}}\right) = \sigma(X_i)$ for all $i \in I$; hence $\sigma(X_i) \subset \sigma(\mathcal{Z}^{\mathcal{E},R})$. Therefore, $\mathcal{A}_I \subset \sigma(\mathcal{Z}^{\mathcal{E},R})$.

(iii) By (ii) and Lemma 14.11, $\mathcal{Z}^{\mathcal{E},R}$ is a π-system that generates $\mathcal{A}$. Hence, the claim follows by Lemma 1.42. $\qquad\square$

Reflection Find an example that shows that in (iii), we cannot simply drop the assumption that there exists a sequence $E_n \uparrow \Omega$ with $\mu(E_n) < \infty$. ♠

> **Takeaways** Consider an arbitrary product of measurable spaces. The product σ-algebra is the smallest σ-algebra such that all coordinate maps are measurable. It is also induced by finite dimensional cylinder sets. For a countable product of Polish spaces, the Borel σ-algebra of the product is the product of the Borel σ-algebras. In either case, a probability measure on the product is uniquely determined by its values on cylinder sets.

Exercise 14.1.1 Show that

$$\bigotimes_{i \in I} \mathcal{A}_i = \bigcup_{J \subset I \text{ countable}} \mathcal{Z}_J. \tag{14.4}$$

Hint: Show that the right-hand side is a σ-algebra. ♣

14.2 Finite Products and Transition Kernels

Consider now the situation of finitely many σ-finite measure spaces $(\Omega_i, \mathcal{A}_i, \mu_i)$, $i = 1, \ldots, n$, where $n \in \mathbb{N}$.

Lemma 14.13 *Let* $A \in \mathcal{A}_1 \otimes \mathcal{A}_2$ *and let* $f : \Omega_1 \times \Omega_2 \to \overline{\mathbb{R}}$ *be an* $\mathcal{A}_1 \otimes \mathcal{A}_2$-*measurable map. Then, for all* $\tilde{\omega}_1 \in \Omega_1$ *and* $\tilde{\omega}_2 \in \Omega_2$,

$$A_{\tilde{\omega}_1} := \{\omega_2 \in \Omega_2 : (\tilde{\omega}_1, \omega_2) \in A\} \in \mathcal{A}_2,$$

$$A_{\tilde{\omega}_2} := \{\omega_1 \in \Omega_1 : (\omega_1, \tilde{\omega}_2) \in A\} \in \mathcal{A}_1,$$

$$f_{\tilde{\omega}_1} : \Omega_2 \to \overline{\mathbb{R}}, \quad \omega_2 \mapsto f(\tilde{\omega}_1, \omega_2) \quad \text{is } \mathcal{A}_2\text{-measurable},$$

$$f_{\tilde{\omega}_2} : \Omega_1 \to \overline{\mathbb{R}}, \quad \omega_1 \mapsto f(\omega_1, \tilde{\omega}_2) \quad \text{is } \mathcal{A}_1\text{-measurable}.$$

Proof For $\tilde{\omega}_1$, define the embedding map $i : \Omega_2 \to \Omega_1 \times \Omega_2$ by $i(\omega_2) = (\tilde{\omega}_1, \omega_2)$. Note that $X_1 \circ i$ is constantly $\tilde{\omega}_1$ (and hence $\mathcal{A}_1$-measurable), and $X_2 \circ i = \mathrm{id}_{\Omega_2}$ (and hence $\mathcal{A}_2$-measurable). Thus, by Corollary 1.82, the map i is measurable with respect to $\mathcal{A}_2 - (\mathcal{A}_1 \otimes \mathcal{A}_2)$. Hence $A_{\tilde{\omega}_1} = i^{-1}(A) \in \mathcal{A}_2$ and $f_{\tilde{\omega}_1} = f \circ i$ is measurable with respect to $\mathcal{A}_2$. $\square$

The following theorem generalizes Theorem 1.61.

Theorem 14.14 (Finite product measures)   *There exists a unique σ-finite measure μ on $\mathcal{A} := \bigotimes_{i=1}^{n} \mathcal{A}_i$ such that*

$$\mu(A_1 \times \cdots \times A_n) = \prod_{i=1}^{n} \mu_i(A_i) \quad for \ A_i \in \mathcal{A}_i, \ i = 1, \dots, n. \tag{14.5}$$

$\bigotimes_{i=1}^{n} \mu_i := \mu_1 \otimes \cdots \otimes \mu_n := \mu$ *is called the **product measure** of the μ_i.*

If all spaces involved equal $(\Omega_0, \mathcal{A}_0, \mu_0)$, then we write $\mu_0^{\otimes n} := \bigotimes_{i=1}^{n} \mu_0$.

Proof Let $\tilde{\mu}$ be the restriction of μ to $\mathcal{Z}^R$. Evidently, $\tilde{\mu}(\emptyset) = 0$, and it is simple to check that $\tilde{\mu}$ is σ-finite. Let $A^1, A^2, \dots \in \mathcal{Z}^R$ be pairwise disjoint and let $A \in \mathcal{Z}^R$ with $A \subset \bigcup_{k=1}^{\infty} A^k$. Then, by the monotone convergence theorem,

$$\tilde{\mu}(A) = \int \mu_1(d\omega_1) \cdots \int \mu_n(d\omega_n) \, \mathbb{1}_A((\omega_1, \dots, \omega_n))$$

$$\leq \int \mu_1(d\omega_1) \cdots \int \mu_n(d\omega_n) \sum_{k=1}^{\infty} \mathbb{1}_{A^k}((\omega_1, \dots, \omega_n)) = \sum_{k=1}^{\infty} \tilde{\mu}(A^k).$$

In particular, if $A = A^1 \uplus A^2$, one similarly gets $\tilde{\mu}(A) = \tilde{\mu}(A^1) + \tilde{\mu}(A^2)$. Hence $\tilde{\mu}$ is a σ-finite, additive, σ-subadditive set function on the semiring $\mathcal{Z}^R$ with $\tilde{\mu}(\emptyset) = 0$. By the measure extension theorem (Theorem 1.53), $\tilde{\mu}$ can be uniquely extended to a σ-finite measure on $\mathcal{A} = \sigma(\mathcal{Z}^R)$. $\square$

Example 14.15 For $i = 1, \dots, n$, let $(\Omega_i, \mathcal{A}_i, \mathbf{P}_i)$ be a probability space. On the space $(\Omega, \mathcal{A}, \mathbf{P}) := \left(\bigtimes_{i=1}^{n} \Omega_i, \bigotimes_{i=1}^{n} \mathcal{A}_i, \bigotimes_{i=1}^{n} \mathbf{P}_i \right)$, the coordinate maps $X_i : \Omega \to \Omega_i$ are independent with distribution $\mathbf{P}_{X_i} = \mathbf{P}_i$. $\Diamond$

In order to formulate Fubini's theorem rigorously, we need the following definition.

Definition 14.16 Let $(\Omega, \mathcal{A}, \mu)$ a measure space and $(\Omega', \mathcal{A}')$ a measurable space. Let $N \in \mathcal{A}$ with $\mu(N) = 0$. A map $f : \Omega \backslash N \to \Omega'$ is called a μ-almost everywhere defined and measurable map from $(\Omega, \mathcal{A})$ to $(\Omega', \mathcal{A}')$, if $f^{-1}(\mathcal{A}') \subset \mathcal{A}$.

Remark 14.17 Let $g, h : \Omega \to \overline{\mathbb{R}}$ be measurable finite almost everywhere. Then $g - h$ is almost everywhere defined and measurable. In particular, this holds if g and h are integrable. $\Diamond$

Remark 14.18 Assume that f is almost everywhere defined and measurable (with null set N) and takes values in $\bar{\mathbb{R}}$. Define $f'(\omega) = 0$ for $\omega \in N$ and $f'(\omega) = f(\omega)$ else. Then f' is measurable (and everywhere defined). If f' is integrable, then we can define the integral $\int f \, d\mu := \int f' d\mu$. $\Diamond$

Theorem 14.19 (Fubini) *Let $(\Omega_i, \mathcal{A}_i, \mu_i)$ be σ-finite measure spaces, $i = 1, 2$. Let $f : \Omega_1 \times \Omega_2 \to \bar{\mathbb{R}}$ be measurable with respect to $\mathcal{A}_1 \otimes \mathcal{A}_2$. If $f \geq 0$ or $f \in \mathcal{L}^1(\mu_1 \otimes \mu_2)$, then*

$$
\omega_1 \mapsto \int f(\omega_1, \omega_2) \, \mu_2(d\omega_2) \quad \text{is } \mu_1\text{-a.e. defined and } \mathcal{A}_1\text{-measurable,}
$$

$$
\omega_2 \mapsto \int f(\omega_1, \omega_2) \, \mu_1(d\omega_1) \quad \text{is } \mu_2\text{-a.e. defined and } \mathcal{A}_2\text{-measurable,}
$$

(14.6)

and

$$
\int_{\Omega_1 \times \Omega_2} f \, d(\mu_1 \otimes \mu_2) = \int_{\Omega_1} \left(\int_{\Omega_2} f(\omega_1, \omega_2) \, \mu_2(d\omega_2) \right) \mu_1(d\omega_1)
$$

$$
= \int_{\Omega_2} \left(\int_{\Omega_1} f(\omega_1, \omega_2) \, \mu_1(d\omega_1) \right) \mu_2(d\omega_2).
$$

(14.7)

Proof The proof follows the usual procedure of stepwise approximations, starting with an indicator function.

We first show the statement for nonnegative f. There are functions $h_i : \Omega_i \to (0, \infty)$ such that $\int h_i \, d\mu_i < \infty$, $i = 1, 2$ (see Lemma 6.23). Now write

$$
f(\omega_1, \omega_2) = \frac{f(\omega_1, \omega_2)}{h_1(\omega_1) h_2(\omega_2)} \cdot h_1(\omega_1) h_2(\omega_2).
$$

Hence, it is enough to show the statement for the finite measures $\tilde{\mu}_i := h_i \mu_i$ instead of μ_i, $i = 1, 2$.

Now assume that μ_1 and μ_2 are finite measures. First let $f = \mathbb{1}_A$ for $A = A_1 \times A_2$ with $A_1 \in \mathcal{A}_1$ and $A_2 \in \mathcal{A}_2$. Then (14.6) and (14.7) hold trivially.

Now consider the set $\mathcal{G} \subset \mathcal{A}_1 \otimes \mathcal{A}_2$ such that $A \in \mathcal{G}$ if and only if (14.6) and (14.7) hold for $f = \mathbb{1}_A$. It is easy to see that $\mathcal{G}$ is a Dynkin system. In fact, $\Omega_1 \times \Omega_2 \in \mathcal{G}$ is trivial. If $A \in \mathcal{G}$ and $f = \mathbb{1}_{\Omega_1 \times \Omega_2 \setminus A}$, then clearly

$$
\omega_i \mapsto \int f(\omega_1, \omega_2) \, \mu_{3-i}(d\omega_{3-i})
$$

$$
= \mu_{3-i}(\Omega_{3-i}) - \int \mathbb{1}_A(\omega_1, \omega_2) \, \mu_{3-i}(d\omega_{3-i}) \quad \text{is } \mathcal{A}_i\text{-measurable,}
$$

for $i = 1, 2$. Similarly, (14.7) holds. Hence, $\mathcal{G}$ is stable under complements.

Finally, for pairwise disjoint sets $A_1, A_2, \ldots \in \mathcal{G}$ and $A := A_1 \cup A_2 \cup \ldots$, we have

$$\omega_i \mapsto \int \mathbb{1}_A(\omega_1, \omega_2)\, \mu_{3-i}(d\omega_{3-i})$$

$$= \sum_{n=1}^{\infty} \int \mathbb{1}_{A_n}(\omega_1, \omega_2)\, \mu_{3-i}(d\omega_{3-i}) \quad \text{is } \mathcal{A}_i\text{-measurable,}$$

for $i = 1, 2$. Similarly, (14.7) holds for $f = \mathbb{1}_A$. Hence $A \in \mathcal{G}$.

Now, $\mathcal{G}$ is a Dynkin system that contains a $\cap$-stable generator of $\mathcal{A}_1 \otimes \mathcal{A}_2$ (namely, the cylinder sets $A_1 \times A_2$, $A_1 \in \mathcal{A}_1$, $A_2 \in \mathcal{A}_2$). Hence $\mathcal{G} = \mathcal{A}_1 \otimes \mathcal{A}_2$ by Dynkin's π-λ theorem (Theorem 1.19).

We have thus shown that (14.6) and (14.7) hold for $f = \mathbb{1}_A$ for all $A \in \mathcal{A}_1 \otimes \mathcal{A}_2$. Building finite sums, (14.6) and (14.7) also hold if f is a *simple function*.

Consider now $f \geq 0$. Then, by Theorem 1.96, there exists a sequence of simple functions $(f_n)_{n \in \mathbb{N}}$ with $f_n \uparrow f$. By the monotone convergence theorem (Theorem 4.20), (14.6) and (14.7) also hold for this f.

Now let $f \in \mathcal{L}^1(\mu_1 \otimes \mu_2)$. Then $f = f^+ - f^-$ with $f^+, f^- \geq 0$ being integrable functions. Since (14.6) and (14.7) hold for f^- and f^+, by Remark 14.17 and 14.18 this is true for f also.   □

Reflection Come up with an example of a measurable function f such that the integrals on the right hand side of (14.7) both exist but do not coincide. Which condition of the theorem would be violated? ♠

In Definition 2.32, we defined the convolution of two real probability measures μ and ν as the distribution of the sum of two independent random variables with distributions μ and ν, respectively. As a simple application of Fubini's theorem, we can give a new definition for the convolution of, more generally, finite measures on $\mathbb{R}^n$. Of course, for real probability measures, it coincides with the old definition. If the measures have Lebesgue densities, then we obtain an explicit formula for the density of the convolution.

Let X and Y be $\mathbb{R}^n$-valued random variables with densities f_X and f_Y. That is, $f_X, f_Y : \mathbb{R}^n \to [0, \infty]$ are measurable and integrable with respect to n-dimensional Lebesgue measure λ^n and, for all $x \in \mathbb{R}^n$,

$$\mathbf{P}[X \leq x] = \int_{(-\infty, x]} f_X(t)\, \lambda^n(dt) \quad \text{and} \quad \mathbf{P}[Y \leq x] = \int_{(-\infty, x]} f_Y(t)\, \lambda^n(dt).$$

Here $(-\infty, x] = \{y \in \mathbb{R}^n : y_i \leq x_i \text{ for } i = 1, \ldots, n\}$ (compare (1.5)).

Definition 14.20 Let $n \in \mathbb{N}$. For two Lebesgue integrable maps $f, g : \mathbb{R}^n \to [0, \infty]$, define the **convolution** $f * g : \mathbb{R}^n \to [0, \infty]$ by

$$(f * g)(x) = \int_{\mathbb{R}^n} f(y)\, g(x - y)\, \lambda^n(dy).$$

For two finite measures $\mu, \nu \in \mathcal{M}_f(\mathbb{R}^n)$, define the convolution $\mu * \nu \in \mathcal{M}_f(\mathbb{R}^n)$ by

$$(\mu * \nu)((-\infty, x]) = \int \int \mathbb{1}_{A_x}(u, v)\, \mu(du)\, \nu(dv),$$

where $A_x := \{(u, v) \in \mathbb{R}^n \times \mathbb{R}^n : u + v \le x\}$.

Lemma 14.21 *The map $f * g$ is measurable and we have $f * g = g * f$ and*

$$\int_{\mathbb{R}^n} (f * g)\, d\lambda^n = \left(\int_{\mathbb{R}^n} f\, d\lambda^n \right) \left(\int_{\mathbb{R}^n} g\, d\lambda^n \right).$$

Furthermore, $\mu * \nu = \nu * \mu$ and $(\mu * \nu)(\mathbb{R}^n) = \mu(\mathbb{R}^n)\, \nu(\mathbb{R}^n)$.

Proof The claims follow immediately from Fubini's theorem.   □

Theorem 14.22 (Convolution of n-dimensional measures)

(i) *If X and Y are independent $\mathbb{R}^n$-valued random variables with densities f_X and f_Y, then $X + Y$ has density $f_X * f_Y$.*

(ii) *If $\mu = f\lambda^n$ and $\nu = g\lambda^n$ are finite measures with Lebesgue densities f and g, then $\mu * \nu = (f * g)\lambda^n$.*

Proof

(i) Let $x \in \mathbb{R}^n$ and $A := \{(u, v) \in \mathbb{R}^n \times \mathbb{R}^n : u + v \le x\}$. Repeated application of Fubini's theorem and the translation invariance of λ^n yields

$$\mathbf{P}[X + Y \le x] = \mathbf{P}[(X, Y) \in A]$$

$$= \int_{\mathbb{R}^n \times \mathbb{R}^n} \mathbb{1}_A(u, v)\, f_X(u)\, f_Y(v)\, \left(\lambda^n\right)^{\otimes 2} (d(u, v))$$

$$= \int_{\mathbb{R}^n} \left(\int_{\mathbb{R}^n} \mathbb{1}_A(u, v)\, f_X(u)\, \lambda^n(du) \right) f_Y(v)\, \lambda^n(dv)$$

$$= \int_{\mathbb{R}^n} \left(\int_{(-\infty, x-v]} f_X(u)\, \lambda^n(du) \right) f_Y(v)\, \lambda^n(dv)$$

$$= \int_{\mathbb{R}^n} \left(\int_{(-\infty, x]} f_X(u - v)\, \lambda^n(du) \right) f_Y(v)\, \lambda^n(dv)$$

$$= \int_{(-\infty, x]} \left(\int_{\mathbb{R}^n} f_X(u - v)\, f_Y(v)\, \lambda^n(dv) \right) \lambda^n(du)$$

$$= \int_{(-\infty, x]} (f_X * f_Y)\, d\lambda^n.$$

(ii) In (i), replace μ by $\mathbf{P}_X$ and ν by $\mathbf{P}_Y$. The claim is immediate.   □

We come next to a concept that generalizes the notion of product measures and points in the direction of the example from the introduction to this chapter.

Recall the definition of a transition kernel from Definition 8.25.

Lemma 14.23 *Let κ be a finite transition kernel from $(\Omega_1, \mathcal{A}_1)$ to $(\Omega_2, \mathcal{A}_2)$ and let $f : \Omega_1 \times \Omega_2 \to [0, \infty]$ be measurable with respect to $\mathcal{A}_1 \otimes \mathcal{A}_2 - \mathcal{B}([0, \infty])$. Then the map*

$$I_f : \Omega_1 \to [0, \infty],$$

$$\omega_1 \mapsto \int f(\omega_1, \omega_2)\, \kappa(\omega_1, d\omega_2)$$

is well-defined and $\mathcal{A}_1$-measurable.

Proof By Lemma 14.13, for every $\omega_1 \in \Omega_1$, the map f_{ω_1} is measurable with respect to $\mathcal{A}_2$; hence $I_f(\omega_1) = \int f_{\omega_1}(\omega_2)\, \kappa(\omega_1, d\omega_2)$ is well-defined. Hence, it remains to show measurability of I_f.

If $g = \mathbb{1}_{A_1 \times A_2}$ for some $A_1 \in \mathcal{A}_1$ and $A_2 \in \mathcal{A}_2$, then clearly $I_g(\omega_1) = \mathbb{1}_{A_1}(\omega_1)\kappa(\omega_1, A_2)$ is measurable. Now let

$$\mathcal{D} = \{A \in \mathcal{A}_1 \otimes \mathcal{A}_2 : I_{\mathbb{1}_A} \text{ is } \mathcal{A}_1\text{-measurable}\}.$$

We show that $\mathcal{D}$ is a λ-system:

  (i) Evidently, $\Omega_1 \times \Omega_2 \in \mathcal{D}$.
 (ii) If $A, B \in \mathcal{D}$ with $A \subset B$, then $I_{\mathbb{1}_{B \setminus A}} = I_{\mathbb{1}_B} - I_{\mathbb{1}_A}$ is measurable, where we used the fact that κ is finite; hence $B \setminus A \in \mathcal{D}$.
(iii) If $A_1, A_2, \ldots \in \mathcal{D}$ are pairwise disjoint and $A := \bigcup_{n=1}^{\infty} A_n$, then $I_{\mathbb{1}_A} = \sum_{n=1}^{\infty} I_{\mathbb{1}_{A_n}}$ is measurable; hence $A \in \mathcal{D}$.

Summarising, $\mathcal{D}$ is a λ-system that contains a π-system that generates $\mathcal{A}_1 \otimes \mathcal{A}_2$ (namely, the rectangles). Hence, by the π–λ theorem (Theorem 1.19), $\mathcal{D} = \mathcal{A}_1 \otimes \mathcal{A}_2$. Hence $I_{\mathbb{1}_A}$ is measurable for all $A \in \mathcal{A}_1 \otimes \mathcal{A}_2$. We infer that I_g is measurable for any simple function g. Now let $(f_n)_{n \in \mathbb{N}}$ be a sequence of simple functions with $f_n \uparrow f$. For any fixed $\omega_1 \in \Omega_1$, by the monotone convergence theorem, $I_f(\omega_1) = \lim_{n \to \infty} I_{f_n}(\omega_1)$. As a limit of measurable functions, I_f is measurable. $\square$

Remark 14.24 In the following, we often write $\int \kappa(\omega_1, d\omega_2)\, f(\omega_1, \omega_2)$ instead of $\int f(\omega_1, \omega_2)\, \kappa(\omega_1, d\omega_2)$ since for multiple integrals this notation allows us to write the integrator closer to the corresponding integral sign. $\Diamond$

Theorem 14.25 *Let $(\Omega_i, \mathcal{A}_i)$, $i = 0, 1, 2$, be measurable spaces. Let κ_1 be a finite transition kernel from $(\Omega_0, \mathcal{A}_0)$ to $(\Omega_1, \mathcal{A}_1)$ and let κ_2 be a finite transition kernel from $(\Omega_0 \times \Omega_1, \mathcal{A}_0 \otimes \mathcal{A}_1)$ to $(\Omega_2, \mathcal{A}_2)$. Then the map*

$$\kappa_1 \otimes \kappa_2 : \Omega_0 \times (\mathcal{A}_1 \otimes \mathcal{A}_2) \to [0, \infty),$$

$$(\omega_0, A) \mapsto \int_{\Omega_1} \kappa_1(\omega_0, d\omega_1) \int_{\Omega_2} \kappa_2((\omega_0, \omega_1), d\omega_2)\, \mathbb{1}_A((\omega_1, \omega_2))$$

is well-defined and is a σ-finite (but not necessarily a finite) transition kernel from $(\Omega_0, \mathcal{A}_0)$ *to* $(\Omega_1 \times \Omega_2, \mathcal{A}_1 \otimes \mathcal{A}_2)$. *If* κ_1 *and* κ_2 *are (sub)stochastic, then* $\kappa_1 \otimes \kappa_2$ *is (sub)stochastic. We call* $\kappa_1 \otimes \kappa_2$ *the **product** of* κ_1 *and* κ_2.

If κ_2 *is a kernel from* $(\Omega_1, \mathcal{A}_1)$ *to* $(\Omega_2, \mathcal{A}_2)$, *then we define the product* $\kappa_1 \otimes \kappa_2$ *similarly by formally understanding* κ_2 *as a kernel from* $(\Omega_0 \times \Omega_1, \mathcal{A}_0 \otimes \mathcal{A}_1)$ *to* $(\Omega_2, \mathcal{A}_2)$ *that does not depend on the* Ω_0-*coordinate.*

Proof Let $A \in \mathcal{A}_1 \otimes \mathcal{A}_2$. By Lemma 14.23, the map

$$g_A : (\omega_0, \omega_1) \mapsto \int \kappa_2((\omega_0, \omega_1), d\omega_2) \, \mathbb{1}_A(\omega_1, \omega_2)$$

is well-defined and $\mathcal{A}_0 \otimes \mathcal{A}_1$-measurable. Thus, again by Lemma 14.23, the map

$$\omega_0 \mapsto \kappa_1 \otimes \kappa_2(\omega_0, A) = \int \kappa_1(\omega_0, d\omega_1) \, g_A(\omega_0, \omega_1)$$

is well-defined and $\mathcal{A}_0$-measurable. For fixed ω_0, by the monotone convergence theorem, the map $A \mapsto \kappa_1 \otimes \kappa_2(\omega_0, A)$ is σ-additive and thus a measure.

For $\omega_0 \in \Omega_0$ and $n \in \mathbb{N}$, let $A_{\omega_0, n} := \{\omega_1 \in \Omega_1 : \kappa_2((\omega_0, \omega_1), \Omega_2) < n\}$. Since κ_2 is finite, we have $\bigcup_{n \geq 1} A_{\omega_0, n} = \Omega_1$ for all $\omega_0 \in \Omega_0$. Furthermore, $\kappa_1 \otimes \kappa_2(\omega_0, A_n \times \Omega_2) \leq n \cdot \kappa_1(\omega_0, A_n) < \infty$. Hence $\kappa_1 \otimes \kappa_2(\omega_0, \cdot)$ is σ-finite and is thus a transition kernel.

The supplement is trivial. $\qquad\qquad\qquad\qquad\qquad\qquad\qquad\qquad\qquad\qquad\qquad$ $\square$

Corollary 14.26 (Products via kernels) *Let* $(\Omega_1, \mathcal{A}_1, \mu)$ *be a finite measure space, let* $(\Omega_2, \mathcal{A}_2)$ *be a measurable space and let* κ *be a finite transition kernel from* Ω_1 *to* Ω_2. *Then there exists a unique* σ-*finite measure* $\mu \otimes \kappa$ *on* $(\Omega_1 \times \Omega_2, \mathcal{A}_1 \otimes \mathcal{A}_2)$ *with*

$$\mu \otimes \kappa(A_1 \times A_2) = \int_{A_1} \kappa(\omega_1, A_2) \, \mu(d\omega_1) \quad \text{for all } A_1 \in \mathcal{A}_1, \ A_2 \in \mathcal{A}_2.$$

If κ *is stochastic and if* μ *is a probability measure, then* $\mu \otimes \kappa$ *is a probability measure.*

Proof Apply Theorem 14.25 with $\kappa_2 = \kappa$ and $\kappa_1(\omega_0, \cdot) = \mu$. $\qquad\qquad\qquad$ $\square$

Corollary 14.27 *Let* $n \in \mathbb{N}$ *and let* $(\Omega_i, \mathcal{A}_i)$, $i = 0, \dots, n$, *be measurable spaces. For* $i = 1, \dots, n$, *let* κ_i *be a substochastic kernel from* $\left(\underset{k=0}{\overset{i-1}{\times}} \Omega_k, \ \underset{k=0}{\overset{i-1}{\bigotimes}} \mathcal{A}_k \right)$ *to* $(\Omega_i, \mathcal{A}_i)$ *or from* $(\Omega_{i-1}, \mathcal{A}_{i-1})$ *to* $(\Omega_i, \mathcal{A}_i)$. *Then the recursion* $\kappa_1 \otimes \cdots \otimes \kappa_i := (\kappa_1 \otimes \cdots \otimes \kappa_{i-1}) \otimes \kappa_i$ *for any* $i = 1, \dots, n$ *defines a substochastic kernel* $\underset{k=1}{\overset{i}{\bigotimes}} \kappa_k := \kappa_1 \otimes \cdots \otimes \kappa_i$ *from* $(\Omega_0, \mathcal{A}_0)$ *to* $\left(\underset{k=1}{\overset{i}{\times}} \Omega_k, \ \underset{k=1}{\overset{i}{\bigotimes}} \mathcal{A}_k \right)$. *If all* κ_k *are stochastic, then all* $\underset{k=1}{\overset{i}{\bigotimes}} \kappa_k$ *are stochastic.*

If μ is a finite measure on $(\Omega_0, \mathcal{A}_0)$, then $\mu_i := \mu \otimes \bigotimes_{k=1}^{i} \kappa_k$ is a finite measure on

$\left(\underset{k=0}{\overset{i}{\times}} \Omega_k, \overset{i}{\underset{k=0}{\bigotimes}} \mathcal{A}_k \right)$. *If μ is a probability measure and if every κ_i is stochastic, then μ_i is a probability measure.*

Proof The claims follow inductively by Theorem 14.25.   □

Definition 14.28 (Composition of kernels) Let $(\Omega_i, \mathcal{A}_i)$ be measurable spaces, $i = 0, 1, 2$, and let κ_i be a substochastic kernel from $(\Omega_{i-1}, \mathcal{A}_{i-1})$ to $(\Omega_i, \mathcal{A}_i)$, $i = 1, 2$. Define the **composition** of κ_1 and κ_2 by

$$\kappa_1 \cdot \kappa_2 : \Omega_0 \times \mathcal{A}_2 \to [0, \infty),$$

$$(\omega_0, A_2) \mapsto \int_{\Omega_1} \kappa_1(\omega_0, d\omega_1) \kappa_2(\omega_1, A_2).$$

Theorem 14.29 *If we denote by $\pi_2 : \Omega_1 \times \Omega_2 \to \Omega_2$ the projection to the second coordinate, then*

$$(\kappa_1 \cdot \kappa_2)(\omega_0, A_2) = (\kappa_1 \otimes \kappa_2)\big(\omega_0, \pi_2^{-1}(A_2)\big) \quad \text{for all } A_2 \in \mathcal{A}_2.$$

In particular, the composition $\kappa_1 \cdot \kappa_2$ is a (sub)stochastic kernel from $(\Omega_0, \mathcal{A}_0)$ to $(\Omega_2, \mathcal{A}_2)$.

Proof This is obvious.   □

Lemma 14.30 (Kernels and convolution) *Let μ and ν be probability measures on $\mathbb{R}^d$ and define the kernels $\kappa_i : (\mathbb{R}^d, \mathcal{B}(\mathbb{R}^d)) \to (\mathbb{R}^d, \mathcal{B}(\mathbb{R}^d))$, $i = 1, 2$, by $\kappa_1(x, dy) = \mu(dy)$ and $\kappa_2(y, dz) = (\delta_y * \nu)(dz)$. Then $\kappa_1 \cdot \kappa_2 = \mu * \nu$.*

Proof This is trivial.   □

Theorem 14.31 (Kernels and convolution) *Assume $X_1, X_2, \ldots$ are independent $\mathbb{R}^d$-valued random variables with distributions $\mu_i := \mathbf{P}_{X_i}$, $i = 1, \ldots, n$. Let $S_k := X_1 + \ldots + X_k$ for $k = 1, \ldots, n$, and define stochastic kernels from $\mathbb{R}^d$ to $\mathbb{R}^d$ by $\kappa_k(x, \cdot) = \delta_x * \mu_k$ for $k = 1, \ldots, n$. Then*

$$\left(\bigotimes_{k=1}^{n} \kappa_k \right)(0, \cdot) = \mathbf{P}_{(S_1, \ldots, S_n)}. \tag{14.8}$$

Proof For $k = 1, \ldots, n$, define the measurable bijection $\varphi_k : (\mathbb{R}^d)^k \to (\mathbb{R}^d)^k$ by

$$\varphi_k(x_1, \ldots, x_k) = (x_1, x_1 + x_2, \ldots, x_1 + \ldots + x_k).$$

Evidently, $\mathcal{B}((\mathbb{R}^d)^n) = \sigma\left(\varphi_n(A_1 \times \cdots \times A_n) : A_1, \ldots, A_n \in \mathcal{B}(\mathbb{R}^d)\right)$. Hence, it is enough to show (14.8) for sets of this type. That is, it is enough to show that

$$\left(\bigotimes_{k=1}^n \kappa_k\right)(0, \varphi_n(A_1 \times \cdots \times A_n)) = \mathbf{P}_{(S_1,\ldots,S_n)}(\varphi_n(A_1 \times \cdots \times A_n)) = \prod_{k=1}^n \mu_k(A_k).$$

For $n = 1$, this is clear. By definition, $\kappa_n(y_{n-1}, y_{n-1} + A_n) = \mu_n(A_n)$. Inductively, we get

$$\left(\bigotimes_{k=1}^n \kappa_k\right)(0, \varphi_n(A_1 \times \cdots \times A_n))$$

$$= \int_{\varphi_{n-1}(A_1 \times \cdots \times A_{n-1})} \left(\bigotimes_{k=1}^{n-1} \kappa_k\right)(0, d(y_1, \ldots, y_{n-1})) \kappa_n\left(y_{ne-1}, y_{n-1} + A_n\right)$$

$$= \left(\prod_{k=1}^{n-1} \mu_k(A_k)\right) \mu_n(A_n). \qquad \square$$

Theorem 14.32 (Fubini for transition kernels) *Let $(\Omega_i, \mathcal{A}_i)$ be measurable spaces, $i = 1, 2$. Let μ be a finite measure on $(\Omega_1, \mathcal{A}_1)$ and let κ be a finite transition kernel from Ω_1 to Ω_2. Assume that $f : \Omega_1 \times \Omega_2 \to \overline{\mathbb{R}}$ is measurable with respect to $\mathcal{A}_1 \otimes \mathcal{A}_2$. If $f \geq 0$ or $f \in \mathcal{L}^1(\mu \otimes \kappa)$, then*

$$\int_{\Omega_1 \times \Omega_2} f \, d(\mu \otimes \kappa) = \int_{\Omega_1} \left(\int_{\Omega_2} f(\omega_1, \omega_2) \kappa(\omega_1, d\omega_2)\right) \mu(d\omega_1). \qquad (14.9)$$

Proof For $f = \mathbb{1}_{A_1 \times A_2}$ with $A_1 \in \mathcal{A}_1$ and $A_2 \in \mathcal{A}_2$, the statement is true by definition. For general f, apply the usual approximation argument as in Theorem 14.19. $\qquad \square$

Example 14.33 We come back to the example from the beginning of this chapter. Let $n \in \mathbb{N}$ and let $(\Omega_2, \mathcal{A}_2) = (\{0, 1\}^n, (2^{\{0,1\}})^{\otimes n})$ be the space of n-fold coin tossing. For any $p \in [0, 1]$, define

$$P_p = (\text{Ber}_p)^{\otimes n} = \left((1 - p)\delta_0 + p\delta_1\right)^{\otimes n}.$$

P_p is that probability measure on $(\Omega_2, \mathcal{A}_2)$ under which the coordinate maps Y_i are independent Bernoulli random variables with success probability p.

Further, let $\Omega_1 = [0, 1]$, let $\mathcal{A}_1 = \mathcal{B}([0, 1])$ be the Borel σ-algebra on Ω_1 and let $\mu = \mathcal{U}_{[0,1]}$ be the uniform distribution on $[0, 1]$. Then the identity map $X : \Omega_1 \to [0, 1]$ is a random variable on $(\Omega_1, \mathcal{A}_1, \mu)$ that is uniformly distributed on $[0, 1]$.

Finally, consider the stochastic kernel κ from Ω_1 to Ω_2, defined by

$$\kappa(\omega_1, A_2) = P_{\omega_1}(A_2).$$

If we let $\Omega = \Omega_1 \times \Omega_2$, $\mathcal{A} = \mathcal{A}_1 \otimes \mathcal{A}_2$ and $\mathbf{P} = \mu \otimes \kappa$, then X and $Y_1, \ldots, Y_n$ describe precisely the random variables on $(\Omega, \mathcal{A}, \mathbf{P})$ from the beginning of this chapter. $\Diamond$

Remark 14.34 The procedure can be extended to n-stage experiments. Let $(\Omega_i, \mathcal{A}_i)$ be the measurable space of the ith experiment, $i = 0, \ldots, n - 1$. Let P_0 be a probability measure on $(\Omega_0, \mathcal{A}_0)$. Assume that for $i = 1, \ldots, n-1$, the distribution on $(\Omega_i, \mathcal{A}_i)$ depends on $(\omega_1, \ldots, \omega_{i-1})$ and is given by a stochastic kernel κ_i from $\Omega_0 \times \cdots \times \Omega_{i-1}$ to Ω_i. The whole n-stage experiment is then described by the coordinate maps on the probability space $\left(\underset{i=0}{\overset{n-1}{\times}} \Omega_i, \, \underset{i=0}{\overset{n-1}{\bigotimes}} \mathcal{A}_i, \, P_0 \otimes \underset{i=1}{\overset{n-1}{\bigotimes}} \kappa_i \right)$. $\Diamond$

Takeaways For finite products of measurable spaces, we define the product measure. The integral of an integrable function on the product space can be computed by successive integration (in arbitrary order) over the individual coordinates (Fubini's theorem).

Depending on the outcome of a random experiment, we choose the distribution of a second random experiment (in a measurable way). The corresponding map is called a stochastic kernel. By concatenation of stochastic kernels we construct multi-step random experiments. This procedure leads to a generalisation of the concept of a product measure.

Exercise 14.2.1 Show the following convolution formulas.

  (i) Normal distribution: $\mathcal{N}_{\mu_1, \sigma_1^2} * \mathcal{N}_{\mu_2, \sigma_2^2} = \mathcal{N}_{\mu_1 + \mu_2, \sigma_1^2 + \sigma_2^2}$ for all $\mu_1, \mu_2 \in \mathbb{R}$ and $\sigma_1^2, \sigma_2^2 > 0$.
 (ii) Gamma distribution: $\Gamma_{\theta, r} * \Gamma_{\theta, s} = \Gamma_{\theta, r+s}$ for all $\theta, r, s > 0$.
(iii) Cauchy distribution: $\mathrm{Cau}_r * \mathrm{Cau}_s = \mathrm{Cau}_{r+s}$ for all $r, s > 0$. ♣

Exercise 14.2.2 (Hilbert–Schmidt operator) Let $(\Omega_i, \mathcal{A}_i, \mu_i)$, $i = 1, 2$, be σ-finite measure spaces and let $a : \Omega_1 \times \Omega_2 \to \mathbb{R}$ be measurable with

$$\int \mu_1(dt_1) \int \mu_2(dt_2) \, a(t_1, t_2)^2 < \infty.$$

For $f \in \mathcal{L}^2(\mu_1)$, define

$$(Af)(t_2) = \int a(t_1, t_2) f(t_1) \, \mu_1(dt_1).$$

Show that A is a continuous linear operator from $\mathcal{L}^2(\mu_1)$ to $\mathcal{L}^2(\mu_2)$. ♣

Exercise 14.2.3 (Partial integration) Let F_μ and F_ν be the distribution functions of locally finite measures μ and ν on $\mathbb{R}$. For $x \in \mathbb{R}$, define the left-sided limit

$F(x-) = \sup_{y<x} F(y)$ and the jump height $\Delta F(x) = F(x) - F(x-)$. Show that, for $a < b$,

$$\int_{(a,b]} F_\mu \, dv = F_\mu(b)F_v(b) - F_\mu(a)F_v(a) - \int_{(a,b]} F_v(x-)\mu(dx)$$

$$= F_\mu(b)F_v(b) - F_\mu(a)F_v(a) - \int_{(a,b]} F_v \, d\mu + \sum_{a<x\leq b} \Delta F_\mu(x)\, \Delta F_v(x). \quad \clubsuit$$

(14.10)

14.3 Kolmogorov's Extension Theorem

In the previous section, we saw how we can implement n-stage experiments on a probability space. In this section, we first show how to implement countably many successive experiments on one probability space (Ionescu–Tulcea's theorem). Thereafter we also construct probability measures on products of uncountably many spaces (Kolmogorov's extension theorem).

Let $(\Omega_i, \mathcal{A}_i)$, $i \in \mathbb{N}_0$, be measurable spaces and let P_0 be a probability measure on $(\Omega_0, \mathcal{A}_0)$. Let $\Omega^i := \bigtimes_{k=0}^{i} \Omega_k$ and $\mathcal{A}^i = \bigotimes_{k=0}^{i} \mathcal{A}_k$ and

$$\Omega := \bigtimes_{k=0}^{\infty} \Omega_k \quad \text{and} \quad \mathcal{A} = \bigotimes_{k=0}^{\infty} \mathcal{A}_k.$$

For every $i \in \mathbb{N}$, let κ_i be a stochastic kernel from $(\Omega^{i-1}, \mathcal{A}^{i-1})$ to $(\Omega_i, \mathcal{A}_i)$. In Corollary 14.27, we defined inductively probability measures $P_i = P_0 \otimes \bigotimes_{k=1}^{i} \kappa_k$ on $(\Omega^i, \mathcal{A}^i)$. By construction, for $i, j \geq k$ and $A \in \mathcal{A}^k$, we had

$$P_i(A \times \Omega_{k+1} \times \cdots \times \Omega_i) = P_j(A \times \Omega_{k+1} \times \cdots \times \Omega_j). \tag{14.11}$$

Now we want to define a probability measure P on $(\Omega, \mathcal{A})$ such that for $k \in \mathbb{N}_0$ and $A \in \mathcal{A}^k$

$$P\left(A \times \bigtimes_{i=k+1}^{\infty} \Omega_i\right) = P_k(A). \tag{14.12}$$

Theorem 14.35 (Ionescu–Tulcea) *There is a uniquely determined probability measure P on $(\Omega, \mathcal{A})$ such that (14.12) holds.*

Proof Uniqueness is clear since the finite-dimensional rectangular cylinders form a π-system that generates $\mathcal{A}$. It remains to show the existence of that measure.

We use (14.12) to define a set function P on cylinder sets. Clearly, P is additive and is hence a content. If we can show that P is $\emptyset$-continuous, then P is a premeasure (by Theorem 1.36) and thus by Carathéodory's theorem (Theorem 1.41) can be extended uniquely to a measure on $\mathcal{A}$.

Hence, let $A_0 \supset A_1 \supset A_2 \supset \ldots$ be a sequence in $\mathcal{Z}$ with $\alpha := \inf_{n \in \mathbb{N}_0} P(A_n) > 0$. It is enough to show that $\bigcap_{n=0}^{\infty} A_n \neq \emptyset$. Without loss of generality, we can assume that $A_n = A'_n \times \bigtimes_{k=n+1}^{\infty} \Omega_k$ for certain $A'_n \in \mathcal{A}^n$. For $n \geq m$, define

$$h_{m,n}(\omega_0, \ldots, \omega_m) := \left(\delta_{(\omega_0, \ldots, \omega_m)} \otimes \bigotimes_{k=m+1}^{n} \kappa_k \right) (A'_n)$$

and $h_m := \inf_{n \geq m} h_{m,n}$. Inductively, we show that for every $i \in \mathbb{N}_0$, there exists a $\varrho_i \in \Omega_i$ such that

$$h_m(\varrho_0, \ldots, \varrho_m) \geq \alpha. \tag{14.13}$$

Since $A'_{n+1} \subset A'_n \times \Omega_{n+1}$, we have

$$h_{m,n+1}(\omega_0, \ldots, \omega_m) = \left(\delta_{(\omega_0, \ldots, \omega_m)} \otimes \bigotimes_{k=m+1}^{n+1} \kappa_k \right) (A'_{n+1})$$

$$\leq \left(\delta_{(\omega_0, \ldots, \omega_m)} \otimes \bigotimes_{k=m+1}^{n+1} \kappa_k \right) (A'_n \times \Omega_{n+1})$$

$$= \left(\delta_{(\omega_0, \ldots, \omega_m)} \otimes \bigotimes_{k=m+1}^{n} \kappa_k \right) (A'_n) = h_{m,n}(\omega_0, \ldots, \omega_m).$$

Hence $h_{m,n} \downarrow h_m$ for $n \to \infty$ and by the monotone convergence theorem,

$$\int h_m \, dP_m = \inf_{n \geq m} \int h_{m,n} \, dP_m = \inf_{n \in \mathbb{N}_0} P_n(A'_n) = \alpha,$$

whence we have (14.13) for $m = 0$. Now assume that (14.13) holds for $m \in \mathbb{N}_0$. Then

$$\int h_{m+1}(\varrho_0, \ldots, \varrho_m, \omega_{m+1}) \, \kappa_{m+1}\big((\varrho_0, \ldots, \varrho_m), d\omega_{m+1}\big)$$

$$= \inf_{n \geq m+1} \int h_{m+1,n}(\varrho_0, \ldots, \varrho_m, \omega_{m+1}) \, \kappa_{m+1}\big((\varrho_0, \ldots, \varrho_m), d\omega_{m+1}\big)$$

$$= h_m(\varrho_0, \ldots, \varrho_m) \geq \alpha.$$

Hence (14.13) holds for $m + 1$.

Let $\varrho := (\varrho_0, \varrho_1, \ldots) \in \Omega$. By construction,

$$\alpha \leq h_{m,m}(\varrho_0, \ldots, \varrho_m) = \mathbb{1}_{A'_m}(\varrho_0, \ldots, \varrho_m),$$

hence $\varrho \in A_m$ for all $m \in \mathbb{N}_0$ and thus $\bigcap_{i=0}^{\infty} A_i \neq \emptyset$.   □

Corollary 14.36 (Product measure) *For every $n \in \mathbb{N}_0$, let P_n be a probability measure on $(\Omega_n, \mathcal{A}_n)$. Then there exists a uniquely determined probability measure P on $(\Omega, \mathcal{A})$ with*

$$P\left(A_0 \times \cdots \times A_n \times \overset{\infty}{\underset{i=n+1}{\LARGE\times}} \Omega_i \right) = \prod_{k=0}^{n} P_k(A_k)$$

for $A_i \in \mathcal{A}_i$, $i = 0, \ldots, n$ and $n \in \mathbb{N}_0$.

$\bigotimes_{i=0}^{\infty} P_i := P$ *is called the product of the measures $P_0, P_1, \ldots$. Under P, the coordinate maps $(X_i)_{i \in \mathbb{N}_0}$ are independent.*

Proof This follows by Ionescu–Tulcea's theorem with $\kappa_i((\omega_0, \ldots, \omega_{i-1}), \cdot) = P_i$.
                                                                                                □

We want to make a statement similar to that of Ionescu–Tulcea's theorem; however, without the assumption that the measures P_k are defined *a priori* by kernels. Before we formulate the theorem, we generalize the consistency condition (14.11). Recall that for $L \subset J \subset I$, $X_L^J : \Omega_J \longrightarrow \Omega_L$ denotes the canonical projection.

Definition 14.37 A family $(P_J, \ J \subset I$ finite$)$ of probability measures on the space $(\Omega_J, \mathcal{A}_J)$ is called **consistent** if

$$P_L = P_J \circ \left(X_L^J\right)^{-1} \qquad \text{for all } L \subset J \subset I \text{ finite.}$$

Recall that $\Omega = \underset{i \in I}{\times} \Omega_i$ and $\mathcal{A} = \underset{i \in I}{\bigotimes} \mathcal{A}_i$. Let P be a probability measure on $(\Omega, \mathcal{A})$. Since $X_L = X_L^J \circ X_J$, the family $(P_J := P \circ X_J^{-1}, J \subset I$ finite$)$ is consistent. Thus, consistency is a necessary condition for the existence of a measure P on the product space with $P_J := P \circ X_J^{-1}$. If all the measurable spaces are Borel spaces (recall Definition 8.35), for example $\mathbb{R}^d$, $\mathbb{Z}^d$, $C([0, 1])$ or more general Polish spaces, then this condition is also sufficient. We formulate this statement first for a countable index set.

Theorem 14.38 *Let I be countable and let $(\Omega_i, \mathcal{A}_i)$ be Borel spaces for all $i \in I$. Let $(P_J, \ J \subset I$ finite$)$ be a consistent family of probability measure. Then there exists a unique probability measure P on $(\Omega, \mathcal{A})$ with $P_J = P \circ X_J^{-1}$ for all finite $J \subset I$.*

Proof Without loss of generality, assume $I = \mathbb{N}_0$. Let $P_n := P_{\{0,\ldots,n\}}$, $\Omega^n :=$ $\Omega_{\{0,\ldots,n\}}$ and $\mathcal{A}^n := \mathcal{A}_{\{0,\ldots,n\}}$. It is easy to check that finite products of Borel spaces are again Borel spaces; hence $(\Omega^n, \mathcal{A}^n)$ is Borel for all $n \in \mathbb{N}_0$.

Let $\mathcal{F} := \{A \times \Omega_{n+1} : A \in \mathcal{A}^n\}$, $Y : \Omega^{n+1} \to \Omega_{n+1}$, $(\omega_0, \ldots, \omega_{n+1}) \mapsto \omega_{n+1}$ and $Z : \Omega^{n+1} \to \Omega^n$, $(\omega_0, \ldots, \omega_{n+1}) \mapsto (\omega_0, \ldots, \omega_n)$. By Theorem 8.37 (with $\Omega = \Omega^{n+1}$, $\mathcal{A} = \mathcal{A}^{n+1}$ and $E = \Omega_{n+1}$), there is a stochastic kernel κ'_{n+1} from $(\Omega^{n+1}, \mathcal{F})$ to $(\Omega_{n+1}, \mathcal{A}_{n+1})$ such that κ'_{n+1} is a regular conditional distribution of Y given $\mathcal{F}$ (under the probability measure P_{n+1}). Hence, for $A \in \mathcal{A}^n$ and $B \in \mathcal{A}_{n+1}$, we have (compare (8.11))

$$P_{n+1}(A \times B) = \int \mathbb{1}_B(Y)\, \mathbb{1}_{A \times \Omega_{n+1}}\, dP_{n+1} = \int \kappa'_{n+1}(\,\cdot\,, B)\, \mathbb{1}_{A \times \Omega_{n+1}}\, dP_{n+1}.$$

Since $\kappa'_{n+1}(\,\cdot\,, B)$ is $\mathcal{F}$-measurable, there is a stochastic kernel κ_{n+1} from $(\Omega^n, \mathcal{A}^n)$ to $(\Omega_{n+1}, \mathcal{A}_{n+1})$ such that

$$\kappa_{n+1}\big((\omega_0, \ldots, \omega_n), \,\cdot\,\big) = \kappa'_{n+1}\big((\omega_0, \ldots, \omega_{n+1}), \,\cdot\,\big) \qquad \text{for all } \omega_0, \ldots, \omega_{n+1}.$$

Hence

$$\kappa'_{n+1}(\,\cdot\,, B) = \kappa_{n+1}(Z(\,\cdot\,), B) \quad \text{and} \quad \mathbb{1}_{A \times \Omega_{n+1}} = \mathbb{1}_A(Z).$$

We infer that

$$P_{n+1}(A \times B) = \int \kappa_{n+1}(Z, B)\, \mathbb{1}_A(Z)\, dP_{n+1}$$

$$= \int \kappa_{n+1}(\,\cdot\,, B)\, \mathbb{1}_A\, d\big(P_{n+1} \circ Z^{-1}\big)$$

$$= \int_A \kappa_{n+1}(\,\cdot\,, B)\, dP_n.$$

Note that in the last equality we used the fact that $(P_n)_{n \in \mathbb{N}_0}$ is a projective family. By Corollary 14.26, we get $P_{n+1} = P_n \otimes \kappa_{n+1}$. Recursively, we obtain $P_n = P_0 \otimes \bigotimes_{k=1}^n \kappa_k$ for all $n \in \mathbb{N}$. Using Theorem 14.35, this yields the claim. $\qquad\square$

The last step in our construction is to replace the countable index set I by an arbitrary index set.

Theorem 14.39 (Kolmogorov's extension theorem) *Let I be an arbitrary index set and let $(\Omega_i, \mathcal{A}_i)$ be Borel spaces, $i \in I$. Let $(P_J,\ J \subset I$ finite) be a consistent family of probability measures. Then there exists a unique probability measure P on $(\Omega, \mathcal{A})$ with $P_J = P \circ X_J^{-1}$ for every finite $J \subset I$. P is called the **projective limit** and will be denoted by $P =: \lim_{J \uparrow I} P_J$.*

Proof For countable $J \subset I$, by Theorem 14.38, there is a unique probability measure P_J on $(\Omega_J, \mathcal{A}_J)$ with $P_J \circ (X_K^J)^{-1} = P_K$ for finite $K \subset J$. By defining $\tilde{P}_J(X_J^{-1}(A_J)) := P_J(A_J)$ for $A_J \in \mathcal{A}_J$, we get a probability measure $\tilde{P}_J$ on $(\Omega, \sigma(X_J))$.

Let $J, J' \subset I$ be countable and let $A \in \sigma(X_J) \cap \sigma(X_{J'}) \cap \mathcal{Z}$ be a $\sigma(X_J) \cap \sigma(X_{J'})$-measurable cylinder with a finite base. Then there exists a finite $K \subset J \cap J'$ and $A_K \in \mathcal{A}_K$ with $A = X_K^{-1}(A_K)$. Hence $\tilde{P}_J(A) = P_K(A_K) = \tilde{P}_{J'}(A)$. Moreover, by Theorem 14.12, $\tilde{P}_J(A) = P_K(A_K) = \tilde{P}_{J'}(A)$ for all $A \in \sigma(X_J) \cap \sigma(X_{J'})$. Now, by Exercise 14.1.1, for any $A \in \mathcal{A}$, there is a countable $J \subset I$ with $A \in \sigma(X_J)$. Hence, independently of the choice of J, we can uniquely define a set function P on $\mathcal{A}$ by $P(A) = \tilde{P}_J(A)$. It remains to show that P is a probability measure. Evidently, $P(\Omega) = 1$. If $A_1, A_2, \ldots \in \mathcal{A}$ are pairwise disjoint and $A := \bigcup_{n=1}^\infty A_n$, then for any $n \in \mathbb{N}$, there is a countable $J_n \subset I$ with $A_n \in \sigma(X_{J_n})$. Define $J = \bigcup_{n \in \mathbb{N}} J_n$. Then each A_n is in $\sigma(X_J)$; thus $A \in \sigma(X_J)$. Therefore,

$$P(A) = \tilde{P}_J(A) = \sum_{n=1}^\infty \tilde{P}_J(A_n) = \sum_{n=1}^\infty P(A_n).$$

This shows that P is a probability measure. $\qquad\square$

Example 14.40 Let $\big((\Omega_i, \tau_i), \ i \in I\big)$ be an arbitrary family of Polish spaces (recall from Theorem 8.36 that Polish spaces are also Borel spaces). Let $\mathcal{A}_i = \sigma(\tau_i)$ and let P_i be an arbitrary probability measure on $(\Omega_i, \mathcal{A}_i)$ for every $i \in I$. For finite $J \subset I$, let $P_J := \bigotimes_{j \in J} P_j$ be the product measure of the P_j, $j \in J$. Evidently, the family $(P_J, \ J \subset I$ finite) is consistent. We call

$$P = \bigotimes_{i \in I} P_i := \lim_{J \uparrow I} P_J$$

the **product measure** on $(\Omega, \mathcal{A})$. Under P, all coordinate maps X_j are independent. $\Diamond$

Example 14.41 (Pólya's urn model) (Compare Example 12.29.) In an urn there are initially k red and $n - k$ blue balls. At each step, one ball is drawn at random and is returned to the urn with an *additional* ball of the same color. Hence, at time $i \in \mathbb{N}_0$ there are $n + i$ balls in the urn. The random number of red balls is denoted by X_i.

For a more formal description, let $n \in \mathbb{N}$ and $k \in \{0, \ldots, n\}$. Let $I = \mathbb{N}_0$, $\Omega_i = \{0, \ldots, n+i\}$, $i \in \mathbb{N}$. Let $P_0[\{k\}] = 1$, and define the stochastic kernels κ_i from Ω_i to Ω_{i+1} by

$$\kappa_i(x_i, \{x_{i+1}\}) = \begin{cases} \frac{x_i}{n+i}, & \text{if } x_{i+1} = x_i + 1, \\[2mm] 1 - \frac{x_i}{n+i}, & \text{if } x_{i+1} = x_i, \\[2mm] 0, & \text{else.} \end{cases}$$

Now let $P_{i+1} = P_i \otimes \kappa_i$. Under the measure $\mathbf{P} = \varprojlim_{i \to \infty} P_i$, the projections $(X_i, i \in \mathbb{N}_0)$ describe Pólya's urn model. $\Diamond$

Takeaways Consider an infinite product of Borel measure spaces. A family of probability measures on all finite products of these spaces is called projective if the image measures under projections coincide. By Kolmogorov's theorem, for a projective family, there exists a probability measure on the infinite product space such that the projections coincide with the original measures we started with. This is a universal construction of a probability measure on the product space and it allows in a very flexible way to construct the large probability spaces needed for the definition of stochastic processes. As a particularly simple application, we get an infinite product measure such that all coordinate maps are independent with a desired distribution.

14.4   Markov Semigroups

Definition 14.42 Let E be a Polish space. Let $I \subset \mathbb{R}$ be a nonempty index set and let $(\kappa_{s,t} : s, t \in I, s < t)$ be a family of stochastic kernels from E to E. We say that the family is **consistent** if $\kappa_{r,s} \cdot \kappa_{s,t} = \kappa_{r,t}$ for any choice of $r, s, t \in I$ with $r < s < t$.

Definition 14.43 Let E be a Polish space. Let $I \subset [0, \infty)$ be an additive semigroup (for example, $I = \mathbb{N}_0$ or $I = [0, \infty)$). A family $(\kappa_t : t \in I)$ of stochastic kernels is called a **semigroup of stochastic kernels**, or a **Markov semigroup**, if

$$\kappa_0(\omega, \cdot) = \delta_\omega \quad \text{for all } \omega \in E \tag{14.14}$$

and if it satisfies the **Chapman–Kolmogorov equation**:

$$\kappa_s \cdot \kappa_t = \kappa_{s+t} \quad \text{for all } s, t \in I. \tag{14.15}$$

Indeed, $(\{\kappa_t : t \in I\}, \cdot)$ is a semigroup in the algebraic sense and the map $t \to \kappa_t$ is a homomorphism of semigroups. In particular, the kernels commute in the sense that $\kappa_s \cdot \kappa_t = \kappa_t \cdot \kappa_s$ for all $s, t \in I$.

Lemma 14.44 If $(\kappa_t : t \in I)$ is a Markov semigroup, then the family of kernels, defined by $\tilde{\kappa}_{s,t} := \kappa_{t-s}$ for $t > s$, is consistent.

Proof This is trivial. $\qquad\qquad\qquad\qquad\qquad\qquad\qquad\qquad\qquad\qquad\square$

Theorem 14.45 (Kernel via a consistent family of kernels) *Let $I \subset [0, \infty)$ with $0 \in I$ and let $(\kappa_{s,t} : s, t \in I, \ s < t)$ be a consistent family of stochastic kernels on the Polish space E. Then there exists a kernel κ from $(E, \mathcal{B}(E))$ to $(E^I, \mathcal{B}(E)^{\otimes I})$ such that, for all $x \in E$ and for any choice of finitely many numbers $0 = j_0 < j_1 < j_2 < \ldots < j_n$ from I, and with the notation $J := \{j_0, \ldots, j_n\}$, we have*

$$\kappa(x, \cdot) \circ X_J^{-1} = \left(\delta_x \otimes \bigotimes_{k=0}^{n-1} \kappa_{j_k, j_{k+1}} \right). \tag{14.16}$$

Proof First we show that, for fixed $x \in E$, (14.16) defines a probability measure $\kappa(x, \cdot)$. Define the family $(P_J : J \subset I$ finite, $0 \in J)$ by $P_J := \delta_x \otimes \bigotimes_{k=0}^{n-1} \kappa_{j_k, j_{k+1}}$. By Kolmogorov's extension theorem, it is enough to show that this family is consistent. In fact, if for $0 \notin J \subset I$ finite, we define P_J as the projection of $P_{J \cup \{0\}}$ to E^J, then the family $(P_J : J \subset I$ finite$)$ is projective. Hence, let $0 \in L \subset J \subset I$ with $J \subset I$ finite. We have to show that $P_J \circ (X_L^J)^{-1} = P_L$. We may assume that $L = J \setminus \{j_l\}$ for some $l = 1, \ldots, n$. The general case can be inferred inductively.

First consider $l = n$. Let $A_{j_0}, \ldots, A_{j_{n-1}} \in \mathcal{B}(E)$ and $A := \times_{j \in L} A_j$. Then

$$P_J \circ (X_L^J)^{-1}(A) = P_J(A \times E) = P_L \otimes \kappa_{j_{n-1}, j_n}(A \times E)$$

$$= \int_A P_L\big(d(\omega_0, \ldots, \omega_{n-1})\big)\kappa_{j_{n-1}, j_n}(\omega_{n-1}, E) = P_L(A).$$

Now let $l \in \{1, \ldots, n-1\}$. For all $j \in L$, let $A_j \in \mathcal{B}(E)$ and $A_{j_l} := E$. Define $A := \times_{j \in L} A_j$, and abbreviate $A' = \times_{k=0}^{l-1} A_{j_k}$ and $P' = \delta_x \otimes \bigotimes_{k=0}^{l-2} \kappa_{j_k, j_{k+1}}$. For $i = 0, \ldots, n-1$, let

$$f_i(\omega_i) = \left(\bigotimes_{k=i}^{n-1} \kappa_{j_k, j_{k+1}} \right) (\omega_i, A_{j_{i+1}} \times \cdots \times A_{j_n}).$$

By assumption and using Fubini's theorem, we get

$$f_{l-1}(\omega_{l-1}) = \int_E \kappa_{j_{l-1}, j_l}(\omega_{l-1}, d\omega_l) \int_{A_{j_{l+1}}} \kappa_{j_l, j_{l+1}}(\omega_l, d\omega_{l+1}) f_{l+1}(\omega_{l+1})$$

$$= \int_{A_{j_{l+1}}} \kappa_{j_{l-1}, j_{l+1}}(\omega_{l-1}, d\omega_{l+1}) f_{l+1}(\omega_{l+1}).$$

This implies

$$P_J \circ (X_L^J)^{-1}(A) = \int_{A'} P'(d(\omega_0, \ldots, \omega_{l-1})) \, f_{l-1}(\omega_{l-1})$$

$$= \int_{A'} P'(d(\omega_0, \ldots, \omega_{l-1})) \int_{A_{j_{l+1}}} \kappa_{j_{l-1}, j_{l+1}}(\omega_{l-1}, d\omega_{l+1}) \, f_{l+1}(\omega_{l+1})$$

$$= P_L(A).$$

It remains to show that κ is a stochastic kernel. That is, we have to show that $x \mapsto \kappa(x, A)$ is measurable with respect to $\mathcal{B}(E) - \mathcal{B}(E)^{\otimes I}$. By Remark 8.26, it suffices to check this for rectangular cylinders with a finite base $A \in \mathcal{Z}^R$ since $\mathcal{Z}^R$ is a π-system that generates $\mathcal{B}(E)^{\otimes I}$. Hence, let $0 = t_0 < t_1 < \ldots < t_n$ and $B_0, \ldots, B_n \in \mathcal{B}(E)$ as well as $A = \bigcap_{i=0}^n X_{t_i}^{-1}(B_i)$. However, by Corollary 14.27, the following map is measurable,

$$x \mapsto \mathbf{P}_x[A] = \left(\delta_x \otimes \bigotimes_{i=0}^{n-1} \kappa_{t_i, t_{i+1}} \right) \left(\underset{i=0}{\overset{n}{\times}} B_i \right).$$

$\square$

Corollary 14.46 (Measures by consistent families of kernels) *Under the assumptions of Theorem 14.45, for every probability measure μ on E, there exists a unique probability measure $\mathbf{P}_\mu$ on $\left(E^I, \mathcal{B}(E)^{\otimes I} \right)$ with the following property: For any choice of finitely many numbers $0 = j_0 < j_1 < j_2 < \ldots < j_n$ from I, and letting $J := \{j_0, \ldots, j_n\}$, we have $\mathbf{P}_\mu \circ X_J^{-1} = \mu \otimes \bigotimes_{k=0}^{n-1} \kappa_{j_k, j_{k+1}}$.*

Proof Take $\mathbf{P}_\mu = \int \mu(dx) \, \kappa(x, \cdot)$.
$\square$

As a simple conclusion of Lemma 14.44 and Theorem 14.45, we get the following statement that we formulate separately because it will play a central role later.

Corollary 14.47 (Measures via Markov semigroups) *Let $(\kappa_t : t \in I)$ be a Markov semigroup on the Polish space E. Then there exists a unique stochastic kernel κ from $(E, \mathcal{B}(E))$ to $(E^I, \mathcal{B}(E)^{\otimes I})$ with the property: For all $x \in E$ and for any choice of finitely many numbers $0 = t_0 < t_1 < t_2 < \ldots < t_n$ from I, and letting $J := \{t_0, \ldots, t_n\}$, we have*

$$\kappa(x, \cdot) \circ X_J^{-1} = \left(\delta_x \otimes \bigotimes_{k=0}^{n-1} \kappa_{t_{k+1} - t_k} \right). \tag{14.17}$$

For any probability measure μ on E, there exists a unique probability measure $\mathbf{P}_\mu$ on $\left(E^I, \mathcal{B}(E)^{\otimes I} \right)$ with the property: For any choice of finitely many numbers $0 = t_0 < t_1 < t_2 < \ldots < t_n$ from I, and letting $J := \{t_0, \ldots, t_n\}$, we have $\mathbf{P}_\mu \circ X_J^{-1} = \mu \otimes \bigotimes_{k=0}^{n-1} \kappa_{t_{k+1} - t_k}$. We denote $\mathbf{P}_x = \mathbf{P}_{\delta_x} = \kappa(x, \cdot)$ for $x \in E$.

Example 14.48 (Independent normally distributed increments) Let $I = [0, \infty)$ and $\Omega_i = \mathbb{R}$, $i \in [0, \infty)$, equipped with the Borel σ-algebra $\mathcal{B} = \mathcal{B}(\mathbb{R})$. Further, let $\Omega = \mathbb{R}^{[0,\infty)}$, $\mathcal{A} = \mathcal{B}^{\otimes[0,\infty)}$ and let X_t be the coordinate map for $t \in [0, \infty)$. In the sense of Definition 14.6, $X = (X_t)_{t\geq 0}$ is thus the canonical process on $(\Omega, \mathcal{A})$.

We construct a probability measure $\mathbf{P}$ on $(\Omega, \mathcal{A})$ such that the stochastic process X has independent, stationary, normally distributed increments (recall Definition 9.7). That is, it should hold that

$$(X_{t_i} - X_{t_{i-1}})_{i=1,\ldots,n} \text{ is independent } \quad \text{for all } 0 =: t_0 < t_1 < \ldots < t_n, \quad (14.18)$$

$$\mathbf{P}_{X_t-X_s} = \mathcal{N}_{0,t-s} \quad \text{for all } t > s. \quad (14.19)$$

To this end, define stochastic kernels $\kappa_t(x, dy) := \delta_x * \mathcal{N}_{0,t}(dy)$ for $t \in [0, \infty)$ where $\mathcal{N}_{0,0} = \delta_0$. By Lemma 14.30, the Chapman–Kolmogorov equation holds since (compare Exercise 14.2.1(i))

$$\kappa_s \cdot \kappa_t(x, dy) = \delta_x * (\mathcal{N}_{0,s} * \mathcal{N}_{0,t})(dy) = \delta_x * \mathcal{N}_{0,s+t}(dy) = \kappa_{s+t}(x, dy).$$

Let $P_0 = \delta_0$ and let $\mathbf{P}$ be the unique probability measure on Ω corresponding to P_0 and $(\kappa_t : t \geq 0)$ according to Corollary 14.47. By Theorem 14.31, the equations (14.18) and (14.19) hold.

With $(X_t)_{t\geq 0}$, we have almost constructed the so-called **Brownian motion**. In addition to the properties we required here, Brownian motion has continuous *paths*; that is, the maps $t \mapsto X_t$ are almost surely continuous. Note that at this point it is not even clear that the paths are measurable maps. We will have some work to do to establish continuity of the paths, and we will come back to this in Chap. 21. ◇

The construction in the preceding example does not depend on the details of the normal distribution but only on the validity of the convolution equation

$$\mathcal{N}_{0,s+t} = \mathcal{N}_{0,s} * \mathcal{N}_{0,t}.$$

Hence, in (14.19) we can replace the normal distribution by *any* parameterized family of distributions $(\nu_t, t \geq 0)$ with the property $\nu_{t+s} = \nu_t * \nu_s$. Examples include the Gamma distribution $\nu_t = \Gamma_{\theta,t}$ (for fixed parameter $\theta > 0$), the Poisson distribution $\nu_t = \text{Poi}_t$, the negative binomial distribution $\nu_t = b^-_{t,p}$ (for fixed $p \in (0, 1]$), the Cauchy distribution $\nu_t = \text{Cau}_t$ and others (compare Theorem 15.13 and Corollary 15.14). We establish the result in a theorem.

Definition 14.49 (Convolution semigroup) Let $I \subset [0, \infty)$ be a semigroup. A family $\nu = (\nu_t : t \in I)$ of probability distributions on $\mathbb{R}^d$ is called a **convolution semigroup** if $\nu_{s+t} = \nu_s * \nu_t$ holds for all $s, t \in I$.

If $I = [0, \infty)$ and if in addition $\nu_t \xrightarrow{t \to 0} \delta_0$, then the convolution semigroup is called **continuous** (in the sense of weak convergence).

If $d = 1$ and $\nu_t((-\infty, 0)) = 0$ for all $t \in I$, then ν is called a nonnegative convolution semigroup.

For the following theorem, compare Definition 9.7.

Theorem 14.50 *For any convolution semigroup $(\nu_t : t \in I)$ and any $x \in \mathbb{R}^d$, there exists a probability measure $\mathbf{P}_x$ on the product space $(\Omega, \mathcal{A}) = \big((\mathbb{R}^d)^I, \mathcal{B}(\mathbb{R}^d)^{\otimes I}\big)$ such that the canonical process $(X_t)_{t \in I}$ is a stochastic process with $\mathbf{P}_x[X_0 = x] = 1$, with stationary independent increments and with $\mathbf{P}_x \circ (X_t - X_s)^{-1} = \nu_{t-s}$ for $t > s$. On the other hand, every stochastic process $(X_t)_{t \in I}$ (on an arbitrary probability space $(\Omega, \mathcal{A}, \mathbf{P})$) with stationary independent increments defines a convolution semigroup by $\nu_t = \mathbf{P} \circ (X_t - X_0)^{-1}$ for all $t \in I$.*

Takeaways For a Markov process, the conditional distribution at time t given the full history until some time $s < t$ is a function of the state at time s only and can be described by a stochastic kernel $\kappa_{s,t}$. On the other hand, if a family of stochastic kernels $\kappa_{s,t}$, $s < t$, fulfills the a minimal consistency condition (the Chapman-Kolmogorov equation), then Kolmogorov's extension theorem allows to construct a probability space and a Markov process on it that fits to these kernels. Convolution semigroups are a special application and yield real valued processes with independent and stationary increments.

Exercise 14.4.1 Assume that $(\nu_t : t \geq 0)$ is a continuous convolution semigroup. Show that $\nu_t = \lim_{s \to t} \nu_s$ for all $t > 0$. ♣

Exercise 14.4.2 Assume that $(\nu_t : t \geq 0)$ is a convolution semigroup. Show that $\nu_{t/n} \xrightarrow{n \to \infty} \delta_0$. ♣

Exercise 14.4.3 Show that a nonnegative convolution semigroup is continuous. ♣

Exercise 14.4.4 Show that a continuous real convolution semigroup $(\nu_t : t \geq 0)$ with $\nu_t((-\infty, 0)) = 0$ for some $t > 0$ is nonnegative. ♣

Exercise 14.4.5 Use the methods developed in this section to construct a stochastic process $(X_t)_{t \geq 0}$ with independent and stationary Poisson-distributed increments. Furthermore, show that such a process can be constructed in such a way that almost surely the map $t \mapsto X_t$ is monotone increasing and right continuous. (Compare Sect. 5.5.) ♣

Chapter 15
Characteristic Functions and the Central Limit Theorem

The main goal of this chapter is the central limit theorem (CLT) for sums of independent random variables (Theorem 15.38) and for independent arrays of random variables (Lindeberg–Feller theorem, Theorem 15.44). For the latter, we prove only that one of the two implications (Lindeberg's theorem) that is of interest in the applications.

The ideal tools for the treatment of central limit theorems are so-called characteristic functions; that is, Fourier transforms of probability measures. We start with a more general treatment of classes of test functions that are suitable to characterize weak convergence and then study Fourier transforms in greater detail. The subsequent section proves the CLT for real-valued random variables by means of characteristic functions. In the fifth section, we prove a multidimensional version of the CLT.

15.1 Separating Classes of Functions

Let (E, d) be a metric space with Borel σ-algebra $\mathcal{E} = \mathcal{B}(E)$.

Denote by $\mathbb{C} = \{u + iv : u, v \in \mathbb{R}\}$ the field of complex numbers. Let

$$\mathrm{Re}(u + iv) = u \quad \text{and} \quad \mathrm{Im}(u + iv) = v$$

denote the real part and the imaginary part, respectively, of $z = u + iv \in \mathbb{C}$. Let $\bar{z} = u - iv$ be the complex conjugate of z and $|z| = \sqrt{u^2 + v^2}$ its modulus. A prominent role will be played by the complex exponential function $\exp : \mathbb{C} \to \mathbb{C}$, which can be defined either by Euler's formula $\exp(z) = \exp(u)\big(\cos(v) + i\sin(v)\big)$ or by the power series $\exp(z) = \sum_{n=0}^{\infty} z^n/n!$. It is well-known that $\exp(z_1 + z_2) =$

© The Editor(s) (if applicable) and The Author(s), under exclusive license to Springer Nature Switzerland AG 2020
A. Klenke, *Probability Theory*, Universitext,
https://doi.org/10.1007/978-3-030-56402-5_15

$\exp(z_1) \cdot \exp(z_2)$. Note that $\text{Re}(z) = (z + \bar{z})/2$ and $\text{Im}(z) = (z - \bar{z})/2i$ imply

$$\cos(x) = \frac{e^{ix} + e^{-ix}}{2} \quad \text{and} \quad \sin(x) = \frac{e^{ix} - e^{-ix}}{2i} \quad \text{for all } x \in \mathbb{R}.$$

A map $f : E \to \mathbb{C}$ is measurable if and only if $\text{Re}(f)$ and $\text{Im}(f)$ are measurable (see Theorem 1.90 with $\mathbb{C} \cong \mathbb{R}^2$). In particular, any continuous function $E \to \mathbb{C}$ is measurable. If $\mu \in \mathcal{M}(E)$, then we define

$$\int f \, d\mu := \int \text{Re}(f) \, d\mu + i \int \text{Im}(f) \, d\mu$$

if both integrals exist and are finite. Let $C_b(E; \mathbb{C})$ denote the Banach space of continuous, bounded, complex-valued functions on E equipped with the supremum norm $\|f\|_\infty = \sup\{|f(x)| : x \in E\}$. We call $\mathcal{C} \subset C_b(E; \mathbb{C})$ a separating class for $\mathcal{M}_f(E)$ if for any two measures $\mu, \nu \in \mathcal{M}_f(E)$ with $\mu \neq \nu$, there is an $f \in \mathcal{C}$ such that $\int f \, d\mu \neq \int f \, d\nu$. The analogue of Theorem 13.34 holds for $\mathcal{C} \subset C_b(E; \mathbb{C})$.

Definition 15.1 *Let $\mathbb{K} = \mathbb{R}$ or $\mathbb{K} = \mathbb{C}$. A subset $\mathcal{C} \subset C_b(E; \mathbb{K})$ is called an **algebra** if*

(i) $1 \in \mathcal{C}$,
(ii) if $f, g \in \mathcal{C}$, then $f \cdot g$ and $f + g$ are in $\mathcal{C}$, and
(iii) if $f \in \mathcal{C}$ and $\alpha \in \mathbb{K}$, then (αf) is in $\mathcal{C}$.

*We say that $\mathcal{C}$ **separates points** if for any two points $x, y \in E$ with $x \neq y$, there is an $f \in \mathcal{C}$ with $f(x) \neq f(y)$.*

Theorem 15.2 (Stone–Weierstraß) *Let E be a compact Hausdorff space. Let $\mathbb{K} = \mathbb{R}$ or $\mathbb{K} = \mathbb{C}$. Let $\mathcal{C} \subset C_b(E; \mathbb{K})$ be an algebra that separates points. If $\mathbb{K} = \mathbb{C}$, then in addition assume that $\mathcal{C}$ is closed under complex conjugation (that is, if $f \in \mathcal{C}$, then the complex conjugate function $\bar{f}$ is also in $\mathcal{C}$).*

Then $\mathcal{C}$ is dense in $C_b(E; \mathbb{K})$ with respect to the supremum norm.

Proof We follow the exposition in Dieudonné [34, Chapter VII.3]. First consider the case $\mathbb{K} = \mathbb{R}$. We proceed in several steps.

Step 1. By Weierstraß's approximation theorem (Example 5.15), there is a sequence $(p_n)_{n \in \mathbb{N}}$ of polynomials that approach the map $[0, 1] \to [0, 1]$, $t \mapsto \sqrt{t}$ uniformly. If $f \in \mathcal{C}$, then also

$$|f| = \|f\|_\infty \lim_{n \to \infty} p_n\left(f^2 / \|f\|_\infty^2\right)$$

is in the closure $\overline{\mathcal{C}}$ of $\mathcal{C}$ in $C_b(E; \mathbb{R})$.

Step 2. Applying Step 1 to the algebra $\overline{\mathcal{C}}$ yields that, for all $f, g \in \overline{\mathcal{C}}$,

$$f \vee g = \frac{1}{2}(f + g + |f - g|) \quad \text{and} \quad f \wedge g = \frac{1}{2}(f + g - |f - g|)$$

are also in $\overline{\mathcal{C}}$.

Step 3. We show that for any $f \in C_b(E; \mathbb{R})$, any $x \in E$ and any $\varepsilon > 0$, there exists a $g_x \in \overline{C}$ with $g_x(x) = f(x)$ and $g_x(y) \leq f(y) + \varepsilon$ for all $y \in E$. As C separates points, for any $z \in E \setminus \{x\}$, there exists an $H_z \in C$ with $H_z(z) \neq H_z(x) = 0$. For such z, define $h_z \in C$ by

$$h_z(y) = f(x) + \frac{f(z) - f(x)}{H_z(z)} H_z(y) \qquad \text{for all } y \in E.$$

In addition, define $h_x := f$. Then $h_z(x) = f(x)$ and $h_z(z) = f(z)$ for all $z \in E$. Since f and h_z are continuous, for any $z \in E$, there exists an open neighborhood $U_z \ni z$ with $h_z(y) \leq f(y) + \varepsilon$ for all $y \in U_z$. We construct a finite covering $U_{z_1}, \ldots, U_{z_n}$ of E consisting of such neighborhoods and define $g_x = \min(h_{z_1}, \ldots, h_{z_n})$. By Step 2, we have $g_x \in \overline{C}$.

Step 4. Let $f \in C_b(E; \mathbb{R})$, $\varepsilon > 0$ and, for any $x \in E$, let g_x be as in Step 3. As f and g_x are continuous, for any $x \in E$, there exists an open neighborhood $V_x \ni x$ with $g_x(y) \geq f(y) - \varepsilon$ for any $y \in V_x$. We construct a finite covering $V_{x_1}, \ldots, V_{x_n}$ of E and define $g := \max(g_{x_1}, \ldots, g_{x_n})$. Then $g \in \overline{C}$ by Step 2 and $\|g - f\|_\infty < \varepsilon$ by construction. Letting $\varepsilon \downarrow 0$, we get $\overline{C} = C_b(E; \mathbb{R})$.

Step 5. Now consider $\mathbb{K} = \mathbb{C}$. If $f \in C$, then by assumption $\mathrm{Re}(f) = (f + \bar{f})/2$ and $\mathrm{Im}(f) = (f - \bar{f})/2i$ are in C. In particular, $C_0 := \{\mathrm{Re}(f) : f \in C\} \subset C$ is a real algebra that, by assumption, separates points and contains the constant functions. Hence C_0 is dense in $C_b(E; \mathbb{R})$. Since $C = C_0 + iC_0$, C is dense in $C_b(E; \mathbb{C})$. $\qquad \square$

Corollary 15.3 *Let E be a compact metric space. Let $\mathbb{K} = \mathbb{R}$ or $\mathbb{K} = \mathbb{C}$. Let $C \subset C_b(E; \mathbb{K})$ be a family that separates points; that is, stable under multiplication and that contains 1. If $\mathbb{K} = \mathbb{C}$, then in addition assume that C is closed under complex conjugation.*

Then C is a separating family for $\mathcal{M}_f(E)$.

Proof Let $\mu_1, \mu_2 \in \mathcal{M}_f(E)$ with $\int g \, d\mu_1 = \int g \, d\mu_2$ for all $g \in C$. Let C' be the algebra of finite linear combinations of elements of C. By linearity of the integral, $\int g \, d\mu_1 = \int g \, d\mu_2$ for all $g \in C'$.

For any $f \in C_b(E, \mathbb{R})$ and any $\varepsilon > 0$, by the Stone–Weierstraß theorem, there exists a $g \in C'$ with $\|f - g\|_\infty < \varepsilon$. By the triangle inequality,

$$\left| \int f \, d\mu_1 - \int f \, d\mu_2 \right| \leq \left| \int f \, d\mu_1 - \int g \, d\mu_1 \right| + \left| \int g \, d\mu_1 - \int g \, d\mu_2 \right|$$

$$+ \left| \int g \, d\mu_2 - \int f \, d\mu_2 \right|$$

$$\leq \varepsilon \, (\mu_1(E) + \mu_2(E)).$$

Letting $\varepsilon \downarrow 0$, we get equality of the integrals and hence $\mu_1 = \mu_2$ (by Theorem 13.11). $\qquad \square$

Reflection In the Stone-Weierstraß theorem for the case $\mathbb{K} = \mathbb{C}$, it is assumed that the algebra is closed under complex conjugation. Find an example that shows that this condition cannot be dropped. ♠♠

The following theorems are simple consequences of Corollary 15.3.

Theorem 15.4 *The distribution of a bounded real random variable X is characterized by its moments.*

Proof Without loss of generality, we can assume that X takes values in $E := [0, 1]$. For $n \in \mathbb{N}$, define the map $f_n : [0, 1] \to [0, 1]$ by $f_n : x \mapsto x^n$. Further, let $f_0 \equiv 1$. The family $\mathcal{C} = \{f_n, \, n \in \mathbb{N}_0\}$ separates points and is closed under multiplication; hence it is a separating class for $\mathcal{M}_f(E)$. Thus $\mathbf{P}_X$ is uniquely determined by its moments $\mathbf{E}[X^n] = \int x^n \, \mathbf{P}_X(dx), \, n \in \mathbb{N}$. □

Example 15.5 (due to [73]) In the preceding theorem, we cannot simply drop the assumption that X is bounded without making other assumptions (see Corollary 15.33). Even if all moments exist, the distribution of X is, in general, not uniquely determined by its moments. As an example consider $X := \exp(Y)$, where $Y \sim \mathcal{N}_{0,1}$. The distribution of X is called the **log-normal distribution**. For every $n \in \mathbb{N}$, nY is distributed as the sum of n^2 independent, standard normally distributed random variables $nY \overset{D}{=} Y_1 + \ldots + Y_{n^2}$. Hence, for $n \in \mathbb{N}$,

$$\mathbf{E}[X^n] = \mathbf{E}[e^{nY}] = \mathbf{E}[e^{Y_1 + \ldots + Y_{n^2}}] = \prod_{i=1}^{n^2} \mathbf{E}[e^{Y_i}] = \mathbf{E}[e^Y]^{n^2}$$

$$= \left(\int_{-\infty}^{\infty} (2\pi)^{-1/2} \, e^y \, e^{-y^2/2} \, dy \right)^{n^2} = e^{n^2/2}. \tag{15.1}$$

We construct a whole family of distributions with the same moments as X. By the transformation formula for densities (Theorem 1.101), the distribution of X has the density

$$f(x) = \frac{1}{\sqrt{2\pi}} x^{-1} \exp\left(-\frac{1}{2} \log(x)^2 \right) \quad \text{for } x > 0.$$

For $\alpha \in [-1, 1]$, define probability densities f_α on $(0, \infty)$ by

$$f_\alpha(x) = f(x)\big(1 + \alpha \sin(2\pi \log(x))\big).$$

In order to show that f_α is a density and has the same moments as f, it is enough to show that, for all $n \in \mathbb{N}_0$,

$$m(n) := \int_0^{\infty} x^n \, f(x) \, \sin(2\pi \log(x)) \, dx = 0.$$

With the substitution $y = \log(x) - n$, we get (note that $\sin(2\pi(y+n)) = \sin(2\pi y)$)

$$m(n) = \int_{-\infty}^{\infty} e^{yn+n^2} (2\pi)^{-1/2} e^{-(y+n)^2/2} \sin(2\pi(y+n)) \, dy$$

$$= (2\pi)^{-1/2} e^{n^2/2} \int_{-\infty}^{\infty} e^{-y^2/2} \sin(2\pi y) \, dy = 0,$$

where the last equality holds since the integrand is an odd function. ◊

Theorem 15.6 (Laplace transform) *A finite measure μ on $[0, \infty)$ is characterized by its Laplace transform*

$$\mathcal{L}_\mu(\lambda) := \int e^{-\lambda x} \mu(dx) \quad \text{for } \lambda \geq 0.$$

Proof We face the problem that the space $[0, \infty)$ is not compact by passing to the one-point compactification $E = [0, \infty]$. For $\lambda \geq 0$, define the continuous function $f_\lambda : [0, \infty] \to [0, 1]$ by $f_\lambda(x) = e^{-\lambda x}$ if $x < \infty$ and $f_\lambda(\infty) = \lim_{x \to \infty} e^{-\lambda x}$. Then $\mathcal{C} = \{f_\lambda, \; \lambda \geq 0\}$ separates points, $f_0 = 1 \in \mathcal{C}$ and $f_\mu \cdot f_\lambda = f_{\mu+\lambda} \in \mathcal{C}$. By Corollary 15.3, $\mathcal{C}$ is a separating class for $\mathcal{M}_f([0, \infty])$ and thus also for $\mathcal{M}_f([0, \infty))$. □

Remark 15.7 Let X and Y be independent nonnegative random variables with Laplace transforms $\mathcal{L}_X := \mathcal{L}_{\mathbf{P}_X}$ and $\mathcal{L}_Y := \mathcal{L}_{\mathbf{P}_Y}$, respectively. Further, let $a \geq 0$ and $b \geq 0$. Then we clearly have $\mathcal{L}_{aX+b}(t) = e^{-bt} \mathcal{L}_X(at)$ and $\mathcal{L}_{X+Y}(t) = \mathcal{L}_X(t) \cdot \mathcal{L}_Y(t)$ for $t \geq 0$. ◊

Reflection Check the statements of the preceding remark! ♠

Definition 15.8 *For $\mu \in \mathcal{M}_f(\mathbb{R}^d)$, define the map $\varphi_\mu : \mathbb{R}^d \to \mathbb{C}$ by*

$$\varphi_\mu(t) := \int e^{i\langle t,x\rangle} \mu(dx).$$

*φ_μ is called the **characteristic function** of μ.*

Theorem 15.9 (Characteristic function) *A finite measure $\mu \in \mathcal{M}_f(\mathbb{R}^d)$ is characterized by its characteristic function.*

Proof Let $\mu_1, \mu_2 \in \mathcal{M}_f(\mathbb{R}^d)$ with $\varphi_{\mu_1}(t) = \varphi_{\mu_2}(t)$ for all $t \in \mathbb{R}^d$. By Theorem 13.11(ii), $C_c(\mathbb{R}^d)$ is a separating class for $\mathcal{M}_f(\mathbb{R}^d)$. Hence, it is enough to show that $\int f \, d\mu_1 = \int f \, d\mu_2$ for all $f \in C_c(\mathbb{R}^d)$.

Let $f : \mathbb{R}^d \to \mathbb{R}$ be continuous with compact support and let $\varepsilon > 0$. Assume that $K > 0$ is large enough such that $f(x) = 0$ for $x \notin (-K/2, K/2)^d$ and such that $\mu_i(\mathbb{R}^d \setminus (-K, K)^d) < \varepsilon$, $i = 1, 2$. Consider the *torus* $E := \mathbb{R}^d/(2K\mathbb{Z}^d)$ and define $\tilde{f} : E \to \mathbb{R}$ by

$$\tilde{f}(x + 2K\mathbb{Z}^d) = f(x) \quad \text{for } x \in [-K, K)^d.$$

Since the support of f is contained in $(-K, K)^d$, $\tilde{f}$ is continuous.

For $m \in \mathbb{Z}^d$ define

$$g_m : \mathbb{R}^d \to \mathbb{C}, \qquad x \mapsto \exp\left(i \langle \pi m / K, x \rangle\right).$$

Let $\mathcal{C}$ be the algebra of finite linear combinations of the g_m. For $g \in \mathcal{C}$, we have $g(x) = g(x + 2Kn)$ for all $x \in \mathbb{R}^d$ and $n \in \mathbb{Z}^d$. Hence, the map

$$\tilde{g} : E \to \mathbb{C}, \qquad \tilde{g}(x + 2K\mathbb{Z}^d) = g(x)$$

is well-defined, continuous and bounded. Furthermore, $\tilde{\mathcal{C}} := \{\tilde{g} : g \in \mathcal{C}\} \subset C_b(E; \mathbb{C})$ is an algebra that separates points and is closed under complex conjugation. As E is compact, by the Stone–Weierstraß theorem, there is a $g \in \mathcal{C}$ such that $\|\tilde{g} - \tilde{f}\|_\infty < \varepsilon$. We infer

$$\left\|(f - g)\mathbb{1}_{[-K,K]^d}\right\|_\infty < \varepsilon$$

and

$$\left\|(f - g)\mathbb{1}_{\mathbb{R}^d \setminus [-K,K]^d}\right\|_\infty \leq \|g\|_\infty = \|\tilde{g}\|_\infty \leq \|\tilde{f}\|_\infty + \varepsilon = \|f\|_\infty + \varepsilon.$$

By assumption of the theorem, $\int g \, d\mu_1 = \int g \, d\mu_2$. Hence, using the triangle inequality, we conclude

$$\left| \int f \, d\mu_1 - \int f \, d\mu_2 \right| \leq \int |f - g| \, d\mu_1 + \int |f - g| \, d\mu_2$$

$$\leq \varepsilon\left(2\|f\|_\infty + 2\varepsilon + \mu_1(\mathbb{R}^d) + \mu_2(\mathbb{R}^d)\right).$$

As $\varepsilon > 0$ was arbitrary, the integrals coincide.   □

Corollary 15.10 *A finite measure μ on $\mathbb{Z}^d$ is uniquely determined by the values*

$$\varphi_\mu(t) = \int e^{i \langle t, x \rangle} \, \mu(dx), \qquad t \in [-\pi, \pi)^d.$$

Proof This is obvious since $\varphi_\mu(t + 2\pi k) = \varphi_\mu(t)$ for all $k \in \mathbb{Z}^d$.   □

While the preceding corollary only yields an abstract uniqueness statement, we will profit also from an explicit inversion formula for Fourier transforms.

Theorem 15.11 (Discrete Fourier inversion formula) *Let $\mu \in \mathcal{M}_f(\mathbb{Z}^d)$ with characteristic function φ_μ. Then, for every $x \in \mathbb{Z}^d$,*

$$\mu(\{x\}) = (2\pi)^{-d} \int_{[-\pi,\pi)^d} e^{-i \langle t, x \rangle} \, \varphi_\mu(t) \, dt.$$

Proof By the dominated convergence theorem,

$$
\int_{[-\pi,\pi)^d} e^{-i\langle t,x\rangle} \varphi_\mu(t)\, dt = \int_{[-\pi,\pi)^d} e^{-i\langle t,x\rangle} \left(\lim_{n\to\infty} \sum_{|y|\le n} e^{i\langle t,y\rangle} \mu(\{y\}) \right) dt
$$

$$
= \lim_{n\to\infty} \int_{[-\pi,\pi)^d} e^{-i\langle t,x\rangle} \sum_{|y|\le n} e^{i\langle t,y\rangle} \mu(\{y\})\, dt
$$

$$
= \sum_{y\in\mathbb{Z}^d} \mu(\{y\}) \int_{[-\pi,\pi)^d} e^{i\langle t,y-x\rangle}\, dt.
$$

The claim follows since, for $y \in \mathbb{Z}^d$,

$$
\int_{[-\pi,\pi)^d} e^{i\langle t,y-x\rangle}\, dt = \begin{cases} (2\pi)^d, & \text{if } x = y, \\ 0, & \text{else.} \end{cases} \qquad \Box
$$

Similar inversion formulas hold for measures μ on $\mathbb{R}^d$. Particularly simple is the case where μ possesses an integrable density $f := \frac{d\mu}{d\lambda}$ with respect to d-dimensional Lebesgue measure λ. In this case, we have the Fourier inversion formula,

$$
f(x) = (2\pi)^{-d} \int_{\mathbb{R}^d} e^{-i\langle t,x\rangle} \varphi_\mu(t)\, \lambda(dt). \tag{15.2}
$$

Furthermore, by Plancherel's theorem, $f \in \mathcal{L}^2(\lambda)$ if and only if $\varphi_\mu \in \mathcal{L}^2(\lambda)$. In this case, $\|f\|_2 = \|\varphi\|_2/(2\pi)^{d/2}$.

Since we will not need these statements in the following, we only refer to the standard literature (e.g., [174, Chapter VI.2] or [54, Theorem XV.3.3 and Equation (XV.3.8)]).

Takeaways A priori, checking equality of two measures by computing integrals or checking weak convergence of a sequence of measures requires to consider a huge class of test functions. The Stone-Weierstraß theorem and its corollaries allow to boil down the class of test functions to a tractable size. In fact, in many cases, it is enough to consider moments, Laplace transforms or characteristic functions. There is some freedom in the choice of the class of test functions so it can be adapted to the individual problem. For example, characteristic functions work well with sums of independent random variables.

Exercise 15.1.1 Show that, in the Stone–Weierstraß theorem, compactness of E is essential. *Hint:* Let $E = \mathbb{R}$ and use the fact that $C_b(\mathbb{R}) = C_b(\mathbb{R}; \mathbb{R})$ is not separable. Construct a countable algebra $\mathcal{C} \subset C_b(\mathbb{R})$ that separates points. ♣

Exercise 15.1.2 Let $d \in \mathbb{N}$ and let μ be a finite measure on $[0, \infty)^d$. Show that μ is characterized by its Laplace transform $\mathcal{L}_\mu(\lambda) = \int e^{-\langle \lambda, x \rangle} \mu(dx)$, $\lambda \in [0, \infty)^d$. ♣

Exercise 15.1.3 Let $n \in \mathbb{N}$ and let $X_1, \ldots, X_n$ be i.i.d. exponentially distributed random variables with parameter 1. Finally, let $Y_1, \ldots, Y_n$ be independent exponentially distributed random variables with $\mathbf{P}_{Y_k} = \exp_k$. That is, $(Y_1, \ldots, Y_n) \overset{\mathcal{D}}{=} (X_1, X_2/2, X_3/3, \ldots, X_n/n)$. Show that

$$\max\{X_1, \ldots, X_n\} \overset{\mathcal{D}}{=} Y_1 + \ldots + Y_n.$$

Hint:

(i) Compute the Laplace transforms $\mathcal{L}_{Y_1}, \ldots, \mathcal{L}_{Y_n}$ and use Remark 15.7 to compute $\mathcal{L}_{Y_1 + \ldots + Y_n}$.

(ii) Use the explicit formula for the Laplace transform $M := \max\{X_1, \ldots, X_n\}$ from Remark 2.24 to compute the Laplace transform

$$\mathcal{L}_M(t) = \frac{n!}{(t+1)(t+2) \cdots (t+n)} \qquad \text{for } t \geq 0.$$

(iii) Use the uniqueness theorem for Laplace transforms.

An alternative strategy of proof is: Sort the values of the X_i by size $M = X_{(1)} > X_{(2)} > \ldots > X_{(n)}$. Check that $X_{(n)} \overset{\mathcal{D}}{=} Y_n$. Show that the conditional distribution $\mathcal{L}\left[\left(X^{(1)} - X^{(n)}, \ldots, X^{(n-1)} - X^{(n)}\right) \mid X^{(n)}\right]$ does not depend on $X^{(n)}$ and that it equals the (unconditional) distribution of the ordered values of $X_1, \ldots, X_{n-1}$. By an iteration procedure, show the even stronger statement

$$\left(X_{(n)}, X_{(n-1)}, \ldots, X_{(1)}\right) \overset{\mathcal{D}}{=} \left(Y_n, Y_{n-1} + Y_n, \ldots, Y_1 + Y_2 + \ldots + Y_n\right).$$

We will use this Exercise later in Example 17.27. ♣

Exercise 15.1.4 Show that, under the assumptions of Theorem 15.11, **Plancherel's equation** holds:

$$\sum_{x \in \mathbb{Z}^d} \mu(\{x\})^2 = (2\pi)^{-d} \int_{[-\pi, \pi)^d} |\varphi_\mu(t)|^2 \, dt. ♣$$

Exercise 15.1.5 (Mellin transform) Let X be a nonnegative real random variable. For $s \geq 0$, define the Mellin transform of $\mathbf{P}_X$ by

$$m_X(s) = \mathbf{E}[X^s]$$

(with values in $[0, \infty]$).

Assume there is an $\varepsilon_0 > 0$ with $m_X(\varepsilon_0) < \infty$ (respectively $m_X(-\varepsilon_0) < \infty$). Show that, for any $\varepsilon > 0$, the distribution $\mathbf{P}_X$ is characterized by the values $m_X(s)$ (respectively $m_X(-s)$), $s \in [0, \varepsilon]$.
Hint: For continuous $f : [0, \infty) \to [0, \infty)$, let

$$\phi_f(z) = \int_0^\infty t^{z-1} f(t)\, dt$$

for those $z \in \mathbb{C}$ for which the integral is well-defined. By a standard result of complex analysis if $\phi_f(s) < \infty$ for an $s > 1$, then ϕ_f is holomorphic in $\{z \in \mathbb{C} : \mathrm{Re}(z) \in (1, s)\}$ (and is thus uniquely determined by the values $\phi_f(r)$, $r \in (1, 1+\varepsilon)$ for any $\varepsilon > 0$). Furthermore, for all $r \in (1, s)$,

$$f(t) = \frac{1}{2\pi i} \int_{-\infty}^\infty t^{-(r+i\rho)} \phi_f(r + i\rho)\, d\rho.$$

(i) Conclude the statement for X with a continuous density.
(ii) For $\delta > 0$, let $Y_\delta \sim \mathcal{U}_{[1-\delta,1]}$ be independent of X. Show that XY_δ has a continuous density.
(iii) Compute m_{XY_δ}, and show that $m_{XY_\delta} \to m_X$ for $\delta \downarrow 0$.
(iv) Show that $XY_\delta \implies X$ for $\delta \downarrow 0$. ♣

Exercise 15.1.6 Let X, Y, Z be independent nonnegative random variables such that $\mathbf{P}[Z > 0] > 0$ and such that the Mellin transform $m_{XZ}(s)$ is finite for some $s > 0$.

Show that if $XZ \overset{\mathcal{D}}{=} YZ$ holds, then $X \overset{\mathcal{D}}{=} Y$. ♣

Exercise 15.1.7 Let μ be a probability measure on $\mathbb{R}$ with integrable characteristic function φ_μ and hence $\varphi_\mu \in \mathcal{L}^1(\lambda)$, where λ is the Lebesgue measure on $\mathbb{R}$. Show that μ is absolutely continuous with bounded continuous density $f = \frac{d\mu}{d\lambda}$ given by

$$f(x) = \frac{1}{2\pi} \int_{-\infty}^\infty e^{-itx} \varphi_\mu(t)\, dt \qquad \text{for all } x \in \mathbb{R}.$$

Hint: Show this first for the normal distribution $\mathcal{N}_{0,\varepsilon}$, $\varepsilon > 0$. Then show that $\mu * \mathcal{N}_{0,\varepsilon}$ is absolutely continuous with density f_ε, which converges pointwise to f (as $\varepsilon \to 0$). ♣

Exercise 15.1.8 Let (Ω, τ) be a separable topological space that satisfies the $T_{3\frac{1}{2}}$ separation axiom: For any closed set $A \subset \Omega$ and any point $x \in \Omega \setminus A$, there exists a continuous function $f : \Omega \to [0, 1]$ with $f(x) = 0$ and $f(y) = 1$ for all $y \in A$. (Note in particular that every metric space is a $T_{3\frac{1}{2}}$-space.)

Show that $\sigma(C_b(\Omega)) = \mathcal{B}(\Omega)$; that is, the Borel σ-algebra is generated by the bounded continuous functions $\Omega \to \mathbb{R}$. ♣

15.2   Characteristic Functions: Examples

Recall that $\mathrm{Re}(z)$ is the real part of $z \in \mathbb{C}$. We collect some simple properties of characteristic functions.

Lemma 15.12 *Let X be a random variable with values in $\mathbb{R}^d$ and characteristic function $\varphi_X(t) = \mathbf{E}[e^{i\langle t, X \rangle}]$. Then:*

(i) $|\varphi_X(t)| \le 1$ for all $t \in \mathbb{R}^d$ and $\varphi_X(0) = 1$.
(ii) $\varphi_{aX+b}(t) = \varphi_X(at)\, e^{i\langle b, t \rangle}$ for all $a \in \mathbb{R}$ and $b \in \mathbb{R}^d$.
(iii) $\mathbf{P}_X = \mathbf{P}_{-X}$ if and only if φ is real-valued.
(iv) If X and Y are independent, then $\varphi_{X+Y} = \varphi_X \cdot \varphi_Y$.
(v) $0 \le 1 - \mathrm{Re}(\varphi_X(2t)) \le 4(1 - \mathrm{Re}(\varphi_X(t)))$ for all $t \in \mathbb{R}^d$.

Proof

(i)   and (ii) are trivial.
(iii) $\overline{\varphi_X(t)} = \varphi_X(-t) = \varphi_{-X}(t)$.
(iv) As $e^{i\langle t, X \rangle}$ and $e^{i\langle t, Y \rangle}$ are independent random variables, we have

$$\varphi_{X+Y}(t) = \mathbf{E}[e^{i\langle t, X \rangle} \cdot e^{i\langle t, Y \rangle}] = \mathbf{E}[e^{i\langle t, X \rangle}]\, \mathbf{E}[e^{i\langle t, Y \rangle}] = \varphi_X(t)\, \varphi_Y(t).$$

(v) By the addition theorem for trigonometric functions,

$$1 - \cos(\langle 2t, X \rangle) = 2\big(1 - (\cos(\langle t, X \rangle))^2\big) \le 4\big(1 - \cos(\langle t, X \rangle)\big).$$

Now take the expectations of both sides.   □

In the next theorem, we collect the characteristic functions for some of the most important distributions.

Theorem 15.13 (Characteristic functions of some distributions)   *For some distributions P with density $x \mapsto f(x)$ on $\mathbb{R}$ or weights $P(\{k\})$, $k \in \mathbb{N}_0$, we state the characteristic function $\varphi(t)$ explicitly:*

Distribution Name Symbol	Parameter	on	Density / Weights	Char. fct. $\varphi(t)$
normal $\mathcal{N}_{\mu,\sigma^2}$	$\mu \in \mathbb{R}\ \sigma^2 > 0$	$\mathbb{R}$	$\frac{1}{\sqrt{2\pi\sigma^2}}\exp\left(-\frac{(x-\mu)^2}{2\sigma^2}\right)$	$e^{i\mu t}\cdot e^{-\sigma^2 t^2/2}$
uniform $\mathcal{U}_{[0,a]}$	$a>0$	$[0,a]$	$1/a$	$\frac{e^{iat}-1}{iat}$
uniform $\mathcal{U}_{[-a,a]}$	$a>0$	$[-a,a]$	$1/2a$	$\frac{\sin(at)}{at}$
triangle Tri_a	$a>0$	$[-a,a]$	$\frac{1}{a}\left(1-|x|/a\right)^+$	$2\frac{1-\cos(at)}{a^2 t^2}$
N.N.	$a>0$	$\mathbb{R}$	$\frac{1}{\pi}\frac{1-\cos(ax)}{ax^2}$	$\left(1-|t|/a\right)^+$
Gamma $\Gamma_{\theta,r}$	$\theta>0\ r>0$	$[0,\infty)$	$\frac{\theta^r}{\Gamma(r)}x^{r-1}\,e^{-\theta x}$	$\left(1-it/\theta\right)^{-r}$
exponential $\exp_\theta$	$\theta>0$	$0,\infty)$	$\theta\,e^{-\theta x}$	$\frac{\theta}{\theta-it}$
two-sided exponential $\exp_\theta^2$	$\theta>0$	$\mathbb{R}$	$\frac{\theta}{2}\,e^{-\theta|x|}$	$\frac{1}{1+(t/\theta)^2}$
Cauchy Cau_a	$a>0$	$\mathbb{R}$	$\frac{1}{a\pi}\frac{1}{1+(x/a)^2}$	$e^{-a|t|}$
binomial $b_{n,p}$	$n\in\mathbb{N}\ p\in[0,1]$	$\{0,\ldots,n\}$	$\binom{n}{k}p^k(1-p)^{n-k}$	$\left((1-p)+pe^{it}\right)^n$
negative binomial $b^-_{r,p}$	$r>0\ p\in(0,1]$	$\mathbb{N}_0$	$\binom{-r}{k}(-1)^k p^r(1-p)^k$	$\left(\dfrac{p}{1-(1-p)e^{it}}\right)^r$
Poisson Poi_λ	$\lambda>0$	$\mathbb{N}_0$	$e^{-\lambda}\,\frac{\lambda^k}{k!}$	$\exp\left(\lambda(e^{it}-1)\right)$

Proof

(i) (**Normal distribution**) By Lemma 15.12, it is enough to consider the case $\mu = 0$ and $\sigma^2 = 1$. By virtue of the differentiation lemma (Theorem 6.28) and using partial integration, we get

$$\frac{d}{dt}\varphi(t) = \frac{1}{\sqrt{2\pi}}\int_{-\infty}^{\infty} e^{itx}\, ix\, e^{-x^2/2}\, dx = -t\,\varphi(t).$$

This linear differential equation with initial value $\varphi(0) = 1$ has the unique solution $\varphi(t) = e^{-t^2/2}$.

(ii) (**Uniform distribution**) This is immediate.

(iii) (**Triangle distribution**) Note that $\mathrm{Tri}_a = \mathcal{U}_{[-a/2,a/2]} * \mathcal{U}_{[-a/2,a/2]}$; hence

$$\varphi_{\mathrm{Tri}_a}(t) = \varphi_{\mathcal{U}_{[-a/2,a/2]}}(t)^2 = 4\,\frac{\sin(at/2)^2}{a^2 t^2} = 2\,\frac{1 - \cos(at)}{a^2 t^2}.$$

Here we used the fact that by the addition theorem for trigonometric functions

$$1 - \cos(x) = \sin(x/2)^2 + \cos(x/2)^2 - \cos(x) = 2\sin(x/2)^2.$$

(iv) (**N.N.**) This can either be computed directly or can be deduced from (iii) by using the Fourier inversion formula (equation (15.2)).

(v) (**Gamma distribution**) Again it suffices to consider the case $\theta = 1$. For $0 \le b < c \le \infty$ and $t \in \mathbb{R}$, let $\gamma_{b,c,t}$ be the linear path in $\mathbb{C}$ from $b - ibt$ to $c - ict$, let $\delta_{b,t}$ be the linear path from b to $b - ibt$ and let $\epsilon_{c,t}$ be the linear path from $c - ict$ to c. Substituting $z = (1 - it)x$, we get

$$\varphi(t) = \frac{1}{\Gamma(r)}\int_0^{\infty} x^{r-1} e^{-x} e^{itx}\, dx = \frac{(1-it)^{-r}}{\Gamma(r)}\int_{\gamma_{0,\infty,t}} z^{r-1}\, e^{-z}\, dz.$$

Hence, it suffices to show that $\int_{\gamma_{0,\infty,t}} z^{r-1}\exp(-z)\,dz = \Gamma(r)$.

The function $z \mapsto z^{r-1}\exp(-z)$ is holomorphic in the right complex plane. Hence, by the residue theorem for $0 < b < c < \infty$,

$$\int_b^c x^{r-1}\exp(-x)\,dx = \int_{\gamma_{b,c,t}} z^{r-1}\exp(-z)\,dz$$

$$+ \int_{\delta_{b,t}} z^{r-1}\exp(-z)\,dz + \int_{\epsilon_{c,t}} z^{r-1}\exp(-z)\,dz.$$

Recall that $\int_0^{\infty} x^{r-1}\exp(-x)\,dx =: \Gamma(r)$. Hence, it is enough to show that the integrals along $\delta_{b,t}$ and $\epsilon_{c,t}$ vanish if $b \to 0$ and $c \to \infty$.

However, $|z^{r-1} \exp(-z)| \le (1+t^2)^{(r-1)/2} b^{r-1} \exp(-b)$ for $z \in \delta_{b,t}$. As the path $\delta_{b,t}$ has length $b\,|t|$, we get the estimate

$$\left| \int_{\delta_{b,t}} z^{r-1} e^{-z}\, dz \right| \le b^r e^{-b} (1+t^2)^{r/2} \longrightarrow 0 \quad \text{for } b \to 0.$$

Similarly,

$$\left| \int_{\epsilon_{c,t}} z^{r-1} e^{-z}\, dz \right| \le c^r e^{-c} (1+t^2)^{r/2} \longrightarrow 0 \quad \text{for } c \to \infty.$$

(vi) **(Exponential distribution)** This follows from (v) since $\exp_\theta = \Gamma_{\theta,1}$.

(vii) **(Two-sided exponential distribution)** If X and Y are independent $\exp_\theta$-distributed random variables, then it is easy to check that $X - Y \sim \exp_\theta^2$. Hence

$$\varphi_{\exp_\theta^2}(t) = \varphi_{\exp_\theta}(t)\, \varphi_{\exp_\theta}(-t) = \frac{1}{1 - it/\theta}\, \frac{1}{1 + it/\theta} = \frac{1}{1 + (t/\theta)^2}.$$

(viii) **(Cauchy distribution)** This can either be computed directly using residue calculus or can be inferred from the statement for the two-sided exponential distribution by the Fourier inversion formula (equation (15.2)).

(ix) **(Binomial distribution)** By the binomial theorem,

$$\varphi(t) = \sum_{k=0}^{n} \binom{n}{k} (1-p)^{n-k} (pe^{it})^k = (1 - p + pe^{it})^n.$$

(x) **(Negative binomial distribution)** By the generalized binomial theorem (Lemma 3.5), for all $x \in \mathbb{C}$ with $|x| < 1$,

$$(1 - x)^{-r} = \sum_{k=0}^{\infty} \binom{-r}{k} (-x)^k.$$

Using this formula with $x = (1 - p)\, e^{it}$ gives the claim.

(xi) **(Poisson distribution)** Clearly, $\varphi_{\mathrm{Poi}_\lambda}(t) = \sum_{n=0}^{\infty} e^{-\lambda}\, \frac{(\lambda e^{it})^n}{n!} = e^{\lambda(e^{it}-1)}$. ☐

Corollary 15.14 *The following convolution formulas hold.*

(i) $\mathcal{N}_{\mu_1,\sigma_1^2} * \mathcal{N}_{\mu_2,\sigma_2^2} = \mathcal{N}_{\mu_1+\mu_2,\sigma_1^2+\sigma_2^2}$ *for* $\mu_1, \mu_2 \in \mathbb{R}$ *and* $\sigma_1^2, \sigma_2^2 > 0$.

(ii) $\Gamma_{\theta,r} * \Gamma_{\theta,s} = \Gamma_{\theta,r+s}$ *for* $\theta, r, s > 0$.

(iii) $\mathrm{Cau}_a * \mathrm{Cau}_b = \mathrm{Cau}_{a+b}$ *for* $a, b > 0$.

(iv) $b_{m,p} * b_{n,p} = b_{m+n,p}$ *for* $m, n \in \mathbb{N}$ *and* $p \in [0, 1]$.

(v) $b_{r,p}^- * b_{s,p}^- = b_{r+s,p}^-$ *for* $r, s > 0$ *and* $p \in (0, 1]$.
(vi) $\mathrm{Poi}_\lambda * \mathrm{Poi}_\mu = \mathrm{Poi}_{\lambda+\mu}$ *for* $\lambda, \mu \geq 0$.

Proof This follows by Theorem 15.13 and by $\varphi_{\mu*\nu} = \varphi_\mu \, \varphi_\nu$ (Lemma 15.12).   □

The following theorem gives two simple procedures for calculating the characteristic functions of compound distributions.

Theorem 15.15

(i) *Let* $\mu_1, \mu_2, \ldots \in \mathcal{M}_f(\mathbb{R}^d)$ *and let* $p_1, p_2, \ldots$ *be nonnegative numbers with*

$$\sum_{n=1}^{\infty} p_n \mu_n(\mathbb{R}^d) < \infty. \text{ Then the measure } \mu := \sum_{n=1}^{\infty} p_n \mu_n \in \mathcal{M}_f(\mathbb{R}^d) \text{ has}$$

characteristic function

$$\varphi_\mu = \sum_{n=1}^{\infty} p_n \, \varphi_{\mu_n}. \tag{15.3}$$

(ii) *Let* $N, X_1, X_2, \ldots$ *be independent random variables. Assume* $X_1, X_2, \ldots$
are identically distributed on $\mathbb{R}^d$ *with characteristic function* φ_X. *Assume*
N takes values in $\mathbb{N}_0$ *and has the probability generating function* f_N. *Then*
$Y := \sum_{n=1}^{N} X_n$ *has the characteristic function* $\varphi_Y(t) = f_N(\varphi_X(t))$.

(iii) *In particular, if we let* $N \sim \mathrm{Poi}_\lambda$ *in (ii), then* $\varphi_Y(t) = \exp(\lambda(\varphi_X(t) - 1))$.

Proof

(i) Define $\nu_n = \sum_{k=1}^{n} p_k \mu_k$. By the linearity of the integral, $\varphi_{\nu_n} = \sum_{k=1}^{n} p_k \varphi_{\mu_k}$. By assumption, $\mu = \text{w-lim}_{n\to\infty} \nu_n$; hence also $\varphi_\mu(t) = \lim_{n\to\infty} \varphi_{\nu_n}(t)$.

(ii) Clearly,

$$\varphi_Y(t) = \sum_{n=0}^{\infty} \mathbf{P}[N = n] \, \mathbf{E}\left[e^{i\langle t, X_1 + \ldots + X_n\rangle}\right]$$

$$= \sum_{n=0}^{\infty} \mathbf{P}[N = n] \, \varphi_X(t)^n = f_N(\varphi_X(t)).$$

(iii) In this special case, $f_N(z) = e^{\lambda(z-1)}$ for $z \in \mathbb{C}$ with $|z| \leq 1$.   □

Example 15.16 Let $n \in \mathbb{N}$, and assume that the points $0 = a_0 < a_1 < \ldots < a_n$ and $1 = y_0 > y_1 > \ldots > y_n = 0$ are given. Let $\varphi : \mathbb{R} \to [0, \infty)$ have the properties that

- $\varphi(a_k) = y_k$ for all $k = 0, \ldots, n$ and φ is linearly interpolated between the points a_k,
- $\varphi(x) = 0$ for $|x| > a_n$, and
- φ is even (that is, $\varphi(x) = \varphi(-x)$).

Assume in addition that the y_k are chosen such that φ is convex on $[0, \infty)$. This is equivalent to the condition that $m_1 \leq m_2 \leq \ldots \leq m_n \leq 0$, where $m_k := \frac{y_k - y_{k-1}}{a_k - a_{k-1}}$ is the slope on the kth interval. We want to show that φ is the characteristic function of a probability measure $\mu \in \mathcal{M}_1(\mathbb{R})$.

Define $p_k = a_k(m_{k+1} - m_k)$ for $k = 1, \ldots, n$.

Let $\mu_k \in \mathcal{M}_1(\mathbb{R})$ be the distribution on $\mathbb{R}$ with density $\frac{1}{\pi} \frac{1-\cos(a_k \pi)}{a_k x^2}$. By Theorem 15.13, μ_k has the characteristic function $\varphi_{\mu_k}(t) = \left(1 - \frac{|t|}{a_k}\right)^+$. The characteristic function φ_μ of $\mu := \sum_{k=1}^n p_k \mu_k$ is then

$$\varphi_\mu(t) = \sum_{k=1}^n p_k(1 - |t|/a_k)^+.$$

This is a continuous, symmetric, real function with $\varphi_\mu(0) = 1$. It is linear on each of the intervals $[a_{k-1}, a_k]$. See Fig. 15.1 for an example with $n = 4$. By partial

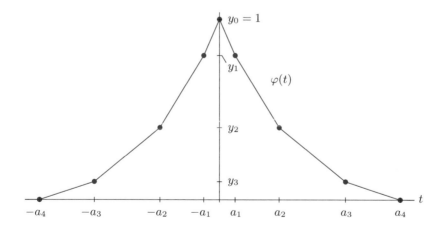

Fig. 15.1 The characteristic function φ from Example 15.16 with $n = 4$.

summation, for all $k = 1, \ldots, n$ (since $m_{n+1} = 0$),

$$\varphi_\mu(a_l) = \sum_{k=1}^{n} a_k(m_{k+1} - m_k)\left(1 - \frac{a_l}{a_k}\right)^+ = \sum_{k=l}^{n} (a_k - a_l)(m_{k+1} - m_k)$$

$$= \left[(a_n - a_l)m_{n+1} - (a_l - a_l)m_l\right] - \sum_{k=l+1}^{n} (a_k - a_{k-1})m_k$$

$$= - \sum_{k=l+1}^{n} (y_k - y_{k-1}) = y_l = \varphi(a_l).$$

Hence $\varphi_\mu = \varphi$. $\Diamond$

Example 15.17 Define the function $\varphi : \mathbb{R} \to [-1, 1]$ for $t \in [-\pi, \pi)$ by $\varphi(t) = 1 - 2|t|/\pi$, and assume φ is periodic (with period 2π). By the discrete Fourier inversion formula (Theorem 15.11), φ is the characteristic function of the probability measure $\mu \in \mathcal{M}_1(\mathbb{Z})$ with $\mu(\{x\}) = (2\pi)^{-1} \int_{-\pi}^{\pi} \cos(tx)\,\varphi(t)\,dt$. In fact, in order that μ be a measure (not only a signed measure), we still have to show that all of the masses $\mu(\{x\})$ are nonnegative. Clearly, $\mu(\{0\}) = 0$. For $x \in \mathbb{Z} \setminus \{0\}$, use partial integration to compute the integral,

$$\int_{-\pi}^{\pi} \cos(tx)\,\varphi(t)\,dt = 2 \int_{0}^{\pi} \cos(tx)\,(1 - 2t/\pi)\,dt$$

$$= \frac{4}{x}\left(1 - \frac{2}{\pi}\right)\sin(\pi x) - \frac{4}{x}\sin(0) + \frac{4}{\pi x}\int_{0}^{\pi}\sin(tx)\,dt$$

$$= \frac{4}{\pi x^2}(1 - \cos(\pi x)).$$

Summing up, we have

$$\mu(\{x\}) = \begin{cases} \frac{4}{\pi^2 x^2}, & \text{if } x \text{ is odd,} \\ 0, & \text{else.} \end{cases}$$

Since $\mu(\mathbb{Z}) = \varphi(0) = 1$, μ is indeed a probability measure. $\Diamond$

Example 15.18 Define the function $\psi : \mathbb{R} \to [0, 1]$ for $t \in [-\pi/2, \pi/2)$ by $\psi(t) = 1 - 2|t|/\pi$. Assume ψ is periodic with period π. If φ is the characteristic function of the measure μ from the previous example, then clearly $\psi(t) = |\varphi(t)|$. On the other hand, $\psi(t) = \frac{1}{2} + \frac{1}{2}\varphi(2t)$. By Theorem 15.15 and Lemma 15.12(ii), we infer that ψ is the characteristic function of the measure ν with $\nu(A) = \frac{1}{2}\delta_0(A) + \frac{1}{2}\mu(A/2)$

for $A \subset \mathbb{R}$. Hence,

$$
\nu(\{x\}) = \begin{cases}
\frac{1}{2}, & \text{if } x = 0, \\
\frac{8}{\pi^2 x^2}, & \text{if } \frac{x}{2} \in \mathbb{Z} \text{ is odd,} \\
0, & \text{else.}
\end{cases}
$$

◊

Example 15.19 Let $\varphi(t) = (1 - 2|t|/\pi)^+$ be the characteristic function of the distribution "N.N." from Theorem 15.13 (with $a = \pi/2$) and let ψ be the characteristic function from the preceding example. Note that $\varphi(t) = \psi(t)$ for $|t| \leq \pi/2$ and $\varphi(t) = 0$ for $|t| > \pi/2$; hence $\varphi^2 = \varphi \cdot \psi$. Now let X, Y, Z be independent real random variables with characteristic functions $\varphi_X = \varphi_Y = \varphi$ and $\varphi_Z = \psi$. Then $\varphi_X \varphi_Y = \varphi_X \varphi_Z$; hence $X + Y \overset{\mathcal{D}}{=} X + Z$. However, the distributions of Y and Z do not coincide. ◊

Takeaways In order to compute characteristic functions, in many cases it is enough to have a table of characteristic functions for some repertoire of standard distributions and to know how characteristic functions transform under linear maps and independent sums. We have studied both in this section. It is of a certain theoretical interest and will be needed later that triangle functions and sums of triangle functions are characteristic functions.

Exercise 15.2.1 Let φ be the characteristic function of the d-dimensional random variable X. Assume that $\varphi(t) = 1$ for some $t \neq 0$. Show that $\mathbf{P}[X \in H_t] = 1$, where

$$
H_t = \{x \in \mathbb{R}^d : \langle x, t \rangle \in 2\pi \mathbb{Z}\}
$$
$$
= \{y + z \cdot (2\pi t/\|t\|_2^2) : z \in \mathbb{Z},\ y \in \mathbb{R}^d \text{ with } \langle y, t \rangle = 0\}.
$$

Infer that $\varphi(t + s) = \varphi(s)$ for all $s \in \mathbb{R}^d$. ♣

Exercise 15.2.2 Show that there are real random variables X, X' and Y, Y' with the properties (i) $X \overset{\mathcal{D}}{=} X'$ and $Y \overset{\mathcal{D}}{=} Y'$, (ii) X' and Y' are independent, (iii) $X + Y \overset{\mathcal{D}}{=} X' + Y'$, and (iv) X and Y are not independent. ♣

Exercise 15.2.3 Let X be a real random variable with characteristic function φ. X is called **lattice distributed** if there are $a, d \in \mathbb{R}$ such that $\mathbf{P}[X \in a + d\mathbb{Z}] = 1$. Show that X is lattice distributed if and only if there exists a $u \neq 0$ such that $|\varphi(u)| = 1$. ♣

Exercise 15.2.4 Let X be a real random variable with characteristic function φ. Assume that there is a sequence $(t_n)_{n \in \mathbb{N}}$ of real numbers such that $|t_n| \downarrow 0$ and $|\varphi(t_n)| = 1$ for any n. Show that there exists a $b \in \mathbb{R}$ such that $X = b$ almost surely. If in addition, $\varphi(t_n) = 1$ for all n, then $X = 0$ almost surely. ♣

15.3   Lévy's Continuity Theorem

The main statement of this section is Lévy's continuity theorem (Theorem 15.24). Roughly speaking, it says that a sequence of characteristic functions converges pointwise to a continuous function if and only if the limiting function is a characteristic function and the corresponding probability measures converge weakly. We prepare for the proof of this theorem by assembling some analytic tools.

Lemma 15.20 Let $\mu \in \mathcal{M}_1(\mathbb{R}^d)$ with characteristic function φ. Then

$$|\varphi(t) - \varphi(s)|^2 \le 2\big(1 - \operatorname{Re}(\varphi(t-s))\big) \quad \text{for all } s, t \in \mathbb{R}^d.$$

Proof By the Cauchy–Schwarz inequality,

$$|\varphi(t) - \varphi(s)|^2 = \left| \int_{\mathbb{R}^d} e^{i \langle t, x \rangle} - e^{i \langle s, x \rangle} \, \mu(dx) \right|^2$$

$$= \left| \int_{\mathbb{R}^d} \big(e^{i \langle t-s, x \rangle} - 1\big) e^{i \langle s, x \rangle} \, \mu(dx) \right|^2$$

$$\le \int_{\mathbb{R}^d} \big| e^{i \langle t-s, x \rangle} - 1 \big|^2 \, \mu(dx) \cdot \int_{\mathbb{R}^d} \big| e^{i \langle s, x \rangle} \big|^2 \, \mu(dx)$$

$$= \int_{\mathbb{R}^d} \big(e^{i \langle t-s, x \rangle} - 1\big)\big(e^{-i \langle t-s, x \rangle} - 1\big) \, \mu(dx)$$

$$= 2\big(1 - \operatorname{Re}(\varphi(t-s))\big). \qquad \square$$

Definition 15.21 Let (E, d) be a metric space. A family $(f_i, \ i \in I)$ of maps $E \to \mathbb{R}$ is called **uniformly equicontinuous** if, for every $\varepsilon > 0$, there exists a $\delta > 0$ such that $|f_i(t) - f_i(s)| < \varepsilon$ for all $i \in I$ and all $s, t \in E$ with $d(s, t) < \delta$.

Theorem 15.22 If $\mathcal{F} \subset \mathcal{M}_1(\mathbb{R}^d)$ is a tight family, then $\{\varphi_\mu : \ \mu \in \mathcal{F}\}$ is uniformly equicontinuous. In particular, every characteristic function is uniformly continuous.

Proof We have to show that, for every $\varepsilon > 0$, there exists a $\delta > 0$ such that, for all $t \in \mathbb{R}^d$, all $s \in \mathbb{R}^d$ with $|t - s| < \delta$ and all $\mu \in \mathcal{F}$, we have $|\varphi_\mu(t) - \varphi_\mu(s)| < \varepsilon$.

As $\mathcal{F}$ is tight, there exists an $N \in \mathbb{N}$ with $\mu([-N, N]^d) > 1 - \varepsilon^2/6$ for all $\mu \in \mathcal{F}$. Furthermore, there exists a $\delta > 0$ such that, for $x \in [-N, N]^d$ and $u \in \mathbb{R}^d$ with $|u| < \delta$, we have $\left|1 - e^{i\langle u, x \rangle}\right| < \varepsilon^2/6$. Hence we get for all $\mu \in \mathcal{F}$

$$
\begin{aligned}
1 - \operatorname{Re}(\varphi_\mu(u)) &\leq \int_{\mathbb{R}^d} \left|1 - e^{i\langle u, x \rangle}\right| \mu(dx) \\
&\leq \frac{\varepsilon^2}{3} + \int_{[-N, N]^d} \left|1 - e^{i\langle u, x \rangle}\right| \mu(dx) \\
&\leq \frac{\varepsilon^2}{3} + \frac{\varepsilon^2}{6} = \frac{\varepsilon^2}{2}.
\end{aligned}
$$

Thus, for $|t - s| < \delta$ by Lemma 15.20, $|\varphi_\mu(t) - \varphi_\mu(s)| \leq \varepsilon$. □

Lemma 15.23 *Let (E, d) be a metric space and let $f, f_1, f_2, \ldots$ be maps $E \to \mathbb{R}$ with $f_n \overset{n \to \infty}{\longrightarrow} f$ pointwise. If $(f_n)_{n \in \mathbb{N}}$ is uniformly equicontinuous, then f is uniformly continuous and $(f_n)_{n \in \mathbb{N}}$ converges to f uniformly on compact sets; that is, for every compact set $K \subset E$, we have $\sup_{s \in K} |f_n(s) - f(s)| \overset{n \to \infty}{\longrightarrow} 0$.*

Proof Fix $\varepsilon > 0$, and choose $\delta > 0$ such that $|f_n(t) - f_n(s)| < \varepsilon$ for all $n \in \mathbb{N}$ and all $s, t \in E$ with $d(s, t) < \delta$. For these s, t, we thus have

$$
|f(s) - f(t)| = \lim_{n \to \infty} |f_n(s) - f_n(t)| \leq \varepsilon.
$$

Hence, f is uniformly continuous.

Now let $K \subset E$ be compact. As compact sets are totally bounded, there exists an $N \in \mathbb{N}$ and points $t_1, \ldots, t_N \in K$ with $K \subset \bigcup_{i=1}^N B_\delta(t_i)$. Choose $n_0 \in \mathbb{N}$ large enough that $|f_n(t_i) - f(t_i)| \leq \varepsilon$ for all $i = 1, \ldots, N$ and $n \geq n_0$.

Now let $s \in K$ and $n \geq n_0$. Choose a t_i with $d(s, t_i) < \delta$. Then

$$
|f_n(s) - f(s)| \leq |f_n(s) - f_n(t_i)| + |f_n(t_i) - f(t_i)| + |f(t_i) - f(s)| \leq 3\varepsilon.
$$

As $\varepsilon > 0$ was arbitrary, we infer that $f_n \overset{n \to \infty}{\longrightarrow} f$ uniformly on K. □

A map $f : \mathbb{R}^d \to \mathbb{R}$ is called **partially continuous** at $x = (x_1, \ldots, x_d)$ if, for any $i = 1, \ldots, d$, the map $y_i \mapsto f(x_1, \ldots, x_{i-1}, y_i, x_{i+1}, \ldots, x_d)$ is continuous at $y_i = x_i$.

Theorem 15.24 (Lévy's continuity theorem) *Let $P, P_1, P_2, \ldots \in \mathcal{M}_1(\mathbb{R}^d)$ with characteristic functions $\varphi, \varphi_1, \varphi_2, \ldots$.*

(i) *If $P = \text{w-}\lim_{n \to \infty} P_n$, then $\varphi_n \overset{n \to \infty}{\longrightarrow} \varphi$ uniformly on compact sets.*

(ii) *If $\varphi_n \overset{n \to \infty}{\longrightarrow} f$ pointwise for some $f : \mathbb{R}^d \to \mathbb{C}$ that is partially continuous at 0, then there exists a probability measure Q such that $\varphi_Q = f$ and $Q = \text{w-}\lim_{n \to \infty} P_n$.*

Proof

(i) By the definition of weak convergence, we have $\varphi_n \overset{n\to\infty}{\longrightarrow} \varphi$ pointwise. As the family $(P_n)_{n\in\mathbb{N}}$ is tight, by Theorem 15.22, $(\varphi_n)_{n\in\mathbb{N}}$ is uniformly equicontinuous. By Lemma 15.23, this implies uniform convergence on compact sets.

(ii) By Theorem 13.34, it is enough to show that the sequence $(P_n)_{n\in\mathbb{N}}$ is tight. For this purpose, it suffices to show that, for every $k = 1, \ldots, n$, the sequence $(P_n^k)_{n\in\mathbb{N}}$ of kth marginal distributions is tight. (Here $P_n^k := P_n \circ \pi_k^{-1}$, where $\pi_k : \mathbb{R}^d \to \mathbb{R}$ is the projection on the kth coordinate.) Let e_k be the kth unit vector in $\mathbb{R}^d$. Then $\varphi_{P_n^k}(t) = \varphi_n(t\, e_k)$ is the characteristic function of P_n^k. By assumption, $\varphi_{P_n^k} \overset{n\to\infty}{\longrightarrow} f_k$ pointwise for some function f_k that is continuous at 0. We have thus reduced the problem to the one-dimensional situation and will henceforth assume $d = 1$.

As $\varphi_n(0) = 1$ for all $n \in \mathbb{N}$, we have $f(0) = 1$. Define the map $h : \mathbb{R} \to [0, \infty)$ by $h(x) = 1 - \sin(x)/x$ for $x \neq 0$ and $h(0) = 0$. Clearly, h is continuously differentiable on $\mathbb{R}$. It is easy to see that $\alpha := \inf\{h(x) : |x| \geq 1\} = 1 - \sin(1) > 0$. Now, for $K > 0$, compute (using Markov's inequality and Fubini's theorem)

$$P_n\big([-K, K]^c\big) \leq \alpha^{-1} \int_{[-K,K]^c} h(x/K)\, P_n(dx)$$

$$\leq \alpha^{-1} \int_{\mathbb{R}} h(x/K)\, P_n(dx)$$

$$= \alpha^{-1} \int_{\mathbb{R}} \left(\int_0^1 \big(1 - \cos(tx/K)\big)\, dt \right) P_n(dx)$$

$$= \alpha^{-1} \int_0^1 \left(\int_{\mathbb{R}} \big(1 - \cos(tx/K)\big)\, P_n(dx) \right) dt$$

$$= \alpha^{-1} \int_0^1 \big(1 - \mathrm{Re}(\varphi_n(t/K))\big)\, dt.$$

Using dominated convergence, we conclude that

$$\limsup_{n\to\infty} P_n([-K, K]^c) \leq \alpha^{-1} \limsup_{n\to\infty} \int_0^1 \big(1 - \mathrm{Re}(\varphi_n(t/K))\big)\, dt$$

$$= \alpha^{-1} \int_0^1 \left(\lim_{n\to\infty} \big(1 - \mathrm{Re}(\varphi_n(t/K))\big) \right) dt$$

$$= \alpha^{-1} \int_0^1 \big(1 - \mathrm{Re}(f(t/K))\big)\, dt.$$

As f is continuous and $f(0) = 1$, the last integral converges to 0 for $K \to \infty$. Hence $(P_n)_{n\in\mathbb{N}}$ is tight.   □

Reflection Find an example of a pointwise convergent sequence of characteristic functions (φ_n) such that the limiting function φ is not continuous (at 0). ♠

Applying Lévy's continuity theorem to Example 15.16, we get a theorem of Pólya.

Theorem 15.25 (Pólya) *Let* $f : \mathbb{R} \to [0, 1]$ *be continuous and even with* $f(0) = 1$. *Assume that* f *is convex on* $[0, \infty)$. *Then* f *is the characteristic function of a probability measure.*

Proof Define f_n by $f_n(k/n) := f(k/n)$ for $k = 0, \dots, n^2$, and assume f_n is linearly interpolated between these points. Furthermore, let f_n be constant to the right of n and for $x < 0$, define $f_n(x) = f_n(-x)$. This is an approximation of f on $[0, \infty)$ by convex and piecewise linear functions. By Example 15.16, every f_n is a characteristic function of a probability measure μ_n. Clearly, $f_n \stackrel{n\to\infty}{\longrightarrow} f$ pointwise; hence f is the characteristic function of a probability measure $\mu = \text{w-lim}_{n\to\infty} \mu_n$ on $\mathbb{R}$. □

Corollary 15.26 *For every* $\alpha \in (0, 1]$ *and* $r > 0$, $\varphi_{\alpha,r}(t) = e^{-|r\,t|^\alpha}$ *is the characteristic function of a symmetric probability measure* $\mu_{\alpha,r}$ *on* $\mathbb{R}$.

Remark 15.27 In fact, $\varphi_{\alpha,r}$ is a characteristic function for every $\alpha \in (0, 2]$ ($\alpha = 2$ corresponds to the normal distribution), see Sect. 16.2. The distributions $\mu_{\alpha,r}$ are the so-called α-**stable** distributions (see Definition 16.20): If $X_1, X_2, \dots, X_n$ are independent and $\mu_{\alpha,a}$-distributed, then $\varphi_{X_1+\dots+X_n}(t) = \varphi_X(t)^n = \varphi_X(n^{1/\alpha}t)$; hence $X_1 + \dots + X_n \stackrel{D}{=} n^{1/\alpha}X_1$. ◊

The Stone–Weierstraß theorem implies that a characteristic function determines a probability distribution uniquely. Pólya's theorem gives a sufficient condition for a symmetric real function to be a characteristic function. Clearly, that condition is not necessary, as, for example, the normal distribution does not fulfill it. For general education we present Bochner's theorem that formulates a necessary and sufficient condition for a function $\varphi : \mathbb{R}^d \to \mathbb{C}$ to be the characteristic function of a probability measure.

Definition 15.28 *A function* $f : \mathbb{R}^d \to \mathbb{C}$ *is called **positive semidefinite** if, for all* $n \in \mathbb{N}$, *all* $t_1, \dots, t_n \in \mathbb{R}^d$ *and all* $y_1, \dots, y_n \in \mathbb{C}$, *we have*

$$\sum_{k,l=1}^{n} y_k \, \bar{y}_l \, f(t_k - t_l) \geq 0,$$

in other words, if the matrix $(f(t_k - t_l))_{k,l=1,\dots,n}$ *is positive semidefinite.*

Lemma 15.29 *If* $\mu \in \mathcal{M}_f(\mathbb{R}^d)$ *has characteristic function* φ, *then* φ *is positive semidefinite.*

Proof We have

$$
\sum_{k,l=1}^{n} y_k \, \bar{y}_l \, \varphi(t_k - t_l) = \sum_{k,l=1}^{n} y_k \, \bar{y}_l \int e^{ix(t_k - t_l)} \, \mu(dx)
$$

$$
= \int \sum_{k,l=1}^{n} y_k \, e^{ixt_k} \, \overline{y_l \, e^{ixt_l}} \, \mu(dx)
$$

$$
= \int \left| \sum_{k=1}^{n} y_k \, e^{ixt_k} \right|^2 \mu(dx) \geq 0. \qquad \square
$$

In the case $d = 1$, the following theorem goes back to Bochner (1932).

Theorem 15.30 (Bochner) *A continuous function* $\varphi : \mathbb{R}^d \to \mathbb{C}$ *is the characteristic function of a probability distribution on* $\mathbb{R}^d$ *if and only if* φ *is positive semidefinite and* $\varphi(0) = 1$.

The statement still holds if $\mathbb{R}^d$ *is replaced by a locally compact Abelian group.*

Proof For the case $d = 1$ see [19, §20, Theorem 23] or [54, Chapter XIX.2, page 622]. For the general case, see, e.g., [71, page 293, Theorem 33.3]. $\qquad \square$

> **Takeaways** In order to check weak convergence of a sequence of probability measures, it is enough to show tightness and pointwise convergence of the characteristic functions. If the limiting function is continuous at 0, then by Levy's theorem, tightness and hence weak convergence are automatic.

Exercise 15.3.1 (Compare [50] and [4]) Show that there exist two exchangeable sequences $X = (X_n)_{n\in\mathbb{N}}$ and $Y = (Y_n)_{n\in\mathbb{N}}$ of real random variables with $\mathbf{P}_X \neq \mathbf{P}_Y$ but such that

$$
\sum_{k=1}^{n} X_k \overset{\mathcal{D}}{=} \sum_{k=1}^{n} Y_k \quad \text{for all } n \in \mathbb{N}. \tag{15.4}
$$

Hint:

(i) Define the characteristic functions (see Theorem 15.13) $\varphi_1(t) = \frac{1}{1+t^2}$ and $\varphi_2(t) = (1 - t/2)^+$. Use Pólya's theorem to show that

$$
\psi_1(t) := \begin{cases} \varphi_1(t), & \text{if } |t| \leq 1, \\ \varphi_2(t), & \text{if } |t| > 1, \end{cases}
$$

and

$$\psi_2(t) := \begin{cases} \varphi_2(t), & \text{if } |t| \leq 1, \\ \varphi_1(t), & \text{if } |t| > 1, \end{cases}$$

are characteristic functions of probability distributions on $\mathbb{R}$.

(ii) Define independent random variables $X_{n,i}$, $Y_{n,i}$, $n \in \mathbb{N}$, $i = 1, 2$, and Θ_n, $n \in \mathbb{N}$ such that $X_{n,i}$ has characteristic function φ_i, $Y_{n,i}$ has characteristic function ψ_i and $\mathbf{P}[\Theta_n = 1] = \mathbf{P}[\Theta_n = -1] = \frac{1}{2}$. Define $X_n = X_{n,\Theta_n}$ and $Y_n = Y_{n,\Theta_n}$. Show that (15.4) holds.

(iii) Determine $\mathbf{E}[e^{i t_1 X_1 + i t_2 X_2}]$ and $\mathbf{E}[e^{i t_1 Y_1 + i t_2 Y_2}]$ for $t_1 = \frac{1}{2}$ and $t_2 = 2$. Conclude that $(X_1, X_2) \overset{\mathcal{D}}{\neq} (Y_1, Y_2)$ and thus $\mathbf{P}_X \neq \mathbf{P}_Y$. ♣

Exercise 15.3.2 Show that for any $\delta > 0$ and $\varepsilon > 0$, there is a $C < \infty$ such that for any $\mu \in \mathcal{M}_1(\mathbb{R})$ with characteristic function φ, we have

$$\mu([-\delta, \delta]^c) \leq C \int_0^\varepsilon (1 - \mathrm{Re}(\varphi(t)))\, dt.$$

For $\varepsilon\delta \leq 3$ one can choose $C = 12/\delta^2\varepsilon^3$. *Hint:* Proceed as in the proof of Lévy's continuity theorem. ♣

Exercise 15.3.3 Let $(\mu_n)_{n\in\mathbb{N}}$ be a sequence of probability measures on $\mathbb{R}$ and denote by $(\varphi_n)_{n\in\mathbb{N}}$ the corresponding characteristic functions. Assume that we have $\varphi_n(t) \overset{n\to\infty}{\longrightarrow} 1$ for t in a neighborhood of 0. Use Exercise 15.3.2 to show that $\mu_n \overset{n\to\infty}{\longrightarrow} \delta_0$. ♣

Exercise 15.3.4 (Continuity theorem for Laplace transforms) Let $(\mu_n)_{n\in\mathbb{N}}$ be a sequence of probability measures on $[0, \infty)$ and let

$$\psi_n(t) = \int e^{-tx} \mu_n(dx) \quad \text{for } t \geq 0, \quad n \in \mathbb{N}$$

be the Laplace transforms. We assume that there exists a map $\psi : [0, \infty) \to [0, 1]$ that is continuous in 0 and such that $\psi_n \overset{n\to\infty}{\longrightarrow} \psi$ pointwise.

Show that ψ is the Laplace transform of a probability measure μ on $[0, \infty)$ and that $\mu_n \overset{n\to\infty}{\longrightarrow} \mu$ weakly. ♣

15.4 Characteristic Functions and Moments

We want to study the connection between the moments of a real random variable X and the derivatives of its characteristic function φ_X. We start with a simple lemma.

Lemma 15.31 *For $t \in \mathbb{R}$ and $n \in \mathbb{N}$, we have*

$$\left| e^{it} - 1 - \frac{it}{1!} - \ldots - \frac{(it)^{n-1}}{(n-1)!} \right| \leq \frac{|t|^n}{n!}.$$

Proof As the nth derivative of e^{it} has modulus 1, this follows by Taylor's formula.
□

Theorem 15.32 (Moments and differentiability) *Let X be a real random variable with characteristic function φ.*

(i) *If $\mathbf{E}[|X|^n] < \infty$, then φ is n-times continuously differentiable with derivatives*

$$\varphi^{(k)}(t) = \mathbf{E}\big[(iX)^k e^{itX} \big] \quad \text{for } k = 0, \ldots, n.$$

(ii) *In particular, if $\mathbf{E}[X^2] < \infty$, then*

$$\varphi(t) = 1 + it\, \mathbf{E}[X] - \frac{1}{2} t^2\, \mathbf{E}[X^2] + \varepsilon(t)\, t^2$$

with $\varepsilon(t) \to 0$ for $t \to 0$.

(iii) *Let $h \in \mathbb{R}$. If $\lim\limits_{n \to \infty} \frac{|h|^n \mathbf{E}[|X|^n]}{n!} = 0$, then, for every $t \in \mathbb{R}$,*

$$\varphi(t + h) = \sum_{k=0}^{\infty} \frac{(ih)^k}{k!} \mathbf{E}\Big[e^{itX} X^k \Big].$$

In particular, this holds if $\mathbf{E}\big[e^{|hX|} \big] < \infty$.

Proof

(i) For $t \in \mathbb{R}$, $h \in \mathbb{R} \setminus \{0\}$ and $k \in \{1, \ldots, n\}$, define

$$Y_k(t, h, x) = k!\, h^{-k} e^{itx} \left(e^{ihx} - \sum_{l=0}^{k-1} \frac{(ihx)^l}{l!} \right).$$

Then

$$\mathbf{E}[Y_k(t, h, X)] = k!\, h^{-k} \left(\varphi(t + h) - \varphi(t) - \sum_{l=1}^{k-1} \mathbf{E}\big[e^{itX} (iX)^l \big] \frac{h^l}{l!} \right).$$

If the limit $\varphi_k(t) := \lim_{h \to 0} \mathbf{E}[Y_k(t, h, X)]$ exists, then φ is k-times differentiable at t with $\varphi^{(k)}(t) = \varphi_k(t)$.

However (by Lemma 15.31 with $n = k + 1$), $Y_k(t, h, x) \xrightarrow{h \to 0} (ix)^k e^{itx}$ for all $x \in \mathbb{R}$ and (by Lemma 15.31 with $n = k$) $|Y_k(t, h, x)| \leq |x|^k$. As

$\mathbf{E}[|X|^k] < \infty$ by assumption, the dominated convergence theorem implies

$$\mathbf{E}[Y_k(t, h, X)] \xrightarrow{h \to 0} \mathbf{E}[(iX)^k e^{itX}] = \varphi^{(k)}(t).$$

Applying the continuity lemma (Theorem 6.27) yields that $\varphi^{(k)}$ is continuous.

(ii) This is a direct consequence of (i).

(iii) By assumption,

$$\left| \varphi(t+h) - \sum_{k=0}^{n-1} \frac{(ih)^k}{k!} \mathbf{E}[e^{itX} X^k] \right| = \frac{|h|^n}{n!} |\mathbf{E}[Y_n(t, h, X)]|$$

$$\leq \frac{|h|^n \, \mathbf{E}[|X|^n]}{n!} \xrightarrow{n \to \infty} 0. \qquad \square$$

Corollary 15.33 (Method of moments) *Let X be a real random variable with*

$$\alpha := \limsup_{n \to \infty} \frac{1}{n} \mathbf{E}[|X|^n]^{1/n} < \infty.$$

Then the characteristic function φ of X is analytic and the distribution of X is uniquely determined by the moments $\mathbf{E}[X^n]$, $n \in \mathbb{N}$. In particular, this holds if $\mathbf{E}[e^{t|X|}] < \infty$ for some $t > 0$.

Proof By Stirling's formula,

$$\lim_{n \to \infty} \frac{1}{n!} n^n e^{-n} \sqrt{2\pi n} = 1.$$

Thus, for $|h| < 1/(3\alpha)$,

$$\limsup_{n \to \infty} \mathbf{E}[|X|^n] \cdot |h|^n/n! = \limsup_{n \to \infty} (2\pi n)^{-1/2} \left(\mathbf{E}[|X|^n]^{1/n} \cdot |h| \cdot e/n \right)^n$$

$$\leq \limsup_{n \to \infty} (2\pi n)^{-1/2} (e/3)^n = 0.$$

Hence the characteristic function can be expanded about any point $t \in \mathbb{R}$ in a power series with radius of convergence at least $1/(3\alpha)$. In particular, it is analytic and is hence determined by the coefficients of its power series about $t = 0$; that is, by the moments of X. $\qquad \square$

Example 15.34

(i) Let $X \sim \mathcal{N}_{\mu,\sigma^2}$. Then, for every $t \in \mathbb{R}$,

$$\mathbf{E}[e^{tX}] = (2\pi\sigma^2)^{-1/2} \int_{-\infty}^{\infty} e^{tx} e^{-(x-\mu)^2/2\sigma^2}\, dx$$

$$= e^{\mu t + t^2\sigma^2/2}(2\pi\sigma^2)^{-1/2} \int_{-\infty}^{\infty} e^{-(x-\mu-t\sigma^2)^2/2\sigma^2}\, dx$$

$$= e^{\mu t + t^2\sigma^2/2} < \infty.$$

Hence the distribution of X is characterized by its moments. The characteristic function $\varphi(t) = e^{i\mu t}\, e^{-\sigma^2 t^2/2}$ that we get by the above calculation with t replaced by it is indeed analytic. (ii) Let X be exponentially distributed with parameter $\theta > 0$. Then, for $t \in (0, \theta)$,

$$\mathbf{E}[e^{tX}] = \theta \int_{0}^{\infty} e^{tx}\, e^{-\theta x}\, dx = \frac{\theta}{\theta - t} < \infty.$$

Hence the distribution of X is characterized by its moments. The above calculation with t replaced by it yields $\varphi(t) = \theta/(\theta - it)$, and this function is indeed analytic. The fact that in the complex plane φ has a singularity at $t = -i\theta$ implies that the power series of φ about 0 has radius of convergence θ. In particular, this implies that not all exponential moments are finite. This is reflected by the above calculation that shows that, for $t \geq \theta$, the exponential moments are infinite.

(iii) Let X be log-normally distributed (see Example 15.5). Then $\mathbf{E}[X^n] = e^{n^2/2}$. In particular, here $\alpha = \infty$. In fact, in Example 15.5, we saw that here the moments do not determine the distribution of X.

(iv) If X takes values in $\mathbb{N}_0$ and if $\beta := \limsup_{n\to\infty} \mathbf{E}[X^n]^{1/n} < 1$, then by Hadamard's criterion $\psi_X(z) := \sum_{k=1}^{\infty} \mathbf{P}[X = k]\, z^k < \infty$ for $|z| < 1/\beta$. In particular, the probability generating function X is characterized by its derivatives $\psi_X^{(n)}(1)$, $n \in \mathbb{N}$, and thus by the moments of X. Compare Theorem 3.2(iii). $\Diamond$

Theorem 15.35 *Let X be a real random variable and let φ be its characteristic function. Let $n \in \mathbb{N}$, and assume that φ is $2n$-times differentiable at 0 with derivative $\varphi^{(2n)}(0)$. Then $\mathbf{E}[X^{2n}] = (-1)^n\, \varphi^{(2n)}(0) < \infty$.*

Proof We carry out the proof by induction on $n \in \mathbb{N}_0$. For $n = 0$, the claim is trivially true. Now, let $n \in \mathbb{N}$, and assume φ is $2n$-times (not necessarily continuously) differentiable at 0. Define $u(t) = \mathrm{Re}(\varphi(t))$. Then u is also $2n$-times differentiable at 0 and $u^{(2k-1)}(0) = 0$ for $k = 1, \ldots, n$ since u is even. Since $\varphi^{(2n)}(0)$ exists, $\varphi^{(2n-1)}$ is continuous at 0 and $\varphi^{(2n-1)}(t)$ exists for all $t \in (-\varepsilon, \varepsilon)$ for some $\varepsilon > 0$. Furthermore, $\varphi^{(k)}$ exists in $(-\varepsilon, \varepsilon)$ and is continuous on $(-\varepsilon, \varepsilon)$ for any $k = 0, \ldots, 2n - 2$. By Taylor's formula, for every $t \in (-\varepsilon, \varepsilon)$,

$$
\left| u(t) - \sum_{k=0}^{n-1} u^{(2k)}(0) \frac{t^{2k}}{(2k)!} \right| \leq \frac{|t|^{2n-1}}{(2n-1)!} \sup_{\theta \in (0,1]} \left| u^{(2n-1)}(\theta t) \right|. \tag{15.5}
$$

Define a continuous function $f_n : \mathbb{R} \to [0, \infty)$ by $f_n(0) = 1$ and

$$
f_n(x) = (-1)^n (2n)! \, x^{-2n} \left[\cos(x) - \sum_{k=0}^{n-1} (-1)^k \frac{x^{2k}}{(2k)!} \right] \quad \text{for } x \neq 0.
$$

By the induction hypothesis, $E[X^{2k}] = (-1)^k u^{(2k)}(0)$ for all $k = 1, \ldots, n-1$. Using (15.5), we infer

$$
E\big[f_n(tX) X^{2n}\big] \leq \frac{2n}{|t|} \sup_{\theta \in (0,1]} |u^{(2n-1)}(\theta t)| \leq g_n(t) := 2n \sup_{\theta \in (0,1]} \frac{|u^{(2n-1)}(\theta t)|}{\theta \, |t|}.
$$

Now Fatou's lemma implies

$$
E\big[X^{2n}\big] = E\big[f_n(0) X^{2n}\big] \leq \liminf_{t \to 0} E\big[f_n(tX) X^{2n}\big]
$$

$$
\leq \liminf_{t \to 0} g_n(t) = 2n \left| u^{(2n)}(0) \right| < \infty.
$$

By Theorem 15.32, this implies $E[X^{2n}] = (-1)^n u^{(2n)}(0) = (-1)^n \varphi^{(2n)}(0)$.   □

Remark 15.36 For odd moments, the statement of the theorem may fail (see, e.g., Exercise 15.4.4 for the first moment). Indeed, φ is differentiable at 0 with derivative $i\,m$ for some $m \in \mathbb{R}$ if and only if $x \, P[|X| > x] \xrightarrow{x \to \infty} 0$ and $E[X \, \mathbb{1}_{\{|X| \leq x\}}] \xrightarrow{x \to \infty} m$. (See [54, Chapter XVII.2a, page 565].) ◊

Takeaways A random variable with finite nth moment possesses a characteristic function that is n-times differentiable. The moments can be read off from the derivatives at 0. If all moments exist and do not grow too quickly, then the moments determine the distribution.

Exercise 15.4.1 Let X and Y be nonnegative random variables with

$$
\limsup_{n \to \infty} \frac{1}{n} E[|X|^n]^{1/n} < \infty, \qquad \limsup_{n \to \infty} \frac{1}{n} E[|Y|^n]^{1/n} < \infty,
$$

and

$$E[X^m Y^n] = E[X^m]E[Y^n] \quad \text{for all } m, n \in \mathbb{N}_0.$$

Show that X and Y are independent.

Hint: Consider the random variable Y with respect to the probability measure $X^m P[\cdot]/E[X^m]$, and use Corollary 15.33 to show that

$$E[X^m \mathbb{1}_A(Y)]/E[X^m] = P[Y \in A] \quad \text{for all } A \in \mathcal{B}(R) \text{ and } m \in \mathbb{N}_0.$$

Now apply Corollary 15.33 to the random variable X with respect to the probability measure $P[\cdot \mid Y \in A]$. ♣

Exercise 15.4.2 Let $r, s > 0$ and let $Z \sim \Gamma_{1,r+s}$ and $B \sim \beta_{r,s}$ be independent (see Example 1.107). Use Exercise 15.4.1 to show that the random variables $X := BZ$ and $Y := (1 - B)Z$ are independent with $X \sim \Gamma_{1,r}$ and $Y \sim \Gamma_{1,s}$. ♣

Exercise 15.4.3 Show that, for $\alpha > 2$, the function $\phi_\alpha(t) = e^{-|t|^\alpha}$ is not a characteristic function.

(*Hint:* Assume the contrary and show that the corresponding random variable would have variance zero.) ♣

Exercise 15.4.4 Let $X_1, X_2, \ldots$ be i.i.d. real random variables with characteristic function φ. Show the following.

(i) If φ is differentiable at 0, then $\varphi'(0) = i\,m$ for some $m \in \mathbb{R}$.
(ii) φ is differentiable at 0 with $\varphi'(0) = i\,m$ if and only if $(X_1 + \ldots + X_n)/n \overset{n \to \infty}{\longrightarrow} m$ in probability.
(iii) Assume that φ is differentiable at 0 and that $X_1 \geq 0$ almost surely. Then $E[X_1] = -i\,\varphi'(0) < \infty$. *Hint:* Use (ii) and the law of large numbers.
(iv) The distribution of X_1 can be chosen such that φ is differentiable at 0 but $E[|X_1|] = \infty$. ♣

Exercise 15.4.5 Let $X_1, X_2, \ldots$ be real random variables. For $r > 0$ let $M_r(X_n) = E[|X_n|^r]$ be the rth absolute moment. For $k \in \mathbb{N}$ let $m_k(X_n) = E[X_n^k]$ be the kth moment if $M_k(X_n) < \infty$.

(i) Assume that X is a real random variable and that $(X_{n_l})_{l \in \mathbb{N}}$ is a subsequence such that

$$P_{X_{n_l}} \overset{l \to \infty}{\Longrightarrow} P_X \quad \text{weakly.}$$

Assume further that there is an $r > 0$ such that $\sup_{n \in \mathbb{N}} M_r(X_n) < \infty$. Show that for any $k \in \mathbb{N} \cap (0, r)$ and $s \in (0, r)$ we have $M_s(X) < \infty$ as well as

$$M_s(X_{n_l}) \overset{l \to \infty}{\longrightarrow} M_s(X) \quad \text{and} \quad m_k(X_{n_l}) \overset{l \to \infty}{\longrightarrow} m_k(X).$$

(ii) Assume that for any $k \in \mathbb{N}$ the limit

$$m_k := \lim_{n \to \infty} m_k(X_n)$$

exists and is finite (note that finitely many of the $m_k(X_n)$ may be undefined for any k.) Show that there exists a real random variable X with $m_k = m_k(X)$ for all $k \in \mathbb{N}$ and a subsequence $(X_{n_l})_{l \in \mathbb{N}}$ such that

$$\mathbf{P}_{X_{n_l}} \overset{l \to \infty}{\longrightarrow} \mathbf{P}_X \text{ weakly.}$$

(iii) Show the theorem of Fréchet–Shohat: If in (ii) the distribution of X is determined by its moments $m_k(X)$, $k \in \mathbb{N}$ (see Corollary 15.33), then

$$\mathbf{P}_{X_n} \overset{n \to \infty}{\longrightarrow} \mathbf{P}_X \text{ weakly.} \quad \clubsuit$$

Exercise 15.4.6 Let $X_1, X_2, \ldots$ be i.i.d. real random variables with $\mathbf{E}[X_1] = 0$ and $\mathbf{E}[|X_1|^k] < \infty$ for all $k \in \mathbb{N}$.

(i) Show that there exist finite numbers $(d_k)_{k \in \mathbb{N}}$ (depending on the distribution $\mathbf{P}_{X_1}$) such that for any $k, n \in \mathbb{N}$ we have

$$\left| \mathbf{E}\left[(X_1 + \ldots + X_n)^{2k-1} \right] \right| \leq d_{2k-1} \, n^{k-1}$$

and

$$\left| \mathbf{E}\left[(X_1 + \ldots + X_n)^{2k} \right] - \frac{(2k)!}{2^k \, k!} \mathbf{E}\left[X_1^2 \right]^k n^k \right| \leq d_{2k} \, n^{k-1}.$$

Hint: Expand the bracket expression, sort the terms by the different mixed moments and compute by combinatorial means the number of each type of summand. The number of summands of the type $\mathbf{E}[X_{l_1}^2 \cdots X_{l_k}^2]$ (for different $l_1, \ldots, l_k$) is of particular importance.

(ii) Let $Y \sim \mathcal{N}_{0,1}$. Use Theorem 15.32(i) to show that for any $k \in \mathbb{N}$ we have

$$\mathbf{E}\left[Y^{2k-1} \right] = 0 \quad \text{and} \quad \mathbf{E}\left[Y^{2k} \right] = \frac{(2k)!}{2^k \, k!}.$$

(iii) Let $S_n^* = (X_1 + \ldots + X_n)/\sqrt{n \, \mathbf{Var}[X_1]}$. Use Exercise 15.4.5 to infer the statement of the central limit theorem (compare Theorem 15.38)

$$\mathbf{P}_{S_n^*} \overset{n \to \infty}{\longrightarrow} \mathcal{N}_{0,1} \text{ weakly.} \quad \clubsuit$$

15.5   The Central Limit Theorem

In the strong law of large numbers, we saw that, for large n, the order of magnitude of the sum $S_n = X_1 + \ldots + X_n$ of i.i.d. integrable random variables is $n \cdot \mathbf{E}[X_1]$. Of course, for any n, the actual value of S_n will sometimes be smaller than $n \cdot \mathbf{E}[X_1]$ and sometimes larger. In the central limit theorem (CLT), we study the size and shape of the *typical fluctuations* around $n \cdot \mathbf{E}[X_1]$ in the case where the X_i have a finite variance.

We prepare for the proof of the CLT with a lemma.

Lemma 15.37 *Let $X_1, X_2, \ldots$ be i.i.d. real random variables with $\mathbf{E}[X_1] = \mu$ and $\mathrm{Var}[X_1] = \sigma^2 \in (0, \infty)$. Let*

$$
S_n^* := \frac{1}{\sqrt{n\sigma^2}} \sum_{k=1}^{n} (X_k - \mu)
$$

be the normalized nth partial sum. Then

$$
\lim_{n \to \infty} \varphi_{S_n^*}(t) = e^{-t^2/2} \quad \text{for all } t \in \mathbb{R}.
$$

Proof Let $\varphi = \varphi_{X_k - \mu}$. Then, by Theorem 15.32(ii),

$$
\varphi(t) = 1 - \frac{\sigma^2}{2} t^2 + \varepsilon(t) t^2,
$$

where the error term $\varepsilon(t)$ goes to 0 if $t \to 0$. By Lemma 15.12(iv) and (ii),

$$
\varphi_{S_n^*}(t) = \varphi\left(\frac{t}{\sqrt{n\sigma^2}}\right)^n.
$$

Now $\left(1 - \frac{t^2}{2n}\right)^n \overset{n \to \infty}{\longrightarrow} e^{-t^2/2}$ and

$$
\left| \left(1 - \frac{t^2}{2n}\right)^n - \varphi\left(\frac{t}{\sqrt{n\sigma^2}}\right)^n \right| \leq n \left| 1 - \frac{t^2}{2n} - \varphi\left(\frac{t}{\sqrt{n\sigma^2}}\right) \right|
$$

$$
\leq n \frac{t^2}{n\sigma^2} \left| \varepsilon\left(\frac{t}{\sqrt{n\sigma^2}}\right) \right| \overset{n \to \infty}{\longrightarrow} 0.
$$

(Note that $|u^n - v^n| \leq |u - v| \cdot n \cdot \max(|u|, |v|)^{n-1}$ for all $u, v \in \mathbb{C}$.) □

Theorem 15.38 (Central limit theorem (CLT)) *Let $X_1, X_2, \ldots$ be i.i.d. real random variables with $\mu := \mathbf{E}[X_1] \in \mathbb{R}$ and $\sigma^2 := \mathbf{Var}[X_1] \in (0, \infty)$. For $n \in \mathbb{N}$, let $S_n^* := \frac{1}{\sqrt{\sigma^2 n}} \sum_{i=1}^{n} (X_i - \mu)$. Then*

$$\mathbf{P}_{S_n^*} \overset{n \to \infty}{\longrightarrow} \mathcal{N}_{0,1} \quad \text{weakly.}$$

For $-\infty \le a < b \le +\infty$, we have $\lim_{n \to \infty} \mathbf{P}[S_n^* \in [a, b]] = \frac{1}{\sqrt{2\pi}} \int_a^b e^{-x^2/2} \, dx$.

Proof By Lemma 15.37 and Lévy's continuity theorem (Theorem 15.24), $(\mathbf{P}_{S_n^*})$ converges to the distribution with characteristic function $\varphi(t) = e^{-t^2/2}$. By Theorem 15.13(i), this is $\mathcal{N}_{0,1}$. The additional claim follows by the Portemanteau theorem (Theorem 13.16) since $\mathcal{N}_{0,1}$ has a density; hence $\mathcal{N}_{0,1}(\partial[a, b]) = 0$. □

Remark 15.39 If we prefer to avoid the continuity theorem, we could argue as follows: For every $K > 0$ and $n \in \mathbb{N}$, we have $\mathbf{P}[|S_n^*| > K] \le \mathbf{Var}[S_n^*]/K^2 = 1/K^2$; hence the sequence $(\mathbf{P}_{S_n^*})$ is tight. As characteristic functions determine distributions, the claim follows by Theorem 13.34. ◊

We want to weaken the assumption in Theorem 15.38 that the random variables are identically distributed. In fact, we can even take a different set of summands for every n. The essential assumptions are that the summands are independent, each summand contributes only a little to the sum and the sum is centered and has variance 1.

Definition 15.40 *For every $n \in \mathbb{N}$, let $k_n \in \mathbb{N}$ and let $X_{n,1}, \ldots, X_{n,k_n}$ be real random variables. We say that $(X_{n,l}) = (X_{n,l}, \, l = 1, \ldots, k_n, \, n \in \mathbb{N})$ is an **array of random variables**. Its row sum is denoted by $S_n = X_{n,1} + \ldots + X_{n,k_n}$. The array is called*

- *independent if, for every $n \in \mathbb{N}$, the family $(X_{n,l})_{l=1,\ldots,k_n}$ is independent,*
- *centered if $X_{n,l} \in \mathcal{L}^1(\mathbf{P})$ and $\mathbf{E}[X_{n,l}] = 0$ for all n and l, and*
- *normed if $X_{n,l} \in \mathcal{L}^2(\mathbf{P})$ and $\sum_{l=1}^{k_n} \mathbf{Var}[X_{n,l}] = 1$ for all $n \in \mathbb{N}$.*

*A centered array is called a **null array** if its individual components are asymptotically negligible in the sense that, for all $\varepsilon > 0$,*

$$\lim_{n \to \infty} \max_{1 \le l \le k_n} \mathbf{P}[|X_{n,l}| > \varepsilon] = 0.$$

Definition 15.41 *A centered array of random variables $(X_{n,l})$ with $X_{n,l} \in \mathcal{L}^2(\mathbf{P})$ for every $n \in \mathbb{N}$ and $l = 1, \ldots, k_n$ is said to satisfy the **Lindeberg condition** if, for all $\varepsilon > 0$,*

$$L_n(\varepsilon) := \frac{1}{\mathbf{Var}[S_n]} \sum_{l=1}^{k_n} \mathbf{E}\left[X_{n,l}^2 \, \mathbb{1}_{\left\{ X_{n,l}^2 > \varepsilon^2 \, \mathbf{Var}[S_n] \right\}} \right] \overset{n \to \infty}{\longrightarrow} 0. \tag{15.6}$$

*The array fulfills the **Lyapunov condition** if there exists a $\delta > 0$ such that*

$$\lim_{n \to \infty} \frac{1}{\mathbf{Var}[S_n]^{1+(\delta/2)}} \sum_{l=1}^{k_n} \mathbf{E}\big[|X_{n,l}|^{2+\delta}\big] = 0. \tag{15.7}$$

Lemma 15.42 *The Lyapunov condition implies the Lindeberg condition.*

Proof For $x \in \mathbb{R}$, we have $x^2 \mathbb{1}_{\{|x|>\varepsilon'\}} \leq (\varepsilon')^{-\delta} |x|^{2+\delta} \mathbb{1}_{\{|x|>\varepsilon'\}} \leq (\varepsilon')^{-\delta} |x|^{2+\delta}$. Letting $\varepsilon' := \varepsilon \sqrt{\mathbf{Var}[S_n]}$, we get

$$L_n(\varepsilon) \leq \varepsilon^{-\delta} \frac{1}{\mathbf{Var}[S_n]^{1+(\delta/2)}} \sum_{l=1}^{k_n} \mathbf{E}\big[|X_{n,l}|^{2+\delta}\big].$$

$\square$

Example 15.43 Let $(Y_n)_{n \in \mathbb{N}}$ be i.i.d. with $\mathbf{E}[Y_n] = 0$ and $\mathbf{Var}[Y_n] = 1$. Let $k_n = n$ and $X_{n,l} = \frac{Y_l}{\sqrt{n}}$. Then $(X_{n,l})$ is independent, centered and normed. Clearly, $\mathbf{P}[|X_{n,l}| > \varepsilon] = \mathbf{P}[|Y_1| > \sqrt{\varepsilon n}] \overset{n \to \infty}{\longrightarrow} 0$; hence $(X_{n,l})$ is a null array. Furthermore, $L_n(\varepsilon) = \mathbf{E}\big[Y_1^2 \mathbb{1}_{\{|Y_1|>\varepsilon\sqrt{n}\}}\big] \overset{n \to \infty}{\longrightarrow} 0$; hence $(X_{n,l})$ satisfies the Lindeberg condition.
 If $Y_1 \in \mathcal{L}^{2+\delta}(\mathbf{P})$ for some $\delta > 0$, then

$$\sum_{l=1}^{n} \mathbf{E}\big[|X_{n,l}|^{2+\delta}\big] = n^{-(\delta/2)} \mathbf{E}\big[|Y_1|^{2+\delta}\big] \overset{n \to \infty}{\longrightarrow} 0.$$

In this case, $(X_{n,l})$ also satisfies the Lyapunov condition. $\Diamond$

The following theorem is due to Lindeberg (1922, see [108]) for the implication (i) $\Longrightarrow$ (ii) and is attributed to Feller (1935 and 1937, see [51, 52]) for the converse implication (ii) $\Longrightarrow$ (i). As most applications only need (i) $\Longrightarrow$ (ii), we only prove that implication. For a proof of (ii) $\Longrightarrow$ (i) see, e.g., [155, Theorem III.4.3].

Theorem 15.44 (Central limit theorem of Lindeberg–Feller) *Let $(X_{n,l})$ be an independent centered and normed array of real random variables. For every $n \in \mathbb{N}$, let $S_n = X_{n,1} + \ldots + X_{n,k_n}$. Then the following are equivalent.*

 (i) The Lindeberg condition holds.
 (ii) $(X_{n,l})$ is a null array and $\mathbf{P}_{S_n} \overset{n \to \infty}{\longrightarrow} \mathcal{N}_{0,1}$.

We prepare for the proof of Lindeberg's theorem with a couple of lemmas.

Lemma 15.45 *If (i) of Theorem 15.44 holds, then $(X_{n,l})$ is a null array.*

Proof For $\varepsilon > 0$, by Chebyshev's inequality,

$$\sum_{l=1}^{k_n} \mathbf{P}\big[|X_{n,l}| > \varepsilon\big] \leq \varepsilon^{-2} \sum_{l=1}^{k_n} \mathbf{E}\big[X_{n,l}^2 \mathbb{1}_{\{|X_{n,l}|>\varepsilon\}}\big] = \varepsilon^{-2} L_n(\varepsilon) \overset{n \to \infty}{\longrightarrow} 0. \qquad \square$$

In the following, $\varphi_{n,l}$ and φ_n will always denote the characteristic functions of $X_{n,l}$ and S_n.

Lemma 15.46 *For every $n \in \mathbb{N}$ and $t \in \mathbb{R}$, we have* $\displaystyle\sum_{l=1}^{k_n}\left|1 - \varphi_{n,l}(t)\right| \le \frac{t^2}{2}.$

Proof For every $x \in \mathbb{R}$, we have $\left|e^{itx} - 1 - itx\right| \le \frac{t^2 x^2}{2}$. Since $\mathbf{E}[X_{n,l}] = 0$,

$$\sum_{l=1}^{k_n}\left|\varphi_{n,l}(t) - 1\right| = \sum_{l=1}^{k_n}\left|\mathbf{E}[e^{itX_{n,l}} - 1]\right|$$

$$\le \sum_{l=1}^{k_n}\mathbf{E}\left[\left|e^{itX_{n,l}} - itX_{n,l} - 1\right|\right] + \left|\mathbf{E}[itX_{n,l}]\right|$$

$$\le \sum_{l=1}^{k_n}\frac{t^2}{2}\mathbf{E}[X_{n,l}^2] = \frac{t^2}{2}.$$

$\square$

Lemma 15.47 *If (i) of Theorem 15.44 holds, then*

$$\lim_{n\to\infty}\left|\log\varphi_n(t) - \sum_{l=1}^{k_n}\mathbf{E}[e^{itX_{n,l}} - 1]\right| = 0.$$

Proof Let $m_n := \max_{l=1,\dots,k_n}\left|\varphi_{n,l}(t) - 1\right|$. Note that, for all $\varepsilon > 0$,

$$\left|e^{itx} - 1\right| \le \begin{cases} 2x^2/\varepsilon^2, & \text{if } |x| > \varepsilon, \\ \varepsilon|t|, & \text{if } |x| \le \varepsilon. \end{cases}$$

This implies

$$\left|\varphi_{n,l}(t) - 1\right| \le \mathbf{E}\left[\left|e^{itX_{n,l}} - 1\right|\mathbb{1}_{\{|X_{n,l}|\le\varepsilon\}}\right] + \mathbf{E}\left[\left|e^{itX_{n,l}} - 1\right|\mathbb{1}_{\{|X_{n,l}|>\varepsilon\}}\right]$$

$$\le \varepsilon t + 2\varepsilon^{-2}\mathbf{E}\left[X_{n,l}^2\mathbb{1}_{\{|X_{n,l}|>\varepsilon\}}\right].$$

Hence, for all $\varepsilon > 0$,

$$\limsup_{n\to\infty} m_n \le \limsup_{n\to\infty}\left(\varepsilon t + 2\varepsilon^{-2}L_n(\varepsilon)\right) = \varepsilon t,$$

and thus $\lim_{n\to\infty} m_n = 0$. Now $|\log(x) - (x-1)| \leq |x-1|^2$ for all $x \in \mathbb{C}$ with $|x-1| \leq \frac{1}{2}$. If n is sufficiently large that $m_n < \frac{1}{2}$, then

$$\left| \log \varphi_n(t) - \sum_{l=1}^{k_n} \mathbf{E}[e^{itX_{n,l}} - 1] \right| = \left| \sum_{l=1}^{k_n} \left(\log(\varphi_{n,l}(t)) - (\varphi_{n,l}(t) - 1) \right) \right|$$

$$\leq \sum_{l=1}^{k_n} |\varphi_{n,l}(t) - 1|^2$$

$$\leq m_n \sum_{l=1}^{k_n} |\varphi_{n,l}(t) - 1|$$

$$\leq \frac{1}{2} m_n t^2 \qquad \text{(by Lemma 15.46)}$$

$$\longrightarrow 0 \quad \text{for } n \to \infty. \qquad\qquad \square$$

In order to work with the concepts of weak convergence in this proof, we introduce the function

$$f_t(x) := \begin{cases} \dfrac{1}{x^2} \left(e^{itx} - 1 - itx \right), & \text{if } x \neq 0, \\[2mm] -\dfrac{t^2}{2}, & \text{if } x = 0, \end{cases} \qquad (15.8)$$

as well as the measures $\nu_n \in \mathcal{M}_f(\mathbb{R})$, $n \in \mathbb{N}$,

$$\nu_n(dx) := \sum_{l=1}^{k_n} x^2 \, \mathbf{P}_{X_{n,l}}(dx).$$

Lemma 15.48 *For every $t \in \mathbb{R}$, we have $f_t \in C_b(\mathbb{R})$.*

Proof Clearly, f_t is continuous on $\mathbb{R} \setminus \{0\}$. On the other hand, for $|x| \geq 1$, we have $|f_t(x)| \leq |e^{itx}| + 1 + |t/x| \leq 2 + |t|$. It remains to show that f_t is continuous at 0. This will imply that f_t is bounded also on the compact set $[-1, 1]$. Taylor's theorem (Lemma 15.31) yields

$$e^{itx} - 1 - itx = -\frac{t^2 x^2}{2} + R(tx)$$

with $|R(tx)| \leq \frac{1}{6}|tx|^3$. Hence, for fixed t, we have

$$\lim_{0 \neq x \to 0} f_t(x) = -\frac{t^2}{2} + \lim_{0 \neq x \to 0} \frac{R(tx)}{x^2} = -\frac{t^2}{2} = f_t(0).$$

$\square$

Lemma 15.49 *If (i) of Theorem 15.44 holds, then* $\nu_n \overset{n \to \infty}{\longrightarrow} \delta_0$ *weakly.*

Proof For every $n \in \mathbb{N}$, we have $\nu_n \in \mathcal{M}_1(\mathbb{R})$ since

$$\nu_n(\mathbb{R}) = \sum_{l=1}^{k_n} \int x^2 \, \mathbf{P}_{X_{n,l}}(dx) = \sum_{l=1}^{k_n} \mathrm{Var}[X_{n,l}] = 1.$$

However, for $\varepsilon > 0$, we have $\nu_n([-\varepsilon, \varepsilon]^c) = L_n(\varepsilon) \overset{n \to \infty}{\longrightarrow} 0$; hence $\nu_n \overset{n \to \infty}{\longrightarrow} \delta_0$. $\square$

Lemma 15.50 *If (i) of Theorem 15.44 holds, then*

$$\int f_t \, d\nu_n \overset{n \to \infty}{\longrightarrow} -\frac{t^2}{2}.$$

Proof By Lemma 15.48, we have $f_t \in C_b(\mathbb{R})$. Furthermore, by Lemma 15.49 we have $\nu_n \overset{n \to \infty}{\longrightarrow} \delta_0$. In other words, we have $\int f_t \, d\nu_n \overset{n \to \infty}{\longrightarrow} f_t(0) = -t^2/2$. $\square$

Proof of Theorem 15.44

"(i) $\Longrightarrow$ (ii)"   We have to show that $\lim_{n \to \infty} \log \varphi_n(t) = -\frac{t^2}{2}$ for every $t \in \mathbb{R}$. By Lemma 15.47, this is equivalent to

$$\lim_{n \to \infty} \sum_{l=1}^{k_n} \big(\varphi_{n,l}(t) - 1\big) = -\frac{t^2}{2}.$$

Now $f_t(x) x^2 + itx = e^{itx} - 1$ and $\int itx \, \mathbf{P}_{X_{n,l}}(dx) = it\mathbf{E}[X_{n,l}] = 0$, since the array $(X_{n,l})$ is centered. Hence, we get

$$\sum_{l=1}^{k_n} \big(\varphi_{n,l}(t) - 1\big) = \sum_{l=1}^{k_n} \int \Big(f_t(x) x^2 + itx\Big) \mathbf{P}_{X_{n,l}}(dx)$$

$$= \sum_{l=1}^{k_n} \int f_t(x) x^2 \, \mathbf{P}_{X_{n,l}}(dx)$$

$$= \int f_t \, d\nu_n$$

$$\overset{n\to\infty}{\longrightarrow} -\frac{t^2}{2} \qquad \text{(by Lemma 15.50).} \qquad \square$$

As an application of the Lindeberg–Feller theorem, we give the so-called **three-series theorem**, which is due to Kolmogorov.

Theorem 15.51 (Kolmogorov's three-series theorem) *Let $X_1, X_2, \ldots$ be independent real random variables. Let $K > 0$ and $Y_n := X_n \mathbb{1}_{\{|X_n| \leq K\}}$ for all $n \in \mathbb{N}$.*

The series $\sum_{n=1}^{\infty} X_n$ converges almost surely if and only if each of the following three conditions holds:

(i) $\displaystyle\sum_{n=1}^{\infty} \mathbf{P}[|X_n| > K] < \infty$.

(ii) $\displaystyle\sum_{n=1}^{\infty} \mathbf{E}[Y_n]$ *converges.*

(iii) $\displaystyle\sum_{n=1}^{\infty} \mathbf{Var}[Y_n] < \infty$.

Proof " $\Longleftarrow$ " Assume that (i), (ii) and (iii) hold. By Exercise 6.1.4, since (iii) holds, the series $\sum_{n=1}^{\infty} (Y_n - \mathbf{E}[Y_n])$ converges a.s. As (ii) holds, $\sum_{n=1}^{\infty} Y_n$ converges almost surely. By the Borel–Cantelli lemma, there exists an $N = N(\omega)$ such that $|X_n| \leq K$; hence $X_n = Y_n$ for all $n \geq N$. Hence $\sum_{n=1}^{\infty} X_n = \sum_{n=1}^{N-1} X_n + \sum_{n=N}^{\infty} Y_n$ converges a.s.

" $\Longrightarrow$ " Assume that $\sum_{n=1}^{\infty} X_n$ converges a.s. Clearly, this implies (i) (otherwise, by the Borel–Cantelli lemma, $|X_n| > K$ infinitely often, contradicting the assumption).

We assume that (iii) does not hold and produce a contradiction. To this end, let $\sigma_n^2 = \sum_{k=1}^{n} \mathbf{Var}[Y_k]$ and define an array $(X_{n,l}; \, l = 1, \ldots, n, \, n \in \mathbb{N})$ by $X_{n,l} = (Y_l - \mathbf{E}[Y_l])/\sigma_n$. This array is independent, centered and normed. Since $\sigma_n^2 \overset{n\to\infty}{\longrightarrow} \infty$, for every $\varepsilon > 0$ and for sufficiently large $n \in \mathbb{N}$, we have $2K < \varepsilon\sigma_n$; thus $|X_{n,l}| < \varepsilon$ for all $l = 1, \ldots, n$. This implies $L_n(\varepsilon) \overset{n\to\infty}{\longrightarrow} 0$, where $L_n(\varepsilon) = \sum_{l=1}^{n} \mathbf{E}[X_{n,l}^2 \mathbb{1}_{\{|X_{n,l}|>\varepsilon\}}]$ is the quantity of the Lindeberg condition (see (15.6)). By the Lindeberg–Feller theorem, we then get $S_n := X_{n,1} + \ldots + X_{n,n} \overset{n\to\infty}{\Longrightarrow} \mathcal{N}_{0,1}$. As shown in the first part of this proof, almost sure convergence of $\sum_{n=1}^{\infty} X_n$ and (i) imply that

$$\sum_{n=1}^{\infty} Y_n \qquad \text{converges almost surely.} \qquad (15.9)$$

In particular, $T_n := (Y_1 + \ldots + Y_n)/\sigma_n \overset{n\to\infty}{\Longrightarrow} 0$. Thus, by Slutzky's theorem, we also have $(S_n - T_n) \overset{n\to\infty}{\Longrightarrow} \mathcal{N}_{0,1}$. On the other hand, for all $n \in \mathbb{N}$, the difference $S_n - T_n$ is deterministic, contradicting the assumption that (iii) does not hold.

Now that we have established (iii), by Exercise 6.1.4, we see that $\sum_{n=1}^{\infty}(Y_n - \mathbf{E}[Y_n])$ converges almost surely. Together with (15.9), we conclude (ii). □

As a supplement, we cite a statement about the speed of convergence in the central limit theorem (see, e.g., [155, Chapter III, §11] for a proof). With different bounds (instead of 0.8), the statement was found independently by Berry [10] and Esseen [46].

Theorem 15.52 (Berry–Esseen) *Let* $X_1, X_2, \ldots$ *be independent and identically distributed with* $\mathbf{E}[X_1] = 0$, $\mathbf{E}[X_1^2] = \sigma^2 \in (0, \infty)$ *and* $\gamma := \mathbf{E}[|X_1|^3] < \infty$. *Let* $S_n^* := \frac{1}{\sqrt{n\sigma^2}}(X_1 + \cdots + X_n)$ *and let* $\Phi : x \mapsto \frac{1}{\sqrt{2\pi}}\int_{-\infty}^{x} e^{-t^2/2}dt$ *be the distribution function of the standard normal distribution. Then, for all* $n \in \mathbb{N}$,

$$\sup_{x\in\mathbb{R}} \left|\mathbf{P}\left[S_n^* \leq x\right] - \Phi(x)\right| \leq \frac{0.8\,\gamma}{\sigma^3\sqrt{n}}.$$

Example 15.53 Let $\alpha \in (0, 1)$. Consider the distribution μ_α on $\mathbb{R}$ with density

$$f_\alpha(x) = \frac{1}{2\alpha}|x|^{-1-1/\alpha}\,\mathbb{1}_{\{|x|\geq 1\}}.$$

Let $X_1, X_2, \ldots$, be i.i.d. random variables with distribution μ_α. Then $\mathbf{E}[X_1] = 0$ and $\sigma^2 := \mathbf{Var}[X_1] = 1/(1 - 2\alpha) < \infty$ if $\alpha < 1/2$. Let F_n denote the distribution function of S_n^* and F_Φ the distribution function of the standard normal distribution. The closer F_n and F_Φ are, the closer lie the points $(F_\Phi^{-1}(t), F_n^{-1}(t))$ on the diagonal $\{(x, x) : x \in \mathbb{R}\}$. A graphical representation of the points $(F_\Phi^{-1}(t), F_n^{-1}(t)), t \in \mathbb{R}$ is called **Q-Q-plot** or quantile-quantile-plot.

As α approaches $1/2$, the distribution μ_α has less and less moments. Hence we expect the convergence in the central limit theorem to be slower. For fixed n, we expect the deviation of F_n from F_Φ to be larger for larger α. The graphs in Fig. 15.2 illustrate this. ◊

Takeaways If a random variable is the sum of many independent centred random variables, each of which takes mainly small values, then its distribution is close to a normal distribution. The Feller-Lindeberg theorem makes rigorous sense of the expressions "mainly small values" and "close to a normal distribution" and formulates the precise statement.

Exercise 15.5.1 The argument of Remark 15.39 is more direct than the argument with Lévy's continuity theorem but is less robust: Give a sequence $X_1, X_2, \ldots$ of

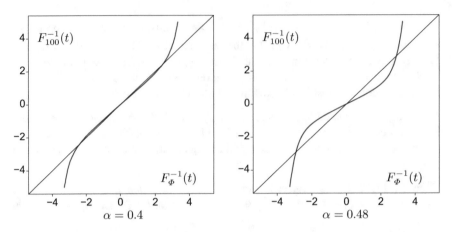

Fig. 15.2 Q-Q-plots for S_{100}^* from Example 15.53 with $\alpha = 0.4$ (left) and $\alpha = 0.48$ (right). The abscissa shows the quantiles of the standard normal distribution. For convenience, also the diagonal is drawn.

independent real random variables with $\mathbf{E}[|X_n|] = \infty$ for all $n \in \mathbb{N}$ but such that

$$\frac{X_1 + \ldots + X_n}{\sqrt{n}} \stackrel{n \to \infty}{\Longrightarrow} \mathcal{N}_{0,1}. \quad \clubsuit$$

Exercise 15.5.2 Let $Y_1, Y_2, \ldots$ be i.i.d. with $\mathbf{E}[Y_i] = 0$ and $\mathbf{E}[Y_i^2] = 1$. Let $Z_1, Z_2, \ldots$ be independent random variables (and independent of $Y_1, Y_2, \ldots$) with

$$\mathbf{P}[Z_i = i] = \mathbf{P}[Z_i = -i] = \frac{1}{2}\big(1 - \mathbf{P}[Z_i = 0]\big) = \frac{1}{2}\frac{1}{i^2}.$$

For $i, n \in \mathbb{N}$, define $X_i := Y_i + Z_i$ and $S_n = X_1 + \ldots + X_n$.

Show that $n^{-1/2} S_n \stackrel{n \to \infty}{\Longrightarrow} \mathcal{N}_{0,1}$ but that $(X_i)_{i \in \mathbb{N}}$ does not satisfy the Lindeberg condition.

Hint: Do not try a direct computation! $\clubsuit$

Exercise 15.5.3 Let $X_1, X_2, \ldots$ be i.i.d. random variables with density

$$f(x) = \frac{1}{|x|^3}\, \mathbb{1}_{\mathbb{R}\setminus[-1,1]}(x).$$

Then $\mathbf{E}[X_1^2] = \infty$ but there are numbers $A_1, A_2, \ldots$, such that

$$\frac{X_1 + \ldots + X_n}{A_n} \stackrel{n \to \infty}{\Longrightarrow} \mathcal{N}_{0,1}.$$

Determine one such sequence $(A_n)_{n \in \mathbb{N}}$ explicitly. $\clubsuit$

15.6   Multidimensional Central Limit Theorem

We come to a multidimensional version of the CLT.

Definition 15.54   *Let C be a (strictly) positive definite symmetric real $d \times d$ matrix and let $\mu \in \mathbb{R}^d$. A random vector $X = (X_1, \ldots, X_d)^T$ is called d-**dimensional normally distributed** with expectation μ and covariance matrix C if X has the density*

$$f_{\mu,C}(x) = \frac{1}{\sqrt{(2\pi)^d \det(C)}} \exp\left(-\frac{1}{2}\langle x - \mu, \ C^{-1}(x - \mu)\rangle\right) \qquad (15.10)$$

for $x \in \mathbb{R}^d$. In this case, we write $X \sim \mathcal{N}_{\mu,C}$.

Theorem 15.55   *Let $\mu \in \mathbb{R}^d$ and let C be a real positive definite symmetric $d \times d$ matrix. If $X \sim \mathcal{N}_{\mu,C}$, then the following statements hold.*

   (i) $\mathbf{E}[X_i] = \mu_i$ *for all $i = 1, \ldots, d$.*
   (ii) $\mathbf{Cov}[X_i, X_j] = C_{i,j}$ *for all $i, j = 1, \ldots, d$.*
   (iii) $\langle \lambda, X \rangle \sim \mathcal{N}_{\langle \lambda,\mu\rangle, \langle \lambda, C\lambda\rangle}$ *for every $\lambda \in \mathbb{R}^d$.*
   (iv) $\varphi(t) := \mathbf{E}[e^{i\langle t, X\rangle}] = e^{i\langle t,\mu\rangle} e^{-\frac{1}{2}\langle t, Ct\rangle}$ *for every $t \in \mathbb{R}^d$.*

   Moreover, $X \sim \mathcal{N}_{\mu,C} \iff (iii) \iff (iv)$.

Proof   (i) and (ii) follow by simple computations. The same is true for (iii) and (iv). The implication (iii) $\Longrightarrow$ (iv) is straightforward as the characteristic function φ uniquely determines the distribution of X by Theorem 15.9. $\square$

Remark 15.56   For one-dimensional normal distributions, it is natural to define the degenerate normal distribution by $\mathcal{N}_{\mu,0} := \delta_\mu$. For the multidimensional situation, there are various possibilities for degeneracy depending on the size of the kernel of C. If C is only positive semidefinite (and symmetric, of course), we define $\mathcal{N}_{\mu,C}$ as that distribution on $\mathbb{R}^n$ with characteristic function $\varphi(t) = e^{i\langle t,\mu\rangle} e^{-\frac{1}{2}\langle t, Ct\rangle}$. $\Diamond$

Theorem 15.57 (Cramér–Wold device)   *Let $X_n = (X_{n,1}, \ldots, X_{n,d})^T \in \mathbb{R}^d$, $n \in \mathbb{N}$, be random vectors. Then, the following are equivalent:*

   (i) There is a random vector X such that $X_n \overset{n\to\infty}{\Longrightarrow} X$.
   (ii) For any $\lambda \in \mathbb{R}^d$, there is a random variable X^λ such that $\langle \lambda, X_n\rangle \overset{n\to\infty}{\Longrightarrow} X^\lambda$.

   If (i) and (ii) hold, then $X^\lambda \overset{\mathcal{D}}{=} \langle \lambda, X\rangle$ for all $\lambda \in \mathbb{R}^d$.

Proof   Assume (i). Let $\lambda \in \mathbb{R}^d$ and $s \in \mathbb{R}$. The map $\mathbb{R}^d \to \mathbb{C}$, $x \mapsto e^{i\,s\langle\lambda, x\rangle}$ is continuous and bounded; hence we have $\mathbf{E}[e^{i\,s\langle\lambda, X_n\rangle}] \overset{n\to\infty}{\longrightarrow} \mathbf{E}[e^{i\,s\langle\lambda, X_\infty\rangle}]$. Thus (ii) holds with $X^\lambda := \langle \lambda, X\rangle$.

   Now assume (ii). Then $(\mathbf{P}_{X_{n,l}})_{n\in\mathbb{N}}$ is tight for every $l = 1, \ldots, d$. Hence $(\mathbf{P}_{X_n})_{n\in\mathbb{N}}$ is tight and thus relatively sequentially compact (Prohorov's theorem). For any weak limit point Q for $(\mathbf{P}_{X_n})_{n \in \mathbb{N}}$ and for any $\lambda \in \mathbb{R}^d$, we have

$$\int Q(dx)\, e^{i \langle \lambda, x \rangle} = \mathbf{E}\big[e^{i X^\lambda}\big].$$

Hence the limit point Q is unique and thus $(\mathbf{P}_{X_n})_{n \in \mathbb{N}}$ converges weakly to Q. That is, (i) holds.

If (ii) holds, then the distributions of the limiting random variables X^λ are uniquely determined and by what we have shown already, $X^\lambda = \langle \lambda, X \rangle$ is one possible choice. Thus $X^\lambda \overset{\mathcal{D}}{=} \langle \lambda, X \rangle$.   □

Theorem 15.58 (Central limit theorem in $\mathbb{R}^d$) *Let* $(X_n)_{n \in \mathbb{N}}$ *be i.i.d. random vectors with* $\mathbf{E}[X_{n,i}] = 0$ *and* $\mathbf{E}[X_{n,i} X_{n,j}] = C_{ij}$, $i, j = 1, \ldots, d$. *Let* $S_n^* := \frac{X_1 + \ldots + X_n}{\sqrt{n}}$. *Then*

$$\mathbf{P}_{S_n^*} \overset{n \to \infty}{\longrightarrow} \mathcal{N}_{0,C} \text{ weakly.}$$

Proof Let $\lambda \in \mathbb{R}^d$. Define $X_n^\lambda = \langle \lambda, X_n \rangle$, $S_n^\lambda = \langle \lambda, S_n^* \rangle$ and $S_\infty \sim \mathcal{N}_{0,C}$. Then $\mathbf{E}[X_n^\lambda] = 0$ and $\mathbf{Var}[X_n^\lambda] = \langle \lambda, C\lambda \rangle$. By the one-dimensional central limit theorem, we have $\mathbf{P}_{S_n^\lambda} \overset{n \to \infty}{\longrightarrow} \mathcal{N}_{0, \langle \lambda, C\lambda \rangle} = \mathbf{P}_{\langle \lambda, S_\infty \rangle}$. By Theorem 15.57, this yields the claim.   □

Takeaways For vector-valued random variables to converge, it is enough that the projections to one-dimensional subspaces converge (Cramér-Wold). We use this to conclude a central limit theorem for multi-dimensional independent and identically distributed random variables.

Exercise 15.6.1 Let $\mu \in \mathbb{R}^d$, let C be a symmetric positive semidefinite real $d \times d$ matrix and let $X \sim \mathcal{N}_{\mu,C}$ (in the sense of Remark 15.56). Show that $AX \sim \mathcal{N}_{A\mu, ACA^T}$ for every $m \in \mathbb{N}$ and every real $m \times d$ matrix A. ♣

Exercise 15.6.2 (Cholesky factorization)   Let C be a positive definite symmetric real $d \times d$ matrix. Then there exists a real $d \times d$ matrix $A = (a_{kl})$ with $A \cdot A^T = C$. The matrix A can be chosen to be lower triangular. Let $W := (W_1, \ldots, W_d)^T$, where $W_1, \ldots, W_d$ are independent and $\mathcal{N}_{0,1}$-distributed. Define $X := AW + \mu$. Show that $X \sim \mathcal{N}_{\mu,C}$. ♣

Chapter 16
Infinitely Divisible Distributions

For every n, the normal distribution $\mathcal{N}_{\mu,\sigma^2}$ is the nth convolution power of a probability measure (namely, of $\mathcal{N}_{\mu/n,\sigma^2/n}$). This property is called infinite divisibility and is shared by other probability distributions such as the Poisson distribution and the Gamma distribution. In the first section, we study which probability measures on $\mathbb{R}$ are infinitely divisible and give an exhaustive description of this class of distributions by means of the Lévy–Khinchin formula.

Unlike the Poisson distribution, the normal distribution is the limit of *rescaled* sums of i.i.d. random variables (central limit theorem). In the second section, we investigate briefly which subclass of the infinitely divisible measures on $\mathbb{R}$ shares this property.

16.1 Lévy–Khinchin Formula

For the sake of brevity, in this section, we use the shorthand "CFP" for "characteristic function of a probability measure on $\mathbb{R}$".

Definition 16.1 *A measure $\mu \in \mathcal{M}_1(\mathbb{R})$ is called **infinitely divisible** if, for every $n \in \mathbb{N}$, there is a $\mu_n \in \mathcal{M}_1(\mathbb{R})$ such that $\mu_n^{*n} = \mu$. Analogously, a CFP φ is called infinitely divisible if, for every $n \in \mathbb{N}$, there is a CFP φ_n such that $\varphi = \varphi_n^n$. A real random variable X is called infinitely divisible if, for every $n \in \mathbb{N}$, there exist i.i.d. random variables $X_{n,1}, \ldots, X_{n,n}$ such that $X \overset{\mathcal{D}}{=} X_{n,1} + \ldots + X_{n,n}$.*

Manifestly, all three notions of infinite divisibility are equivalent, and we will use them synonymously. Note that the uniqueness of μ_n and φ_n, respectively, is by no means evident. Indeed, n-fold divisibility alone does not imply uniqueness of the nth convolution root $\mu^{*1/n} := \mu_n$ or of φ_n, respectively. As an example for even n, choose a real-valued CFP φ for which $|\varphi| \neq \varphi$ is also a CFP (see Examples 15.17 and 15.18). Then $\varphi^n = |\varphi|^n$ is n-fold divisible; however, the factors are not unique.

By virtue of Lévy's continuity theorem, one can show that (see Exercise 16.1.2) $\varphi(t) \neq 0$ for all $t \in \mathbb{R}$ if φ is infinitely divisible. The probabilistic meaning of this fact is that as a continuous function $\log(\varphi(t))$ is uniquely defined and thus there exists only one continuous function $\varphi^{1/n} = \exp(\log(\varphi)/n)$. The nth convolution roots are thus unique if the distribution is *infinitely* divisible.

Example 16.2

(i) δ_x is infinitely divisible with $\delta_{x/n}^{*n} = \delta_x$ for every $n \in \mathbb{N}$.

(ii) The normal distribution is infinitely divisible with $\mathcal{N}_{m,\sigma^2} = \mathcal{N}_{m/n,\sigma^2/n}^{*n}$.

(iii) The Cauchy distribution Cau_a with density $x \mapsto (a\pi)^{-1} (1 + (x/a)^2)^{-1}$ is infinitely divisible with $\mathrm{Cau}_a = \mathrm{Cau}_{a/n}^{*n}$. Indeed, Cau_a has CFP $\varphi_a(t) = e^{-a|t|}$; hence $\varphi_{a/n}^n = \varphi_a$.

(iv) Every symmetric stable distribution with index $\alpha \in (0, 2]$ and scale parameter $\gamma > 0$ (that is, the distribution with CFP $\varphi_{\alpha,\gamma}(t) = e^{-|\gamma t|^\alpha}$) is infinitely divisible. Indeed, $\varphi_{\alpha,\gamma/n^{1/\alpha}}^n = \varphi_{\alpha,\gamma}$. (To be precise, we have shown only for $\alpha \in (0, 1]$ (in Corollary 15.26) and for $\alpha = 2$ (normal distribution) that $\varphi_{\alpha,\gamma}$ is in fact a CFP. In Sect. 16.2, we will show that this is true for all $\alpha \in (0, 2]$. For $\alpha > 2$, $\varphi_{\alpha,\gamma}$ is not a CFP, see Exercise 15.4.3.)

(v) The Gamma distribution $\Gamma_{\theta,r}$ with CFP $\varphi_{\theta,r}(t) = \exp(r\psi_\theta(t))$, where $\psi_\theta(t) = \log(1 - it/\theta)$, is infinitely divisible with $\Gamma_{\theta,r} = \Gamma_{\theta,r/n}^{*n}$.

(vi) The Poisson distribution is infinitely divisible with $\mathrm{Poi}_\lambda = \mathrm{Poi}_{\lambda/n}^{*n}$.

(vii) The negative binomial distribution $b_{r,p}^-(\{k\}) = \binom{-r}{k}(-1)^k p^r (1-p)^k$, $k \in \mathbb{N}_0$, with parameters $r > 0$ and $p \in (0, 1)$, is infinitely divisible with $b_{r,p}^- = (b_{r/n,p}^-)^{*n}$. Indeed, $\varphi_{r,p}(t) = e^{r\psi_p(t)}$, where

$$\psi_p(t) = \log(p) - \log\left(1 - (1 - p)e^{it}\right).$$

(viii) Let X and Y be independent with $X \sim \mathcal{N}_{0,\sigma^2}$ and $Y \sim \Gamma_{\theta,r}$, where $\sigma^2, \theta, r > 0$. It can be shown that the random variable $Z := X/\sqrt{Y}$ is infinitely divisible (see [65] or [131]). In particular, Student's t-distribution with $k \in \mathbb{N}$ degrees of freedom is infinitely divisible (this is the case where $\sigma^2 = 1$ and $\theta = r = k/2$).

(ix) The binomial distribution $b_{n,p}$ with parameters $n \in \mathbb{N}$ and $p \in (0, 1)$ is *not* infinitely divisible (why?).

(x) Somewhat more generally, there is no nontrivial infinitely divisible distribution that is concentrated on a bounded interval. ◊

A main goal of this section is to show that every infinitely divisible distribution can be composed of three generic ones:

- the Dirac measures δ_x with $x \in \mathbb{R}$,
- the normal distributions $\mathcal{N}_{\mu,\sigma^2}$ with $\mu \in \mathbb{R}$ and $\sigma^2 > 0$, and
- (limits of) convolutions of Poisson distributions.

As convolutions of Poisson distributions play a special role, we will consider them separately.

If $\nu \in \mathcal{M}_1(\mathbb{R})$ with CFP φ_ν and if $\lambda > 0$, then one can easily check that $\varphi(t) = \exp(\lambda(\varphi_\nu(t) - 1))$ is the CFP of $\mu_\lambda = \sum_{k=0}^{\infty} e^{-\lambda} \frac{\lambda^k}{k!} \nu^{*k}$. Hence, formally we can write $\mu_\lambda = e^{*\lambda(\nu - \delta_0)}$. Indeed, μ_λ is infinitely divisible with $\mu_\lambda = \mu_{\lambda/n}^{*n}$. We want to combine the two parameters λ and ν into one parameter $\lambda\nu$. For $\nu \in \mathcal{M}_f(\mathbb{R})$, we can define $\nu^{*n} = \nu(\mathbb{R})^n (\nu/\nu(\mathbb{R}))^{*n}$ (and $\nu^{*n} = 0$ if $\nu = 0$). In both cases, let $\nu^{*0} := \delta_0$. Hence we make the following definition.

Definition 16.3 *The **compound Poisson distribution** with intensity measure $\nu \in \mathcal{M}_f(\mathbb{R})$ is the following probability measure on $\mathbb{R}$:*

$$\mathrm{CPoi}_\nu := e^{*(\nu - \nu(\mathbb{R})\delta_0)} := e^{-\nu(\mathbb{R})} \sum_{n=0}^{\infty} \frac{\nu^{*n}}{n!}.$$

The CFP of CPoi_ν is given by

$$\varphi_\nu(t) = \exp\left(\int (e^{itx} - 1)\, \nu(dx)\right). \tag{16.1}$$

In particular, $\mathrm{CPoi}_{\mu+\nu} = \mathrm{CPoi}_\mu * \mathrm{CPoi}_\nu$; hence CPoi_ν is infinitely divisible.

Example 16.4 For every measurable set $A \subset \mathbb{R} \setminus \{0\}$ and every $r > 0$,

$$r^{-1}\mathrm{CPoi}_{r\nu}(A) = e^{-r\nu(\mathbb{R})}\nu(A) + e^{-r\nu(\mathbb{R})} \sum_{k=2}^{\infty} \frac{r^{k-1}\nu^{*k}(A)}{k!} \xrightarrow{r\downarrow 0} \nu(A).$$

We use this in order to show that $b_{r,p}^- = \mathrm{CPoi}_{r\nu}$ for some $\nu \in \mathcal{M}_f(\mathbb{N})$. To this end, for $k \in \mathbb{N}$, we compute

$$r^{-1}b_{r,p}^-(\{k\}) = \frac{r(r+1)\cdots(r+k-1)}{r\,k!}\, p^r(1-p)^k \xrightarrow{r\downarrow 0} \frac{(1-p)^k}{k}.$$

If we had $b_{r,p}^- = \mathrm{CPoi}_{r\nu}$ for some $\nu \in \mathcal{M}_f(\mathbb{N})$, then we would have $\nu(\{k\}) = (1-p)^k/k$. We compute the CFP of $\mathrm{CPoi}_{r\nu}$ for this ν,

$$\varphi_{r\nu}(t) = \exp\left(r\sum_{k=1}^{\infty} \frac{((1-p)e^{it})^k - (1-p)^k}{k}\right) = p^r\left(1 - (1-p)e^{it}\right)^{-r}.$$

However, this is the CFP of $b_{r,p}^-$; hence indeed $b_{r,p}^- = \mathrm{CPoi}_{r\nu}$. $\Diamond$

Not every infinitely divisible distribution is of the type CPoi_ν, however we have the following theorem.

Theorem 16.5 *A probability measure μ on $\mathbb{R}$ is infinitely divisible if and only if there is a sequence $(\nu_n)_{n\in\mathbb{N}}$ in $\mathcal{M}_f(\mathbb{R}\setminus\{0\})$ such that $\mathrm{CPoi}_{\nu_n} \xrightarrow{n\to\infty} \mu$.*

If μ is infinitely divisible and $\mu_n \in \mathcal{M}_1(\mathbb{R})$ is such that $\mu_n^{*n} = \mu$ for all $n \in \mathbb{N}$, then $\nu_n = \mathbb{1}_{\mathbb{R}\setminus\{0\}} n\mu_n$ is a possible choice.

Since every CPoi_{ν_n} is infinitely divisible, on the one hand we have to show that this property is preserved under weak limits. On the other hand, we show that, for infinitely divisible μ, the sequence $\nu_n = \mathbb{1}_{\mathbb{R}\setminus\{0\}} n\mu^{*1/n}$ does the trick. We prepare for the proof of Theorem 16.5 with a further theorem.

Theorem 16.6 *Let $(\varphi_n)_{n\in\mathbb{N}}$ be a sequence of CFPs. Then the following are equivalent.*

(i) *For every $t \in \mathbb{R}$, the limit $\varphi(t) = \lim_{n\to\infty} \varphi_n^n(t)$ exists and φ is continuous at 0.*

(ii) *For every $t \in \mathbb{R}$, the limit $\psi(t) = \lim_{n\to\infty} n(\varphi_n(t)-1)$ exists and ψ is continuous at 0.*

If (i) and (ii) hold, then $\varphi = e^\psi$ is a CFP.

Proof The proof is based on a Taylor expansion of the logarithm,

$$|\log(z) - (z-1)| \le |z-1|^2 \quad \text{for } z \in \mathbb{C} \text{ with } |z-1| < \frac{1}{2}.$$

As an immediate consequence, we get

$$\frac{1}{2}|z-1| \le |\log(z)| \le \frac{3}{2}|z-1| \quad \text{for } z \in \mathbb{C} \text{ with } |z-1| < \frac{1}{2}.$$

In particular, for $(z_n)_{n\in\mathbb{N}}$ in $\mathbb{C}$,

$$\limsup_{n\to\infty} n\,|z_n-1| < \infty \iff \limsup_{n\to\infty} |n\log(z_n)| < \infty, \tag{16.2}$$

and $\lim_{n\to\infty} n(z_n-1) = \lim_{n\to\infty} n\log(z_n)$ if one of the limits exists.

Applying this to $z_n = \varphi_n(t)$, we see that (ii) implies (i). On the other hand, (i) implies (ii) if $\liminf_{n\to\infty} n\log(|\varphi_n(t)|) > -\infty$ and hence if $\varphi(t) \ne 0$ for all $t \in \mathbb{R}$.

Since φ is continuous at 0 and since $\varphi(0) = 1$, there is an $\varepsilon > 0$ with $|\varphi(t)| > \frac{1}{2}$ for all $t \in [-\varepsilon, \varepsilon]$. Since φ and φ_n are CFPs, $|\varphi|^2$ and $|\varphi_n|^2$ are also CFPs. Thus, since $|\varphi_n(t)|^{2n}$ converges to $|\varphi(t)|^2$ pointwise, Lévy's continuity theorem implies uniform convergence on compact sets. Now apply (16.2) with $z_n = |\varphi_n(t)|^2$. Thus $(n(1 - |\varphi_n(t)|^2))_{n\in\mathbb{N}}$ is bounded for $t \in [-\varepsilon, \varepsilon]$. Hence, by Lemma 15.12(v), $n(1 - |\varphi_n(2t)|^2) \le 4n(1 - |\varphi_n(t)|^2)$ also is bounded; thus

$$|\varphi(2t)|^2 \ge \liminf_{n\to\infty} \exp(4n(|\varphi_n(t)|^2 - 1)) = (|\varphi(t)|^2)^4.$$

Inductively, we get $|\varphi(t)| \geq 2^{-(4^k)}$ for $|t| \leq 2^k \varepsilon$. Hence there is a $\gamma > 0$ such that

$$|\varphi(t)| > \frac{1}{2} e^{-\gamma t^2} \qquad \text{for all } t \in \mathbb{R}. \tag{16.3}$$

If (i) and (ii) hold, then

$$\log \varphi(t) = \lim_{n \to \infty} n \log(\varphi_n(t)) = \lim_{n \to \infty} n(\varphi_n(t) - 1) = \psi(t).$$

By Lévy's continuity theorem, as a continuous limit of CFPs, φ is a CFP. □

Corollary 16.7 *If the conditions of Theorem 16.6 hold, then φ^r is a CFP for every $r > 0$. In particular, $\varphi = (\varphi^{1/n})^n$ is infinitely divisible.*

Proof If φ_n is the CFP of $\mu_n \in \mathcal{M}_1(\mathbb{R})$, then $e^{rn(\varphi_n - 1)}$ is the CFP of $\text{CPoi}_{rn\mu_n}$. Being a limit of CFPs that is continuous at 0, by Lévy's continuity theorem, $\varphi^r = e^{r\psi} = \lim_{n \to \infty} e^{rn(\varphi_n - 1)}$ is a CFP. Letting $r = \frac{1}{n}$, we get that $\varphi = (\varphi^{1/n})^n$ is infinitely divisible. □

Corollary 16.8 *Let $\varphi : \mathbb{R} \to \mathbb{C}$ be continuous at 0. φ is an infinitely divisible CFP if and only if there is a sequence $(\varphi_n)_{n \in \mathbb{N}}$ of CFPs such that $\varphi_n^n(t) \to \varphi(t)$ for all $t \in \mathbb{R}$.*

Proof One implication has been shown already in Corollary 16.7. Hence, let φ be an infinitely divisible CFP. Then $\varphi_n = \varphi^{1/n}$ serves the purpose. □

Corollary 16.9 *If $(\mu_n)_{n \in \mathbb{N}}$ is a (weakly) convergent sequence of infinitely divisible probability measures on $\mathbb{R}$, then $\mu = \lim_{n \to \infty} \mu_n$ is infinitely divisible.*

Proof Apply Theorem 16.6, where φ_n is the CFP of $\mu_n^{*1/n}$. □

Corollary 16.10 *If $\mu \in \mathcal{M}_1(\mathbb{R})$ is infinitely divisible, then there exists a continuous convolution semigroup $(\mu_t)_{t \geq 0}$ with $\mu_1 = \mu$ and a stochastic process $(X_t)_{t \geq 0}$ with independent, stationary increments $X_t - X_s \sim \mu_{t-s}$ for $t > s$.*

Proof Let φ be the CFP of μ. The existence of the convolution semigroup follows by Corollaries 16.8 and 16.7 if we define μ_r by φ^r. Since $\varphi^r(t) \neq 0$ for all $t \in \mathbb{R}$, we have $\varphi^r \to 1$ for $r \to 0$ and thus the semigroup is continuous. Finally, Theorem 14.50 implies the existence of the process X. □

Corollary 16.11 *If φ is an infinitely divisible CFP, then there exists a $\gamma > 0$ with $|\varphi(t)| \geq \frac{1}{2} e^{-\gamma t^2}$ for all $t \in \mathbb{R}$. In particular, for $\alpha > 2$, $t \mapsto e^{-|t|^\alpha}$ is not a CFP.*

Proof This is a direct consequence of (16.3). □

Proof (of Theorem 16.5) As every CPoi_{ν_n} is infinitely divisible, by Corollary 16.9, the weak limit is also infinitely divisible.

Now let μ be infinitely divisible with CFP φ. For $n \in \mathbb{N}$ choose $\mu_n \in \mathcal{M}_1(\mathbb{R})$ such that $\mu_n^{*n} = \mu$ and let φ_n the CFP of μ_n. Then $\varphi_n^n = \varphi$. By Theorem 16.6, we have $e^{n(\varphi_n-1)} \overset{n\to\infty}{\longrightarrow} \varphi$. As $e^{n(\varphi_n-1)}$ is the CFP of $\mathrm{CPoi}_{n\mu_n}$, we infer $\mathrm{CPoi}_{n\mu_n} \overset{n\to\infty}{\longrightarrow}$ μ. Now let $\nu_n = \mathbb{1}_{\mathbb{R}\setminus\{0\}} n\mu_n$. Then $\mathrm{CPoi}_{\nu_n} = \mathrm{CPoi}_{n\mu_n} \overset{n\to\infty}{\longrightarrow} \mu$. $\qquad\square$

Without proof, we quote the following strengthening of Corollary 16.8 that relies on a finer analysis using the arguments from the proof of Theorem 16.6.

Theorem 16.12 Let $(\varphi_{n,l}; l = 1, \dots, k_n, n \in \mathbb{N})$ be an array of CFPs with the property

$$\sup_{L>0} \limsup_{n\to\infty} \sup_{t\in[-L,L]} \sup_{l=1,\dots,k_n} |\varphi_{n,l}(t) - 1| = 0. \qquad (16.4)$$

Assume that, for every $t \in \mathbb{R}$, the limit $\varphi(t) := \lim_{n\to\infty} \prod_{l=1}^{k_n} \varphi_{n,l}(t)$ exists and that φ is continuous at 0. Then φ is an infinitely divisible CFP.

Proof See, e.g., [54, Chapter XV.7]. $\qquad\square$

In the special case where for every n, the individual $\varphi_{n,l}$ are equal and where $k_n \overset{n\to\infty}{\longrightarrow} \infty$, equation (16.4) holds automatically if the product converges to a continuous function. Thus, the theorem is in fact an improvement of Corollary 16.8.

The benefit of this theorem will become clear through the following observation. Let $(X_{n,l}; l = 1, \dots, k_n, n \in \mathbb{N})$ be an array of real random variables with CFPs $\varphi_{n,l}$. This array is a null array if and only if (16.4) holds. In fact, if $\mathbf{P}[|X_{n,l}| > \varepsilon] < \delta$, then we have $|\varphi_{n,l}(t) - 1| \leq 2\varepsilon + \delta$ for all $t \in [-1/\varepsilon, 1/\varepsilon]$. Hence (16.4) holds if the array $(X_{n,l})$ is a null array. On the other hand, (16.4) implies $\varphi_{n,l_n} \overset{n\to\infty}{\longrightarrow} 1$ for every sequence (l_n) with $l_n \leq k_n$. Hence $X_{n,l_n} \overset{n\to\infty}{\longrightarrow} 0$ in probability.

From these considerations and from Theorem 16.12, we conclude the following theorem.

Theorem 16.13 Let $(X_{n,l}; l = 1, \dots, k_n, n \in \mathbb{N})$ be an independent null array of real random variables. If there exists a random variable S with

$$X_{n,1} + \dots + X_{n,k_n} \overset{n\to\infty}{\Longrightarrow} S,$$

then S is infinitely divisible.

As a direct application of Theorem 16.5, we give a complete description of the class of infinitely divisible probability measures on $[0, \infty)$ in terms of their Laplace transforms. The following theorem is of independent interest. Here, however, it is primarily used to provide familiarity with the techniques that will be needed for the more challenging classification of the infinitely divisible probability measures on $\mathbb{R}$.

Theorem 16.14 (Lévy–Khinchin formula on $[0, \infty)$) Let $\mu \in \mathcal{M}_1([0, \infty))$ and let $u : [0, \infty) \to [0, \infty)$, $t \mapsto -\log \int e^{-tx} \mu(dx)$ be the log-Laplace transform μ. μ is infinitely divisible if and only if there exists an $\alpha \geq 0$ and a σ-finite measure

$\nu \in \mathcal{M}((0, \infty))$ *with*

$$\int (1 \wedge x) \, \nu(dx) < \infty \tag{16.5}$$

and such that

$$u(t) = \alpha t + \int \left(1 - e^{-tx}\right) \nu(dx) \quad \text{for } t \geq 0. \tag{16.6}$$

In this case, the pair (α, ν) is unique. ν is called the canonical measure or Lévy measure of μ, and α is called the deterministic part.

Proof "$\Longrightarrow$ "   First assume μ is infinitely divisible. The case $\mu = \delta_0$ is trivial. Now let $\mu \neq \delta_0$; hence $u(1) > 0$.

For $n \in \mathbb{N}$, there exists a $\mu_n \in \mathcal{M}_1(\mathbb{R})$ such that $\mu_n^{*n} = \mu$. Clearly, we have $\mu_n((-\infty, 0))^n \leq \mu((-\infty, 0)) = 0$. Hence μ_n is supported by $[0, \infty)$. Define $\nu_n = n\mu_n \in \mathcal{M}_f([0, \infty))$. By Theorem 16.5, we have $\mathrm{CPoi}_{\nu_n} \overset{n\to\infty}{\longrightarrow} \mu$.

If we define $u_n(t) := \int (1 - e^{-tx}) \, \nu_n(dx)$, then (as in (16.1)) $u_n(t) \overset{n\to\infty}{\longrightarrow} u(t)$ for all $t \geq 0$. In particular, $u_n(1) > 0$ for sufficiently large n. Define $\tilde{\nu}_n \in \mathcal{M}_1([0, \infty))$ by $\tilde{\nu}_n(dx) := \frac{1-e^{-x}}{u_n(1)} \nu_n(dx)$. Hence, for all $t \geq 0$,

$$\int e^{-tx} \, \tilde{\nu}_n(dx) = \frac{u_n(t+1) - u_n(t)}{u_n(1)} \overset{n\to\infty}{\longrightarrow} \frac{u(t+1) - u(t)}{u(1)}.$$

The right hand side is continuous and hence a variation of Lévy's continuity theorem for Laplace transforms (compare Exercise 15.3.4) yields that the weak limit $\tilde{\nu} := \text{w-lim } \tilde{\nu}_n$ (in $\mathcal{M}_1([0, \infty))$) exists and is uniquely determined by u. Let $\alpha := \tilde{\nu}(\{0\}) \, u(1)$ and define $\nu \in \mathcal{M}((0, \infty))$ by

$$\nu(dx) = u(1)(1 - e^{-x})^{-1} \mathbb{1}_{(0,\infty)}(x) \, \tilde{\nu}(dx).$$

Since $1 \wedge x \leq 2(1 - e^{-x})$ for all $x \geq 0$, clearly

$$\int (1 \wedge x) \, \nu(dx) \leq 2 \int (1 - e^{-x}) \, \nu(dx) \leq 2u(1) < \infty.$$

For all $t \geq 0$, the function (compare (15.8))

$$f_t : [0, \infty) \to [0, \infty), \quad x \mapsto \begin{cases} \frac{1-e^{-tx}}{1-e^{-x}}, & \text{if } x > 0, \\ t, & \text{if } x = 0, \end{cases}$$

is continuous and bounded (by $t \vee 1$). Hence we have

$$u(t) = \lim_{n \to \infty} u_n(t) = \lim_{n \to \infty} u_n(1) \int f_t \, d\tilde{v}_n$$

$$= u(1) \int f_t \, d\tilde{v} = \alpha t + \int (1 - e^{-tx}) \, v(dx).$$

"$\Longleftarrow$"   Now assume that α and v are given. Define the intervals $I_0 = [1, \infty)$ and $I_k = [1/(k+1), 1/k)$ for $k \in \mathbb{N}$. Let $X_0, X_1, \dots$ be independent random variables with $\mathbf{P}_{X_k} = \mathrm{CPoi}_{(v|_{I_k})}$ for $k = 0, 1, \dots$, and let $X := \alpha + \sum_{k=0}^{\infty} X_k$. For every $k \in \mathbb{N}$, we have $\mathbf{E}[X_k] = \int_{I_k} x \, v(dx)$; hence $\sum_{k=1}^{\infty} \mathbf{E}[X_k] = \int_{(0,1)} x \, v(dx) < \infty$. Thus $X < \infty$ almost surely and $\alpha + \sum_{k=0}^{n} X_k \stackrel{n \to \infty}{\Longrightarrow} X$. Therefore,

$$-\log \mathbf{E}[e^{-tX}] = \alpha t - \sum_{k=0}^{\infty} \log \mathbf{E}[e^{-tX_k}] = \alpha t + \int (1 - e^{-tx}) \, v(dx).$$

"Uniqueness" For $x, t > 0$, we have

$$0 \leq t^{-1} (1 - e^{-tx}) \leq x \wedge \frac{1}{t} \stackrel{t \to \infty}{\longrightarrow} 0.$$

By the dominated convergence theorem with dominating function $1 \wedge x$ for $t \geq 1$, we infer

$$t^{-1} \int (1 - e^{-tx}) \, v(dx) \stackrel{t \to \infty}{\longrightarrow} 0.$$

Thus we can compute α as

$$\alpha = \lim_{t \to \infty} u(t)/t.$$

Define the map $h : (0, \infty) \to (0, 1)$ by

$$h(x) = 1 - \frac{1 - e^{-x}}{x}.$$

Note that $h(x)/(1 \wedge x) \leq 1$ for all $x > 0$. Hence $\tilde{v} := hv$ is a finite measure and $v = h^{-1}\tilde{v}$ is uniquely defined by $\tilde{v}$. Let

$$\bar{u}(t) := \frac{\alpha}{2} + u(t) - \int_t^{t+1} u(s) \, ds$$

$$= \int e^{-tx} \left(1 - \int_0^1 e^{-sx} ds\right) v(dx) = \int e^{-tx} \tilde{v}(dx).$$

That is, $\bar{u}$ is the Laplace transform of $\tilde{v}$ which determines $\tilde{v}$ and v uniquely. $\square$

Example 16.15 For an infinitely divisible distribution μ on $[0, \infty)$, we can compute the Lévy measure ν by the vague limit

$$\nu = \underset{n \to \infty}{\text{v-lim}} \, n\mu^{*1/n}\big|_{(0,\infty)}. \tag{16.7}$$

Often α is also easy to obtain (e.g., via the representation from Exercise 16.1.3). For example, for the Gamma distribution, we get $\alpha = 0$ and

$$n\Gamma_{\theta,1/n}(A) = \frac{\theta^{1/n}}{\Gamma(1/n)/n} \int_A x^{(1/n)-1} e^{-\theta x} \, dx \overset{n \to \infty}{\longrightarrow} \int_A x^{-1} e^{-\theta x} \, dx,$$

hence $\nu(dx) = x^{-1} e^{-\theta x} \, dx$. $\Diamond$

For infinitely divisible distributions on $\mathbb{R}$, we would like to obtain a description similar to that in the preceding theorem. However, an infinitely divisible real random variable X is not simply the difference of two infinitely divisible nonnegative random variables, as the normal distribution shows. In addition, we have more freedom if, as in the last proof, we want to express X as a sum of independent random variables X_k.

Hence we define the real random variable X as the sum of independent random variables,

$$X = b + X^N + X_0 + \sum_{k=1}^{\infty} (X_k - \alpha_k), \tag{16.8}$$

where $b \in \mathbb{R}$, $X^N = \mathcal{N}_{0,\sigma^2}$ for some $\sigma^2 \geq 0$ and $\mathbf{P}_{X_k} = \text{CPoi}_{\nu_k}$ with intensity measure ν_k that is concentrated on $I_k := (-1/k, -1/(k+1)] \cup [1/(k+1), 1/k)$ (with the convention $1/0 = \infty$), $k \in \mathbb{N}_0$. Furthermore, $\alpha_k = \mathbf{E}[X_k] = \int x \, \nu_k(dx)$ for $k \geq 1$. In order for the series to converge almost surely, it is sufficient (and also necessary, as a simple application of Kolmogorov's three-series theorem shows) that

$$\sum_{k=1}^{\infty} \mathbf{Var}[X_k] < \infty. \tag{16.9}$$

(In contrast to the situation in Theorem 16.14, here it is not necessary to have $\sum_{k=1}^{\infty} \mathbf{E}[|X_k - \alpha_k|] < \infty$. This allows for greater freedom in the choice of ν than in the case of nonnegative random variables.) Now $\mathbf{Var}[X_k] = \int x^2 \, \nu_k(dx)$. Hence, if we let $\nu = \sum_{k=0}^{\infty} \nu_k$, then (16.9) is equivalent to $\int_{(-1,1)} x^2 \, \nu(dx) < \infty$. As ν_0 is always finite, this in turn is equivalent to $\int (x^2 \wedge 1) \, \nu(dx) < \infty$.

Definition 16.16 *A σ-finite measure v on $\mathbb{R}$ is called a **canonical measure** if $v(\{0\}) = 0$ and*

$$\int \left(x^2 \wedge 1\right) v(dx) < \infty. \tag{16.10}$$

*If $\sigma^2 \geq 0$ and $b \in \mathbb{R}$, then (σ^2, b, v) is called a **canonical triple**.*

To every canonical triple, by (16.8) there corresponds an infinitely divisible random variable. Define

$$\psi_0(t) \;=\; \log \mathbf{E}\!\left[e^{it X_0}\right] \;=\; \int_{I_0} \left(e^{itx} - 1\right) v(dx).$$

For $k \in \mathbb{N}$, let

$$\psi_k(t) \;=\; \log \mathbf{E}\!\left[e^{it(X_k - \alpha_k)}\right] \;=\; \int_{I_k} \left(e^{itx} - 1 - itx\right) v(dx).$$

(Note that it is not a priori clear, that the logarithm in the two equations above is well-defined. In general, the expectation could be zero for some t. However, the right hand sides are well-defined and by (16.1) are the exponentials of the left hand sides. Hence the logarithms could be applied on both sides. This also shows that the expectations on the left hand side never equal zero. Hence

$$\psi(t) \;:=\; \log \mathbf{E}\!\left[e^{it X}\right] \;=\; -\frac{\sigma^2}{2} t^2 + ibt + \sum_{k=0}^{\infty} \psi_k(t)$$

satisfies the Lévy–Khinchin formula

$$\psi(t) \;=\; -\frac{\sigma^2}{2} t^2 + ibt + \int \left(e^{itx} - 1 - itx \, \mathbb{1}_{\{|x|<1\}}\right) v(dx). \tag{16.11}$$

Theorem 16.17 (Lévy–Khinchin formula) *Let $\mu \in \mathcal{M}_1(\mathbb{R})$ and*

$$\psi(t) \;:=\; \log \int e^{itx} \, \mu(dx).$$

μ is infinitely divisible if and only if ψ is well-defined and there exists a canonical triple (σ^2, b, v) such that (16.11) holds. By (16.11), this triple is uniquely determined.

Again, v is called the Lévy measure of μ, σ^2 is called the Gaussian coefficient and b is called the centering constant.

Proof We have shown already that via (16.11) every canonical triple (σ^2, b, ν) corresponds to an infinitely divisible distribution μ. It remains to show:

  (i) A canonical triple is uniquely determined by (16.11).
 (ii) For every infinitely divisible distribution, there exists a canonical triple such that (16.11) holds.

(i) Uniqueness. For $\varepsilon \in [0, 1/2)$ and $t \geq 0$, define

$$g_{t,\varepsilon}(x) = e^{itx} - 1 - itx \, \mathbb{1}_{\{|x| < 1 - \varepsilon\}}.$$

For $\varepsilon = 0$, this is the function in the Lévy-Khinchin formula (16.11). However, since we will work with weak convergence and since $g_{t,\varepsilon}$ is discontinuous at $x = -(1 - \varepsilon)$ and $x = 1 - \varepsilon$, we will later need to adjust the parameter ε in such a way that ν does not have atoms at these points of discontinuity. Finally, we will let $\varepsilon \to 0$.

Obviously, we have $|g_{t,\varepsilon}(x)| \leq 2 + |t|$. Hence, for $x \neq 0$,

$$\left| \frac{g_{t,\varepsilon}(x)}{t^2(1 \wedge x^2)} \right| \leq \frac{2 + |t|}{t^2} \frac{1}{1 \wedge x^2} \xrightarrow{t \to \infty} 0. \tag{16.12}$$

For $x \in (-1/2, 1/2)$, by Lemma 15.31, we have

$$\left| g_{t,\varepsilon}(x) \right| = \left| e^{itx} - 1 - itx \right| \leq (tx)^2.$$

For $t \geq 1$, $\varepsilon \in [0, 1/2)$ and $x \neq 0$, we thus have (note that $(2 + |t|)/t^2 \leq 3$)

$$\left| \frac{g_{t,\varepsilon}(x)}{t^2(1 \wedge x^2)} \right| \leq 12. \tag{16.13}$$

Since (16.10) holds, by the dominated convergence theorem,

$$\lim_{t \to \infty} \frac{\psi(t)}{t^2} = -\frac{\sigma^2}{2} + \lim_{t \to \infty} \frac{ib}{t} + \lim_{t \to \infty} \int_{-\infty}^{\infty} \left(\frac{g_t(x)}{t^2(1 \wedge x^2)} \right) (1 \wedge x^2) \nu(dx) \tag{16.14}$$

$$= -\frac{\sigma^2}{2}.$$

This implies the uniqueness of σ^2. Thus we can and will assume $\sigma^2 = 0$ in the following. Define

$$\overline{\psi}(t) = \psi(t) - \frac{1}{2} \int_{t-1}^{t+1} \psi(s) \, ds. \tag{16.15}$$

Then

$$\overline{\psi}(t) = \int_{\mathbb{R}} e^{itx} \left(1 - \frac{1}{2} \int_{-1}^{1} e^{isx}\, ds\right) v(dx) = \int e^{itx} h(x)\, v(dx), \qquad (16.16)$$

where $h(x) = 1 - \frac{\sin(x)}{x}$ for $x \neq 0$ and $h(0) = 0$. Define $\hat{h}(x) = h(x)/(1 \wedge x^2)$ for $x \neq 0$ and $\hat{h}(0) = 1/6$. Clearly, h and $\hat{h}$ are bounded and continuous and

$$\frac{1}{7} < 1 - \sin(1) \leq \hat{h}(x) \leq \frac{3}{2} \qquad \text{for any } x \in \mathbb{R}. \qquad (16.17)$$

$\overline{\psi}$ is the characteristic function of $\tilde{v} \in \mathcal{M}_f(\mathbb{R})$, where $\tilde{v}(dx) = h(x)v(dx)$. Hence $\tilde{v}$ is uniquely determined by ψ. Since $v(dx) = (\mathbb{1}_{\{x\neq0\}}/h(x))\tilde{v}(dx)$, v is also uniquely determined by ψ. Now the number b is the difference of the remaining terms.

(ii) **Existence of a canonical triple.** Let μ be infinitely divisible and let

$$\psi(t) = \log \int e^{itx}\, \mu(dx).$$

Clearly, $\mathrm{Im}(\psi)$ is odd and $\mathrm{Re}(\psi(t)) \leq 0$ for all $t \in \mathbb{R}$. Hence $\overline{\psi}(0) \geq 0$ (with $\overline{\psi}$ from (16.15)) and $\overline{\psi}(0) = 0$ if $\mathrm{Re}\psi(t) = $ for all $t \in [-1, t]$. By Exercise 15.2.4, this is the case if and only if $\mu = \delta_b$ for some $b \in \mathbb{R}$. In this case, $(0, b, 0)$ is the corresponding canonical triple.

Now assume $\overline{\psi}(0) > 0$. By Theorem 16.5, there exists a sequence $(v_n)_{n\in\mathbb{N}}$ in $\mathcal{M}_f(\mathbb{R})$ with $\mathrm{CPoi}_{v_n} \overset{n\to\infty}{\longrightarrow} \mu$ and $v_n(\{0\}) = 0$ for any $n \in \mathbb{N}$. Define

$$b_{n,\varepsilon} = \int x\, \mathbb{1}_{\{|x|<1-\varepsilon\}}\, v_n(dx) \qquad \text{for } \varepsilon \in [0, 1/2).$$

Then, by (16.1) and with $g_{t,\varepsilon}$ from (i) (and for any $\varepsilon \in [0, 1/2)$),

$$\psi_n(t) := \log \int e^{itx}\, \mathrm{CPoi}_{v_n}(dx) = \int (e^{itx} - 1)\, v_n(dx) = \int g_{t,\varepsilon}\, dv_n + ib_{n,\varepsilon}t.$$

As in (16.16), we have

$$\overline{\psi}_n(t) := \psi_n(t) - \frac{1}{2} \int_{t-1}^{t+1} \psi_n(s)\, ds = \int e^{itx} h(x)\, v_n(dx).$$

As $\psi_n \overset{n\to\infty}{\longrightarrow} \psi$ converges uniformly on compact sets (Theorem 15.24(i)), and since ψ is continuous and thus locally bounded, we have $\overline{\psi}_n \overset{n\to\infty}{\longrightarrow} \overline{\psi}$ pointwise. Therefore,

$$\int e^{itx} h(x)\, v_n(dx) \overset{n\to\infty}{\longrightarrow} \overline{\psi}(t). \tag{16.18}$$

In particular, $\overline{\psi}_n(0) > 0$ for large n. If we let $\tilde{v}_n(dx) = (h(x)/\overline{\psi}_n(0))v_n(dx) \in \mathcal{M}_1(\mathbb{R})$, then $\int e^{itx}\tilde{v}_n(dx) \overset{n\to\infty}{\longrightarrow} \overline{\psi}(t)/\overline{\psi}(0)$ and the right-hand side is continuous. Hence, by Lévy's continuity theorem, there is a $\tilde{v} \in \mathcal{M}_1(\mathbb{R})$ with $\tilde{v}_n \overset{n\to\infty}{\longrightarrow} \tilde{v}$ and

$$\overline{\psi}(t) = \overline{\psi}(0) \int e^{itx} \tilde{v}(dx).$$

Let $\sigma^2 := -6\,\overline{\psi}(0)\,\tilde{v}(\{0\})$ and define a canonical measure v by

$$v(dx) = \frac{\overline{\psi}(0)}{h(x)} \mathbb{1}_{\{x\neq 0\}}\, \tilde{v}(dx).$$

For any $t \in \mathbb{R}$ and $\varepsilon \in [0, 1/2)$, the map (compare (15.8))

$$f_{t,\varepsilon} : \mathbb{R} \to \mathbb{C}, \quad x \mapsto \begin{cases} \frac{g_{t,\varepsilon}(x)}{h(x)}, & \text{if } x \neq 0, \\ -3t^2, & \text{if } x = 0, \end{cases}$$

is continuous except for the points $|x| = 1 - \varepsilon$ and is bounded (by $84\,t^2$, see (16.13) and (16.17)). Since $v((-1/2, 1/2)^c) < \infty$, the set of ε such that $v(\{1 - \varepsilon\} \cup \{-1 + \varepsilon\}) = 0$ is dense in $[0, 1/2]$. Let $(\varepsilon_k)_{k\in\mathbb{N}}$ be a sequence in $[0, 1/2]$ such that $\varepsilon_k \downarrow 0$ and $v(\{1 - \varepsilon_k\} \cup \{-1 + \varepsilon_k\}) = 0$. Fix $k \in \mathbb{N}$. By the Portemanteau Theorem (Theorem 13.16(iii)), we infer

$$\int g_{t,\varepsilon_k}\, dv_n = \overline{\psi}_n(0) \int f_{t,\varepsilon_k}\, d\tilde{v}_n \overset{n\to\infty}{\longrightarrow} \overline{\psi}(0) \int f_{t,\varepsilon_k}\, d\tilde{v} = -\frac{\sigma^2}{2}t^2 + \int g_{t,\varepsilon_k}\, dv.$$

Hence also the limit

$$it\, b_{\varepsilon_k} := \lim_{n\to\infty} it\, b_{n,\varepsilon_k} = \lim_{n\to\infty}\left(\psi_n(t) - \int g_{t,\varepsilon_k}\, dv_n\right) = \psi(t) + \frac{\sigma^2}{2}t^2 - \int g_{t,\varepsilon_k}\, dv$$

exists and we have

$$\psi(t) = -\frac{\sigma^2}{2}t^2 + ib_{\varepsilon_k}t + \int g_{t,\varepsilon_k}\, dv.$$

As $|g_{t,\varepsilon_k}(x)| \leq 12t^2(1 \wedge x^2)$, and since $g_{t,\varepsilon_k}(x) \overset{k \to \infty}{\longrightarrow} g_{t,0}(x)$, the dominated convergence theorem yields

$$\int g_{t,0}\, dv = \lim_{k \to \infty} \int g_{t,\varepsilon_k}\, dv.$$

Also, since $v((-1/2, 1/2)^c) < \infty$, the limit

$$b := \lim_{k \to \infty} b_{\varepsilon_k}$$

exists and we have

$$\psi(t) = -\frac{\sigma^2}{2} t^2 + ibt + \int g_{t,0}\, dv. \qquad\qquad \square$$

Remark 16.18 There are many versions of the Lévy–Khinchin formula

$$\psi(t) = -\frac{\sigma^2}{2} t^2 + ibt + \int \left(e^{itx} - 1 - it\, f(x)\right) v(dx)$$

that differ in the function $it\, f(x)$ that is subtracted for the centering in the integral. We chose $f(x) = x\, \mathbb{1}_{\{|x|<1\}}$ since this fits best to the construction with the random variables X_k. However, for a given canonical measure v, any function $\tilde{f}$ for which $\int |f - \tilde{f}|\, dv < \infty$ holds is possible; that is, every $\tilde{f}$ for which $|f(x) - \tilde{f}(x)|/(1 \wedge x^2)$ is bounded. One common function is, e.g., $\tilde{f}(x) = \sin(x)$. The Lévy measure and the Gaussian coefficient σ^2 do not change but the b differs:

$$\tilde{b} - b = \int \left(\tilde{f} - f\right) dv.$$

If v is a measure that is concentrated on $(0, \infty)$ and such that $\int (1 \wedge x)\, v(dx) < \infty$ holds, then this f is integrable with respect to v and can thus be replaced by $\tilde{f} = 0$. Hence we recover Theorem 16.14 as a special case. However, condition (16.10) is weaker than $\int (1 \wedge x)\, v(dx) < \infty$ and thus describes a larger class of measures than is considered in Theorem 16.14. This implies that to a canonical triple $(b, 0, v)$ with $v((-\infty, 0)) = 0$ and $\int (1 \wedge x)\, v(dx) = \infty$, there corresponds an infinitely divisible probability distribution μ that is not concentrated on $[0, \infty)$, no matter how b is chosen. $\lozenge$

Reflection In the proof of Theorem 16.17, why was it necessary to introduce the $\varepsilon > 0$? How can we get rid of this technicality by replacing the function $f(x) = x\, \mathbb{1}_{\{|x|<1\}}$ by $\tilde{f}(x) = \sin(x)$, as in Remark 16.18? What are the problems that come with this approach? ♠♠

For a given infinitely divisible distribution μ, we can compute the canonical measure ν as the vague limit

$$\nu = \operatorname*{v\text{-}lim}_{n\to\infty} n\mu^{*1/n}\big|_{\mathbb{R}\setminus\{0\}}. \tag{16.19}$$

Example 16.19 For the Cauchy distribution Cau_a with $\psi(t) = -a\,|t|$, by symmetry, we get $b = 0$ and, by (16.14), $\sigma^2 = -2\lim_{t\to\infty}\psi(t)/t^2 = 0$. Finally, if $A \subset \mathbb{R}$ with $(-\varepsilon, \varepsilon) \cap A = \emptyset$ for some $\varepsilon > 0$, then

$$n\,\mathrm{Cau}_{1/n}(A) = \frac{1}{\pi}\int_A \frac{n^2}{1+(nx)^2}\,dx \xrightarrow{n\to\infty} \frac{1}{\pi}\int_A \frac{1}{x^2}\,dx.$$

Hence Cau_1 has the canonical triple $\big(0, 0, (\pi x^2)^{-1}dx\big)$. $\Diamond$

> **Takeaways** An infinitely divisible random variable can be written as a sum of arbitrarily many independent and identically distributed random variables. On the other hand, it can also be split into three characteristic parts: a deterministic number, a normally distributed random variable and a mixture of Poisson jumps of various sizes (Lévy-Khinchin formula). Every infinitely divisible distribution is a weak limit of compound Poisson distributions.

Exercise 16.1.1 Use a variance argument to show that an infinitely divisible distribution that is concentrated on a bounded interval is a Dirac measure. ♣

Exercise 16.1.2 Let φ be infinitely divisible, and for every $n \in \mathbb{N}$, let φ_n be a CFP with $\varphi_n^n = \varphi$. Use Lévy's continuity theorem to show that $\varphi_n \xrightarrow{n\to\infty} 1$ uniformly on compact sets $\varphi_n \xrightarrow{n\to\infty} 1$. Conclude that $\varphi(t) \neq 0$ for all $t \in \mathbb{R}$. ♣

Exercise 16.1.3 Under the conditions of Theorem 16.14, show that

$$\alpha = \sup\big\{x \geq 0 : \mu([0, x)) = 0\big\}. \quad ♣$$

16.2 Stable Distributions

A distribution μ on the real numbers is called stable if for any $n \in \mathbb{N}$, the n-fold convolution μ^{*n} equals μ up to an affine linear transformation. Hence stability can be interpreted as self-similarity. We first show that the class of stable distributions is rather simple and can easily be parameterized. Then we quote results which say that stable distributions are exactly those distributions that occur as limits of sums of i.i.d. random variables.

Symmetric Stable Distributions

For $\alpha \in (0, 2)$, let

$$\theta_\alpha := \int_{\mathbb{R}} (1 - \cos(x)) |x|^{-\alpha-1} dx = \begin{cases} -2\Gamma(-\alpha)\cos(\alpha\pi/2), & \text{if } \alpha \neq 1, \\ \pi, & \text{if } \alpha = 1. \end{cases}$$

(Note that the integral diverges for $\alpha \in \mathbb{R} \setminus (0, 2)$). Then $\nu_\alpha(dx) = \theta_\alpha^{-1} |x|^{-\alpha-1} dx$ is a canonical measure since

$$\int (1 \wedge x^2) \, \nu_\alpha(dx) = 2\theta_\alpha^{-1}(\alpha^{-1} + (2-\alpha)^{-1}) < \infty.$$

Let ψ_α be the logarithm of the characteristic function that corresponds to the infinitely divisible measure μ_α with canonical triple $(0, 0, \nu_\alpha)$. By the Lévy–Khinchin formula, we have

$$\psi_\alpha(t) = \int_{-\infty}^{\infty} \left(e^{itx} - 1 - itx \, \mathbb{1}_{\{|x|<1\}}\right) \theta_\alpha^{-1} |x|^{-\alpha-1} dx$$

$$= -\theta_\alpha^{-1} \int_{-\infty}^{\infty} \left(1 - \cos(tx)\right) |x|^{-\alpha-1} dx$$

$$= -|t|^\alpha.$$

Hence $\varphi_\alpha(t) := e^{-|t|^\alpha}$ is the characteristic function of the infinitely divisible measure μ_α, which is called the symmetric **stable distribution** with index α. The name is due to the fact that, for i.i.d. random variables $X_1, X_2, \ldots$ that are μ_α-distributed, we have

$$X_1 + \ldots + X_n \overset{\mathcal{D}}{=} n^{1/\alpha} X_n \quad \text{for all } n \in \mathbb{N}. \tag{16.20}$$

General Stable Distributions

Motivated by equation (16.20), we present a somewhat more general notion of stability of a distribution.

Definition 16.20 (Stable distribution) *Let $\mu \in \mathcal{M}_1(\mathbb{R})$ be a probability distribution on the real numbers that is not concentrated in one point. Assume that $X_1, X_2, \ldots$ are i.i.d. random variables with distribution μ. The distribution μ is* *said to be* **stable in the broad sense** *if there exist nonnegative numbers* $a_1, a_2, \ldots$ *and real numbers* $d_1, d_2, \ldots$ *such that*

$$X_1 + \ldots + X_n \overset{\mathcal{D}}{=} a_n X_1 + d_n \quad \text{for all } n \in \mathbb{N}. \tag{16.21}$$

μ *is called* **stable** *(in the strict sense), if (16.21) holds with* $d_1 = d_2 = \ldots = 0$.

μ *is called stable in the broad sense with index* $\alpha \in (0, 2]$, *if (16.21) holds with* $a_n = n^{1/\alpha}$, $n \in \mathbb{N}$. *It is called stable (in the strict sense) with index* $\alpha \in (0, 2]$, *if in addition, we can choose* $d_1 = d_2 = \ldots = 0$.

Remark 16.21 If μ is stable in the broad sense, then it is infinitely divisible. $\Diamond$

Theorem 16.22 *Let* μ *be stable in the broad sense.*

 (i) There is an $\alpha \in (0, 2]$ *such that* μ *is stable in the broad sense with index* α.
 (ii) If $\alpha = 2$, *then* μ *is a normal distribution.*
(iii) If $\alpha \in (0, 2)$, *then the Lévy measure* ν *of* μ *has the density*

$$\frac{\nu(dx)}{dx} = \begin{cases} c^- \, (-x)^{-\alpha-1}, & \text{if } x < 0, \\ c^+ x^{-\alpha-1}, & \text{if } x > 0, \end{cases} \tag{16.22}$$

for some $c^-, c^+ \geq 0$, $c^- + c^+ > 0$.
 (iv) If $\alpha \neq 1$, *then there exists a* $b \in \mathbb{R}$ *such that* $\mu * \delta_{-b}$ *is stable with index* α.
  (v) If $\alpha = 1$, *then* $d_n = (c^+ - c^-) n \log(n)$, $n \in \mathbb{N}$. *If* $c^- = c^+$, *then* μ *is a Cauchy distribution.*

Remark 16.23 If μ is infinitely divisible with Lévy measure ν given by (16.22), then $\psi(t) := \log \int e^{itx} \, \mu(dx)$ is given by

$$\psi(t) = \begin{cases} |t|^\alpha \Gamma(-\alpha)\left[(c^+ + c^-) \cos\left(\frac{\pi\alpha}{2}\right) - i \, \text{sign}(t) \, (c^+ - c^-) \sin\left(\frac{\pi\alpha}{2}\right)\right], & \alpha \neq 1, \\ -|t|(c^+ + c^-)\left[\frac{\pi}{2} + i \, \text{sign}(t)(c^+ - c^-) \log(|t|)\right], & \alpha = 1. \end{cases} \tag{16.23}$$

$\Diamond$

Lemma 16.24 *Let* μ *be infinitely divisible with canonical triple* (σ^2, b, ν); *that is, with log-characteristic function* $\psi(t) := \log\left(\int e^{itx} \mu(dx)\right)$ *given by*

$$\psi(t) = -\frac{\sigma^2}{2}t^2 + ibt + \int \left(e^{itx} - 1 - itx \, \mathbb{1}_{\{|x|<1\}}\right) \nu(dx).$$

Further, let $a > 0$, $d \in \mathbb{R}$, $n \in \mathbb{N}$ *and let* $X, X_1, \ldots, X_n$ *be i.i.d. random variables with distribution* μ.

 (i) The canonical triple of $X_1 + \ldots + X_n$ *is* $(n\sigma^2, nb, n\nu)$.

(ii) *The canonical triple of* $aX + d$ *is* $(a^2\sigma^2, \tilde{b}, v \circ m_a^{-1})$, *where* $m_a : \mathbb{R} \to \mathbb{R}$, $x \mapsto ax$ *is the multiplication by* a *and*

$$\tilde{b} := ab + d + a \int (\mathbb{1}_{\{|x|<1/a\}} - \mathbb{1}_{\{|x|<1\}}) x \, v(dx). \tag{16.24}$$

Proof

(i) The log-characteristic function of $X_1 + \ldots + X_n$ is $n\psi$.

(ii) The log-characteristic function of $aX + d$ is

$$\psi_{aX+d}(t) = \psi(at) + idt$$

$$= -\frac{a^2\sigma^2}{2}t^2 + i(ab + d)t + \int \left(e^{iatx} - 1 - iatx \, \mathbb{1}_{\{|x|<1\}}\right) v(dx)$$

$$= -\frac{a^2\sigma^2}{2}t^2 + i\tilde{b}t + \int \left(e^{iatx} - 1 - iatx \, \mathbb{1}_{\{|x|<1/a\}}\right) v(dx)$$

$$= -\frac{a^2\sigma^2}{2}t^2 + i\tilde{b}t + \int \left(e^{itx} - 1 - itx \, \mathbb{1}_{\{|x|<1\}}\right) v \circ m_a^{-1}(dx). \qquad \square$$

Lemma 16.25 (Scaling of the canonical triple) *Under the assumptions of Theorem 16.22, let* (σ^2, b, v) *be the canonical triple of* μ.

(i) *We have*

$$(a_n^2 - n)\sigma^2 = 0 \quad \text{for all } n \in \mathbb{N} \tag{16.25}$$

and (with m_{a_n} *as in Lemma 16.24)*

$$nv = v \circ m_{a_n}^{-1} \quad \text{for all } n \in \mathbb{N}. \tag{16.26}$$

(ii) *If* $v = 0$, *then* $a_n = n^{1/2}$ *for all* $n \in \mathbb{N}$ *and*

$$d_n = b \left(n - n^{1/2}\right). \tag{16.27}$$

(iii) *Assume that* $\alpha \in (0, 2)$, $a_n = n^{1/\alpha}$, *and that* v *is given by (16.22). Then we have*

$$d_n = \left(b + \frac{c^+ - c^-}{\alpha - 1}\right)\left(n - n^{1/\alpha}\right) \quad \text{if } \alpha \neq 1, \tag{16.28}$$

and

$$d_n = (c^+ - c^-) n \log(n) \quad \text{if } \alpha = 1. \tag{16.29}$$

Proof

(i) Let $(a_n^2\sigma^2, \tilde{b}_n, \nu \circ m_{a_n}^{-1})$ be the canonical triple of $a_n X + d_n$ as determined in the preceding lemma and let $(n\sigma^2, nb, n\nu)$ be the canonical triple of $X_1 + \ldots + X_n$. By (16.21) and due to the uniqueness of the canonical triple (Theorem 16.17), we infer $a_n^2\sigma^2 = n\sigma^2$, $\tilde{b}_n = nb$ and $\nu \circ m_{a_n}^{-1} = n\nu$.

(ii) If $\nu = 0$, then $\sigma^2 > 0$, since by assumption, μ is not concentrated in one point. Hence, by (16.25), we get $a_n = n^{1/2}$. By virtue of Lemma 16.24(ii), we have $nb = \tilde{b}_n = bn^{1/2} + d_n$ and thus (16.27) holds.

(iii) Using (16.24), we compute $\tilde{b}_n$ more explicitly:

$$nb = \tilde{b}_n = bn^{1/\alpha} + d_n - n^{1/\alpha} \int \mathbb{1}_{\{n^{-1/\alpha} \le |x| < 1\}} x \, \nu(dx)$$

$$= bn^{1/\alpha} + d_n - n^{1/\alpha}(c^+ - c^-) \int_{n^{-1/\alpha}}^{1} x^{-\alpha} \, dx$$

$$= bn^{1/\alpha} + d_n - (c^+ - c^-) \begin{cases} (1-\alpha)^{-1}(n^{1/\alpha} - n), & \text{if } \alpha \ne 1, \\ n\log(n), & \text{if } \alpha = 1. \end{cases}$$

Rearranging terms yields (16.28) and (16.29).   □

Proof (of Theorem 16.22) We distinguish the cases $\liminf_{n\to\infty} a_n n^{-1/2} < \infty$ and "$= \infty$".

Case 1.   Assume that $\liminf_{n\to\infty} a_n n^{-1/2} < \infty$. Let $C \in [1, \infty)$ and let $(n_k)_{k\in\mathbb{N}}$ be a subsequence such that $a_{n_k} n_k^{-1/2} \le C$ for any $k \in \mathbb{N}$. Then for any $x \in \mathbb{R} \setminus \{0\}$, we have

$$C^2 \ge n_k^{-1}(1 \vee a_{n_k}^2) \ge \frac{n_k^{-1}(1 \wedge a_{n_k}^2 x^2)}{1 \wedge x^2} \xrightarrow{k\to\infty} 0.$$

Using (16.26) and (16.10), the dominated convergence theorem yields

$$\int_{-\infty}^{\infty} (1 \wedge x^2) \, \nu(dx) = \int_{-\infty}^{\infty} \frac{n_k^{-1}(1 \wedge a_{n_k}^2 x^2)}{1 \wedge x^2} (1 \wedge x^2) \, \nu(dx) \xrightarrow{k\to\infty} 0.$$

That is, we have $\nu = 0$. By Lemma 16.25(ii), we see that $\mu * \delta_{-b}$ is stable with index 2. This shows (ii).

Case 2.   Assume that

$$a_n n^{-1/2} \xrightarrow{n\to\infty} \infty. \tag{16.30}$$

By (16.25), we have $\sigma^2 = 0$ and hence $\nu \neq 0$. We define the function

$$F(x) = \begin{cases} \nu([x, \infty)), & \text{if } x > 0, \\ \nu((-\infty, x]), & \text{if } x < 0. \end{cases}$$

Since we have $\nu \neq 0$, there is an $x_0 \in \mathbb{R} \setminus \{0\}$ such that $F(x_0) > 0$. By symmetry, we may assume that $x_0 > 0$. Using (16.26), we infer

$$n\, F(x) = F(x/a_n) \quad \text{for any } x \in \mathbb{R} \setminus \{0\},\ n \in \mathbb{N},$$

and thus

$$F\left(\left(\frac{a_{n+1}}{a_n}\right)^k x_0\right) = \left(\frac{n}{n+1}\right)^k F(x_0) \quad \text{for any } k \in \mathbb{Z}.$$

We can rephrase this as

$$F(x) = (x/x_0)^{-\alpha_n} F(x_0) \quad \text{for any } x \in \left\{(a_{n+1}/a_n)^k x_0 : k \in \mathbb{Z}\right\},$$

where $\alpha_n := \log((n+1)/n)/\log(a_{n+1}/a_n)$. Since F is monotone decreasing and since $F(x) \xrightarrow{x \to \infty} 0$, we have $\alpha_n > 0$ for all $n \in \mathbb{N}$, and

$$\left(\frac{m}{m+1}\right)\left(\frac{x}{x_0}\right)^{-\alpha_m} \leq \frac{F(x)}{F(x_0)} \leq \left(\frac{n+1}{n}\right)\left(\frac{x}{x_0}\right)^{-\alpha_n} \quad \text{for } x > 0,\ m, n \in \mathbb{N}.$$

Letting $x \to \infty$, we obtain $\alpha_m \geq \alpha_n$. By symmetry, we also get $\alpha_m \leq \alpha_n$. Hence, we define $\alpha := \alpha_1 > 0$ and get $a_n = n^{1/\alpha}$ for all $n \in \mathbb{N}$ (note that (16.21) implies $a_1 = 1$). By the assumption (16.30), we have $\alpha < 2$. This shows (i).

We have $F(1) = x_0^\alpha F(x_0) > 0$ and $F(x) = x^{-\alpha} F(1)$ for all $x > 0$. Similarly, we get $F(x) = (-x)^{-\alpha} F(-1)$ for $x < 0$ (with the same $\alpha \in (0, 2)$ since it is determined by the sequence $(a_n)_{n \in \mathbb{N}}$). Defining $c^+ = \alpha\, \nu([1, \infty))$ and $c^- := \alpha \nu((-\infty, -1])$, we get (16.22) and thus (iii) and (i).

The statements (iv) and (v) are immediate consequences of Lemma 16.25. $\square$

Convergence to Stable Distributions

To complete the picture, we cite theorems from [54, Chapter XVII.5] (see also [62] and [128]) that state that only stable distributions occur as limiting distributions of rescaled sums of i.i.d. random variables $X_1, X_2, \ldots$.

In the following, let $X, X_1, X_2, \ldots$ be i.i.d. random variables and for $n \in \mathbb{N}$, let $S_n = X_1 + \ldots + X_n$.

Definition 16.26 (Domain of attraction) *Let* $\mu \in \mathcal{M}_1(\mathbb{R})$ *be nontrivial. The* **domain of attraction** $\mathrm{Dom}(\mu) \subset \mathcal{M}_1(\mathbb{R})$ *is the set of all distributions* $\mathbf{P}_X$ *with the property that there exist sequences of numbers* $(a_n)_{n\in\mathbb{N}}$ *and* $(d_n)_{n\in\mathbb{N}}$ *with*

$$\frac{S_n - d_n}{a_n} \overset{n\to\infty}{\Longrightarrow} \mu.$$

If μ *is stable (in the broader sense) with index* $\alpha \in (0, 2]$, *then* $\mathbf{P}_X$ *is said to be in the* **domain of normal attraction** *if we can choose* $a_n = n^{1/\alpha}$.

Theorem 16.27 *Let* $\mu \in \mathcal{M}_1(\mathbb{R})$ *be nontrivial. Then* $\mathrm{Dom}(\mu) \neq \emptyset$ *if and only if* μ *is stable (in the broader sense). In this case,* $\mu \in \mathrm{Dom}(\mu)$.

In the following, an important role is played by the function

$$U(x) := \mathbf{E}\big[X^2 \, \mathbb{1}_{\{|X|\le x\}}\big]. \tag{16.31}$$

A function $H : (0, \infty) \to (0, \infty)$ is called **slowly varying** at ∞ if

$$\lim_{x\to\infty} \frac{H(\gamma x)}{H(x)} = 1 \quad \text{for all } \gamma > 0.$$

In the following, we assume that there exists an $\alpha \in (0, 2]$ such that

$$U(x)\, x^{\alpha-2} \text{ is slowly varying at } \infty. \tag{16.32}$$

Theorem 16.28

(i) *If* $\mathbf{P}_X$ *is in the domain of attraction of some distribution, then there exists an* $\alpha \in (0, 2]$ *such that (16.32) holds.*

(ii) *In the case* $\alpha = 2$, *we have: If* $\mathbf{P}_X$ *is not concentrated at one point, then (16.32) implies that* $\mathbf{P}_X$ *is in the domain of attraction of some distribution.*

(iii) *In the case* $\alpha \in (0, 2)$, *we have:* $\mathbf{P}_X$ *is in the domain of attraction of some distribution if and only if (16.32) holds and the limit*

$$p := \lim_{x\to\infty} \frac{\mathbf{P}[X \ge x]}{\mathbf{P}[|X| \ge x]} \quad \text{exists.} \tag{16.33}$$

Theorem 16.29 *Let* $\mathbf{P}_X$ *be in the domain of attraction of an* α-*stable distribution (that is, assume that condition (ii) or (iii) of Theorem 16.28 holds), and assume that* $(a_n)_{n\in\mathbb{N}}$ *is such that*

$$C := \lim_{n\to\infty} \frac{n\, U(a_n)}{a_n^2} \in (0, \infty)$$

exists. Further, let μ *be the stable distribution with index* α *whose characteristic function is given by (16.23) with* $c^+ = Cp$ *and* $c^- = C(1 - p)$.

(i)   *In the case $\alpha \in (0, 1)$, let $b_n \equiv 0$.*
(ii) *In the case $\alpha = 2$ and $\mathbf{Var}[X] < \infty$, let $\mathbf{E}[X] = 0$.*
(iii) *In the case $\alpha \in (1, 2]$, let $d_n = n\,\mathbf{E}[X]$ for all $n \in \mathbb{N}$.*
(iv) *In the case $\alpha = 1$, let $d_n = n\,a_n\,\mathbf{E}[\sin(X/a_n)]$ for all $n \in \mathbb{N}$.*

Then

$$\frac{S_n - d_n}{a_n} \overset{n \to \infty}{\Longrightarrow} \mu.$$

Corollary 16.30 *If $\mathbf{P}_X$ is in the domain of attraction of a stable distribution with index α, then $\mathbf{E}\big[|X|^\beta\big] < \infty$ for all $\beta \in (0, \alpha)$ and $\mathbf{E}\big[|X|^\beta\big] = \infty$ if $\beta > \alpha$ and $\alpha < 2$.*

Takeaways A random variable X is called infinitely divisible if for any $n \in \mathbb{N}$ it can be written as a sum of n independent and identically distributed random variables. It is called *stable* if each of the summands has the same distribution as $b_n + X/a_n$ for some sequences (a_n) and (b_n). We have seen that in this case, we must have $a_n = n^{1/\alpha}$ for some $\alpha \in (0, 2]$. A stable distribution is characterised by its index α and a skewness parameter (and, of course, a scale parameter); see Remark 16.23.

Exercise 16.2.1 Let μ be an α-stable distribution and let φ be its characteristic function.

(i)   Show by a direct computation using only the definition of stability that $|\varphi(t) - 1| \leq C|t|^\alpha$ for t close to 0 (for some $C < \infty$).
(ii) Use Exercise 15.3.2 to infer that $\mu = \delta_0$ if $\alpha > 2$.
(iii) Modify the argument in order to show that for $\alpha > 2$, the α-stable distributions in the broad sense are also necessarily trivial. ♣

Exercise 16.2.2 Show that the distribution on $\mathbb{R}$ with density $f(x) = \dfrac{1 - \cos(x)}{\pi x^2}$ is not infinitely divisible. ♣

Exercise 16.2.3 Let Φ be the distribution function of the standard normal distribution $\mathcal{N}_{0,1}$ and let $F : \mathbb{R} \to [0, 1]$ be defined by

$$F(x) = \begin{cases} 2\left(1 - \Phi\left(x^{-1/2}\right)\right), & \text{if } x > 0, \\[2mm] 0, & \text{else.} \end{cases}$$

Show the following.

(i) F is the distribution function of a $\frac{1}{2}$-stable distribution.
(ii) If $X_1, X_2, \ldots$ are i.i.d. with distribution function F, then $\frac{1}{n}\sum_{k=0}^{n} X_k$ diverges almost surely for $n \to \infty$.

Hint: Compute the density of F, and show that the Laplace transform is given by $\lambda \mapsto e^{-\sqrt{2\lambda}}$. ♣

Exercise 16.2.4 Which of the following distributions is in the domain of attraction of a stable distribution and for which parameter?

(i) The distribution on $\mathbb{R}$ with density

$$f(x) = \begin{cases} \varrho\,\frac{1}{1+\alpha}\,|x|^\alpha, & \text{if } x < -1, \\ (1-\varrho)\,\frac{1}{1+\beta}\,x^\beta, & \text{if } x > 1, \\ 0, & \text{else.} \end{cases}$$

Here $\alpha, \beta < -1$ and $\varrho \in [0, 1]$.
(ii) The exponential distribution $\exp_\theta$ for $\theta > 0$.
(iii) The distribution on $\mathbb{N}$ with weights $c\,n^\alpha$ if n is even and $c\,n^\beta$ if n is odd. Here $\alpha, \beta < -1$ and $c = (2^\alpha \zeta(-\alpha) + (1 - 2^\beta)\zeta(-\beta))^{-1}$ (ζ is the Riemann zeta function) is the normalization constant. ♣

Chapter 17
Markov Chains

In spite of their simplicity, Markov processes with countable state space (and discrete time) are interesting mathematical objects with which a variety of real-world phenomena can be modeled. We give an introduction to the basic concepts and then study certain examples in more detail. The connection with discrete potential theory will be investigated later, in Chap. 19. Some readers might prefer to skip the somewhat technical construction of general Markov processes in Sect. 17.1.

There is a vast literature on Markov chains. For further reading, see, e.g., [21, 26, 64, 66, 91, 116, 123, 124, 144, 153].

17.1 Definitions and Construction

In the following, E is always a Polish space with Borel σ-algebra $\mathcal{B}(E)$, $I \subset \mathbb{R}$ and $(X_t)_{t \in I}$ is an E-valued stochastic process. We assume that $(\mathcal{F}_t)_{t \in I} = \mathbb{F} = \sigma(X)$ is the filtration generated by X.

Definition 17.1 *We say that X has the **Markov property** (MP) if, for every $A \in \mathcal{B}(E)$ and all $s, t \in I$ with $s \leq t$,*

$$\mathbf{P}\big[X_t \in A \,\big|\, \mathcal{F}_s\big] = \mathbf{P}\big[X_t \in A \,\big|\, X_s\big].$$

Remark 17.2 If E is a countable space, then X has the Markov property if and only if, for all $n \in \mathbb{N}$, all $s_1 < \ldots < s_n < t$ and all $i_1, \ldots, i_n, i \in E$ with $\mathbf{P}[X_{s_1} = i_1, \ldots, X_{s_n} = i_n] > 0$, we have

$$\mathbf{P}\big[X_t = i \,\big|\, X_{s_1} = i_1, \ldots, X_{s_n} = i_n\big] = \mathbf{P}\big[X_t = i \,\big|\, X_{s_n} = i_n\big]. \tag{17.1}$$

A. Klenke, *Probability Theory*, Universitext,
https://doi.org/10.1007/978-3-030-56402-5_17

In fact, (17.1) clearly implies the Markov property. On the other hand, if X has the Markov property, then (see (8.6)) $\mathbf{P}[X_t = i \mid X_{S_n}](\omega) = \mathbf{P}[X_t = i \mid X_{S_n} = i_n]$ for almost all $\omega \in \{X_{S_n} = i_n\}$. Hence, for $A := \{X_{S_1} = i_1, \ldots, X_{S_n} = i_n\}$ (using the Markov property in the second equation),

$$\mathbf{P}\big[X_t = i,\, X_{S_1} = i_1, \ldots, X_{S_n} = i_n\big]$$

$$= \mathbf{E}\big[\mathbf{E}[\mathbb{1}_{\{X_t=i\}} \mid \mathcal{F}_{S_n}]\mathbb{1}_A\big] \;=\; \mathbf{E}\big[\mathbf{E}[\mathbb{1}_{\{X_t=i\}} \mid X_{S_n}]\mathbb{1}_A\big]$$

$$= \mathbf{E}\big[\mathbf{P}[X_t = i \mid X_{S_n} = i_n]\mathbb{1}_A\big] \;=\; \mathbf{P}\big[X_t = i \mid X_{S_n} = i_n\big]\mathbf{P}[A].$$

Dividing both sides by $\mathbf{P}[A]$ yields (17.1). $\Diamond$

Definition 17.3 *Let $I \subset [0, \infty)$ be closed under addition and assume $0 \in I$. A stochastic process $X = (X_t)_{t \in I}$ is called a time-homogeneous **Markov process** with distributions $(\mathbf{P}_x)_{x \in E}$ on the space $(\Omega, \mathcal{A})$ if:*

(i) *For every $x \in E$, X is a stochastic process on the probability space $(\Omega, \mathcal{A}, \mathbf{P}_x)$ with $\mathbf{P}_x[X_0 = x] = 1$.*

(ii) *The map $\kappa : E \times \mathcal{B}(E)^{\otimes I} \to [0, 1]$, $(x, B) \mapsto \mathbf{P}_x[X \in B]$ is a stochastic kernel.*

(iii) *X has the time-homogeneous **Markov property** (MP): For every $A \in \mathcal{B}(E)$, every $x \in E$ and all $s, t \in I$, we have*

$$\mathbf{P}_x\big[X_{t+s} \in A \mid \mathcal{F}_s\big] = \kappa_t(X_s, A) \qquad \mathbf{P}_x\text{-a.s.}$$

*Here, for every $t \in I$, the **transition kernel** $\kappa_t : E \times \mathcal{B}(E) \to [0, 1]$ is the stochastic kernel defined for $x \in E$ and $A \in \mathcal{B}(E)$ by*

$$\kappa_t(x, A) := \kappa\big(x, \{y \in E^I : y(t) \in A\}\big) = \mathbf{P}_x[X_t \in A].$$

*The family $(\kappa_t(x, A),\, t \in I,\, x \in E,\, A \in \mathcal{B}(E))$ is also called the family of **transition probabilities** of X.*

We write $\mathbf{E}_x$ for expectation with respect to $\mathbf{P}_x$, $\mathcal{L}_x[X] = \mathbf{P}_x$ and $\mathcal{L}_x[X \mid \mathcal{F}] = \mathbf{P}_x[X \in \cdot \mid \mathcal{F}]$ (for a regular conditional distribution of X given $\mathcal{F}$).

*If E is countable, then X is called a **discrete Markov process**.*

*In the special case $I = \mathbb{N}_0$, X is called a **Markov chain**. In this case, κ_n is called the family of n-step transition probabilities.*

Remark 17.4 We will see that the existence of the transition kernels (κ_t) implies the existence of the kernel κ. Thus, a time-homogeneous Markov process is simply a stochastic process with the Markov property and for which the transition probabilities are time-homogeneous. Although it is sometimes convenient to allow also time-inhomogeneous Markov processes, for a wide range of applications it is sufficient to consider time-homogeneous Markov processes. We will not go into the details but will henceforth assume that all Markov processes are time-homogeneous. ◊

In the following, we will use the somewhat sloppy notation $\mathbf{P}_{X_s}[X \in \cdot] := \kappa(X_s, \cdot)$. That is, we understand X_s as the initial value of a *second* Markov process with the same distributions $(\mathbf{P}_x)_{x \in E}$.

Example 17.5 Let $Y_1, Y_2, \ldots$ be i.i.d. $\mathbb{R}^d$-valued random variables and let

$$S_n^x = x + \sum_{i=1}^{n} Y_i \quad \text{for } x \in \mathbb{R}^d \quad \text{and} \quad n \in \mathbb{N}_0.$$

Define probability measures $\mathbf{P}_x$ on $((\mathbb{R}^d)^{\mathbb{N}_0}, (\mathcal{B}(\mathbb{R}^d))^{\otimes \mathbb{N}_0})$ by $\mathbf{P}_x = \mathbf{P} \circ (S^x)^{-1}$. Then the canonical process $X_n : (\mathbb{R}^d)^{\mathbb{N}_0} \to \mathbb{R}^d$ is a Markov chain with distributions $(\mathbf{P}_x)_{x \in \mathbb{R}^d}$. The process X is called a random walk on $\mathbb{R}^d$ with initial value x. ◊

Example 17.6 In the previous example, it is simple to pass to continuous time; that is, $I = [0, \infty)$. To this end, let $(\nu_t)_{t \geq 0}$ be a convolution semigroup on $\mathbb{R}^d$ and let $\kappa_t(x, dy) = \delta_x * \nu_t(dy)$. In Theorem 14.50, for every $x \in \mathbb{R}^d$, we constructed a measure $\mathbf{P}_x$ on $((\mathbb{R}^d)^{[0,\infty)}, \mathcal{B}(\mathbb{R}^d)^{\otimes[0,\infty)})$ with

$$\mathbf{P}_x \circ (X_0, X_{t_1}, \ldots, X_{t_n})^{-1} = \delta_x \otimes \bigotimes_{i=0}^{n-1} \kappa_{t_{i+1}-t_i}$$

for any choice of finitely many points $0 = t_0 < t_1 < \ldots < t_n$. It is easy to check that the map $\kappa : \mathbb{R}^d \times \mathcal{B}(\mathbb{R}^d)^{\otimes[0,\infty)}, (x, A) \mapsto \mathbf{P}_x[A]$ is a stochastic kernel. The time-homogeneous Markov property is immediate from the fact that the increments are independent and stationary. ◊

Example 17.7 (See Example 9.5 and Theorem 5.36) Let $\theta > 0$ and $\nu_t^{\theta}(\{k\}) = e^{-\theta t} \frac{t^k \theta^k}{k!}$, $k \in \mathbb{N}_0$, the convolution semigroup of the Poisson distribution. The Markov process X on $\mathbb{N}_0$ with this semigroup is called a **Poisson process** with (jump) rate θ. ◊

As in Example 17.6, we will construct a Markov process for a more general Markov semigroup of stochastic kernels.

Theorem 17.8 Let $I \subset [0, \infty)$ be closed under addition and let $(\kappa_t)_{t \in I}$ be a Markov semigroup of stochastic kernels from E to E. Then there is a measurable space $(\Omega, \mathcal{A})$ and a Markov process $((X_t)_{t \in I}, (\mathbf{P}_x)_{x \in E})$ on $(\Omega, \mathcal{A})$ with transition probabilities

$$\mathbf{P}_x[X_t \in A] = \kappa_t(x, A) \quad \text{for all } x \in E, A \in \mathcal{B}(E), t \in I. \tag{17.2}$$

Conversely, for every Markov process X, Equation (17.2) defines a semigroup of stochastic kernels. By (17.2), the finite-dimensional distributions of X are uniquely determined.

Proof "$\Longrightarrow$" We construct X as a canonical process. Let $\Omega = E^{[0,\infty)}$ and $\mathcal{A} = \mathcal{B}(E)^{\otimes[0,\infty)}$. Further, let X_t be the projection on the tth coordinate. For $x \in E$, define (see Corollary 14.46) on $(\Omega, \mathcal{A})$ the probability measure $\mathbf{P}_x$ such that, for finitely many time points $0 = t_0 < t_1 < \ldots < t_n$, we have

$$\mathbf{P}_x \circ (X_{t_0}, \ldots, X_{t_n})^{-1} = \delta_x \otimes \bigotimes_{i=0}^{n-1} \kappa_{t_{i+1}-t_i}.$$

Then

$$\mathbf{P}_x\big[X_{t_0} \in A_0, \ldots, X_{t_n} \in A_n\big]$$
$$= \int_{A_{n-1}} \mathbf{P}_x\big[X_{t_0} \in A_0, \ldots, X_{t_{n-2}} \in A_{n-2}, X_{t_{n-1}} \in dx_{n-1}\big]$$
$$\kappa_{t_n-t_{n-1}}(x_{n-1}, A_n);$$

hence $\mathbf{P}_x[X_{t_n} \in A_n | \mathcal{F}_{t_{n-1}}] = \kappa_{t_n-t_{n-1}}(X_{t_{n-1}}, A_n)$. Thus X is recognized as a Markov process. Furthermore, we have $\mathbf{P}_x[X_t \in A] = (\delta_x \cdot \kappa_t)(A) = \kappa_t(x, A)$.
"$\Longleftarrow$" Now let $(X, (\mathbf{P}_x)_{x \in E})$ be a Markov process. Then a stochastic kernel κ_t is defined by

$$\kappa_t(x, A) := \mathbf{P}_x [X_t \in A] \quad \text{for all } x \in E, \ A \in \mathcal{B}(E), \ t \in I.$$

By the Markov property, we have

$$\kappa_{t+s}(x, A) = \mathbf{P}_x [X_{t+s} \in A] = \mathbf{E}_x \big[\mathbf{P}_{X_s} [X_t \in A]\big]$$
$$= \int \mathbf{P}_x [X_s \in dy] \mathbf{P}_y [X_t \in A]$$
$$= \int \kappa_s(x, dy)\kappa_t(y, A) = (\kappa_s \cdot \kappa_t)(x, A).$$

Hence $(\kappa_t)_{t \in I}$ is a Markov semigroup. $\square$

Theorem 17.9 *A stochastic process $X = (X_t)_{t \in I}$ is a Markov process if and only if there exists a stochastic kernel $\kappa : E \times \mathcal{B}(E)^{\otimes I} \to [0, 1]$ such that, for every bounded $\mathcal{B}(E)^{\otimes I} - \mathcal{B}(\mathbb{R})$-measurable function $f : E^I \to \mathbb{R}$ and for every $s \geq 0$ and $x \in E$, we have*

$$\mathbf{E}_x \big[f ((X_{t+s})_{t \in I}) \big| \mathcal{F}_s\big] = \mathbf{E}_{X_s} [f(X)] := \int_{E^I} \kappa(X_s, dy) f(y). \tag{17.3}$$

Proof "$\Longleftarrow$" The time-homogeneous Markov property follows by (17.3) with the function $f(y) = \mathbb{1}_A(y(t))$ since $\mathbf{P}_{X_s}[X_t \in A] = \mathbf{P}_x[X_{t+s} \in A | \mathcal{F}_s] = \kappa_t(X_s, A)$.

"$\Longrightarrow$" By the usual approximation arguments, it is enough to consider functions f that depend only on finitely many coordinates $0 \leq t_1 \leq t_2 \leq \cdots \leq t_n$. We perform the proof by induction on n.

For $n = 1$ and f an indicator function, this is the (time-homogeneous) Markov property. For general measurable f, the statement follows by the usual approximation arguments.

Now assume the claim is proved for $n \in \mathbb{N}$. Again it suffices to assume that f is an indicator function of the type $f(x) = \mathbb{1}_{B_1 \times \cdots \times B_{n+1}}(x_{t_1}, \ldots, x_{t_{n+1}})$ (with $B_1, \ldots, B_{n+1} \in \mathcal{B}(E)$). Using the Markov property (third and fifth equalities in the following equation) and the induction hypothesis (fourth equality), we get

$$\mathbf{E}_x\left[f\left((X_{t+s})_{t\geq 0}\right) \big| \mathcal{F}_s \right]$$

$$= \mathbf{E}_x\left[\mathbf{E}_x\left[f\left((X_{t+s})_{t\geq 0}\right) \big| \mathcal{F}_{t_n+s} \right] \big| \mathcal{F}_s \right]$$

$$= \mathbf{E}_x\left[\mathbf{E}_x\left[\mathbb{1}_{\{X_{t_{n+1}+s} \in B_{n+1}\}} \big| \mathcal{F}_{t_n+s} \right] \mathbb{1}_{B_1}(X_{t_1+s}) \cdots \mathbb{1}_{B_n}(X_{t_n+s}) \big| \mathcal{F}_s \right]$$

$$= \mathbf{E}_x\left[\mathbf{P}_{X_{t_n+s}}\left[X_{t_{n+1}-t_n} \in B_{n+1} \right] \mathbb{1}_{B_1}(X_{t_1+s}) \cdots \mathbb{1}_{B_n}(X_{t_n+s}) \big| \mathcal{F}_s \right]$$

$$= \mathbf{E}_{X_s}\left[\mathbf{P}_{X_{t_n}}\left[X_{t_{n+1}-t_n} \in B_{n+1} \right] \mathbb{1}_{B_1}(X_{t_1}) \cdots \mathbb{1}_{B_n}(X_{t_n}) \right]$$

$$= \mathbf{E}_{X_s}\left[\mathbf{P}_{X_0}\left[X_{t_{n+1}} \in B_{n+1} \big| \mathcal{F}_{t_n} \right] \mathbb{1}_{B_1}(X_{t_1}) \cdots \mathbb{1}_{B_n}(X_{t_n}) \right]$$

$$= \mathbf{E}_{X_s}\left[\mathbf{P}_{X_0}\left[X_{t_1} \in B_1, \ldots, X_{t_{n+1}} \in B_{n+1} \big| \mathcal{F}_{t_n} \right] \right]$$

$$= \mathbf{E}_{X_s}\left[f(X) \right]. \qquad \square$$

Corollary 17.10 *A stochastic process $(X_n)_{n\in\mathbb{N}_0}$ is a Markov chain if and only if*

$$\mathcal{L}_x\left[(X_{n+k})_{n\in\mathbb{N}_0} \big| \mathcal{F}_k \right] = \mathcal{L}_{X_k}\left[(X_n)_{n\in\mathbb{N}_0} \right] \quad \text{for every } k \in \mathbb{N}_0. \qquad (17.4)$$

Proof If the conditional distributions exist, then, by Theorem 17.9, the equation (17.4) is equivalent to X being a Markov chain. Hence we only have to show that the conditional distributions exist.

Since E is Polish, $E^{\mathbb{N}_0}$ is also Polish and we have $\mathcal{B}(E^{\mathbb{N}_0}) = \mathcal{B}(E)^{\otimes \mathbb{N}_0}$ (see Theorem 14.8). Hence, by Theorem 8.37, there exists a regular conditional distribution of $(X_{n+k})_{n\in\mathbb{N}_0}$ given $\mathcal{F}_k$. $\qquad \square$

Theorem 17.11 *Let $I = \mathbb{N}_0$. If $(X_n)_{n\in\mathbb{N}_0}$ is a stochastic process with distributions $(\mathbf{P}_x, \ x \in E)$, then the Markov property in Definition 17.3(iii) is implied by the existence of a stochastic kernel $\kappa_1 : E \times \mathcal{B}(E) \to [0, 1]$ with the property that for every $A \in \mathcal{B}(E)$, every $x \in E$ and every $s \in I$, we have*

$$\mathbf{P}_x\left[X_{s+1} \in A \,\middle|\, \mathcal{F}_s\right] = \kappa_1(X_s, A). \tag{17.5}$$

In this case, the n-step transition kernels κ_n can be computed inductively by

$$\kappa_n = \kappa_{n-1} \cdot \kappa_1 = \int_E \kappa_{n-1}(\,\cdot\,, dx)\,\kappa_1(x, \,\cdot\,).$$

In particular, the family $(\kappa_n)_{n\in\mathbb{N}}$ is a Markov semigroup and the distribution X is uniquely determined by κ_1.

Proof In Theorem 17.9, let $t_i = i$ for every $i \in \mathbb{N}_0$. For the proof of that theorem, only (17.5) was needed. $\square$

The (time-homogeneous) Markov property of a process means that, for fixed time t, the future (after t) depends on the past (before t) only via the present (that is, via the value X_t). We can generalize this concept by allowing random times τ instead of fixed times t.

Definition 17.12 *Let $I \subset [0, \infty)$ be closed under addition. A Markov process $(X_t)_{t\in I}$ with distributions $(\mathbf{P}_x, \ x \in E)$ has the **strong Markov property** if, for every a.s. finite stopping time τ, every bounded $\mathcal{B}(E)^{\otimes I} - \mathcal{B}(\mathbb{R})$ measurable function $f : E^I \to \mathbb{R}$ and every $x \in E$, we have*

$$\mathbf{E}_x\left[f\left((X_{\tau+t})_{t\in I}\right)\middle|\mathcal{F}_\tau\right] = \mathbf{E}_{X_\tau}[f(X)] := \int_{E^I} \kappa(X_\tau, dy)\,f(y). \tag{17.6}$$

Remark 17.13 If I is countable, then the strong Markov property holds if and only if, for every almost surely finite stopping time τ, we have

$$\mathcal{L}_x\left[(X_{\tau+t})_{t\in I}\,\middle|\,\mathcal{F}_\tau\right] = \mathcal{L}_{X_\tau}\left[(X_t)_{t\in I}\right] := \kappa(X_\tau, \,\cdot\,). \tag{17.7}$$

This follows just as in Corollary 17.10. $\Diamond$

Most Markov processes one encounters have the strong Markov property. In particular, for countable time sets, the strong Markov property follows from the Markov property. For continuous time, however, in general, some work has to be done to establish the strong Markov property.

Theorem 17.14 *If $I \subset [0, \infty)$ is countable and closed under addition, then every Markov process $(X_n)_{n\in I}$ with distributions $(\mathbf{P}_x)_{x\in E}$ has the strong Markov property.*

Proof Let $f : E^I \to \mathbb{R}$ be measurable and bounded. Then, for every $s \in I$, the random variable $\mathbb{1}_{\{\tau=s\}} \mathbf{E}_x\left[f\left((X_{s+t})_{t\in I}\right)\middle|\mathcal{F}_\tau\right]$ is measurable with respect to $\mathcal{F}_s$.

Using the tower property of the conditional expectation and Theorem 17.9 in the third equality, we thus get

$$\mathbf{E}_x\left[f\left((X_{\tau+t})_{t\in I}\right)\,\middle|\,\mathcal{F}_\tau\right] = \sum_{s\in I} \mathbb{1}_{\{\tau=s\}}\,\mathbf{E}_x\left[f\left((X_{s+t})_{t\in I}\right)\,\middle|\,\mathcal{F}_\tau\right]$$

$$= \sum_{s\in I}\mathbf{E}_x\left[\mathbb{1}_{\{\tau=s\}}\,\mathbf{E}_x\left[f\left((X_{s+t})_{t\in I}\right)\,\middle|\,\mathcal{F}_s\right]\,\middle|\,\mathcal{F}_\tau\right]$$

$$= \sum_{s\in I}\mathbf{E}_x\left[\mathbb{1}_{\{\tau=s\}}\,\mathbf{E}_{X_s}\left[f\left((X_t)_{t\in I}\right)\right]\,\middle|\,\mathcal{F}_\tau\right]$$

$$= \mathbf{E}_{X_\tau}\left[f\left((X_t)_{t\in I}\right)\right]. \qquad\qquad\square$$

As a simple application of the strong Markov property, we show the reflection principle for random walks.

Theorem 17.15 (Reflection principle) *Let $Y_1, Y_2, \dots$ be i.i.d. real random variables with symmetric distribution $\mathcal{L}[Y_1] = \mathcal{L}[-Y_1]$. Define $X_0 = 0$ and $X_n := Y_1 + \dots + Y_n$ for $n \in \mathbb{N}$. Then, for every $n \in \mathbb{N}_0$ and $a > 0$,*

$$\mathbf{P}\left[\sup_{m\le n} X_m \ge a\right] \le 2\,\mathbf{P}[X_n \ge a] - \mathbf{P}[X_n = a]. \qquad (17.8)$$

If we have $\mathbf{P}[Y_1 \in \{-1, 0, 1\}] = 1$, then for $a \in \mathbb{N}$ equality holds in (17.8).

Proof Let $a > 0$ and $n \in \mathbb{N}$. Define the time of first excess of a (truncated at $(n + 1)$),

$$\tau := \inf\{m \ge 0 : X_m \ge a\} \wedge (n + 1).$$

Then τ is a bounded stopping time and

$$\sup_{m\le n} X_m \ge a \quad\Longleftrightarrow\quad \tau \le n.$$

Let $f(m, X) = \mathbb{1}_{\{m\le n\}}\left(\mathbb{1}_{\{X_{n-m}>a\}} + \tfrac{1}{2}\mathbb{1}_{\{X_{n-m}=a\}}\right)$. Then

$$f\left(\tau, (X_{\tau+m})_{m\in\mathbb{N}_0}\right) = \mathbb{1}_{\{\tau\le n\}}\left(\mathbb{1}_{\{X_n>a\}} + \tfrac{1}{2}\mathbb{1}_{\{X_n=a\}}\right).$$

The strong Markov property of X yields

$$\mathbf{E}_0\left[f\left(\tau, (X_{\tau+m})_{m\ge 0}\right)\,\middle|\,\mathcal{F}_\tau\right] = \varphi\left(\tau, X_\tau\right),$$

where $\varphi(m, x) = \mathbf{E}_x[f(m, X)]$. (Recall that $\mathbf{E}_x$ denotes the expectation for X if $X_0 = x$.)

Due to the symmetry of Y_i, we have

$$
\varphi(m, x) \begin{cases} \geq \frac{1}{2}, & \text{if } m \leq n \text{ and } x \geq a, \\ = \frac{1}{2}, & \text{if } m \leq n \text{ and } x = a, \\ = 0, & \text{if } m > n. \end{cases}
$$

Hence

$$
\{\tau \leq n\} = \{\tau \leq n\} \cap \{X_\tau \geq a\} \subset \left\{\varphi(\tau, X_\tau) \geq \frac{1}{2}\right\} \cap \{\tau \leq n\}
$$

$$
= \{\varphi(\tau, X_\tau) > 0\} \cap \{\tau \leq n\}.
$$

Now (17.8) is implied by

$$
\mathbf{P}[X_n > a] + \frac{1}{2} \mathbf{P}[X_n = a] = \mathbf{E}\left[f\left(\tau, (X_{\tau+m})_{m \geq 0}\right)\right]
$$

$$
(17.9)
$$

$$
= \mathbf{E}_0\left[\varphi(\tau, X_\tau) \mathbb{1}_{\{\tau \leq n\}}\right] \geq \frac{1}{2} \mathbf{P}_0\left[\tau \leq n\right].
$$

Now assume $\mathbf{P}[Y_1 \in \{-1, 0, 1\}] = 1$ and $a \in \mathbb{N}$. Then $X_\tau = a$ if $\tau \leq n$. Hence

$$
\{\varphi(\tau, X_\tau) > 0\} \cap \{\tau \leq n\} = \left\{\varphi(\tau, X_\tau) = \frac{1}{2}\right\} \cap \{\tau \leq n\}.
$$

Thus, in the last step of (17.9), equality holds and hence also in (17.8). □

Reflection Find an example for strict inequality in (17.8). ♠

Reflection Consider the case $\mathbf{P}[Y_1 = 1] = \mathbf{P}[Y_1 = -1] = \frac{1}{2}$. Then we have equality in (17.8). In fact, in this case, the reflection principle can be derived also in an elementary way via a bijection that changes the signs of those Y_i with $i > \tau$. Each path of the process X of partial sums that ends above a corresponds to a unique path that reaches a but ends below a.

Try and fill the details in this argument. ♠

> **Takeaways** For Markov processes, the future depends upon the information up to a given time only via the state *at* this very time. If the time set is countable, this property can be generalized to random (stopping) times and is then called strong Markov property. Markov processes can be characterised by their transition probabilities (stochastic kernels).

Exercise 17.1.1 Let $I \subset \mathbb{R}$ and let $X = (X_t)_{t \in I}$ be a stochastic process. For $t \in I$, define the σ-algebras that code the past before t and the future beginning with t by

$$\mathcal{F}_{\leq t} := \sigma(X_s : s \in I, \, s \leq t) \quad \text{and} \quad \mathcal{F}_{\geq t} := \sigma(X_s : s \in I, \, s \geq t).$$

Show that X has the Markov property if and only if, for every $t \in I$, the σ-algebras $\mathcal{F}_{\leq t}$ and $\mathcal{F}_{\geq t}$ are independent given $\sigma(X_t)$ (compare Definition 12.20).

In other words, a process has the (possibly time-inhomogeneous) Markov property if and only if past and future are independent given the present. ♣

17.2 Discrete Markov Chains: Examples

Let E be countable and $I = \mathbb{N}_0$. By Definition 17.3, a Markov process $X = (X_n)_{n \in \mathbb{N}_0}$ on E is a discrete Markov chain (or Markov chain with discrete state space).

If X is a discrete Markov chain, then $(\mathbf{P}_x)_{x \in E}$ is determined by the **transition matrix**

$$p = (p(x, y))_{x, y \in E} := (\mathbf{P}_x[X_1 = y])_{x, y \in E}.$$

The n-step transition probabilities

$$p^{(n)}(x, y) := \mathbf{P}_x[X_n = y]$$

can be computed as the n-fold matrix product

$$p^{(n)}(x, y) = p^n(x, y),$$

where

$$p^n(x, y) = \sum_{z \in E} p^{n-1}(x, z) p(z, y)$$

and where $p^0 = I$ is the unit matrix.

By induction, we get the **Chapman–Kolmogorov equation** (see (14.15)) for all $m, n \in \mathbb{N}_0$ and $x, y \in E$,

$$p^{(m+n)}(x, y) = \sum_{z \in E} p^{(m)}(x, z)\, p^{(n)}(z, y). \tag{17.10}$$

Definition 17.16 *A matrix* $(p(x, y))_{x,y \in E}$ *with nonnegative entries and with*

$$\sum_{y \in E} p(x, y) = 1 \quad \text{for all } x \in E$$

is called a **stochastic matrix** *on* E.

A stochastic matrix is essentially a stochastic kernel from E to E. In Theorem 17.8 we saw that, for the semigroup of kernels $(p^n)_{n \in \mathbb{N}}$, there exists a unique discrete Markov chain whose transition probabilities are given by p. The arguments we gave there were rather abstract. Here we give a construction for X that could actually be used to implement a computer simulation of X.

Let $(R_n)_{n \in \mathbb{N}_0}$ be an independent family of random variables with values in E^E and with the property

$$\mathbf{P}[R_n(x) = y] = p(x, y) \quad \text{for all } x, y \in E. \tag{17.11}$$

For example, choose $(R_n(x),\ x \in E,\ n \in \mathbb{N})$ as an independent family of random variables with values in E and distributions

$$\mathbf{P}[R_n(x) = y] = p(x, y) \quad \text{for all } x, y \in E \text{ and } n \in \mathbb{N}_0.$$

Note, however, that in (17.11) we have *required* neither independence of the random variables $(R_n(x),\ x \in E)$ nor that all R_n had the same distribution. Only the one-dimensional marginal distributions are determined. In fact, in many applications it is useful to have subtle dependence structures in order to *couple* Markov chains with different initial chains. We pick up this thread again in Sect. 18.2.

For $x \in E$, define

$$X_0^x = x \quad \text{and} \quad X_n^x = R_n(X_{n-1}^x) \quad \text{for } n \in \mathbb{N}.$$

Finally, let $\mathbf{P}_x := \mathcal{L}[X^x]$ be the distribution of X^x. Recall that this is a probability measure on the space of sequences $(E^{\mathbb{N}_0}, \mathcal{B}(E)^{\otimes \mathbb{N}_0})$.

Theorem 17.17

(i) *With respect to the distribution* $(\mathbf{P}_x)_{x \in E}$, *the canonical process* X *on* $(E^{\mathbb{N}_0}, \mathcal{B}(E)^{\otimes \mathbb{N}_0})$ *is a Markov chain with transition matrix* p.

(ii) *In particular, to any stochastic matrix* p, *there corresponds a unique discrete Markov chain* X *with transition probabilities* p.

Proof "(ii)"   This follows from (i) since Theorem 17.11 yields uniqueness of X.
"(i)"   For $n \in \mathbb{N}_0$ and $x, y, z \in E$, by construction,

$$\mathbf{P}_x[X_{n+1} = z \,|\, \mathcal{F}_n, \, X_n = y] = \mathbf{P}\big[X_{n+1}^x = z \,\big|\, \sigma\big(R_m, \, m \leq n\big), \, X_n^x = y\big]$$
$$= \mathbf{P}\big[R_{n+1}(X_n^x) = z \,\big|\, \sigma\big(R_m, \, m \leq n\big), \, X_n^x = y\big]$$
$$= \mathbf{P}\big[R_{n+1}(y) = z\big]$$
$$= p(y, z).$$

Hence, by Theorem 17.11, X is a Markov chain with transition matrix p. $\square$

Example 17.18 (Random walk on $\mathbb{Z}$)   Let $E = \mathbb{Z}$, and assume

$$p(x, y) = p(0, y - x) \quad \text{for all } x, y \in \mathbb{Z}.$$

In this case, we say that p is **translation invariant**. A discrete Markov chain X with transition matrix p is a random walk on $\mathbb{Z}$. Indeed, $X_n \overset{\mathcal{D}}{=} X_0 + Z_1 + \ldots + Z_n$, where $(Z_n)_{n \in \mathbb{N}}$ are i.i.d. with $\mathbf{P}[Z_n = x] = p(0, x)$.

The R_n that we introduced in the explicit construction are given by $R_n(x) := x + Z_n$. $\lozenge$

Example 17.19 (Computer simulation)   Consider the situation where the state space $E = \{1, \ldots, k\}$ is finite. The aim is to simulate a Markov chain X with transition matrix p on a computer. Assume that the computer provides a random number generator that generates an i.i.d. sequence $(U_n)_{n \in \mathbb{N}}$ of random variables that are uniformly distributed on $[0, 1]$. (Of course, this is wishful thinking. But modern random number generators produce sequences that for many purposes are close enough to really random sequences.)

Define $r(i, 0) = 0$, $r(i, j) = p(i, 1) + \ldots + p(i, j)$ for $i, j \in E$, and define Y_n by

$$R_n(i) = j \quad \Longleftrightarrow \quad U_n \in [r(i, j-1), r(i, j)).$$

Then, by construction, $\mathbf{P}[R_n(i) = j] = r(i, j) - r(i, j-1) = p(i, j)$. $\lozenge$

Example 17.20 (Branching process as a Markov chain)   We want to understand the Galton–Watson branching process (see Definition 3.9) as a Markov chain on $E = \mathbb{N}_0$.

To this end, let $(q_k)_{k \in \mathbb{N}_0}$ be a probability vector, the offspring distribution of one individual. Define $q_k^{*0} = \mathbb{1}_{\{0\}}(k)$ and

$$q_k^{*n} = \sum_{l=0}^{k} q_{k-l}^{*(n-1)} q_l \quad \text{for } n \in \mathbb{N}$$

as the n-fold convolutions of q. Hence, for n individuals, q_k^{*n} is the probability to have exactly k offspring. Finally, define the matrix p by $p(x, y) = q_y^{*x}$ for $x, y \in \mathbb{N}_0$.

Now let $(Y_{n,i}, n \in \mathbb{N}_0, i \in \mathbb{N}_0)$ be i.i.d. with $\mathbf{P}[Y_{n,i} = k] = q_k$. For $x \in \mathbb{N}_0$, define the branching process X with x ancestors and offspring distribution q by $X_0 = x$ and $X_n := \sum_{i=1}^{X_{n-1}} Y_{n-1,i}$. In order to show that X is a Markov chain, we compute

$$\mathbf{P}[X_n = x_n \mid X_0 = x, X_1 = x_1, \ldots, X_{n-1} = x_{n-1}]$$
$$= \mathbf{P}[Y_{n-1,1} + \ldots + Y_{n-1,x_{n-1}} = x_n]$$
$$= \mathbf{P}_{Y_{1,1}}^{*x_{n-1}}(\{x_n\}) = q_{x_n}^{*x_{n-1}} = p(x_{n-1}, x_n).$$

Hence X is a Markov chain on $\mathbb{N}_0$ with transition matrix p. $\Diamond$

Example 17.21 (Wright's evolution model) In population genetics, Wright's evolution model [172] describes the hereditary transmission of a genetic trait with two possible specifications (say A and B); for example, resistance/no resistance to a specific antibiotic. It is assumed that the population has a constant size of $N \in \mathbb{N}$ individuals and the generations change at discrete times and do not overlap. Furthermore, for simplicity, the individuals are assumed to be **haploid**; that is, cells bear only one copy of each chromosome (like certain protozoans do) and not two copies (as in mammals).

Here we consider the case where none of the traits is favored by selection. Hence, it is assumed that each individual of the new generation chooses independently and uniformly at random one individual of the preceding generation as ancestor and becomes a perfect clone of that. Thus, if the number of individuals of type A in the current generation is $k \in \{0, \ldots, N\}$, then in the new generation it will be random and binomially distributed with parameters N and k/N.

The gene frequencies k/N in this model can be described by a Markov chain X on $E = \{0, 1/N, \ldots, (N-1)/N, 1\}$ with transition matrix $p(x, y) = b_{N,x}(\{Ny\})$. Note that X is a (bounded) martingale. Hence, by the martingale convergence theorem (Theorem 11.7), X converges $\mathbf{P}_x$-almost surely to a random variable X_∞ with $\mathbf{E}_x[X_\infty] = \mathbf{E}_x[X_0] = x$. As with the voter model (see Example 11.16) that is closely related to Wright's model, we can argue that the limit X_∞ can take only the stable values 0 and 1. That is, $\mathbf{P}_x[\lim_{n \to \infty} X_n = 1] = x = 1 - \mathbf{P}_x[\lim_{n \to \infty} X_n = 0]$. $\Diamond$

Example 17.22 (Discrete Moran model) In contrast to Wright's model, the Moran model also allows overlapping generations. The situation is similar to that of Wright's model; however, now in each time step, only (exactly) one individual gets replaced by a new one, whose type is chosen at random from the whole population.

As the new and the old types of the replaced individual are independent, as a model for the gene frequencies, we obtain a Markov chain X on $E = \{0, \frac{1}{N}, \ldots, 1\}$ with transition matrix

$$p(x, y) = \begin{cases} x(1 - x), & \text{if } y = x + 1/N, \\ x^2 + (1 - x)^2, & \text{if } y = x, \\ x(1 - x), & \text{if } y = x - 1/N, \\ 0, & \text{else.} \end{cases}$$

Here also, X is a bounded martingale and we can compute the square variation process,

$$\langle X \rangle_n = \sum_{i=1}^{n} \mathbf{E}\left[(X_i - X_{i-1})^2 \,\middle|\, X_{i-1} \right] = \frac{2}{N^2} \sum_{i=0}^{n-1} X_i (1 - X_i). \tag{17.12}$$

◇

> **Takeaways** A Markov process indexed by the natural numbers is called a Markov chain. Markov chains with discrete state spaces give rise to many interesting probabilistic examples. The transition probabilities are given by stochastic matrices.

Exercise 17.2.1 (Discrete martingale problem) Let $E \subset \mathbb{R}$ be countable and let X be a Markov chain on E with transition matrix p and with the property that, for any x, there are at most three choices for the next step; that is, there exists a set $A_x \subset E$ of cardinality 3 with $p(x, y) = 0$ for all $y \in E \setminus A_x$. Let $d(x) := \sum_{y \in E} (y - x) \, p(x, y)$ for $x \in E$.

(i) Show that $M_n := X_n - \sum_{k=0}^{n-1} d(X_k)$ defines a martingale M with square variation process $\langle M \rangle_n = \sum_{i=0}^{n-1} f(X_i)$ for a unique function $f : E \to [0, \infty)$.
(ii) Show that the transition matrix p is uniquely determined by f and d.
(iii) For the Moran model (Example 17.22), use the explicit form (17.12) of the square variation process to compute the transition matrix. ♣

17.3   Discrete Markov Processes in Continuous Time

Let E be countable and let $(X_t)_{t \in [0,\infty)}$ be a Markov process on E with transition probabilities $p_t(x, y) = \mathbf{P}_x[X_t = y]$ (for $x, y \in E$). (Some authors call such a process a Markov chain in continuous time.)

Let $x, y \in E$ with $x \neq y$. We say that X jumps *with rate* $q(x, y)$ from x to y if the following limit exists:

$$q(x, y) := \lim_{t \downarrow 0} \frac{1}{t} \mathbf{P}_x[X_t = y].$$

Henceforth we assume that the limit $q(x, y)$ exists for all $y \neq x$ and that

$$\sum_{y \neq x} q(x, y) < \infty \quad \text{for all } x \in E. \tag{17.13}$$

Then we define

$$q(x, x) = -\sum_{y \neq x} q(x, y). \tag{17.14}$$

Finally we assume that (which is equivalent to exchangeability of the limit and the sum over $y \neq x$ in the display preceding (17.13))

$$\lim_{t \downarrow 0} \frac{1}{t} \left(\mathbf{P}_x[X_t = y] - \mathbb{1}_{\{x=y\}} \right) = q(x, y) \quad \text{for all } x, y \in E. \tag{17.15}$$

Definition 17.23 *If* (17.13), (17.14) *and* (17.15) *hold, then* q *is called the **Q-matrix** of X. Sometimes q is also called the **generator** of the semigroup* $(p_t)_{t \geq 0}$.

Example 17.24 (Poisson process) The Poisson process with rate $\alpha > 0$ (compare Sect. 5.5) has the Q-matrix $q(x, y) = \alpha(\mathbb{1}_{\{y=x+1\}} - \mathbb{1}_{\{y=x\}})$. ◊

Theorem 17.25 *Let q be an $E \times E$ matrix such that $q(x, y) \geq 0$ for all $x, y \in E$ with $x \neq y$. Assume that* (17.13) *and* (17.14) *hold and that*

$$\lambda := \sup_{x \in E} |q(x, x)| < \infty. \tag{17.16}$$

Then q is the Q-matrix of a unique Markov process X.

Intuitively, (17.15) suggests that we define $p_t = e^{tq}$ in a suitable sense. Then, formally, $q = \frac{d}{dt} p_t \big|_{t=0}$. The following proof shows that this formal argument can be made rigorous.

Proof Let I be the unit matrix on E. Define

$$p(x, y) = \frac{1}{\lambda} q(x, y) + I(x, y) \quad \text{for } x, y \in E,$$

if $\lambda > 0$ and $p = I$ otherwise. Then p is a stochastic matrix and $q = \lambda(p - I)$. Let $\big((Y_n)_{n \in \mathbb{N}_0}, (\mathbf{P}_x^Y)_{x \in E}\big)$ be a discrete Markov chain with transition matrix p and let $\big((T_t)_{t \geq 0}, (\mathbf{P}_n^T)_{n \in \mathbb{N}_0}\big)$ be a Poisson process with rate λ. Let $X_t := Y_{T_t}$ and $\mathbf{P}_x = \mathbf{P}_x^Y \otimes \mathbf{P}_0^T$. Then $\mathfrak{X} := \big((X_t)_{t \geq 0}, (\mathbf{P}_x)_{x \in E}\big)$ is a Markov process and

$$p_t(x, y) := \mathbf{P}_x[X_t = y] = \sum_{n=0}^{\infty} \mathbf{P}_0^T[T_t = n] \, \mathbf{P}_x^Y[Y_n = y]$$

$$= e^{-\lambda t} \sum_{n=0}^{\infty} \frac{\lambda^n t^n}{n!} p^n(x, y).$$

This power series (in t) converges everywhere (note that as a linear operator, p has finite norm $\|p\|_\infty \leq 1$) to the matrix exponential function $e^{\lambda t p}(x, y)$. Furthermore,

$$p_t(x, y) = e^{-\lambda t} e^{\lambda t p}(x, y) = e^{\lambda t(p - I)}(x, y) = e^{t q}(x, y).$$

Differentiating the power series termwise yields $\frac{d}{dt} p_t(x, y)\big|_{t=0} = q(x, y)$. Hence $\mathfrak{X}$ is the required Markov process.

Now assume that $(\widetilde{p}_t)_{t \geq 0}$ are the transition probabilities of another Markov process $\widetilde{\mathfrak{X}}$ with the same generator q; that is, with

$$\lim_{s \downarrow 0} \frac{1}{s} \big(\widetilde{p}_s(x, y) - I(x, y)\big) = q(x, y).$$

It is easy to check that

$$\lim_{s \downarrow 0} \frac{1}{s} \big(p_{t+s}(x, y) - p_t(x, y)\big) = (q \cdot p_t)(x, y).$$

That is, we have $(d/dt) p_t(x, y) = q \, p_t(x, y)$. Similarly, we get $(d/dt)\widetilde{p}_t = q \, \widetilde{p}_t(x, y)$. Hence also,

$$p_t(x, y) - \widetilde{p}_t(x, y) = \int_0^t \big(q(p_s - \widetilde{p}_s)\big)(x, y) \, ds.$$

If we let $r_s = p_s - \tilde{p}_s$, then $\|r_s\|_\infty \le 2$ and $\|q\|_\infty \le 2\lambda$; hence

$$\sup_{s \le t} \|r_s\|_\infty \le \sup_{s \le t} \int_0^s \|qr_u\|_\infty \, du$$

$$\le \|q\|_\infty \sup_{s \le t} \int_0^s \|r_u\|_\infty \, du \le 2\lambda t \sup_{s \le t} \|r_s\|_\infty.$$

For $t < 1/2\lambda$, this implies $r_t = 0$. Assuming $r_s = 0$ for all $s \le (n-1)t$, we conclude

$$\sup_{s \le nt} \|r_s\|_\infty \le 2\lambda \int_{(n-1)t}^{nt} \|r_u\|_\infty \, du \le 2\lambda t \sup_{s \le nt} \|r_s\|_\infty = 0,$$

hence $r_s = 0$ for all $s \le nt$. By induction, we get $r_s = 0$ hence $\tilde{p}_s = p_s$ for all $s \ge 0$. $\square$

Remark 17.26 The condition (17.16) cannot be dropped easily, as the following example shows. Let $E = \mathbb{N}$ and

$$q(x, y) = \begin{cases} x^2, & \text{if } y = x + 1, \\ -x^2, & \text{if } y = x, \\ 0, & \text{else.} \end{cases}$$

We construct explicitly a candidate X for a Markov process with Q-matrix q. Let $T_1, T_2, \ldots$ be independent, exponentially distributed random variables with $\mathbf{P}_{T_n} = \exp_{n^2}$. Define $S_n = T_1 + \ldots + T_{n-1}$ and $X_t = \sup\{n \in \mathbb{N}_0 : S_n \le t\}$. Then, at any time, X makes at most one step to the right. Furthermore, due to the lack of memory of the exponential distribution (see Exercise 8.1.1),

$$\mathbf{P}[X_{t+s} \ge n + 1 \,|\, X_t = n]$$

$$= \mathbf{P}[S_{n+1} \le t + s \,|\, S_n \le t, \, S_{n+1} > t]$$

$$= \mathbf{P}[T_n \le s + t - S_n \,|\, S_n \le t, \, T_n > t - S_n] = \mathbf{P}[T_n \le s]$$

$$= 1 - \exp(-n^2 s).$$

Therefore,

$$\lim_{s \downarrow 0} s^{-1} \mathbf{P}[X_{t+s} = n + 1 \,|\, X_t = n] = n^2$$

and

$$\lim_{s \downarrow 0} s^{-1} \left(\mathbf{P}[X_{t+s} = n \,|\, X_t = n] - 1 \right) = -n^2;$$

hence

$$\lim_{s \downarrow 0} s^{-1} \left(\mathbf{P}[X_{t+s} = m \mid X_t = n] - I(m, n) \right) = q(m, n) \quad \text{for all } m, n \in \mathbb{N}.$$

Let

$$\tau^n = \inf\{t \geq 0 : X_t = n\} = S_n \quad \text{for } n \in \mathbb{N}.$$

Then $\mathbf{E}_1[\tau^n] = \sum_{k=1}^{n-1} \frac{1}{k^2}$. By monotone convergence, $\mathbf{E}_1\left[\sup_{n \in \mathbb{N}} \tau^n \right] < \infty$. That is, in finite time, X exceeds all levels. We say that X *explodes.* ◇

Example 17.27 (Yule process)   We consider an example that resembles the preceding one at first glance. However, the qualitative behaviour will be quite different. Let $E = \mathbb{N}$ and let $X = (X_t)_{t \geq 0}$ be a Markov process on E with Q-matrix

$$q(x, y) = \begin{cases} x, & \text{if } y = x + 1, \\ -x, & \text{if } y = x, \\ 0, & \text{else.} \end{cases}$$

This process can be regarded as a Galton-Watson branching process in continuous time. Each of the x individuals has an exponentially distributed lifetime with parameter 1. When a particle dies, it has two offspring. Each of the offspring gets its own independent exponential lifetime. Due to the lack-of-memory property of the exponential distribution, also the remaining lifetimes of the other $x - 1$ individuals are independent and exponentially distributed with parameter 1. As the minimum of x independent $\exp_1$-distributed random variables is $\exp_x$-distributed, the waiting time for the next branching event is $\exp_x$-distributed. This property is reflected by the assumption that the jump rate $q(x, x + 1)$ equals x.

For $n \in \mathbb{N}_0$ and $t \geq 0$, define the probability

$$f_n(t) := \mathbf{P}_1[X_t > n].$$

Clearly, we have $f_n(0) = 0$ for all $n \in \mathbb{N}$. Now X jumps from n to $n + 1$ at rate n. Hence, we have

$$\frac{d}{dt} f_n(t) = n \, \mathbf{P}_1[X_t = n] = n \, (f_{n-1}(t) - f_n(t)).$$

Note that $f_0(t) = 1$ for all $t \geq 0$ and hence $f_1'(t) = 1 - f_1(t)$. The unique solution of this differential equation is $f_1(t) = 1 - e^{-t}$. By induction, we get

$$\mathbf{P}_1[X_t > n] = f_n(t) = (1 - e^{-t})^n \quad \text{for all } n \in \mathbb{N}, \ t \geq 0.$$

In particular, we see that X_t is finite for all t. Now we determine the asymptotic growth rate of X. For $x > 0$, we have

$$\mathbf{P}_1[e^{-t}X_t > x] = \mathbf{P}[X_t > e^t x] = (1 - e^{-t})^{\lfloor e^t x \rfloor} \overset{t \to \infty}{\longrightarrow} e^{-x}.$$

That is, $e^{-t}X_t$ converges in distribution to a random variable W with $W \sim \exp_1$.

We can use the jump times of X for an alternative derivation. Similarly as in Remark 17.28, we define independent random variables $T_1, T_2, \ldots$ with $T_n \sim \exp_n$. Now let $S_n := T_1 + \ldots + T_{n-1}$ and define X by

$$X_t := \sup\{n \in \mathbb{N} : S_n \leq t\} \quad \text{for all } t \geq 0.$$

Let $Z_1, Z_2, \ldots$ be independent exponentially distributed random variables with parameter 1. From Exercise 15.1.3, we know that

$$S_{n+1} \overset{\mathcal{D}}{=} \max\{Z_1, \ldots, Z_n\}.$$

Hence, we have

$$\mathbf{P}[X_t > n] = \mathbf{P}[S_{n+1} \leq t] = \mathbf{P}[Z_1 \leq t]^n = (1 - e^{-t})^n. \quad \Diamond$$

Example 17.28 (A variant of Pólya's urn model) Consider a variant of Pólya's urn model with black and red balls (compare Example 12.29). In contrast to the original model, we do not simply add *one* ball of the same color as the ball that we return. Rather, the number of balls that we add varies from time to time. More precisely, the kth ball of a given color will be returned together with r_k more balls of the same color. The numbers $r_1, r_2, \ldots \in \mathbb{N}$ are parameters of the model. In particular, the case $1 = r_1 = r_2 = \ldots$ is the classical Pólya's urn model. Let

$$X_n := \begin{cases} 1, & \text{if the } n\text{th ball is black,} \\ 0, & \text{else.} \end{cases}$$

For the classical model, we saw (Example 12.29) that the fraction of black balls in the urn converges a.s. to a Beta-distributed random variable Z. Furthermore, given Z, the sequence $X_1, X_2, \ldots$ is independent and Ber_Z-distributed. A similar statement holds for the case where $r = r_1 = r_2 = \ldots$ for some $r \in \mathbb{N}$. Indeed, here only the parameters of the Beta distribution change. In particular (as the Beta distribution is continuous and, in particular, does not have atoms at 0 or 1), almost surely we draw infinitely many balls of each color. Formally, $\mathbf{P}[B] = 0$ where B is the event where there is one color of which only finitely many balls are drawn.

The situation changes when the numbers r_k grow quickly as $k \to \infty$. Assume that in the beginning there is one black and one red ball in the urn. Denote by $w_n = 1 + \sum_{k=1}^{n} r_k$ the total number of balls of a given color after n balls of that color have been drawn already ($n \in \mathbb{N}_0$).

For illustration, first consider the extreme situation where w_n grows very quickly; for example, $w_n = 2^n$ for every $n \in \mathbb{N}$. Denote by

$$S_n = 2(X_1 + \ldots + X_n) - n$$

the number of black balls drawn in the first n steps minus the number of red balls drawn in these steps. Then, for every $n \in \mathbb{N}_0$,

$$\mathbf{P}[X_{n+1} = 1 \,|\, S_n] = \frac{2^{S_n}}{1 + 2^{S_n}} \quad \text{and} \quad \mathbf{P}[X_{n+1} = 0 \,|\, S_n] = \frac{2^{-S_n}}{1 + 2^{-S_n}}.$$

We conclude that $(Z_n)_{n \in \mathbb{N}_0} := (|S_n|)_{n \in \mathbb{N}_0}$ is a Markov chain on $\mathbb{N}_0$ with transition matrix

$$p(z, z') = \begin{cases} 2^z/(1 + 2^z), & \text{if } z' = z + 1 > 1, \\ 1, & \text{if } z' = z + 1 = 1, \\ 1/(1 + 2^z), & \text{if } z' = z - 1 \geq 0, \\ 0, & \text{else.} \end{cases}$$

The event B from above can be written as

$$B = \big\{ Z_{n+1} < Z_n \text{ only finitely often} \big\}.$$

Let $A = \big\{ Z_{n+1} > Z_n \text{ for all } n \in \mathbb{N}_0 \big\}$ denote the event where Z *flees directly to* ∞ and let $\tau_z = \inf\{n \in \mathbb{N}_0 : Z_n \geq z\}$. Evidently,

$$\mathbf{P}_z[A] = \prod_{z'=z}^{\infty} p(z', z'+1) \geq 1 - \sum_{z'=z}^{\infty} \frac{1}{1 + 2^{z'}} \geq 1 - 2^{1-z}.$$

It is easy to check that $\mathbf{P}_0[\tau_z < \infty] = 1$ for all $z \in \mathbb{N}_0$. Using the strong Markov property, we get that, for all $z \in \mathbb{N}_0$,

$$\mathbf{P}_0[B] \geq \mathbf{P}_0[Z_{n+1} > Z_n \text{ for all } n \geq \tau_z] = \mathbf{P}_z[A] \geq 1 - 2^{1-z},$$

and thus $\mathbf{P}_0[B] = 1$. In prose, almost surely eventually only balls of one color will be drawn.

This example was a bit extreme. In order to find a necessary and sufficient condition on the growth of (w_n), we need more subtle methods that appeal to the above example of the explosion of a Markov process.

We will show that $\mathbf{P}[B] = 1$ if and only if $\sum_{n=0}^{\infty} \frac{1}{w_n} < \infty$. To this end, consider independent random variables $T_1^s, T_1^r, T_2^s, T_2^r, \ldots$ with $\mathbf{P}_{T_n^r} = \mathbf{P}_{T_n^s} = \exp_{w_{n-1}}$. Let $T_{\infty}^r = \sum_{n=1}^{\infty} T_n^r$ and $T_{\infty}^s = \sum_{n=1}^{\infty} T_n^s$. Clearly, $\mathbf{E}[T_{\infty}^r] = \sum_{n=0}^{\infty} 1/w_n < \infty$; hence, in particular, $\mathbf{P}[T_{\infty}^r < \infty] = 1$. The corresponding statement holds for T_{∞}^s.

Note that T_∞^r and T_∞^s are independent and have densities (since T_1^r and T_1^s have densities); hence we have $\mathbf{P}[T_\infty^r = T_\infty^s] = 0$.

Now let

$$R_t := \sup \left\{ n \in \mathbb{N} : T_1^r + \ldots + T_{n-1}^r \leq t \right\}$$

and

$$S_t := \sup \left\{ n \in \mathbb{N} : T_1^s + \ldots + T_{n-1}^s \leq t \right\}.$$

Let $R := \{T_1^r + \ldots + T_n^r, \ n \in \mathbb{N}\}$ and let $S := \{T_1^s + \ldots + T_n^s, \ n \in \mathbb{N}\}$ be the jump times of (R_t) and (S_t). Define $U := R \cup S = \{u_1, u_2, \ldots\}$, where $u_1 < u_2 < \ldots$. Let

$$X_n = \begin{cases} 1, & \text{if } u_n \in S, \\ 0, & \text{else.} \end{cases}$$

Let $L_n = x_1 + \ldots + x_n$. Then

$$\mathbf{P}[X_{n+1} = 1 \,|\, X_1 = x_1, \ldots, X_n = x_n]$$
$$= \mathbf{P}\left[u_{n+1} \in S \,\middle|\, (u_k \in S \iff x_k = 1) \text{ for every } k \leq n\right]$$
$$= \mathbf{P}\left[T_1^s + \ldots + T_{L_n+1}^s < T_1^r + \ldots + T_{n-L_n+1}^r \right.$$
$$\left. \,\middle|\, T_1^s + \ldots + T_{L_n+1}^s > T_1^r + \ldots + T_{n-L_n}^r\right]$$
$$= \mathbf{P}\left[T_{L_n+1}^s < T_{n-L_n+1}^r\right] = \frac{w_{L_n}}{w_{L_n} + w_{n-L_n}}.$$

Hence $(X_n)_{n \in \mathbb{N}_0}$ is our generalized urn model with weights $(w_n)_{n \in \mathbb{N}_0}$. Consider now the event B^c where infinitely many balls of each color are drawn. Evidently, $\{X_n = 1 \text{ infinitely often}\} = \{\sup S = \sup U\}$ and $\{X_n = 0 \text{ infinitely often}\} = \{\sup R = \sup U\}$. Since $\sup S = T_\infty^s$ and $\sup R = T_\infty^r$, we thus have $\mathbf{P}[B^c] = \mathbf{P}[T_\infty^r = T_\infty^s] = 0$. $\Diamond$

Takeaways A Markov processes in continuous time and with discrete state space can be described by its jump rates (q-matrix). If the jump rates are bounded, then the process can be constructed as a Markov chain at random times given by a Poisson clock.

Exercise 17.3.1 Consider the Yule process X from Example 17.27. Show that $W_t := e^{-t}X_t, \ t \geq 0$, is a martingale. Conclude that $(W_n)_{n \in \mathbb{N}}$ converges almost surely an in L^1 to a random variable W that is exponentially distributed with parameter 1. (In fact, using Exercise 21.4.2, it can even be shown that $(W_t)_{t \geq 0}$ converges to W almost surely and in L^1.) ♣

Exercise 17.3.2 Let $r, s, R, S \in \mathbb{N}$. Consider the generalized version of Pólya's urn model $(X_n)_{n \in \mathbb{N}_0}$ with $r_k = r$ and $s_k = s$ for all $k \in \mathbb{N}$. Assume that in the beginning there are R red balls and S black balls in the urn. Show that the fraction of black balls converges almost surely to a random variable Z with a Beta distribution and determine the parameters. Show that $(X_n)_{n \in \mathbb{N}_0}$ is i.i.d. given Z and $X_i \sim \mathrm{Ber}_Z$ for all $i \in \mathbb{N}_0$. ♣

Exercise 17.3.3 Show that, almost surely, infinitely many balls of each color are drawn if $\displaystyle\sum_{n=0}^{\infty} \frac{1}{w_n} = \infty$. ♣

17.4   Discrete Markov Chains: Recurrence and Transience

In the following, let $X = (X_n)_{n \in \mathbb{N}_0}$ be a Markov chain on the countable space E with transition matrix p.

Definition 17.29 *For any $x \in E$, let $\tau_x := \tau_x^1 := \inf\{n > 0 : X_n = x\}$ and*

$$\tau_x^k = \inf\left\{n > \tau_x^{k-1} : X_n = x\right\} \quad \text{for } k \in \mathbb{N}, \, k \geq 2.$$

τ_x^k *is the kth **entrance time** of X for x. For $x, y \in E$, let*

$$F(x, y) := \mathbf{P}_x[\tau_y^1 < \infty] = \mathbf{P}_x\Big[\text{there is an } n \geq 1 \text{ with } X_n = y\Big]$$

be the probability of ever going from x to y. In particular, $F(x, x)$ is the return probability (after the first jump) from x to x.

Note that $\tau_x^1 > 0$ even if we start the chain at $X_0 = x$.

Theorem 17.30 *For all $x, y \in E$ and $k \in \mathbb{N}$, we have*

$$\mathbf{P}_x\left[\tau_y^k < \infty\right] = F(x, y)\, F(y, y)^{k-1}.$$

Proof We carry out the proof by induction on k. For $k = 1$, the claim is true by definition. Now let $k \geq 2$. Using the strong Markov property of X (see Theorem 17.14), we get

$$\mathbf{P}_x\left[\tau_y^k < \infty\right] = \mathbf{E}_x\left[\mathbf{P}_x\left[\tau_y^k < \infty \,\big|\, \mathcal{F}_{\tau_y^{k-1}}\right] \mathbb{1}_{\{\tau_y^{k-1} < \infty\}}\right]$$

$$= \mathbf{E}_x\left[F(y, y) \cdot \mathbb{1}_{\{\tau_y^{k-1} < \infty\}}\right]$$

$$= F(y, y) \cdot F(x, y)\, F(y, y)^{k-2} = F(x, y)\, F(y, y)^{k-1}. \qquad \square$$

Definition 17.31 *A state $x \in E$ is called*

- **recurrent** *if $F(x, x) = 1$,*
- **positive recurrent** *if $\mathbf{E}_x[\tau_x^1] < \infty$,*
- **null recurrent** *if x is recurrent but not positive recurrent,*
- **transient** *if $F(x, x) < 1$, and*
- **absorbing** *if $p(x, x) = 1$.*

The Markov chain X is called (positive/null) recurrent if every state $x \in E$ is (positive/null) recurrent and is called transient if every recurrent state is absorbing.

Remark 17.32 Clearly, we have:
"absorbing" $\Longrightarrow$ "positive recurrent" $\Longrightarrow$ "recurrent". $\lozenge$

Example 17.33

(i) In Fig. 17.1, the state 2 is absorbing. If it does not get trapped in 2, the chain will eventually jump from 5 to 6 and will not return after that. Hence 1, 3, 4 and 5 are transient. The states 6, 7 and 8 are positive recurrent. One can show (see Exercise 17.6.1) that $\mathbf{E}_6[\tau_6] = \frac{17}{4}$, $\mathbf{E}_7[\tau_7] = \frac{17}{5}$ and $\mathbf{E}_8[\tau_8] = \frac{17}{8}$.

(ii) The chain in Fig. 17.2 has a drift to the right if $r > \frac{1}{2}$. Hence, in this case, every state is transient. On the other hand, if $r \in (0, \frac{1}{2})$, then the chain has a drift to the left (except at the point 0) and hence visits every state again and again. Thus the chain is recurrent. With a little thought, one can show (see Exercise 17.6.4) that in this case, the chain is actually positive recurrent and in the remaining case $r = \frac{1}{2}$ it is null recurrent. $\lozenge$

Definition 17.34 *Denote by $N(y) = \sum_{n=0}^{\infty} \mathbb{1}_{\{X_n=y\}}$ the total number of visits of X to y and by*

$$G(x, y) = \mathbf{E}_x[N(y)] = \sum_{n=0}^{\infty} p^n(x, y)$$

*the **Green function** of X.*

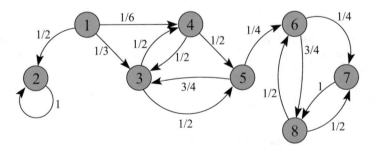

Fig. 17.1 Markov chain with eight states. The numbers are the transition probabilities for the corresponding arrows. State 2 is absorbing, the states 1, 3, 4 and 5 are transient and the states 6, 7 and 8 are (positive) recurrent.

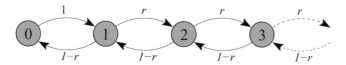

Fig. 17.2 Markov chain on $\mathbb{N}_0$ with parameter $r \in (0, 1)$. The chain is positive recurrent if $r \in (0, 1/2)$, null recurrent if $r = 1/2$ and transient if $r \in (1/2, 1)$.

Reflection The Green function G is defined by a geometric series. Hence, formally we should have $G = (I - p)^{-1}$ with I the unit matrix. In which situations can this formalism be justified? ♠♠

Theorem 17.35

(i) *For all $x, y \in E$, we have (with the convention $1/0 := \infty$, $0/0 := 0$ and $0 \cdot \infty := 0$)*

$$
G(x, y) =
\begin{cases}
\dfrac{F(x, y)}{1 - F(y, y)}, & \text{if } x \neq y \\[3mm]
\dfrac{1}{1 - F(y, y)}, & \text{if } x = y
\end{cases}
= F(x, y)\, G(y, y) + \mathbb{1}_{\{x=y\}}.
$$

$$(17.17)$$

(ii) *A non-absorbing state $x \in E$ is recurrent if and only if $G(x, x) = \infty$.*

Proof (ii) follows by (i). Hence, it remains to show (17.17). By Theorem 17.30, we have

$$
G(x, y) = \mathbf{E}_x[N(y)] = \sum_{k=1}^{\infty} \mathbf{P}_x[N(y) \geq k]
$$

$$
= \mathbb{1}_{\{x=y\}} + \sum_{k=1}^{\infty} \mathbf{P}_x\left[\tau_y^k < \infty\right] = \mathbb{1}_{\{x=y\}} + \sum_{k=1}^{\infty} F(x, y)\, F(y, y)^{k-1}
$$

$$
=
\begin{cases}
\dfrac{F(x, y)}{1 - F(y, y)}, & \text{if } x \neq y, \\[3mm]
\dfrac{1}{1 - F(x, x)}, & \text{if } x = y.
\end{cases}
$$

The second equality in (17.17) is an immediate consequence. □

Theorem 17.36 *If x is recurrent and $F(x, y) > 0$, then y is also recurrent, and $F(x, y) = F(y, x) = 1$.*

Proof Let $x, y \in E$, $x \neq y$, be such that $F(x, y) > 0$. Then there is a $k \in \mathbb{N}$ and states $x_1, \ldots, x_k \in E$ with $x_k = y$ and $x_i \neq x$ for all $i = 1, \ldots, k$ and such that

$$\mathbf{P}_x\left[X_i = x_i \text{ for all } i = 1, \dots, k\right] > 0.$$

In particular, $p^k(x, y) > 0$. By the Markov property, we have

$$1 - F(x, x) = \mathbf{P}_x\left[\tau_x^1 = \infty\right] \geq \mathbf{P}_x\left[X_1 = x_1, \dots, X_k = x_k, \ \tau_x^1 = \infty\right]$$

$$= \mathbf{P}_x\left[X_1 = x_1, \dots, X_k = x_k\right] \cdot \mathbf{P}_y\left[\tau_x^1 = \infty\right]$$

$$= \mathbf{P}_x\left[X_1 = x_1, \dots, X_k = x_k\right](1 - F(y, x)).$$

If now $F(x, x) = 1$, then also $F(y, x) = 1$. Since $F(y, x) > 0$, there exists an $l \in \mathbb{N}$ with $p^l(y, x) > 0$. Hence, for $n \in \mathbb{N}_0$,

$$p^{l+n+k}(y, y) \geq p^l(y, x) \, p^n(x, x) \, p^k(x, y).$$

If x is recurrent, then we conclude that

$$G(y, y) \ \geq \ \sum_{n=0}^{\infty} p^{l+n+k}(y, y) \ \geq \ p^l(y, x) p^k(x, y) G(x, x) \ = \ \infty$$

and hence also that y is recurrent. Changing the roles of x and y in the above argument, we get $F(x, y) = 1$.   □

Definition 17.37 *A discrete Markov chain is called*

- **irreducible** *if* $F(x, y) > 0$ *for all* $x, y \in E$, *or equivalently* $G(x, y) > 0$, *and*
- **weakly irreducible** *if* $F(x, y) + F(y, x) > 0$ *for all* $x, y \in E$.

Theorem 17.38 *An irreducible discrete Markov chain is either recurrent or transient. If* $|E| \geq 2$, *then there is no absorbing state.*

Proof This follows directly from Theorem 17.36.   □

Theorem 17.39 *If* E *is finite and* X *is irreducible, then* X *is recurrent.*

Proof Evidently, for all $x \in E$,

$$\sum_{y \in E} G(x, y) \ = \ \sum_{n=0}^{\infty} \sum_{y \in E} p^n(x, y) \ = \ \sum_{n=0}^{\infty} 1 \ = \ \infty.$$

As E is finite, there is a $y \in E$ with $G(x, y) = \infty$. Since $F(y, x) > 0$, there exists a $k \in \mathbb{N}$ with $p^k(y, x) > 0$. Therefore, since $p^{n+k}(x, x) \geq p^n(x, y)\, p^k(y, x)$, we have

$$G(x, x) \geq \sum_{n=0}^{\infty} p^n(x, y)\, p^k(y, x) = p^k(y, x)\, G(x, y) = \infty. \qquad \square$$

Takeaways A state of a Markov chain is called recurrent if the chain returns to it almost surely. Otherwise it is called transient. A recurrent state is called positive recurrent if the expected time of return is finite; otherwise it is null recurrent. For irreducible chains all states are in the same class: either positive recurrent, null recurrent or transient. A state is recurrent if and only if the expected number of visits (Green function) is infinite.

Exercise 17.4.1 Let x be positive recurrent and let $F(x, y) > 0$. Show that y is also positive recurrent. ♣

17.5 Application: Recurrence and Transience of Random Walks

In this section, we study recurrence and transience of random walks on the D-dimensional integer lattice $\mathbb{Z}^D$, $D = 1, 2, \ldots$. A more exhaustive investigation can be found in Spitzer's book [159].

Consider first the simplest situation of symmetric simple random walk X on $\mathbb{Z}^D$. That is, at each step, X jumps to any of its $2D$ neighbors with the same probability $1/2D$. Hence, in terms of the Markov chain notation, we have $E = \mathbb{Z}^D$ and

$$p(x, y) = \begin{cases} \frac{1}{2D}, & \text{if } |x - y| = 1, \\ 0, & \text{else.} \end{cases}$$

Is this random walk recurrent or transient?

The central limit theorem suggests that

$$p^n(0, 0) \approx C_D\, n^{-D/2} \quad \text{as } n \to \infty$$

for some constant C_D that depends on the dimension D. However, first we have to exclude the case where n is odd since here clearly $p^n(0, 0) = 0$. Thus let $Y_1, Y_2, \ldots$ be independent $\mathbb{Z}^D$-valued random variables with $\mathbf{P}[Y_i = x] = p^2(0, x)$. Then $X_{2n} \stackrel{D}{=} S_n := Y_1 + \ldots + Y_n$ for $n \in \mathbb{N}_0$; hence $G(0, 0) = \sum_{n=0}^{\infty} \mathbf{P}[S_n = 0]$.

Clearly, $Y_1 = (Y_1^1, \ldots, Y_1^D)$ has covariance matrix $C_{i,j} := \mathbf{E}[Y_1^i \cdot Y_1^j] = \frac{2}{D} \mathbb{1}_{\{i=j\}}$.
By the local central limit theorem (see, e.g., [20, pages 224ff] for a one-dimensional
version of that theorem or Exercise 17.5.1 for an analytic derivation), we have

$$n^{D/2} p^{2n}(0,0) = n^{D/2} \mathbf{P}[S_n = 0] \overset{n \to \infty}{\longrightarrow} 2 (4\pi/D)^{-D/2}. \tag{17.18}$$

Now $\sum_{n=1}^{\infty} n^{-\alpha} < \infty$ if and only if $\alpha > 1$. Hence $G(0,0) < \infty$ if and only if
$D > 2$. We have thus shown the following theorem of Pólya [134].

Theorem 17.40 (Pólya [134]) *Symmetric simple random walk on $\mathbb{Z}^D$ is recurrent
if and only if $D \leq 2$.*

The procedure we used here to derive Pólya's theorem has the disadvantage that it
relies on the local central limit theorem, which we have not proved (and will not).
Hence we will consider different methods of proof that yield further insight into the
problem.

Consider first the one-dimensional simple random walk that with probability p
jumps one step to the right and with probability $1 - p$ jumps one step to the left.
Then

$$G(0,0) = \sum_{n=0}^{\infty} \binom{2n}{n} (p(1-p))^n = \sum_{n=0}^{\infty} \binom{-1/2}{n} (-4p(1-p))^n.$$

Using the generalized binomial theorem (see Lemma 3.5), we get (since we have
$(1 - 4p(1-p))^{1/2} = |2p - 1|$)

$$G(0,0) = \begin{cases} \frac{1}{|2p-1|}, & \text{if } p \neq \frac{1}{2}, \\ \infty, & \text{if } p = \frac{1}{2}. \end{cases} \tag{17.19}$$

Thus, simple random walk on $\mathbb{Z}$ is recurrent if and only if it is symmetric; that is, if
$p = \frac{1}{2}$.

Of course, transience in the case $p \neq \frac{1}{2}$ could also be deduced directly from the
strong law of large numbers since $\lim_{n\to\infty} \frac{1}{n} X_n = \mathbf{E}_0[X_1] = 2p - 1$ almost surely.
In fact, this argument is even more robust since it uses only that the single steps of
X have an expectation that is not zero.

Consider now the situation where X does not necessarily jump to one of its
nearest neighbors but where we still have $\mathbf{E}_0[|X_1|] < \infty$ and $\mathbf{E}_0[X_1] = 0$. The
strong law of large numbers does not yield recurrence immediately and we have to
do some work:

By the Markov property, for every $N \in \mathbb{N}$ and every $y \neq x$,

$$G_N(x,y) := \sum_{k=0}^{N} \mathbf{P}_x[X_k = y] = \sum_{k=0}^{N} \mathbf{P}_x[\tau_y^1 = k] \sum_{l=0}^{N-k} \mathbf{P}_y[X_l = y] \leq G_N(y,y).$$

This implies for all $L \in \mathbb{N}$

$$G_N(0, 0) \geq \frac{1}{2L+1} \sum_{|y| \leq L} G_N(0, y)$$

$$= \frac{1}{2L+1} \sum_{k=0}^{N} \sum_{|y| \leq L} p^k(0, y)$$

$$\geq \frac{1}{2L+1} \sum_{k=1}^{N} \sum_{y : |y/k| \leq L/N} p^k(0, y).$$

By the weak law of large numbers, we have $\liminf_{k \to \infty} \sum_{|y| \leq \varepsilon k} p^k(0, y) = 1$ for every $\varepsilon > 0$. Hence, letting $L = \varepsilon N$, we get

$$\liminf_{N \to \infty} G_N(0, 0) \geq \frac{1}{2\varepsilon} \quad \text{for every } \varepsilon > 0.$$

Thus $G(0, 0) = \infty$, which shows that X is recurrent.

We summarize the discussion in a theorem.

Theorem 17.41 *A random walk on $\mathbb{Z}$ with $\sum_{x=-\infty}^{\infty} |x| \, p(0, x) < \infty$ is recurrent if and only if $\sum_{x=-\infty}^{\infty} x \, p(0, x) = 0$.*

Now what about symmetric simple random walk in dimension $D = 2$ or in higher dimensions? In order that the random walk be at the origin after $2n$ steps, it must perform k_i steps in the ith direction and k_i steps in the opposite direction for some numbers $k_1, \ldots, k_D \in \mathbb{N}_0$ with $k_1 + \ldots + k_D = n$. We thus get

$$p^{2n}(0, 0) = (2D)^{-2n} \sum_{k_1 + \ldots + k_D = n} \binom{2n}{k_1, k_1, \ldots, k_D, k_D}, \tag{17.20}$$

where $\binom{N}{l_1, \ldots, l_r} = \frac{N!}{l_1! \cdots l_r!}$ is the multinomial coefficient. In particular, for $D = 2$,

$$p^{2n}(0, 0) = 4^{-2n} \sum_{k=0}^{n} \frac{(2n)!}{(k!)^2((n-k)!)^2}$$

$$= 4^{-2n} \binom{2n}{n} \sum_{k=0}^{n} \binom{n}{k} \binom{n}{n-k} = \left(2^{-2n} \binom{2n}{n} \right)^2.$$

Note that in the last step, we used a simple combinatorial identity that follows, e.g., by the convolution formula $(b_{n,p} * b_{n,p})(\{n\}) = b_{2n,p}(\{n\})$. Now, by Stirling's formula,

$$\lim_{n\to\infty} \sqrt{n}\, 2^{-2n} \binom{2n}{n} = \frac{1}{\sqrt{\pi}},$$

hence $\lim_{n\to\infty} np^{2n}(0,0) = \frac{1}{\pi}$. In particular, we have $\sum_{n=1}^{\infty} p^{2n}(0,0) = \infty$. That is, two-dimensional symmetric simple random walk is recurrent.

For $D \geq 3$, the sum over the multinomial coefficients cannot be computed in a satisfactory way. However, it is not too hard to give an estimate that shows that there exists a $c = c_D$ such that $p^{2n}(0,0) \leq cn^{-D/2}$, which implies $G(0,0) \leq c\sum_{n=1}^{\infty} n^{-D/2} < \infty$ (see, e.g., [53, page 361] or [59, Example 6.31]). Here, however, we follow a different route.

Things would be easy if the individual coordinates of the chain were *independent* one-dimensional random walks. In this case, the probability that at time $2n$ all coordinates are zero would be the Dth power of the probability that the first coordinate is zero. For one coordinate, however, which moves only with probability $1/D$ and thus has variance $1/D$, the probability of being back at the origin at time $2n$ is approximately $(n\pi/D)^{-1/2}$. Up to a factor, we would thus get (17.18) without using the multidimensional local central limit theorem.

An elegant way to decouple the coordinates is to pass from discrete time to continuous time in such a way that the individual coordinates become independent but such that the Green function remains unchanged.

We give the details. Let $(T_t^i)_{t\geq0}$, $i = 1, \ldots, D$ be independent Poisson processes with rate $1/D$. Let $Z^1, \ldots, Z^D$ be independent (and independent of the Poisson processes) symmetric simple random walks on $\mathbb{Z}$. Define $T := T^1 + \ldots + T^D$, $Y_t^i := Z_{T_t^i}^i$ for $i = 1, \ldots, D$ and let $Y_t = (Y_t^1, \ldots, Y_t^D)$. Then Y is a Markov chain in continuous time with Q-matrix $q(x,y) = p(x,y) - \mathbb{1}_{\{x=y\}}$. As T is a Poisson process with rate 1, $(X_{T_t})_{t\geq0}$ is also a Markov process with Q-matrix q. It follows that $(X_{T_t})_{t\geq0} \overset{D}{=} (Y_t)_{t\geq0}$. We now compute

$$G_Y := \int_0^\infty \mathbf{P}_0[Y_t = 0]\, dt = \int_0^\infty \sum_{n=0}^\infty \mathbf{P}_0[X_{2n} = 0,\, T_t = 2n]\, dt$$

$$= \sum_{n=0}^\infty p^{2n}(0,0) \int_0^\infty e^{-t}\frac{t^{2n}}{(2n)!}\, dt = G(0,0).$$

The two processes $(X_n)_{n\in\mathbb{N}_0}$ and $(Y_t)_{t\in[0,\infty)}$ thus have the same Green function. As the coordinates of Y are independent, we have

$$G_Y = \int_0^\infty \mathbf{P}_0[Y_t^1 = 0]^D\, dt.$$

Hence we only have to compute the asymptotics of $\mathbf{P}_0[Y_t^1 = 0]$ for large t. We can argue as follows. By the law of large numbers, we have $T_t^1 \approx t/D$ for large t. Furthermore, $\mathbf{P}_0[Y_t^1$ is even$] \approx \frac{1}{2}$. Hence we have, with $n_t = \lfloor t/2D \rfloor$ for $t \to \infty$ (compare Exercise 17.5.2),

$$\mathbf{P}_0[Y_t^1 = 0] \sim \frac{1}{2}\mathbf{P}[Z_{2n_t}^1 = 0] = \frac{1}{2}\binom{2n_t}{n_t}4^{-n_t} \sim (2\pi/D)^{-1/2}\, t^{-1/2}. \qquad (17.21)$$

Since $\int_1^\infty t^{-\alpha}\, dt < \infty$ if and only if $\alpha > 1$, we also have $G_Y < \infty$ if and only if $D > 2$. However, this is the statement of Pólya's theorem.

Finally, we present a third method of studying recurrence and transience of random walks that does not rely on the Euclidean properties of the integer lattice but rather on the Fourier inversion formula.

First consider a general (discrete time) irreducible random walk with transition matrix p on $\mathbb{Z}^D$. By $\phi(t) = \sum_{x\in\mathbb{Z}^D} e^{i\langle t,x\rangle}p(0,x)$ denote the characteristic function of a single transition. The convolution of the transition probabilities translates into powers of the characteristic function; hence

$$\phi^n(t) = \sum_{x\in\mathbb{Z}^D} e^{i\langle t,x\rangle}p^n(0,x).$$

By the Fourier inversion formula (Theorem 15.11), we recover the n-step transition probabilities from ϕ^n by

$$p^n(0,x) = (2\pi)^{-D}\int_{[-\pi,\pi)^D} e^{-i\langle t,x\rangle}\,\phi^n(t)\,dt.$$

In particular, for $\lambda \in (0,1)$,

$$R_\lambda := \sum_{n=0}^\infty \lambda^n p^n(0,0)$$

$$= (2\pi)^{-D}\sum_{n=0}^\infty \int_{[-\pi,\pi)^D}\lambda^n\,\phi^n(t)\,dt$$

$$= (2\pi)^{-D}\int_{[-\pi,\pi)^D}\frac{1}{1-\lambda\phi(t)}\,dt.$$

$$= (2\pi)^{-D}\int_{[-\pi,\pi)^D}\mathrm{Re}\left(\frac{1}{1-\lambda\phi(t)}\right)dt.$$

Now $G(0, 0) = \lim_{\lambda \uparrow 1} R_\lambda$ and hence

$$X \text{ is recurrent} \iff \lim_{\lambda \uparrow 1} \int_{[-\pi, \pi)^D} \text{Re}\left(\frac{1}{1 - \lambda \phi(t)}\right) dt = \infty. \tag{17.22}$$

If we had $\phi(t) = 1$ for some $t \in (-2\pi, 2\pi)^D \setminus \{0\}$, then we would have $\phi^n(t) = 1$ for every $n \in \mathbb{N}$ and hence, by Exercise 15.2.1, $\mathbf{P}_0[\langle X_n, t/(2\pi)\rangle \in \mathbb{Z}] = 1$. Thus X would not be irreducible contradicting the assumption. Due to the continuity of ϕ for all $\varepsilon > 0$, we thus have

$$\inf\{|\phi(t) - 1| : t \in [-\pi, \pi)^D \setminus (-\varepsilon, \varepsilon)^D\} > 0.$$

We summarize the discussion in a theorem due to Chung and Fuchs [27].

Theorem 17.42 (Chung–Fuchs [27]) *An irreducible random walk on $\mathbb{Z}^D$ with characteristic function ϕ is recurrent if and only if, for every $\varepsilon > 0$,*

$$\lim_{\lambda \uparrow 1} \int_{(-\varepsilon, \varepsilon)^D} \text{Re}\left(\frac{1}{1 - \lambda \phi(t)}\right) dt = \infty. \tag{17.23}$$

Now consider symmetric simple random walk. Here $\phi(t) = \frac{1}{D}\sum_{i=1}^D \cos(t_i)$. Expanding the cosine function in a Taylor series, we get $\cos(t_i) = 1 - \frac{1}{2}t_i^2 + O(t_i^4)$; hence $1 - \phi(t) = \frac{1}{2D}\|t\|_2^2 + O(\|t\|_2^4)$. We infer that X is recurrent if and only if $\int_{\|t\|_2 < \varepsilon} \|t\|_2^{-2} dt = \infty$. We compute this integral in polar coordinates (with C_D the surface of the unit sphere in $\mathbb{R}^D$):

$$\int_{\|t\|_2 < \varepsilon} \|t\|_2^{-2} dt = C_D \int_0^\varepsilon r^{D-1} r^{-2} dr = \infty \iff D \leq 2.$$

Hence, X is recurrent if and only if $D \leq 2$.

In Sect. 19.3, we will encounter a further method of proving Pólya's theorem that has a completely different structure and that is based on the connection between Markov chains and electrical networks.

In fact, the Chung–Fuchs theorem can be used to compute the numerical values of the Green function $G_D(0, 0)$ of symmetric simple random walk on $\mathbb{Z}^D$ if we compute numerically the so-called **Watson integral**

$$G_D(0, 0) = (2\pi)^{-D} \int_{[-\pi, \pi)^D} \frac{D}{D - (\cos(x_1) + \ldots + \cos(x_D))} dx. \tag{17.24}$$

For this purpose, we follow [80] (where there are further refinements of the method) to transform the D-fold integral into a double integral. Denote by

$$I_0(t) := \frac{1}{\pi} \int_0^\pi e^{t \cos(\theta)} \, d\theta$$

the so-called modified Bessel function of the first kind. Using the identity $\frac{1}{\lambda} = \int_0^\infty e^{-\lambda t} \, dt$ for the integrand and applying Fubini's theorem, we get

$$G_D(0,0) = \frac{D}{(2\pi)^D} \int_0^\infty e^{-Dt} \left(\int_{[-\pi,\pi)^D} e^{t(\cos(x_1)+...+\cos(x_D))} \, dx \right) dt$$

and thus

$$G_D(0,0) = D \int_0^\infty e^{-Dt} I_0(t)^D \, dt. \tag{17.25}$$

The right-hand side of (17.25) can quickly be computed numerically with great accuracy (see Table 17.1).

For the case $D = 3$, Watson [169] found the expression

Table 17.1 Green function $G_D(0,0)$ and return probability $F_D(0,0)$ of simple symmetric random walk on $\mathbb{Z}^D$. The numerical computations are based on (17.25).

D	$G_D(0,0)$	$F_D(0,0)$
2	∞	1
3	1.51638605915	0.34053732955
4	1.23946712185	0.19320167322
5	1.15630812484	0.13517860982
6	1.11696337322	0.10471549562
7	1.09390631559	0.08584493411
8	1.07864701202	0.07291264996
9	1.06774608638	0.06344774965
10	1.05954374789	0.05619753597
11	1.05313615291	0.05045515982
12	1.04798637482	0.04578912090
13	1.04375406289	0.04191989708
14	1.04021240323	0.03865787709
15	1.03720412092	0.03586962312
16	1.03461657857	0.03345836447
17	1.03236691238	0.03135214040
18	1.03039276285	0.02949628913
19	1.02864627888	0.02784852234
20	1.02709011674	0.02637559869

$$G_3(0,0) = 12 \frac{18 + 12\sqrt{2} - 10\sqrt{3} - 7\sqrt{6}}{\pi^2} \, K\Big((2 - \sqrt{3})(\sqrt{3} - \sqrt{2})\Big)^2,$$

where $K(m) = \int_0^1 \big((1 - t^2)(1 - mt^2)\big)^{-1/2} \, dt$ is the complete elliptic integral of the first kind with modulus $m \in (-1, 1)$. This in turn can be expressed as a (quickly convergent) series

$$K(m) = \frac{\pi}{2}\left(1 + \sum_{n=1}^{\infty} \left(\frac{(2n)!}{4^n (n!)^2}\right)^2 m^2\right).$$

Glasser and Zucker [61] found an expression as a product of four Gamma functions,

$$G_3(0,0) = \frac{\sqrt{6}}{32\pi^3} \, \Gamma\left(\frac{1}{24}\right)\Gamma\left(\frac{5}{24}\right)\Gamma\left(\frac{7}{24}\right)\Gamma\left(\frac{11}{24}\right) = 1.5163860591519780181\ldots$$

Takeaways Symmetric nearest neighbour random walk on the D-dimensional integer lattice is recurrent if $D \leq 2$ and is transient if $D > 2$. A (one-dimensional) random walk on the integers is recurrent if and only if the jump distribution is centred. The (numerical) computation of the Green function is possible via the Fourier inversion formula.

Exercise 17.5.1 For $n \in \mathbb{N}_0$, let p^n be the matrix of n-step transition probabilities of simple symmetric random walk on $\mathbb{Z}^D$. For $n \in \mathbb{N}$, derive the formula (see Theorem 15.11)

$$p^{2n}(0,0) = (2\pi)^{-D} \int_{[-\pi,\pi)^D} D^{-2n}\big(\cos(t_1) + \ldots + \cos(t_D)\big)^{2n} \, dt.$$

By a suitable bound for the integral, conclude the convergence $n^{D/2} p^{2n}(0,0) \xrightarrow{n \to \infty} 2(4\pi/D)^{-D/2}$ (see (17.18)). ♣

Exercise 17.5.2 Show (17.21) formally. ♣

Exercise 17.5.3 Use Theorem 17.42 to show that a random walk on $\mathbb{Z}^2$ with $\sum_{x \in \mathbb{Z}^2} x \, p(0,x) = 0$ is recurrent if $\sum_{x \in \mathbb{Z}^2} \|x\|_2^2 \, p(0,x) < \infty$. ♣

Exercise 17.5.4 Use Theorem 17.42 to show that, for $D \geq 3$ every irreducible random walk on $\mathbb{Z}^D$ is transient ♣

Exercise 17.5.5 Show (17.25) for $G_D(0,0)$ directly with the $p^{2n}(0,0)$ from (17.20) and using the representation of $I_0(t)$ as the series $I_0(t) = \sum_{k=0}^{\infty}(k!)^{-2} (t/2)^k$. ♣

17.6   Invariant Distributions

In the following, let p be a stochastic matrix on the discrete space E and let $(X_n)_{n \in \mathbb{N}_0}$ be a corresponding Markov chain.

This section is devoted to the question: Which distributions are preserved under the dynamics of the Markov chain? Of course, often the chain will not stay put in a specific state but the *distribution* of the random state of the chain might nevertheless be the same for all times. If such an invariant distribution exists, we will see in Chap. 18 that under rather weak conditions, the distribution of a Markov chain (started in an arbitrary state) converges in the large time limit to such an invariant distribution.

Definition 17.43 *If μ is a measure on E and $f : E \to \mathbb{R}$ is a map, then we write $\mu p(\{x\}) = \sum_{y \in E} \mu(\{y\}) p(y, x)$ and $pf(x) = \sum_{y \in E} p(x, y) f(y)$ if the sums converge.*

Definition 17.44

(i) *A σ-finite measure μ on E is called an **invariant measure** if*

$$\mu p = \mu.$$

*A probability measure that is an invariant measure is called an **invariant distribution**. Denote by $\mathcal{I}$ the set of invariant distributions.*

(ii) *A function $f : E \to \mathbb{R}$ is called **subharmonic** if pf exists and if $f \leq pf$. f is called superharmonic if $f \geq pf$ and harmonic if $f = pf$.*

Remark 17.45 In the terminology of linear algebra, an invariant measure is a left eigenvector of p corresponding to the eigenvalue 1. A harmonic function is a right eigenvector corresponding to the eigenvalue 1. ◊

Lemma 17.46 *If f is bounded and (sub-, super-) harmonic, then $(f(X_n))_{n \in \mathbb{N}_0}$ is a (sub-, super-) martingale with respect to the filtration $\mathbb{F} = \sigma(X)$ generated by X.*

Proof Let f be bounded and subharmonic. Then

$$\mathbf{E}_x[f(X_n) \,|\, \mathcal{F}_{n-1}] = \mathbf{E}_{X_{n-1}}[f(X_1)] = \sum_{y \in E} p(X_{n-1}, y) f(y)$$

$$= pf(X_{n-1}) \geq f(X_{n-1}). \qquad \square$$

Theorem 17.47 *If any point is transient, then an invariant distribution does not exist.*

Proof By assumption, $G(x, y) = \sum_{n=0}^{\infty} p^n(x, y) < \infty$ for all $x, y \in E$; hence $p^n(x, y) \xrightarrow{n \to \infty} 0$. For every probability measure μ on E, we thus have that $\mu p^n(\{x\}) \xrightarrow{n \to \infty} 0$. If μ was invariant, however, then we would have $\mu p^n(\{x\}) = \mu(\{x\})$ for all $n \in \mathbb{N}$. $\qquad \square$

Theorem 17.48 *Let x be a recurrent state and let $\tau_x^1 = \inf\{n \geq 1 : X_n = x\}$.*
Then one invariant measure μ_x is defined by

$$\mu_x(\{y\}) = \mathbf{E}_x\left[\sum_{n=0}^{\tau_x^1-1} \mathbb{1}_{\{X_n=y\}}\right] = \sum_{n=0}^{\infty} \mathbf{P}_x\left[X_n = y, \ \tau_x^1 > n\right].$$

Proof First we have to show that $\mu_x(\{y\}) < \infty$ for all $y \in E$. For $y = x$, clearly
$\mu_x(\{x\}) = 1$. For $y \neq x$ and $F(x, y) = 0$, we have $\mu_x(\{y\}) = 0$. Now let $y \neq x$ and
$F(x, y) > 0$. As x is recurrent, we have $F(x, y) = F(y, x) = 1$ and y is recurrent
(Theorem 17.36). Let

$$\widehat{F}(x, y) = \mathbf{P}_x\left[\tau_x^1 > \tau_y^1\right].$$

Then $\widehat{F}(x, y) > 0$ (otherwise y would not be visited). Changing the roles of x and
y, we also get $\widehat{F}(y, x) > 0$.

By the strong Markov property (Theorem 17.14), we have

$$\mathbf{E}_y\left[\sum_{n=0}^{\tau_x^1-1} \mathbb{1}_{\{X_n=y\}}\right] = 1 + \mathbf{E}_y\left[\sum_{n=\tau_y^1}^{\tau_x^1-1} \mathbb{1}_{\{X_n=y\}}; \ \tau_x^1 > \tau_y^1\right]$$

$$= 1 + \left(1 - \widehat{F}(y, x)\right)\mathbf{E}_y\left[\sum_{n=0}^{\tau_x^1-1} \mathbb{1}_{\{X_n=y\}}\right].$$

Hence,

$$\mathbf{E}_y\left[\sum_{n=0}^{\tau_x^1-1} \mathbb{1}_{\{X_n=y\}}\right] = \frac{1}{\widehat{F}(y, x)}.$$

Therefore,

$$\mu_x(\{y\}) = \mathbf{E}_x\left[\sum_{n=0}^{\tau_x^1-1} \mathbb{1}_{\{X_n=y\}}\right] = \mathbf{E}_x\left[\sum_{n=\tau_y^1}^{\tau_x^1-1} \mathbb{1}_{\{X_n=y\}}; \ \tau_x^1 > \tau_y^1\right] = \frac{\widehat{F}(x, y)}{\widehat{F}(y, x)} < \infty.$$

Define $\overline{p}_n(x, y) = \mathbf{P}_x\left[X_n = y; \ \tau_x^1 > n\right]$. Then, for every $z \in E$,

$$\mu_x \, p(\{z\}) = \sum_{y \in E} \mu_x(\{y\}) \, p(y, z) = \sum_{n=0}^{\infty} \sum_{y \in E} \overline{p}_n(x, y) \, p(y, z).$$

Case 1: $x \neq z$. In this case,

$$\sum_{y \in E} \overline{p}_n(x, y) p(y, z) = \sum_{y \in E} \mathbf{P}_x \left[X_n = y, \tau_x^1 > n, X_{n+1} = z \right]$$

$$= \mathbf{P}_x \left[\tau_x^1 > n + 1; X_{n+1} = z \right] = \overline{p}_{n+1}(x, z).$$

Hence (since $\overline{p}_0(x, z) = 0$)

$$\mu_x \, p(\{z\}) = \sum_{n=0}^{\infty} \overline{p}_{n+1}(x, z) = \sum_{n=1}^{\infty} \overline{p}_n(x, z) = \sum_{n=0}^{\infty} \overline{p}_n(x, z) = \mu_x(\{z\}).$$

Case 2: $x = z$. In this case, we have

$$\sum_{y \in E} \overline{p}_n(x, y) p(y, x) = \sum_{y \in E} \mathbf{P}_x \left[X_n = y; \tau_x^1 > n; X_{n+1} = x \right] = \mathbf{P}_x \left[\tau_x^1 = n + 1 \right].$$

Thus (since $\mathbf{P}_x \left[\tau_x^1 = 0 \right] = 0$)

$$\mu_x \, p(\{x\}) = \sum_{n=0}^{\infty} \mathbf{P}_x \left[\tau_x^1 = n + 1 \right] = 1 = \mu_x(\{x\}). \qquad \square$$

Corollary 17.49 *If X is positive recurrent, then $\pi := \dfrac{\mu_x}{\mathbf{E}_x \left[\tau_x^1 \right]}$ is an invariant distribution for any $x \in E$.*

Theorem 17.50 *If X is irreducible, then X has at most one invariant distribution.*

Remark 17.51

(i) One could in fact show that if X is irreducible and recurrent, then an invariant measure of X is unique up to a multiplicative factor. However, the proof is a little more involved. Since we will not need the statement here, we leave its proof as an exercise (compare Exercise 17.6.6; see also [39, Section 6.5]).

(ii) For transient X, there can be more than one invariant measure. For example, consider the asymmetric random walk on $\mathbb{Z}$ that jumps one step to the right with probability r and one step to the left with probability $1 - r$ (for some $r \in (0, 1)$). The invariant measures are the nonnegative linear combinations of the measures μ_1 and μ_2 given by $\mu_1(\{x\}) \equiv 1$ and $\mu_2(\{x\}) = (r/(1 - r))^x$, $x \in \mathbb{Z}$. X is transient if and only if $r \neq 1/2$, in which case we have $\mu_1 \neq \mu_2$. $\diamondsuit$

Proof Let π and ν be invariant distributions. Choose an arbitrary probability vector $(g_n)_{n \in \mathbb{N}}$ with $g_n > 0$ for all $n \in \mathbb{N}$. Define the stochastic matrix $\widetilde{p}(x, y) = \sum_{n=1}^{\infty} g_n \, p^n(x, y)$. Then $\widetilde{p}(x, y) > 0$ for all $x, y \in E$ and $\pi \widetilde{p} = \pi$ as well as $\nu \widetilde{p} = \nu$.

Consider now the signed measure $\mu = \pi - \nu$. We have $\mu \widetilde{p} = \mu$. If we had $\mu \neq 0$, then there would exist (since $\mu(E) = 0$) points $x_1, x_2 \in E$ with $\mu(\{x_1\}) > 0$ and $\mu(\{x_2\}) < 0$. Clearly, for every $y \in E$, this would imply $\left| \mu(\{x_1\}) \widetilde{p}(x_1, y) + \mu(\{x_2\}) \widetilde{p}(x_2, y) \right| < \left| \mu(\{x_1\}) \widetilde{p}(x_1, y) \right| + \left| \mu(\{x_2\}) \widetilde{p}(x_2, y) \right|$; hence

$$\| \mu \widetilde{p} \|_{TV} = \sum_{y \in E} \left| \sum_{x \in E} \mu(\{x\}) \widetilde{p}(x, y) \right|$$

$$< \sum_{y \in E} \sum_{x \in E} |\mu(\{x\})| \, \widetilde{p}(x, y) = \sum_{x \in E} |\mu(\{x\})| = \| \mu \|_{TV}.$$

Since this is a contradiction, we conclude that $\mu = 0$. $\square$

Recall that $\mathcal{I}$ is the set of invariant distributions of X.

Theorem 17.52 *Let X be irreducible. X is positive recurrent if and only if $\mathcal{I} \neq \emptyset$. In this case, $\mathcal{I} = \{\pi\}$ with*

$$\pi(\{x\}) = \frac{1}{\mathbf{E}_x[\tau_x^!]} > 0 \quad \text{for all } x \in E.$$

Proof If X is positive recurrent, then $\mathcal{I} \neq \emptyset$ by Corollary 17.49. Now let $\mathcal{I} \neq \emptyset$ and $\pi \in \mathcal{I}$. As X is irreducible, we have $\pi(\{x\}) > 0$ for all $x \in E$. Let $\mathbf{P}_\pi = \sum_{x \in E} \pi(\{x\}) \mathbf{P}_x$. Fix an $x \in E$ and for $n \in \mathbb{N}_0$, let

$$\sigma_x^n = \sup \{m \leq n : X_m = x\} \in \mathbb{N}_0 \cup \{-\infty\}$$

be the time of last entrance in x before time n. (Note that this is not a stopping time.) By the Markov property, for all $k \leq n$,

$$\mathbf{P}_\pi \left[\sigma_x^n = k \right] = \mathbf{P}_\pi \left[X_k = x, \, X_{k+1} \neq x, \ldots, X_n \neq x \right]$$

$$= \mathbf{P}_\pi \left[X_{k+1} \neq x, \ldots, X_n \neq x \, | \, X_k = x \right] \mathbf{P}_\pi [X_k = x]$$

$$= \pi(\{x\}) \, \mathbf{P}_x \left[X_1, \ldots, X_{n-k} \neq x \right]$$

$$= \pi(\{x\}) \, \mathbf{P}_x \left[\tau_x^! \geq n - k + 1 \right].$$

Hence, for every $n \in \mathbb{N}_0$ (since $\mathbf{P}_y\left[\tau_x^1 < \infty\right] = 1$ for all $y \in E$),

$$1 = \sum_{k=0}^{n} \mathbf{P}_\pi\left[\sigma_x^n = k\right] + \mathbf{P}_\pi\left[\sigma_x^n = -\infty\right]$$

$$= \pi(\{x\}) \sum_{k=0}^{n} \mathbf{P}_x\left[\tau_x^1 \geq n - k + 1\right] + \mathbf{P}_\pi\left[\tau_x^1 \geq n + 1\right]$$

$$\xrightarrow{n \to \infty} \pi(\{x\}) \sum_{k=1}^{\infty} \mathbf{P}_x\left[\tau_x^1 \geq k\right] = \pi(\{x\}) \, \mathbf{E}_x\left[\tau_x^1\right].$$

Therefore, $\mathbf{E}_x\left[\tau_x^1\right] = \frac{1}{\pi(\{x\})} < \infty$, and thus X is positive recurrent. $\square$

Example 17.53 Let $(p_x)_{x \in \mathbb{N}_0}$ be numbers in $(0, 1]$ and let X be an irreducible Markov chain on $\mathbb{N}_0$ with transition matrix

$$p(x, y) = \begin{cases} p_x, & \text{if } y = x + 1, \\ 1 - p_x, & \text{if } y = 0, \\ 0, & \text{else.} \end{cases}$$

If μ is an invariant measure, then the equations for $\mu p = \mu$ read

$$\mu(\{n\}) = p_{n-1} \, \mu(\{n - 1\}) \quad \text{for } n \in \mathbb{N},$$

$$\mu(\{0\}) = \sum_{n=0}^{\infty} \mu(\{n\})(1 - p_n).$$

Hence we get

$$\mu(\{n\}) = \mu(\{0\}) \prod_{k=0}^{n-1} p_k$$

and (note that the sum is a telescope sum)

$$\mu(\{0\}) = \mu(\{0\}) \sum_{n=0}^{\infty} (1 - p_n) \prod_{k=0}^{n-1} p_k = \mu(\{0\}) \left(1 - \prod_{n=0}^{\infty} p_n\right).$$

Hence there exists a nontrivial invariant measure μ (that is, $\mu(\{0\})$ can be chosen strictly positive) if and only if $\prod_{n=0}^{\infty} p_n = 0$. This, however, is true if and only if $\sum_{n=0}^{\infty}(1 - p_n) = \infty$. Using a Borel–Cantelli argument, it is not hard to show that this is exactly the condition for recurrence of X.

If $\mu \neq 0$, then μ is a finite measure if and only if

$$M := \sum_{n=0}^{\infty} \prod_{k=0}^{n-1} p_k < \infty.$$

Hence X is positive recurrent if and only if $M < \infty$. In fact, it is not hard to show that M is the expected time to return to 0; hence the criterion for positive recurrence could also be deduced by Theorem 17.52.

A necessary condition for $M < \infty$ is of course that the series $\sum_{n=0}^{\infty}(1 - p_n)$ diverge; that is, that X is recurrent. One sufficient condition for $M < \infty$ is

$$\sum_{n=0}^{\infty} \exp\left(-\sum_{k=0}^{n-1}(1 - p_k)\right) < \infty. \quad \Diamond$$

Takeaways An irreducible Markov chain possesses an invariant distribution if and only if it is positive recurrent. The invariant distribution is unique and is given as the reciprocals of the expected return times.

Exercise 17.6.1 Consider the Markov chain from Fig. 17.1 (page 412). Determine the set of all invariant distributions. Show that the states 6, 7 and 8 are positive recurrent and compute the expected first entrance times

$$\mathbf{E}_6[\tau_6] = \frac{17}{4}, \qquad \mathbf{E}_7[\tau_7] = \frac{17}{5} \quad \text{and} \quad \mathbf{E}_8[\tau_8] = \frac{17}{8}. \quad \clubsuit$$

Exercise 17.6.2 Let $X = (X_t)_{t \geq 0}$ be a Markov chain on E in continuous time with Q-matrix q. Show that a probability measure π on E is an invariant distribution for X if and only if $\sum_{x \in E} \pi(\{x\})q(x, y) = 0$ for all $y \in E$. $\clubsuit$

Exercise 17.6.3 Let G be a countable Abelian group and let p be the transition matrix of an irreducible random walk X on G. That is, we have $p(hg, hf) = p(g, f)$ for all $h, g, f \in G$. (This generalizes the notion of a random walk on $\mathbb{Z}^D$.) Use Theorem 17.52 to show that X is positive recurrent if and only if G is finite. $\clubsuit$

Exercise 17.6.4 Let $r \in [0, 1]$ and let X be the Markov chain on $\mathbb{N}_0$ with transition matrix (see Fig. 17.2 on page 413)

$$p(x, y) = \begin{cases} 1, & \text{if } x = 0 \text{ and } y = 1, \\ r, & \text{if } y = x + 1 \geq 2, \\ 1 - r, & \text{if } y = x - 1, \\ 0, & \text{else.} \end{cases}$$

Compute the invariant measure and show the following using Theorem 17.52:

(i) If $r \in \left(0, \frac{1}{2}\right)$, then X is positive recurrent.
(ii) If $r = \frac{1}{2}$, then X is null recurrent.
(iii) If $r \in \{0\} \cup \left(\frac{1}{2}, 1\right]$, then X is transient. ♣

Exercise 17.6.5

(i) Use a direct argument to show that the Markov chain in Example 17.53 is recurrent if and only if $\sum_{n=0}^{\infty}(1 - p_n) = \infty$.
(ii) Show that the expected time to return to 0 is M and infer that the chain is positive recurrent if and only if $M < \infty$.
(iii) Give examples of sequences $(p_x)_{x \in \mathbb{N}_0}$ such that the chain is (a) transient, (b) null recurrent, (c) positive recurrent, and (d) positive recurrent but

$$\sum_{n=0}^{\infty} \exp\left(-\sum_{k=0}^{n-1}(1 - p_k)\right) = \infty. \quad ♣$$

Exercise 17.6.6 Let X be irreducible and recurrent. Show that, as claimed in Remark 17.51, the invariant measure is unique up to constant multiples.
Hint: Let $\pi \neq 0$ be an invariant measure for X and abbreviate $\mathbf{P}_\pi = \sum_{x \in E} \pi(\{x\})\mathbf{P}_x$ (note that, in general, this need not be a finite measure). Let $x, y \in E$ with $x \neq y$ and deduce by induction that

$$\pi(\{y\}) = \mathbf{P}_\pi\left[\tau_x^1 \geq n, X_0 \neq x, X_n = y\right] + \sum_{k=1}^{n} \mathbf{P}_\pi\left[\tau_x^1 \geq k, X_0 = x, X_k = y\right].$$

Infer that

$$\pi(\{y\}) \geq \sum_{k=1}^{\infty} \mathbf{P}_\pi\left[\tau_x^1 \geq k, X_0 = x, X_k = y\right] = \pi(\{x\})\mu_x(\{y\}),$$

where μ_x is the invariant measure defined in Theorem 17.48. Now use the fact that $\pi p^n = \pi$ and $\mu_x p^n = \mu_x$ for all $n \in \mathbb{N}$ to conclude that even $\pi(\{y\}) = \pi(\{x\})\mu_x(\{y\})$ holds. ♣

17.7 Stochastic Ordering and Coupling

In many situations, for the comparison of two distributions, it is helpful to construct a product space such that the two distributions are the marginal distributions but are not necessarily independent. We first introduce the abstract principle of such *couplings* and then give some examples.

There are many concepts to order probability measures on $\mathbb{R}$ or $\mathbb{R}^d$ such that the "larger" one has a greater preference for large values than the "smaller" one. As one of the most prominent orders we present here the so-called stochastic order and illustrate its connection with couplings. As an excuse for presenting this section in a chapter on Markov chains, we fill finally use a simple Markov chain in order to prove a theorem on the stochastic order of binomial distributions.

Definition 17.54 *Let $(E_1, \mathcal{E}_1, \mu_1)$ and $(E_2, \mathcal{E}_2, \mu_2)$ be probability spaces. A probability measure μ on $(E_1 \times E_2, \mathcal{E}_1 \otimes \mathcal{E}_2)$ with $\mu(\cdot \times E_2) = \mu_1$ and $\mu(E_1 \times \cdot) = \mu_2$ is called a **coupling** of μ_1 and μ_2.*

Clearly, the product measure $\mu = \mu_1 \otimes \mu_2$ is a coupling, but in many situations there are more interesting ones.

Example 17.55 Let X be a real random variable and let $f, g : \mathbb{R} \to \mathbb{R}$ be monotone increasing functions with $\mathbf{E}[f(X)^2] < \infty$ and $\mathbf{E}[g(X)^2] < \infty$. We want to show that the random variables $f(X)$ and $g(X)$ are nonnegatively correlated.

To this end, let Y be an **independent copy** of X; that is, a random variable with $\mathbf{P}_Y = \mathbf{P}_X$ that is independent of X. Note that $\mathbf{E}[f(X)] = \mathbf{E}[f(Y)]$ and $\mathbf{E}[g(X)] = \mathbf{E}[g(Y)]$. For all numbers $x, y \in \mathbb{R}$, we have $(f(x) - f(y))(g(x) - g(y)) \geq 0$. Hence

$$
\begin{aligned}
0 \leq \mathbf{E}\big[\big(f(X) - f(Y)\big)\big(g(X) - g(Y)\big)\big] \\
= \mathbf{E}[f(X)g(X)] - \mathbf{E}[f(X)]\,\mathbf{E}[g(Y)] + \mathbf{E}[f(Y)g(Y)] - \mathbf{E}[f(Y)]\,\mathbf{E}[g(X)] \\
= 2\,\mathbf{Cov}[f(X), g(X)]. \quad \Diamond \qquad\qquad (17.26)
\end{aligned}
$$

Example 17.56 Let (E, ϱ) be a Polish space. For two probability measures P and Q on $(E, \mathcal{B}(E))$, denote by $K(P, Q) \subset \mathcal{M}_1(E \times E)$ the set of all couplings of P and Q. The so-called **Wasserstein metric** on $\mathcal{M}_1(E)$ is defined by

$$
d_W(P, Q) := \inf\left\{ \int \varrho(x, y)\,\varphi(d(x, y)) : \varphi \in K(P, Q) \right\}. \qquad (17.27)
$$

It can be shown that (this is the Kantorovich–Rubinstein theorem [84]; see also [37, pages 420ff])

$$
d_W(P, Q) = \sup\left\{ \int f\,d(P - Q) : f \in \mathrm{Lip}_1(E; \mathbb{R}) \right\}. \qquad (17.28)
$$

Compare this representation of the Wasserstein metric with that of the total variation norm,

$$
\|P - Q\|_{TV} = \sup\left\{ \int f\,d(P - Q) : f \in \mathcal{L}^\infty(E) \text{ with } \|f\|_\infty \leq 1 \right\}. \qquad (17.29)
$$

In fact, we can also give a definition for the total variation in terms of a coupling: Let $D := \{(x, x) : x \in E\}$ be the diagonal in $E \times E$. Then

$$\|P - Q\|_{TV} = \inf \{\varphi((E \times E) \setminus D) : \varphi \in K(P, Q)\}. \tag{17.30}$$

See [60] for a comparison of different metrics on $\mathcal{M}_1(E)$.◊

As an example of a more involved coupling, we quote the following theorem that is due to Skorohod.

Theorem 17.57 (Skorohod coupling) *Let $\mu, \mu_1, \mu_2, \dots$ be probability measures on a Polish space E with $\mu_n \overset{n \to \infty}{\longrightarrow} \mu$. Then there exists a probability space $(\Omega, \mathcal{A}, \mathbf{P})$ with random variables $X, X_1, X_2, \dots$ with $\mathbf{P}_X = \mu$ and $\mathbf{P}_{X_n} = \mu_n$ for every $n \in \mathbb{N}$ such that $X_n \overset{n \to \infty}{\longrightarrow} X$ almost surely.*

Proof See, e.g., [83, page 79]. □

Reflection For real valued random variables, a Skorohod coupling can be constructed explicitly using the distribution functions. How does this work in detail? ♠

We now come to the concept of stochastic order.

Definition 17.58 *Let $\mu_1, \mu_2 \in \mathcal{M}_1(\mathbb{R}^d)$. We write $\mu_1 \leq_{st} \mu_2$ if*

$$\int f \, d\mu_1 \leq \int f \, d\mu_2$$

*for every monotone increasing bounded function $f : \mathbb{R}^d \to \mathbb{R}$. In this case, we say that μ_2 is **stochastically larger** than μ_1.*

Evidently, $\leq_{st}$ is a partial order on $\mathcal{M}_1(\mathbb{R}^d)$. The stochastic order belongs to the class of so-called integral orders that are defined by the requirement that the integrals with respect to a certain class of functions (here: monotone increasing and bounded) are ordered. Other classes of functions that are often considered are convex functions or indicator functions on lower or upper orthants.

Let F_1 and F_2 be the distribution functions of μ_1 and μ_2. Clearly, $\mu_1 \leq_{st} \mu_2$ implies $F_1(x) \geq F_2(x)$ for all $x \in \mathbb{R}^d$. If $d = 1$, then both statements are equivalent. However, for $d \geq 2$, the condition $F_1 \geq F_2$ is weaker than $\mu_1 \leq_{st} \mu_2$. For example, consider $d = 2$ and

$$\mu_1 = \frac{1}{2}\delta_{(0,0)} + \frac{1}{2}\delta_{(1,1)} \quad \text{and} \quad \mu_2 = \frac{1}{2}\delta_{(1,0)} + \frac{1}{2}\delta_{(0,1)}.$$

The partial order defined by the comparison of the distribution functions is called (lower) orthant order.

For a survey on different orders of probability measures, see, e.g., [120].

The following theorem was shown by Strassen [161] in larger generality for integral orders.

Theorem 17.59 (Strassen's theorem) *Let*

$$L := \{(x_1, x_2) \in \mathbb{R}^d \times \mathbb{R}^d : x_1 \leq x_2\}.$$

Then $\mu_1 \leq_{st} \mu_2$ if and only if there is a coupling φ of μ_1 and μ_2 with $\varphi(L) = 1$.

Proof Let φ be such a coupling. For monotone increasing bounded $f : \mathbb{R}^d \to \mathbb{R}$, we have $f(x_1) - f(x_2) \leq 0$ for every $x = (x_1, x_2) \in L$; hence $\int f \, d\mu_1 - \int f \, d\mu_2 = \int_L \big(f(x_1) - f(x_2)\big) \varphi(dx) \leq 0$ and thus $\mu_1 \leq_{st} \mu_2$.

Now assume $\mu_1 \leq_{st} \mu_2$. We only consider the case $d = 1$ (see [120, Thm. 3.3.5] for $d \geq 2$). Here $F((x_1, x_2)) := \min(F_1(x_1), F_2(x_2))$ defines a distribution function on $\mathbb{R} \times \mathbb{R}$ (see Exercise 1.5.5) that corresponds to a coupling φ with $\varphi(L) = 1$. A somewhat more explicit representation can be obtained using random variables. Let U be a random variable that is uniformly distributed on $(0, 1)$. Then

$$X_i := F_i^{-1}(U) := \inf\{x \in \mathbb{R} : F_i(x) \geq U\}$$

is a real random variable with distribution μ_i (see proof of Theorem 1.104). Clearly, we have $X_1 \leq X_2$ almost surely; that is, $\mathbf{P}[(X_1, X_2) \in L] = 1$. Evidently, the distribution function of (X_1, X_2) is F.   □

While Strassen's theorem yields the existence of an abstract coupling, in many examples a natural coupling can be established and used as a tool for proving, e.g., stochastic orders.

Example 17.60 Let $n \in \mathbb{N}$ and $0 \leq p_1 \leq p_2 \leq 1$. Let $Y_1, \ldots, Y_n$ be independent random variables that are uniformly distributed on $[0, 1]$. Define $X_i = \#\{k \leq n : Y_k \leq p_i\}$, $i = 1, 2$. Then $X_i \sim b_{n,p_i}$ and $X_1 \leq X_2$ almost surely. This coupling shows that $b_{n,p_1} \leq_{st} b_{n,p_2}$.

An even simpler coupling can be used to show that $b_{m,p} \leq_{st} b_{n,p}$ for $m \leq n$ and $p \in [0, 1]$. ◇

Theorem 17.61 *Let $n_1, n_2 \in \mathbb{N}$ and $p_1, p_2 \in (0, 1)$. We have $b_{n_1,p_1} \leq_{st} b_{n_2,p_2}$ if and only if*

$$(1 - p_1)^{n_1} \geq (1 - p_2)^{n_2} \tag{17.31}$$

and

$$n_1 \leq n_2. \tag{17.32}$$

Proof (The proof follows the exposition in [100, Section 3]) Since $b_{n_i,p_i}(\{0\}) = (1 - p_i)^{n_i}$, conditions (17.31) and (17.32) are clearly necessary for $b_{n_1,p_1} \leq_{st} b_{n_2,p_2}$. Hence we only have to show sufficiency of the two conditions.

Assume that (17.31) and (17.32) hold. By Example 17.60, it is enough to consider the smallest p_2 that fulfills (17.31). Hence we assume $(1 - p_1)^{n_1} = (1 - p_2)^{n_2}$. Define $\lambda := -n_1 \log(1 - p_1) = -n_2 \log(1 - p_2)$. We will construct a binomially distributed random variable by throwing a Poi_λ-distributed number T of balls in n_i boxes and count the number of nonempty boxes. More precisely, let $T \sim \mathrm{Poi}_\lambda$ and let $X_1, X_2, \ldots$ be independent and uniformly distributed on $[0, 1]$ and independent of T. For $n \in \mathbb{N}$, $t \in \mathbb{N}_0$ and $l = 1, \ldots, n$, define

$$M_{n,t,l} = \#\big\{s \le t : X_s \in ((l - 1)/n, \, l/n]\big\}$$

and the number of nonempty boxes after t balls are thrown:

$$N_{n,t} := \sum_{l=1}^{n} \mathbb{1}_{\{M_{n,t,l} > 0\}}.$$

By Theorem 5.35, the random variables $M_{n,T,1}, \ldots, M_{n,T,n}$ are independent and $\mathrm{Poi}_{\lambda/n}$-distributed. In particular, we have

$$\mathbf{P}[M_{n_i,T,l} > 0] = 1 - e^{-\lambda/n_i} = p_i$$

and thus $N_{n_i,T} \sim b_{n_i,p_i}$, $i = 1, 2$. Hence it suffices to show that $N_{n_1,T} \le_{\mathrm{st}} N_{n_2,T}$. For this in turn it is enough to show

$$N_{n_1,t} \le_{\mathrm{st}} N_{n_2,t} \qquad \text{for all } t \in \mathbb{N}_0. \tag{17.33}$$

In fact, let $f : \{0, \ldots, n\} \to \mathbb{R}$ be monotone increasing. Then

$$\mathbf{E}[f(N_{n_1,T})] = \sum_{t=0}^{\infty} \mathbf{E}[f(N_{n_1,t})]\, \mathbf{P}[T = t]$$

$$\le \sum_{t=0}^{\infty} \mathbf{E}[f(N_{n_2,t})]\, \mathbf{P}[T = t] = \mathbf{E}[f(N_{n_2,T})].$$

We use an induction argument to show (17.33). For $t = 0$, the claim holds trivially. Now assume that (17.33) holds for some given $t \in \mathbb{N}_0$. We are now at the point to use a Markov chain. Note that (for fixed n), $(N_{n,t})_{t=0,1,\ldots}$ is a Markov chain with state space $\{0, \ldots, n\}$ and transition matrix

$$p_n(k, l) = \begin{cases} k/n, & \text{if } l = k, \\ 1 - k/n, & \text{if } l = k + 1, \\ 0, & \text{otherwise.} \end{cases}$$

We define for $k, l = 0, \ldots, n$

$$
h_{n,l}(k) = \sum_{j=l}^{n} p_n(k, j) = \begin{cases} 0, & \text{if } k < l - 1, \\ 1 - k/n, & \text{if } k = l - 1, \\ 1, & \text{if } k > l - 1. \end{cases}
$$

Then $\mathbf{P}[N_{n,t+1} \geq l] = \mathbf{E}[h_{n,l}(N_{n,t})]$ and $h_{n,l}(k)$ is monotone increasing both in k and in n. Hence by the induction hypothesis, we have

$$
\mathbf{P}[N_{n_1,t+1} \geq l] = \mathbf{E}[h_{n_1,l}(N_{n_1,t})] \leq \mathbf{E}[h_{n_1,l}(N_{n_2,t})]
$$
$$
\leq \mathbf{E}[h_{n_2,l}(N_{n_2,t})] = \mathbf{P}[N_{n_2,t+1} \geq l].
$$

We conclude that $N_{n_1,t+1} \leq_{\mathrm{st}} N_{n_2,t+1}$ which completes the induction and the proof of the theorem. □

> **Takeaways** A coupling is a probability measure on a product space with given marginals. In many situations it is desirable to have a coupling with additional properties like all the mass lies above the diagonal. The latter property can be achieved if the marginals are stochastically ordered. Optimal couplings can be used also for the definitions of a metric on probability measures such as the Wasserstein metric, for example. Using Markov chains we construct a coupling to prove a theorem on the stochastic ordering of binomial distributions.

Exercise 17.7.1 Use an elementary direct coupling argument to show the claim of Theorem 17.61 for the case $n_2/n_1 \in \mathbb{N}$. ♣

Exercise 17.7.2 For the Poisson distribution, show that

$$
\mathrm{Poi}_{\lambda_1} \leq_{\mathrm{st}} \mathrm{Poi}_{\lambda_2} \iff \lambda_1 \leq \lambda_2. \quad ♣
$$

Exercise 17.7.3 Let $n \in \mathbb{N}$, $p \in (0, 1)$ and $\lambda > 0$. Show that

$$
b_{n,p} \leq_{\mathrm{st}} \mathrm{Poi}_{\lambda} \iff (1 - p)^n \geq e^{-\lambda}. \quad ♣
$$

Chapter 18
Convergence of Markov Chains

We consider a Markov chain X with invariant distribution π and investigate conditions under which the distribution of X_n converges to π for $n \to \infty$. Essentially it is necessary and sufficient that the state space of the chain cannot be decomposed into subspaces

- that the chain does not leave
- or that are visited by the chain periodically; e.g., only for odd n or only for even n.

In the first case, the chain would be called *reducible*, and in the second case, it would be *periodic*.

We study periodicity of Markov chains in the first section. In the second section, we prove the convergence theorem. The third section is devoted to applications of the convergence theorem to computer simulations with the so-called Monte Carlo method. In the last section, we describe the speed of convergence to the equilibrium by means of the spectrum of the transition matrix.

18.1   Periodicity of Markov Chains

We study the conditions under which a positive recurrent Markov chain X on the countable space E (and with transition matrix p), started in an arbitrary $\mu \in \mathcal{M}_1(E)$, converges in distribution to an invariant distribution π; that is, $\mu p^n \overset{n\to\infty}{\longrightarrow} \pi$. Clearly, it is necessary that π be the *unique* invariant distribution; that is, up to a factor π it is the unique left eigenvector of p for the eigenvalue 1. As shown in Theorem 17.50, for this uniqueness it is sufficient that the chain be irreducible.

In order for $\mu p^n \overset{n\to\infty}{\longrightarrow} \pi$ to hold for every $\mu \in \mathcal{M}_1(E)$, a certain contraction property of p is necessary. Manifestly, 1 is the largest (absolute value of an)

© The Editor(s) (if applicable) and The Author(s), under exclusive license to Springer Nature Switzerland AG 2020
A. Klenke, *Probability Theory*, Universitext,
https://doi.org/10.1007/978-3-030-56402-5_18

eigenvalue of p. However, p is sufficiently contractive only if the multiplicity of the eigenvalue 1 is exactly 1 and if there are no further (possibly complex-valued) eigenvalues of modulus 1.

For the latter property, it is not sufficient that the chain be irreducible. For example, consider on $E = \{0, \ldots, N-1\}$ the Markov chain with transition matrix $p(x, y) = \mathbb{1}_{\{y=x+1(\mathrm{mod}\ N)\}}$. The eigenvalue 1 has the multiplicity 1. However, all complex Nth roots of unity $e^{2\pi i k/N}$, $k = 0, \ldots, N-1$, are eigenvalues of modulus 1. Clearly, the uniform distribution on E is invariant but $\lim_{n\to\infty} \delta_x p^n$ does not exist for any $x \in E$. In fact, every point is visited periodically after N steps. In order to obtain criteria for the convergence of Markov chains, we thus have to understand periodicity first. Thereafter, for irreducible *aperiodic* chains, we state the convergence theorem.

If $m, n \in \mathbb{N}$, then write $m | n$ if m is a divisor of n; that is, if $\frac{n}{m} \in \mathbb{N}$. If $M \subset \mathbb{N}$, then denote by $\gcd(M)$ the greatest common divisor of all $n \in M$. In the following, let X be a Markov chain on the countable space E with transition matrix p.

Definition 18.1

(i) For $x, y \in E$, define

$$N(x, y) := \{n \in \mathbb{N}_0 : p^n(x, y) > 0\}.$$

For any $x \in E$, $d_x := \gcd(N(x, x))$ is called the **period** of the state x.
(ii) If $d_x = d_y$ for all $x, y \in E$, then $d := d_x$ is called the period of X.
(iii) If $d_x = 1$ for all $x \in E$, then X is called **aperiodic**.

See Figs. 18.1 and 18.2 for illustrations of aperiodic and periodic Markov chains.

Lemma 18.2 *For any $x \in E$, there exists an $n_x \in \mathbb{N}$ with*

$$p^{nd_x}(x, x) > 0 \quad \text{for all } n \geq n_x. \tag{18.1}$$

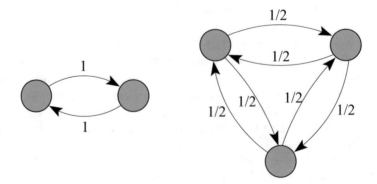

Fig. 18.1 The left Markov chain is periodic with period 2, and the right Markov chain is aperiodic.

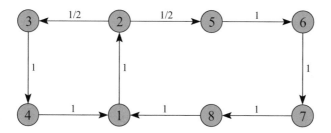

Fig. 18.2 Here $N(8, 8) = \{6, 10, 12, 14, 16, \ldots\}$; hence $d_8 := \gcd(\{6, 10, 12, \ldots\}) = 2$ and $n_8 = 5$. The chain thus has period 2. However, $n_1 = 2$ and $n_4 = 4$.

Proof Let $k_1, \ldots, k_r \in N(x, x)$ with $\gcd(\{k_1, \ldots, k_r\}) = d_x$. Then, for all $m_1, \ldots, m_r \in \mathbb{N}_0$, we also have $\sum_{i=1}^{r} k_i m_i \in N(x, x)$. Basic number theory then yields that, for every $n \geq n_x := r \cdot \prod_{i=1}^{r}(k_i/d_x)$, there are numbers $m_1, \ldots, m_r \in \mathbb{N}_0$ with $n\, d_x = \sum_{i=1}^{r} k_i m_i$. Hence (18.1) holds. $\qquad\square$

The problem of finding the *smallest* number N such that any $n\, d_x$, $n \geq N$ can be written as a nonnegative integer linear combination of $k_1, \ldots, k_r$ is called the *Frobenius problem*. The general solution is unknown; however, for the case $r = 2$, Sylvester [163] showed that $N = (k_1/d_x - 1)(k_2/d_x - 1)$ is minimal. In the general case, for N, the upper bound $2 \max\{k_i : i = 1, \ldots, r\}^2/(rd_x^2)$ is known; see, e.g., [45].

Lemma 18.3 *Let X be irreducible. Then the following statements hold.*

(i) $d := d_x = d_y$ *for all $x, y \in E$.*
(ii) *For all $x, y \in E$, there exist $n_{x,y} \in \mathbb{N}$ and $L_{x,y} \in \{0, \ldots, d-1\}$ such that*

$$nd + L_{x,y} \in N(x, y) \qquad \text{for all } n \geq n_{x,y}. \tag{18.2}$$

$L_{x,y}$ is uniquely determined, and we have

$$L_{x,y} + L_{y,z} + L_{z,x} = 0 \;(\mathrm{mod}\; d) \quad \text{for all } x, y, z \in E. \tag{18.3}$$

Proof

(i) Let $m, n \in \mathbb{N}_0$ with $p^m(x, y) > 0$ and $p^n(y, z) > 0$. Then

$$p^{m+n}(x, z) \geq p^m(x, y)\, p^n(y, z) > 0.$$

Hence we have

$$N(x, y) + N(y, z) := \{m + n : m \in N(x, y),\, n \in N(y, z)\} \subset N(x, z). \tag{18.4}$$

If, in particular, $m \in N(x, y)$, $n \in N(y, x)$ and $k \geq n_y$, then $kd_y \in N(y, y)$; hence $m + kd_y \in N(x, y)$ and $m + n + kd_y \in N(x, x)$. Therefore, $d_x | (m + n + kd_y)$ for every $k \geq n_y$; hence $d_x | d_y$. Similarly, we get $d_y | d_x$; hence $d_x = d_y$.

(ii) Let $m \in N(x, y)$. Then $m + kd \in N(x, y)$ for every $k \geq n_x$. Hence (18.2) holds with

$$n_{x,y} := n_x + \left\lfloor \frac{m}{d} \right\rfloor \quad \text{and} \quad L_{x,y} := m - d \left\lfloor \frac{m}{d} \right\rfloor.$$

Owing to (18.4), we have

$$(n_{x,y} + n_{y,z})d + L_{x,y} + L_{y,z} \in N(x, z).$$

Together with $z = x$, it follows that $d | (L_{x,y} + L_{y,x})$. Hence the value of $L_{x,y}$ is unique in $\{0, \ldots, d-1\}$ and $L_{x,y} = -L_{y,x} \pmod{d}$. For general z, we infer that $d | (L_{x,y} + L_{y,z} + L_{z,x})$; hence (18.3).   □

Theorem 18.4 *Let X be irreducible with period d. Then there exists a disjoint decomposition of the state space*

$$E = \biguplus_{i=0}^{d-1} E_i \tag{18.5}$$

with the property

$$p(x, y) > 0 \ and \ x \in E_i \quad \Longrightarrow \quad y \in E_{i+1 \, (\text{mod} \, d)}. \tag{18.6}$$

This decomposition is unique up to cyclic permutations.

See Fig. 18.3 for an illustration of the state space decomposition of a periodic Markov chain.

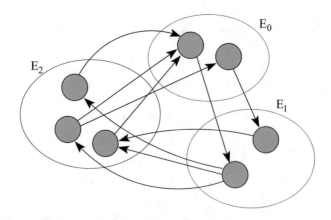

Fig. 18.3 State space decomposition of a Markov chain with period $d = 3$.

Property (18.6) says that X visits the E_i one after the other (see Fig. 18.3 or 18.2, where $d = 2$, $E_0 = \{1, 3, 5, 7\}$ and $E_1 = \{2, 4, 6, 8\}$). Somewhat more formally, we could write: If $x \in E_i$ for some i, then $\mathbf{P}_x[X_n \in E_{i+n \pmod d}] = 1$.

Proof

Existence $x_0 \in E$ and let

$$E_i := \{y \in E : L_{x_0, y} = i\} \quad \text{for } i = 0, \ldots, d - 1.$$

Clearly, (18.5) holds. Let $i \in \{0, \ldots, d-1\}$ and $x \in E_i$. If $y \in E$ with $p(x, y) > 0$, then $L_{x,y} = 1$ and hence $L_{x_0, y} = L_{x_0, x} + L_{x, y} = i + 1 \pmod d$.

Uniqueness Let $(\widetilde{E}_i, \ i = 0, \ldots, \widetilde{d} - 1)$ be another decomposition that satisfies (18.5) and (18.6). Without loss of generality, assume $E_0 \cap \widetilde{E}_0 \neq \emptyset$ (otherwise permute the $\widetilde{E}_i$ cyclically until this holds). Fix an arbitrary $x_0 \in E_0 \cap \widetilde{E}_0$. By assumption, $p(x_0, y) > 0$ now implies $y \in E_1$ and $y \in \widetilde{E}_1$; hence $y \in E_1 \cap \widetilde{E}_1$. Inductively, we get that $p^{nd+i}(x, y) > 0$ implies $y \in E_i \cap \widetilde{E}_i$ (for all $n \in \mathbb{N}$ and $i = 0, \ldots, d - 1$).

However, since the chain is irreducible, for every $y \in E$, there exist numbers $n(y)$ and $i(y)$ such that $p^{n(y) d + i(y)}(x_0, y) > 0$; hence $y \in E_{i(y)} \cap \widetilde{E}_{i(y)}$. Therefore, we have $E_i = \widetilde{E}_i$ for every $i = 0, \ldots, d - 1$. $\qquad\square$

Takeaways Assume that a Markov chain can return to a given state only at times that are a multiple of some natural number d and assume that d is the largest number with this property. Then d is said to be the period of that state. For irreducible chains, all states have the same period. For example, for nearest neighbour random walk on the integers, every state has period $d = 2$. If we have $d = 1$ for every state, then the Markov chain is called aperiodic. For periodic chains, the state space decomposes into d subspaces that can be entered at specific times (mod d) only. In this sense, aperiodicity has the flavour of an irreducibility condition which is needed in order that two independent chains started in arbitrary states could meet each other.

18.2 Coupling and Convergence Theorem

Our goal is to use a coupling of two discrete Markov chains that are started in different distributions μ and ν in order to show the convergence theorem for Markov chains.

In the following, let E be a countable space and let p be a stochastic matrix on E. Recall the definition of a general coupling of two probability measures from Definition 17.54.

Definition 18.5 *A bivariate process* $((X_n, Y_n))_{n \in \mathbb{N}_0}$ *with values in* $E \times E$ *is called a coupling if* $(X_n)_{n \in \mathbb{N}_0}$ *and* $(Y_n)_{n \in \mathbb{N}_0}$ *are Markov chains, each with transition matrix p.*

A coupling is called successful if $\mathbf{P}_{(x,y)}\left[\bigcup_{m \geq n}\{X_m \neq Y_m\}\right] \overset{n \to \infty}{\longrightarrow} 0$ *for all* $x, y \in E$.

Of course, two independent chains form a coupling, though maybe not the most interesting one.

Example 18.6 (Independent coalescence) The most important coupling is Markov chains that run independently until they coalesce: Let X and Y be independent chains with transition matrix p until they first meet. After that, the chains run together. We call this coupling the **independent coalescent**. The transition matrix is

$$\bar{p}\big((x_1, y_1), (x_2, y_2)\big) = \begin{cases} p(x_1, x_2) \cdot p(y_1, y_2), & \text{if } x_1 \neq y_1, \\ p(x_1, x_2), & \text{if } x_1 = y_1, \ x_2 = y_2, \\ 0, & \text{if } x_1 = y_1, \ x_2 \neq y_2. \end{cases}$$

Denote by $\tau := \inf\{n \in \mathbb{N}_0 : X_n = Y_n\}$ the time of coalescence. We can construct the coupling using two independent chains $\tilde{X}$ and $\tilde{Y}$ by defining $X := \tilde{X}$, $\tilde{\tau} := \inf\{n \in \mathbb{N}_0 : \tilde{X}_n = \tilde{Y}_n\}$ and

$$Y_n := \begin{cases} \tilde{Y}_n, & \text{if } n < \tilde{\tau}, \\ X_n, & \text{if } n \geq \tilde{\tau}. \end{cases}$$

Instead of checking by a direct computation that this process (X, Y) is indeed a coupling with transition matrix $\bar{p}$, consider the construction of Markov chains from Theorem 17.17: Let $(R_n(x) : n \in \mathbb{N}_0, x \in E)$ be independent random variables with distribution $\mathbf{P}[R_n(x_1) = x_2] = p(x_1, x_2)$, and let $\tilde{R}_n((x_1, y_1)) = (R_n(x_1), R_n(y_1))$. Then $(\tilde{R}_n)_{n \in \mathbb{N}_0}$ is independent and we have

$$\mathbf{P}\big[\tilde{R}_n((x_1, y_1)) = (x_2, y_2)\big] = \bar{p}\big((x_1, y_1), (x_2, y_2)\big).$$

As we saw in Theorem 17.17, by $X_{n+1} := R_n(X_n)$ and $Y_{n+1} := R_n(Y_n)$, two Markov chains X and Y are defined with transition matrix p. On the other hand, we have $(X_{n+1}, Y_{n+1}) = \tilde{R}_n((X_n, Y_n))$. Hence the bivariate process is indeed a coupling with transition matrix $\bar{p}$. $\Diamond$

Example 18.7 Let $E = \mathbb{Z}$ and $p(x, y) = 1/3$ if $|x-y| \leq 1$ and 0 otherwise. Clearly, p is the transition matrix of an aperiodic recurrent random walk on $\mathbb{Z}$. We will show that we can obtain a successful coupling by coalescing independent chains.

Accordingly, let $\tilde{X}$ and $\tilde{Y}$ be independent random walks with transition matrix p. Then the difference chain $(Z_n)_{n \in \mathbb{N}_0} := (\tilde{X}_n - \tilde{Y}_n)_{n \in \mathbb{N}_0}$ is a symmetric random walk with finite expectation and hence recurrent. Furthermore, Z is irreducible. For any two points $x, y \in \mathbb{Z}$, we thus have

$$\mathbf{P}_{(x,y)}[\tilde{\tau} < \infty] = \mathbf{P}_{x-y}[Z_n = 0 \text{ for some } n \in \mathbb{N}_0] = 1.$$

Therefore, X and Y coalesce almost surely. $\Diamond$

Recurrence, irreducibility and aperiodicity alone are not sufficient for the independent coalescence coupling to be successful. In Exercise 18.2.4, an example is studied that shows that spacial homogeneity cannot easily be dropped if we want to have a successful coupling. Dropping the assumption of recurrence is easier, as the following theorem shows.

Theorem 18.8 *Let X be an arbitrary aperiodic and irreducible random walk on $\mathbb{Z}^d$ with transition matrix p. Then there exists a successful coupling (X, Y).*

Proof
Step 1.   First, consider the case where $p(0, x) = 3^{-d}$ for all $x \in \{-1, 0, 1\}^d$. The individual coordinates $X^{(1)}, \ldots, X^{(d)}$ of X are independent random walks on $\mathbb{Z}$ with transition probabilities $\mathbf{P}_0[X_1^{(i)} = x_i] = 1/3$ for $x_i = -1, 0, 1$. By Example 18.7, we can construct independent successful couplings $(X^{(i)}, Y^{(i)})$, $i = 1, \ldots, d$, with merging times $\tau^{(i)}$. Define $Y = (Y^{(1)}, \ldots, Y^{(d)})$ and $\tau = \max\{\tau^{(1)}, \ldots, \tau^{(d)}\} < \infty$. Then (X, Y) is a successful coupling and $X_n = Y_n$ for $n \geq \tau$.
Step 2.   Now, consider the case where

$$\lambda := 3^d \min\{p(0, x) : x \in \{-1, 0, 1\}^d\} > 0.$$

If $\lambda = 1$, then the condition of Step 1 is fulfilled and we are done. Hence, we assume that $\lambda \in (0, 1)$. We define the transition matrix $\hat{p}$ on $\mathbb{Z}^d$ by $\hat{p}(x, y) = 3^{-d}$ for $y - x \in \{-1, 0, 1\}^d$. Note that also $\check{p} := (p - \lambda\hat{p})/(1 - \lambda)$ is the transition matrix of a random walk on $\mathbb{Z}^d$ and that

$$p = \lambda\hat{p} + (1 - \lambda)\check{p}.$$

Let $\hat{X}$ and $\check{X}$ be independent random walks with transition matrices $\hat{p}$ and $\check{p}$, respectively. Assume that $\hat{X}_0 = X_0$ and $\check{X}_0 = 0$. Furthermore, let $Z_1, Z_2, \ldots$ be i.i.d. Bernoulli random variables with parameter λ that are independent of $\hat{X}$ and $\check{X}$. Define $S_n := Z_1 + \ldots + Z_n$ for $n \in \mathbb{N}$ and

$$X_n := \hat{X}_{S_n} + \check{X}_{n-S_n}.$$

That is, in each time step, a coin flip decides whether X makes a jump according to the matrix $\hat{p}$ or $\check{p}$. Hence X is a random walk with transition matrix p.

By Step 1, there exists a successful coupling $(\hat{X}, \hat{Y})$ such that $\hat{Y}$ is independent of $\check{X}$ and $Z_1, Z_2, \ldots$. Consequently,

$$Y_n := \hat{Y}_{S_n} + \check{X}_{n-S_n}, \quad n \in \mathbb{N},$$

is also a random walk with transition matrix p. Since we have $S_n \to \infty$ almost surely, the coupling (X, Y) is successful.

Step 3. Finally, we consider the general situation. Since X is irreducible and aperiodic, by Lemma 18.3(ii), there exists an $N \in \mathbb{N}$, such that the N-step transition matrix fulfills

$$p^N(0, x) > 0 \quad \text{for all } x \in \{-1, 0, 1\}^d.$$

Hence, the random walk $X' = (X'_n)_{n \in \mathbb{N}} := (X_{nN})_{n \in \mathbb{N}}$ fulfills the condition from Step 2. Let (X', Y') be the coupling that was constructed in Step 2 and let

$$\tau := \inf\{n \in \mathbb{N}_0 : X'_m = Y'_m \text{ for all } m \geq n\}.$$

Then Y' is a random walk with transition matrix p^N. For $n \in \mathbb{N}_0$, define $Y_{nN} := Y'_n$. It remains to close the gaps between the points $\{0, N, 2N, \ldots\}$ in such a way that Y is a random walk and (X, Y) is a successful coupling.

Let $(U^{x,y,n} : x, y \in \mathbb{Z}^d, n \in \mathbb{N}_0)$ be an independent family of $(\mathbb{Z}^d)^{N-1}$-valued random variables $U^{x,y,n} = (U_1^{x,y,n}, \ldots, U_{N-1}^{x,y,n})$ such that

$$\mathbf{P}[(X_1, \ldots, X_{N-1}) \in \cdot \mid X_0 = x, X_N = y] = \mathbf{P}_{U^{x,y,n}}$$

for all $x, y \in \mathbb{Z}^d$ with $p^N(x, y) > 0$ and for all $n \in \mathbb{N}_0$. We further assume that the $U^{x,y,n}$ are independent of X and Y'. For $k \in \{nN + 1, \ldots, (n+1)N - 1\}$, define

$$Y_k := \begin{cases} U_{k-nN}^{Y'_n, Y'_{n+1}, n}, & \text{if } n < \tau, \\ X_k, & \text{else.} \end{cases}$$

It is easy to check that Y is indeed a random walk with transition matrix p. By construction, the coupling (X, Y) is successful. $\square$

Theorem 18.9 *Let X be a Markov chain on E with transition matrix p. If there exists a successful coupling, then every bounded harmonic function is constant.*

Proof Let $f : E \to \mathbb{R}$ be bounded and harmonic; hence $pf = f$. Let $x, y \in E$, and let (X, Y) be a successful coupling. By Lemma 17.46, $(f(X_n))_{n \in \mathbb{N}_0}$ and $(f(Y_n))_{n \in \mathbb{N}_0}$ are martingales; hence we have

$$|f(x) - f(y)| = |\mathbf{E}_{(x,y)}[f(X_n) - f(Y_n)]| \leq 2\|f\|_\infty \mathbf{P}_{(x,y)}[X_n \neq Y_n] \xrightarrow{n \to \infty} 0. \quad \square$$

Corollary 18.10 *If X is an irreducible random walk on $\mathbb{Z}^d$, then every bounded harmonic function is constant.*

This statement holds more generally if we replace $\mathbb{Z}^d$ by a locally compact Abelian group. In that form, the theorem goes back to Choquet and Deny [24], see also [144].

Proof Let p be the transition matrix of X. Let $\bar{X}$ be a Markov chain with transition matrix $\bar{p}(x, y) = \frac{1}{2}p(x, y) + \frac{1}{2}\mathbb{1}_{\{x\}}(y)$. Clearly, X and $\bar{X}$ have the same harmonic functions. Now $\bar{X}$ is an aperiodic irreducible random walk; hence, by Theorem 18.8, there is a successful coupling for all initial states. ☐

Reflection Consider the random walk on the integers with transition matrix given by $p(k, k + 1) = r$ and $p(k, k - 1) = 1 - r$ for some $r \in [0, 1]$. The bounded harmonic functions are constant, but what are the *unbounded* harmonic functions? Be careful, the cases $r = \frac{1}{2}$ and $r \neq \frac{1}{2}$ are different. ♠

Theorem 18.11 *Let p be the transition matrix of an irreducible, positive recurrent, aperiodic Markov chain on E. Then the independent coalescent chain is a successful coupling.*

Proof Let $\tilde{X}$ and $\tilde{Y}$ be two independent Markov chains on E, each with transition matrix p. Then the bivariate Markov chain $Z := ((\tilde{X}_n, \tilde{Y}_n))_{n \in \mathbb{N}_0}$ has the transition matrix $\tilde{p}$ defined by

$$\tilde{p}\big((x_1, y_1), (x_2, y_2)\big) = p(x_1, x_2) \cdot p(y_1, y_2).$$

We first show that the matrix $\tilde{p}$ is irreducible. Only here do we need aperiodicity of p. Accordingly, fix $(x_1, y_1), (x_2, y_2) \in E \times E$. Then, by Lemma 18.2, there exists an $m_0 \in \mathbb{N}$ such that

$$p^n(x_1, x_2) > 0 \quad \text{and} \quad p^n(y_1, y_2) > 0 \quad \text{for all } n \geq m_0.$$

For $n \geq m_0$, we thus have $\tilde{p}^n\big((x_1, y_1), (x_2, y_2)\big) > 0$. Hence $\tilde{p}$ is irreducible.

Now define the stopping time τ of the first entrance of $(\tilde{X}, \tilde{Y})$ into the diagonal $D := \{(x, x) : x \in E\}$ by $\tau := \inf\{n \in \mathbb{N}_0 : \tilde{X}_n = \tilde{Y}_n\}$. Let π be the invariant distribution of $\tilde{X}$. Then, clearly, the product measure $\pi \otimes \pi \in \mathcal{M}_1(E \times E)$ is an (and then *the*) invariant distribution of $(\tilde{X}, \tilde{Y})$. Thus $(\tilde{X}, \tilde{Y})$ is positive recurrent (hence, in particular, recurrent) by Theorem 17.52. Therefore, $\mathbf{P}_{(x,y)}[\tau < \infty] = 1$ for all initial points $(x, y) \in E \times E$ of Z. ☐

Theorem 18.12 *Let X be a Markov chain with transition matrix p such that there exists a successful coupling. Then $\|(\mu - \nu)p^n\|_{TV} \xrightarrow{n \to \infty} 0$ for all $\mu, \nu \in \mathcal{M}_1(E)$.*

If X is aperiodic and positive recurrent with invariant distribution π, then we have $\|\mathcal{L}_\mu[X_n] - \pi\|_{TV} \xrightarrow{n \to \infty} 0$ for all $\mu \in \mathcal{M}_1(E)$.

Proof It is enough to consider the case $\mu = \delta_x$, $\nu = \delta_y$ for some $x, y \in E$. Summation over x and y yields the general case. Let $(X_n, Y_n)_{n \in \mathbb{N}_0}$ be a successful coupling. Then

$$\left\| (\delta_x - \delta_y) p^n \right\|_{TV} \leq 2 \mathbf{P}_{(x,y)}[X_n \neq Y_n] \overset{n \to \infty}{\longrightarrow} 0. \qquad \square$$

We summarize the connection between aperiodicity and convergence of distributions of X in the following theorem.

Theorem 18.13 (Convergence of Markov chains) *Let X be an irreducible, positive recurrent Markov chain on E with invariant distribution π. Then the following are equivalent:*

 (i) X is aperiodic.
 (ii) For every $x \in E$, we have

$$\left\| \mathcal{L}_x[X_n] - \pi \right\|_{TV} \overset{n \to \infty}{\longrightarrow} 0. \tag{18.7}$$

(iii) Equation (18.7) holds for some $x \in E$.
(iv) For every $\mu \in \mathcal{M}_1(E)$, we have $\left\| \mu p^n - \pi \right\|_{TV} \overset{n \to \infty}{\longrightarrow} 0$.

Proof The implications (iv) $\iff$ (ii) $\implies$ (iii) are evident. The implication (i) $\implies$ (ii) was shown in Theorem 18.12. Hence we only show (iii) $\implies$ (i). **"(iii) $\implies$ (i)"** Assume that (i) does not hold. If X has period $d \geq 2$, and if $n \in \mathbb{N}$ is not a multiple of d, then, by Theorem 17.52,

$$\left\| \delta_x p^n - \pi \right\|_{TV} \geq |p^n(x, x) - \pi(\{x\})| = \pi(\{x\}) > 0.$$

Thus, for every $x \in E$, we have $\limsup\limits_{n \to \infty} \left\| \delta_x p^n - \pi \right\|_{TV} > 0$. Therefore, (iii) does not hold. $\qquad \square$

Takeaways For two examples of aperiodic and irreducible Markov chains, we have constructed a coupling such that two chains meet almost surely: Random walks on the d-dimensional integer lattice and positive recurrent Markov chains. In both cases, we infer that bounded harmonic functions are constant. In the latter case, we also get convergence of the Markov chain to the invariant distribution (in distribution).

Exercise 18.2.1 Let d_P be the Prohorov metric (see (13.4) and Exercise 13.2.1). Show that $d_P(P, Q) \leq \sqrt{d_W(P, Q)}$ for all $P, Q \in \mathcal{M}_1(E)$. If E has a finite diameter $\text{diam}(E)$, then $d_W(P, Q) \leq (\text{diam}(E) + 1) d_P(P, Q)$ for all $P, Q \in \mathcal{M}_1(E)$. ♣

Exercise 18.2.2 Consider the bivariate process (X, Y) that was constructed from $\tilde{X}$ and $\tilde{Y}$ in Example 18.6. Show that (X, Y) is a coupling with transition matrix $\bar{p}$. ♣

Exercise 18.2.3 Let X be an arbitrary aperiodic irreducible recurrent random walk on $\mathbb{Z}^d$. Show that, for any two starting points, the independent coalescent coupling is successful.

Hint: Show that the difference of two independent recurrent random walks is a recurrent random walk. ♣

Exercise 18.2.4 Let X be a Markov chain on $\mathbb{Z}^2$ with transition matrix

$$
p((x_1, x_2), (y_1, y_2)) = \begin{cases} \frac{1}{4}, & \text{if } x_1 = 0, \|y - x\|_2 = 1, \\ \frac{1}{4}, & \text{if } x_1 \neq 0 \text{ and } y_1 = x_1 \pm 1, \ x_2 = y_2, \\ \frac{1}{2}, & \text{if } x_1 \neq 0 \text{ and } y_1 = x_1, \ x_2 = y_2, \\ 0, & \text{else.} \end{cases}
$$

Intuitively, this is the symmetric simple random walk whose vertical transitions are all blocked away from the vertical axis. Show that X is null recurrent, irreducible and aperiodic and that independent coalescence does not give a successful coupling. ♣

18.3 Markov Chain Monte Carlo Method

Let E be a finite set and let $\pi \in \mathcal{M}_1(E)$ with $\pi(x) := \pi(\{x\}) > 0$ for every $x \in E$. We consider the problem of sampling a random variable Y with distribution π on a computer. For example, this is a relevant problem if E is a very large set and if sums of the type $\sum_{x \in E} f(x)\pi(x)$ have to be approximated numerically by the estimator $n^{-1} \sum_{i=1}^n f(Y_i)$ (see Example 5.21).

Assume that our computer has a random number generator that provides realizations of i.i.d. random variables $U_1, U_2, \ldots$ that are uniformly distributed on $[0, 1]$. In order for the problem to be interesting, assume also that the distribution π cannot be constructed directly too easily.

Metropolis Algorithm

We have seen already in Example 17.19 how to simulate a Markov chain on a computer. Now the idea is to construct a Markov chain X whose distribution converges to π in the long run. If we simulate such a chain and let it run long enough this should give a sample that is distributed approximately like π. The chain should be designed so that at each step, only a small number of transitions are possible in order to ensure that the procedure described in Example 17.19 works efficiently. (Of course, the chain with transition matrix $p(x, y) = \pi(y)$ converges to π, but this does not help a lot.) This method of producing (approximately) π-distributed samples and using them to estimate expected values of functions of interest is called the **Markov chain Monte Carlo method** or, briefly, **MCMC** (see [15, 112, 119]).

Let q be the transition matrix of an arbitrary irreducible Markov chain on E (with $q(x, y) = 0$ for most $y \in E$). We use this to construct the Metropolis matrix (see [70, 114]).

Definition 18.14 *Define a stochastic matrix p on E by*

$$
p(x, y) = \begin{cases} q(x, y) \min\left(1, \frac{\pi(y)q(y,x)}{\pi(x)q(x,y)}\right), & \text{if } x \neq y, \ q(x, y) > 0, \\ \qquad\qquad\qquad\qquad 0, & \text{if } x \neq y, \ q(x, y) = 0, \\ \qquad\quad 1 - \sum_{z \neq x} p(x, z), & \text{if } x = y. \end{cases}
$$

*p is called the **Metropolis matrix** of q and π.*

Note that p is **reversible** (see Sect. 19.2); that is, for all $x, y \in E$, we have

$$
\pi(x)\, p(x, y) = \pi(y)\, p(y, x). \tag{18.8}
$$

In particular, π is invariant (check this!). We thus obtain the following theorem.

Theorem 18.15 *Assume that q is irreducible and that for any $x, y \in E$, we have $q(x, y) > 0$ if and only if $q(y, x) > 0$. Then the Metropolis matrix p of q and π is irreducible with unique invariant distribution π. If, in addition, q is aperiodic, or q is not reversible with respect to π, then p is aperiodic.*

In order to simulate a chain X that converges to π, we take a reference chain with transition matrix q and use the **Metropolis algorithm**: If the chain with transition matrix q proposes a transition from the present state x to state y, then we accept this proposal with probability

$$
\frac{\pi(y)\, q(y, x)}{\pi(x)\, q(x, y)} \wedge 1.
$$

Otherwise the chain X stays at x.

In the definition of p, the distribution π appears only in terms of the quotients $\pi(y)/\pi(x)$. In many cases of interest, these quotients are easy to compute even though $\pi(x)$ and $\pi(y)$ are not. We illustrate this with an example.

Example 18.16 (Ising model) The Ising model (pronounced like the English word "easing") is a thermodynamical (and quantum mechanical) model for ferromagnetism in crystals. It makes the following assumptions:

- Atoms are placed at the sites of a lattice Λ (for example, $\Lambda = \{0, \ldots, N-1\}^2$).

- Each atom $i \in \Lambda$ has a magnetic spin $x(i) \in \{-1, 1\}$ that either points upwards ($x(i) = +1$) or downwards ($x(i) = -1$).
- Neighboring atoms interact.
- Due to thermic fluctuations, the state of the system is random and distributed according to the so-called **Boltzmann distribution** π on the state space $E := \{-1, 1\}^{\Lambda}$. A parameter of this distribution is the inverse temperature $\beta = \frac{1}{T} \geq 0$ (with T the absolute temperature).

Define the local energy that describes the energy level of a single atom at $i \in \Lambda$ as a function H^i of the state x of the whole system,

$$H^i(x) = \frac{1}{2} \sum_{j \in \Lambda : i \sim j} \mathbb{1}_{\{x(i) \neq x(j)\}}.$$

Here $i \sim j$ indicates that i and j are neighbors in Λ (that is, coordinate-wise mod N, we also speak of *periodic boundary conditions*). The total energy (or Hamilton function) of the system in state x is the sum of the individual energies,

$$H(x) = \sum_{i \in \Lambda} H^i(x) = \sum_{i \sim j} \mathbb{1}_{\{x(i) \neq x(j)\}}.$$

The Boltzmann distribution π on $E := \{-1, 1\}^{\Lambda}$ for the inverse temperature $\beta \geq 0$ is defined by

$$\pi(x) = Z_{\beta}^{-1} \exp(-\beta H(x)),$$

where the **partition sum** $Z_{\beta} = \sum_{x \in E} \exp(-\beta H(x))$ is the normalising constant such that π is a probability measure.

Macroscopically, the individual spins cannot be observed but the average magnetization can; that is, the modulus of the average of all spins,

$$m_{\Lambda}(\beta) = \sum_{x \in E} \pi(x) \left| \frac{1}{\#\Lambda} \sum_{i \in \Lambda} x(i) \right|.$$

If we consider a very large system, then we are close to the so-called thermodynamic limit

$$m(\beta) := \lim_{\Lambda \uparrow \mathbb{Z}^d} m_{\Lambda}(\beta).$$

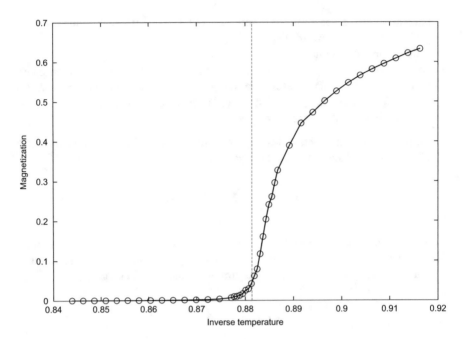

Fig. 18.4 Computer simulation of the magnetization curve of the Ising model on a 1000×1000 grid. The dashed vertical line indicates the critical inverse temperature.

Using a contour argument, as for percolation (see [127]), one can show that (for $d \geq 2$) there exists a critical value $\beta_c = \beta_c(d) \in (0, \infty)$ such that

$$m(\beta) \begin{cases} > 0, & \text{if } \beta > \beta_c, \\ = 0, & \text{if } \beta < \beta_c. \end{cases} \tag{18.9}$$

See Fig. 18.4 for a computer simulation of the curve $\beta \mapsto m(\beta)$.

For a similar model, the Weiss ferromagnet, we will prove in Example 23.20 the existence of such a **phase transition**. In the physical literature, $T_c := 1/\beta_c$ is called the **Curie temperature** for spontaneous magnetization. This is a material-dependent constant (chromium bromide (CrBr) 37 Kelvin, nickel 645 K, iron 1017 K, cobalt 1404 K). Below the Curie temperature, these materials are magnetic, and above it they are not. Below the critical temperature, the magnetization increases with decreasing temperature. We will see in a computer simulation that the Ising model displays this critical temperature effect.

If $x \in E$, then denote by $x^{i,\sigma}$ the state in which at site i the spin is changed to $\sigma \in \{-1, +1\}$; that is,

$$x^{i,\sigma}(j) = \begin{cases} \sigma, & \text{if } j = i, \\ x(j), & \text{if } j \neq i. \end{cases}$$

Furthermore, define the state x^i in which the spin at i is reversed, $x^i := x^{i,-x(i)}$. As reference chain, we choose a chain with transition probabilities

$$q(x, y) = \begin{cases} \frac{1}{\#\Lambda}, & \text{if } y = x^i \text{ for some } i \in \Lambda, \\ 0, & \text{else.} \end{cases}$$

In words, we choose a random site $i \in \Lambda$ (uniformly on Λ) and invert the spin at that site. Clearly, q is irreducible.

The Metropolis algorithm for this chain accepts the proposal of the reference chain with probability 1 if $\pi(x^i) \geq \pi(x)$. Otherwise the proposal is accepted only with probability $\pi(x^i)/\pi(x)$. However, now

$$H(x^i) - H(x) = \sum_{j:\, j\sim i} \mathbb{1}_{\{x(j)\neq -x(i)\}} - \sum_{j:\, j\sim i} \mathbb{1}_{\{x(j)\neq x(i)\}}$$

$$= -2 \sum_{j:\, j\sim i} \left(\mathbb{1}_{\{x(j)\neq x(i)\}} - \frac{1}{2} \right).$$

Hence $\pi(x^i)/\pi(x) = \exp\left(-2\beta \sum_{j\sim i} \left(\mathbb{1}_{\{x(j)=x(i)\}} - \frac{1}{2} \right) \right)$, and this expression is easy to compute as it depends only on the $2d$ neighboring spins and, in particular, does not require knowledge of the value of Z_β. We thus obtain the Metropolis transition matrix

$$p(x, y) = \begin{cases} \frac{1}{\#\Lambda}\left(1 \wedge \exp\left[2\beta \sum_{j:\, j\sim i} \left(\mathbb{1}_{\{x(j)\neq x(i)\}} - \frac{1}{2} \right) \right] \right), & \text{if } y = x^i \text{ for some } i \in \Lambda, \\ 1 - \sum_{i\in\Lambda} p(x, x^i), & \text{if } x = y, \\ 0, & \text{else.} \end{cases}$$

For a practical simulation use the computer's random number generator to produce independent random variables $I_1, I_2, \dots$ and $U_1, U_2, \dots$ with $I_n \sim \mathcal{U}_\Lambda$ and $U_n \sim \mathcal{U}_{[0,1]}$. Then define

$$F_n(x) = \begin{cases} x^{I_n}, & \text{if } U_n \leq \exp\left[2\beta \sum_{j:\, j\sim i} \left(\mathbb{1}_{\{x(j)\neq x(i)\}} - \frac{1}{2} \right) \right], \\ x, & \text{else,} \end{cases}$$

and define the Markov chain $(X_n)_{n\in\mathbb{N}}$ by $X_n = F_n(X_{n-1})$ for $n \in \mathbb{N}$. See Figs. 18.5 and 18.6 for computer simulations of equilibrium states and metastable states of the Ising model. $\Diamond$

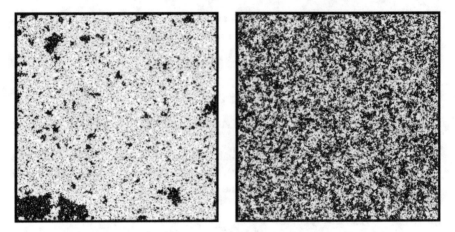

Fig. 18.5 Equilibrium states of the Ising model on an 800×800 grid (black dot = spin +1). Left side: below the critical temperature ($\beta > \beta_c$); Right side: above the critical temperature.

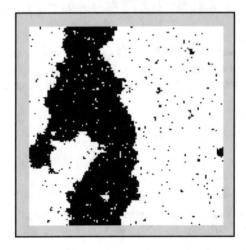

Fig. 18.6 Ising model (150×150 grid) below the critical temperature. Even after a long time, the computer simulation does not produce the equilibrium state but rather so-called metastable states, in which the Weiss domains are clearly visible.

Gibbs Sampler

We consider a situation where, as in the above example, a state consists of many components $x = (x_i)_{i \in \Lambda} \in E$ and where Λ is a finite set. As an alternative to the Metropolis chain, we consider a different procedure to establish a Markov chain with a given invariant distribution. For the so-called **Gibbs sampler** or *heat bath* *algorithm*, the idea is to adapt the state *locally* to the stationary distribution. If x is a state and $i \in \Lambda$, then define

$$x_{-i} := \{y \in E : y(j) = x(j) \text{ for } j \neq i\}.$$

Definition 18.17 (Gibbs sampler) *Let $q \in \mathcal{M}_1(\Lambda)$ with $q(i) > 0$ for every $i \in \Lambda$. The transition matrix p on E with*

$$p(x, y) = \begin{cases} q_i \dfrac{\pi(x^{i,\sigma})}{\pi(x_{-i})}, & \text{if } y = x^{i,\sigma} \text{ for some } i \in \Lambda, \\ 0, & \text{else,} \end{cases}$$

*is called a **Gibbs sampler** for the invariant distribution π.*

Verbally, each step of the chain with transition matrix p can be described by the following instructions.

(1) Choose a random coordinate I according to some distribution $(q_i)_{i \in \Lambda}$.
(2) With probability $\pi(x^{I,\sigma})/\pi(x_{-I})$, replace x by $x^{I,\sigma}$.

If $I = i$, then the new state has the distribution $\mathcal{L}(X|X_{-i} = x_{-i})$, where X is a random variable with distribution π. Note that, for the Gibbs sampler also it is enough to know the values of the distribution π only up to the normalising constant. (In a more general framework, the Gibbs sampler and the Metropolis algorithm can be understood as special cases of one and the same method.) For states x and y that differ only in the ith coordinate, we have (since $x_{-i} = y_{-i}$)

$$\pi(x)\, p(x, y) = \pi(x)\, q_i \frac{\pi(y)}{\pi(x_{-i})} = \pi(y)\, q_i \frac{\pi(x)}{\pi(y_{-i})} = \pi(y)\, p(y, x).$$

Thus the Gibbs sampler is a reversible Markov chain with invariant measure π. Irreducibility of the Gibbs sampler, however, has to be checked for each case.

Example 18.18 (Ising model) In the Ising model described above, we have $x_{-i} = \{x^{i,-1}, x^{i,+1}\}$. Hence, for $i \in \Lambda$ and $\sigma \in \{-1, +1\}$,

$$\pi(x^{i,\sigma}|x_{-i}) = \frac{\pi(x^{i,\sigma})}{\pi(\{x^{i,-1}, x^{i,+1}\})}$$

$$= \frac{e^{-\beta H(x^{i,\sigma})}}{e^{-\beta H(x^{i,-1})} + e^{-\beta H(x^{i,+1})}}$$

$$= \left(1 + \exp\left[\beta\big(H(x^{i,\sigma}) - H(x^{i,-\sigma})\big)\right]\right)^{-1}$$

$$= \left(1 + \exp\left[2\beta \sum_{j:\, j \sim i} (\mathbb{1}_{\{x(j) \neq \sigma\}} - \tfrac{1}{2})\right]\right)^{-1}.$$

The Gibbs sampler for the Ising model is thus the Markov chain $(X_n)_{n \in \mathbb{N}_0}$ with values in $E = \{-1, 1\}^\Lambda$ and with transition matrix

$$
p(x, y) = \begin{cases} \frac{1}{\#\Lambda}\left(1 + \exp\left[2\beta \sum_{j:\, j \sim i} (\mathbb{1}_{\{x(j) \neq x(i)\}} - \frac{1}{2})\right]\right)^{-1}, & \text{if } y = x^i \text{ for some } i \in \Lambda, \\ 0, & \text{otherwise.} \end{cases}
$$

$\Diamond$

Perfect Sampling

The MCMC method as described above is based on hope: We let the chain run for a long time and hope that its distribution is close to the invariant distribution. Even if we can compute the speed of convergence (and in many cases, this is not trivial, we come back to this point in Sect. 18.4), the distribution will never be *exactly* the invariant distribution.

Although this flaw might seem inevitable in the MCMC method, it is in fact, at least theoretically, possible to use a very similar method that allows *perfect sampling* according to the invariant distribution π, even if we do not know anything about the speed of convergence. The idea is simple. Assume that $F_1, F_2, \ldots$ are i.i.d. random maps $E \to E$ with $\mathbf{P}[F(x) = y] = p(x, y)$ for all $x, y \in E$. We have seen how to construct the Markov chain X with initial value $X_0 = x$ by defining $X_n = F_n \circ F_{n-1} \circ \cdots \circ F_1(x)$.

Note that $F_1^n(x) := F_1 \circ \ldots \circ F_n(x) \overset{\mathcal{D}}{=} F_n \circ \ldots \circ F_1(x)$. Hence we have

$$
\mathbf{P}[F_1^n(x) = y] \overset{n \to \infty}{\longrightarrow} \pi(y) \quad \text{for every } y.
$$

However, if F_1^n turns out to be a constant map (e.g., $F_1^n \equiv x^*$ for some random x^*), then we will also have $F_1^m \equiv x^*$ for all $m \geq n$. If by some clever choice of the distribution of F_n one can ensure that the stopping time $T := \inf\{n \in \mathbb{N} : F_1^n \text{ is constant}\}$ is almost surely finite (and this is always possible), then we will have $\mathbf{P}[F_1^T(x) = y] = \pi(y)$ for all $x, y \in E$. A simple algorithm for this method is the following.

(1) Let $F \leftarrow \mathrm{id}_E$ and $n \leftarrow 0$.
(2) Let $n \leftarrow n + 1$. Generate F_n and let $F \leftarrow F \circ F_n$.
(3) If F is not a constant map, then go to (2).
(4) Output $F(*)$.

This method is called **coupling from the past** and goes back to Propp and Wilson [138] (see also [55, 56, 92, 137, 139, 171]). David Wilson has nice simulations and a survey of the current research on his web site http://www.dbwilson.com/. A nice survey on MCMC methods including coupling from the past is [66].

For a practical implementation, there are two main problems: (1) The full map F_n has to be generated and has to be composed with F. The computer time needed for this is at least of the order of the size of the space E. (2) Checking if F is constant needs computer time of the same order of magnitude. Consequently, the method can be efficiently implemented only if there is more structure. For example, assume that E is partially ordered with a smallest element $\underline{0}$ and a largest element $\underline{1}$ (like the Ising model). Further, assume that the maps F_n can be chosen to be almost surely monotone increasing. In this case, it is enough to compute at each step $F(\underline{0})$ and $F(\underline{1})$ since F is constant if the values coincide.

> **Takeaways** In order to draw random samples (approximately) according to a given distribution, it is sometimes feasible to simulate a suitable Markov chain that converges to this distribution as its invariant measure. The Metropolis algorithm constitutes a universal tool for the construction of such a Markov chain. The Gibbs sampler is a more specific algorithm often helpful in statistical mechanics. The technique of coupling from the past allows for drawing *exactly* according to the desired distribution.

18.4 Speed of Convergence

So far we have ignored the question of the speed of convergence of the distribution $\mathbf{P}_{X_n}$ to π. For practical purposes, however, this is often the most interesting question. We do not intend to go into the details and we only briefly touch upon the topic. Without loss of generality, assume $E = \{1, \ldots, N\}$. If p is reversible (Equation (18.8)), then $f \mapsto pf$ defines a symmetric linear operator on $L^2(E, \pi)$ (exercise!). All eigenvalues $\lambda_1, \ldots, \lambda_N$ (listed according to the corresponding multiplicity) are real and have modulus at most 1 since p is stochastic. Thus we can arrange the eigenvalues by decreasing modulus $\lambda_1 = 1 \geq |\lambda_2| \geq \ldots \geq |\lambda_N|$. If p is irreducible and aperiodic, then $|\lambda_2| < 1$. Let $\mu_1 = \pi, \mu_2, \ldots, \mu_N$ be an orthonormal basis of left eigenvectors for the eigenvalues $\lambda_1, \ldots, \lambda_N$. Then, for every probability measure $\mu = \alpha_1 \mu_1 + \ldots + \alpha_N \mu_N$, we have $\mu p^n = \sum_{i=1}^N \lambda_i^n \alpha_i \mu_i \xrightarrow{n \to \infty} \alpha_1 \pi$. Since for each $n \in \mathbb{N}$, the left hand side is a probability measure, we have $\alpha_1 = 1$ and

$$\|\mu p^n - \pi\|_{TV} \leq C |\lambda_2|^n \tag{18.10}$$

for a constant C (that does not depend on μ). A similar formula holds if p is not reversible; however, with a correction term of order at most n^{V-1}. Here, V is the size of the largest Jordan block square matrix for the eigenvalue λ_2 in the

Jordan canonical form of p. In particular, V is no larger than the multiplicity of the eigenvalue with second largest modulus.

The speed of convergence is thus exponential with a rate that is determined by the **spectral gap** $1 - |\lambda_2|$ of the second largest eigenvalue of p. In practice, for a large space E, computing the spectral gap is often extremely difficult.

Reflection Why have we restricted ourselves to aperiodic Markov chains? What is λ_2 in the periodic case? For example, consider $E = \{0, \ldots, N-1\}$ and the transition matrix given by $p(k, k + 1 \pmod{N}) = 1$ for all k. ♠

Example 18.19 Let $r \in (0, 1)$ and $N \in \mathbb{N}$, $N \geq 2$. Further, let $E = \{0, \ldots, N - 1\}$. We consider the transition matrix

$$p(i, j) = \begin{cases} r, & \text{if } j = i + 1 \pmod{N}, \\ 1 - r, & \text{if } j = i - 1 \pmod{N}, \\ 0, & \text{else.} \end{cases}$$

p is the transition matrix of simple (asymmetric) random walk on the discrete torus $\mathbb{Z}/(N)$, which with probability r makes a jump to the right and with probability $1 - r$ makes a jump to the left. Clearly, p is irreducible, and p is aperiodic if and only if N is odd. Furthermore, the uniform distribution $\mathcal{U}_E$ is the unique invariant distribution.

Case 1: N odd.   Let $\theta_k = e^{2\pi i k/N}$, $k = 0, \ldots, N - 1$, be the Nth roots of unity and let the corresponding (right) eigenvectors be

$$x^k := \left(\theta_k^0, \theta_k^1, \ldots, \theta_k^{N-1} \right).$$

It is easy to check that p has the eigenvalues

$$\lambda_k := r\,\theta_k + (1 - r)\,\overline{\theta}_k = \cos\left(\tfrac{2\pi k}{N}\right) + (2r - 1)\,i\,\sin\left(\tfrac{2\pi k}{N}\right), \quad k = 0, \ldots, N - 1.$$

The moduli of the eigenvalues are given by $|\lambda_k| = f(2\pi k/N)$, where

$$f(\vartheta) = \sqrt{1 - 4r(1 - r)\sin(\vartheta)^2} \quad \text{for } \vartheta \in \mathbb{R}.$$

Since N is odd, $|\lambda_k|$ is maximal (except for $k = 0$) for $k = \frac{N-1}{2}$ and for $k = \frac{N+1}{2}$. For these k, $|\lambda_k|$ equals $\gamma := \sqrt{1 - 4r(1 - r)\sin(\pi/N)^2}$. Since all eigenvalues are different, every eigenvalue has multiplicity 1. Hence there is a constant $C < \infty$ such that

$$\|\mu p^n - \mathcal{U}_E\|_{TV} \leq C\gamma^n \quad \text{for all } n \in \mathbb{N}, \ \mu \in \mathcal{M}_1(E).$$

Case 2: N even. In this case, p is not aperiodic. Nevertheless, the eigenvalues and eigenvectors are of the same form as in Case 1. In order to get an aperiodic chain, for $\varepsilon > 0$, define the transition matrix

$$p_\varepsilon := (1 - \varepsilon)p + \varepsilon I,$$

where I is the unit matrix on E. p_ε describes the random walk on E that with probability ε does not move and with probability $1 - \varepsilon$ makes a jump according to p. Clearly, p_ε is irreducible and aperiodic. The eigenvalues are

$$\lambda_{\varepsilon,k} = (1 - \varepsilon)\lambda_k + \varepsilon, \quad k = 0, \ldots, N - 1,$$

and the corresponding eigenvectors are the x^k from above. Evidently, $\lambda_{\varepsilon,0} = 1$, and if $\varepsilon > 0$ is very small, then $\lambda_{\varepsilon,N/2} = 2\varepsilon - 1$ is the eigenvalue with the second largest modulus. For larger values of ε, we have $|\lambda_{\varepsilon,1}| > |\lambda_{\varepsilon,N/2}|$. More precisely, if we let

$$\varepsilon_0 := \frac{(1 - (2r - 1)^2) \sin(2\pi/N)^2}{(1 - (2r - 1)^2) \sin(2\pi/N)^2 + 2\cos(2\pi/N)},$$

then the eigenvalue with the second largest modulus has modulus

$$\gamma_\varepsilon = |\lambda_{\varepsilon,N/2}| = 1 - 2\varepsilon, \quad \text{if } \varepsilon \le \varepsilon_0,$$

or

$$\gamma_\varepsilon = |\lambda_{\varepsilon,1}|$$

$$= \sqrt{\left((1 - \varepsilon)\cos\left(\frac{2\pi}{N}\right) + \varepsilon\right)^2 + \left((1 - \varepsilon)(2r - 1)\sin\left(\frac{2\pi}{N}\right)\right)^2}, \quad \text{if } \varepsilon \ge \varepsilon_0.$$

It is easy to check that $\varepsilon \mapsto |\lambda_{\varepsilon,N/2}|$ is monotone decreasing and that $\varepsilon \mapsto |\lambda_{\varepsilon,1}|$ is monotone increasing. Hence γ_ε is minimal for $\varepsilon = \varepsilon_0$.

Hence there is a $C < \infty$ with

$$\|\mu p_\varepsilon^n - \mathcal{U}_E\|_{TV} \le C\gamma_\varepsilon^n \quad \text{for all } n \in \mathbb{N}, \ \mu \in \mathcal{M}_1(E),$$

and the best speed of convergence (in this class of transition matrices) can be obtained by choosing $\varepsilon = \varepsilon_0$. $\Diamond$

Example 18.20 (Gambler's ruin) We consider the gambler's ruin problem from Example 10.19 with the probability of a gain $r \in (0, 1)$. Here the state space is $E = \{0, \dots, N\}$, and the transition matrix is of the form

$$p(i, j) = \begin{cases} r, & \text{if } j = i + 1 \in \{2, \dots, N\}, \\ 1 - r, & \text{if } j = i - 1 \in \{0, \dots, N - 2\}, \\ 1, & \text{if } j = i \in \{0, N\}, \\ 0, & \text{else.} \end{cases}$$

This transition matrix is not irreducible; rather it has two absorbing states 0 and N. In Example 10.19 (Equation (10.7)) for the case $r \neq \frac{1}{2}$, and Example 10.16 for the case $r = \frac{1}{2}$, it was shown that, for every $\mu \in \mathcal{M}_1(E)$,

$$\mu p^n \overset{n \to \infty}{\longrightarrow} (1 - m(\mu))\delta_0 + m(\mu)\delta_N. \tag{18.11}$$

Here $m(\mu) = \int p_N(x)\,\mu(dx)$, where the probability $p_N(x)$ that the chain, if started at x, hits N is given by

$$p_N(x) = \begin{cases} \dfrac{1 - \left(\frac{1-r}{r}\right)^x}{1 - \left(\frac{1-r}{r}\right)^N}, & \text{if } r \neq \frac{1}{2}, \\[4mm] \dfrac{x}{N}, & \text{if } r = \frac{1}{2}. \end{cases}$$

How quick is the convergence in (18.11)? Here also the convergence has exponential speed and the rate is determined by the second largest eigenvalue of p.

Hence we have to compute the spectrum of p. Clearly, $x^0 = (1, 0, \dots, 0)$ and $x^N = (0, \dots, 0, 1)$ are left eigenvectors for the eigenvalue 1. In order for $x = (x_0, \dots, x_N)$ to be a left eigenvector for the eigenvalue λ, the following equations have to hold:

$$\lambda x_k = r x_{k-1} + (1 - r)x_{k+1} \quad \text{for } k = 2, \dots, N - 2, \tag{18.12}$$

and

$$\lambda x_{N-1} = r x_{N-2}. \tag{18.13}$$

If (18.12) and (18.13) hold for $x_1, \dots, x_{N-1}$, then we define $x_0 := \frac{1-r}{\lambda-1}x_1$ and $x_N := \frac{r}{\lambda-1}x_{N-1}$ and get that in fact $xr = \lambda x$. We make the ansatz

$$\lambda = (1 - r)\rho(\theta + \bar{\theta}) \quad \text{and} \quad x_k = \rho^k(\theta^k - \bar{\theta}^k) \quad \text{for } k = 1, \dots, N - 1,$$

where

$$\rho = \sqrt{r/(1-r)} \quad \text{and} \quad \theta \in \mathbb{C} \setminus \{-1, +1\} \quad \text{with } |\theta| = 1.$$

Thus we have $\theta\bar{\theta} = 1$ and $(1-r)\rho^{k+1} = r\rho^{k-1}$. Therefore, for every $k = 2, \ldots, N-1$,

$$
\begin{aligned}
\lambda x_k &= (1-r)\,\rho^{k+1}(\theta^k - \bar{\theta}^k)(\theta + \bar{\theta}) \\
&= (1-r)\,\rho^{k+1}\big[(\theta^{k+1} - \bar{\theta}^{k+1}) + \theta\bar{\theta}\,(\theta^{k-1} - \bar{\theta}^{k-1})\big] \\
&= r\,\rho^{k-1}(\theta^{k-1} - \bar{\theta}^{k-1}) + (1-r)\,\rho^{k+1}(\theta^{k+1} - \bar{\theta}^{k+1}) \\
&= r\,x_{k-1} + (1-r)\,x_{k+1}.
\end{aligned}
$$

That is, (18.12) holds. The same computation with $k = N-1$ shows that (18.13) holds if and only if $\theta^N - \bar{\theta}^N = 0$; that is, if $\theta^{2N} = 1$. In all, then, for θ, we get $N-1$ different values (note that the complex conjugates of the values considered here lead to the same values λ_n),

$$\theta_n = e^{(n/N)\pi i} \quad \text{for } n = 1, \ldots, N-1.$$

The corresponding eigenvalues are

$$\lambda_n = \sigma \cos\left(\frac{n\,\pi}{N}\right) \quad \text{for } n = 1, \ldots, N-1.$$

Here the variance of the individual random walk step is

$$\sigma^2 := 4r(1-r). \tag{18.14}$$

As all eigenvalues are real, the corresponding eigenvectors are given by

$$x_k^n = 2\left(\frac{r}{1-r}\right)^{n/2} \sin\left(\frac{n\,\pi}{N}\right), \quad k = 1, \ldots, N-1.$$

The second largest modulus of an eigenvalue is $|\lambda_n| = \sigma \cos\left(\frac{\pi}{N}\right)$ if $n = 1$ or $n = N-1$. Thus there exists a $C > 0$ such that, for every $\mu \in \mathcal{M}_1(E)$, we have

$$\mu p^n(\{1, \ldots, N-1\}) \le C\left(\sigma \cos\left(\frac{\pi}{N}\right)\right)^n \quad \text{for every } n \in \mathbb{N}.$$

In other words, the probability that the game has not finished up to the nth round is at most $C\left(\sigma \cos(\pi/N)\right)^n$.

An alternative approach to the eigenvalues can be made via the roots of the characteristic polynomial

$$\chi_N(x) = \det(p - xI), \qquad x \in \mathbb{R}.$$

Clearly, $\chi_1(x) = (1 - x)^2$ and $\chi_2(x) = -x(1 - x)^2$. Using Laplace's expansion formula for the determinant (elimination of rows and columns), we get the recursion

$$\chi_N(x) = -x \, \chi_{N-1}(x) - r(1 - r) \, \chi_{N-2}(x). \tag{18.15}$$

The solution is (check this!)

$$\chi_N(x) = (-1)^{N-1} (\sigma/2)^{N-1} (1 - x)^2 \, U_{N-1}(x/\sigma), \tag{18.16}$$

where

$$U_m(x) := \sum_{k=0}^{\lfloor m/2 \rfloor} (-1)^k \binom{m - k}{k} (2x)^{m-2k}$$

denotes the so-called mth **Chebyshev polynomial** of the second kind.

Using de Moivre's formula, one can show that, for $x \in (-\sigma, \sigma)$,

$$\chi_N(x) = (-1)^{N-1} (\sigma/2)^{N-1} (1 - x)^2 \, \frac{\sin\left(N \, \arccos\left(x/\sigma\right)\right)}{\sqrt{1 - (x/\sigma)^2}}$$

$$= (1 - x)^2 \prod_{k=1}^{N-1} \left(\sigma \cos\left(\frac{\pi k}{N}\right) - x\right). \tag{18.17}$$

Apart from the double zero at 1, we get the zeros

$$\sigma \cos\left(\pi k / N\right), \quad k = 1, \ldots, N - 1. \quad \Diamond$$

Takeaways The speed at which a Markov chain converges towards its invariant distribution is determined by the spectral gap of its transition matrix. For two examples we could compute the spectral gap explicitly.

Exercise 18.4.1 Show (18.16). ♣

Exercise 18.4.2 Show (18.17). ♣

Exercise 18.4.3 Let $v(dx) = \frac{2}{\pi}\sqrt{1 - x^2}\,\mathbb{1}_{[-1,1]}(x)\,dx$. Show that the Chebyshev polynomials of the second kind are orthonormal with respect to v; that is,

$$\int U_m U_n\,dv = \mathbb{1}_{\{m=n\}}.\quad\clubsuit$$

Exercise 18.4.4 Let $E = \{1, 2, 3\}$ and $p = \begin{pmatrix} 1/2 & 1/3 & 1/6 \\ 1/3 & 1/3 & 1/3 \\ 0 & 3/4 & 1/4 \end{pmatrix}$. Compute the invariant distribution and the exponential rate of convergence. $\clubsuit$

Exercise 18.4.5 Let $E = \{0, \ldots, N - 1\}$, $r \in (0, 1)$ and

$$p(i, j) = \begin{cases} r, & \text{if } j = i + 1 \pmod{N}, \\ 1 - r, & \text{if } j = i \pmod{N}, \\ 0, & \text{else.} \end{cases}$$

Show that p is the transition matrix of an irreducible, aperiodic random walk and compute the invariant distribution and the exponential rate of convergence. $\clubsuit$

Exercise 18.4.6 Let $N \in \mathbb{N}$ and let $E = \{0, 1\}^N$ denote the N-dimensional hypercube. That is, two points $x, y \in E$ are connected by an edge if they differ in exactly one coordinate. Let p be the transition matrix of the random walk on E that stays put with probability $\varepsilon > 0$ and that with probability $1 - \varepsilon$ makes a jump to a randomly (uniformly) chosen neighboring site.

Describe p formally and show that p is aperiodic and irreducible. Compute the invariant distribution and the exponential rate of convergence. $\clubsuit$

Chapter 19
Markov Chains and Electrical Networks

We consider symmetric simple random walk on $\mathbb{Z}^2$. By Pólya's theorem (Theorem 17.40), this random walk is recurrent. However, is this still true if we remove a single edge from the lattice $\mathbb{L}^2$ of $\mathbb{Z}^2$? Intuitively, such a small local change should not make a difference for a global phenomenon such as recurrence. However, the computations used in Sect. 17.5 to prove recurrence are not very robust and would need a substantial improvement in order to cope with even a small change. The situation becomes even more puzzling if we restrict the random walk to, e.g., the upper half plane $\{(x, y) : x \in \mathbb{Z}, y \in \mathbb{N}_0\}$ of $\mathbb{Z}^2$. Is this random walk recurrent? Or consider **bond percolation** on $\mathbb{Z}^2$. Fix a parameter $p \in [0, 1]$ and independently declare any edge of $\mathbb{L}^2$ *open* with probability p and *closed* with probability $1 - p$. At a second stage, start a random walk on the random subgraph of open edges. At each step, the walker chooses one of the adjacent open edges at random (with equal probability) and traverses it. For $p > \frac{1}{2}$, there exists a unique infinite connected component of open edges (Theorem 2.47). The question that we answer at the end of this chapter is: Is a random walk on the infinite open cluster recurrent or transient?

The aim of this chapter is to establish a connection between certain Markov chains and electrical networks. This connection

- in some cases allows us to distinguish between recurrence and transience by means of easily computable quantities, and
- in other cases provides a comparison criterion that says that if a random walk on a graph is recurrent, then a random walk on any connected subgraph is recurrent. Any of the questions raised above can be answered using this comparison technique.

Some of the material of this chapter is taken from [110] and [36].

19.1   Harmonic Functions

In this chapter, E is always a countable set and X is a discrete Markov chain on E with transition matrix p and Green function G. Recall that $F(x, y)$ is the probability of hitting y at least once when starting at x. Compare Sect. 17.4, in particular, Definitions 17.29 and 17.34.

Definition 19.1 Let $A \subset E$. A function $f : E \to \mathbb{R}$ is called **harmonic** on $E \setminus A$ if $pf(x) = \sum_{y \in E} p(x, y) f(y)$ exists and if $pf(x) = f(x)$ for all $x \in E \setminus A$.

Theorem 19.2 (Superposition principle) Assume f and g are harmonic on $E \setminus A$ and let $\alpha, \beta \in \mathbb{R}$. Then $\alpha f + \beta g$ is also harmonic on $E \setminus A$.

Proof This is trivial.   □

Example 19.3 Let X be transient and let $a \in E$ be a transient state (that is, a is not absorbing). Then $f(x) := G(x, a)$ is harmonic on $E \setminus \{a\}$: For $x \neq a$, we have

$$pf(x) = p \sum_{n=0}^{\infty} p^n(x, a) = \sum_{n=1}^{\infty} p^n(x, a) = G(x, a) - \mathbb{1}_{\{a\}}(x) = G(x, a). \quad \Diamond$$

Example 19.4 For $x \in E$, let $\tau_x := \inf\{n > 0 : X_n = x\}$. For $A \subset E$, let

$$\tau := \tau_A := \inf_{x \in A} \tau_x$$

be the stopping time of the first entrance to A. Assume that A is chosen so that $\mathbf{P}_x[\tau_A < \infty] = 1$ for every $x \in E$. Let $g : A \to \mathbb{R}$ be a bounded function. Define

$$f(x) := \begin{cases} g(x), & \text{if } x \in A, \\ \mathbf{E}_x[g(X_\tau)], & \text{if } x \in E \setminus A. \end{cases} \tag{19.1}$$

Then f is harmonic on $E \setminus A$. We give two proofs for this statement.
1. Proof.   By the Markov property, for $x \notin A$ and $y \in E$,

$$\mathbf{E}_x\big[g(X_\tau)\,\big|\,X_1 = y\big] = \begin{cases} g(y), & \text{if } y \in A \\ \mathbf{E}_y[g(X_\tau)], & \text{if } y \in E \setminus A \end{cases} = f(y).$$

Hence, for $x \in E \setminus A$,

$$f(x) = \mathbf{E}_x[g(X_\tau)] = \sum_{y \in E} \mathbf{E}_x\big[g(X_\tau); \ X_1 = y\big]$$

$$= \sum_{y \in E} p(x, y)\, \mathbf{E}_x\big[g(X_\tau)\,\big|\,X_1 = y\big] = \sum_{y \in E} p(x, y)\, f(y) = pf(x).$$

2. Proof. We change the Markov chain by adjoining a cemetery state Δ. That is, the new state space is $\tilde{E} = E \cup \{\Delta\}$ and the transition matrix is

$$
\tilde{p}(x, y) = \begin{cases} p(x, y), & \text{if } x \in E \setminus A, \ y \neq \Delta, \\ 0, & \text{if } x \in E \setminus A, \ y = \Delta, \\ 1, & \text{if } x \in A \cup \{\Delta\}, \ y = \Delta. \end{cases} \tag{19.2}
$$

The corresponding Markov chain $\tilde{X}$ is transient, and Δ is the only absorbing state. Furthermore, we have $pf = f$ on $E \setminus A$ if and only if $\tilde{p}f = f$ on $E \setminus A$. Since $\tilde{G}(y, y) = 1$ for all $y \in A$, we have (compare Theorem 17.35)

$$
\mathbf{P}_x[X_\tau = y] = \mathbf{P}_x[\tilde{\tau}_y < \infty] = \tilde{F}(x, y) = \tilde{G}(x, y) \quad \text{for all } x \in E \setminus A, \ y \in A.
$$

Now $x \mapsto \tilde{G}(x, y)$ is harmonic on $E \setminus A$. Hence, by the superposition principle,

$$
f(x) = \sum_{y \in A} \tilde{G}(x, y) \, g(y) \tag{19.3}
$$

is harmonic on $E \setminus A$. Due to the analogy of (19.3) to Green's formula in continuous space potential theory, the function $\tilde{G}$ is called the **Green function** for the equation $(p - I)f = 0$ on $E \setminus A$. $\Diamond$

Definition 19.5 *The system of equations*

$$
\begin{aligned}
(p - I)f(x) &= 0, & \text{for } x \in E \setminus A, \\
f(x) &= g(x), & \text{for } x \in A,
\end{aligned} \tag{19.4}
$$

*is called the **Dirichlet problem** on $E \setminus A$ with respect to $p - I$ and with boundary value g on A.*

We have shown the existence of solutions of the Dirichlet problem in Example 19.4. In order to show uniqueness (under certain conditions) we first derive the maximum principle for harmonic functions.

If $p = I$ then any function f that coincides with g on A is a solution of the Dirichlet problem. However, even in less extreme situations the solution of (19.4) may be ambiguous. This is the case if $E \setminus A$ decomposes into domains between which the chain that is stopped in A cannot change.

In order to describe formally the irreducibility condition that we have to impose, we introduce the transition matrix p_A of the chain stopped upon reaching A by

$$
p_A(x, y) := \begin{cases} p(x, y), & \text{if } x \notin A, \\ \mathbb{1}_{\{x=y\}}, & \text{if } x \in A. \end{cases}
$$

Further, define F_A for p_A similarly as F was defined for p. Finally, for $x \in E$ let

$$S_A^n(x) = \{y \in E : (p_A)^n(x, y) > 0\}, \quad \text{for } n \in \mathbb{N}_0$$

and

$$S_A(x) = \bigcup_{n=0}^{\infty} S_A^n(x) = \{y \in E : F_A(x, y) > 0\}.$$

Theorem 19.6 (Maximum principle) *Let f be a harmonic function on $E \setminus A$.*

(i) *If there exists an $x_0 \in E \setminus A$ such that*

$$f(x_0) = \sup f(S_A(x_0)), \tag{19.5}$$

then $f(y) = f(x_0)$ for any $y \in S_A(x_0)$.

(ii) *In particular, if $F_A(x, y) > 0$ for all $x, y \in E \setminus A$, and if there is an $x_0 \in E \setminus A$ such that $f(x_0) = \sup f(E)$, then $f(x_0) = f(y)$ for any $y \in E \setminus A$.*

Proof

(i) Let $m := \sup f(S_A(x_0))$. As f is harmonic on $E \setminus A$, we have $p_A f = f$ on E. Hence, for any $n \in \mathbb{N}$,

$$f(x_0) = (p_A)^n f(x_0) = \sum_{y \in S_A^n(x_0)} p_A^n(x_0, y) f(y) \le m$$

with equality if and only if $f(y) = m$ for all $y \in S_A^n(x_0)$. Since (19.5) implies equality, we infer $f(x_0) = f(y)$ for all $y \in S_A(x_0)$.

(ii) This is a direct consequence of (i) since $S_A(x) \supset E \setminus A$ for any $x \in E \setminus A$. □

Theorem 19.7 (Uniqueness of harmonic functions) *Assume that $F(x, y) > 0$ for all $x, y \in E$. Let $A \subset E$ be such that $A \ne \emptyset$ and $E \setminus A$ is finite. Assume that f_1 and f_2 are harmonic on $E \setminus A$. If $f_1 = f_2$ on A, then $f_1 = f_2$.*

In other words, the Dirichlet problem (19.4) has a unique solution given by (19.3) (or equivalently by (19.1)).

Proof By the superposition principle, $f := f_1 - f_2$ is harmonic on $E \setminus A$ with $f\big|_A \equiv 0$.

We will show $f \le 0$. Then, by symmetry, also $f \ge 0$ and hence $f \equiv 0$. To this end, we assume that there exists an $x \in E$ such that $f(x) > 0$ and deduce a contradiction.

Since $f\big|_A \equiv 0$ and since $E \setminus A$ is finite, there is an $x_0 \in E \setminus A$ such that $f(x_0) = \max f(E) \ge f(x) > 0$.

Since $F(x, y) > 0$ for all $x, y \in E$, we have

$$n_0 := \min \{n \in \mathbb{N}_0 : p^n(x_0, y) > 0 \text{ for some } y \in A\} < \infty.$$

Clearly, we have $p^{n_0}(x_0, y) = (p_A)^{n_0}(x_0, y)$ for all $y \in A$. Hence, there exists a $y \in A$ such that $(p_A)^{n_0}(x_0, y) > 0$, i.e., $y \in S_A(x_0)$. By Theorem 19.6, this implies $f(x_0) = f(y) = 0$ contradicting the assumption. $\qquad\square$

Takeaways Let E be countable and let p be a stochastic matrix on E. Let $A \subset E$. A function f is called harmonic on $G := E \setminus A$ if $(p - I)f = 0$ holds on G. If the values of f on A are prescribed, then we say that f solves a Dirichlet problem. If p is irreducible on G and f is not constant, then f does not assume its maximum in G. As a consequence, we get uniqueness of the solution of the Dirichlet problem.

Exercise 19.1.1 Let $\overline{p}$ be the substochastic $E \times E$ matrix that is given by $\overline{p}(x, y) = \tilde{p}(x, y)$, $x, y \in E$ (with $\tilde{p}$ as in (19.2)). Hence $\overline{p}(x, y) = p(x, y) \mathbb{1}_{x \in E \setminus A}$. Let I be the unit matrix on E.

(i) Show that $I - \overline{p}$ is invertible.
(ii) Define $\overline{G} := (I - \overline{p})^{-1}$. Show that $\overline{G}(x, y) = \tilde{G}(x, y)$ for all $x, y \in E \setminus A$ and that $\overline{G}(x, y) = \mathbb{1}_{\{x = y\}}$ if $x \in A$. In particular,

$$\overline{G}(x, y) = \mathbf{P}_x[X_{\tau_A} = y] \quad \text{for } x \in E \setminus A \text{ and } y \in A. \quad \clubsuit$$

19.2 Reversible Markov Chains

Definition 19.8 *The Markov chain X is called **reversible** with respect to the measure π if*

$$\pi(\{x\}) \, p(x, y) = \pi(\{y\}) \, p(y, x) \quad \text{for all } x, y \in E. \tag{19.6}$$

*Equation (19.6) is sometimes called the equation of **detailed balance**. X is called reversible if there is a π with respect to which X is reversible.*

Remark 19.9 If X is reversible with respect to π, then π is an invariant measure for X since

$$\pi \, p(\{x\}) = \sum_{y \in E} \pi(\{y\}) \, p(y, x) = \sum_{y \in E} \pi(\{x\}) \, p(x, y) = \pi(\{x\}).$$

If X is irreducible and recurrent, then, by Remark 17.51, π is thus unique up to constant multiples. $\Diamond$

Reflection Let $X = (X_n)_{n \in \mathbb{Z}}$ be a reversible Markov chain (with transition matrix p) with respect to π and assume that $\mathbf{P}_{X_0} = \pi$. Check that $X'_n := X_{-n}$, $n \in \mathbb{Z}$, is also a Markov chain with transition matrix p. This justifies the term *reversible*. ♠

Example 19.10 Let (E, K) be a graph with vertex set (or set of nodes) E and with edge set K (see page 73). By $\langle x, y \rangle = \langle y, x \rangle \in K$, denote an (undirected) edge that connects x with y. Let $C := (C(x, y), x, y \in E)$ be a family of weights with $C(x, y) = C(y, x) \geq 0$ for all $x, y \in E$ and

$$C(x) := \sum_{y \in E} C(x, y) < \infty \quad \text{for all } x \in E.$$

If we define $p(x, y) := \frac{C(x,y)}{C(x)}$ for all $x, y \in E$, then X is reversible with respect to $\pi(\{x\}) = C(x)$. In fact,

$$\pi(\{x\}) \, p(x, y) = C(x) \frac{C(x, y)}{C(x)} = C(x, y)$$

$$= C(y, x) = C(y) \frac{C(y, x)}{C(y)} = \pi(\{y\}) \, p(y, x). \quad \Diamond \quad (19.7)$$

Definition 19.11 *Let (E, K), C and X be as in Example 19.10. Then X is called a random walk on E with weights C. In particular, if $C(x, y) = \mathbb{1}_{\{\langle x, y \rangle \in K\}}$, then X is called a **simple random walk** on (E, K).*

Thus the random walk with weights C is reversible. However, the converse is also true.

Theorem 19.12 *If X is a reversible Markov chain and if π is an invariant measure, then X is a random walk on E with weights $C(x, y) = p(x, y) \, \pi(\{x\})$. If X is irreducible and recurrent, then π and hence C are unique up to a factor.*

Proof This is obvious.   □

Takeaways A reversible Markov chain is stationary and fulfills an even stronger equilibrium condition: The condition of detailed balance says that on average the Markov chain jumps from x to y as often as it jumps from y to x (for all x and y).

Exercise 19.2.1 Show that p is reversible with respect to π if and only if the linear map $L^2(\pi) \to L^2(\pi)$, $f \mapsto pf$ is self-adjoint. ♣

Exercise 19.2.2 Let $\beta > 0$, $K \in \mathbb{N}$ and $W_1, \ldots, W_K \in \mathbb{R}$. Define

$$p(i, j) := \frac{1}{Z} \exp(-\beta W_j) \quad \text{for all } i, j = 1, \ldots, K,$$

where $Z := \sum_{j=1}^{K} \exp(-\beta W_j)$ is the normalising constant.

Assume that in K (enumerated) urns there are a total of N indistinguishable balls. At each step, choose one of the N balls uniformly at random. If i is the number of the urn from which the ball is drawn, then with probability $p(i, j)$ move the ball to the urn with number j.

(i) Give a formal description of this process as a Markov chain.
(ii) Determine the invariant distribution π and show that the chain is reversible with respect to π. ♣

19.3 Finite Electrical Networks

An electrical network (E, C) consists of a set E of sites (the electrical contacts) and wires between pairs of sites. The **conductance** of the wire that connects the points $x \in E$ and $y \in E \setminus \{x\}$ is denoted by $C(x, y) \in [0, \infty)$. If $C(x, y) = 0$, then we could just as well assume that there is no wire connecting x and y. By symmetry, we have $C(x, y) = C(y, x)$ for all x and y. Denote by

$$R(x, y) = \frac{1}{C(x, y)} \in (0, \infty]$$

the **resistance** of the connection $\langle x, y \rangle$. A particular case is that of a graph (E, K) where all edges have the same conductance, say 1; that is, $C(x, y) = \mathbb{1}_{\{\langle x, y \rangle \in K\}}$. The corresponding network (E, C) will be called the **unit network** on (E, K).

In the remainder of this section, assume that (E, C) is a **finite** electrical network.

Now let $A \subset E$. At the points $x_0 \in A$, we apply the voltages $u(x_0)$ (e.g., using batteries). What is the voltage $u(x)$ at $x \in E \setminus A$?

Definition 19.13 *A map* $I : E \times E \rightarrow \mathbb{R}$ *is called a* **flow** *on* $E \setminus A$ *if it is antisymmetric (that is,* $I(x, y) = -I(y, x)$*) and if it obeys* **Kirchhoff's rule***:*

$$I(x) = 0, \quad \text{for } x \in E \setminus A,$$
$$I(A) = 0. \tag{19.8}$$

Here we denoted

$$I(x) := \sum_{y \in E} I(x, y) \quad \text{and} \quad I(A) := \sum_{x \in A} I(x).$$

Definition 19.14 *A flow* $I : E \times E \to \mathbb{R}$ *on* $E \setminus A$ *is called a **current flow** if there exists a function* $u : E \to \mathbb{R}$ *with respect to which **Ohm's rule** is fulfilled:*

$$I(x, y) = \frac{u(x) - u(y)}{R(x, y)} \quad \text{for all } x, y \in E, \ x \neq y.$$

In this case, $I(x, y)$ is called the flow from x to y and $u(x)$ is called the electrical potential (or voltage) at x.

Reflection Give an example of a current that is not an electrical current. ♦

Theorem 19.15 *An electrical potential u in (E, C) is a harmonic function on $E \setminus A$:*

$$u(x) = \sum_{y \in E} \frac{1}{C(x)} C(x, y) u(y) \quad \text{for all } x \in E \setminus A.$$

In particular, if the network is irreducible, an electrical potential is uniquely determined by the values on A.

Proof By Ohm's rule and Kirchhoff's rule,

$$u(x) - \sum_{y \in E} \frac{C(x, y)}{C(x)} u(y) = \sum_{y \in E} \frac{C(x, y)}{C(x)} (u(x) - u(y)) = \frac{1}{C(x)} \sum_{y \in E} I(x, y) = 0.$$

Hence u is harmonic for the stochastic matrix $p(x, y) = C(x, y)/C(x)$. The claim follows by the uniqueness theorem for harmonic functions (Theorem 19.7). □

Corollary 19.16 *Let X be a Markov chain on E with edge weights C. Then $u(x) = \mathbf{E}_x[u(X_{\tau_A})]$.*

Assume $A = \{x_0, x_1\}$ where $x_0 \neq x_1$, and $u(x_0) = 0$, $u(x_1) = 1$. Then $I(x_1)$ is the total flow *into* the network and $-I(x_0)$ is the total flow *out of* the network. Kirchhoff's rule says that the flow is divergence-free and that the flows into and out of the network are equal. In other words, the net flow is $I(x_0) + I(x_1) = 0$.

Recall that, by Ohm's rule, the resistance of a wire is the quotient of the potential difference and the current flow. Hence we define the *effective resistance* between x_0 and x_1 as

$$R_{\text{eff}}(x_0 \leftrightarrow x_1) = \frac{u(x_1) - u(x_0)}{I(x_1)} = \frac{1}{I(x_1)} = -\frac{1}{I(x_0)}.$$

Correspondingly, the *effective conductance* is $C_{\text{eff}}(x_0 \leftrightarrow x_1) = R_{\text{eff}}(x_0 \leftrightarrow x_1)^{-1}$. As I and u are uniquely determined by x_0, x_1 and C, the quantities $C_{\text{eff}}(x_0 \leftrightarrow x_1)$ and $R_{\text{eff}}(x_0 \leftrightarrow x_1)$ are well-defined and can be computed from C.

Consider now two sets $A_0, A_1 \subset E$ with $A_0 \cap A_1 = \emptyset$, $A_0, A_1 \neq \emptyset$. Define $u(x) = 0$ for every $x \in A_0$ and $u(x) = 1$ for every $x \in A_1$. Let I be the corresponding current flow. In a manner similar to the above, we make the following definition.

Definition 19.17 *We call* $C_{\text{eff}}(A_0 \leftrightarrow A_1) := I(A_1)$ *the **effective conductance** between A_0 and A_1 and $R_{\text{eff}}(A_0 \leftrightarrow A_1) := \frac{1}{I(A_1)}$ the **effective resistance** between A_0 and A_1.*

Example 19.18

(i) Let $E = \{0, 1, 2\}$ with $C(0, 2) = 0$, and $A_0 = \{x_0\} = \{0\}$, $A_1 = \{x_1\} = \{2\}$. Define $u(0) = 0$ and $u(2) = 1$. Then (with $p(x, y) = C(x, y)/C(x)$),

$$u(1) = 1 \cdot p(1, 2) + 0 \cdot p(1, 0) = \frac{C(1, 2)}{C(1, 2) + C(1, 0)} = \frac{R(1, 0)}{R(1, 0) + R(1, 2)}$$

$$= \frac{R_{\text{eff}}(1 \leftrightarrow 0)}{R_{\text{eff}}(1 \leftrightarrow 0) + R_{\text{eff}}(1 \leftrightarrow 2)}.$$

The total current flow is

$$I(\{2\}) = u(1) \, C(0, 1) = \frac{1}{R(0, 1) + R(1, 2)} = \frac{1}{\frac{1}{C(0,1)} + \frac{1}{C(1,2)}}.$$

Hence we have $R_{\text{eff}}(0 \leftrightarrow 2) = \frac{1}{I(\{2\})} = R(0, 1) + R(1, 2)$ and $C_{\text{eff}}(0 \leftrightarrow 2) = \left(C(0, 1)^{-1} + C(1, 2)^{-1}\right)^{-1}$.

(ii) **(Series connection (see Fig.19.1))**   Let $n \in \mathbb{N}$, $n \geq 2$ and $E = \{0, \dots, n\}$ with conductances $C(k - 1, k) > 0$ and $C(k, l) = 0$ if $|k - l| > 1$. By Kirchhoff's rule, we have $I(l, l + 1) = -I(x_1)$ for any $l = 0, \dots, n - 1$. By Ohm's rule, we get $u(1) = u(0) + I(x_1) \, R(0, 1)$, $u(2) = u(1) + I(x_1) \, R(1, 2)$

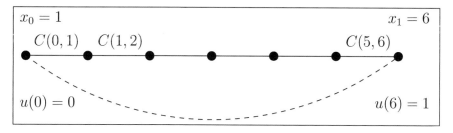

Fig. 19.1 Series connection of six resistors. The effective resistance is $R_{\text{eff}}(0 \leftrightarrow 6) = R(0, 1) + \dots + R(5, 6)$.

and so on, yielding

$$u(k) - u(0) = I(x_1) \sum_{l=0}^{k-1} R(l, l+1).$$

Hence

$$R_{\text{eff}}(0 \leftrightarrow k) = \frac{u(k) - u(0)}{I(x_1)} = \sum_{l=0}^{k-1} R(l, l+1).$$

By symmetry, we also have

$$R_{\text{eff}}(k \leftrightarrow n) = \sum_{l=k}^{n-1} R(l, l+1)$$

and thus $R_{\text{eff}}(0 \leftrightarrow n) = R_{\text{eff}}(0 \leftrightarrow k) + R_{\text{eff}}(k \leftrightarrow n)$.
Finally, for $k \in \{1, \ldots, n-1\}$, we get

$$u(k) = \frac{R_{\text{eff}}(0 \leftrightarrow k)}{R_{\text{eff}}(0 \leftrightarrow k) + R_{\text{eff}}(k \leftrightarrow n)}.$$

Note that this yields the ruin probability of the corresponding Markov chain X on $\{0, \ldots, n\}$,

$$\mathbf{P}_k[\tau_n < \tau_0] = u(k) = \frac{R_{\text{eff}}(0 \leftrightarrow k)}{R_{\text{eff}}(0 \leftrightarrow n)} = \sum_{l=0}^{k-1} R(l, l+1) \Big/ \sum_{l=0}^{n-1} R(l, l+1).$$

$$(19.9)$$

(iii) **(Parallel connection (see Fig.19.2))** Let $E = \{0, 1\}$. We extend the model a little by allowing for more than one wire to connect 0 and 1. Denote the conductances of these wires by $C_1, \ldots, C_n$. Then, by Ohm's rule, the current flow along the ith wire is $I_i = \frac{u(1) - u(0)}{R_i} = \frac{1}{R_i}$. Hence the total current is $I = \sum_{i=1}^{n} \frac{1}{R_i}$ and thus we have

$$C_{\text{eff}}(0 \leftrightarrow 1) = \sum_{i=1}^{n} C_i \quad \text{and} \quad R_{\text{eff}}(0 \leftrightarrow 1) = \left(\sum_{i=1}^{n} \frac{1}{R_i} \right)^{-1}. \quad \Diamond$$

In each of the three preceding examples, the effective resistance is a monotone function of the individual resistances. This is more than just coincidence.

Theorem 19.19 (Rayleigh's monotonicity principle) *Let (E, C) and (E, C') be electrical networks with $C(x, y) \geq C'(x, y)$ for all $x, y \in E$.*
Then, for $A_0, A_1 \subset E$ with $A_0, A_1 \neq \emptyset$ and $A_0 \cap A_1 = \emptyset$,

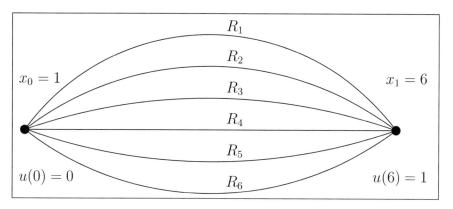

Fig. 19.2 Parallel connection of six resistors. The effective resistance is $R_{\text{eff}}(0 \leftrightarrow 1) = (R_1^{-1} + \ldots + R_6^{-1})^{-1}$.

$$C_{\text{eff}}(A_0 \leftrightarrow A_1) \geq C'_{\text{eff}}(A_0 \leftrightarrow A_1).$$

The remainder of this section is devoted to the proof of this theorem. We will need a theorem on conservation of energy and Thomson's principle (also called Dirichlet's principle) on the minimization of the energy dissipation.

Theorem 19.20 (Conservation of energy) *Let $A = A_0 \cup A_1$, and let I be a flow on $E \setminus A$ (but not necessarily a current flow; that is, Kirchhoff's rule holds but Ohm's rule need not). Further, let $w : E \to \mathbb{R}$ be a function that is constant both on A_0 and on A_1: $w\big|_{A_0} \equiv: w_0$ and $w\big|_{A_1} \equiv: w_1$. Then*

$$(w_1 - w_0)I(A_1) = \frac{1}{2} \sum_{x,y \in E} (w(x) - w(y)) \, I(x, y).$$

Note that this is a discrete version of Gauß's integral theorem for (wI). In fact, Kirchhoff's rule says that I is divergence-free on $E \setminus A$.

Proof We compute

$$\sum_{x,y \in E} (w(x) - w(y))I(x, y) = \sum_{x \in E} \left(w(x) \sum_{y \in E} I(x, y) \right) - \sum_{y \in E} \left(w(y) \sum_{x \in E} I(x, y) \right)$$

$$= \sum_{x \in A} \left(w(x) \sum_{y \in E} I(x, y) \right) - \sum_{y \in A} \left(w(y) \sum_{x \in E} I(x, y) \right)$$

$$= w_0 I(A_0) + w_1 I(A_1) - w_0(-I(A_0)) - w_1(-I(A_1))$$

$$= 2(w_1 - w_0)I(A_1). \qquad \square$$

Definition 19.21 *Let I be a flow on $E \setminus A$. Denote by*

$$L_I := L_I^C := \frac{1}{2} \sum_{x,y \in E} I(x,y)^2 R(x,y)$$

*the **energy dissipation** of I in the network (E, C).*

Theorem 19.22 (Thomson's (or Dirichlet's) principle of minimization of energy dissipation) *Let I and J be unit flows from A_1 to A_0 (that is, $I(A_1) = J(A_1) = 1$). Assume in addition that I is a current flow (that is, it satisfies Ohm's rule with some potential u that is constant both on A_0 and on A_1). Then*

$$L_I \leq L_J$$

with equality if and only if $I = J$. In particular, the unit current flow is uniquely determined.

Proof Let $D = J - I \not\equiv 0$ be the difference of the flows. Then clearly $D(A_0) = D(A_1) = 0$. We infer

$$\sum_{x,y \in E} J(x,y)^2 R(x,y)$$

$$= \sum_{x,y \in E} \big(I(x,y) + D(x,y)\big)^2 R(x,y)$$

$$= \sum_{x,y \in E} \big(I(x,y)^2 + D(x,y)^2\big) R(x,y) + 2 \sum_{x,y \in E} I(x,y) D(x,y) R(x,y)$$

$$= \sum_{x,y \in E} \big(I(x,y)^2 + D(x,y)^2\big) R(x,y) + 2 \sum_{x,y \in E} \big(u(x) - u(y)\big) D(x,y).$$

By the principle of conservation of energy, the last term equals

$$2 \sum_{x,y \in E} \big(u(x) - u(y)\big) D(x,y) = 4D(A_1)(u_1 - u_0) = 0.$$

Therefore (since $D \not\equiv 0$),

$$L_J = L_I + \frac{1}{2} \sum_{x,y \in E} D(x,y)^2 R(x,y) > L_I. \qquad \square$$

Proof *(Rayleigh's monotonicity principle, Theorem 19.19)* Let I and I' be the unit current flows from A_1 to A_0 with respect to C and C', respectively. By Thomson's principle, the principle of conservation of energy and the assumption $R(x, y) \leq R'(x, y)$ for all $x, y \in E$, we have

$$
\begin{aligned}
R_{\text{eff}}(A_0 \leftrightarrow A_1) &= \frac{u(1) - u(0)}{I(A_1)} = u(1) - u(0) \\
&= \frac{1}{2} \sum_{x,y \in E} I(x, y)^2 R(x, y) \\
&\leq \frac{1}{2} \sum_{x,y \in E} I'(x, y)^2 R(x, y) \leq \frac{1}{2} \sum_{x,y \in E} I'(x, y)^2 R'(x, y) \\
&= u'(1) - u'(0) = R'_{\text{eff}}(A_0 \leftrightarrow A_1).
\end{aligned}
$$
$\square$

Takeaways Let us imagine the edges of some graph as resistors in an electrical network. We can compute the effective resistances in parallel and serial connections. The electrical current between two contacts (vertices in the graph) minimises the electrical power among all unit currents. We use this to infer uniqueness of the electrical current. As a consequence we get a formal proof for the intuitive fact that increasing the resistance along an individual bond (or even removing the bond which is the same as increasing the resistance to infinity) increases the effective resistance between any two given points.

19.4   Recurrence and Transience

We consider the situation where E is countable and $A_1 = \{x_1\}$ for some $x_1 \in E$. Let X be a random walk on E with weights $C = (C(x, y), x, y \in E)$ and hence with transition probabilities $p(x, y) = C(x, y)/C(x)$ (compare Definition 19.11).

The main goal of this section is to express the probability $1 - F(x_1, x_1)$ that the random walk never returns to x_1 in terms of effective resistances in the network. In order to apply the results on *finite* electrical networks from the last section, we henceforth assume that $A_0 \subset E$ is such that $E \setminus A_0$ is finite. We will obtain $1 - F(x_1, x_1)$ as the limit of the probability that a random walk started at x_1 hits A_0 before returning to x_1 as $A_0 \downarrow \emptyset$.

Let $u = u_{x_1, A_0}$ be the unique potential function on E with $u(x_1) = 1$ and $u(x) = 0$ for any $x \in A_0$. By Theorem 19.7, u is harmonic and can be written as

$$u_{x_1, A_0}(x) = \mathbf{E}_x \left[\mathbb{1}_{\{X_{\tau_{A_0 \cup \{x_1\}}} = x_1\}} \right]$$

$$= \mathbf{P}_x \left[\tau_{x_1} < \tau_{A_0} \right] \qquad \text{for every } x \in E \setminus (A_0 \cup \{x_1\}).$$

Hence the current flow I with respect to u satisfies

$$-I(A_0) = I(x_1) = \sum_{x \in E} I(x_1, x) = \sum_{x \in E} \left(u(x_1) - u(x) \right) C(x_1, x)$$

$$= C(x_1) \sum_{x \in E} \left(1 - u(x) \right) p(x_1, x)$$

$$= C(x_1) \left(\sum_{x \notin A_0 \cup \{x_1\}} p(x_1, x) \, \mathbf{P}_x \left[\tau_{A_0} < \tau_{x_1} \right] + \sum_{x \in A_0} p(x_1, x) \right)$$

$$= C(x_1) \, \mathbf{P}_{x_1} \left[\tau_{A_0} < \tau_{x_1} \right].$$

Therefore,

$$p_F(x_1, A_0) := \mathbf{P}_{x_1} \left[\tau_{A_0} < \tau_{x_1} \right]$$

$$= \frac{C_{\text{eff}}(x_1 \leftrightarrow A_0)}{C(x_1)} = \frac{1}{C(x_1)} \frac{1}{R_{\text{eff}}(x_1 \leftrightarrow A_0)}. \qquad (19.10)$$

Definition 19.23 *We denote the **escape probability** of x_1 by*

$$p_F(x_1) = \mathbf{P}_{x_1} \left[\tau_{x_1} = \infty \right] = 1 - F(x_1, x_1).$$

We denote the effective conductance from x_1 to ∞ by

$$C_{\text{eff}}(x_1 \leftrightarrow \infty) := C(x_1) \inf \left\{ p_F(x_1, A_0) : A_0 \subset E \text{ with } |E \setminus A_0| < \infty, \ A_0 \not\ni x_1 \right\}.$$

Lemma 19.24 *For any decreasing sequence $A_0^n \downarrow \emptyset$ such that $|E \setminus A_0^n| < \infty$ and $x_1 \notin A_0^n$ for all $n \in \mathbb{N}$, we have*

$$C_{\text{eff}}(x_1 \leftrightarrow \infty) = \lim_{n \to \infty} C_{\text{eff}}(x_1 \leftrightarrow A_0^n).$$

Proof This is obvious since

$$C_{\text{eff}}(x_1 \leftrightarrow \infty) = C(x_1) \inf \left\{ p_F(x_1, A_0) : |E \setminus A_0| < \infty, \ A_0 \not\ni x_1 \right\} \qquad (19.11)$$

and since $p_F(x_1, A_0)$ is monotone decreasing in A_0. $\qquad \square$

Theorem 19.25 *We have*

$$p_F(x_1) = \frac{1}{C(x_1)} \, C_{\text{eff}}(x_1 \leftrightarrow \infty). \tag{19.12}$$

In particular,

$$x_1 \text{ is recurrent} \quad \Longleftrightarrow \quad C_{\text{eff}}(x_1 \leftrightarrow \infty) = 0 \quad \Longleftrightarrow \quad R_{\text{eff}}(x_1 \leftrightarrow \infty) = \infty.$$

Proof Let $A_0^n \downarrow \emptyset$ be a decreasing sequence such that $|E \setminus A_0^n| < \infty$ and $x_1 \notin A_0^n$ for all $n \in \mathbb{N}$. Define $F_n := \{\tau_{A_0^n} < \tau_{x_1}\}$. For every $M \in \mathbb{N}$, we have

$$\mathbf{P}_{x_1}[\tau_{A_0^n} \leq M] \leq \sum_{k=0}^{M} \mathbf{P}_{x_1}[X_k \in A_0^n] \xrightarrow{n \to \infty} 0.$$

Hence $\tau_{A_0^n} \uparrow \infty$ almost surely, and thus $F_n \downarrow \{\tau_{x_1} = \infty\}$ (up to a null set). We conclude

$$\frac{1}{C(x_1)} \, C_{\text{eff}}(x_1 \leftrightarrow \infty) = \lim_{n \to \infty} \mathbf{P}_{x_1}[F_n] = \mathbf{P}_{x_1}[\tau_{x_1} = \infty] = p_F(x_1). \qquad \square$$

Example 19.26 Symmetric simple random walk on $E = \mathbb{Z}$ is recurrent. Here $C(x, y) = \mathbb{1}_{\{|x-y|=1\}}$. The effective resistance from 0 to ∞ can be computed by the formulas for parallel and sequence connections,

$$R_{\text{eff}}(0 \leftrightarrow \infty) = \frac{1}{2} \sum_{i=0}^{\infty} R(i, i+1) = \infty. \quad \Diamond$$

Example 19.27 Asymmetric simple random walk on $E = \mathbb{Z}$ with $p(x, x+1) = p \in (\frac{1}{2}, 1)$, $p(x, x-1) = 1 - p$ is transient. Here one choice (and thus up to multiples the unique choice) for the conductances is

$$C(x, x+1) = \left(\frac{p}{1-p}\right)^x \quad \text{for } x \in \mathbb{Z},$$

and $C(x, y) = 0$ if $|x - y| > 1$. By the monotonicity principle, the effective resistance from 0 to ∞ can be bounded by

$$R_{\text{eff}}(0 \leftrightarrow \infty) = \lim_{n \to \infty} R_{\text{eff}}(0 \leftrightarrow \{-n, n\})$$

$$\leq \lim_{n \to \infty} R_{\text{eff}}(0 \leftrightarrow n)$$

$$= \sum_{n=0}^{\infty} \left(\frac{1-p}{p}\right)^n = \frac{p}{2p-1} < \infty. \quad \Diamond$$

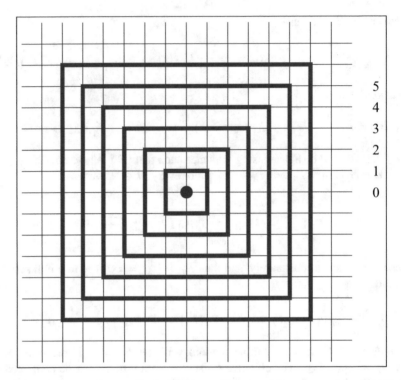

Fig. 19.3 Electrical network on $\mathbb{Z}^2$. The bold lines are *superconductors*. The nth and the $(n+1)$th superconductors are connected by $4(2n+1)$ edges.

Reflection Check that nearest neighbour random walk on an infinite binary tree (see Fig. 2.3) is transient. ♠

Example 19.28 Symmetric simple random walk on $E = \mathbb{Z}^2$ is recurrent. Here again $C(x, y) = \mathbb{1}_{\{|x-y|=1\}}$. Let $B_n = \{-n, \ldots, n\}^2$ and $\partial B_n = B_n \setminus B_{n-1}$. We construct a network C' with greater conductances by adding ring-shaped *superconductors* along ∂B. (See Figs. 19.3 and 19.4 for illustrations.) That is, we replace $C(x, y)$ by

$$
C'(x, y) = \begin{cases} \infty, & \text{if } x, y \in \partial B_n \text{ for some } n \in \mathbb{N}, \\ C(x, y), & \text{else.} \end{cases}
$$

Then $R'_{\text{eff}}(B_n \leftrightarrow B_n^c) = \frac{1}{4(2n+1)}$ (note that there are $4(2n+1)$ edges that connect B_n with B_n^c), and thus

$$
R'_{\text{eff}}(0 \leftrightarrow \infty) = \sum_{n=0}^{\infty} \frac{1}{4(2n+1)} = \infty.
$$

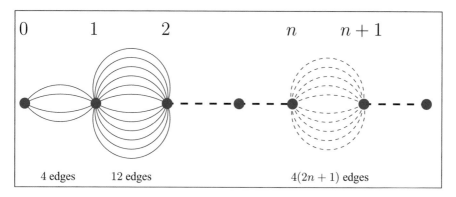

Fig. 19.4 Effective network after adding superconductors to $\mathbb{Z}^2$. The ring-shaped superconductors have melted down to single points.

By the monotonicity principle, we thus have $R_{\text{eff}}(0 \leftrightarrow \infty) \geq R'_{\text{eff}}(0 \leftrightarrow \infty) = \infty$.
◊

Example 19.29 Let (E, K) be an arbitrary connected subgraph of the square lattice $(\mathbb{Z}^2, \mathbb{L}^2)$. Then simple random walk on (E, K) (see Definition 19.11) is recurrent. Indeed, by the monotonicity principle, we have

$$R_{\text{eff}}^{(E,K)}(0 \leftrightarrow \infty) \geq R_{\text{eff}}^{(\mathbb{Z}^2, \mathbb{L}^2)}(0 \leftrightarrow \infty) = \infty. \quad ◊$$

We formulate the method used in the foregoing examples as a theorem.

Theorem 19.30 *Let C and C' be edge weights on E with $C'(x, y) \leq C(x, y)$ for all $x, y \in E$. If the Markov chain X with weights C is recurrent, then the Markov chain X' with weights C' is also recurrent.*

In particular, consider a graph (E, K) and a subgraph (E', K'). If simple random walk on (E, K) is recurrent, then so is simple random walk on (E', K').

Proof This follows from Theorem 19.25 and Rayleigh's monotonicity principle (Theorem 19.19). □

Example 19.31 Symmetric simple random walk on $\mathbb{Z}^3$ is transient. In order to prove this, we construct a subgraph for which we can compute $R'_{\text{eff}}(0 \leftrightarrow \infty) < \infty$.

Sketch We consider the set of all infinite paths starting at 0 and that

- begin by taking one step in the x-direction, the y-direction or the z-direction,
- continue by choosing a possibly different direction x, y or z and make *two* steps in that direction, and
- at the nth stage choose a direction x, y or z and take 2^{n+1} steps in that direction.

For example, by $xyyxxxxzzzzzzzz\ldots$ we denote the path that starts with one step in direction x, then chooses y, then x, then z and so on. Note that after two

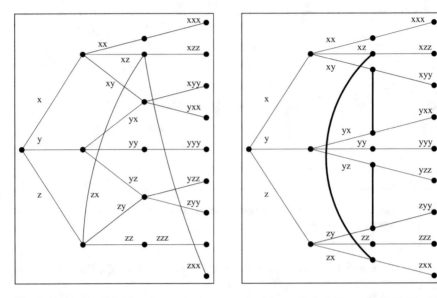

Fig. 19.5 Scheme of the first three steps (two stages) of the graph from Example 19.31. The left figure shows the actual edges where, e.g., xyy indicates that the first step is in direction x, the second step is in direction y and then the third step is necessarily also in direction y. In the right figure, the nodes at the ends of xz/zx, xy/yx and yz/zy are split into two nodes and then connected by a superconductor (bold line). If we remove the superconductors from the network, we end up with the network of Fig. 19.6 whose effective resistance $R'_{\text{eff}}(0 \leftrightarrow \infty)$ is not smaller than that of $\mathbb{Z}^3$. (If at the root we apply a voltage of 1 and at the points to the right the voltage 0, then by symmetry no current flows through the superconductors. Thus, in fact, the network is *equivalent* to that in Fig. 19.6.)

paths follow different directions for the first time, they will not have any common *edge* again, though some of the *nodes* can be visited by both paths.

Consider the electrical network with unit resistors. Apply a voltage of 1 at the origin and 0 at the endpoints of the paths at the nth stage. By symmetry, the potential at a given node depends only on the distance (length of the shortest path) from the origin. We thus obtain an equivalent network if we replace multiply used nodes by multiple nodes (see Fig. 19.5). Thus we obtain a tree-shaped network: For any $n \in \mathbb{N}_0$, after 2^n steps each path splits into three (see Fig. 19.6). The 3^n paths leading from the nodes of the nth generation to those of the $(n+1)$th generation are disjoint paths, each of length 2^{n-1}. If $B(n)$ denotes the set of points up to the nth generation, then

$$R'_{\text{eff}}\big(0 \leftrightarrow B(n+1)^c\big) = \sum_{k=0}^{n-1} R'_{\text{eff}}\big(B(k) \leftrightarrow B(k)^c\big) = \sum_{k=0}^{n-1} 2^k \, 3^{-k}.$$

Therefore, $R'_{\text{eff}}(0 \leftrightarrow \infty) = \frac{1}{3} \sum_{k=0}^{\infty} \left(\frac{2}{3}\right)^k = 1 < \infty$. On this tree, random walk is transient. Hence, by Theorem 19.30, random walk on $\mathbb{Z}^3$ is also transient. $\Diamond$

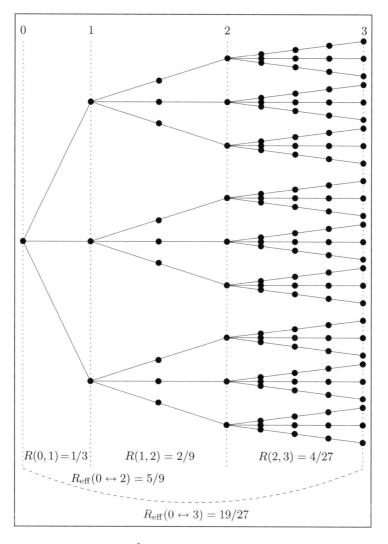

Fig. 19.6 A tree as a subgraph of $\mathbb{Z}^3$ on which random walk is still transient.

Takeaways A reversible Markov chain escapes from a point x to infinity (that is, it never returns to x) with positive probability if and only if in the corresponding electrical network, the effective resistance between x and infinity is finite. We use this idea to give a different proof for the fact that simple random walk on $\mathbb{Z}^d$ is recurrent if and only if $d \leq 2$. Furthermore, and more importantly, we use Rayleigh's monotonicity principle to show that if a random on a graph is recurrent, then it is also recurrent on any subgraph.

Exercise 19.4.1 Consider the electrical network on $\mathbb{Z}^d$ with unit resistors between neighboring points. Let X be a symmetric simple random walk on $\mathbb{Z}^d$. Finally, fix two arbitrary neighboring points $x_0, x_1 \in \mathbb{Z}^d$. Show the following:

(i) The effective conductance between x_0 and x_1 is $C_{\text{eff}}(x_0 \leftrightarrow x_1) = d$.

(ii) If $d \leq 2$, then $\mathbf{P}_{x_0}[\tau_{x_1} < \tau_{x_0}] = \frac{1}{2}$.

(iii) If $d \geq 3$, then $\mathbf{P}_{x_0}[\tau_{x_1} < \tau_{x_0} \mid \tau_{x_0} \wedge \tau_{x_1} < \infty] = \frac{1}{2}$. ♣

19.5   Network Reduction

Example 19.32 Consider a random walk on the graph in Fig. 19.7 that starts at x and at each step jumps to one of its neighbors at random with equal probability. What is the probability P that this Markov chain visits 1 before it visits 0?

We can regard the graph as an electrical network with unit resistors at each edge, voltage 0 at 0 and voltage 1 at 1. Then P equals the voltage at point x:

$$P = u(x).$$

In order to compute $u(x)$, we replace the network step by step by simpler networks such that the effective resistances between 0, 1, and x remain unchanged. Hence in each step the voltage $u(x)$ at point x does not change. ◊

Reduced Network

Assume that we have already reduced the network to a network with the three points 0, 1 and x and with resistors between these points $R'(0, 1)$, $R'(0, x)$ and $R'(1, x)$. See Fig. 19.8.

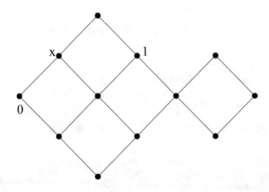

Fig. 19.7 Initial situation.

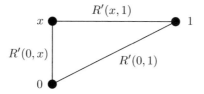

Fig. 19.8 Reduced network with three nodes.

Clearly, we have

$$P = u(x) = \frac{R'(0, x)}{R'(0, x) + R'(1, x)}. \tag{19.13}$$

If we knew the effective resistances $R_{\text{eff}}(0 \leftrightarrow x)$, $R_{\text{eff}}(1 \leftrightarrow x)$ and $R_{\text{eff}}(0 \leftrightarrow 1)$, we could avoid the hassle of reducing the network and we could compute $u(x)$ directly. In order to derive the formula for $u(x)$, we make the following observations. In the reduced network, the effective resistances are easy to compute: If $\{a, b, c\} = \{0, 1, x\}$, then

$$R_{\text{eff}}(a \leftrightarrow b) = \left(\frac{1}{R'(a, b)} + \frac{1}{R'(a, c) + R'(b, c)} \right)^{-1}. \tag{19.14}$$

Solving these three equations for $R'(0, 1)$, $R'(0, x)$ and $R'(1, x)$ and plugging the values into (19.13) yields

$$P = u(x) = \frac{R_{\text{eff}}(0 \leftrightarrow 1) + R_{\text{eff}}(0 \leftrightarrow x) - R_{\text{eff}}(x \leftrightarrow 1)}{2 R_{\text{eff}}(0 \leftrightarrow 1)}. \tag{19.15}$$

In particular, in the case $R'(0, 1) = \infty$ (or equivalently $R_{\text{eff}}(0 \leftrightarrow 1) = R_{\text{eff}}(0 \leftrightarrow x) + R_{\text{eff}}(x \leftrightarrow 1)$), we have $R_{\text{eff}}(0 \leftrightarrow x) = R'(0, x)$ and $R_{\text{eff}}(1 \leftrightarrow x) = R'(1, x)$, hence

$$u(x) = \frac{R_{\text{eff}}(0 \leftrightarrow x)}{R_{\text{eff}}(0 \leftrightarrow x) + R_{\text{eff}}(x \leftrightarrow 1)}. \tag{19.16}$$

Since we always have $u(x) \in [0, 1]$, rearranging the terms yields (again in the general situation)

$$R_{\text{eff}}(1 \leftrightarrow x) \leq R_{\text{eff}}(0 \leftrightarrow 1) + R_{\text{eff}}(0 \leftrightarrow x). \tag{19.17}$$

This is the triangle inequality for the effective resistances and it shows that the effective resistance is a metric in any electrical network.

Step-by-Step Reduction of the Network

Having seen how to compute $u(x)$ from the effective resistances, we now turn to the systematic computation of these effective resistances. Later we will come back to the introductory example and make the computations explicit.

There are four elementary transformations for the reduction of an electrical network:

1. **Deletion of loops.**   The three points on the very right of the graph form a loop that can be deleted from the network without changing any of the remaining voltages. In particular, any edge that directly connects 0 to 1 can be deleted.
2. **Joining serial edges.**   If two (or more) edges are in a row such that the nodes along them do not have any further adjacent edges, this sequence of edges can be substituted by a single edge whose resistance is the sum of the resistances of the single edges (see Fig. 19.1).
3. **Joining parallel edges.**   Two (or more) edges with resistances $R_1, \ldots, R_n$ that connect the same two nodes can by replaced by a single edge with resistance $R = (R_1^{-1} + \ldots + R_n^{-1})^{-1}$ (see Fig. 19.2).
4. **Star–triangle transformation (see Exercise 19.5.1).**   The star-shaped part of a network (left in Fig. 19.9) is equivalent to the triangle-shaped part (right in Fig. 19.9) if the resistances $R_1, R_2, R_3, \tilde{R}_1, \tilde{R}_2, \tilde{R}_3$ satisfy the condition

$$R_i \tilde{R}_i = \delta \quad \text{for any } i = 1, 2, 3, \tag{19.18}$$

where

$$\delta = R_1 R_2 R_3 \left(R_1^{-1} + R_2^{-1} + R_3^{-1} \right) = \frac{\tilde{R}_1 \tilde{R}_2 \tilde{R}_3}{\tilde{R}_1 + \tilde{R}_2 + \tilde{R}_3}.$$

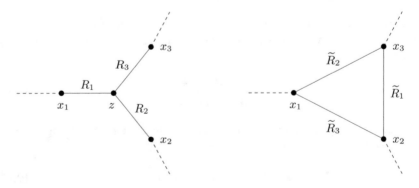

Fig. 19.9 Star–triangle transformation.

Application to Example 19.32

With the four transformations at hand, we solve the problem of Example 19.32. Assume that initially all edges have resistance 1. In the figures we label each edge with its resistance if it differs (in the course of the reduction) from 1.

Step 1. Delete the loop at the right-hand side (left in Fig. 19.10).

Step 2. Replace the series on top, bottom and right by edges with resistance 2 (right in Fig. 19.10).

Step 3. Use the star-triangle transformation to remove the lower left node (left in Fig. 19.11). Here $R_1 = 1$, $R_2 = 2$, $R_3 = 1$, $\delta = 5$, $\widetilde{R}_1 = \delta/R_1 = 5$, $\widetilde{R}_2 = \delta/R_2 = 5/2$ and $\widetilde{R}_3 = \delta/R_3 = 5$.

Step 4. Replace the parallel edges with resistances $R_1 = 5$ and $R_2 = 1$ by one edge with $R = (\frac{1}{5} + 1)^{-1} = \frac{5}{6}$ (right in Fig. 19.11).

Step 5. Use the star-triangle transformation to remove the lower right node (left in Fig. 19.12). Here $R_1 = 5$, $R_2 = 2$, $R_3 = \frac{5}{6}$, $\delta = 95/6$, $\widetilde{R}_1 = \delta/R_1 = 19/6$, $\widetilde{R}_2 = \delta/R_2 = 95/12$ and $\widetilde{R}_3 = \delta/R_3 = 19$.

Step 6. Replace the parallel edges by edges with resistances $(\frac{12}{95} + \frac{2}{5})^{-1} = \frac{19}{10}$ and $(\frac{6}{19} + 1)^{-1} = \frac{19}{25}$, respectively (right in Fig. 19.12).

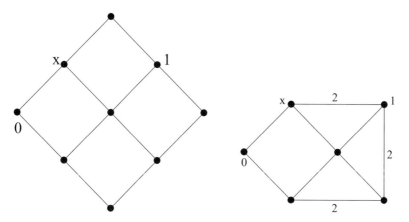

Fig. 19.10 Steps 1 and 2.

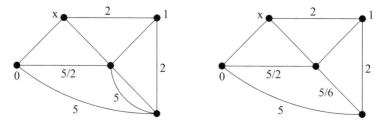

Fig. 19.11 Steps 3 and 4.

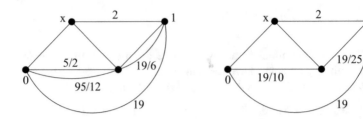

Fig. 19.12 Steps 5 and 6.

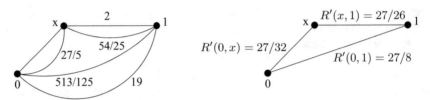

Fig. 19.13 Steps 7 and 8.

Step 7. Use the star-triangle transformation to remove the lower right node (left in Fig. 19.13). Here $R_1 = \frac{19}{10}$, $R_2 = \frac{19}{25}$, $R_3 = 1$, $\delta = \frac{513}{125}$, $\widetilde{R}_1 = \delta/R_1 = \frac{54}{25}$, $\widetilde{R}_2 = \delta/R_2 = \frac{27}{5}$ and $\widetilde{R}_3 = \delta/R_3 = \frac{513}{125}$.

Step 8. Replace the three pairs of parallel edges by single edges with resistances $(\frac{5}{27} + 1)^{-1} = \frac{27}{32}$, $(\frac{25}{54} + \frac{1}{2})^{-1} = \frac{27}{26}$ and $(\frac{1}{19} + \frac{125}{513})^{-1} = \frac{27}{8}$, respectively.

In the reduced network, we have the resistances $R'(0, x) = \frac{27}{32}$ and $R'(x, 1) = \frac{27}{26}$. Using (19.13), the probability that the random walk visits 1 before 0 is

$$P = \frac{\frac{27}{32}}{\frac{27}{32} + \frac{27}{26}} = \frac{13}{29}.$$

Using the values of $R'(0, x)$, $R'(1, x)$ and $R'(0, 1)$ and equation (19.14), we compute the effective resistances in the reduced network (and hence in the original network):

$$R_{\text{eff}}(0 \leftrightarrow x) = \left(\frac{32}{27} + \frac{1}{\frac{27}{8} + \frac{27}{26}} \right)^{-1} = \frac{17}{24},$$

$$R_{\text{eff}}(1 \leftrightarrow x) = \left(\frac{26}{27} + \frac{1}{\frac{27}{32} + \frac{27}{8}} \right)^{-1} = \frac{5}{6},$$

$$R_{\text{eff}}(0 \leftrightarrow 1) = \left(\frac{8}{27} + \frac{1}{\frac{27}{26} + \frac{27}{32}} \right)^{-1} = \frac{29}{24}.$$

Using (19.15) we can use the values to compute $u(x)$:

$$P = u(x) = \frac{\frac{29}{24} + \frac{17}{24} - \frac{5}{6}}{2 \cdot \frac{29}{24}} = \frac{13}{29}.$$

Clearly, the latter computation is more complicated than using the resistances R' from the reduced network directly. However, it has the advantage that it can be performed without going through all the network reduction steps if, for some reason, we know the effective resistances already. For example, we could buy resistors in an electronic market, solder the network and measure the resistances with a multimeter.
◊

Alternative Solution

A different approach to solving the problem of Example 19.32 is to use linear algebra instead of network reduction. It is a matter of taste as to which solution is preferable. First generate the transition matrix p of the Markov chain. To this end, enumerate the nodes of the graph from 1 to 12 as in Fig. 19.14. The chain starts at 2, and we want to compute the probability that it visits 3 before 5.

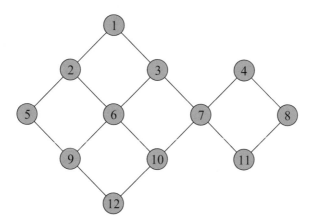

Fig. 19.14 Graph with enumerated nodes.

Generate the matrix $\overline{p}$ of the chain that is killed at 3 and at 5 and compute $\overline{G} = (I - \overline{p})^{-1}$. By Exercise 19.1.1 (with $A = \{3, 5\}$, $x = 2$ and $y = 3$), the probability of visiting 3 before 5 is $P = \overline{G}(2, 3) = \frac{13}{29}$.

$$
\overline{p} := \begin{pmatrix}
0 & \frac{1}{2} & \frac{1}{2} & 0 & 0 & 0 & 0 & 0 & 0 & 0 & 0 & 0 \\
\frac{1}{3} & 0 & 0 & 0 & \frac{1}{3} & \frac{1}{3} & 0 & 0 & 0 & 0 & 0 & 0 \\
0 & 0 & 0 & 0 & 0 & 0 & 0 & 0 & 0 & 0 & 0 & 0 \\
0 & 0 & 0 & 0 & 0 & 0 & \frac{1}{2} & \frac{1}{2} & 0 & 0 & 0 & 0 \\
0 & 0 & 0 & 0 & 0 & 0 & 0 & 0 & 0 & 0 & 0 & 0 \\
0 & \frac{1}{4} & \frac{1}{4} & 0 & 0 & 0 & 0 & 0 & \frac{1}{4} & \frac{1}{4} & 0 & 0 \\
0 & 0 & \frac{1}{4} & \frac{1}{4} & 0 & 0 & 0 & 0 & 0 & \frac{1}{4} & \frac{1}{4} & 0 \\
0 & 0 & 0 & \frac{1}{2} & 0 & 0 & 0 & 0 & 0 & 0 & \frac{1}{2} & 0 \\
0 & 0 & 0 & 0 & \frac{1}{3} & \frac{1}{3} & 0 & 0 & 0 & 0 & 0 & \frac{1}{3} \\
0 & 0 & 0 & 0 & 0 & \frac{1}{3} & \frac{1}{3} & 0 & 0 & 0 & 0 & \frac{1}{3} \\
0 & 0 & 0 & 0 & 0 & 0 & \frac{1}{2} & \frac{1}{2} & 0 & 0 & 0 & 0 \\
0 & 0 & 0 & 0 & 0 & 0 & 0 & 0 & \frac{1}{2} & \frac{1}{2} & 0 & 0
\end{pmatrix},
$$

$$
\overline{G} := (I - \overline{p})^{-1} = \begin{pmatrix}
\frac{143}{116} & \frac{81}{116} & \frac{21}{29} & \frac{3}{58} & \frac{8}{29} & \frac{19}{58} & \frac{3}{29} & \frac{3}{58} & \frac{15}{116} & \frac{9}{58} & \frac{3}{58} & \frac{11}{116} \\
\frac{27}{58} & \frac{81}{58} & \frac{13}{29} & \frac{3}{29} & \frac{16}{29} & \frac{19}{29} & \frac{6}{29} & \frac{3}{29} & \frac{15}{58} & \frac{9}{29} & \frac{3}{29} & \frac{11}{58} \\
0 & 0 & 1 & 0 & 0 & 0 & 0 & 0 & 0 & 0 & 0 & 0 \\
\frac{3}{58} & \frac{9}{58} & \frac{24}{29} & \frac{165}{58} & \frac{5}{29} & \frac{15}{29} & \frac{78}{29} & \frac{68}{29} & \frac{21}{58} & \frac{30}{29} & \frac{107}{58} & \frac{27}{58} \\
0 & 0 & 0 & 0 & 1 & 0 & 0 & 0 & 0 & 0 & 0 & 0 \\
\frac{19}{116} & \frac{57}{116} & \frac{18}{29} & \frac{15}{58} & \frac{11}{29} & \frac{95}{58} & \frac{15}{29} & \frac{15}{58} & \frac{75}{116} & \frac{45}{58} & \frac{15}{58} & \frac{55}{116} \\
\frac{3}{58} & \frac{9}{58} & \frac{24}{29} & \frac{39}{29} & \frac{5}{29} & \frac{15}{29} & \frac{78}{29} & \frac{39}{29} & \frac{21}{58} & \frac{30}{29} & \frac{39}{29} & \frac{27}{58} \\
\frac{3}{58} & \frac{9}{58} & \frac{24}{29} & \frac{68}{29} & \frac{5}{29} & \frac{15}{29} & \frac{78}{29} & \frac{97}{29} & \frac{21}{58} & \frac{30}{29} & \frac{68}{29} & \frac{27}{58} \\
\frac{5}{58} & \frac{15}{58} & \frac{11}{29} & \frac{7}{29} & \frac{18}{29} & \frac{25}{29} & \frac{14}{29} & \frac{7}{29} & \frac{93}{58} & \frac{21}{29} & \frac{7}{29} & \frac{45}{58} \\
\frac{3}{29} & \frac{9}{29} & \frac{19}{29} & \frac{20}{29} & \frac{10}{29} & \frac{30}{29} & \frac{40}{29} & \frac{20}{29} & \frac{21}{29} & \frac{60}{29} & \frac{20}{29} & \frac{27}{29} \\
\frac{3}{58} & \frac{9}{58} & \frac{24}{29} & \frac{107}{58} & \frac{5}{29} & \frac{15}{29} & \frac{78}{29} & \frac{68}{29} & \frac{21}{58} & \frac{30}{29} & \frac{165}{58} & \frac{27}{58} \\
\frac{11}{116} & \frac{33}{116} & \frac{15}{29} & \frac{27}{58} & \frac{14}{29} & \frac{55}{58} & \frac{27}{29} & \frac{27}{58} & \frac{135}{116} & \frac{81}{58} & \frac{27}{58} & \frac{215}{116}
\end{pmatrix}.
$$

Takeaways There are essentially two possibilities to systematically determine the effective resistance between two points in an electrical network. Firstly, one can solve the system of linear equations belonging to the Markov chain killed in the two points. Secondly, the network can be reduced in a series of elementary steps: Resolving parallel connections, resolving serial connections and resolving intermediate points using the star-triangle transformation. While solving the linear equations is a simple job for a computer, network reduction can give insights into the structure of the problem and can lead to general formulas also for similar networks.

Exercise 19.5.1 Show the validity of the star-triangle transformation. ♣

Exercise 19.5.2 Consider a random walk on the honeycomb graph shown below. Show that if the walk starts at x, then the probability of visiting 1 before 0 is $\frac{8}{17}$ using

(i) the method of network reduction, and
(ii) the method of matrix inversion. ♣

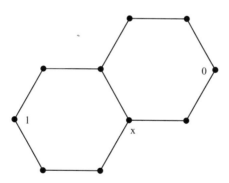

Exercise 19.5.3 Consider the graph of Fig. 19.15.

(i) For the effective conductance between a and z, show that $C_{\text{eff}}(a \longleftrightarrow z) = \sqrt{3}$.
(ii) For a random walk started at a, show that the probability $\mathbf{P}_a[\tau_z < \tau_a]$ of visiting z before returning to a is $\mathbf{P}_a[\tau_z < \tau_a] = 1/\sqrt{3}$. ♣

Exercise 19.5.4 For the graph of Fig. 19.16, determine $C_{\text{eff}}(a \longleftrightarrow z)$ and $\mathbf{P}_a[\tau_z < \tau_a]$. (This is simpler than in Exercise 19.5.3!) ♣

Exercise 19.5.5 For a random walk on the graph of Fig. 19.17, determine the probability $\mathbf{P}_a[\tau_z < \tau_a]$. ♣

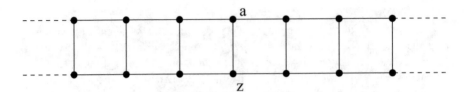

Fig. 19.15 Simple ladder graph

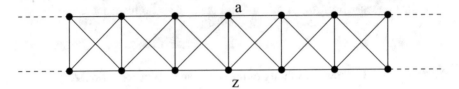

Fig. 19.16 Crossed ladder graph

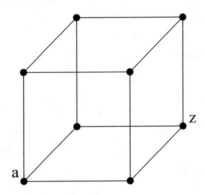

Fig. 19.17 Random walk on a hypercube.

19.6   Random Walk in a Random Environment

(Compare [143, 175] and [76, 77, 96].) Consider a Markov chain X on $\mathbb{Z}$ that at each step makes a jump either to the left (with probability w_i^-) or to the right (with probability w_i^+) if X is at $i \in \mathbb{Z}$. Hence, let $w_i^- \in (0, 1)$ and $w_i^+ := 1 - w_i^-$ for $i \in \mathbb{Z}$. Then X is the Markov chain with transition matrix

$$
p_w(i, j) = \begin{cases} w_i^-, & \text{if } j = i - 1, \\ w_i^+, & \text{if } j = i + 1, \\ 0, & \text{else.} \end{cases}
$$

We consider $(w_i^-)_{i \in \mathbb{Z}}$ as an environment in which X walks and later choose the environment at random.

In order to describe X in terms of conductances of an electrical network, we define $\varrho_i := w_i^- / w_i^+$ for $i \in \mathbb{Z}$. Let $C_w(i, j) := 0$ if $|i - j| \neq 1$ and

$$
C_w(i + 1, i) := C_w(i, i + 1) := \begin{cases} \prod_{k=0}^{i} \varrho_k^{-1}, & \text{if } i \geq 0, \\ \prod_{k=i}^{-1} \varrho_k, & \text{if } i < 0. \end{cases}
$$

With this definition,

$$
\frac{C_w(i, i + 1)}{C_w(i)} = \frac{1}{\varrho_i + 1} = w_i^+ \quad \text{and} \quad \frac{C_w(i, i - 1)}{C_w(i)} = \frac{\varrho_i}{\varrho_i + 1} = w_i^-.
$$

Hence the transition probabilities p_w are indeed described by the C_w. Let

$$
R_w^+ := \sum_{i=0}^{\infty} R_w(i, i + 1) = \sum_{i=0}^{\infty} \frac{1}{C_w(i, i + 1)} = \sum_{i=0}^{\infty} \prod_{k=0}^{i} \varrho_k
$$

and

$$
R_w^- := \sum_{i=0}^{\infty} R_w(-i, -i - 1) = \sum_{i=0}^{\infty} \frac{1}{C_w(-i, -i - 1)} = \sum_{i=1}^{\infty} \prod_{k=-i}^{1} \varrho_k^{-1}.
$$

Note that R_w^+ and R_w^- are the effective resistances from 0 to $+\infty$ and from 0 to $-\infty$, respectively. Hence

$$
R_{w,\text{eff}}(0 \leftrightarrow \infty) = \frac{1}{\frac{1}{R_w^-} + \frac{1}{R_w^+}}
$$

is finite if and only if $R_w^- < \infty$ or $R_w^+ < \infty$. Therefore, by Theorem 19.25,

$$
X \text{ is transient} \quad \Longleftrightarrow \quad R_w^- < \infty \text{ or } R_w^+ < \infty. \tag{19.19}
$$

If X is transient, in which direction does it get lost?

Theorem 19.33 *(i) If $R_w^- < \infty$ or $R_w^+ < \infty$, then (agreeing on $\frac{\infty}{\infty} = 1$)*

$$
\mathbf{P}_0\big[X_n \overset{n\to\infty}{\longrightarrow} -\infty\big] = \frac{R_w^+}{R_w^- + R_w^+} \quad \text{and} \quad \mathbf{P}_0\big[X_n \overset{n\to\infty}{\longrightarrow} +\infty\big] = \frac{R_w^-}{R_w^- + R_w^+}.
$$

(ii) If $R_w^- = \infty$ and $R_w^+ = \infty$, then $\liminf\limits_{n\to\infty} X_n = -\infty$ and $\limsup\limits_{n\to\infty} X_n = \infty$ almost surely.

Proof

(i) Let $\tau_N := \inf\{n \in \mathbb{N}_0 : X_n \in \{-N, N\}\}$. As X is transient, we have $\mathbf{P}_0[\tau_N < \infty] = 1$ and (as in (19.9))

$$\mathbf{P}_0[X_{\tau_N} = -N] = \frac{R_{w,\mathrm{eff}}(0 \leftrightarrow N)}{R_{w,\mathrm{eff}}(-N \leftrightarrow N)} = \frac{R_{w,\mathrm{eff}}(0 \leftrightarrow N)}{R_{w,\mathrm{eff}}(0 \leftrightarrow -N) + R_{w,\mathrm{eff}}(0 \leftrightarrow N)}.$$

Again, since X is transient, we infer

$$\begin{aligned}
\mathbf{P}_0[X_n \overset{n\to\infty}{\longrightarrow} -\infty] &= \mathbf{P}\big[\sup\{X_n : n \in \mathbb{N}_0\} < \infty\big] \\
&= \lim_{N\to\infty} \mathbf{P}\big[\sup\{X_n : n \in \mathbb{N}_0\} < N\big] \\
&\leq \limsup_{N\to\infty} \mathbf{P}\big[X_{\tau_N} = -N\big] \\
&= \frac{R_w^+}{R_w^- + R_w^+}.
\end{aligned}$$

By symmetry (and since X is transient), we get

$$\mathbf{P}_0[X_n \overset{n\to\infty}{\longrightarrow} -\infty] = 1 - \mathbf{P}_0[X_n \overset{n\to\infty}{\longrightarrow} \infty] \geq 1 - \frac{R_w^-}{R_w^- + R_w^+} = \frac{R_w^+}{R_w^- + R_w^+}.$$

(ii) If $R_w^- = R_w^+ = \infty$, then X is recurrent and hence *every* point is visited infinitely often. That is, $\limsup_{n\to\infty} X_n = \infty$ and $\liminf_{n\to\infty} X_n = -\infty$ a.s.   □

We now consider the situation where the sequence $w = (w_i^-)_{i\in\mathbb{Z}}$ is also random. That is, we consider a two-stage experiment: At the first stage we choose a realization of i.i.d. random variables $W = (W_i^-)_{i\in\mathbb{Z}}$ on $(0, 1)$ and let $W_i^+ := 1 - W_i^-$. At the second stage, given W, we construct a Markov chain X on $\mathbb{Z}$ with transition matrix

$$p_W(i, j) = \begin{cases} W_i^-, & \text{if } j = i - 1, \\ W_i^+, & \text{if } j = i + 1, \\ 0, & \text{else.} \end{cases}$$

Note that X is a Markov chain only given W; that is, under the probability measure $\mathbf{P}[X \in \cdot \,|\, W]$. However, it is not a Markov chain with respect to the so-called annealed measure $\mathbf{P}[X \in \cdot\,]$. In fact, if W is unknown, observing X gives an increasing amount of information on the true realization of W. This is precisely what memory is and is thus in contrast with the Markov property of X.

Definition 19.34 *The process X is called a **random walk in the random environment** W.*

We are now in the position to prove a theorem of Solomon [158]. Let $\varrho_i :=$ W_i^- / W_i^+ for $i \in \mathbb{Z}$ and R_W^- and R_W^+ be defined as above.

Theorem 19.35 (Solomon [158]) *Assume that* $\mathbf{E}[|\log(\varrho_0)|] < \infty$.

(i) *If* $\mathbf{E}[\log(\varrho_0)] < 0$, *then* $X_n \overset{n\to\infty}{\longrightarrow} \infty$ *a.s.*

(ii) *If* $\mathbf{E}[\log(\varrho_0)] > 0$, *then* $X_n \overset{n\to\infty}{\longrightarrow} -\infty$ *a.s.*

(iii) *If* $\mathbf{E}[\log(\varrho_0)] = 0$, *then* $\liminf\limits_{n\to\infty} X_n = -\infty$ *and* $\limsup\limits_{n\to\infty} X_n = \infty$ *a.s.*

Proof (i) **and** (ii) By symmetry, it is enough to show (ii). Hence, let $c :=$ $\mathbf{E}[\log(\varrho_0)] > 0$. By the strong law of large numbers, there is an $n_0^- = n_0^-(\omega)$ with

$$\prod_{k=-n}^{1} \varrho_k^{-1} = \exp\left(-\sum_{k=-n}^{1} \log(\varrho_k)\right) < e^{-cn/2} \quad \text{for all } n \geq n_0^-.$$

Therefore,

$$R_W^- = \sum_{n=1}^{\infty} \prod_{k=-n}^{1} \varrho_k^{-1} \leq \sum_{n=1}^{n_0^- - 1} \prod_{k=-n}^{1} \varrho_k^{-1} + \sum_{n=n_0^-}^{\infty} e^{-cn/2} < \infty \quad \text{a.s.}$$

Similarly, there is an $n_0^+ = n_0^+(\omega)$ with

$$\prod_{k=0}^{n} \varrho_k > e^{cn/2} \quad \text{for all } n \geq n_0^+.$$

We conclude

$$R_W^+ = \sum_{n=0}^{\infty} \prod_{k=0}^{n} \varrho_k \geq \sum_{n=0}^{n_0^+ - 1} \prod_{k=0}^{n} \varrho_k + \sum_{n=n_0^+}^{\infty} e^{cn/2} = \infty \quad \text{a.s.}$$

Now, by Theorem 19.33, we get $X_n \overset{n\to\infty}{\longrightarrow} -\infty$ almost surely.

(iii) In order to show $R_W^- = R_W^+ = \infty$ almost surely, it is enough to show $\limsup_{n\to\infty} \sum_{k=0}^{n} \log(\varrho_k) > -\infty$ and $\limsup_{n\to\infty} \sum_{k=-n}^{1} \log(\varrho_k^{-1}) > -\infty$ almost surely if $\mathbf{E}[\log(\varrho_0)] = 0$. If $\log(\varrho_0)$ has a finite variance, this follows by the central limit theorem. In the general case, it follows by Theorem 20.21. □

Reflection In the situation of Theorem 19.35, come up with an example such that $\mathbf{E}[W_i^+ - W_i^-] > 0$ but still $X_n \overset{n\to\infty}{\longrightarrow} -\infty$ holds. ♠

Takeaways One-dimensional random walks in random environment can be regarded as random walks in a network of random conductances. This is a specialty of dimension one that makes it easy to check if the random walk is transient or recurrent.

Exercise 19.6.1 Consider the situation of Theorem 19.35 but with the random walk restricted to $\mathbb{N}_0$. To this end, change the walk so that whenever it attempts to make a step from 0 to -1, it simply stays in 0. Show that this random walk in a random environment is

- a.s. transient if $\mathbf{E}[\log(\varrho_0)] < 0$,
- a.s. null recurrent if $\mathbf{E}[\log(\varrho_0)] = 0$, and
- a.s. positive recurrent if $\mathbf{E}[\log(\varrho_0)] > 0$. ♣

Chapter 20
Ergodic Theory

Laws of large numbers, e.g., for i.i.d. random variables $X_1, X_2, \ldots$, state that the sequence of averages converges a.s. to the expected value, $n^{-1} \sum_{i=1}^{n} X_i \overset{n \to \infty}{\longrightarrow}$ $E[X_1]$. Hence averaging over one realization of many random variables is equivalent to averaging over all possible realizations of one random variable. In the terminology of statistical physics this means that the *time average*, or path (Greek: *odos*) average, equals the *space average*. The "space" in "space average" is the probability space in mathematical terminology, and in physics it is considered the space of admissible states with a certain energy (Greek: *ergon*). Combining the Greek words gives rise to the name *ergodic theory*, which studies laws of large numbers for possibly dependent, but stationary, random variables.

For further reading, see, for example [103] or [88].

20.1 Definitions

Definition 20.1 *Let* $I \subset \mathbb{R}$ *be a set that is closed under addition (for us the important examples are* $I = \mathbb{N}_0$, $I = \mathbb{N}$, $I = \mathbb{Z}$, $I = \mathbb{R}$, $I = [0, \infty)$, $I = \mathbb{Z}^d$ *and so on). A stochastic process* $X = (X_t)_{t \in I}$ *is called* **stationary** *if*

$$\mathcal{L}[(X_{t+s})_{t \in I}] = \mathcal{L}[(X_t)_{t \in I}] \quad \text{for all } s \in I. \tag{20.1}$$

Remark 20.2 If $I = \mathbb{N}_0$, $I = \mathbb{N}$ or $I = \mathbb{Z}$, then (20.1) is equivalent to

$$\mathcal{L}[(X_{n+1})_{n \in I}] = \mathcal{L}[(X_n)_{n \in I}]. \quad \Diamond$$

© The Editor(s) (if applicable) and The Author(s), under exclusive license to Springer Nature Switzerland AG 2020
A. Klenke, *Probability Theory*, Universitext,
https://doi.org/10.1007/978-3-030-56402-5_20

Example 20.3

(i) If $X = (X_t)_{t \in I}$ is i.i.d., then X is stationary. If only $\mathbf{P}_{X_t} = \mathbf{P}_{X_0}$ holds for every $t \in I$ (without the independence), then in general, X is not stationary. For example, consider $I = \mathbb{N}_0$ and $X_1 = X_2 = X_3 = \ldots$ but $X_0 \neq X_1$. Then X is not stationary.

(ii) Let X be a Markov chain with invariant distribution π. If $\mathcal{L}[X_0] = \pi$, then X is stationary.

(iii) Let $(Y_n)_{n \in \mathbb{Z}}$ be i.i.d. real random variables and let $c_1, \ldots, c_k \in \mathbb{R}$. Then

$$X_n := \sum_{l=1}^{k} c_l \, Y_{n-l}, \qquad n \in \mathbb{Z},$$

defines a stationary process X that is called the **moving average** with weights $(c_1, \ldots, c_k)$. In fact, X is stationary if only Y is stationary. $\Diamond$

Lemma 20.4 *If $(X_n)_{n \in \mathbb{N}_0}$ is stationary, then X can be extended to a stationary process $(\widetilde{X}_n)_{n \in \mathbb{Z}}$.*

Proof Let $\widetilde{X}$ be the canonical process on $\Omega = E^{\mathbb{Z}}$. For $n \in \mathbb{N}$, define a probability measure $\widetilde{\mathbf{P}}^{\{-n, -n+1, \ldots\}} \in \mathcal{M}_1\big(E^{\{-n, -n+1, \ldots\}}\big)$ by

$$\widetilde{\mathbf{P}}^{\{-n, -n+1, \ldots\}}\big[\widetilde{X}_{-n} \in A_{-n}, \widetilde{X}_{-n+1} \in A_{-n+1}, \ldots\big]$$

$$= \mathbf{P}\big[X_0 \in A_{-n}, X_1 \in A_{-n+1}, \ldots\big].$$

Then $\{-n, -n+1, \ldots\} \uparrow \mathbb{Z}$ and $\big(\widetilde{\mathbf{P}}^{\{-n, -n+1, \ldots\}}, n \in \mathbb{N}\big)$ is a consistent family. By the Ionescu–Tulcea theorem (Theorem 14.35), the projective limit $\widetilde{\mathbf{P}} := \lim_{n \to \infty} \widetilde{\mathbf{P}}^{\{-n, -n+1, \ldots\}}$ exists. By construction, $\widetilde{X}$ is stationary with respect to $\widetilde{\mathbf{P}}$ and

$$\widetilde{\mathbf{P}} \circ \big((\widetilde{X}_n)_{n \in \mathbb{N}_0}\big)^{-1} = \mathbf{P} \circ \big((X_n)_{n \in \mathbb{N}_0}\big)^{-1}. \qquad \square$$

In the following, assume that $(\Omega, \mathcal{A}, \mathbf{P})$ is a probability space and $\tau : \Omega \to \Omega$ is a measurable map.

Definition 20.5 *An event $A \in \mathcal{A}$ is called* **invariant** *if $\tau^{-1}(A) = A$ and* **quasi-invariant** *if $\mathbb{1}_{\tau^{-1}(A)} = \mathbb{1}_A$ $\mathbf{P}$-a.s. Denote the σ-algebra of* **invariant events** *by*

$$\mathcal{I} = \big\{ A \in \mathcal{A} : \tau^{-1}(A) = A \big\}.$$

Recall that a σ-algebra $\mathcal{I}$ is called **P**-trivial if $\mathbf{P}[A] \in \{0, 1\}$ for every $A \in \mathcal{I}$.

Definition 20.6

(i) τ is called **measure-preserving** if

$$\mathbf{P}\big[\tau^{-1}(A)\big] = \mathbf{P}[A] \quad \text{for all } A \in \mathcal{A}.$$

In this case, $(\Omega, \mathcal{A}, \mathbf{P}, \tau)$ is called a **measure-preserving dynamical system**.
(ii) If τ is measure-preserving and $\mathcal{I}$ is **P**-trivial, then $(\Omega, \mathcal{A}, \mathbf{P}, \tau)$ is called **ergodic**.

Lemma 20.7

(i) A measurable map $f : (\Omega, \mathcal{A}) \to (\mathbb{R}, \mathcal{B}(\mathbb{R}))$ is $\mathcal{I}$-measurable if and only if $f \circ \tau = f$.
(ii) $(\Omega, \mathcal{A}, \mathbf{P}, \tau)$ is ergodic if and only if any $\mathcal{I}$-measurable $f : (\Omega, \mathcal{I}) \to (\mathbb{R}, \mathcal{B}(\mathbb{R}))$ is **P**-almost surely constant.

Proof

(i) The statement is obvious if $f = \mathbb{1}_A$ is an indicator function. The general case, can be inferred by the usual approximation arguments (see Theorem 1.96((i))).
(ii) " $\Longrightarrow$ " Assume that $(\Omega, \mathcal{A}, \mathbf{P}, \tau)$ is ergodic. Then, for any $c \in \mathbb{R}$, we have $f^{-1}((c, \infty)) \in \mathcal{I}$ and thus $\mathbf{P}[f^{-1}((c, \infty))] \in \{0, 1\}$. We conclude that

$$f = \inf\big\{c \in \mathbb{R} : \mathbf{P}\big[f^{-1}((c, \infty))\big] = 0\big\} \quad \mathbf{P}\text{-a.s.}$$

" $\Longleftarrow$ " Assume any $\mathcal{I}$-measurable map is **P**-a.s. constant. If $A \in \mathcal{I}$, then $\mathbb{1}_A$ is $\mathcal{I}$-measurable and hence **P**-a.s. equals either 0 or 1. Thus $\mathbf{P}[A] \in \{0, 1\}$. $\square$

Example 20.8 Let $n \in \mathbb{N} \setminus \{1\}$, let $\Omega = \mathbb{Z}/(n)$, let $\mathcal{A} = 2^{\Omega}$ and let **P** be the uniform distribution on Ω. Let $r \in \{1, \dots, n\}$ and

$$\tau : \Omega \to \Omega, \quad x \mapsto x + r \pmod{n}.$$

Then τ is measure-preserving. If $d = \gcd(n, r)$ and

$$A_i = \big\{i, \tau(i), \tau^2(i), \dots, \tau^{n-1}(i)\big\} = i + \langle r \rangle \quad \text{for } i = 0, \dots, d-1,$$

then $A_0, \dots, A_{d-1}$ are the disjoint coset classes of the normal subgroup $\langle r \rangle \trianglelefteq \Omega$. Hence we have $A_i \in \mathcal{I}$ for $i = 0, \dots, d-1$, and each $A \in \mathcal{I}$ is a union of certain A_i's. Hence we have

$$(\Omega, \mathcal{A}, \mathbf{P}, \tau) \text{ is ergodic} \quad \Longleftrightarrow \quad \gcd(r, n) = 1. \quad \Diamond$$

Example 20.9 (Rotation) Let $\Omega = [0, 1)$, let $\mathcal{A} = \mathcal{B}(\Omega)$ and let $\mathbf{P} = \lambda$ be the Lebesgue measure. Let $r \in (0, 1)$ and $\tau_r(x) = x + r \pmod{1}$. Clearly, $(\Omega, \mathcal{A}, \mathbf{P}, \tau_r)$ is a measure-preserving dynamical system. We will show

$$(\Omega, \mathcal{A}, \mathbf{P}, \tau_r) \text{ is ergodic} \quad \Longleftrightarrow \quad r \text{ is irrational.}$$

Let $f : [0, 1) \to \mathbb{R}$ be an $\mathcal{I}$-measurable function. Without loss of generality, we assume that f is bounded and thus square integrable. Hence f can be expanded in a Fourier series

$$f(x) = \sum_{n=-\infty}^{\infty} c_n e^{2\pi i n x} \quad \text{for } \mathbf{P}\text{-a.a. } x.$$

This series converges in L^2, and the sequence of square summable coefficients $(c_n)_{n \in \mathbb{Z}}$ is unique (compare Exercise 7.3.1 with $c_n = (-i/2)a_n + (1/2)b_n$ and $c_{-n} = (i/2)a_n + (1/2)b_n$ for $n \in \mathbb{N}$ as well as $c_0 = b_0$). Now we compute

$$(f \circ \tau_r)(x) = \sum_{n=-\infty}^{\infty} \left(c_n e^{2\pi i n r} \right) e^{2\pi i n x} \quad \text{a.e.}$$

By Lemma 20.7, f is $\mathcal{I}$-measurable if and only if $f = f \circ \tau_r$; that is, if and only if

$$c_n = c_n e^{2\pi i n r} \quad \text{for all } n \in \mathbb{Z}.$$

If r is irrational, this implies $c_n = 0$ for $n \neq 0$, and thus f is almost surely constant. Therefore, $(\Omega, \mathcal{A}, \mathbf{P}, \tau_r)$ is ergodic.

On the other hand, if r is rational, then there exists some $n \in \mathbb{Z} \setminus \{0\}$ with $e^{2\pi i n r} = e^{-2\pi i n r} = 1$. Hence $x \mapsto e^{2\pi i n x} + e^{-2\pi i n x} = 2\cos(2\pi n x)$ is $\mathcal{I}$-measurable but not a.s. constant. Thus, in this case $(\Omega, \mathcal{A}, \mathbf{P}, \tau_r)$ is not ergodic. $\Diamond$

Example 20.10 Let $X = (X_n)_{n \in \mathbb{N}_0}$ be a stochastic process with values in a Polish space E. Without loss of generality, we may assume that X is the canonical process on the probability space $(\Omega, \mathcal{A}, \mathbf{P}) = \left(E^{\mathbb{N}_0}, \mathcal{B}(E)^{\otimes \mathbb{N}_0}, \mathbf{P} \right)$. Define the **shift** operator

$$\tau : \Omega \to \Omega, \quad (\omega_n)_{n \in \mathbb{N}_0} \mapsto (\omega_{n+1})_{n \in \mathbb{N}_0}.$$

Then $X_n(\omega) = X_0(\tau^n(\omega))$. Hence X is stationary if and only if $(\Omega, \mathcal{A}, \mathbf{P}, \tau)$ is a measure-preserving dynamical system. $\Diamond$

Definition 20.11 *The stochastic process X (from Example 20.10) is called **ergodic** if $(\Omega, \mathcal{A}, \mathbf{P}, \tau)$ is ergodic.*

Example 20.12 Let $(X_n)_{n \in \mathbb{N}_0}$ be i.i.d. and let $X_n(\omega) = X_0(\tau^n(\omega))$. If $A \in \mathcal{I}$, then, for every $n \in \mathbb{N}$,

$$A = \tau^{-n}(A) = \{\omega : \tau^n(\omega) \in A\} \in \sigma(X_n, X_{n+1}, \ldots).$$

Hence, if we let $\mathcal{T}$ be the tail σ-algebra of $(X_n)_{n \in \mathbb{N}}$ (see Definition 2.34), then

$$\mathcal{I} \subset \mathcal{T} = \bigcap_{n=1}^{\infty} \sigma(X_n, X_{n+1}, \ldots).$$

By Kolmogorov's 0-1 law (Theorem 2.37), $\mathcal{T}$ is **P**-trivial. Hence $\mathcal{I}$ is also **P**-trivial and therefore $(X_n)_{n \in \mathbb{N}_0}$ is ergodic. $\Diamond$

Takeaways A measure preserving dynamical system consists of a probability space $(\Omega, \mathcal{A}, \mathbf{P})$ and a measure preserving map $\tau : \Omega \to \Omega$. It is called ergodic if there are no nontrivial (w.r.t. **P**) invariant sets. Examples for measure preserving dynamical systems are stationary stochastic processes, i.i.d. random variables, rotations and so on.

Exercise 20.1.1 Let G be a finite group of measure-preserving measurable maps on $(\Omega, \mathcal{A}, \mathbf{P})$ and let $\mathcal{A}_0 := \{A \in \mathcal{A} : g(A) = A \text{ for all } g \in G\}$.
   Show that, for every $X \in \mathcal{L}^1(\mathbf{P})$, we have

$$\mathbf{E}[X \mid \mathcal{A}_0] = \frac{1}{\#G} \sum_{g \in G} X \circ g. \quad \clubsuit$$

20.2 Ergodic Theorems

In this section, $(\Omega, \mathcal{A}, \mathbf{P}, \tau)$ always denotes a measure-preserving dynamical system. Further, let $f : \Omega \to \mathbb{R}$ be measurable and

$$X_n(\omega) = f \circ \tau^n(\omega) \quad \text{for all } n \in \mathbb{N}_0.$$

Hence $X = (X_n)_{n \in \mathbb{N}_0}$ is a stationary real-valued stochastic process. Let

$$S_n = \sum_{k=0}^{n-1} X_k$$

denote the nth partial sum. Ergodic theorems are laws of large numbers for $(S_n)_{n \in \mathbb{N}}$. We start with a preliminary lemma.

Lemma 20.13 (Hopf's maximal-ergodic lemma) *Let $X_0 \in \mathcal{L}^1(\mathbf{P})$. Define $M_n = \max\{0, S_1, \ldots, S_n\}$, $n \in \mathbb{N}$. Then*

$$\mathbf{E}\left[X_0 \, \mathbb{1}_{\{M_n > 0\}}\right] \geq 0 \quad \text{for every } n \in \mathbb{N}.$$

Proof For $k \leq n$, we have $M_n(\tau(\omega)) \geq S_k(\tau(\omega))$. Hence

$$X_0 + M_n \circ \tau \;\geq\; X_0 + S_k \circ \tau \;=\; S_{k+1}.$$

Thus $X_0 \geq S_{k+1} - M_n \circ \tau$ for $k = 1, \ldots, n$. Manifestly, $S_1 = X_0$ and $M_n \circ \tau \geq 0$ and hence also (for $k = 0$) $X_0 \geq S_1 - M_n \circ \tau$. Therefore,

$$X_0 \;\geq\; \max\{S_1, \ldots, S_n\} - M_n \circ \tau. \tag{20.2}$$

Furthermore, we have

$$\{M_n > 0\}^c \;\subset\; \{M_n = 0\} \cap \{M_n \circ \tau \geq 0\} \;\subset\; \{M_n - M_n \circ \tau \leq 0\}. \tag{20.3}$$

By (20.2) and (20.3), and since τ is measure-preserving, we conclude that

$$\begin{aligned}
\mathbf{E}\left[X_0 \, \mathbb{1}_{\{M_n > 0\}}\right] &\geq \mathbf{E}\left[(\max\{S_1, \ldots, S_n\} - M_n \circ \tau)\, \mathbb{1}_{\{M_n > 0\}}\right] \\
&= \mathbf{E}\left[(M_n - M_n \circ \tau)\, \mathbb{1}_{\{M_n > 0\}}\right] \\
&\geq \mathbf{E}\left[M_n - M_n \circ \tau\right] = \mathbf{E}[M_n] - \mathbf{E}[M_n] = 0. \qquad \square
\end{aligned}$$

Theorem 20.14 (Individual ergodic theorem, Birkhoff [16]) *Let $f = X_0 \in \mathcal{L}^1(\mathbf{P})$. Then*

$$\frac{1}{n}\sum_{k=0}^{n-1} X_k = \frac{1}{n}\sum_{k=0}^{n-1} f \circ \tau^k \;\overset{n\to\infty}{\longrightarrow}\; \mathbf{E}[X_0 \,|\, \mathcal{I}] \quad \mathbf{P}\text{-a.s.}$$

In particular, if τ is ergodic, then $\dfrac{1}{n}\sum\limits_{k=0}^{n-1} X_k \;\overset{n\to\infty}{\longrightarrow}\; \mathbf{E}[X_0]$ *$\mathbf{P}$-a.s.*

Proof If τ is ergodic, then $\mathbf{E}[X_0 \,|\, \mathcal{I}] = \mathbf{E}[X_0]$ and the supplement is a consequence of the first statement.

Consider now the general case. By Lemma 20.7, we have $\mathbf{E}[X_0 \,|\, \mathcal{I}] \circ \tau = \mathbf{E}[X_0 \,|\, \mathcal{I}]$ $\mathbf{P}$-a.s. Hence, by passing to $\widetilde{X}_n := X_n - \mathbf{E}[X_0 \,|\, \mathcal{I}]$, without loss of generality, we can assume $\mathbf{E}[X_0 \,|\, \mathcal{I}] = 0$. Define

$$Z := \limsup_{n\to\infty} \frac{1}{n} S_n.$$

Let $\varepsilon > 0$ and $F := \{Z > \varepsilon\}$. We have to show that $\mathbf{P}[F] = 0$. From this we infer $\mathbf{P}[Z > 0] = 0$ and similarly (with $-X$ instead of X) also $\liminf\limits_{n \to \infty} \frac{1}{n} S_n \geq 0$ almost surely. Hence $\frac{1}{n} S_n \overset{n \to \infty}{\longrightarrow} 0$ a.s.

Evidently, $Z \circ \tau = Z$; hence $F \in \mathcal{I}$. Define

$$X_n^\varepsilon := (X_n - \varepsilon)\,\mathbb{1}_F, \qquad\qquad S_n^\varepsilon := X_0^\varepsilon + \ldots + X_{n-1}^\varepsilon,$$

$$M_n^\varepsilon := \max\{0,\, S_1^\varepsilon, \ldots, S_n^\varepsilon\}, \qquad F_n := \{M_n^\varepsilon > 0\}.$$

Then $F_1 \subset F_2 \subset \ldots$ and

$$\bigcup_{n=1}^\infty F_n = \left\{\sup_{k \in \mathbb{N}} \frac{1}{k} S_k^\varepsilon > 0\right\} = \left\{\sup_{k \in \mathbb{N}} \frac{1}{k} S_k > \varepsilon\right\} \cap F = F,$$

hence $F_n \uparrow F$. Dominated convergence yields $\mathbf{E}\left[X_0^\varepsilon \mathbb{1}_{F_n}\right] \overset{n \to \infty}{\longrightarrow} \mathbf{E}\left[X_0^\varepsilon\right]$.

By the maximal-ergodic lemma (applied to X^ε), we have $\mathbf{E}\left[X_0^\varepsilon \mathbb{1}_{F_n}\right] \geq 0$; hence

$$0 \leq \mathbf{E}\left[X_0^\varepsilon\right] = \mathbf{E}\left[(X_0 - \varepsilon)\,\mathbb{1}_F\right] = \mathbf{E}\left[\mathbf{E}[X_0 | \mathcal{I}]\,\mathbb{1}_F\right] - \varepsilon \mathbf{P}[F] = -\varepsilon \mathbf{P}[F].$$

We conclude that $\mathbf{P}[F] = 0$. □

As a consequence, we obtain the statistical ergodic theorem, or L^p-ergodic theorem, that was found by von Neumann in 1931 right before Birkhoff proved his ergodic theorem, but was published only later in [122]. Before we formulate it, we state one more lemma.

Lemma 20.15 *Let $p \geq 1$ and let $X_0, X_1, \ldots$ be identically distributed, real random variables with $\mathbf{E}[|X_0|^p] < \infty$. Define $Y_n := \left|\frac{1}{n} \sum\limits_{k=0}^{n-1} X_k\right|^p$ for $n \in \mathbb{N}$. Then $(Y_n)_{n \in \mathbb{N}}$ is uniformly integrable.*

Proof Evidently, the singleton $\{|X_0|^p\}$ is uniformly integrable. Hence, by Theorem 6.19, there exists a monotone increasing convex map $f : [0, \infty) \to [0, \infty)$ with $\frac{f(x)}{x} \to \infty$ for $x \to \infty$ and $C := \mathbf{E}[f(|X_0|^p)] < \infty$. Again, by Theorem 6.19, it is enough to show that $\mathbf{E}[f(Y_n)] \leq C$ for every $n \in \mathbb{N}$. By Jensen's inequality (for $x \mapsto |x|^p$), we have

$$Y_n \leq \frac{1}{n} \sum_{k=0}^{n-1} |X_k|^p.$$

Again, by Jensen's inequality (now applied to f), we get that

$$f(Y_n) \leq f\left(\frac{1}{n}\sum_{k=0}^{n-1}|X_k|^p\right) \leq \frac{1}{n}\sum_{k=0}^{n-1}f(|X_k|^p).$$

Hence $\mathbf{E}[f(Y_n)] \leq \frac{1}{n}\sum_{k=0}^{n-1}\mathbf{E}[f(|X_k|^p)] = C.$ □

Theorem 20.16 (L^p-**ergodic theorem, von Neumann (1931)**) *Let* $(\Omega, \mathcal{A}, \mathbf{P}, \tau)$ *be a measure-preserving dynamical system, $p \geq 1$, $X_0 \in \mathcal{L}^p(\mathbf{P})$ and $X_n = X_0 \circ \tau^n$. Then*

$$\frac{1}{n}\sum_{k=0}^{n-1}X_k \stackrel{n\to\infty}{\longrightarrow} \mathbf{E}[X_0|\mathcal{I}] \quad in \; L^p(\mathbf{P}).$$

In particular, if τ is ergodic, then $\frac{1}{n}\sum_{k=0}^{n-1}X_k \stackrel{n\to\infty}{\longrightarrow} \mathbf{E}[X_0]$ *in* $L^p(\mathbf{P})$.

Proof Define

$$Y_n := \left|\frac{1}{n}\sum_{k=0}^{n-1}X_k - \mathbf{E}[X_0|\mathcal{I}]\right|^p \quad \text{for every } n \in \mathbb{N}.$$

By Lemma 20.15, $(Y_n)_{n\in\mathbb{N}}$ is uniformly integrable, and by Birkhoff's ergodic theorem, we have $Y_n \stackrel{n\to\infty}{\longrightarrow} 0$ almost surely. By Theorem 6.25, we thus have $\lim_{n\to\infty}\mathbf{E}[Y_n] = 0$.

If τ is ergodic, then $\mathbf{E}[X_0|\mathcal{I}] = \mathbf{E}[X_0]$. □

Takeaways For ergodic dynamical systems, averages over ω and averages over trajectories coincide (ergodic theorems). In general dynamical systems, a similar statement is true if we replace the average over ω by the conditional expectation given the σ-algebra of invariant events.

20.3 Examples

Example 20.17 Let $(X, (\mathbf{P}_x)_{x\in E})$ be a positive recurrent, irreducible Markov chain on the countable space E. Let π be the invariant distribution of X. Then $\pi(\{x\}) > 0$ for every $x \in E$. Define $\mathbf{P}_\pi = \sum_{x\in E}\pi(\{x\})\mathbf{P}_x$. Then X is stationary on $(\Omega, \mathcal{A}, \mathbf{P}_\pi)$. Denote the shift by τ; that is, $X_n = X_0 \circ \tau^n$.

Now let $A \in \mathcal{I}$ be invariant. Then $A \in \mathcal{T} = \bigcap\limits_{n=1}^{\infty} \sigma(X_n, X_{n+1}, \ldots)$. By the strong Markov property, for every finite stopping time σ (recall that $\mathcal{F}_\sigma$ is the σ-algebra of the σ-past),

$$\mathbf{P}_\pi[X \in A \mid \mathcal{F}_\sigma] = \mathbf{P}_{X_\sigma}[X \in A]. \tag{20.4}$$

Indeed, we have $\{X \in A\} = \{X \in \tau^{-n}(A)\} = \{(X_n, X_{n+1}, \ldots) \in A\}$. For $B \in \mathcal{F}_\sigma$, using the Markov property (in the third line), we get

$$\mathbf{E}_\pi\left[\mathbb{1}_{\{X \in B\}}\, \mathbb{1}_{\{X \in A\}}\right] = \sum_{n=0}^{\infty} \sum_{x \in E} \mathbf{P}_\pi\left[X \in B,\ \sigma = n,\ X_n = x,\ X \in A\right]$$

$$= \sum_{n=0}^{\infty} \sum_{x \in E} \mathbf{P}_\pi\left[X \in B,\ \sigma = n,\ X_n = x,\ X \circ \tau^n \in A\right]$$

$$= \sum_{n=0}^{\infty} \sum_{x \in E} \mathbf{P}_\pi\left[X \in B,\ \sigma = n,\ X_n = x\right] \mathbf{P}_x[X \in A]$$

$$= \mathbf{E}_\pi\left[\mathbb{1}_{\{X \in B\}}\, \mathbf{P}_{X_\sigma}[X \in A]\right].$$

In particular, if $x \in E$ and $\sigma_x = \inf\{n \in \mathbb{N}_0 : X_n = x\}$, then $\sigma_x < \infty$ since X is recurrent and irreducible. By (20.4), we conclude that, for every $x \in E$,

$$\mathbf{P}_\pi[X \in A] \;=\; \mathbf{E}_\pi\left[\mathbf{P}_x[X \in A]\right] \;=\; \mathbf{P}_x[X \in A].$$

Hence $\mathbf{P}_{X_n}[X \in A] = \mathbf{P}_\pi[X \in A]$ almost surely and thus (with $\sigma = n$ in (20.4))

$$\mathbf{P}_\pi[X \in A \mid X_0, \ldots, X_n] \;=\; \mathbf{P}_{X_n}[X \in A] \;=\; \mathbf{P}_\pi[X \in A].$$

Now $A \in \mathcal{I} \subset \sigma(X_1, X_2, \ldots)$; hence

$$\mathbf{P}_\pi[X \in A \mid X_0, \ldots, X_n] \overset{n \to \infty}{\longrightarrow} \mathbf{P}_\pi[X \in A \mid \sigma(X_0, X_1, \ldots)] \;=\; \mathbb{1}_{\{X \in A\}}.$$

This implies $\mathbf{P}_\pi[X \in A] \in \{0, 1\}$. Hence X is ergodic.

Birkhoff's ergodic theorem now implies that, for every $x \in E$,

$$\frac{1}{n} \sum_{k=0}^{n-1} \mathbb{1}_{\{X_k = x\}} \overset{n \to \infty}{\longrightarrow} \pi(\{x\}) \quad \mathbf{P}_\pi\text{-a.s.}$$

In this sense, $\pi(\{x\})$ is the average time X spends in x in the long run. $\Diamond$

Example 20.18 Let P and Q be probability measures on the measurable space $(\Omega, \mathcal{A})$, and let $(\Omega, \mathcal{A}, P, \tau)$ and $(\Omega, \mathcal{A}, Q, \tau)$ be ergodic. Then either $P = Q$ or $P \perp Q$. Indeed, if $P \neq Q$, then there exists an f with $|f| \leq 1$ and $\int f \, dP \neq \int f \, dQ$. However, by Birkhoff's ergodic theorem,

$$\frac{1}{n} \sum_{k=0}^{n-1} f \circ \tau^k \overset{n \to \infty}{\longrightarrow} \begin{cases} \int f \, dP & P\text{-a.s.,} \\ \int f \, dQ & Q\text{-a.s.} \end{cases}$$

If we define $A := \left\{ \frac{1}{n} \sum_{k=0}^{n-1} f \circ \tau^k \overset{n \to \infty}{\longrightarrow} \int f \, dP \right\}$, then $P(A) = 1$ and $Q(A) = 0$. Thus $P \perp Q$. $\Diamond$

> **Takeaways** The ergodic theorem yields a law of large numbers for the occupation times of a positive recurrent Markov chain. Furthermore, it shows that two ergodic measures are either equal or mutually singular.

Exercise 20.3.1 Let $(\Omega, \mathcal{A})$ be a measurable space and let $\tau : \Omega \to \Omega$ be a measurable map.

(i) Show that the set $\mathcal{M} := \{ \mu \in \mathcal{M}_1(\Omega) : \mu \circ \tau^{-1} = \mu \}$ of τ-invariant measures is convex.
(ii) An element μ of $\mathcal{M}$ is called *extremal* if $\mu = \lambda \mu_1 + (1 - \lambda) \mu_2$ for some $\mu_1, \mu_2 \in \mathcal{M}$ and $\lambda \in (0, 1)$ implies $\mu = \mu_1 = \mu_2$. Show that $\mu \in \mathcal{M}$ is extremal if and only if τ is ergodic with respect to μ. ♣

Exercise 20.3.2 Let $p = 2, 3, 5, 6, 7, 10, \ldots$ be square-free (that is, there is no number $r = 2, 3, 4, \ldots$, whose square is a divisor of p) and let $q \in \{2, 3, \ldots, p-1\}$. For every $n \in \mathbb{N}$, let a_n be the leading digit of the p-adic expansion of q^n.

Show the following version of Benford's law: For every $d \in \{1, \ldots, p - 1\}$,

$$\frac{1}{n} \#\{ i \leq n : a_i = d \} \overset{n \to \infty}{\longrightarrow} \frac{\log(d + 1) - \log(d)}{\log(p)}. \quad ♣$$

20.4   Application: Recurrence of Random Walks

Let $(X_n)_{n \in \mathbb{N}}$ be a stationary process with values in $\mathbb{R}^d$. Define $S_n := \sum_{k=1}^{n} X_k$ for $n \in \mathbb{N}_0$. Further, let

$$R_n = \#\{S_1, \ldots, S_n\}$$

denote the **range** of S; that is, the number of distinct points visited by S up to time n. Finally, let $A := \{S_n \neq 0$ for every $n \in \mathbb{N}\}$ be the event of an "escape" from 0.

Theorem 20.19 *We have* $\lim\limits_{n\to\infty} \dfrac{1}{n} R_n = \mathbf{P}[A \,|\, \mathcal{I}]$ *almost surely.*

Proof Let X be the canonical process on $(\Omega, \mathcal{A}, \mathbf{P}) = \big((\mathbb{R}^d)^{\mathbb{N}}, \mathcal{B}(\mathbb{R}^d)^{\otimes \mathbb{N}}, \mathbf{P}\big)$ and let $\tau : \Omega \to \Omega$ be the shift; that is, $X_n = X_0 \circ \tau^n$.
Evidently,

$$R_n = \#\big\{k \leq n : S_l \neq S_k \text{ for all } l \in \{k+1, \ldots, n\}\big\}$$

$$\geq \#\big\{k \leq n : S_l \neq S_k \text{ for all } l > k\big\}$$

$$= \sum_{k=1}^{n} \mathbb{1}_A \circ \tau^k.$$

Birkhoff's ergodic theorem yields

$$\liminf_{n\to\infty} \frac{1}{n} R_n \geq \mathbf{P}[A \,|\, \mathcal{I}] \quad \text{a.s.} \tag{20.5}$$

For the converse inequality, consider $A_m = \{S_l \neq 0 \text{ for } l = 1, \ldots, m\}$. Then, for every $n \geq m$,

$$R_n \leq m + \#\big\{k \leq n - m : S_l \neq S_k \text{ for all } l \in \{k+1, \ldots, n\}\big\}$$

$$\leq m + \#\big\{k \leq n - m : S_l \neq S_k \text{ for all } l \in \{k+1, \ldots, k+m\}\big\}$$

$$= m + \sum_{k=1}^{n-m} \mathbb{1}_{A_m} \circ \tau^k.$$

Again, by the ergodic theorem,

$$\limsup_{n\to\infty} \frac{1}{n} R_n \leq \mathbf{P}[A_m \,|\, \mathcal{I}] \quad \text{a.s.} \tag{20.6}$$

Since $A_m \downarrow A$ and $\mathbf{P}[A_m \,|\, \mathcal{I}] \overset{m\to\infty}{\longrightarrow} \mathbf{P}[A \,|\, \mathcal{I}]$ almost surely (by Theorem 8.14(8.14)), the claim follows from (20.5) and (20.6). $\qquad\square$

Theorem 20.20 *Let* $X = (X_n)_{n\in\mathbb{N}}$ *be an integer-valued, integrable, stationary process with the property* $\mathbf{E}[X_1 \,|\, \mathcal{I}] = 0$ *a.s. Let* $S_n = X_1 + \ldots + X_n$, $n \in \mathbb{N}$. *Then*

$$\mathbf{P}\big[S_n = 0 \text{ for infinitely many } n \in \mathbb{N}\big] = 1.$$

In particular, a random walk on $\mathbb{Z}$ *with centered increments is recurrent (Chung–Fuchs theorem, compare Theorem 17.41).*

Proof Define $A = \{S_n \neq 0 \text{ for all } n \in \mathbb{N}\}$.

Step 1. We show $\mathbf{P}[A] = 0$. (If X is i.i.d., then S is a Markov chain, and this implies immediately that 0 is recurrent. Only for the more general case of stationary X do we need an additional argument.) By the ergodic theorem, we have $\frac{1}{n} S_n \overset{n \to \infty}{\longrightarrow}$ $\mathbf{E}[X_1 \,|\, \mathcal{I}] = 0$ a.s. Thus, for every $m \in \mathbb{N}$,

$$\limsup_{n \to \infty} \left(\frac{1}{n} \max_{k=1,\ldots,n} |S_k| \right) = \limsup_{n \to \infty} \left(\frac{1}{n} \max_{k=m,\ldots,n} |S_k| \right)$$

$$\leq \max_{k \geq m} \frac{|S_k|}{k} \overset{m \to \infty}{\longrightarrow} 0.$$

Therefore,

$$\lim_{n \to \infty} \left(\frac{1}{n} \max_{k=1,\ldots,n} S_k \right) = \lim_{n \to \infty} \left(\frac{1}{n} \min_{k=1,\ldots,n} S_k \right) = 0.$$

Now $R_n \leq 1 + \left(\max_{k=1,\ldots,n} S_k \right) - \left(\min_{k=1,\ldots,n} S_k \right)$; hence $\frac{1}{n} R_n \overset{n \to \infty}{\longrightarrow} 0$. By Theorem 20.19, this implies $\mathbf{P}[A] = 0$.

Step 2. Define $\sigma_n := \inf\{m \in \mathbb{N} : S_{m+n} = S_n\}$, $B_n := \{\sigma_n < \infty\}$ for $n \in \mathbb{N}_0$ and

$$B := \bigcap_{n=0}^{\infty} B_n.$$

Since $\{\sigma_0 = \infty\} = A$, we have $\mathbf{P}[\sigma_0 < \infty] = 1$. By stationarity, $\mathbf{P}[\sigma_n < \infty] = 1$ for every $n \in \mathbb{N}_0$; hence $\mathbf{P}[B] = 1$.

Let $\tau_0 = 0$ and inductively define $\tau_{n+1} = \tau_n + \sigma_{\tau_n}$ for $n \in \mathbb{N}_0$. Then τ_n is the time of the nth return of S to 0. On B we have $\tau_n < \infty$ for every $n \in \mathbb{N}_0$ and hence

$$\mathbf{P}\big[S_n = 0 \text{ infinitely often}\big] = \mathbf{P}\big[\tau_n < \infty \text{ for all } n \in \mathbb{N}\big] \geq \mathbf{P}[B] = 1. \qquad \square$$

If in Theorem 20.20 the random variables X_n are not integer-valued, then there is no hope that $S_n = 0$ for any $n \in \mathbb{N}$ with positive probability. On the other hand, in this case, there is also some kind of recurrence property, namely $S_n/n \overset{n \to \infty}{\longrightarrow} 0$ almost surely by the ergodic theorem. Note, however, that this does not exclude the possibility that $S_n \overset{n \to \infty}{\longrightarrow} \infty$ with positive probability; for instance, if S_n grows like $\sqrt{n}$. The next theorem shows that if the X_n are integrable, then the process of partial sums can go to infinity only with a linear speed.

Theorem 20.21 *Let* $(X_n)_{n \in \mathbb{N}}$ *be an integrable ergodic process and define* $S_n = X_1 + \ldots + X_n$ *for* $n \in \mathbb{N}_0$. *Then the following statements are equivalent.*

(i) $S_n \overset{n\to\infty}{\longrightarrow} \infty$ almost surely.

(ii) $\mathbf{P}\left[S_n \overset{n\to\infty}{\longrightarrow} \infty \right] > 0.$

(iii) $\lim\limits_{n\to\infty} \dfrac{S_n}{n} = \mathbf{E}[X_1] > 0$ almost surely.

If the random variables $X_1, X_2, \ldots$ are i.i.d. with $\mathbf{E}[X_1] = 0$ and $\mathbf{P}[X_1 = 0] < 1$, then $\liminf_{n\to\infty} S_n = -\infty$ and $\limsup_{n\to\infty} S_n = \infty$ almost surely.

Proof "(i) $\Longleftrightarrow$ (ii)"   Clearly, $B := \{S_n \overset{n\to\infty}{\longrightarrow} \infty\}$ is an invariant event and thus has probability either 0 or 1.

"(iii) $\Longrightarrow$ (i)"   This is trivial.

"(i) $\Longrightarrow$ (iii)"   The equality follows by the individual ergodic theorem. Hence, it is enough to show that $\liminf_{n\to\infty} S_n/n > 0$ almost surely.

Note that $\mathbf{P}[B] = 1$. For $n \in \mathbb{N}_0$ and $\varepsilon > 0$, let

$$A_n^\varepsilon := \big\{ S_m > S_n + \varepsilon \ \text{ for all } \ m \geq n + 1 \big\} \cap B.$$

Let $S^- := \inf\{S_n : n \in \mathbb{N}_0\}$. By assumption (i), we have $S^- > -\infty$ almost surely and $\tau := \sup\{n \in \mathbb{N}_0 : S_n = S^-\}$ is finite almost surely. Hence there is an $N \in \mathbb{N}$ with $\mathbf{P}[\tau < N] \geq \frac{1}{2}$. Therefore,

$$\mathbf{P}\left[\bigcup_{n=0}^{N-1} A_n^0 \right] = \mathbf{P}[\tau < N] \geq \frac{1}{2}.$$

Since $A_n^\varepsilon \uparrow A_n^0$ for $\varepsilon \downarrow 0$, there is an $\varepsilon > 0$ with $p := \mathbf{P}[A_0^\varepsilon] \geq \frac{1}{4N} > 0$. As $(X_n)_{n\in\mathbb{N}}$ is ergodic, $\big(\mathbb{1}_{A_n^\varepsilon}\big)_{n\in\mathbb{N}_0}$ is also ergodic. By the individual ergodic theorem, we conclude that $\frac{1}{n}\sum_{i=0}^{n-1} \mathbb{1}_{A_i^\varepsilon} \overset{n\to\infty}{\longrightarrow} p$ almost surely. Hence there exists an $n_0 = n_0(\omega)$ such that $\sum_{i=0}^{n-1} \mathbb{1}_{A_i^\varepsilon} \geq \frac{pn}{2}$ for all $n \geq n_0$. This implies $S_n \geq S^- + \frac{pn\varepsilon}{2}$ for $n \geq n_0$ and hence $\liminf_{n\to\infty} S_n/n \geq \frac{p\varepsilon}{2} > 0$.

Now we show the additional statement. Assume that $\mathbf{E}[X_1] = 0$ and $\mathbf{P}[X_1 = 0] < 1$. Hence, there is an $\varepsilon > 0$ such that $\mathbf{P}[X_1 < -2\varepsilon] > \varepsilon$. Let $L := \liminf_{n\to\infty} S_n$. As shown above, we have $\mathbf{P}[L = \infty] = 0$. The event $\{L > -\infty\}$ is invariant and hence has probability 0 or 1. We assume $\mathbf{P}[L > -\infty] = 1$ and construct a contradiction. Inductively, define stopping times $\tau_1 := \inf\{k \in \mathbb{N} : S_k < L + \varepsilon\}$ and

$$\tau_{n+1} := \inf\{k > \tau_n : S_k < L + \varepsilon\} \quad \text{ for } n \in \mathbb{N}.$$

By assumption, we have $\tau_n < \infty$ almost surely for all n. Let $\mathbb{F} = (\mathcal{F}_n)_{n\in\mathbb{N}_0} = \sigma((X_n)_{n\in\mathbb{N}})$ be the filtration generated by $X = (X_n)_{n\in\mathbb{N}}$. Let $\mathcal{F}_{\tau_n}$ be the σ-algebra of τ_n-past. Define the events $A_n := \{X_{\tau_n+1} < -2\varepsilon\}$, $n \in \mathbb{N}$. On A_n, we have $S_{\tau_n+1} < L - \varepsilon$. Clearly, A_n is independent of $\mathcal{F}_{\tau_n}$ and thus

$$\mathbf{P}\big[A_n \,\big|\, \mathcal{F}_{\tau_n}\big] = \mathbf{P}[A_n] > \varepsilon.$$

By the conditional version of the Borel-Cantelli Lemma (see Exercise 11.2.7), we infer

$$\mathbf{P}\Big[\limsup_{n\to\infty} A_n\Big] = 1.$$

This shows that almost surely $S_{\tau_n+1} < L - \varepsilon$ infinitely often and this in turn contradicts the assumption that L be finite.

Consequently, we have $\mathbf{P}[\liminf S_n = -\infty] = 1$. The statement for $\limsup S_n$ is similar.   □

Remark 20.22 It can be shown that Theorem 20.21 holds also without the assumption that the X_n are integrable. See [93]. ◊

Takeaways The set of points visited by a random walk within the first n steps grows with n at a speed that is the probability of no return. Hence, in the recurrent case, the set grows sublinearly. For a random walk on $\mathbb{Z}$ with a finite first moment, this shows that it is recurrent if and only if the increments are centred. Consequently, for such a random walk, one out of three alternatives holds: (i) The random walk goes to ∞ at positive speed. (ii) The random walk goes to $-\infty$ at positive speed. (iii) The random walk oscillates around 0 with a growing amplitude. It is impossible that the random walk would go to ∞ (or $-\infty$) slower than linearly.

20.5   Mixing

Ergodicity provides a weak notion of "independence" or "mixing". At the other end of the scale, the strongest notion is "i.i.d.". Here we are concerned with notions of mixing that lie between these two.

In the following, we always assume that $(\Omega, \mathcal{A}, \mathbf{P}, \tau)$ is a measure-preserving dynamical system and that $X_n := X_0 \circ \tau^n$. We start with a simple observation.

Theorem 20.23 $(\Omega, \mathcal{A}, \mathbf{P}, \tau)$ *is ergodic if and only if, for all $A, B \in \mathcal{A}$,*

$$\lim_{n\to\infty} \frac{1}{n} \sum_{k=0}^{n-1} \mathbf{P}\Big[A \cap \tau^{-k}(B)\Big] = \mathbf{P}[A]\,\mathbf{P}[B]. \tag{20.7}$$

Proof "⟹" Let $(\Omega, \mathcal{A}, \mathbf{P}, \tau)$ be ergodic. Define

$$Y_n := \frac{1}{n} \sum_{k=0}^{n-1} \mathbb{1}_{\tau^{-k}(B)} = \frac{1}{n} \sum_{k=0}^{n-1} \mathbb{1}_B \circ \tau^k.$$

By Birkhoff's ergodic theorem, we have $Y_n \overset{n\to\infty}{\longrightarrow} \mathbf{P}[B]$ almost surely. Hence $Y_n \mathbb{1}_A \overset{n\to\infty}{\longrightarrow} \mathbb{1}_A \mathbf{P}[B]$ almost surely. Dominated convergence yields

$$\frac{1}{n} \sum_{k=0}^{n-1} \mathbf{P}\left[A \cap \tau^{-k}(B)\right] = \mathbf{E}[Y_n \mathbb{1}_A] \overset{n\to\infty}{\longrightarrow} \mathbf{E}[\mathbb{1}_A \mathbf{P}[B]] = \mathbf{P}[A] \mathbf{P}[B].$$

"⟸" Now assume that (20.7) holds. Let $A \in \mathcal{I}$ (recall that $\mathcal{I}$ is the invariant σ-algebra) and $B = A$. Evidently, $A \cap \tau^{-k}(A) = A$ for every $k \in \mathbb{N}_0$. Hence, by (20.7),

$$\mathbf{P}[A] = \frac{1}{n} \sum_{k=0}^{n-1} \mathbf{P}\left[A \cap \tau^{-k}(A)\right] \overset{n\to\infty}{\longrightarrow} \mathbf{P}[A]^2.$$

Thus $\mathbf{P}[A] \in \{0, 1\}$; hence $\mathcal{I}$ is trivial and therefore τ is ergodic. □

We consider a strengthening of (20.7).

Definition 20.24 *A measure-preserving dynamical system* $(\Omega, \mathcal{A}, \mathbf{P}, \tau)$ *is called* **mixing** *if*

$$\lim_{n\to\infty} \mathbf{P}\left[A \cap \tau^{-n}(B)\right] = \mathbf{P}[A] \mathbf{P}[B] \quad \text{for all } A, B \in \mathcal{A}. \tag{20.8}$$

Remark 20.25 Sometimes the mixing property of (20.8) is called **strongly mixing**, in contrast with a **weakly mixing** system $(\Omega, \mathcal{A}, \mathbf{P}, \tau)$, for which we require only

$$\lim_{n\to\infty} \frac{1}{n} \sum_{i=0}^{n-1} \left|\mathbf{P}\left[A \cap \tau^{-i}(B)\right] - \mathbf{P}[A] \mathbf{P}[B]\right| = 0 \quad \text{for all } A, B \in \mathcal{A}.$$

"Strongly mixing" implies "weakly mixing" (see Exercise 20.5.1). On the other hand, there exist weakly mixing systems that are not strongly mixing (see [81]). ◊

Example 20.26 Let $I = \mathbb{N}_0$ or $I = \mathbb{Z}$, and let $(X_n)_{n\in I}$ be an i.i.d. sequence with values in the measurable space $(E, \mathcal{E})$. Hence τ is the shift on the product space $\Omega = E^I, \mathbf{P} = (\mathbf{P}_{X_0})^{\otimes I}$. Let $A, B \in \mathcal{E}^{\otimes I}$. For every $\varepsilon > 0$, there exist events A^ε and B^ε that depend on only finitely many coordinates and such that $\mathbf{P}[A \triangle A^\varepsilon] < \varepsilon$ and $\mathbf{P}[B \triangle B^\varepsilon] < \varepsilon$. Clearly, $\mathbf{P}[\tau^{-n}(A \triangle A^\varepsilon)] < \varepsilon$ and $\mathbf{P}[\tau^{-n}(B \triangle B^\varepsilon)] < \varepsilon$ for every $n \in \mathbb{Z}$. For sufficiently large $|n|$, the sets A^{ε} and $\tau^{-n}(B^{\varepsilon})$ depend on different coordinates and are thus independent. This implies

$$\limsup_{|n| \to \infty} \left| \mathbf{P}[A \cap \tau^{-n}(B)] - \mathbf{P}[A]\,\mathbf{P}[B] \right|$$

$$\leq \limsup_{|n| \to \infty} \left| \mathbf{P}[A^{\varepsilon} \cap \tau^{-n}(B^{\varepsilon})] - \mathbf{P}[A^{\varepsilon}]\,\mathbf{P}[B^{\varepsilon}] \right| + 4\varepsilon = 4\varepsilon.$$

Hence τ is mixing. Letting $A = B \in \mathcal{I}$, we obtain the 0-1 law for invariant events: $\mathbf{P}[A] \in \{0, 1\}$. $\Diamond$

Remark 20.27 Clearly, (20.8) implies (20.7) and hence "mixing" implies "ergodic". The converse implication is false. $\Diamond$

Example 20.28 Let $\Omega = [0, 1)$, $\mathcal{A} = \mathcal{B}([0, 1))$ and let $\mathbf{P} = \lambda$ be the Lebesgue measure on $([0, 1), \mathcal{B}([0, 1)))$. For $r \in [0, 1)$, define $\tau_r : \Omega \to \Omega$ by

$$\tau_r(x) = x + r - \lfloor x + r \rfloor = x + r \pmod 1.$$

If r is irrational, then τ_r is ergodic (Example 20.9). However, τ_r is not mixing: Since r is irrational, there exists a sequence $k_n \uparrow \infty$ such that

$$\tau_r^{k_n}(0) \in \left(\frac{1}{4}, \frac{3}{4} \right) \qquad \text{for } n \in \mathbb{N}.$$

Hence, for $A = \left[0, \frac{1}{4} \right]$, we have $A \cap \tau_r^{-k_n}(A) = \emptyset$. Therefore,

$$\liminf_{n \to \infty} \mathbf{P}\left[A \cap \tau_r^{-n}(A) \right] = 0 \neq \frac{1}{16} = \mathbf{P}[A]^2. \quad \Diamond$$

Reflection Why is τ_r not mixing if r is rational? ♠

Theorem 20.29 *Let X be an irreducible, positive recurrent Markov chain on the countable space E and let π be its invariant distribution. Let $\mathbf{P}_{\pi} = \sum_{x \in E} \pi(\{x\})\,\mathbf{P}_x$. Then:*

(i) X is ergodic (on $(\Omega, \mathcal{A}, \mathbf{P}_{\pi})$).
(ii) X is mixing if and only if X is aperiodic.

Proof

(i) This has been shown already in Example 20.17.
(ii) As X is irreducible, by Theorem 17.52, we have $\pi(\{x\}) > 0$ for every $x \in E$.

" $\Longrightarrow$ " Let X be periodic with period $d \geq 2$. If $n \in \mathbb{N}$ is not a multiple of d, then $p^n(x, x) = 0$. Hence, for $A = B = \{X_0 = x\}$,

$$\liminf_{n\to\infty} \mathbf{P}_\pi[X_0 = x, \ X_n = x] = \liminf_{n\to\infty} \pi(\{x\}) \, p^n(x, x)$$

$$= 0 \neq \pi(\{x\})^2 = \mathbf{P}_\pi[X_0 = x]^2.$$

Thus X is not mixing.

" $\Longleftarrow$ " Let X be aperiodic. In order to simplify the notation, we may assume that X is the canonical process on $E^{\mathbb{N}_0}$. Let $A, B \subset \Omega = E^{\mathbb{N}_0}$ be measurable. For every $\varepsilon > 0$, there exists an $N \in \mathbb{N}$ and a $\tilde{A}^\varepsilon \in E^{\{0,\dots,N\}}$ such that, letting $A^\varepsilon = \tilde{A}^\varepsilon \times E^{\{N+1,N+2,\dots\}}$, we have $\mathbf{P}[A \triangle A^\varepsilon] < \varepsilon$. By the Markov property, for every $n \geq N$,

$$\mathbf{P}_\pi\left[A^\varepsilon \cap \tau^{-n}(B)\right] = \mathbf{P}_\pi\left[(X_0, \dots, X_N) \in \tilde{A}^\varepsilon, \ (X_n, X_{n+1}, \dots) \in B\right]$$

$$= \sum_{x,y\in E} \mathbf{E}_\pi\left[\mathbb{1}_{A^\varepsilon} \mathbb{1}_{\{X_N=x\}} \mathbb{1}_{\{X_n=y\}} (X_n, X_{n+1}, \dots) \in B\right]$$

$$= \sum_{x,y\in E} \mathbf{E}_\pi\left[\mathbb{1}_{A^\varepsilon} \mathbb{1}_{\{X_N=x\}}\right] p^{n-N}(x, y) \mathbf{P}_y[B].$$

By Theorem 18.13, we have $p^{n-N}(x, y) \overset{n\to\infty}{\longrightarrow} \pi(\{y\})$ for all $x, y \in E$. (For periodic X, this is false.) Dominated convergence thus yields

$$\lim_{n\to\infty} \mathbf{P}_\pi\left[A^\varepsilon \cap \tau^{-n}(B)\right] = \sum_{x,y\in E} \mathbf{E}_\pi\left[\mathbb{1}_{A^\varepsilon} \mathbb{1}_{\{X_N=x\}}\right] \pi(\{y\}) \mathbf{P}_y[B]$$

$$= \mathbf{P}_\pi\left[A^\varepsilon\right] \mathbf{P}_\pi[B].$$

Since $\left|\mathbf{P}_\pi\left[A^\varepsilon \cap \tau^{-n}(B)\right] - \mathbf{P}\left[A \cap \tau^{-n}(B)\right]\right| < \varepsilon$, the statement follows by letting $\varepsilon \to 0$. $\qquad\square$

Takeaways Mixing is a concept of independence stronger than ergodicity but weaker than stochastic independence. It allows dependencies between events as long as they wash out when the events are shifted apart. An irreducible and aperiodic positive recurrent Markov chain is mixing. Rotations are not.

Exercise 20.5.1 Show that "strongly mixing" implies "weakly mixing", which in turn implies "ergodic". Give an example of a measure-preserving dynamical system that is ergodic but not weakly mixing. ♣

20.6   Entropy

The entropy $H(\mathbf{P})$ of a probability distribution $\mathbf{P}$ (see Definition 5.25) measures the amount of randomness in this distribution. In fact, the entropy of a delta distribution is zero and for a distribution on n points, the maximal entropy is achieved by the uniform distribution and equals $\log(n)$ (see Exercise 5.3.3). It is natural to use the entropy in order to quantify also the randomness of a dynamical system.

First we consider the situation of a simple shift: Let $\Omega = E^{\mathbb{N}_0}$, where E is a finite set equipped with the product σ-algebra $\mathcal{A} = (2^E)^{\otimes \mathbb{N}_0}$. Let τ be the shift on Ω and let $\mathbf{P}$ be an invariant probability measure. For $n \in \mathbb{N}$, denote by P_n the projection of $\mathbf{P}$ on $E^n = E^{\{0,\dots,n-1\}}$; that is,

$$P_n(\{(e_0, \dots, e_{n-1})\}) = \mathbf{P}\left[\{e_0\} \times \dots \times \{e_{n-1}\} \times E^{\{n,n+1,\dots\}}\right].$$

Denote by h_n the entropy of P_n. By Exercise 5.3.4, the entropy is subadditive:

$$h_{m+n} \leq h_m + h_n \quad \text{for } m, n \in \mathbb{N}.$$

Hence the following limit exists (see Exercise 20.6.2)

$$h := h(\mathbf{P}, \tau) := \lim_{n \to \infty} \frac{1}{n} h_n = \inf_{n \in \mathbb{N}} \frac{1}{n} h_n.$$

Definition 20.30 (Entropy of the simple shift) $h(\mathbf{P}, \tau)$ *is called the **entropy** of the dynamical system* $(\Omega, \mathcal{A}, \mathbf{P}, \tau)$.

Example 20.31 Assume that $\mathbf{P}$ is a product measure with marginals π on E. Then

$$h = H(\pi) = -\sum_{e \in E} \pi(\{e\}) \log(\pi(\{e\})).$$

$\Diamond$

Example 20.32 (Markov chain) Let $(X_n)_{n \in \mathbb{N}_0}$ be a Markov chain on E with transition matrix P and stationary distribution π. Let $(\Omega, \mathcal{A}, \mathbf{P}, \tau)$ be the corresponding dynamical system. For $x = (x_0, \dots, x_{n-1})$ and $0 \leq k < n - 1$, let

$$p(k, x) = \pi(\{x_k\}) P(x_k, x_{k+1}) \cdots P(x_{n-2}, x_{n-1}).$$

Then the entropy of P_n is (using stationarity of π in the third line)

$$H(P_n) = - \sum_{x_0,\dots,x_{n-1}\in E} p(0,x) \log(p(0,x))$$

$$= - \sum_{x_0,\dots,x_{n-1}\in E} p(0,x) \left[\log(\pi(\{x_0\})) + \sum_{k=0}^{n-2} \log(P(x_k, x_{k+1})) \right]$$

$$= H(\pi) - \sum_{k=0}^{n-2} \sum_{x_k,\dots,x_{n-1}} p(k,x) \log(P(x_k, x_{k+1}))$$

$$= H(\pi) - (n-1) \sum_{x_0,x_1\in E} \pi(\{x_0\}) P(x_0, x_1) \log(P(x_0, x_1)).$$

We infer that the entropy of the dynamical system is

$$h(\mathbf{P}, \tau) = - \sum_{x,y\in E} \pi(\{x\}) P(x,y) \log(P(x,y)). \tag{20.9}$$

$\Diamond$

Example 20.33 (Integer rotation) Consider the rotation of Example 20.8. Let $n \in \mathbb{N} \setminus \{1\}$, $E = \mathbb{Z}/(n)$ and let $\mathbf{P}$ be the uniform distribution on Ω. Let $r \in \{1, \dots, n\}$ and

$$\tau : \Omega \to \Omega, \quad x \mapsto x + r \pmod{n}.$$

Clearly, $\tau^{(n)}$ is the identity map, hence $h_n = h_{2n} = \dots$ and thus $h(\mathbf{P}, \tau) = 0$. $\Diamond$

We now come to the situation of the general dynamical system. Let $\mathcal{P}$ be a finite measurable partition of Ω; that is, $\mathcal{P} = \{A_1, \dots, A_k\}$ for certain pairwise disjoint non-empty sets $A_1, \dots, A_k \in \mathcal{A}$ with $\Omega = A_1 \cup \dots \cup A_k$. Denote by $\mathcal{P}_n$ the partition that is generated by the sets $\bigcap_{l=0}^{n-1} \tau^{-l}(A_{i_l})$, $i_1, \dots, i_n \in \{1, \dots, k\}$. We define

$$h_n(\mathbf{P}, \tau; \mathcal{P}) = - \sum_{A\in\mathcal{P}_n} \mathbf{P}[A] \log(\mathbf{P}[A]).$$

Similarly as in the simple shift case, we obtain the subadditivity of (h_n) and thus the existence of

$$h(\mathbf{P}, \tau; \mathcal{P}) := \lim_{n\to\infty} \frac{1}{n} h_n(\mathbf{P}, \tau; \mathcal{P}) = \inf_{n\in\mathbb{N}} \frac{1}{n} h_n(\mathbf{P}, \tau; \mathcal{P}).$$

Definition 20.34 (Kolmogorov–Sinai entropy)   *The **entropy** of a (general) mea-sure-preserving dynamical system $(\Omega, \mathcal{A}, \mathbf{P}, \tau)$ is*

$$h(\mathbf{P}, \tau) = \sup_{\mathcal{P}} h(\mathbf{P}, \tau; \mathcal{P}),$$

where the supremum is taken over all finite measurable partitions of Ω.

Theorem 20.35 (Kolmogorov–Sinai)   *Let $\mathcal{P}$ be a generator of $\mathcal{A}$; that is $\mathcal{A} = \sigma\big(\bigcup_{n \in \mathbb{N}_0} \tau^{-n}(\mathcal{P})\big)$. Then*

$$h(\mathbf{P}, \tau) = h(\mathbf{P}, \tau; \mathcal{P}).$$

Proof   See, e.g., [88, Theorem 3.2.18], [168, Theorem 4.17] or [156].   □

The Kolmogorov–Sinai theorem shows that the entropy that was introduced in Definition 20.30 for simple shifts coincides with the entropy of Definition 20.34; simply take $\mathcal{P} = \{\{e\} \times E^{\mathbb{N}}, \ e \in E\}$ which generates the product σ-algebra on $\Omega = E^{\mathbb{N}_0}$.

Example 20.36 (Rotation)   We come back to the rotation of Example 20.9. Let $\Omega = [0, 1)$, $\mathcal{A} = \mathcal{B}(\Omega)$, $\mathbf{P} = \lambda$ the Lebesgue measure, $r \in (0, 1)$ and $\tau_r(x) = x + r \pmod 1$.

First assume that r is rational. Let $\mathcal{P}$ be an arbitrary finite measurable partition of Ω. Choose $n \in \mathbb{N}$ such that $rn \in \mathbb{N}_0$. As in Example 20.33 we obtain $h_n(\mathbf{P}, \tau_r; \mathcal{P}) = h_{kn}(\mathbf{P}, \tau_r; \mathcal{P})$ for all $k \in \mathbb{N}$, hence $h(\mathbf{P}, \tau_r, \mathcal{P}) = 0$. Concluding, we get $h(\mathbf{P}, \tau_r) = 0$.

Now assume that r is irrational. Choose the partition $\mathcal{P} = \{[0, 1/2), [1/2, 1)\}$. As r is irrational, it is easy to see that $\mathcal{A}$ is generated by $\bigcup_{n \in \mathbb{N}_0} \tau_r^{-n}(\mathcal{P})$. Hence $h(\mathbf{P}, \tau_r) = h(\mathbf{P}, \tau_r, \mathcal{P})$. In order to compute the latter quantity, we first determine the cardinality $\#\mathcal{P}_n$. To this end, consider the map

$$\phi_n : [0, 1) \to \{0, 1\}^n$$

$$x \mapsto \big(\mathbb{1}_{[1/2,1)}(x), \mathbb{1}_{[1/2,1)}(\tau_r(x)), \ldots, \mathbb{1}_{[1/2,1)}(\tau_r^{n-1}(x))\big)$$

Clearly, we have $\#\phi_n([0, 1)) = \#\mathcal{P}_n$. As $x \in [0, 1)$ increases, each coordinate $\mathbb{1}_{[1/2,1)}(\tau_r^k(x))$, $k = 1, \ldots, n - 1$, changes its value exactly twice. Only $\mathbb{1}_{[1/2,1)}(x)$ changes the value exactly once. Summing up, we get $\#\phi_n([0, 1)) \leq 2n$. The maximal entropy of a probability measure on N points is achieved by the uniform distribution and is $\log(N)$. Consequently, $h_n(\mathbf{P}, \tau_r; \mathcal{P}) \leq \log(2n)$. We conclude that

$$h(\mathbf{P}, \tau_r) = h(\mathbf{P}, \tau_r; \mathcal{P}) = 0. \quad \Diamond$$

Takeaways The entropy is an important characteristic of a dynamical system. It measures the amount of new randomness added in each step. For Markov chains, the entropy can be computed explicitly. This is particularly helpful in the context of statistical mechanics when a Markov chain is needed that maximises the entropy under certain constraints.

Exercise 20.6.1 Let $\Omega = [0, 1)$ and $\tau : x \mapsto 2x \pmod 1$. Let $\mathbf{P}$ be the Lebesgue measure on Ω. Determine $h(\mathbf{P}, \tau)$.

Exercise 20.6.2 Let $(a_n)_{n \in \mathbb{N}}$ be a sequence on nonnegative numbers. The sequence is called **subadditive**, if $a_{m+n} \leq a_m + a_n$ for all $m, n \in \mathbb{N}$. Show that the limit $\lim_{n \to \infty} a_n / n$ exists and that

$$\lim_{n \to \infty} \frac{1}{n} a_n = \inf_{n \in \mathbb{N}} \frac{1}{n} a_n. \quad \clubsuit$$

Exercise 20.6.3 Let p_i be the transition matrix of a Markov chain on the countable set E_i with entropy h_i, $i = 1, 2$. Compute the entropy of the bivariate chain on $E_1 \times E_2$ with transition matrix p given by $p((x_1, x_2), (y_1, y_2)) = p_1(x_1, y_1)p_2(x_2, y_2)$. $\clubsuit$

Exercise 20.6.4 Consider a Markov chain on $E = \{1, 2, 3\}$ with transition matrix p.

(i) Which p maximises the entropy?
(ii) Now we set the constraint $p(2, 1) = 0$. Which p maximises the entropy under the constraint? $\clubsuit$

Chapter 21
Brownian Motion

In Example 14.48, we constructed a (canonical) process $(X_t)_{t\in[0,\infty)}$ with independent stationary normally distributed increments. For example, such a process can be used to describe the motion of a particle immersed in water or the change of prices in the stock market. We are now interested in properties of this process X that cannot be described in terms of finite-dimensional distributions but reflect the whole path $t \mapsto X_t$. For example, we want to compute the distribution of the functional $F(X) := \sup_{t\in[0,1]} X_t$. The first problem that has to be resolved is to show that $F(X)$ is a random variable.

In this chapter, we investigate continuity properties of paths of stochastic processes and show how they ensure measurability of some path functionals. Then we construct a version of X that has continuous paths, the so-called *Wiener process* or *Brownian motion*. Without exaggeration, it can be stated that Brownian motion is *the* central object of probability theory.

For further reading, we recommend, e.g., [86, 118, 145, 152].

21.1 Continuous Versions

A priori the paths of a canonical process are of course not continuous since *every* map $[0, \infty) \to \mathbb{R}$ is possible. Hence, it will be important to find out which paths are **P**-almost surely negligible.

Definition 21.1 *Let* X *and* Y *be stochastic processes on* $(\Omega, \mathcal{A}, \mathbf{P})$ *with time set* I *and state space* E. X *and* Y *are called*

(i) **modifications** *or* **versions** *of each other if, for any* $t \in I$, *we have*

$$X_t = Y_t \qquad \textbf{P}\textit{-almost surely,}$$

© The Editor(s) (if applicable) and The Author(s), under exclusive license to Springer Nature Switzerland AG 2020
A. Klenke, *Probability Theory*, Universitext,
https://doi.org/10.1007/978-3-030-56402-5_21

(ii) **indistinguishable** *if there exists an* $N \in \mathcal{A}$ *with* $\mathbf{P}[N] = 0$ *such that*

$$\{X_t \neq Y_t\} \subset N \quad \text{for all } t \in I.$$

Clearly, indistinguishable processes are modifications of each other. Under certain assumptions on the continuity of the paths, however, the two notions coincide.

Reflection Find an example of two stochastic processes that are modifications of each other but that are not indistinguishable. ♠

Definition 21.2 *Let* (E, d) *and* (E', d') *be metric spaces and* $\gamma \in (0, 1]$. *A map* $\varphi : E \to E'$ *is called* **Hölder-continuous** *of order* γ *(briefly, Hölder-γ-continuous) at the point* $r \in E$ *if there exist* $\varepsilon > 0$ *and* $C < \infty$ *such that, for any* $s \in E$ *with* $d(s, r) < \varepsilon$, *we have*

$$d'(\varphi(r), \varphi(s)) \leq C \, d(r, s)^\gamma. \tag{21.1}$$

φ *is called locally Hölder-continuous of order* γ *if, for every* $t \in E$, *there exist* $\varepsilon > 0$ *and* $C = C(t, \varepsilon) > 0$ *such that, for all* $s, r \in E$ *with* $d(s, t) < \varepsilon$ *and* $d(r, t) < \varepsilon$, *the inequality* (21.1) *holds. Finally,* φ *is called Hölder-continuous of order* γ *if there exists a* C *such that* (21.1) *holds for all* $s, r \in E$.

In the case $\gamma = 1$, Hölder continuity is Lipschitz continuity (see Definition 13.8). Furthermore, for $E = \mathbb{R}$ and $\gamma > 1$, every locally Hölder-γ-continuous function is constant. Evidently, a locally Hölder-γ-continuous map is Hölder-γ-continuous at every point. On the other hand, for a function φ that is Hölder-γ-continuous at a given point t, there need not exist an open neighborhood in which φ is continuous. In particular, φ need not be locally Hölder-γ-continuous.

Reflection Let $f : \mathbb{R} \to \mathbb{R}$ be continuously differentiable. Check that f is Hölder-γ-continuous for any $\gamma \in (0, 1]$. If the derivative of f is bounded, then f is also (globally) Hölder-1-continuous, but not necessarily (globally) Hölder-γ-continuous for any $\gamma \in (0, 1)$. ♠

We collect some simple properties of Hölder-continuous functions.

Lemma 21.3 *Let* $I \subset \mathbb{R}$ *and let* $f : I \to \mathbb{R}$ *be locally Hölder-continuous of order* $\gamma \in (0, 1]$. *Then the following statements hold.*

(i) f *is locally Hölder-continuous of order* γ' *for every* $\gamma' \in (0, \gamma)$.
(ii) *If* I *is compact, then* f *is Hölder-continuous.*
(iii) *Let* I *be a bounded interval of length* $T > 0$. *Assume that there exists an* $\varepsilon > 0$ *and an* $C(\varepsilon) < \infty$ *such that, for all* $s, t \in I$ *with* $|t - s| \leq \varepsilon$, *we have*

$$|f(t) - f(s)| \leq C(\varepsilon) \, |t - s|^\gamma.$$

Then f *is Hölder-continuous of order* γ *with constant* $C := C(\varepsilon) \lceil T/\varepsilon \rceil^{1-\gamma}$.

Proof

(i) This is obvious since $|t - s|^\gamma \le |t - s|^{\gamma'}$ for all $s, t \in I$ with $|t - s| \le 1$.
(ii) For $t \in I$ and $\varepsilon > 0$, let $U_\varepsilon(t) := \{s \in I : |s - t| < \varepsilon\}$. For every $t \in I$, choose $\varepsilon(t) > 0$ and $C(t) < \infty$ such that

$$|f(r) - f(s)| \le C(t) \cdot |r - s|^\gamma \quad \text{for all } r, s \in U_t := U_{\varepsilon(t)}(t).$$

There exists a finite subcovering $\mathfrak{U}' = \{U_{t_1}, \ldots, U_{t_n}\}$ of the covering $\mathfrak{U} := \{U_t, t \in I\}$ of I. Let $\varrho > 0$ be a Lebesgue number of the covering $\mathfrak{U}'$; that is, $\varrho > 0$ is such that, for every $t \in I$, there exists a $U \in \mathfrak{U}$ such that $U_\varrho(t) \subset U$. Define

$$\overline{C} := \max \left\{ C(t_1), \ldots, C(t_n), 2\|f\|_\infty \varrho^{-\gamma} \right\}.$$

For $s, t \in I$ with $|t - s| < \varrho$, there is an $i \in \{1, \ldots, n\}$ with $s, t \in U_{t_i}$. By assumption, we have $|f(t) - f(s)| \le C(t_i) |t - s|^\gamma \le \overline{C} |t - s|^\gamma$. Now let $s, t \in I$ with $|s - t| \ge \varrho$. Then

$$|f(t) - f(s)| \le 2\|f\|_\infty \left(\frac{|t - s|}{\varrho} \right)^\gamma \le \overline{C} |t - s|^\gamma.$$

Hence f is Hölder-continuous of order γ with constant $\overline{C}$.
(iii) Let $n = \lceil \frac{T}{\varepsilon} \rceil$. For $s, t \in I$, by assumption, $\frac{|t-s|}{n} \le \varepsilon$ and thus

$$|f(t) - f(s)| \le \sum_{k=1}^n \left| f\left(s + (t - s)\frac{k}{n}\right) - f\left(s + (t - s)\frac{k - 1}{n}\right) \right|$$

$$\le C(\varepsilon) n^{1-\gamma} |t - s|^\gamma = C |t - s|^\gamma. \qquad \square$$

Definition 21.4 (Path properties) *Let $I \subset \mathbb{R}$ and let $X = (X_t, t \in I)$ be a stochastic process on some probability space $(\Omega, \mathcal{A}, \mathbf{P})$ with values in a metric space (E, d). Let $\gamma \in (0, 1]$. For every $\omega \in \Omega$, we say that the map $I \to E$, $t \mapsto X_t(\omega)$ is a **path** of X.*

We say that X has almost surely continuous paths, or briefly that X is a.s. continuous, if for almost all $\omega \in \Omega$, the path $t \mapsto X_t(\omega)$ is continuous. Similarly, we define locally Hölder-γ-continuous paths and so on.

Lemma 21.5 *Let X and Y be modifications of each other. Assume that one of the following conditions holds.*

(i) *I is countable.*
(ii) *$I \subset \mathbb{R}$ is a (possibly unbounded) interval and X and Y are almost surely right continuous.*

Then X and Y are indistinguishable.

Proof Define $N_t := \{X_t \neq Y_t\}$ for $t \in I$ and $\bar{N} = \bigcup_{t \in I} N_t$. By assumption, $\mathbf{P}[N_t] = 0$ for every $t \in I$. We have to show that there exists an $N \in \mathcal{A}$ with $\bar{N} \subset N$ and $\mathbf{P}[N] = 0$.

(i) If I is countable, then $N := \bar{N}$ is measurable and $\mathbf{P}[N] \leq \sum_{t \in I} \mathbf{P}[N_t] = 0$.

(ii) Now let $I \subset \mathbb{R}$ be an interval and let X and Y be almost surely right continuous. Define

$$\bar{R} := \{X \text{ and } Y \text{ are right continuous}\}$$

and choose an $R \in \mathcal{A}$ with $R \subset \bar{R}$ and $\mathbf{P}[R] = 1$. Define

$$\tilde{I} := \begin{cases} \mathbb{Q} \cap I, & \text{if } I \text{ is open to the right,} \\ (\mathbb{Q} \cap I) \cup \max I, & \text{if } I \text{ is closed to the right,} \end{cases}$$

and $\tilde{N} := \bigcup_{r \in \tilde{I}} N_r$. By (i), we have $\mathbf{P}[\tilde{N}] = 0$. Furthermore, for every $t \in I$,

$$N_t \cap R \subset \bigcup_{r \geq t, \, r \in \tilde{I}} (N_r \cap R) \subset \tilde{N}.$$

Hence

$$\bar{N} \subset R^c \cup \bigcup_{t \in I} N_t \subset R^c \cup \tilde{N} =: N,$$

and thus $\mathbf{P}[N] \leq \mathbf{P}[R^c] + \mathbf{P}[\tilde{N}] = 0$. $\square$

We come to the main theorem of this section.

Theorem 21.6 (Kolmogorov–Chentsov) *Let $X = (X_t, \, t \in [0, \infty))$ be a real-valued process. Assume for every $T > 0$, there are numbers $\alpha, \beta, C > 0$ such that*

$$\mathbf{E}\big[|X_t - X_s|^\alpha\big] \leq C |t - s|^{1 + \beta} \quad \text{for all } s, t \in [0, T]. \tag{21.2}$$

Then the following statements hold.

(i) *There is a modification $\tilde{X} = (\tilde{X}_t, \, t \in [0, \infty))$ of X whose paths are locally Hölder-continuous of every order $\gamma \in \left(0, \frac{\beta}{\alpha}\right)$.*

(ii) *Let $\gamma \in \left(0, \frac{\beta}{\alpha}\right)$. For every $\varepsilon > 0$ and $T < \infty$, there exists a number $K < \infty$ that depends only on $\varepsilon, T, \alpha, \beta, C, \gamma$ such that*

$$\mathbf{P}\Big[|\tilde{X}_t - \tilde{X}_s| \leq K \, |t - s|^\gamma, \; s, t \in [0, T]\Big] \geq 1 - \varepsilon. \tag{21.3}$$

Proof

(i) It is enough to show that, for any $T > 0$, the process X on $[0, T]$ has a modification X^T that is locally Hölder-continuous of any order $\gamma \in (0, \beta/\alpha)$. For $S, T > 0$, by Lemma 21.5, two such modifications X^S and X^T are indistinguishable on $[0, S \wedge T]$; hence

$$\Omega_{S,T} := \left\{ \text{there is a } t \in [0, S \wedge T] \text{ with } X_t^T \neq X_t^S \right\}$$

is a null set and thus also $\Omega_\infty := \bigcup_{S,T \in \mathbb{N}} \Omega_{S,T}$ is a null set. Therefore, defining $\tilde{X}_t(\omega) := X_t^t(\omega), t \geq 0$, for $\omega \in \Omega \setminus \Omega_\infty$, we get that $\tilde{X}$ is a locally Hölder-continuous modification of X on $[0, \infty)$.

Without loss of generality, assume $T = 1$. We show that X has a continuous modification on $[0, 1]$. By Markov's inequality, for every $\varepsilon > 0$,

$$\mathbf{P}[|X_t - X_s| \geq \varepsilon] \leq C \varepsilon^{-\alpha} |t - s|^{1+\beta}. \tag{21.4}$$

Hence

$$X_s \xrightarrow{s \to t} X_t \quad \text{in probability.} \tag{21.5}$$

The idea is first to construct $\tilde{X}$ on the dyadic rational numbers and then to extend it continuously to $[0, 1]$. To this end, we will need (21.5). In particular, for $\gamma > 0, n \in \mathbb{N}$ and $k \in \{1, \ldots, 2^n\}$, we have

$$\mathbf{P}\left[\left|X_{k2^{-n}} - X_{(k-1)2^{-n}}\right| \geq 2^{-\gamma n}\right] \leq C 2^{-n(1+\beta-\alpha\gamma)}.$$

Define

$$A_n = A_n(\gamma) := \left\{ \max\left\{|X_{k2^{-n}} - X_{(k-1)2^{-n}}|, \, k \in \{1, \ldots, 2^n\}\right\} \geq 2^{-\gamma n}\right\}$$

and

$$B_n := \bigcup_{m=n}^{\infty} A_m \quad \text{and} \quad N := \limsup_{n \to \infty} A_n = \bigcap_{n=1}^{\infty} B_n.$$

It follows that, for every $n \in \mathbb{N}$,

$$\mathbf{P}[A_n] \leq \sum_{k=1}^{2^n} \mathbf{P}\left[|X_{k2^{-n}} - X_{(k-1)2^{-n}}| \geq 2^{-\gamma n}\right] \leq C 2^{-n(\beta-\alpha\gamma)}.$$

Now fix $\gamma \in (0, \beta/\alpha)$ to obtain

$$\mathbf{P}[B_n] \leq \sum_{m=n}^{\infty} \mathbf{P}[A_m] \leq C \frac{2^{-(\beta-\alpha\gamma)n}}{1 - 2^{\alpha\gamma-\beta}} \xrightarrow{n\to\infty} 0, \tag{21.6}$$

hence $\mathbf{P}[N] = 0$. Now fix $\omega \in \Omega \setminus N$ and choose $n_0 = n_0(\omega)$ such that $\omega \notin \bigcup_{n=n_0}^{\infty} A_n$. Hence

$$\left| X_{k2^{-n}}(\omega) - X_{(k-1)2^{-n}}(\omega) \right| < 2^{-\gamma n} \quad \text{for } k \in \{1, \ldots, 2^n\}, \ n \geq n_0. \tag{21.7}$$

Define the sets of finite dyadic rationals $D_m = \{k2^{-m}, k = 0, \ldots, 2^m\}$, and let $D = \bigcup_{m\in\mathbb{N}} D_m$. Any $t \in D_m$ has a unique dyadic expansion

$$t = \sum_{i=0}^{m} b_i(t) \, 2^{-i} \quad \text{for some } b_i(t) \in \{0, 1\}, \ i = 0, \ldots, m.$$

Let $m \geq n \geq n_0$ and $s, t \in D_m$, $s \leq t$ with $|s - t| \leq 2^{-n}$. Let $u := \max(D_n \cap [0, s])$. Then

$$u \leq s < u + 2^{-n} \quad \text{and} \quad u \leq t < u + 2^{1-n}$$

and hence $b_i(t - u) = b_i(s - u) = 0$ for $i < n$. Define

$$t_l = u + \sum_{i=n}^{l} b_i(t - u) \, 2^{-i} \quad \text{for } l = n - 1, \ldots, m.$$

Then, we have $t_{n-1} = u$ and $t_m = t$. Furthermore, $t_l \in D_l$ for $l = n, \ldots, m$ and

$$t_l - t_{l-1} \leq 2^{-l} \quad \text{for } l = n, \ldots, m.$$

Hence, by (21.7),

$$|X_t(\omega) - X_u(\omega)| \leq \sum_{l=n}^{m} |X_{t_l}(\omega) - X_{t_{l-1}}(\omega)| \leq \sum_{l=n}^{m} 2^{-\gamma l} \leq \frac{2^{-\gamma n}}{1 - 2^{-\gamma}}.$$

Analogously, we obtain $|X_s(\omega) - X_u(\omega)| \leq 2^{-\gamma n}(1 - 2^{-\gamma})^{-1}$, and thus

$$|X_t(\omega) - X_s(\omega)| \leq 2 \frac{2^{-\gamma n}}{1 - 2^{-\gamma}}. \tag{21.8}$$

Define $C_0 = 2^{1+\gamma}(1 - 2^{-\gamma})^{-1} < \infty$. Let $s, t \in D$ with $|s - t| \leq 2^{-n_0}$. By choosing the minimal $n \geq n_0$ such that $|t - s| \geq 2^{-n}$, we obtain by (21.8),

$$|X_t(\omega) - X_s(\omega)| \leq C_0 |t - s|^\gamma. \tag{21.9}$$

As in the proof of Lemma 21.3(iii), we infer (with $K := C_0 2^{(1-\gamma)n_0}$)

$$|X_t(\omega) - X_s(\omega)| \leq K |t - s|^\gamma \qquad \text{for all } s, t \in D. \tag{21.10}$$

In other words, for dyadic rationals D, $X(\omega)$ is (globally) Hölder-γ-continuous. In particular, X is uniformly continuous on D; hence it can be extended to $[0, 1]$. For $t \in D$, define $\tilde{X}_t := X_t$. For $t \in [0, 1] \setminus D$ and $\{s_n, n \in \mathbb{N}\} \subset D$ with $s_n \longrightarrow t$, the sequence $(X_{s_n}(\omega))_{n\in\mathbb{N}}$ is a Cauchy sequence. Hence the limit

$$\tilde{X}_t(\omega) := \lim_{D \ni s \to t} X_s(\omega) \tag{21.11}$$

exists. Furthermore, the statement analogous to (21.10) holds not only for $s, t \in D$:

$$\left|\tilde{X}_t(\omega) - \tilde{X}_s(\omega)\right| \leq K |t - s|^\gamma \qquad \text{for all } s, t \in [0, 1]. \tag{21.12}$$

Hence $\tilde{X}$ is locally Hölder-continuous of order γ. By (21.5) and (21.11), we have $\mathbf{P}[X_t \neq \tilde{X}_t] = 0$ for every $t \in [0, 1]$. Hence $\tilde{X}$ is a modification of X.

(ii) Let $\varepsilon > 0$ and choose $n \in \mathbb{N}$ large enough that (see (21.6))

$$\mathbf{P}[B_n] \leq C \frac{2^{-(\beta-\alpha\gamma)n}}{1 - 2^{\alpha\gamma-\beta}} < \varepsilon.$$

For $\omega \notin B_n$, we conclude that (21.10) holds. However, this is exactly (21.3) with $T = 1$. For general T, the claim follows by linear scaling. $\square$

Remark 21.7 The statement of Theorem 21.6 remains true if X assumes values in some Polish space (E, ϱ) since in the proof we did not make use of the assumption that the range was in $\mathbb{R}$. However, if we change the time set, then the assumptions have to be strengthened: If $(X_t)_{t\in\mathbb{R}^d}$ is a process with values in E, and if, for certain $\alpha, \beta > 0$, all $T > 0$ and some $C < \infty$, we have

$$\mathbf{E}[\varrho(X_t, X_s)^\alpha] \leq C \|t - s\|_2^{d+\beta} \qquad \text{for all } s, t \in [-T, T]^d, \tag{21.13}$$

then for every $\gamma \in (0, \beta/\alpha)$, there is a locally Hölder-γ-continuous version of X. $\Diamond$

Takeaways The distribution of a stochastic process determines only properties that can be described by the values at countably many time points. For example, continuity is not among those properties. Under certain moment conditions, a stochastic process X allows for a modification $\tilde{X}$ that is continuous and equals X with probability 1 at any given time.

Exercise 21.1.1 Show the claim of Remark 21.7. ♣

Exercise 21.1.2 Let $X = (X_t)_{t \geq 0}$ be a real-valued process with continuous paths. Show that, for all $0 \leq a < b$, the map $\omega \mapsto \int_a^b X_t(\omega)\, dt$ is measurable. ♣

Exercise 21.1.3 (Optional sampling/stopping) Let $\mathbb{F}$ be a filtration and let $(X_t)_{t \geq 0}$ be an $\mathbb{F}$-supermartingale with right continuous paths. Let σ and τ be bounded stopping times with $\sigma \leq \tau$. Define $\sigma^n := 2^{-n}\lceil 2^n \sigma \rceil$ and $\tau^n := 2^{-n}\lceil 2^n \tau \rceil$.

 (i) Show that $\mathbf{E}[X_{\tau^m} \mid \mathcal{F}_{\sigma^n}] \overset{n \to \infty}{\longrightarrow} \mathbf{E}[X_{\tau^m} \mid \mathcal{F}_\sigma]$ almost surely and in L^1 as well as $X_{\sigma_n} \overset{n \to \infty}{\longrightarrow} X_\sigma$ almost surely and in L^1.
 (ii) Infer the optional sampling theorem for right continuous supermartingales by using the analogous statement for discrete time (Theorem 10.11); that is, $X_\sigma \geq \mathbf{E}[X_\tau \mid \mathcal{F}_\sigma]$.
 (iii) Show that if Y is adapted, integrable and right continuous, then Y is a martingale if and only if $\mathbf{E}[Y_\tau] = \mathbf{E}[Y_0]$ for every bounded stopping time τ.
 (iv) Assume that X is uniformly integrable and that $\sigma \leq \tau$ are *finite* (not necessarily bounded) stopping times. Show that $X_\sigma \geq \mathbf{E}[X_\tau \mid \mathcal{F}_\sigma]$.
 (v) Now let τ be an arbitrary stopping time. Deduce the optional stopping theorem for right continuous supermartingales: $(X_{\tau \wedge t})_{t \geq 0}$ is a right continuous supermartingale. ♣

Exercise 21.1.4 Let $X = (X_t)_{t \geq 0}$ be a stochastic process on $(\Omega, \mathcal{F}, \mathbf{P})$ with values in the Polish space E and with right continuous paths. Show the following.

 (i) The map $(\omega, t) \mapsto X_t(\omega)$ is measurable with respect to $\mathcal{F} \otimes \mathcal{B}([0, \infty)) - \mathcal{B}(E)$.
 (ii) If in addition X is adapted to the filtration $\mathbb{F}$, then for any $t \geq 0$, the map $\Omega \times [0, t] \to E$, $(\omega, s) \mapsto X_s(\omega)$ is $\mathcal{F}_t \otimes \mathcal{B}([0, t]) - \mathcal{B}(E)$ measurable.
 (iii) If τ is an $\mathbb{F}$-stopping time and X is adapted, then X_τ is an $\mathcal{F}_\tau$-measurable random variable. ♣

21.2   Construction and Path Properties

Definition 21.8 *A real-valued stochastic process $B = (B_t,\ t \in [0, \infty))$ is called a* ***Brownian motion*** *if*

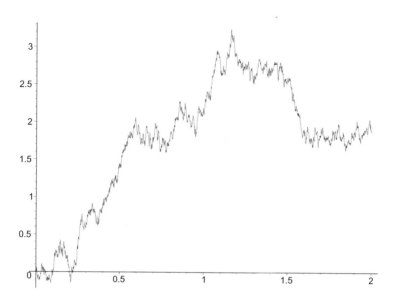

Fig. 21.1   Computer simulation of a Brownian motion.

(i) $B_0 = 0$,
(ii) B has independent, stationary increments (compare Definition 9.7),
(iii) $B_t \sim \mathcal{N}_{0,t}$ for all $t > 0$, and
(iv) $t \mapsto B_t$ is **P**-almost surely continuous.

See Fig. 21.1 for a computer simulation of a Brownian motion.

Theorem 21.9   *There exists a probability space $(\Omega, \mathcal{A}, \mathbf{P})$ and a Brownian motion B on $(\Omega, \mathcal{A}, \mathbf{P})$. The paths of B are a.s. locally Hölder-γ-continuous for every $\gamma < \frac{1}{2}$.*

Proof   As in Example 14.48 or Corollary 16.10 there exists a stochastic process X that fulfills (i), (ii) and (iii). Evidently, $X_t - X_s \overset{\mathcal{D}}{=} \sqrt{t-s}\, X_1 \sim \mathcal{N}_{0,t-s}$ for all $t > s \geq 0$. Thus, for every $n \in \mathbb{N}$, writing $C_n := \mathbf{E}[X_1^{2n}] = \frac{(2n)!}{2^n n!} < \infty$, we have

$$\mathbf{E}\left[(X_t - X_s)^{2n}\right] = \mathbf{E}\left[\left(\sqrt{t-s}\, X_1\right)^{2n}\right] = C_n\, |t - s|^n.$$

Now let $n \geq 2$ and $\gamma \in (0, \frac{n-1}{2n})$. Theorem 21.6 yields the existence of a version B of X that has Hölder-γ-continuous paths. Since all continuous versions of a process are equivalent, B is locally Hölder-γ-continuous for every $\gamma \in (0, \frac{n-1}{2n})$ and every $n \geq 2$ and hence for every $\gamma \in (0, \frac{1}{2})$. $\square$

Recall that a stochastic process $(X_t)_{t \in I}$ is called a **Gaussian process** if, for every $n \in \mathbb{N}$ and for all $t_1, \ldots, t_n \in I$, we have that

$$(X_{t_1}, \ldots, X_{t_n}) \quad \text{is } n\text{-dimensional normally distributed.}$$

X is called **centered** if $\mathbf{E}[X_t] = 0$ for every $t \in I$. The map

$$\Gamma(s, t) := \mathbf{Cov}[X_s, X_t] \quad \text{for } s, t \in I$$

is called the **covariance function** of X.

Remark 21.10 The covariance function determines the finite-dimensional distributions of a centered Gaussian process since a multidimensional normal distribution is determined by the vector of expectations and by the covariance matrix. ◊

Theorem 21.11 *Let $X = (X_t)_{t \in [0, \infty)}$ be a stochastic process. Then the following are equivalent:*

(i) X is a Brownian motion.
(ii) X is a continuous centered Gaussian process with $\mathbf{Cov}[X_s, X_t] = s \wedge t$ for all $s, t \geq 0$.

Proof By Remark 21.10, X is characterized by (ii). Hence, it is enough to show that, for Brownian motion X, we have $\mathbf{Cov}[X_s, X_t] = \min(s, t)$. This is indeed true since for $t > s$, the random variables X_s and $X_t - X_s$ are independent; hence

$$\mathbf{Cov}[X_s, X_t] = \mathbf{Cov}[X_s, X_t - X_s] + \mathbf{Cov}[X_s, X_s] = \mathbf{Var}[X_s] = s. \qquad \square$$

Reflection Let $X = (X_{(s,t)})_{s,t \geq 0}$ be a centred Gaussian process on the time set $[0, \infty)^2$ with covariance function

$$\mathbf{Cov}[X_{(s,t)}, X_{(s',t')}] = (s \wedge s') \cdot (t \wedge t').$$

Check that there exists a continuous modification of X. This modification is called Brownian sheet. ♠♠

Corollary 21.12 (Scaling property of Brownian motion) *If B is a Brownian motion and if $K \neq 0$, then $(K^{-1} B_{K^2 t})_{t \geq 0}$ is also a Brownian motion.*

Example 21.13 Another example of a continuous Gaussian process is the so-called **Brownian bridge** $X = (X_t)_{t \in [0, 1]}$ that is defined by the covariance function $\Gamma(s, t) = s \wedge t - st$. We construct the Brownian bridge as follows.

Let $B = (B_t, t \in [0, 1])$ be a Brownian motion and let

$$X_t := B_t - t B_1.$$

Clearly, X is a centered Gaussian process with continuous paths. We compute the covariance function Γ of X,

$$\begin{aligned}
\Gamma(s, t) = \mathbf{Cov}[X_s, X_t] &= \mathbf{Cov}[B_s - s B_1, B_t - t B_1] \\
&= \mathbf{Cov}[B_s, B_t] - s\,\mathbf{Cov}[B_1, B_t] - t\,\mathbf{Cov}[B_s, B_1] + st\,\mathbf{Cov}[B_1, B_1] \\
&= \min(s, t) - st - st + st = \min(s, t) - st. \quad \Diamond
\end{aligned} \tag{21.14}$$

Theorem 21.14 *Let $(B_t)_{t \geq 0}$ be a Brownian motion and*

$$X_t = \begin{cases} t B_{1/t}, & \text{if } t > 0, \\ 0, & \text{if } t = 0. \end{cases}$$

Then X is a Brownian motion.

Proof Clearly, X is a Gaussian process. For $s, t > 0$, we have

$$\mathbf{Cov}[X_s, X_t] = ts \cdot \mathbf{Cov}[B_{1/s}, B_{1/t}] = ts\,\min\left(s^{-1}, t^{-1}\right) = \min(s, t).$$

Clearly, $t \mapsto X_t$ is continuous at every point $t > 0$. To show continuity at $t = 0$, consider

$$\limsup_{t \downarrow 0} X_t = \limsup_{t \to \infty} \frac{1}{t} B_t$$

$$\leq \limsup_{n \to \infty} \frac{1}{n} B_n + \limsup_{n \to \infty} \frac{1}{n} \sup\left\{ B_t - B_n,\ t \in [n, n+1] \right\}.$$

By the strong law of large numbers, we have $\lim_{n \to \infty} \frac{1}{n} B_n = 0$ a.s. Using a generalization of the reflection principle (Theorem 17.15; see also Theorem 21.19), for $x > 0$, we have (using the abbreviation $B_{[a,b]} := \{B_t : t \in [a, b]\}$)

$$\begin{aligned}
\mathbf{P}\left[\sup B_{[n,n+1]} - B_n > x \right] &= \mathbf{P}\left[\sup B_{[0,1]} > x \right] = 2\,\mathbf{P}[B_1 > x] \\
&= \frac{2}{\sqrt{2\pi}} \int_x^\infty e^{-u^2/2}\, du \leq \frac{1}{x} e^{-x^2/2}.
\end{aligned}$$

In particular, $\sum_{n=1}^{\infty} \mathbf{P}\left[\sup B_{[n,n+1]} - B_n > n^\varepsilon \right] < \infty$ for every $\varepsilon > 0$. By the Borel–Cantelli lemma (Theorem 2.7), we infer

$$\limsup_{n \to \infty} \frac{1}{n} \sup\left\{ B_t - B_n,\ t \in [n, n+1] \right\} = 0 \quad \text{almost surely.}$$

Hence X is also continuous at 0. $\qquad\qquad\qquad\qquad\qquad\qquad\qquad\qquad\Box$

Theorem 21.15 (Blumenthal's 0-1 law, see [18]) *Let B be a Brownian motion and let $\mathbb{F} = (\mathcal{F}_t)_{t \geq 0} = \sigma(B)$ be the filtration generated by B. Further, let $\mathcal{F}_0^+ = \bigcap_{t > 0} \mathcal{F}_t$. Then $\mathcal{F}_0^+$ is a $\mathbf{P}$-trivial σ-algebra.*

Proof Define $Y^n = (B_{2^{-n}+t} - B_{2^{-n}})_{t \in [0, 2^{-n}]}$, $n \in \mathbb{N}$. Then $(Y^n)_{n \in \mathbb{N}}$ is an independent family of random variables (with values in $C([0, 2^{-n}])$). By Kolmogorov's 0-1 law (Theorem 2.37), the tail σ-algebra $\mathcal{T} = \bigcap_{n \in \mathbb{N}} \sigma(Y^m, \, m \geq n)$ is $\mathbf{P}$-trivial. On the other hand, $\sigma(Y^m, \, m \geq n) = \mathcal{F}_{2^{-n+1}}$; hence

$$\mathcal{F}_0^+ = \bigcap_{t > 0} \mathcal{F}_t = \bigcap_{n \in \mathbb{N}} \mathcal{F}_{2^{-n+1}} = \mathcal{T}$$

is $\mathbf{P}$-trivial.   □

Example 21.16 Let B be a Brownian motion. For every $K > 0$, we have

$$\mathbf{P}\big[\inf\{t > 0 : B_t \geq K\sqrt{t}\} = 0\big] = 1. \tag{21.15}$$

To check this, define $A_s := \big\{\inf\{t > 0 : B_t \geq K\sqrt{t}\} \leq s\big\}$ and

$$A := \Big\{\inf\{t > 0 : B_t \geq K\sqrt{t}\} = 0\Big\} = \bigcap_{s > 0} A_s \in \mathcal{F}_0^+.$$

Then $\mathbf{P}[A] \in \{0, 1\}$. By the scaling property of Brownian motion,

$$\mathbf{P}[A] = \inf_{s > 0} \mathbf{P}[A_s] \geq \mathbf{P}[B_1 \geq K] > 0$$

and thus $\mathbf{P}[A] = 1$. ◇

The preceding example shows that, for every $t \geq 0$, almost surely B is not Hölder-$\frac{1}{2}$-continuous at t. Note that the order of quantifiers is subtle. We have *not* shown that almost surely B was not Hölder-$\frac{1}{2}$-continuous at any $t \geq 0$ (however, see Remark 22.4). However, it is not too hard to show the following theorem, which for the case $\gamma = 1$ is due to Paley, Wiener and Zygmund [126]. The proof presented here goes back to an idea of Dvoretzky, Erdös and Kakutani (see [40]).

Theorem 21.17 (Paley–Wiener–Zygmund [126]) *For every $\gamma > \frac{1}{2}$, almost surely the paths of Brownian motion $(B_t)_{t \geq 0}$ are not Hölder-continuous of order γ at any point. In particular, the paths are almost surely nowhere differentiable.*

Proof Let $\gamma > \frac{1}{2}$. It suffices to consider $B = (B_t)_{t \in [0,1]}$. Denote by $H_{\gamma, t}$ the set of maps $[0, 1] \to \mathbb{R}$ that are Hölder-γ-continuous at t and define $H_\gamma := \bigcup_{t \in [0,1)} H_{\gamma, t}$. The aim is to show that almost surely $B \notin H_\gamma$. By translation invariance, this implies that B is nowhere Hölder-γ-continuous a.s. in any of the countably many intervals $[n/2, (n/2) + 1)$, $n \in \mathbb{N}_0$, which implies the claim of the theorem.

If $t \in [0, 1)$ and $w \in H_{\gamma,t}$, then for every $\delta > 0$ there exists a $c = c(\delta, w)$ with the property $|w_s - w_t| \leq c\,|s - t|^\gamma$ for every $s \in [0, 1]$ with $|s - t| < \delta$. Choose a $k \in \mathbb{N}$ with $k > \frac{2}{2\gamma - 1}$. Then, for $n \in \mathbb{N}$ with $n \geq n_0 := \lceil (k + 1)/\delta \rceil$, $i = \lfloor tn \rfloor + 1$ and $l \in \{0, \ldots, k - 1\}$, we get

$$\left| w_{(i+l+1)/n} - w_{(i+l)/n} \right| \leq \left| w_{(i+l+1)/n} - w_t \right| + \left| w_{(i+l)/n} - w_t \right| \leq 2c\,(k + 1)^\gamma n^{-\gamma}.$$

Hence, for $N \geq 2c\,(k + 1)^\gamma$, we have $w \in A_{N,n,i}$, where

$$A_{N,n,i} := \bigcap_{l=0}^{k-1} \left\{ w : \left| w_{(i+l+1)/n} - w_{(i+l)/n} \right| \leq N\,n^{-\gamma} \right\}.$$

Define $A_{N,n} = \bigcup_{i=1}^{n} A_{N,n,i}$, $A_N = \liminf_{n \to \infty} A_{N,n}$ and $A = \bigcup_{N=1}^{\infty} A_N$. Clearly, $H_\gamma \subset A$. Owing to the independence of increments and since the density of the standard normal distribution is bounded by 1, we get

$$\mathbf{P}[B \in A_{N,n,i}] = \mathbf{P}\big[|B_{1/n}| \leq N\,n^{-\gamma}\big]^k = \mathbf{P}\big[|B_1| \leq N\,n^{-\gamma+1/2}\big]^k$$
$$\leq N^k\,n^{k(-\gamma+1/2)}.$$

By the choice of k and since the increments of B are stationary, we have

$$\mathbf{P}\big[B \in A_N\big] = \lim_{n \to \infty} \mathbf{P}\bigg[B \in \bigcap_{m \geq n} A_{N,m}\bigg]$$
$$\leq \limsup_{n \to \infty} \mathbf{P}[B \in A_{N,n}]$$
$$\leq \limsup_{n \to \infty} \sum_{i=1}^{n} \mathbf{P}[B \in A_{N,n,i}]$$
$$\leq \limsup_{n \to \infty} n\,\mathbf{P}[B \in A_{N,n,1}]$$
$$\leq N^k \limsup_{n \to \infty} n^{1+k(-\gamma+1/2)} = 0.$$

Thus $\mathbf{P}[B \in A] = 0$. Therefore, we almost surely have $B \notin H_\gamma$.   □

Takeaways We have used the Kolmogorov-Chentsov theorem to construct Brownian motion as a continuous Gaussian process. The paths are Hölder continuous of any order less than $1/2$, but almost surely they are not Hölder continuous of any order larger than $1/2$ at any point. In particular, the paths are almost surely nowhere differentiable. Any property of Brownian motion that can be checked arbitrarily early after time 0 has either probability 0 or 1.

Exercise 21.2.1 Let B be a Brownian motion and let λ be the Lebesgue measure on $[0, \infty)$.

(i) Compute the expectation and variance of $\int_0^1 B_s \, ds$. (For the measurability of the integral see Exercise 21.1.2.)
(ii) Show that almost surely $\lambda(\{t : B_t = 0\}) = 0$.
(iii) Compute the expectation and variance of

$$\int_0^1 \left(B_t - \int_0^1 B_s \, ds \right)^2 dt. \quad \clubsuit$$

Exercise 21.2.2 Let B be a Brownian motion. Show that $(B_t^2 - t)_{t \geq 0}$ is a martingale. $\clubsuit$

Exercise 21.2.3 Let B be a Brownian motion and $\sigma > 0$. Show that the process $\left(\exp\left(\sigma B_t - \frac{\sigma^2}{2} t \right) \right)_{t \geq 0}$ is a martingale. $\clubsuit$

Exercise 21.2.4 Let B be a Brownian motion, $a < 0 < b$. Define the stopping time $\tau_{a,b} = \inf\{t \geq 0 : B_t \in \{a, b\}\}$.

Show that almost surely $\tau_{a,b} < \infty$ and that $\mathbf{P}[B_{\tau_{a,b}} = b] = -\frac{a}{b-a}$. Furthermore, show (using Exercise 21.2.2) that $\mathbf{E}[\tau_{a,b}] = -ab$. $\clubsuit$

Exercise 21.2.5 Let B be a Brownian motion, $b > 0$ and $\tau_b = \inf\{t \geq 0 : B_t = b\}$. Show the following.

(i) $\mathbf{E}[e^{-\lambda \tau_b}] = e^{-b\sqrt{2\lambda}}$ for $\lambda \geq 0$. (*Hint:* Use Exercise 21.2.3 and the optional sampling theorem.)
(ii) τ_b has a $\frac{1}{2}$-stable distribution with Lévy measure

$$\nu(dx) = \left(b/(\sqrt{2\pi}) \right) x^{-3/2} \mathbb{1}_{\{x > 0\}} \, dx.$$

(iii) The distribution of τ_b has density $f_b(x) = \frac{b}{\sqrt{2\pi}} e^{-b^2/(2x)} x^{-3/2}$. $\clubsuit$

Exercise 21.2.6 Let B be a Brownian motion, $a \in \mathbb{R}$, $b > 0$ and $\tau = \inf\{t \geq 0 : B_t = at + b\}$. For $\lambda \geq 0$, show that

$$\mathbf{E}[e^{-\lambda \tau}] = \exp\left(-ba - b\sqrt{a^2 + 2\lambda} \right).$$

Conclude that $\mathbf{P}[\tau < \infty] = 1 \wedge e^{-2ba}$. ♣

21.3 Strong Markov Property

Denote by $\mathbf{P}_x$ the probability measure such that $B = (B_t)_{t \geq 0}$ is a Brownian motion started at $x \in \mathbb{R}$. To put it differently, under $\mathbf{P}_x$, the process $(B_t - x)_{t \geq 0}$ is a standard Brownian motion. While the (simple) Markov property of $(B, (\mathbf{P}_x)_{x \in \mathbb{R}})$ is evident, it takes some work to check the strong Markov property.

Theorem 21.18 (Strong Markov property) *Brownian motion B with distributions $(\mathbf{P}_x)_{x \in \mathbb{R}}$ has the strong Markov property.*

Proof Let $\mathbb{F} = \sigma(B)$ be the filtration generated by B and let $\tau < \infty$ be an $\mathbb{F}$-stopping time. We have to show that, for every bounded measurable $F : \mathbb{R}^{[0,\infty)} \to \mathbb{R}$, we have:

$$\mathbf{E}_x\left[F\left((B_{t+\tau})_{t \geq 0}\right) \mid \mathcal{F}_\tau\right] = \mathbf{E}_{B_\tau}[F(B)]. \tag{21.16}$$

It is enough to consider continuous bounded functions F that depend on only finitely many coordinates $t_1, \ldots, t_N$ since these functions determine the distribution of $(B_{t+\tau})_{t \geq 0}$. Hence, let $f : \mathbb{R}^N \to \mathbb{R}$ be continuous and bounded and $F(B) = f(B_{t_1}, \ldots, B_{t_N})$. Manifestly, the map

$$x \mapsto \mathbf{E}_x[F(B)] = \mathbf{E}_0[f(B_{t_1} + x, \ldots, B_{t_N} + x)]$$

is continuous and bounded. Now let $\tau^n := 2^{-n}\lfloor 2^n\tau + 1\rfloor$ for $n \in \mathbb{N}$. Then τ^n is a stopping time and $\tau^n \downarrow \tau$; hence $B_{\tau^n} \xrightarrow{n \to \infty} B_\tau$ almost surely. Now every Markov process with countable time set (here all positive rational linear combinations of $1, t_1, \ldots, t_N$) is a strong Markov process (by Theorem 17.14); hence we have

$$\mathbf{E}_x\left[F\left((B_{\tau^n+t})_{t \geq 0}\right) \mid \mathcal{F}_{\tau^n}\right] = \mathbf{E}_x\left[f(B_{\tau^n+t_1}, \ldots, B_{\tau^n+t_N}) \mid \mathcal{F}_{\tau^n}\right]$$

$$= \mathbf{E}_{B_{\tau^n}}\left[f(B_{t_1}, \ldots, B_{t_N})\right] \tag{21.17}$$

$$\xrightarrow{n \to \infty} \mathbf{E}_{B_\tau}\left[f(B_{t_1}, \ldots, B_{t_N})\right] = \mathbf{E}_{B_\tau}[F(B)].$$

As B is right continuous, we have $F\left((B_{\tau^n+t})_{t \geq 0}\right) \xrightarrow{n \to \infty} F\left((B_{\tau+t})_{t \geq 0}\right)$ almost surely and in L^1 and thus

$$\mathbf{E}\left[\left|\mathbf{E}_x\left[F\left((B_{\tau^n+t})_{t \geq 0}\right) \mid \mathcal{F}_{\tau^n}\right] - \mathbf{E}_x\left[F\left((B_{\tau+t})_{t \geq 0}\right) \mid \mathcal{F}_{\tau^n}\right]\right|\right]$$

$$\leq \mathbf{E}_x\left[\left|F\left((B_{\tau^n+t})_{t \geq 0}\right) - F\left((B_{\tau+t})_{t \geq 0}\right)\right|\right] \xrightarrow{n \to \infty} 0. \tag{21.18}$$

Furthermore,

$$\mathcal{F}_{\tau^n} \downarrow \mathcal{F}_{\tau+} := \bigcap_{\sigma > \tau \text{ is a stopping time}} \mathcal{F}_\sigma \supset \mathcal{F}_\tau.$$

In fact, obviously, we have $\mathcal{F}_{\tau_n} \supset \mathcal{F}_{\tau+}$ for all n. On the other hand, for $A \in \bigcap_{n \in \mathbb{N}} \mathcal{F}_{\tau_n}$ and $\sigma > \tau$ a stopping time, we have for all t

$$\mathcal{F}_t \ni A \cap \{\tau_n \leq t\} \cap \{\sigma \leq t\} = A \cap \{(\sigma \vee \tau_n) \leq t\} \uparrow A \cap \{\sigma \leq t\}.$$

Hence, $A \in \mathcal{F}_\sigma \subset \mathcal{F}_{\tau+}$.

By (21.17) and (21.18), using the convergence theorem for backwards martingales (Theorem 12.14), we get that in the sense of L^1-limits

$$\mathbf{E}_{B_\tau}[F(B)] = \lim_{n \to \infty} \mathbf{E}_x\left[F\big((B_{\tau^n+t})_{t \geq 0}\big) \,\big|\, \mathcal{F}_{\tau^n}\right]$$

$$= \lim_{n \to \infty} \mathbf{E}_x\left[F\big((B_{\tau+t})_{t \geq 0}\big) \,\big|\, \mathcal{F}_{\tau^n}\right] = \mathbf{E}_x\left[F\big((B_{\tau+t})_{t \geq 0}\big) \,\big|\, \mathcal{F}_{\tau+}\right].$$

The left-hand side is $\mathcal{F}_\tau$-measurable. The tower property of conditional expectation thus yields (21.16). $\square$

Using the strong Markov property, we show the reflection principle for Brownian motion.

Theorem 21.19 (Reflection principle for Brownian motion) *For every $a > 0$ and $T > 0$,*

$$\mathbf{P}\big[\sup\{B_t : t \in [0, T]\} > a\big] = 2\,\mathbf{P}[B_T > a] \leq \frac{2\sqrt{T}}{\sqrt{2\pi}} \frac{1}{a} e^{-a^2/2T}.$$

Proof By the scaling property of Brownian motion (Corollary 21.12), without loss of generality, we may assume $T = 1$. Let $\tau := \inf\{t \geq 0 : B_t \geq a\} \wedge 1$. By the strong Markov property, $(B_t')_{t \geq 0} := (B_{\tau+t})_{t \geq 0}$ is a Brownian motion started at a and is independent of $\mathcal{F}_\tau$. By symmetry, we have $\mathbf{P}_a[B_{1-\tau}' > a \,|\, \tau < 1] = \frac{1}{2}$; hence

$$\mathbf{P}[B_1 > a] = \mathbf{P}[B_1 > a \,|\, \tau < 1]\,\mathbf{P}[\tau < 1]$$

$$= \mathbf{P}_a[B_{1-\tau} > a]\,\mathbf{P}[\tau < 1] = \frac{1}{2}\,\mathbf{P}[\tau < 1].$$

For the inequality compute

$$\mathbf{P}[B_1 > a] = \frac{1}{\sqrt{2\pi}} \int_a^\infty e^{-x^2/2}\,dx$$

$$\leq \frac{1}{\sqrt{2\pi}} \frac{1}{a} \int_a^\infty x\,e^{-x^2/2}\,dx = \frac{1}{\sqrt{2\pi}} \frac{1}{a} e^{-a^2/2}.\qquad\qquad \square$$

As an application of the reflection principle we derive Paul Lévy's arcsine law [107, page 216] for the last time a Brownian motion visits zero.

Theorem 21.20 (Lévy's arcsine law) *Let $T > 0$ and $\zeta_T := \sup\{t \le T : B_t = 0\}$. Then, for $t \in [0, T]$,*

$$\mathbf{P}\big[\zeta_T \le t\big] = \frac{2}{\pi} \arcsin\left(\sqrt{t/T}\,\right).$$

Proof Without loss of generality, assume $T = 1$ and $\zeta = \zeta_1$. Let $\widetilde{B}$ be a further, independent Brownian motion. By the reflection principle,

$$
\begin{aligned}
\mathbf{P}[\zeta \le t] &= \mathbf{P}\big[B_s \ne 0 \ \text{ for all } \ s \in [t, 1]\big] \\
&= \int_{-\infty}^{\infty} \mathbf{P}\big[B_s \ne 0 \ \text{ for all } \ s \in [t, 1]\,\big|\, B_t = a\big]\,\mathbf{P}[B_t \in da] \\
&= \int_{-\infty}^{\infty} \mathbf{P}_{|a|}\big[\widetilde{B}_s > 0 \ \text{ for all } \ s \in [0, 1 - t]\big]\,\mathbf{P}[B_t \in da] \\
&= \int_{-\infty}^{\infty} \mathbf{P}_0\big[|\widetilde{B}_{1-t}| \le |a|\big]\,\mathbf{P}[B_t \in da] \\
&= \mathbf{P}\big[|\widetilde{B}_{1-t}| \le |B_t|\big].
\end{aligned}
$$

If X and Y are independent and $\mathcal{N}_{0,1}$-distributed, then

$$\big(B_t,\ \widetilde{B}_{1-t}\big) \overset{\mathcal{D}}{=} \big(\sqrt{t}\, X,\ \sqrt{1 - t}\, Y\big).$$

Hence

$$
\begin{aligned}
\mathbf{P}[\zeta \le t] &= \mathbf{P}\big[\sqrt{1 - t}\,|Y| \le \sqrt{t}\,|X|\big] \\
&= \mathbf{P}\big[Y^2 \le t(X^2 + Y^2)\big] \\
&= \frac{1}{2\pi} \int_{-\infty}^{\infty} dx \int_{-\infty}^{\infty} dy\, e^{-(x^2 + y^2)/2}\, \mathbb{1}_{\{y^2 \le t(x^2 + y^2)\}}.
\end{aligned}
$$

Passing to polar coordinates, we obtain

$$\mathbf{P}[\zeta \le t] = \frac{1}{2\pi} \int_0^{\infty} r\, dr\, e^{-r^2/2} \int_0^{2\pi} d\varphi\, \mathbb{1}_{\{\sin(\varphi)^2 \le t\}} = \frac{2}{\pi} \arcsin\left(\sqrt{t}\,\right). \qquad \square$$

> **Takeaways** Brownian motion can be constructed as a strong Markov process. This is the starting point for many conclusions. As examples, we have shown the reflection principle and Lévy's arcsine law.

Exercise 21.3.1 (Hard problem!) Let $\mathbf{P}_x$ be the distribution of Brownian motion started at $x \in \mathbb{R}$. Let $a > 0$ and $\tau = \inf\{t \geq 0 : B_t \in \{0, a\}\}$. Use the reflection principle to show that, for every $x \in (0, a)$,

$$\mathbf{P}_x[\tau > T] = \sum_{n=-\infty}^{\infty} (-1)^n \mathbf{P}_x\big[B_T \in [na, (n+1)a]\big]. \tag{21.19}$$

If f is the density of a probability distribution on $\mathbb{R}$ with characteristic function φ and $\sup_{x \in \mathbb{R}} x^2 f(x) < \infty$, then the Poisson summation formula holds,

$$\sum_{n=-\infty}^{\infty} f(s+n) = \sum_{k=-\infty}^{\infty} \varphi(k) e^{2\pi i s} \quad \text{for all } s \in \mathbb{R}. \tag{21.20}$$

Use (21.19) and (21.20) (compare also (21.38)) to conclude that

$$\mathbf{P}_x[\tau > T] = \frac{4}{\pi} \sum_{k=0}^{\infty} \frac{1}{2k+1} \exp\left(-\frac{(2k+1)^2\pi^2 T}{2a^2}\right) \sin\left(\frac{(2k+1)\pi x}{a}\right). \tag{21.21}$$

♣

21.4   Supplement: Feller Processes

In many situations, a continuous version of a process would be too much to expect, for instance, the Poisson process is generically discontinuous. However, often there is a version with right continuous paths that have left-sided limits. At this point, we only briefly make plausible the existence theorem for such regular versions of processes in the case of so-called Feller semigroups.

Definition 21.21 *Let E be a Polish space. A map $f : [0, \infty) \to E$ is called **RCLL** (right continuous with left limits) or **càdlàg** (continue à droit, limites à gauche) if $f(t) = f(t+) := \lim_{s \downarrow t} f(s)$ for every $t \geq 0$ and if, for every $t > 0$, the left-sided limit $f(t-) := \lim_{s \uparrow t} f(s)$ exists and is finite.*

Definition 21.22 *A filtration $\mathbb{F} = (\mathcal{F}_t)_{t \geq 0}$ is called **right continuous** if $\mathbb{F} = \mathbb{F}^+$, where $\mathcal{F}_t^+ = \bigcap_{s>t} \mathcal{F}_s$. We say that a filtration $\mathbb{F}$ satisfies the **usual conditions** (from the French conditions habituelles) if $\mathbb{F}$ is right continuous and if $\mathcal{F}_0$ is $\mathbf{P}$-complete.*

Remark 21.23 If $\mathbb{F}$ is an arbitrary filtration and $\mathcal{F}_t^{+,*}$ is the completion of $\mathcal{F}_t^+$, then $\mathbb{F}^{+,*}$ satisfies the usual conditions. $\Diamond$

Theorem 21.24 (Doob's regularization) *Let $\mathbb{F}$ be a filtration that satisfies the usual conditions and let $X = (X_t)_{t \geq 0}$ be an $\mathbb{F}$-supermartingale such that $t \mapsto \mathbf{E}[X_t]$ is right continuous. Then there exists a modification $\widetilde{X}$ of X with RCLL paths.*

Proof For $a, b \in \mathbb{Q}^+$, $a < b$ and $I \subset [0, \infty)$, let $U_I^{a,b}$ be the number of upcrossings of $(X_t)_{t \in I}$ over $[a, b]$. By the upcrossing inequality (Lemma 11.3), for every $N > 0$ and every finite set $I \subset [0, N]$, we have $\mathbf{E}[U_I^{a,b}] \leq (\mathbf{E}[|X_N|] + |a|)/(b - a)$. Define $U_N^{a,b} = U_{\mathbb{Q}^+ \cap [0,N]}^{a,b}$. Then $\mathbf{E}[U_N^{a,b}] \leq (\mathbf{E}[|X_N|] + |a|)/(b - a)$. By Exercise 11.1.1, for $\lambda > 0$, we have

$$\lambda \, \mathbf{P}\big[\sup\{|X_t| : t \in \mathbb{Q}^+ \cap [0, N]\} > \lambda \big]$$

$$= \lambda \sup \Big\{ \mathbf{P}\big[\sup\{|X_t| : t \in I\} > \lambda \big] : I \subset \mathbb{Q}^+ \cap [0, N] \text{ finite} \Big\}$$

$$\leq 6 \, \mathbf{E}[|X_0|] + 4 \, \mathbf{E}[|X_N|].$$

Consider the event

$$A := \bigcap_{N \in \mathbb{N}} \Big(\bigcap_{\substack{a,b \in \mathbb{Q}^+ \\ 0 \leq a < b \leq N}} \{U_N^{a,b} < \infty\} \cap \big\{ \sup\{|X_t| : t \in \mathbb{Q}^+ \cap [0, N]\} < \infty \big\} \Big).$$

We have $\mathbf{P}[A] = 1$; hence $A \in \mathcal{F}_t$ for every $t \geq 0$ since $\mathbb{F}$ satisfies the usual conditions. For $\omega \in A$, for every $t \geq 0$, the limit

$$\widetilde{X}_t(\omega) := \lim_{\mathbb{Q}^+ \ni s \downarrow t, \, s > t} X_s(\omega)$$

exists and is RCLL. For $\omega \in A^c$, we define $\widetilde{X}_t(\omega) = 0$. As $\mathbb{F}$ satisfies the usual conditions, $\widetilde{X}$ is $\mathbb{F}$-adapted. As X is a supermartingale, for every N, the family $(X_s)_{s \leq N}$ is uniformly integrable. Hence, by assumption,

$$\mathbf{E}[\widetilde{X}_t] = \lim_{\mathbb{Q}^+ \ni s \downarrow t, \, s > t} \mathbf{E}[X_s] = \mathbf{E}[X_t].$$

However, since X is a supermartingale, for every $s > t$, we have

$$X_t \geq \mathbf{E}[X_s | \mathcal{F}_t] \xrightarrow{\mathbb{Q}^+ \ni s \downarrow t, \, s > t} \mathbf{E}[\widetilde{X}_t | \mathcal{F}_t] = \widetilde{X}_t \quad \text{in } L^1.$$

Therefore, $X_t = \widetilde{X}_t$ almost surely and hence $\widetilde{X}$ is a modification of X. $\qquad \square$

Reflection Why cannot we drop the assumption that $t \mapsto \mathbf{E}[X_t]$ be right continuous? ♠

Corollary 21.25 *Let* $(v_t)_{t \geq 0}$ *be a continuous convolution semigroup and assume that* $\int |x| v_1(dx) < \infty$. *Then there exists a Markov process X with RCLL paths and with independent stationary increments* $\mathbf{P}_{X_t - X_s} = v_{t-s}$ *for all* $t > s$.

Let E be a locally compact Polish space and let $C_0(E)$ be the set of (bounded) continuous functions that vanish at infinity. If κ is a stochastic kernel from E to E and if f is measurable and bounded, then we define $\kappa f(x) = \int \kappa(x, dy) f(y)$.

Definition 21.26 *A Markov semigroup* $(\kappa_t)_{t \geq 0}$ *on E is called a **Feller semigroup** if*

$$f(x) = \lim_{t \to 0} \kappa_t f(x) \quad \text{for all } x \in E, \ f \in C_0(E)$$

and $\kappa_t f \in C_0(E)$ *for every* $f \in C_0(E)$.

Let X be a Markov process with transition kernels $(\kappa_t)_{t \geq 0}$ and with respect to a filtration $\mathbb{F}$ that satisfies the usual conditions.

Let $g \in C_0(E), g \geq 0$. Let $h = \int_0^\infty e^{-t} \kappa_t g \, dt$. Then

$$e^{-s} \kappa_s h = e^{-s} \int_0^\infty e^{-t} \kappa_s \kappa_t g \, dt = \int_s^\infty e^{-t} \kappa_t g \, dt \leq h.$$

Hence $X^g := (e^{-t} h(X_t))_{t \geq 0}$ is an $\mathbb{F}$-supermartingale.

The Feller property and Theorem 21.24 ensure the existence of an RCLL version $\widetilde{X}^g$ of X^g. It takes a little more work to show that there exists a countable set $G \subset C_0(E)$ and a process $\widetilde{X}$ that is uniquely determined by $\widetilde{X}^g, g \in G$, and is an RCLL version of X. See, e.g., [147, Chapter III.7ff].

Let us take a moment's thought and look back at how we derived the strong Markov property of Brownian motion in Sect. 21.3. Indeed, there we needed only right continuity of the paths and a certain continuity of the distribution as a function of the starting point, which is exactly the Feller property. With a little more work, one can show the following theorem (see, e.g., [147, Chapter III.8ff] or [145, Chapter III, Theorem 2.7]).

Theorem 21.27 *Let* $(\kappa_t)_{t \geq 0}$ *be a Feller semigroup on the locally compact Polish space E. Then there exists a strong Markov process* $(X_t)_{t \geq 0}$ *with RCLL paths and transition kernels* $(\kappa_t)_{t \geq 0}$.

*Such a process X is called a **Feller process**.*

Takeaways A Feller semigroup of stochastic kernels is a Markov semigroup with just enough additional regularity such that we can construct an RCLL version of the corresponding Markov process.

Exercise 21.4.1 (Doob's inequality) Let $X = (X_t)_{t\geq 0}$ be a martingale or a nonnegative submartingale with RCLL paths. For $T \geq 0$, let $|X|_T^* = \sup_{t\in[0,T]} |X_t|$. Show Doob's inequalities:

(i) For any $p \geq 1$ and $\lambda > 0$, we have $\lambda^p \, \mathbf{P}\big[|X|_T^* \geq \lambda\big] \leq \mathbf{E}\big[|X_T|^p\big]$.

(ii) For any $p > 1$, we have $\mathbf{E}\big[|X_T|^p\big] \leq \mathbf{E}\big[(|X|_T^*)^p\big] \leq \left(\frac{p}{p-1}\right)^p \mathbf{E}\big[|X_T|^p\big]$.

Construct a counterexample that shows that right continuity of the paths of X is essential. ♣

Exercise 21.4.2 (Martingale convergence theorems) Let X be a stochastic process with RCLL paths. Use Doob's inequality (Exercise 21.4.1) to show that the martingale convergence theorems (a.s. convergence (Theorem 11.4), a.s. and L^1-convergence for uniformly integrable martingales (Theorem 11.7) and the L^p-martingale convergence theorem (Theorem 11.10)) hold for X. ♣

Exercise 21.4.3 Let $p \geq 1$ and let $X^1, X^2, X^3, \ldots$ be L^p-integrable martingales. Assume that, for every $t \geq 0$, there exists an $\widetilde{X}_t \in \mathcal{L}^p(\mathbf{P})$ such that $X_t^n \overset{n\to\infty}{\longrightarrow} \widetilde{X}_t$ in L^p.

(i) Show that $(\widetilde{X}_t)_{t\geq 0}$ is a martingale.
(ii) Use Doob's inequality to show the following. If $p > 1$ and if $X^1, X^2, \ldots$ are a.s. continuous, then there is a continuous martingale X with the following properties: X is a modification of $\widetilde{X}$ and $X_t^n \overset{n\to\infty}{\longrightarrow} X_t$ in L^p for every $t \geq 0$. ♣

Exercise 21.4.4 Let X be a stochastic process with values in a Polish space E and with RCLL paths. Let $\mathbb{F} = \sigma(X)$ be the filtration generated by X and define $\mathbb{F}^+ := (\mathcal{F}_t^+)_{t\geq 0}$ by $\mathcal{F}_t^+ = \bigcap_{s>t} \mathcal{F}_s$. Let $U \subset E$ be open and let $C \subset E$ be closed. For every set $A \subset E$, define $\tau_A := \inf\{t > 0 : X_t \in A\}$. Show the following.

(i) τ_C is an $\mathbb{F}$-stopping time (and an $\mathbb{F}^+$-stopping time).
(ii) τ_U is an $\mathbb{F}^+$-stopping time but in general (even for continuous X) is not an $\mathbb{F}$-stopping time. ♣

Exercise 21.4.5 Show the statement of Remark 21.23. Conclude that if $\mathbb{F}$ is a filtration and if B is a Brownian motion that is an $\mathbb{F}$-martingale, then B is also an $\mathbb{F}^{+,*}$-martingale. ♣

21.5 Construction via L^2-Approximation

We give an alternative construction of Brownian motion by functional analytic means as an L^2-approximation. For simplicity, as the time interval we take $[0, 1]$ instead of $[0, \infty)$.

Let $H = L^2([0, 1])$ be the Hilbert space of square integrable (with respect to Lebesgue measure λ) functions $[0, 1] \to \mathbb{R}$ with inner product

$$\langle f, g \rangle = \int_{[0,1]} f(x)g(x)\,\lambda(dx)$$

and with norm $\|f\| = \sqrt{\langle f, f \rangle}$ (compare Sect. 7.3). Two functions $f, g \in H$ are considered equal if $f = g$ λ-a.e. Let $(b_n)_{n \in \mathbb{N}}$ be an orthonormal basis (ONB) of H; that is, $\langle b_m, b_n \rangle = \mathbb{1}_{\{m=n\}}$ and

$$\lim_{n \to \infty} \left\| f - \sum_{m=1}^{n} \langle f, b_m \rangle b_m \right\| = 0 \quad \text{for all } f \in H.$$

In particular, for every $f \in H$, **Parseval's equation**

$$\|f\|^2 = \sum_{m=1}^{\infty} \langle f, b_m \rangle^2 \tag{21.22}$$

holds and for $f, g \in H$

$$\langle f, g \rangle = \sum_{m=1}^{\infty} \langle f, b_m \rangle \langle g, b_m \rangle. \tag{21.23}$$

Now consider an i.i.d. sequence $(\xi_n)_{n \in \mathbb{N}}$ of $\mathcal{N}_{0,1}$-random variables on some probability space $(\Omega, \mathcal{A}, \mathbf{P})$. For $n \in \mathbb{N}$ and $t \in [0, 1]$, define

$$X_t^n = \int \mathbb{1}_{[0,t]}(s) \left(\sum_{m=1}^{n} \xi_m b_m(s) \right) \lambda(ds) = \sum_{m=1}^{n} \xi_m \langle \mathbb{1}_{[0,t]}, b_m \rangle.$$

Clearly, for $n \geq m$,

$$\mathbf{E}\big[(X_t^m - X_t^n)^2\big] = \mathbf{E}\left[\left(\sum_{k=m+1}^{n} \xi_k \langle \mathbb{1}_{[0,t]}, b_k \rangle \right) \left(\sum_{l=m+1}^{n} \xi_l \langle \mathbb{1}_{[0,t]}, b_l \rangle \right) \right]$$

$$= \sum_{k=m+1}^{n} \langle \mathbb{1}_{[0,t]}, b_k \rangle^2 \leq \sum_{k=m+1}^{\infty} \langle \mathbb{1}_{[0,t]}, b_k \rangle^2.$$

Since $\sum_{k=1}^{\infty} \langle \mathbb{1}_{[0,t]}, b_k \rangle^2 = \|\mathbb{1}_{[0,t]}\|^2 = t < \infty$, we have $X_t^n \in L^2(\mathbf{P})$ and

$$\lim_{m \to \infty} \sup_{n \geq m} \mathbf{E}\big[(X_t^m - X_t^n)^2\big] = 0.$$

Hence $\left(X_t^n\right)_{n\in\mathbb{N}}$ is a Cauchy sequence in $L^2(\mathbf{P})$ and thus (since $L^2(\mathbf{P})$ is complete, see Theorem 7.3) has an L^2-limit X_t. Thus, for $N \in \mathbb{N}$ and $0 \le t_1, \ldots, t_N \le 1$,

$$\lim_{n\to\infty} \mathbf{E}\left[\sum_{i=1}^{N}\left(X_{t_i}^n - X_{t_i}\right)^2\right] = 0.$$

In particular, $\left(X_{t_1}^n, \ldots, X_{t_N}^n\right) \overset{n\to\infty}{\longrightarrow} \left(X_{t_1}, \ldots, X_{t_N}\right)$ in $\mathbf{P}$-probability.

Manifestly, $\left(X_{t_1}^n, \ldots, X_{t_N}^n\right)$ is normally distributed and centered. For $s, t \in [0, 1]$, we have

$$\mathbf{Cov}\left[X_s^n, X_t^n\right] = \mathbf{E}\left[\left(\sum_{k=1}^{n}\xi_k\left\langle\mathbb{1}_{[0,s]}, b_k\right\rangle\right)\left(\sum_{l=1}^{n}\xi_l\left\langle\mathbb{1}_{[0,t]}, b_l\right\rangle\right)\right]$$

$$= \sum_{k,l=1}^{n}\mathbf{E}[\xi_k\xi_l]\left\langle\mathbb{1}_{[0,s]}, b_k\right\rangle\left\langle\mathbb{1}_{[0,t]}, b_l\right\rangle$$

$$= \sum_{k=1}^{n}\left\langle\mathbb{1}_{[0,s]}, b_k\right\rangle\left\langle\mathbb{1}_{[0,t]}, b_k\right\rangle$$

$$\overset{n\to\infty}{\longrightarrow} \left\langle\mathbb{1}_{[0,s]}, \mathbb{1}_{[0,t]}\right\rangle = \min(s, t).$$

Hence $(X_t)_{t\in[0,1]}$ is a centered Gaussian process with

$$\mathbf{Cov}[X_s, X_t] = \min(s, t). \tag{21.24}$$

Lévy Construction of Brownian motion

Up to continuity of paths, X is thus a Brownian motion. A continuous version of X can be obtained via the Kolmogorov–Chentsov theorem (Theorem 21.6). However, by a clever choice of the ONB $(b_n)_{n\in\mathbb{N}}$, we can construct X directly as a continuous process. The **Haar functions** $b_{n,k}$ are one such choice: Let $b_{0,1} \equiv 1$ and for $n \in \mathbb{N}$ and $k = 1, \ldots, 2^{n-1}$, let

$$b_{n,k}(t) = \begin{cases} 2^{(n-1)/2}, & \text{if } \dfrac{2k-2}{2^n} \le t < \dfrac{2k-1}{2^n}, \\ -2^{(n-1)/2}, & \text{if } \dfrac{2k-1}{2^n} \le t < \dfrac{2k}{2^n}, \\ 0, & \text{else.} \end{cases}$$

Then $(b_{n,k})$ is an orthonormal system: $\langle b_{m,k}, b_{n,l}\rangle = \mathbb{1}_{\{(m,k)=(n,l)\}}$. It is easy to check that $(b_{n,k})$ is a basis (exercise!). Define the **Schauder functions** by

$$B_{n,k}(t) = \int_{[0,t]} b_{n,k}(s)\,\lambda(ds) = \langle \mathbb{1}_{[0,t]}, b_{n,k}\rangle.$$

Let $\xi_{0,1}$, $(\xi_{n,k})_{n\in\mathbb{N},\, k=1,\ldots,2^{n-1}}$ be independent and $\mathcal{N}_{0,1}$-distributed. Let

$$X^n := \xi_{0,1}\, B_{0,1} + \sum_{m=1}^{n}\sum_{k=1}^{2^{m-1}} \xi_{m,k}\, B_{m,k},$$

and define $\tilde{X}_t$ as the $L^2(\mathbf{P})$-limit $\tilde{X}_t = L^2 - \lim_{n\to\infty} X_t^n$. See Fig. 21.2 for a computer simulation of X^n, $n = 0, 1, 2, 3, 10$.

Theorem 21.28 (Brownian motion, L^2-approximation) *There is a continuous version X of $\tilde{X}$. X is a Brownian motion and we have*

$$\lim_{n\to\infty} \big\| X^n - X \big\|_\infty = 0 \quad \mathbf{P}\text{-almost surely.} \tag{21.25}$$

Proof By (21.25), we have $X_t = \tilde{X}_t$ a.s. for all $t \in [0, 1]$. As uniform limits of continuous functions are continuous, (21.25) implies that X is continuous. Hence,

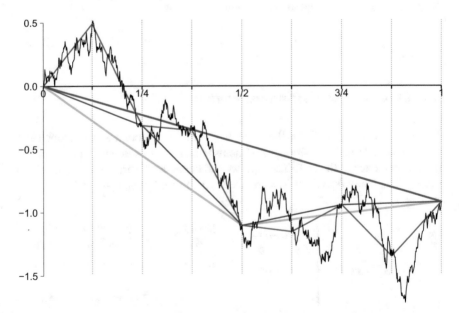

Fig. 21.2 The processes X^n, $n = 0, 1, 2, 3, 10$ of the Lévy construction of Brownian motion.

by (21.24) (and Theorem 21.11), X is a Brownian motion. Therefore, it is enough to prove the existence of an X such that (21.25) holds.

Since $(C([0, 1]), \| \cdot \|_\infty)$ is complete, it suffices to show that **P**-almost surely (X^n) is a Cauchy sequence in $(C([0, 1]), \| \cdot \|_\infty)$. Note that $\|B_{n,k}\|_\infty \le 2^{-(n+1)/2}$ if $n \in \mathbb{N}$ and $B_{n,k} B_{n,l} = 0$ if $k \ne l$. Hence

$$\left\| X^n - X^{n-1} \right\|_\infty \le 2^{-(n+1)/2} \max \left\{ |\xi_{n,k}|, \ k = 1, \ldots, 2^{n-1} \right\}.$$

Therefore,

$$\mathbf{P}\left[\left\| X^n - X^{n-1} \right\|_\infty > 2^{-n/4} \right] \le \sum_{k=1}^{2^{n-1}} \mathbf{P}\left[|\xi_{n,k}| > 2^{(n+2)/4} \right]$$

$$= 2^{n-1} \frac{2}{\sqrt{2\pi}} \int_{2^{(n+2)/4}}^{\infty} e^{-x^2/2} dx$$

$$\le 2^n \exp\left(-2^{n/2} \right).$$

Evidently, $\sum\limits_{n=1}^{\infty} \mathbf{P}[\| X^n - X^{n-1} \|_\infty > 2^{-n/4}] < \infty$; hence, by the Borel–Cantelli lemma,

$$\mathbf{P}\left[\left\| X^n - X^{n-1} \right\|_\infty > 2^{-n/4} \quad \text{only finitely often} \right] = 1.$$

We conclude that $\lim\limits_{n \to \infty} \sup\limits_{m \ge n} \| X^m - X^n \|_\infty = 0$ **P**-almost surely.   □

Brownian Motion and White Noise

The construction of Brownian motion via Haar functions has the advantage that continuity of the paths is straightforward. For some applications, however, a decomposition in trigonometric functions is preferable. Here as the orthonormal basis of $L^2([0, 1])$ we use $b_0 = 1$ and

$$b_n(x) = \sqrt{2} \cos(n\pi x) \quad \text{for } n \in \mathbb{N}.$$

For $t \in [0, 1]$ and $n \in \mathbb{N}_0$, define

$$B_n(t) = \int_0^t b_n(s)\, \lambda(ds);$$

that is, $B_0(t) = t$ and

$$B_n(t) = \frac{\sqrt{2}}{n\,\pi}\,\sin(n\pi\,t) \quad \text{for } n \in \mathbb{N}.$$

Let ξ_n, $n \in \mathbb{N}_0$, be independent standard normally distributed random variables. Define $A_0 = \xi_0$ and

$$A_n := \frac{\sqrt{2}}{\pi\,n}\,\xi_n \quad \text{for } n \in \mathbb{N}.$$

Finally, let

$$X^n := \sum_{k=0}^{n} \xi_k\,B_k;$$

that is,

$$X^n(t) = \xi_0\,t + \sum_{k=1}^{n} A_k\,\sin(k\pi\,t).$$

See Fig. 21.3 for a computer simulation of X^n, $n = 0, 1, 4, 64, 8192$.

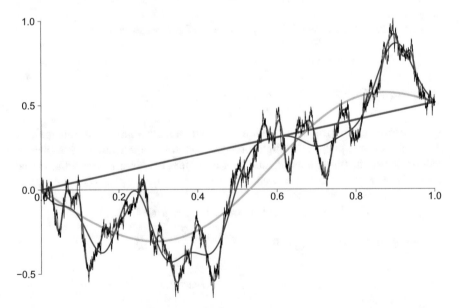

Fig. 21.3 The processes X^n, $n = 0, 1, 4, 64, 8192$ from the Fourier Construction of Brownian motion.

As shown above, the sequence (X^n) converges in $L^2([0, 1])$ towards a process X, which (up to continuity of paths) has all properties of Brownian motion:

$$X_t = \xi_0 t + \sum_{n=1}^{\infty} \frac{\sqrt{2}}{n\pi} \xi_n \sin(n\pi t).$$

This representation of the Brownian motions goes back to Paley and Wiener who also show that along a suitable subsequence the series converges uniformly almost surely and hence the limit X is indeed continuous, see [125, Theorem XLIII, page 148]. The representation is also sometimes called Karhunen–Loève expansion. More precisely, up to the first summand, it is the Karhunen–Loève expansion of the Brownian bridge $(X_t - tX_1)_{t \in [0,1]}$ (see, e.g., [1, Chapter 3.3]).

Taking the formal derivative

$$\dot{X}_t := \frac{d}{dt} X_t = \xi_0 + \sqrt{2} \sum_{n=1}^{\infty} \xi_n \cos(n\pi t)$$

we get independent identically distributed Fourier coefficients for all frequencies. Hence, the formal object $\dot{X}$ is often referred to as **white noise** as opposed to colored noise where the coefficients for the different frequencies have different distributions.

The Fourier basis is not too well suited to showing continuity of paths. For example, the sufficient criterion of absolute summability of coefficients (A_n) fails (see Exercise 21.5.5).

Example 21.29 (Stochastic integral à la Paley–Wiener) Assume that $(\xi_n)_{n \in \mathbb{N}}$ is an i.i.d. sequence of $\mathcal{N}_{0,1}$-distributed random variables. Let $(b_n)_{n \in \mathbb{N}}$ be an orthonormal basis of $L^2([0, 1])$ such that $W_t := \lim_{n \to \infty} \sum_{k=1}^{n} \xi_k \langle \mathbb{1}_{[0,t]}, b_k \rangle$, $t \in [0, 1]$, is a Brownian motion. For $f \in L^2([0, 1])$, define

$$I(f) := \sum_{n=1}^{\infty} \xi_n \langle f, b_n \rangle.$$

By Parseval's equation and the Bienaymé formula, we have

$$\|f\|_2^2 = \sum_{n=1}^{\infty} \langle f, b_n \rangle^2 = \mathbf{Var}\big[I(f)\big] = \mathbf{E}\big[I(f)^2\big].$$

Hence

$$I : L^2([0, 1]) \to L^2(\mathbf{P}), \qquad f \mapsto I(f) \quad \text{is an isometry.} \tag{21.26}$$

We call

$$\int_0^t f(s)\,dW_s := I(f\,\mathbb{1}_{[0,t]}), \quad t \in [0,1], \ f \in L^2([0,1]),$$

the **stochastic integral** of f with respect to W. In the special case of the Fourier basis $b_0(x) = 1$ and $b_n(x) = \sqrt{2}\,\cos(n\pi x)$, $n \in \mathbb{N}$, this construction goes back to Paley and Wiener [125, Theorem XLV, page 154].

The process $X_t := \int_0^t f(s)\,dW_s$, $t \in [0,1]$, is centered Gaussian with covariance function

$$\mathbf{Cov}[X_s, X_t] = \int_0^{s \wedge t} f^2(u)\,du.$$

In fact, it is obvious that X is centered and Gaussian (since it is a limit of the Gaussian processes of partial sums) and has the given covariance function. Furthermore, the existence of a continuous version can be obtained as for Brownian motion by employing the fourth moments of the increments, which for normal random variables can be computed from the variances (compare Theorem 21.9). In the following we will assume for the stochastic integral that such a continuous version is chosen.

In the special case, $f = \sum_{i=1}^n \alpha_i \mathbb{1}_{(t_{i-1}, t_i]}$ for some $n \in \mathbb{N}$ and $0 = t_0 < t_1 < \ldots < t_n$ and $\alpha_1, \ldots, \alpha_n \in \mathbb{R}$, we obtain

$$\int_0^1 f(s)\,dW_s = \sum_{i=1}^n \alpha_i \left(W_{t_i} - W_{t_{i-1}} \right). \quad \Diamond$$

Takeaways Consider an orthonormal basis of the Hilbert space $L^2([0,1])$ and assign to each basis vector an i.i.d. standard normally distributed factor. Now integrate over $[0, t]$ and sum up. The infinite series is a Gaussian process with the same covariance function as Brownian motion. If we choose the orthonormal basis cleverly, then we automatically get a continuous process. As one possible choice for the basis consists of cosine functions, this procedure is known as frequency decomposition of Brownian motion.

Exercise 21.5.1 Use the representation of Brownian motion $(W_t)_{t \in [0,1]}$ as a random linear combination of the Schauder functions $(B_{n,k})$ to show that the Brownian bridge $Y = (Y_t)_{t \in [0,1]} = (W_t - tW_1)_{t \in [0,1]}$ is a continuous, Gaussian process with covariance function $\mathbf{Cov}[Y_t, Y_s] = (s \wedge t) - st$. Further, show that

$$P_Y = \lim_{\varepsilon \downarrow 0} \mathbf{P}\big[W \in \cdot \,|\, W_1 \in (-\varepsilon, \varepsilon)\big]. \quad \clubsuit$$

Exercise 21.5.2 (Compare Example 8.32.) Fix $T \in (0, 1)$. Use an orthonormal basis $b_{0,1}$, $(c_{n,k})$, $(d_{n,k})$ of suitably modified Haar functions (such that the $c_{n,k}$ have support $[0, T]$ and the $d_{n,k}$ have support $[T, 1]$) to show that a regular conditional distribution of W_T given W_1 is defined by

$$\mathbf{P}[W_T \in \cdot \,|\, W_1 = x] = \mathcal{N}_{Tx,T}. \quad \clubsuit$$

Exercise 21.5.3 Define $Y := (Y_t)_{t \in [0,1]}$ by $Y_1 = 0$ and

$$Y_t = (1 - t) \int_0^t (1 - s)^{-1} \, dW_s \quad \text{for } t \in [0, 1).$$

Show that Y is a Brownian bridge.
Hint: Show that Y is a continuous Gaussian process with the correct covariance function. In particular, it has to be shown that $\lim_{t \uparrow 1} Y_t = 0$ almost surely. $\clubsuit$

Exercise 21.5.4 Let $d \in \mathbb{N}$. Use a suitable orthonormal basis on $[0, 1]^d$ to show:

(i) There is a Gaussian process $(W_t)_{t \in [0,1]^d}$ with covariance function

$$\mathbf{Cov}[W_t, W_s] = \prod_{i=1}^{d} (t_i \wedge s_i).$$

(ii) There is a modification of W such that $t \mapsto W_t$ is almost surely continuous (see Remark 21.7).

A process W with properties (i) and (ii) is called a **Brownian sheet**. $\clubsuit$

Exercise 21.5.5 Consider the coefficients $(A_n)_{n \in \mathbb{N}_0}$ of the Fourier basis of the construction of Brownian motion. Show the following statements:

(i) $\sum_{n=0}^{\infty} A_n^2 < \infty$ almost surely.
(ii) $\sum_{n=0}^{\infty} |A_n| = \infty$ almost surely.
(iii) $\sum_{k=0}^{n} A_k$, $n \in \mathbb{N}$ converges almost surely.

Hint: Kolmogorov's three-series theorem (Theorem 15.51). $\clubsuit$

Exercise 21.5.6 Let $t \in (0, 1)$ and $f_0(x) := t$ as well as

$$f_n(x) := \frac{2 \sin(n\pi t)}{n\pi} \cos(n\pi x) \quad \text{for } n \in \mathbb{N}, \; x \in [0, 1].$$

Show that $\sum_{n=0}^{\infty} f_n(x) = \mathbb{1}_{[0,t]}(x)$ for $x \in (0, 1) \setminus \{t\}$. $\clubsuit$

21.6   The Space $C([0, \infty))$

Are functionals that depend on the whole path of a Brownian motion measurable?
For example, is $\sup\{X_t,\ t \in [0, 1]\}$ measurable? For general stochastic processes,
this is false since the supremum depends on more than countably many coordinates.
However, for processes with continuous paths, this is true, as we will show in this
section in a somewhat more general framework.

We may consider Brownian motion as the canonical process on the space $\Omega :=$
$C([0, \infty))$ of continuous paths.

We start by collecting some properties of the space $\Omega = C([0, \infty)) \subset \mathbb{R}^{[0,\infty)}$.
Define the **evaluation map**

$$X_t : \Omega \to \mathbb{R}, \qquad \omega \mapsto \omega(t), \tag{21.27}$$

that is, the restriction of the canonical projection $\mathbb{R}^{[0,\infty)} \to \mathbb{R}$ to Ω.

For $f, g \in C([0, \infty))$ and $n \in \mathbb{N}$, let $d_n(f, g) := \left\| (f - g)\big|_{[0,n]} \right\|_\infty \wedge 1$ and

$$d(f, g) = \sum_{n=1}^{\infty} 2^{-n} d_n(f, g). \tag{21.28}$$

Theorem 21.30 *d is a complete metric on $\Omega := C([0, \infty))$ that induces the
topology of uniform convergence on compact sets. The space (Ω, d) is separable
and hence Polish.*

Proof Clearly, every d_n is a complete metric on $(C([0, n]), \|\cdot\|_\infty)$. Thus, for every
Cauchy sequence (f_N) in (Ω, d) and every $n \in \mathbb{N}$, there exists a $g_n \in \Omega$ with
$d_n(f_N, g_n) \overset{N\to\infty}{\longrightarrow} 0$. Evidently, $g_n(x) = g_m(x)$ for every $x \le m \wedge n$; hence there
exists a $g \in \Omega$ with $g(x) = g_n(x)$ for every $x \le n$ for every $n \in \mathbb{N}$. Hence, clearly,
$d(f_N, g) \overset{N\to\infty}{\longrightarrow} 0$, and thus d is complete.

The set of polynomials with rational coefficients is countable and by the
Weierstraß theorem, it is dense in any $(C([0, n]), \|\cdot\|_\infty)$; hence it is dense in
(Ω, d). $\square$

Theorem 21.31 *With respect to the Borel σ-algebra $\mathcal{B}(\Omega, d)$, the canonical
projections $X_t,\ t \in [0, \infty)$ are measurable. On the other hand, the X_t generate
$\mathcal{B}(\Omega, d)$. Hence*

$$(\mathcal{B}(\mathbb{R}))^{\otimes[0,\infty)}\Big|_{\Omega} = \sigma\big(X_t,\ t \in [0, \infty)\big) = \mathcal{B}(\Omega, d).$$

Proof The first equation holds by definition. For the second one, we must show the
mutual inclusions.

"$\subset$"   Clearly, every $X_t : \Omega \longrightarrow \mathbb{R}$ is continuous and hence $(\mathcal{B}(\Omega, d) - \mathcal{B}(\mathbb{R}))$-
measurable. Thus $\sigma\big(X_t,\ t \in [0, \infty)\big) \subset \mathcal{B}(\Omega, d)$.

"$\supset$" We have to show that open subsets of (Ω, d) are in $\mathcal{A} := (\mathcal{B}(\mathbb{R}))^{\otimes [0, \infty)}$. Since (Ω, d) is separable (Theorem 21.30), every open set is a countable union of ε-balls. Hence it suffices to show that for fixed $\omega_0 \in \Omega$, the map $\omega \mapsto d(\omega_0, \omega)$ is $\mathcal{A}$-measurable. To this end it is enough to show that for any $n \in \mathbb{N}$, the map $\omega \mapsto Y_n(\omega) := d_n(\omega_0, \omega)$ (see (21.28)) is $\mathcal{A}$-measurable. However, the map

$$\omega \mapsto Z_t(\omega) := |X_t(\omega) - X_t(\omega_0)| \wedge 1$$

is $\mathcal{A}$-measurable. Since each ω is continuous, Y_n is a countable supremum

$$Y_n = \sup_{t \in [0, n] \cap \mathbb{Q}} Z_t$$

and is hence $\mathcal{A}$-measurable. $\qquad \square$

In the following, let $\mathcal{A} := \sigma(X_t, \, t \in [0, \infty))$.

Corollary 21.32 *The map* $F_1 : \Omega \to [0, \infty)$, $\omega \mapsto \sup\{\omega(t) : \, t \in [0, 1]\}$ *is* $\mathcal{A}$-*measurable.*

Proof F_1 is continuous with respect to d and hence $\mathcal{B}(\Omega, d)$-measurable. $\qquad \square$

If B is a Brownian motion (on some probability space $(\widetilde{\Omega}, \widetilde{\mathcal{A}}, \widetilde{\mathbf{P}})$), then there exists an $\overline{\Omega} \in \widetilde{\mathcal{A}}$ with $\widetilde{\mathbf{P}}[\, \overline{\Omega}\,] = 1$ and $B(\omega) \in C([0, \infty))$ for every $\omega \in \overline{\Omega}$. Let $\overline{\mathcal{A}} = \widetilde{\mathcal{A}}\big|_{\overline{\Omega}}$ and $\overline{\mathbf{P}} = \widetilde{\mathbf{P}}\big|_{\overline{\mathcal{A}}}$. Then $B : \overline{\Omega} \longrightarrow C([0, \infty))$ is measurable with respect to $(\overline{\mathcal{A}}, \mathcal{A})$. With respect to the image measure $\mathbf{P} = \overline{\mathbf{P}} \circ B^{-1}$ on $\Omega = C([0, \infty))$, the canonical process $X = (X_t, \, t \in [0, \infty))$ on $C([0, \infty))$ is a Brownian motion.

Definition 21.33 *Let* $\mathbf{P}$ *be the probability measure on* $\Omega = C([0, \infty))$ *with respect to which the canonical process* X *is a Brownian motion. Then* $\mathbf{P}$ *is called the* **Wiener measure**. *The triple* $(\Omega, \mathcal{A}, \mathbf{P})$ *is called the* **Wiener space**, *and* X *is called the* **canonical Brownian motion** *or the* **Wiener process**.

Remark 21.34 Sometimes we want a Brownian motion to start not at $X_0 = 0$ but at an arbitrary point x. Denote by $\mathbf{P}_x$ that measure on $C([0, \infty))$ for which $\widetilde{X} = (X_t - x, \, t \in [0, \infty))$ is a Brownian motion (with $\widetilde{X}_0 = 0$). $\Diamond$

Takeaways A continuous stochastic process can be considered as a random variable with values in the space $C([0, \infty))$ of continuous functions. Here we have studied the properties of this space as a topological space and as a measure space.

Exercise 21.6.1 Show that the map

$$F_\infty : \Omega \to [0, \infty], \qquad \omega \mapsto \sup\{\omega(t) : t \in [0, \infty)\},$$

is $\mathcal{A}$-measurable. $\clubsuit$

21.7 Convergence of Probability Measures on $C([0, \infty))$

Let X and $(X^n)_{n \in \mathbb{N}}$ be random variables with values in $C([0, \infty))$ (i.e., continuous stochastic processes) with distributions $\mathbf{P}_X$ and $(\mathbf{P}_{X^n})_{n \in \mathbb{N}}$.

Definition 21.35 *We say that the finite-dimensional distributions of (X^n) converge to those of X if, for every $k \in \mathbb{N}$ and $t_1, \ldots, t_k \in [0, \infty)$, we have*

$$(X^n_{t_1}, \ldots, X^n_{t_k}) \overset{n \to \infty}{\Longrightarrow} (X_{t_1}, \ldots, X_{t_k}).$$

In this case, we write $X^n \overset{n \to \infty}{\underset{fdd}{\Longrightarrow}} X$ *or* $\mathbf{P}_{X^n} \overset{n \to \infty}{\underset{fdd}{\longrightarrow}} \mathbf{P}_X$.

Lemma 21.36 $P_n \overset{n \to \infty}{\underset{fdd}{\longrightarrow}} P$ *and* $P_n \overset{n \to \infty}{\underset{fdd}{\longrightarrow}} Q$ *imply* $P = Q$.

Proof By Theorem 14.12(iii), the finite-dimensional distributions determine P uniquely.   □

Theorem 21.37 *Weak convergence in $\mathcal{M}_1(\Omega, d)$ implies fdd-convergence:*

$$P_n \overset{n \to \infty}{\longrightarrow} P \quad \Longrightarrow \quad P_n \overset{n \to \infty}{\underset{fdd}{\longrightarrow}} P.$$

Proof Let $k \in \mathbb{N}$ and $t_1, \ldots, t_k \in [0, \infty)$. The map

$$\varphi : C([0, \infty)) \to \mathbb{R}^k, \qquad \omega \mapsto (\omega(t_1), \ldots, \omega(t_k))$$

is continuous. By the continuous mapping theorem (Theorem 13.25 on page 287), we have $P_n \circ \varphi^{-1} \overset{n \to \infty}{\longrightarrow} P \circ \varphi^{-1}$; hence $P_n \overset{n \to \infty}{\underset{fdd}{\longrightarrow}} P$.   □

The converse statement in the preceding theorem does not hold. However, we still have the following.

Theorem 21.38 *Let $(P_n)_{n \in \mathbb{N}}$ and P be probability measures on $C([0, \infty))$. Then the following are equivalent:*

(i) $P_n \overset{n \to \infty}{\underset{fdd}{\longrightarrow}} P$ *and* $(P_n)_{n \in \mathbb{N}}$ *is tight.*

(ii) $P_n \overset{n \to \infty}{\longrightarrow} P$ *weakly.*

Proof "(ii) $\Longrightarrow$ (i)" This is a direct consequence of Prohorov's theorem (Theorem 13.29 with $E = C([0, \infty))$).

"(i) $\Longrightarrow$ (ii)" By Prohorov's theorem, $(P_n)_{n \in \mathbb{N}}$ is relatively sequentially compact. Let Q be a limit point for $(P_{n_k})_{k \in \mathbb{N}}$ along some subsequence (n_k). Then $P_{n_k} \overset{fdd}{\longrightarrow} Q, k \to \infty$. By Lemma 21.36, we have $P = Q$.   □

Next we derive a useful criterion for tightness of sets $\{P_n\} \subset \mathcal{M}_1(C([0, \infty)))$. We start by recalling the Arzelà–Ascoli characterization of relatively compact sets in $C([0, \infty))$ (see, e.g., [37, Theorem 2.4.7] or [174, Theorem III.3]).

For $N, \delta > 0$ and $\omega \in C([0, \infty))$, let

$$V^N(\omega, \delta) := \sup \left\{ |\omega(t) - \omega(s)| : |t - s| \leq \delta, \ s, t \leq N \right\}.$$

Theorem 21.39 (Arzelà–Ascoli) *A set $A \subset C([0, \infty))$ is relatively compact if and only if the following two conditions hold.*

(i) $\{\omega(0), \ \omega \in A\} \subset \mathbb{R}$ is bounded.
(ii) For every N, we have $\lim\limits_{\delta \downarrow 0} \sup\limits_{\omega \in A} V^N(\omega, \delta) = 0$.

Theorem 21.40 *A family $(P_i, \ i \in I)$ of probability measures on $C([0, \infty))$ is weakly relatively compact if and only if the following two conditions hold.*

(i) $(P_i \circ X_0^{-1}, \ i \in I)$ is tight; that is, for every $\varepsilon > 0$, there is a $K > 0$ such that

$$P_i\left(\{\omega : |\omega(0)| > K\}\right) \leq \varepsilon \quad \text{for all } i \in I. \tag{21.29}$$

(ii) For all $\eta, \varepsilon > 0$ and $N \in \mathbb{N}$, there is a $\delta > 0$ such that

$$P_i\left(\{\omega : V^N(\omega, \delta) > \eta\}\right) \leq \varepsilon \quad \text{for all } i \in I. \tag{21.30}$$

Proof "$\Longrightarrow$"   By Prohorov's theorem (Theorem 13.29), weak relative compactness of $(P_i, \ i \in I)$ implies tightness of this family. Thus, for every $\varepsilon > 0$, there exists a compact set $A \subset C([0, \infty))$ with $P_i(A) > 1 - \varepsilon$ for every $i \in I$. Using the Arzelà–Ascoli characterization of the compactness of A, we infer (i) and (ii).

"$\Longleftarrow$"   Now assume that (i) and (ii) hold. Then, for $\varepsilon > 0$ and $k, N \in \mathbb{N}$, choose numbers K_ε and $\delta_{N,k,\varepsilon}$ such that

$$\sup_{i \in I} P_i\left(\{\omega : |\omega(0)| > K_\varepsilon\}\right) \leq \frac{\varepsilon}{2}$$

and

$$\sup_{i \in I} P_i\left(\left\{\omega : V^N(\omega, \delta_{N,k,\varepsilon}) > \frac{1}{k}\right\}\right) \leq 2^{-N-k-1}\,\varepsilon.$$

Define

$$C_{N,\varepsilon} = \left\{\omega : |\omega(0)| \leq K_\varepsilon, \ V^N(\omega, \delta_{N,k,\varepsilon}) \leq \frac{1}{k} \ \text{ for all } \ k \in \mathbb{N}\right\}.$$

By the Arzelà–Ascoli theorem, $C_\varepsilon := \bigcap_{N \in \mathbb{N}} C_{N,\varepsilon}$ is relatively compact in $C([0, \infty))$ and we have

$$P_i(C_\varepsilon^c) \leq \frac{\varepsilon}{2} + \sum_{k,N=1}^{\infty} P_i(\{\omega : V^N(\omega, \delta_{N,k,\varepsilon}) > 1/k\}) \leq \varepsilon \quad \text{for all } i \in I.$$

Hence the claim follows.   □

Corollary 21.41 *Let* $(X_i, i \in I)$ *and* $(Y_i, i \in I)$ *be families of random variables in* $C([0, \infty))$. *Assume that* $(\mathbf{P}_{X_i}, i \in I)$ *and* $(\mathbf{P}_{Y_i}, i \in I)$ *are tight. Then* $(\mathbf{P}_{X_i+Y_i}, i \in I)$ *is tight.*

Proof Apply the triangle inequality in order to check (i) and (ii) in the preceding theorem.   □

The following is an important tool to check weak relative compactness.

Theorem 21.42 (Kolmogorov's criterion for weak relative compactness) *Let* $(X^i, i \in I)$ *be a sequence of continuous stochastic processes. Assume that the following conditions are satisfied.*

(i) *The family* $(\mathbf{P}[X_0^i \in \cdot], i \in I)$ *of initial distributions is tight.*
(ii) *For any* $N > 0$ *there are numbers* $C, \alpha, \beta > 0$ *such that, for all* $s, t \in [0, N]$ *and every* $i \in I$, *we have*

$$\mathbf{E}\big[|X_s^i - X_t^i|^\alpha\big] \leq C\,|s - t|^{\beta+1}.$$

Then the family $(\mathbf{P}_{X^i}, i \in I) = (\mathcal{L}[X^i], i \in I)$ *of distributions of* X^i *is weakly relatively compact in* $\mathcal{M}_1(C([0, \infty)))$.

Proof We check the conditions of Theorem 21.40. The first condition of Theorem 21.40 is exactly (i).

Let $N > 0$. By the Kolmogorov–Chentsov theorem (Theorem 21.6(ii)), for $\varepsilon > 0$ and $\gamma \in (0, \beta/\alpha)$, there exists a K such that, for every $i \in I$, we have

$$\mathbf{P}\big[|X_t^i - X_s^i| \leq K\,|t - s|^\gamma \text{ for all } s, t \in [0, N]\big] \geq 1 - \varepsilon.$$

This clearly implies (21.30) with $\delta = (\eta/K)^{1/\gamma}$.   □

Takeaways In many situations, stochastic processes in continuous time are constructed as limits of simpler processes. In order to do, criteria for relative compactness of probability measures on $C([0, \infty))$ are needed. Particularly helpful is a moment criterion that postulates that moments of increments over small intervals decay quickly as the intervals get smaller.

21.8 Donsker's Theorem

Let $Y_1, Y_2, \ldots$ be i.i.d. random variables with $\mathbf{E}[Y_1] = 0$ and $\mathbf{Var}[Y_1] = \sigma^2 > 0$. For $t > 0$, let $S_t^n = \sum_{i=1}^{\lfloor nt \rfloor} Y_i$ and $\widetilde{S}_t^n = \frac{1}{\sqrt{\sigma^2 n}} S_t^n$. By the central limit theorem, for $t > s \geq 0$, we have $\mathcal{L}[\widetilde{S}_t^n - \widetilde{S}_s^n] \overset{n \to \infty}{\longrightarrow} \mathcal{N}_{0, t-s}$.

Let $B = (B_t, t \geq 0)$ be a Brownian motion. Then

$$\mathcal{L}[\widetilde{S}_t^n - \widetilde{S}_s^n] \overset{n \to \infty}{\longrightarrow} \mathcal{L}[B_t - B_s] \quad \text{for any } t > s \geq 0.$$

For $N \in \mathbb{N}$ and $0 = t_0 < t_1 < \ldots < t_N$, the random variables $\widetilde{S}_{t_i}^n - \widetilde{S}_{t_{i-1}}^n$, $i = 1, \ldots, N$, are independent, and hence, we have

$$\mathcal{L}[(\widetilde{S}_{t_1}^n - \widetilde{S}_{t_0}^n, \ldots, \widetilde{S}_{t_N}^n - \widetilde{S}_{t_{N-1}}^n)] \overset{n \to \infty}{\longrightarrow} \mathcal{L}[(B_{t_1} - B_{t_0}, \ldots, B_{t_N} - B_{t_{N-1}})].$$

We infer that

$$\mathcal{L}[(\widetilde{S}_{t_1}^n, \ldots, \widetilde{S}_{t_N}^n)] \overset{n \to \infty}{\longrightarrow} \mathcal{L}[(B_{t_1}, \ldots, B_{t_N})]. \tag{21.31}$$

We now define $\bar{S}^n$ as $\widetilde{S}^n$ but linearly interpolated:

$$\bar{S}_t^n = \frac{1}{\sqrt{\sigma^2 n}} \sum_{i=1}^{\lfloor nt \rfloor} Y_i + \frac{tn - \lfloor tn \rfloor}{\sqrt{\sigma^2 n}} Y_{\lfloor nt \rfloor + 1}. \tag{21.32}$$

Then, for $\varepsilon > 0$,

$$\mathbf{P}[|\widetilde{S}_t^n - \bar{S}_t^n| > \varepsilon] \leq \varepsilon^{-2} \mathbf{E}[(\widetilde{S}_t^n - \bar{S}_t^n)^2] \leq \frac{1}{\varepsilon^2 n} \frac{1}{\sigma^2} \mathbf{E}[Y_1^2] = \frac{1}{\varepsilon^2 n} \overset{n \to \infty}{\longrightarrow} 0.$$

By Slutzky's theorem (Theorem 13.18), we thus have convergence of the finite-dimensional distributions to the Wiener measure $\mathbf{P}_W$:

$$\mathbf{P}_{\bar{S}^n} \overset{n \to \infty}{\underset{\text{fdd}}{\Longrightarrow}} \mathbf{P}_W. \tag{21.33}$$

The aim of this section is to strengthen this convergence statement to weak convergence of probability measures on $C([0, \infty))$. The main theorem of this section is the **functional central limit theorem**, which goes back to Donsker [35]. Theorems of this type are also called **invariance principles** since the limiting distribution is the same for all distributions Y_i with expectation 0 and the same variance.

Theorem 21.43 (Donsker's invariance principle) *In the sense of weak convergence on* $C([0, \infty))$, *the distributions of* $\bar{S}^n$ *converge to the Wiener measure,*

$$\mathcal{L}[\bar{S}^n] \stackrel{n \to \infty}{\Longrightarrow} \mathbf{P}_W. \tag{21.34}$$

Proof Owing to (21.33) and Theorem 21.38, it is enough to show that $(\mathcal{L}[\bar{S}^n], n \in \mathbb{N})$ is tight. To this end, we want to apply Kolmogorov's moment criterion. However, as in the proof of existence of Brownian motion, second moments are not enough; rather we need fourth moments in order that we can choose $\beta > 0$. Hence the strategy is to truncate the Y_i to obtain fourth moments.

For $K > 0$, define

$$Y_i^K := Y_i \, \mathbb{1}_{\{|Y_i| \leq K/2\}} - \mathbf{E}[Y_i \, \mathbb{1}_{\{|Y_i| \leq K/2\}}] \quad \text{and} \quad Z_i^K := Y_i - Y_i^K \quad \text{for } i \in \mathbb{N}.$$

Then $\mathbf{E}[Y_i^K] = \mathbf{E}[Z_i^K] = 0$ and $\mathbf{Var}[Z_i^K] \stackrel{K \to \infty}{\longrightarrow} 0$ as well as $\mathbf{Var}[Y_i^K] \leq \sigma^2$, $i \in \mathbb{N}$. Clearly, $|Y_i^K| \leq K$ for every i. Define

$$T_n^K := \sum_{i=1}^n Y_i^K \quad \text{and} \quad U_n^K := \sum_{i=1}^n Z_i^K \quad \text{for } n \in \mathbb{N}.$$

Let $\bar{T}_t^{K,n}$ and $\bar{U}_t^{K,n}$ be the linearly interpolated versions of

$$\widetilde{T}_t^{K,n} := \frac{1}{\sqrt{\sigma^2 n}} T_{\lfloor nt \rfloor}^K \quad \text{and} \quad \widetilde{U}_t^{K,n} := \frac{1}{\sqrt{\sigma^2 n}} U_{\lfloor nt \rfloor}^K \quad \text{for } t \geq 0.$$

Evidently, $\bar{S}^n = \bar{T}^{K,n} + \bar{U}^{K,n}$. By Corollary 21.41, it is enough to show that, for a sequence $(K_n)_{n \in \mathbb{N}}$ (chosen later), the families $(\mathcal{L}[\bar{U}^{K_n,n}], n \in \mathbb{N})$ and $(\mathcal{L}[\bar{T}^{K_n,n}], n \in \mathbb{N})$ are tight.

We consider first the remainder term. As U^K is a martingale, Doob's inequality (Theorem 11.2) yields

$$\mathbf{P}\left[\sup_{l=1,\ldots,n} |U_l^K| > \varepsilon \sqrt{n} \right] \leq \varepsilon^{-2} \, \mathbf{Var}[Z_1^K] \quad \text{for every } \varepsilon > 0.$$

Now, if $K_n \uparrow \infty$, $n \to \infty$, then for every $N > 0$, we have

$$\mathbf{P}\left[\sup_{t \in [0,N]} |\bar{U}_t^{K_n,n}| > \varepsilon \right] \leq \frac{N}{\varepsilon^2 \sigma^2} \, \mathbf{Var}[Z_1^{K_n}] \stackrel{n \to \infty}{\longrightarrow} 0,$$

hence $\bar{U}^{K_n,n} \stackrel{n \to \infty}{\Longrightarrow} 0$ in $C([0, \infty))$. In particular, $(\mathcal{L}[\bar{U}^{K_n,n}], n \in \mathbb{N})$ is tight.

Next, for $N > 0$ and $s, t \in [0, N]$, we compute the fourth moments of the differences $\bar{T}_{t+s}^{K_n,n} - \bar{T}_s^{K_n,n}$ for the main term. In the following, let $K_n = n^{1/4}$. Fix $n \in \mathbb{N}$. We distinguish two cases:

Case 1: $t < n^{-1}$. Let $k := \lfloor (t + s)n \rfloor$. If $sn \geq k$, then

$$\bar{T}_{t+s}^{K_n,n} - \bar{T}_s^{K_n,n} = \frac{tn}{\sqrt{n\sigma^2}} Y_{k+1}^{K_n}.$$

If $sn < k$, then

$$\bar{T}_{t+s}^{K_n,n} - \bar{T}_s^{K_n,n} = \frac{1}{\sqrt{n\sigma^2}} \left(((t+s)n - k)Y_{k+1}^{K_n} + (k - sn)Y_k^{K_n} \right).$$

In either case, we have

$$\left| \bar{T}_{t+s}^{K_n,n} - \bar{T}_s^{K_n,n} \right| \leq \frac{t\sqrt{n}}{\sigma} \left(|Y_k^{K_n}| + |Y_{k+1}^{K_n}| \right),$$

hence

$$\begin{aligned}
\mathbf{E}\left[(\bar{T}_{t+s}^{K_n,n} - \bar{T}_s^{K_n,n})^4 \right] &\leq \frac{n^2 t^4}{\sigma^4} (2K_n)^2 \, \mathbf{E}\left[(|Y_1^{K_n}| + |Y_2^{K_n}|)^2 \right] \\
&\leq \frac{16 n^{5/2} t^4}{\sigma^4} \, \mathrm{Var}\left[Y_1^{K_n} \right] \leq \frac{16}{\sigma^2} t^{3/2}.
\end{aligned} \tag{21.35}$$

Case 2: $t \geq n^{-1}$. Using the binomial theorem, we get (note that the mixed terms with odd moments vanish since $\mathbf{E}[Y_1^{K_n}] = 0$)

$$\begin{aligned}
\mathbf{E}\left[(T_n^{K_n})^4 \right] &= n \, \mathbf{E}\left[(Y_1^{K_n})^4 \right] + 3n(n-1) \, \mathbf{E}\left[(Y_1^{K_n})^2 \right]^2 \\
&\leq n K_n^2 \sigma^2 + 3n(n-1)\sigma^4.
\end{aligned} \tag{21.36}$$

Note that, for independent real random variables X, Y with $\mathbf{E}[X] = \mathbf{E}[Y] = 0$ and $\mathbf{E}[X^4]$, $\mathbf{E}[Y^4] < \infty$ and for $a \in [-1, 1]$, we have

$$\begin{aligned}
\mathbf{E}\left[(aX + Y)^4 \right] &= a^4 \, \mathbf{E}[X^4] + 6a^2 \, \mathbf{E}[X^2] \, \mathbf{E}[Y^2] + \mathbf{E}[Y^4] \\
&\leq \mathbf{E}[X^4] + 6 \, \mathbf{E}[X^2] \, \mathbf{E}[Y^2] + \mathbf{E}[Y^4] = \mathbf{E}[(X + Y)^4].
\end{aligned}$$

We apply this twice (with $a = (t+s)n - \lfloor (t+s)n \rfloor$ and $a = \lceil sn \rceil - sn$) and obtain (using the rough estimate $\lceil (t+s)n \rceil - \lfloor sn \rfloor \le tn + 2 \le 3tn$) from (21.36) (since $t \le N$)

$$
\mathbf{E}\big[(\bar{T}_{t+s}^{K_n,n} - \bar{T}_s^{K_n,n})^4\big] \le n^{-2}\sigma^{-4}\, \mathbf{E}\big[(T_{\lceil (t+s)n \rceil}^{K_n} - T_{\lfloor sn \rfloor}^{K_n})^4\big]
$$

$$
\begin{aligned}
&= n^{-2}\sigma^{-4}\, \mathbf{E}\big[(T_{\lceil (t+s)n \rceil - \lfloor sn \rfloor}^{K_n})^4\big] \\
&\le \frac{3tn K_n^2}{n^2 \sigma^2} + 18t^2 \;=\; \frac{3}{\sigma^2} tn^{-1/2} + 18t^2 \\
&\le \frac{3}{\sigma^2} t^{3/2} + 18t^2 \;\le\; \left(\frac{3}{\sigma^2} + 18\sqrt{N}\right) t^{3/2}.
\end{aligned} \tag{21.37}
$$

By (21.35) and (21.37), for every $N > 0$, there exists a $C = C(N,\sigma^2)$ such that, for every $n \in \mathbb{N}$ and all $s, t \in [0, N]$, we have

$$
\mathbf{E}\big[(\bar{T}_{t+s}^{K_n,n} - \bar{T}_s^{K_n,n})^4\big] \le C\, t^{3/2}.
$$

Hence, by Kolmogorov's moment criterion (Theorem 21.42 with $\alpha = 4$ and $\beta = 1/2$), $(\mathcal{L}[\bar{T}^{K_n,n}],\, n \in \mathbb{N})$ is tight in $\mathcal{M}_1(C([0,\infty)))$. $\square$

Takeaways Properly rescaled sums of i.i.d. centred random variables with second moments converge to a normally distributed random variable. Using the moment criterion, here we have shown that also the process of partial sums converges and that the limit is Brownian motion. This is Donsker's theorem that is known in the physics literature as the invariance principle.

Exercise 21.8.1 Let $X_1, X_2, \ldots$ be i.i.d. random variables with continuous distribution function F. Let $G_n : [0,1] \to \mathbb{R},\, t \mapsto n^{-1/2} \sum_{i=1}^n \big(\mathbb{1}_{[0,t]}(F(X_i)) - t\big)$ and $M_n := \|G_n\|_\infty$. Further, let $M = \sup_{t \in [0,1]} |B_t|$, where B is a Brownian bridge.

(i)   Show that $\mathbf{E}[G_n(t)] = 0$ and $\mathrm{Cov}[G_n(s), G_n(t)] = s \wedge t - st$ for $s, t \in [0,1]$.
(ii) Show that $\mathbf{E}[(G_n(t) - G_n(s))^4] \le C\big((t-s)^2 + |t-s|/n\big)$ for some $C > 0$.
(iii) Conclude that a suitable continuous version of G_n converges weakly to B. For example, choose

$$
H_n(t) = n^{-1/2} \sum_{i=1}^n \big(h_n(F(X_i) - t) - g_n(t)\big),
$$

where h_n is a suitable smoothed version of $\mathbb{1}_{(-\infty,0]}$, for example, $h_n(s) = 1 - (s/\varepsilon_n \vee 0) \wedge 1$ for some sequence $\varepsilon_n \downarrow 0$, and $g_n(t) := \int_0^1 h_n(t - u)\, du$.

(iv) Finally, show that $M_n \overset{n\to\infty}{\Longrightarrow} M$.

Remark: The distribution of M can be expressed by the Kolmogorov–Smirnov formula ([101] and [157]; see, e.g., [133])

$$P[M > x] = 2 \sum_{n=1}^{\infty} (-1)^{n-1} e^{-2n^2 x^2}. \qquad (21.38)$$

Compare (21.21). Using the statistic M_n, one can test if random variables of a known distribution are independent. Let $X_1, X_2, \ldots$ and $\tilde{X}_1, \tilde{X}_2, \ldots$ be independent random variables with unknown continuous distribution functions F and $\tilde{F}$ and with empirical distribution functions F_n and $\tilde{F}_n$. Further, let

$$D_n := \sup_{t \in \mathbb{R}} |F_n(t) - \tilde{F}_n(t)|.$$

Under the assumption that $F = \tilde{F}$ holds, $\sqrt{n/2}\, D_n$ converges in distribution to M. This fact is the basis for nonparametric tests on the equality of distributions. ♣

21.9 Pathwise Convergence of Branching Processes*

In this section, we investigate the convergence of rescaled Galton–Watson processes (branching processes). As for sums of independent random variables, we first show convergence for a fixed time point to the distribution of a certain limiting process. The next step is to show convergence of finite-dimensional distributions. Finally, using Kolmogorov's moment criterion for tightness, we show convergence in the path space $C([0, \infty))$.

Consider a Galton–Watson process $(Z_n)_{n \in \mathbb{N}_0}$ with geometric offspring distribution

$$p(k) = 2^{-k-1} \quad \text{for } k \in \mathbb{N}_0.$$

That is, let $X_{n,i}$, $n, i \in \mathbb{N}_0$ be i.i.d. random variables on $\mathbb{N}_0$ with $P[X_{n,i} = k] = p(k)$, $k \in \mathbb{N}_0$, and based on the initial state Z_0 define inductively

$$Z_{n+1} = \sum_{i=1}^{Z_n} X_{n,i}.$$

Thus Z is a Markov chain with transition probabilities $p(i, j) = p^{*i}(j)$, where p^{*i} is the ith convolution power of p. In other words, if Z, $Z^1, \ldots, Z^i$ are independent copies of our Galton–Watson process, with $Z_0 = i$ and $Z_0^1 = \ldots = Z_0^i = 1$, then

$$Z \overset{\mathcal{D}}{=} Z^1 + \ldots + Z^i. \qquad (21.39)$$

We consider now the probability generating function of $X_{1,1}$, $\psi^{(1)}(s) := \psi(s) := \mathbf{E}[s^{X_{1,1}}]$, $s \in [0, 1]$. Denote by $\psi^{(n)} := \psi^{(n-1)} \circ \psi$ its nth iterate for $n \in \mathbb{N}$. Then, by Lemma 3.10,

$$\mathbf{E}_i[s^{Z_n}] = \mathbf{E}_1[s^{Z_n}]^i = \left(\psi^{(n)}(s)\right)^i.$$

For the geometric offspring distribution, $\psi^{(n)}$ can be computed explicitly.

Lemma 21.44 *For the branching process with critical geometric offspring distribution, the nth iterate of the probability generating function is*

$$\psi^{(n)}(s) = \frac{n - (n-1)s}{n + 1 - ns}.$$

Proof Compute

$$\psi(s) = \sum_{k=0}^{\infty} 2^{-k-1} s^k = \frac{1}{-s+2}.$$

In order to compute the iterated function, first consider general linear rational functions of the form $f(x) = \frac{ax+b}{cx+d}$. For such f, define the matrix $M_f = \begin{pmatrix} a & b \\ c & d \end{pmatrix}$. For two linear rational functions f and g, we have $M_{f \circ g} = M_f \cdot M_g$. The powers of M are easy to compute:

$$M_\psi = \begin{pmatrix} 0 & 1 \\ -1 & 2 \end{pmatrix}, \quad M_\psi^2 = \begin{pmatrix} -1 & 2 \\ -2 & 3 \end{pmatrix}, \quad M_\psi^3 = \begin{pmatrix} -2 & 3 \\ -3 & 4 \end{pmatrix},$$

and inductively

$$M_\psi^n = \begin{pmatrix} -(n-1) & n \\ -n & n+1 \end{pmatrix}. \qquad \square$$

If we let $s = e^{-\lambda}$, then we get the Laplace transform of Z_n,

$$\mathbf{E}_i[e^{-\lambda Z_n}] = \psi^{(n)}(e^{-\lambda})^i.$$

By Example 6.29, we can compute the moments of Z_n by differentiating the Laplace transform. That is, we obtain the following lemma.

Lemma 21.45 *The moments of Z_n are*

$$\mathbf{E}_i[Z_n^k] = (-1)^k \frac{d^k}{d\lambda^k} \left(\psi^{(n)}(e^{-\lambda})^i\right)\Big|_{\lambda=0}. \tag{21.40}$$

In particular, the first six moments are

$$\mathbf{E}_i[Z_n] = i,$$

$$\mathbf{E}_i[Z_n^2] = 2i\,n + i^2,$$

$$\mathbf{E}_i[Z_n^3] = 6i\,n^2 + 6i^2\,n + i^3,$$

$$\mathbf{E}_i[Z_n^4] = 24i\,n^3 + 36i^2\,n^2 + (12i^3 + 2i)\,n + i^4, \qquad (21.41)$$

$$\mathbf{E}_i[Z_n^5] = 120i\,n^4 + 240i^2\,n^3 + (120i^3 + 30i)\,n^2 + (20i^4 + 10i^2)\,n + i^5,$$

$$\mathbf{E}_i[Z_n^6] = 720i\,n^5 + 1800i^2\,n^4 + (1200i^3 + 360i)\,n^3,$$

$$+ (300i^4 + 240i^2)n^2 + (30i^5 + 30i^3 + 2i)n + i^6.$$

Hence, Z is a martingale, and the first six centered moments are

$$\mathbf{E}_i[(Z_n - i)^2] = 2i\,n,$$

$$\mathbf{E}_i[(Z_n - i)^3] = 6i\,n^2,$$

$$\mathbf{E}_i[(Z_n - i)^4] = 24i\,n^3 + 12i^2\,n^2 + 2i\,n,$$

$$\mathbf{E}_i[(Z_n - i)^5] = 120i\,n^4 + 120i^2\,n^3 + 30i\,n^2,$$

$$\mathbf{E}_i[(Z_n - i)^6] = 720i\,n^5 + 1080i^2\,n^4 + (120i^3 + 360i)\,n^3 + 60i^2n^2 + 2i\,n.$$
$$(21.42)$$

Proof The exact formulas for the first six moments are obtained by tenaciously computing the right-hand side of (21.40).   □

Now consider the following rescaling: Fix $x \geq 0$, start with $Z_0 = \lfloor nx \rfloor$ individuals and consider $\tilde{Z}_t^n := \frac{Z_{\lfloor tn \rfloor}}{n}$ for $t \geq 0$. We abbreviate

$$\mathcal{L}_x[\tilde{Z}^n] := \mathcal{L}_{\lfloor nx \rfloor}\big[(n^{-1} Z_{\lfloor nt \rfloor})_{t \geq 0}\big]. \qquad (21.43)$$

Evidently, $\mathbf{E}_x[\tilde{Z}_t^n] = \frac{\lfloor nx \rfloor}{n} \le x$ for every n; hence $(\mathcal{L}_x[\tilde{Z}_t^n], \, n \in \mathbb{N})$ is tight. By considering Laplace transforms, we obtain that, for every $\lambda \ge 0$, the sequence of distributions converges:

$$
\begin{aligned}
\lim_{n \to \infty} \mathbf{E}_x[e^{-\lambda \tilde{Z}_t^n}] &= \lim_{n \to \infty} \left(\psi^{(\lfloor tn \rfloor)}(e^{-\lambda/n}) \right)^{nx} \\
&= \lim_{n \to \infty} \left(\frac{nt - (nt-1)e^{-\lambda/n}}{nt + 1 - nt\, e^{-\lambda/n}} \right)^{nx} \\
&= \lim_{n \to \infty} \left(1 - \frac{1 - e^{-\lambda/n}}{n(1 - e^{-\lambda/n})t + 1} \right)^{nx} \\
&= \exp\left(- \lim_{n \to \infty} \frac{x\,n(1 - e^{-\lambda/n})}{n(1 - e^{-\lambda/n})t + 1} \right) \\
&= \exp\left(-\frac{\lambda}{\lambda + 1/t}\,(x/t) \right) \quad := \quad \psi_t(\lambda)^x.
\end{aligned}
\tag{21.44}
$$

However, the function ψ_t^x is the Laplace transform of the compound Poisson distribution $\mathrm{CPoi}_{(x/t)\,\mathrm{exp}_{1/t}}$ (see Definition 16.3).

Consider the stochastic kernel $\kappa_t(x, dy) := \mathrm{CPoi}_{(x/t)\,\mathrm{exp}_{1/t}}(dy)$. This is the kernel on $[0, \infty)$ whose Laplace transform is given by

$$
\int_0^\infty \kappa_t(x, dy)\, e^{-\lambda y} = \psi_t(\lambda)^x.
\tag{21.45}
$$

Lemma 21.46 $(\kappa_t)_{t \ge 0}$ *is a Markov semigroup and there exists a Markov process* $(Y_t)_{t \ge 0}$ *with transition kernels* $\mathbf{P}_x[Y_t \in dy] = \kappa_t(x, dy)$.

Proof It suffices to check that the Chapman–Kolmogorov equation $\kappa_t \cdot \kappa_s = \kappa_{s+t}$ holds. We compute the Laplace transform for these kernels. For $\lambda \ge 0$, applying (21.45) twice yields

$$
\begin{aligned}
\int \int \kappa_t(x, dy)\kappa_s(y, dz)\, e^{-\lambda z} &= \int \kappa_t(x, dy) \exp\left(-\frac{\lambda y}{\lambda s + 1} \right) \\
&= \exp\left(-\frac{\frac{\lambda}{\lambda s + 1}}{\frac{\lambda}{\lambda s + 1}t + 1}\, x \right) \\
&= \exp\left(-\frac{\lambda x}{\lambda(t + s) + 1} \right) \\
&= \int \kappa_{t+s}(x, dz)\, e^{-\lambda z}. \qquad \qquad \square
\end{aligned}
$$

Next we show that Y has a continuous version. To this end, we compute some of its moments and then use the Kolmogorov–Chentsov theorem (Theorem 21.6).

Lemma 21.47 *The first k moments of Y_t can be computed by differentiating the Laplace transform,*

$$\mathbf{E}_x[Y_t^k] = (-1)^k \frac{d^k}{d\lambda^k} \left(\psi(\lambda)^x \right) \Big|_{\lambda=0} ,$$

where $\psi_t(\lambda) = \exp\left(-\frac{\lambda}{\lambda t + 1} \right)$. In particular, we have

$$\mathbf{E}_x[Y_t] = x,$$
$$\mathbf{E}_x[Y_t^2] = 2x\,t + x^2,$$
$$\mathbf{E}_x[Y_t^3] = 6x\,t^2 + 6x^2\,t + x^3,$$
$$\mathbf{E}_x[Y_t^4] = 24x\,t^3 + 36x^2\,t^2 + 12x^3\,t + x^4,$$
$$\mathbf{E}_x[Y_t^5] = 120x\,t^4 + 240x^2\,t^3 + 120x^3\,t^2 + 20x^4\,t + x^5,$$
$$\mathbf{E}_x[Y_t^6] = 720x\,t^5 + 1800x^2\,t^4 + 1200x^3\,t^3 + 300x^4\,t^2 + 30x^5\,t + x^6.$$

$$(21.46)$$

Hence Y is a martingale, and the first centered moments are

$$\mathbf{E}_x[(Y_t - x)^2] = 2x\,t,$$
$$\mathbf{E}_x[(Y_t - x)^3] = 6x\,t^2,$$
$$\mathbf{E}_x[(Y_t - x)^4] = 24x\,t^3 + 12x^2\,t^2, \qquad\qquad (21.47)$$
$$\mathbf{E}_x[(Y_t - x)^5] = 120x\,t^4 + 120x^2\,t^3,$$
$$\mathbf{E}_x[(Y_t - x)^6] = 720x\,t^5 + 1080x^2\,t^4 + 120x^3\,t^3.$$

Theorem 21.48 *There is a continuous version of the Markov process Y with transition kernels $(\kappa_t)_{t\geq 0}$ given by (21.45). This version is called **Feller's (continuous) branching diffusion**.*

See Fig. 26.4 for a computer simulation of Feller's branching diffusion.

Proof For fixed $N > 0$ and $s, t \in [0, N]$, we have

$$\mathbf{E}_x\big[(Y_{t+s} - Y_s)^4\big] = \mathbf{E}_x\big[\mathbf{E}_{Y_s}[(Y_t - Y_0)^4]\big] = \mathbf{E}_x\big[24Y_s\,t^3 + 12Y_s^2\,t^2\big]$$
$$= 24x\,t^3 + 12(2sx + x^2)\,t^2 \leq \left(48Nx + 12x^2\right) t^2.$$

Thus Y satisfies the condition of Theorem 21.6 (Kolmogorov–Chentsov) with $\alpha = 4$ and $\beta = 1$. $\square$

Remark 21.49

(i) By using higher moments, it can be shown that the paths of Y are Hölder-continuous of any order $\gamma \in (0, \frac{1}{2})$.

(ii) It can be shown that Y is the (unique strong) solution of the stochastic (Itô-) differential equation (see Examples 26.11 and 26.31)

$$dY_t = \sqrt{2Y_t} \, dW_t, \tag{21.48}$$

where W is a Brownian motion. $\Diamond$

Theorem 21.50 *We have* $\mathcal{L}_x[\tilde{Z}^n] \xrightarrow[\text{fdd}]{n \to \infty} \mathcal{L}_x[Y]$.

Proof As in (21.44) for $0 \leq t_1 \leq t_2$, $\lambda_1, \lambda_2 \geq 0$ and $x \geq 0$, we get

$$\lim_{n \to \infty} \mathbf{E}_x \left[e^{-\left(\lambda_1 \tilde{Z}^n_{t_1} + \lambda_2 \tilde{Z}^n_{t_2}\right)} \right] = \lim_{n \to \infty} \mathbf{E}_x \left[\mathbf{E}_x \left[e^{-\lambda_2 \tilde{Z}^n_{t_2}} \Big| \tilde{Z}^n_{t_1} \right] e^{-\lambda_1 \tilde{Z}^n_{t_1}} \right]$$

$$= \lim_{n \to \infty} \mathbf{E}_x \left[\exp\left(-\frac{\lambda_2}{\lambda_2(t_2 - t_1) + 1} \tilde{Z}^n_{t_1} \right) e^{-\lambda_1 \tilde{Z}^n_{t_1}} \right]$$

$$= \exp\left(-\frac{\left(\frac{\lambda_2}{\lambda_2(t_2-t_1)+1} + \lambda_1 \right) x}{\left(\frac{\lambda_2}{\lambda_2(t_2-t_1)+1} + \lambda_1 \right) t_1 + 1} \right)$$

$$= \mathbf{E}_x \left[\exp(-(\lambda_1 Y_{t_1} + \lambda_2 Y_{t_2})) \right].$$

Hence, we obtain

$$\mathcal{L}_x \left[\lambda_1 \tilde{Z}^n_{t_1} + \lambda_2 \tilde{Z}^n_{t_2} \right] \xrightarrow{n \to \infty} \mathcal{L}_x \left[\lambda_1 Y_{t_1} + \lambda_2 Y_{t_2} \right].$$

Using the Cramér–Wold device (Theorem 15.57), this implies

$$\mathcal{L}_x \left[(\tilde{Z}^n_{t_1}, \tilde{Z}^n_{t_2}) \right] \xrightarrow{n \to \infty} \mathcal{L}_x \left[(Y_{t_1}, Y_{t_2}) \right].$$

Iterating the argument, for every $k \in \mathbb{N}$ and $0 \leq t_1 \leq t_2 \leq \ldots \leq t_k$, we get

$$\mathcal{L}_x \left[(\tilde{Z}^n_{t_i})_{i=1,\ldots,k} \right] \xrightarrow{n \to \infty} \mathcal{L}_x \left[(Y_{t_i})_{i=1,\ldots,k} \right].$$

However, this was the claim. $\square$

The final step is to show convergence in path space. To this end, we have to modify the rescaled processes so that they become continuous. Assume that

$(Z_i^n)_{i \in \mathbb{N}_0}$, $n \in \mathbb{N}$ is a sequence of Galton–Watson processes with $Z_0^n = \lfloor nx \rfloor$. Define the linearly interpolated processes

$$\bar{Z}_t^n := \left(t - n^{-1}\lfloor tn \rfloor\right)\left(Z_{\lfloor tn \rfloor+1}^n - Z_{\lfloor tn \rfloor}^n\right) + \frac{1}{n}Z_{\lfloor tn \rfloor}^n.$$

Theorem 21.51 (Lindvall (1972), see [109]) *As $n \to \infty$, in the sense of weak convergence in $\mathcal{M}_1(C([0, \infty)))$, the rescaled Galton–Watson processes $\bar{Z}^n$ converge to Feller's diffusion Y:*

$$\mathcal{L}_x[\bar{Z}^n] \overset{n \to \infty}{\Longrightarrow} \mathcal{L}_x[Y].$$

Proof We have shown already the convergence of the finite-dimensional distributions. By Theorem 21.38, it is thus enough to show tightness of $(\mathcal{L}_x[\bar{Z}^n], n \in \mathbb{N})$ in $\mathcal{M}_1(C([0, \infty)))$. To this end, we apply Kolmogorov's moment criterion (Theorem 21.42 with $\alpha = 4$ and $\beta = 1$). Hence, for fixed $N > 0$, we compute the fourth moments $\mathbf{E}_x[(\bar{Z}_{t+s}^n - \bar{Z}_s^n)^4]$ for $s, t \in [0, N]$. We distinguish two cases:

Case 1: $t < \frac{1}{n}$. Let $k = \lfloor (t+s)n \rfloor$. First assume that $\lfloor sn \rfloor = k$. Then (by Lemma 21.45)

$$\begin{aligned}
\mathbf{E}_x[(\bar{Z}_{t+s}^n - \bar{Z}_s^n)^4] &= n^{-4}(tn)^4\, \mathbf{E}_{\lfloor nx \rfloor}[(Z_{k+1}^n - Z_k^n)^4] \\
&= t^4\, \mathbf{E}_{\lfloor nx \rfloor}[24Z_k^n + 12(Z_k^n)^2 + 2Z_k^n] \\
&= t^4\left(26\lfloor nx \rfloor + 24\lfloor nx \rfloor k + \lfloor nx \rfloor^2\right) \\
&\leq 26x\, t^3 + 24xs\, t^2 + x^2 t^2 \\
&\leq (50Nx + x^2)\, t^2.
\end{aligned}$$

In the case $\lfloor sn \rfloor = k - 1$, we get a similar estimate. Therefore, there is a constant $C = C(N, x)$ such that

$$\mathbf{E}_x[(\bar{Z}_{s+t}^n - \bar{Z}_s^n)^4] \leq C t^2 \quad \text{for all } s, t \in [0, N] \text{ with } t < \frac{1}{n}. \tag{21.49}$$

Case 2: $t \geq \frac{1}{n}$. Define $k := \lceil (t + s)n \rceil - \lfloor sn \rfloor \leq tn + 1 \leq 2tn$. Then (by Lemma 21.45)

$$
\mathbf{E}_x\left[(\bar{Z}^n_{t+s} - \bar{Z}^n_s)^4\right]
$$

$$
\leq n^{-4} \mathbf{E}_{\lfloor nx \rfloor}\left[(Z^n_{\lfloor (t+s)n \rfloor} - Z^n_{\lfloor sn \rfloor})^4\right]
$$

$$
= n^{-4} \mathbf{E}_{\lfloor nx \rfloor}\left[\mathbf{E}_{Z^n_{\lfloor sn \rfloor}}[(Z^n_k - Z^n_0)^4]\right]
$$

$$
= n^{-4} \mathbf{E}_{\lfloor nx \rfloor}\left[24 Z^n_{\lfloor sn \rfloor} k^3 + 12 (Z^n_{\lfloor sn \rfloor})^2 k^2 + 2 Z^n_{\lfloor sn \rfloor} k\right] \tag{21.50}
$$

$$
\leq n^{-4}\left(24xn(2tn)^3 + (24xn\,sn + 12x^2n^2)(2tn)^2 + 4xtn^2\right)
$$

$$
\leq 192xt^3 + (96xs + 48x^2)t^2 + 4xn^{-1}t^2
$$

$$
\leq (292Nx + 48x^2)\,t^2.
$$

Combining the estimates (21.49) and (21.50), the assumptions of Kolmogorov's moment criterion for tightness (Theorem 21.42) are fulfilled with $\alpha = 4$ and $\beta = 1$. Hence the sequence $(\mathcal{L}_x[\bar{Z}^n],\ n \in \mathbb{N})$ is tight. $\square$

Takeaways Branching processes with very large populations undergo only small relative changes in each generation. By a proper rescaling of time and population size, we get a continuous limiting process (Lindvall's theorem). The details are rather technical as the computation of many moments is necessary.

21.10 Square Variation and Local Martingales

By the Paley–Wiener–Zygmund theorem (Theorem 21.17), the paths $t \mapsto W_t$ of Brownian motion are almost surely nowhere differentiable and hence have locally infinite variation. In particular, the stochastic integral $\int_0^1 f(s)\,dW_s$ that we introduced in Example 21.29 cannot be understood as a Lebesgue–Stieltjes integral. However, as a preparation for the construction of integrals of this type for larger classes of integrands and integrators (in Chap. 25), here we investigate the path properties of Brownian motion and, somewhat more generally, of continuous local martingales in more detail.

Definition 21.52 *Let* $G : [0, \infty) \to \mathbb{R}$ *be a continuous function. For any* $t \geq 0$, *define the* **variation** *up to* t *by*

$$V_t^1(G) := \sup \left\{ \sum_{i=0}^{n-1} |G_{t_{i+1}} - G_{t_i}| : 0 = t_0 \leq t_1 \leq \ldots \leq t_n = t, \ n \in \mathbb{N} \right\}.$$

We say that G *has locally finite variation if* $V_t^1(G) < \infty$ *for all* $t \geq 0$. *We write* $\mathcal{C}_v$ *for the vector space of continuous functions* G *with continuous variation* $t \mapsto V_t^1(G)$.

Remark 21.53 Clearly, $V^1(F + G) \leq V^1(F) + V^1(G)$ and $V^1(\alpha G) = |\alpha| \, V^1(G)$ for all continuous $F, G : [0, \infty) \to \mathbb{R}$ and for all $\alpha \in \mathbb{R}$. Hence $\mathcal{C}_v$ is indeed a vector space. ◊

Remark 21.54

(i) If G is of the form $G_t = \int_0^t f(s) \, ds$ for some locally integrable function f, then we have $G \in \mathcal{C}_v$ with $V_t^1(G) = \int_0^t |f(s)| \, ds$.
(ii) If $G = G^+ - G^-$ is the difference of two continuous monotone increasing functions G^+ and G^-, then

$$V_t^1(G) - V_s^1(G) \leq (G_t^+ - G_s^+) + (G_t^- - G_s^-) \quad \text{for all } t > s, \qquad (21.51)$$

hence we have $G \in \mathcal{C}_v$. In (21.51), equality holds if G^- and G^+ "do not grow on the same sets"; that is, more formally, if G^- and G^+ are the distribution functions of mutually singular measures μ^- and μ^+. The measures μ^- and μ^+ are then the Jordan decomposition of the signed measure $\mu = \mu^+ - \mu^-$ whose distribution function is G. Then the **Lebesgue–Stieltjes integral** is defined by

$$\int_0^t F(s) \, dG_s := \int_{[0,t]} F \, d\mu^+ - \int_{[0,t]} F \, d\mu^-. \qquad (21.52)$$

(iii) If $G \in \mathcal{C}_v$, then clearly

$$G_t^+ := \frac{1}{2} \left(V_t^1(G) + G_t \right) \quad \text{and} \quad G_t^- := \frac{1}{2} \left(V_t^1(G) - G_t \right)$$

is a decomposition of G as in (ii). ◊

The fact that the paths of Brownian motion are nowhere differentiable can be used to infer that the paths have infinite variation. However, there is also a simple direct argument.

Theorem 21.55 *Let* W *be a Brownian motion. Then* $V_t^1(W) = \infty$ *almost surely for every* $t > 0$.

Proof It is enough to consider $t = 1$ and to show

$$Y_n := \sum_{i=1}^{2^n} \left| W_{i2^{-n}} - W_{(i-1)2^{-n}} \right| \xrightarrow{n \to \infty} \infty \quad \text{a.s.} \tag{21.53}$$

We have $\mathbf{E}[Y_n] = 2^{n/2} \mathbf{E}[|W_1|] = 2^{n/2}\sqrt{2/\pi}$ and $\mathbf{Var}[Y_n] = 1 - 2/\pi$. By Chebyshev's inequality,

$$\sum_{n=1}^{\infty} \mathbf{P}\left[Y_n \le \frac{1}{2} 2^{n/2}\sqrt{2/\pi} \right] \le \sum_{n=1}^{\infty} \frac{2\pi - 4}{2^n} = 2\pi - 4 < \infty.$$

Using the Borel–Cantelli lemma, this implies (21.53). □

Evidently, the variation is too crude a measure to quantify essential path properties of Brownian motion. Hence, instead of the increments (in the definition of the variation), we will sum up the (smaller) *squared* increments. For the definition of this *square* variation, more care is needed than in Definition 21.52 for the variation.

Definition 21.56 *A sequence* $\mathcal{P} = (\mathcal{P}^n)_{n \in \mathbb{N}}$ *of countable subsets of* $[0, \infty)$,

$$\mathcal{P}^n := \{t_0, t_1, t_2, \ldots\} \quad \text{with } 0 = t_0 < t_1 < t_2 < \ldots,$$

is called an **admissible partition sequence** *if*

(i) $\mathcal{P}^1 \subset \mathcal{P}^2 \subset \ldots$,
(ii) $\sup \mathcal{P}^n = \infty$ *for every* $n \in \mathbb{N}$, *and*
(iii) *the* **mesh size**

$$|\mathcal{P}^n| := \sup_{t \in \mathcal{P}^n} \min_{s \in \mathcal{P}^n, s \ne t} |s - t|$$

tends to 0 *as* $n \to \infty$.

If $0 \le S < T$, *then define*

$$\mathcal{P}^n_{S,T} := \mathcal{P}^n \cap [S, T) \quad \text{and} \quad \mathcal{P}^n_T := \mathcal{P}^n \cap [0, T).$$

If $t = t_k \in \mathcal{P}^n_T$, *then let* $t' := t_{k+1} \wedge T = \min\{s \in \mathcal{P}^n_T \cup \{T\} : s > t\}$.

Example 21.57 $\mathcal{P}^n = \{k2^{-n} : k = 0, 1, 2, \ldots\}$. ◊

Definition 21.58 *For continuous* $F, G : [0, \infty) \to \mathbb{R}$ *and for* $p \ge 1$, *define the* *p-variation of* G *(along* $\mathcal{P}$*) by*

$$V_T^p(G) := V_T^{\mathcal{P}, p}(G) := \lim_{n \to \infty} \sum_{t \in \mathcal{P}^n_T} \left| G_{t'} - G_t \right|^p \quad \text{for } T \ge 0$$

*if the limit exists. In particular, $\langle G \rangle := V^2(G)$ is called the **square variation** of G. If $T \mapsto V_T^2(G)$ is continuous, then we write $G \in C_{qv} := C_{qv}^{\mathcal{P}}$.*

If, for every $T \geq 0$, the limit

$$V_T^{\mathcal{P},2}(F, G) := \lim_{n \to \infty} \sum_{t \in \mathcal{P}_T^n} (F_{t'} - F_t)(G_{t'} - G_t)$$

*exists, then we call $\langle F, G \rangle := V^2(F, G) := V^{\mathcal{P},2}(F, G)$ the **quadratic covariation** of F and G (along $\mathcal{P}$).*

Remark 21.59 If $p' > p$ and $V_T^p(G) < \infty$, then $V_T^{p'}(G) = 0$. In particular, we have $\langle G \rangle \equiv 0$ if G has locally finite variation. ◊

Remark 21.60 By the triangle inequality, we have

$$\sum_{t \in \mathcal{P}_T^{n+1}} |G_{t'} - G_t| \geq \sum_{t \in \mathcal{P}_T^n} |G_{t'} - G_t| \quad \text{for all } n \in \mathbb{N}, \ T \geq 0.$$

Hence in the case $p = 1$, the limit always exists and coincides with $V^1(G)$ from Definition 21.52 (and is hence independent of the particular choice of $\mathcal{P}$). A similar inequality does not hold for V^2 and thus the limit need not exist or may depend on the choice of $\mathcal{P}$. In the following, we will, however, show that, for a large class of continuous stochastic processes, V^2 exists almost surely along a suitable subsequence of partitions and is almost surely unique. ◊

Remark 21.61

(i) If $\langle F + G \rangle_T$ and $\langle F - G \rangle_T$ exist, then the covariation $\langle F, G \rangle_T$ exists and the **polarization formula** holds:

$$\langle F, G \rangle_T = \frac{1}{4} (\langle F + G \rangle_T - \langle F - G \rangle_T).$$

(ii) If $\langle F \rangle_T$, $\langle G \rangle_T$ and $\langle F, G \rangle_T$ exist, then by the Cauchy–Schwarz inequality, we have for the approximating sums

$$V_T^1(\langle F, G \rangle) \leq \sqrt{\langle F \rangle_T \langle G \rangle_T}. \quad ◊$$

Remark 21.62 If $f \in C^1(\mathbb{R})$ and $G \in C_{qv}$, then (exercise!) in the sense of the Lebesgue–Stieltjes integral

$$\langle f(G) \rangle_T = \int_0^T (f'(G_s))^2 d\langle G \rangle_s. \quad ◊$$

Corollary 21.63 *If F has locally finite square variation and if $\langle G \rangle \equiv 0$ (hence, in particular, if G has locally finite variation), then $\langle F, G \rangle \equiv 0$ and $\langle F + G \rangle = \langle F \rangle$.*

Theorem 21.64 *For Brownian motion W and for every admissible sequence of partitions, we have*

$$\langle W \rangle_T = T \quad \text{for all } T \geq 0 \quad a.s.$$

Proof We prove this only for the case where

$$\sum_{n=1}^{\infty} |\mathcal{P}^n| < \infty. \tag{21.54}$$

For the general case, we only sketch the argument.

Accordingly, assume (21.54). If $\langle W \rangle$ exists, then $T \mapsto \langle W \rangle_T$ is monotone increasing. Hence, it is enough to show that $\langle W \rangle_T$ exists for every $T \in \mathbb{Q}^+ = \mathbb{Q} \cap [0, \infty)$ and that almost surely $\langle W \rangle_T = T$. Since $(\widetilde{W}_t)_{t \geq 0} = \left(T^{-1/2} W_{tT}\right)_{t \geq 0}$ is a Brownian motion, and since $\langle \widetilde{W} \rangle_1 = T^{-1} \langle W \rangle_T$, it is enough to consider the case $T = 1$.

Define

$$Y_n := \sum_{t \in \mathcal{P}_1^n} (W_{t'} - W_t)^2 \quad \text{for all } n \in \mathbb{N}.$$

Then $\mathbf{E}[Y_n] = \sum_{t \in \mathcal{P}_1^n} (t' - t) = 1$ and

$$\mathbf{Var}[Y_n] = \sum_{t \in \mathcal{P}_1^n} \mathbf{Var}\big[(W_{t'} - W_t)^2\big] = 2 \sum_{t \in \mathcal{P}_1^n} (t' - t)^2 \leq 2 |\mathcal{P}^n|.$$

By assumption (21.54), we thus have $\sum_{n=1}^{\infty} \mathbf{Var}[Y_n] \leq 2 \sum_{n=1}^{\infty} |\mathcal{P}^n| < \infty$; hence $Y_n \xrightarrow{n \to \infty} 1$ almost surely.

If we drop the assumption (21.54), then we still have $\mathbf{Var}[Y_n] \xrightarrow{n \to \infty} 0$; hence $Y_n \xrightarrow{n \to \infty} 1$ in probability. However, it is not too hard to show that $(Y_n)_{n \in \mathbb{N}}$ is a backwards martingale (see, e.g., [140, Theorem I.28]) and thus converges almost surely to 1. $\qquad\square$

Corollary 21.65 *If W and $\widetilde{W}$ are independent Brownian motions, then we have* $\langle W, \widetilde{W} \rangle_T = 0$.

Proof The continuous processes $((W + \widetilde{W})/\sqrt{2})$ and $(W - \widetilde{W})/\sqrt{2})$ have independent normally distributed increments. Hence they are Brownian motions. By Remark 21.61(i), we have

$$4\langle W, \widetilde{W} \rangle_T = \langle W + \widetilde{W} \rangle_T - \langle W - \widetilde{W} \rangle_T$$

$$= 2\langle (W + \widetilde{W})/\sqrt{2} \rangle_T - 2\langle (W - \widetilde{W})/\sqrt{2} \rangle_T = 2T - 2T = 0. \quad \square$$

Clearly, $(W_t \, \widetilde{W}_t)_{t \geq 0}$ is a continuous martingale. Now, by Exercise 21.4.2, the process $(W_t^2 - t)_{t \geq 0}$ is also a continuous martingale. Thus, as shown above, the processes $W^2 - \langle W \rangle$ and $W \widetilde{W} - \langle W, \widetilde{W} \rangle$ are martingales. We will see (Theorem 21.70) that the square variation $\langle M(\omega) \rangle$ of a square integrable continuous martingale M always exists (for almost all ω) and that the process $\langle M \rangle$ is uniquely determined by the property that $M^2 - \langle M \rangle$ is a martingale.

In order to obtain a similar statement for continuous martingales that are not square integrable, we make the following definition.

Definition 21.66 (Local martingale) *Let $\mathbb{F}$ be a filtration on $(\Omega, \mathcal{F}, \mathbf{P})$ and let τ be an $\mathbb{F}$-stopping time. An adapted real-valued stochastic process $M = (M_t)_{t \geq 0}$ is called a **local martingale** up to time τ if there exists a sequence $(\tau_n)_{n \in \mathbb{N}}$ of stopping times such that $\tau_n \uparrow \tau$ almost surely and such that, for every $n \in \mathbb{N}$, the stopped process $M^{\tau_n} = (M_{\tau_n \wedge t})_{t \geq 0}$ is a uniformly integrable martingale. Such a sequence $(\tau_n)_{n \in \mathbb{N}}$ is called a **localising sequence** for M. M is called a local martingale if M is a local martingale up to time $\tau \equiv \infty$. Denote by $\mathcal{M}_{loc,c}$ the space of continuous local martingales.*

Remark 21.67 Let M be a continuous adapted process and let τ be a stopping time. Then the following are equivalent:

 (i) M is a local martingale up to time τ.
 (ii) There is a sequence $(\tau_n)_{n \in \mathbb{N}}$ of stopping times with $\tau_n \uparrow \tau$ almost surely and such that every M^{τ_n} is a martingale.
 (iii) There is a sequence $(\tau_n)_{n \in \mathbb{N}}$ of stopping times with $\tau_n \uparrow \tau$ almost surely and such that every M^{τ_n} is a bounded martingale.

Indeed, (iii) $\Longrightarrow$ (i) $\Longrightarrow$ (ii) is trivial. Hence assume that (ii) holds, and define τ'_n by

$$\tau'_n := \inf\{t \geq 0 : |M_t| \geq n\} \quad \text{for all } n \in \mathbb{N}.$$

Since M is continuous, we have $\tau'_n \uparrow \infty$. Hence $(\sigma_n)_{n \in \mathbb{N}} := (\tau_n \wedge \tau'_n)_{n \in \mathbb{N}}$ is a localising sequence for M such that every M^{σ_n} is a bounded martingale. $\Diamond$

Remark 21.68 A bounded local martingale M is a martingale. Indeed, assume that $|M_t| \leq C < \infty$ almost surely for all $t \geq 0$ and that $(\tau_n)_{n \in \mathbb{N}}$ is a localising sequence for M. Let $t > s \geq 0$ and $A \in \mathcal{F}_s$. Then $A \cap \{\tau_n > s\} \in \mathcal{F}_{\tau_n \wedge s}$ and hence

$$\mathbf{E}\big[M_{\tau_n \wedge t} \, \mathbb{1}_{A \cap \{\tau_n > s\}}\big] = \mathbf{E}\big[M_{\tau_n \wedge s} \, \mathbb{1}_{A \cap \{\tau_n > s\}}\big].$$

Since $\tau_n \uparrow \infty$, the dominated convergence theorem (Corollary 6.26) yields

$$\mathbf{E}\big[M_t \, \mathbb{1}_A\big] = \mathbf{E}\big[M_s \, \mathbb{1}_A\big].$$

Hence $\mathbf{E}[M_t \mid \mathcal{F}_s] = M_s$ and thus M is a martingale. $\Diamond$

Example 21.69

(i) Every martingale is a local martingale.

(ii) In Remark 21.68, we saw that bounded local martingales are martingales. On the other hand, even a uniformly integrable local martingale need not be a martingale: Let $W = (W^1, W^2, W^3)$ be a three-dimensional Brownian motion (that is, W^1, W^2 and W^3 are independent Brownian motions) that starts at $W_0 = x \in \mathbb{R}^3 \setminus \{0\}$. Let

$$u(y) = \|y\|^{-1} \quad \text{for } y \in \mathbb{R}^3 \setminus \{0\}.$$

It is easy to check that u is harmonic; that is, $\triangle u(y) = 0$ for all $y \neq 0$. We will see later (Corollary 25.34) that this implies that $M := (u(W_t))_{t \geq 0}$ is a local martingale. Define a localising sequence for M by

$$\tau_n := \inf \big\{ t > 0 : M_t \geq n \big\} = \inf \big\{ t > 0 : \|W_t\| \leq 1/n \big\}, \quad n \in \mathbb{N}.$$

An explicit computation with the three-dimensional normal distribution shows $\mathbf{E}[M_t] \leq t^{-1/2} \overset{t \to \infty}{\longrightarrow} 0$; hence M is integrable but is not a martingale. Since $M_t \overset{t \to \infty}{\longrightarrow} 0$ in L^1, M is uniformly integrable. $\Diamond$

Theorem 21.70 *Let M be a continuous local martingale.*

(i) There exists a unique continuous, monotone increasing, adapted process $\langle M \rangle = (\langle M \rangle_t)_{t \geq 0}$ with $\langle M \rangle_0 = 0$ such that

$$\big(M_t^2 - \langle M \rangle_t \big)_{t \geq 0} \quad \text{is a continuous local martingale.}$$

(ii) If M is a continuous square integrable martingale, then $M^2 - \langle M \rangle$ is a martingale.

(iii) For every admissible sequence of partitions $\mathcal{P} = (\mathcal{P}^n)_{n \in \mathbb{N}}$, we have

$$U_T^n := \sum_{t \in \mathcal{P}_T^n} \big(M_{t'} - M_t \big)^2 \overset{n \to \infty}{\longrightarrow} \langle M \rangle_T \quad \text{in probability} \quad \text{for all } T \geq 0.$$

*The process $\langle M \rangle$ is called the **square variation process** of M.*

Remark 21.71 By possibly passing in (iii) to a subsequence $\mathcal{P}'$ (that might depend on T), we may assume that $U_T^n \overset{n \to \infty}{\longrightarrow} \langle M \rangle_T$ almost surely. Using the diagonal sequence argument, we obtain (as in the proof of Helly's theorem) a sequence of partitions such that $U_T^n \overset{n \to \infty}{\longrightarrow} \langle M \rangle_T$ almost surely for all $T \in \mathbb{Q}^+$. Since both $T \mapsto U_T^n$ and $T \mapsto \langle M \rangle_T$ are monotone and continuous, we get $U_T^n \overset{n \to \infty}{\longrightarrow} \langle M \rangle_T$

for all $T \geq 0$ almost surely. Hence, for this sequence of partitions, the pathwise square variation almost surely equals the square variation process:

$$\langle M(\omega) \rangle = V^2(M(\omega)) = \langle M \rangle(\omega). \quad \Diamond$$

Reflection   In (iii) we only claim stochastic convergence, not almost sure convergence. Can you find an example where the convergence is really not almost sure but only stochastic? ♠♠♠

Proof (of Theorem 21.70) Step 1.   First let $|M_t| \leq C$ almost surely for all $t \geq 0$ for some $C < \infty$. Then, in particular, M is a martingale (by Remark 21.68). Write $U_T^n = M_T^2 - M_0^2 - N_T^n$, where

$$N_T^n = 2 \sum_{t \in \mathcal{P}_T^n} M_t (M_{t'} - M_t), \quad T \geq 0,$$

is a continuous martingale. Assume that we can show that, for every $T \geq 0$, $(U_T^n)_{n \in \mathbb{N}}$ is a Cauchy sequence in $\mathcal{L}^2(\mathbf{P})$. Then also $(N_T^n)_{n \in \mathbb{N}}$ is a Cauchy sequence, and we can define $\tilde{N}_T$ as the L^2-limit of $(N_T^n)_{n \in \mathbb{N}}$. By Exercise 21.4.3, $\tilde{N}$ has a continuous modification N, and we have $N_T^n \overset{n \to \infty}{\longrightarrow} N_T$ in L^2 for all $T \geq 0$. Thus there exists a continuous process $\langle M \rangle$ with

$$U_T^n \overset{n \to \infty}{\longrightarrow} \langle M \rangle_T \quad \text{in } L^2 \quad \text{for all } T \geq 0, \tag{21.55}$$

and $N = M^2 - M_0^2 - \langle M \rangle$ is a continuous martingale.

It remains to show that, for all $T \geq 0$,

$$(U_T^n)_{n \in \mathbb{N}} \text{ is a Cauchy sequence in } L^2. \tag{21.56}$$

For $m \in \mathbb{N}$, let

$$Z_m := \max \left\{ (M_t - M_s)^2 : s \in \mathcal{P}_T^m, t \in \mathcal{P}_{s,s'}^n, n \geq m \right\}.$$

Since M is almost surely uniformly continuous on $[0, T]$, we have $Z_m \overset{m \to \infty}{\longrightarrow} 0$ almost surely. As $Z_m \leq 4C^2$, we infer

$$\mathbf{E}[Z_m^2] \overset{m \to \infty}{\longrightarrow} 0. \tag{21.57}$$

For $n \in \mathbb{N}$ and numbers $a_0, \ldots, a_n$, we have

$$(a_n - a_0)^2 - \sum_{k=0}^{n-1} (a_{k+1} - a_k)^2 = 2 \sum_{k=0}^{n-1} (a_k - a_0)(a_{k+1} - a_k).$$

In the following computation, we apply this to each summand in the outer sum to obtain for $m \in \mathbb{N}$ and $n \geq m$

$$
U_T^m - U_T^n = \sum_{s \in \mathcal{P}_T^m} \left((M_{s'} - M_s)^2 - \sum_{t \in \mathcal{P}_{s,s'}^n} (M_{t'} - M_t)^2 \right)
$$
$$
= 2 \sum_{s \in \mathcal{P}_T^m} \sum_{t \in \mathcal{P}_{s,s'}^n} (M_t - M_s)(M_{t'} - M_t). \tag{21.58}
$$

Since M is a martingale, for $s_1, s_2 \in \mathcal{P}_T^m$ and $t_1 \in \mathcal{P}_{s_1,s_1'}^n$, $t_2 \in \mathcal{P}_{s_2,s_2'}^n$ with $t_1 < t_2$, we have

$$
\mathbf{E}\left[(M_{t_1} - M_{s_1})(M_{t_1'} - M_{t_1})(M_{t_2} - M_{s_2})(M_{t_2'} - M_{t_2}) \right]
$$
$$
= \mathbf{E}\left[(M_{t_1} - M_{s_1})(M_{t_1'} - M_{t_1})(M_{t_2} - M_{s_2}) \, \mathbf{E}\left[M_{t_2'} - M_{t_2} \mid \mathcal{F}_{t_2} \right] \right] = 0.
$$

If we use (21.58) to compute the expectation of $\left(U_T^m - U_T^n \right)^2$, then the mixed terms vanish and we get (using the Cauchy–Schwarz inequality in the third line)

$$
\mathbf{E}\left[(U_T^n - U_T^m)^2 \right] = 4 \, \mathbf{E}\left[\sum_{s \in \mathcal{P}_T^m} \sum_{t \in \mathcal{P}_{s,s'}^n} (M_t - M_s)^2 (M_{t'} - M_t)^2 \right]
$$
$$
\leq 4 \, \mathbf{E}\left[Z_m \sum_{t \in \mathcal{P}_T^n} (M_{t'} - M_t)^2 \right] \tag{21.59}
$$
$$
\leq 4 \, \mathbf{E}[Z_m^2]^{1/2} \, \mathbf{E}\left[\left(\sum_{t \in \mathcal{P}_T^n} (M_{t'} - M_t)^2 \right)^2 \right]^{1/2}.
$$

For the second factor,

$$
\mathbf{E}\left[\left(\sum_{t \in \mathcal{P}_T^n} (M_{t'} - M_t)^2 \right)^2 \right] = \mathbf{E}\left[\sum_{t \in \mathcal{P}_T^n} (M_{t'} - M_t)^4 \right]
$$
$$
+ 2 \, \mathbf{E}\left[\sum_{s \in \mathcal{P}_T^n} (M_{s'} - M_s)^2 \sum_{t \in \mathcal{P}_{s',T}^n} (M_{t'} - M_t)^2 \right]. \tag{21.60}
$$

The first summand in (21.60) is bounded by

$$
4C^2 \, \mathbf{E}\left[\sum_{t \in \mathcal{P}_T^n} (M_{t'} - M_t)^2 \right] = 4C^2 \, \mathbf{E}\left[(M_T - M_0)^2 \right] \leq 16\, C^4.
$$

The second summand in (21.60) equals

$$2\,\mathbf{E}\left[\sum_{s\in\mathcal{P}_T^n}(M_{s'}-M_s)^2\,\mathbf{E}\left[\sum_{t\in\mathcal{P}_{s',T}^n}(M_{t'}-M_t)^2\,\middle|\,\mathcal{F}_{s'}\right]\right]$$

$$=2\,\mathbf{E}\left[\sum_{s\in\mathcal{P}_T^n}(M_{s'}-M_s)^2\,\mathbf{E}\left[(M_T-M_{s'})^2\,\middle|\,\mathcal{F}_{s'}\right]\right]$$

$$\leq 8C^2\,\mathbf{E}\left[(M_T-M_0)^2\right]\leq 32\,C^4.$$

Together with (21.59) and (21.57), we obtain

$$\sup_{n\geq m}\mathbf{E}\left[(U_T^n-U_T^m)^2\right]\leq 16\sqrt{3}\,C^2\,\mathbf{E}[Z_m^2]^{1/2}\overset{m\to\infty}{\longrightarrow}0.$$

This shows (21.56).

Step 2.   Now let $M\in\mathcal{M}_{loc,c}$ and let $(\tau_N)_{N\in\mathbb{N}}$ be a localising sequence such that every M^{τ_N} is a bounded martingale (see Remark 21.67). By Step 1, for $T\geq 0$ and $N\in\mathbb{N}$, we have

$$U_T^{N,n}:=\sum_{t\in\mathcal{P}_T^n}\left(M_{t'}^{\tau_N}-M_t^{\tau_N}\right)^2\overset{n\to\infty}{\Longrightarrow}\langle M^{\tau_N}\rangle_T\ \text{ in }L^2.$$

Since $U_T^{N,n}=U_T^{N+1,n}$ if $T\leq\tau_N$, there is a continuous process U with $U_T^{N,n}\overset{n\to\infty}{\longrightarrow}U_T$ in probability if $T\leq\tau_N$. Thus $\langle M^{\tau_N}\rangle_T=\langle M\rangle_T:=U_T$ if $T\leq\tau_N$. Since $\tau_N\uparrow\infty$ almost surely, for all $T\geq 0$,

$$U_T^n\overset{n\to\infty}{\longrightarrow}\langle M\rangle_T\ \text{ in probability.}$$

As $\left((M_T^{\tau_N})^2-\langle M^{\tau_N}\rangle_T\right)_{T\geq 0}$ is a continuous martingale and since $\langle M^{\tau_N}\rangle=\langle M\rangle^{\tau_N}$, we have $M^2-\langle M\rangle\in\mathcal{M}_{loc,c}$.

Step 3.   It remains to show (ii). Let M be a continuous square integrable martingale and let $(\tau_n)_{n\in\mathbb{N}}$ be a localising sequence for the local martingale $M^2-\langle M\rangle$. Let $T>0$ and let $\tau\leq T$ be a stopping time. Then $M_{\tau_n\wedge\tau}^2\leq\mathbf{E}[M_T^2\,|\,\mathcal{F}_{\tau_n\wedge\tau}]$ since M^2 is a nonnegative submartingale. Hence $(M_{\tau_n\wedge\tau}^2)_{n\in\mathbb{N}}$ is uniformly integrable and thus (using the monotone convergence theorem in the last step)

$$\mathbf{E}[M_\tau^2]=\lim_{n\to\infty}\mathbf{E}[M_{\tau_n\wedge\tau}^2]=\lim_{n\to\infty}\mathbf{E}[\langle M\rangle_{\tau_n\wedge\tau}]+\mathbf{E}[M_0^2]=\mathbf{E}[\langle M\rangle_\tau]+\mathbf{E}[M_0^2].$$

Thus, by the optional sampling theorem, $M^2-\langle M\rangle$ is a martingale.

Step 4 (Uniqueness).   Let A and A' be continuous, monotone increasing, adapted processes with $A_0=A_0'$ such that M^2-A and M^2-A' are local martingales. Then also $N = A - A'$ is a local martingale, and for almost all ω, the path $N(\omega)$ has locally finite variation. Thus $\langle N \rangle \equiv 0$ and hence $N^2 - \langle N \rangle = N^2$ is a continuous local martingale with $N_0 = 0$. Let $(\tau_n)_{n \in \mathbb{N}}$ be a localising sequence for N^2. Then $\mathbf{E}[N^2_{\tau_n \wedge t}] = 0$ for any $n \in \mathbb{N}$ and $t \geq 0$; hence $N^2_{\tau_n \wedge t} = 0$ almost surely and thus $N^2_t = \lim_{n \to \infty} N^2_{\tau_n \wedge t} = 0$ almost surely. We conclude $A = A'$. □

Corollary 21.72 *Let M be a continuous local martingale with $\langle M \rangle \equiv 0$. Then $M_t = M_0$ for all $t \geq 0$ almost surely. In particular, this holds if the paths of M have locally finite variation.*

Corollary 21.73 *Let $M, N \in \mathcal{M}_{loc,c}$. Then there exists a unique continuous adapted process $\langle M, N \rangle$ with almost surely locally finite variation and $\langle M, N \rangle_0 = 0$ such that*

$$MN - \langle M, N \rangle \text{ is a continuous local martingale.}$$

$\langle M, N \rangle$ *is called the* **quadratic covariation process** *of M and N. For every admissible sequence of partitions $\mathcal{P}$ and for every $T \geq 0$, we have*

$$\langle M, N \rangle_T = \lim_{n \to \infty} \sum_{t \in \mathcal{P}^n_T} (M_{t'} - M_t)(N_{t'} - N_t) \text{ in probability.} \qquad (21.61)$$

Proof **Existence** Manifestly, $M + N, M - N \in \mathcal{M}_{loc,c}$. Define

$$\langle M, N \rangle := \frac{1}{4}\big(\langle M + N \rangle - \langle M - N \rangle\big).$$

As the difference of two monotone increasing functions, $\langle M, N \rangle$ has locally finite variation. Using Theorem 21.70(iii), we get (21.61). Furthermore,

$$MN - \langle M, N \rangle = \frac{1}{4}\big((M+N)^2 - \langle M+N \rangle\big) - \frac{1}{4}\big((M-N)^2 - \langle M-N \rangle\big)$$

is a local martingale.

Uniqueness Let A and A' with $A_0 = A'_0 = 0$ be continuous, adapted and with locally finite variation such that $MN - A$ and $MN - A'$ are in $\mathcal{M}_{loc,c}$. Then $A - A' \in \mathcal{M}_{loc,c}$ have locally finite variation; hence $A - A' = 0$. □

Corollary 21.74 *If $M \in \mathcal{M}_{loc,c}$ and A are continuous and adapted with $\langle A \rangle \equiv 0$, then $\langle M + A \rangle = \langle M \rangle$.*

If M is a continuous local martingale up to the stopping time τ, then $M^\tau \in \mathcal{M}_{loc,c}$, and we write $\langle M \rangle_t := \langle M^\tau \rangle_t$ for $t < \tau$.

Theorem 21.75 *Let τ be a stopping time, M be a continuous local martingale up to τ and $\tau_0 < \tau$ a stopping time with $\mathbf{E}[\langle M \rangle_{\tau_0}] < \infty$. Then $\mathbf{E}[M_{\tau_0}] = \mathbf{E}[M_0]$, and M^{τ_0} is an L^2-bounded martingale if $\mathbf{E}[M_0^2] < \infty$.*

Proof Let $\tau_n \uparrow \tau$ be a localising sequence of stopping times for M such that every M^{τ_n} is even a bounded martingale (see Remark 21.67). Then $M^{\tau_0 \wedge \tau_n}$ is also a bounded martingale, and for every $t \geq 0$, we have

$$\mathbf{E}\big[M^2_{\tau_0 \wedge \tau_n \wedge t}\big] = \mathbf{E}\big[M^2_0\big] + \mathbf{E}\big[\langle M \rangle_{\tau_0 \wedge \tau_n \wedge t}\big] \leq \mathbf{E}\big[M^2_0\big] + \mathbf{E}\big[\langle M \rangle_{\tau_0}\big] < \infty. \quad (21.62)$$

Hence $\big((M_{\tau_0 \wedge \tau_n \wedge t}), \ n \in \mathbb{N}, \ t \geq 0\big)$ is bounded in L^2 and is thus uniformly integrable. Therefore, by the optional sampling theorem for uniformly integrable martingales,

$$\mathbf{E}[M_{\tau_0}] = \lim_{n \to \infty} \mathbf{E}[M_{\tau_0 \wedge \tau_n}] = \mathbf{E}[M_0],$$

and, for $t > s$,

$$\begin{aligned}
\mathbf{E}\big[M_t^{\tau_0} \,\big|\, \mathcal{F}_s\big] &= \mathbf{E}\Big[\lim_{n \to \infty} M_t^{\tau_0 \wedge \tau_n} \,\Big|\, \mathcal{F}_s \Big] \\
&= \lim_{n \to \infty} \mathbf{E}\big[M_t^{\tau_0 \wedge \tau_n} \,\big|\, \mathcal{F}_s\big] \\
&= \lim_{n \to \infty} M_s^{\tau_0 \wedge \tau_n} = M_s^{\tau_0}.
\end{aligned}$$

Hence M^{τ_0} is a martingale. $\qquad\square$

Corollary 21.76 *If $M \in \mathcal{M}_{loc,c}$ with $\mathbf{E}[M_0^2] < \infty$ and $\mathbf{E}\big[\langle M \rangle_t\big] < \infty$ for every $t \geq 0$, then M is a square integrable martingale.*

> **Takeaways** Brownian motion is nowhere differentiable and hence has infinite variation. On the other hand, the square variation (defined as limit of sums of squared increments along finer and finer partition sequences) is finite over compact sets. In the general context, we identify the pathwise defined square variation as the variance process of a local martingale M. This is the increasing process $\langle M \rangle$ from the Doob decomposition of M^2 which turns $M^2 - \langle M \rangle$ into a local martingale. Hence, $\langle M \rangle$ is a measure for the random fluctuations of M.

Exercise 21.10.1 Show that the random variables $(Y_n)_{n \in \mathbb{N}}$ from the proof of Theorem 21.64 form a backwards martingale. ♣

Exercise 21.10.2 Let $f : [0, \infty) \to \mathbb{R}$ be continuous and let $X \in \mathcal{C}_{qv}^{\mathcal{P}}$ for the admissible sequence of partitions $\mathcal{P}$. Show that

$$\int_0^T f(s) \, d\langle X \rangle_s = \lim_{n \to \infty} \sum_{t \in \mathcal{P}_T^n} f(t) \big(X_{t'} - X_t \big)^2 \quad \text{for all } T \geq 0. \quad \clubsuit$$

Exercise 21.10.3 Show by a counterexample that if M is a continuous local martingale with $M_0 = 0$ and if τ is a stopping time with $\mathbf{E}\big[\langle M \rangle_\tau \big] = \infty$, then this does not necessarily imply $\mathbf{E}\big[M_\tau^2 \big] = \infty$. $\clubsuit$

Chapter 22
Law of the Iterated Logarithm

For sums of independent random variables we already know two limit theorems: the law of large numbers and the central limit theorem. The law of large numbers describes for large $n \in \mathbb{N}$, the typical behavior, or average value behavior, of sums of n random variables. On the other hand, the central limit theorem quantifies the typical fluctuations about this average value.

In Chap. 23, we will study *atypically* large deviations from the average value in greater detail. The aim of this chapter is to quantify the *typical* fluctuations of the whole process as $n \to \infty$. The main message is: While for fixed time the partial sum S_n deviates by approximately $\sqrt{n}$ from its expected value (central limit theorem), the *maximal* fluctuation up to time n is of order $\sqrt{n \log \log n}$ (Hartman–Wintner theorem, Theorem 22.11).

We start with the simpler task of computing the fluctuations for Brownian motion (Theorem 22.1). After that, we will see how sums of independent centered random variables (with finite variance) can be embedded in a Brownian motion (Skorohod's theorem, Theorem 22.5). This embedding will be used to prove the Hartman–Wintner theorem.

In this chapter, we follow essentially the exposition of [39, Section 8.8].

22.1 Iterated Logarithm for the Brownian Motion

Let $(B_t)_{t \geq 0}$ be a Brownian motion. In Example 21.16, as an application of Blumenthal's 0–1 law, we saw that $\limsup_{t \downarrow 0} B_t/\sqrt{t} = \infty$ a.s. Since by Theorem 21.14, $(t B_{1/t})_{t \geq 0}$ also is a Brownian motion, we get

$$\limsup_{t \to \infty} \frac{B_t}{\sqrt{t}} = \infty \quad \text{a.s.}$$

© The Editor(s) (if applicable) and The Author(s), under exclusive license to Springer Nature Switzerland AG 2020
A. Klenke, *Probability Theory*, Universitext,
https://doi.org/10.1007/978-3-030-56402-5_22

The aim of this section is to replace $\sqrt{t}$ by a function such that the limes superior is finite and nontrivial.

Theorem 22.1 (Law of the iterated logarithm for Brownian motion)

$$\limsup_{t\to\infty} \frac{B_t}{\sqrt{2t\log\log(t)}} = 1 \quad a.s. \tag{22.1}$$

Before proving the theorem, we state an elementary lemma.

Lemma 22.2 *Let $X \sim \mathcal{N}_{0,1}$ be standard normally distributed. Then, for any $x > 0$,*

$$\frac{1}{\sqrt{2\pi}} \frac{1}{x+\frac{1}{x}} e^{-x^2/2} \leq \mathbf{P}[X \geq x] \leq \frac{1}{\sqrt{2\pi}} \frac{1}{x} e^{-x^2/2}. \tag{22.2}$$

Proof Let $\varphi(t) = \frac{1}{\sqrt{2\pi}} e^{-t^2/2}$ be the density of the standard normal distribution. Partial integration yields the second inequality in (22.2),

$$\mathbf{P}[X \geq x] = \int_x^\infty \frac{1}{t}(t\varphi(t))\, dt = -\frac{1}{t}\varphi(t)\Big|_x^\infty - \int_x^\infty \frac{1}{t^2}\varphi(t)\, dt \leq \frac{1}{x}\varphi(x).$$

Similarly, we get

$$\mathbf{P}[X \geq x] \geq \frac{1}{x}\varphi(x) - \frac{1}{x^2}\int_x^\infty \varphi(t)\, dt = \frac{1}{x}\varphi(x) - \frac{1}{x^2}\mathbf{P}[X \geq x].$$

This implies the first inequality in (22.2).   □

Proof of Theorem 22.1
Step 1. "$\leq$" Let $\alpha > 1$, and define $t_n = \alpha^n$ for $n \in \mathbb{N}$. Later, we let $\alpha \downarrow 1$. Define $f(t) = 2\alpha^2 \log\log t$. Then by the reflection principle (Theorem 21.19) and using the abbreviation $B_{[a,b]} := \{B_t : t \in [a,b]\}$, we obtain

$$\mathbf{P}\left[\sup B_{[t_n,t_{n+1}]} > \sqrt{t_n f(t_n)}\right] \leq \mathbf{P}\left[t_{n+1}^{-1/2} \sup B_{[0,t_{n+1}]} > \sqrt{f(t_n)/\alpha}\right]$$

$$= \mathbf{P}\left[\sup B_{[0,1]} > \sqrt{f(t_n)/\alpha}\right]$$

$$\leq \sqrt{\frac{\alpha}{f(t_n)}}\, e^{-f(t_n)/2\alpha} \tag{22.3}$$

$$= (\log\alpha)^{-\alpha} \sqrt{\frac{\alpha}{f(t_n)}}\, n^{-\alpha}$$

$$\leq n^{-\alpha} \quad \text{for large enough } n.$$

In the next to last step, we used

$$\frac{f(t_n)}{2\alpha} = \alpha\big(\log(n\log\alpha)\big) = \alpha\log n + \alpha\log\log\alpha.$$

Since $\alpha > 1$, the right-hand side of (22.3) is summable in n:

$$\sum_{n=1}^{\infty}\mathbf{P}\Big[\sup B_{[t_n, t_{n+1}]} > \sqrt{t_n f(t_n)}\Big] < \infty.$$

The Borel–Cantelli lemma (Theorem 2.7) then yields (note that $t \mapsto \sqrt{tf(t)}$ is monotone increasing)

$$\limsup_{t\to\infty}\frac{B_t}{\sqrt{tf(t)}} \leq 1 \quad \text{a.s.}$$

Now let $\alpha \downarrow 1$ to obtain

$$\limsup_{t\to\infty}\frac{B_t}{\sqrt{2t\log\log t}} \leq 1 \quad \text{a.s.} \tag{22.4}$$

Step 2. "$\geq$"   Here we show the other inequality in (22.1). To this end, we let $\alpha \to \infty$. Let $\beta := \frac{\alpha}{\alpha-1} > 1$ and $g(t) = \frac{2}{\beta^2}\log\log t$. Choose n_0 large enough that $\beta g(t_n) \geq 1$ for all $n \geq n_0$. Then, by Brownian scaling (note that $t_n - t_{n-1} = \frac{1}{\beta}t_n$) and (22.2) (since $(x + \frac{1}{x})^{-1} \geq \frac{1}{2}\frac{1}{x}$ for $x = (\beta g(t_n))^{1/2} \geq 1$),

$$\mathbf{P}\Big[B_{t_n} - B_{t_{n-1}} > \sqrt{t_n g(t_n)}\Big] = \mathbf{P}\Big[B_1 > \sqrt{\beta g(t_n)}\Big]$$

$$\geq \frac{1}{\sqrt{2\pi}}\frac{1}{2}\frac{1}{\sqrt{\beta g(t_n)}}\, e^{-\beta g(t_n)/2}$$

$$= \frac{1}{\sqrt{2\pi}}\frac{1}{2}(\log\alpha)^{-1/\beta}\frac{1}{\sqrt{\beta g(t_n)}}\, n^{-1/\beta}.$$

If $\varepsilon \in (0, 1 - 1/\beta)$, then, for sufficiently large $n \in \mathbb{N}$, the right-hand side of the above equation is $\geq n^{-\varepsilon}n^{-1/\beta} \geq n^{-1}$. Hence

$$\sum_{n=2}^{\infty}\mathbf{P}\Big[B_{t_n} - B_{t_{n-1}} > \sqrt{t_n g(t_n)}\Big] = \infty.$$

The events are independent and hence the Borel–Cantelli lemma yields

$$\mathbf{P}\Big[B_{t_n} - B_{t_{n-1}} > \sqrt{t_n g(t_n)} \quad \text{for infinitely many } n\Big] = 1. \tag{22.5}$$

Since $\dfrac{t_n \log \log t_n}{t_{n-1} \log \log t_{n-1}} \xrightarrow{n \to \infty} \alpha$, (22.4) and symmetry of Brownian motion imply that, for $\varepsilon > 0$,

$$B_{t_{n-1}} > -(1+\varepsilon)\alpha^{-1/2}\sqrt{2t_n \log \log t_n} \quad \text{for almost all } n \in \mathbb{N} \quad \text{a.s.} \quad (22.6)$$

From (22.5) and (22.6), it follows that

$$\limsup_{n \to \infty} \frac{B_{t_n}}{\sqrt{2t_n \log \log t_n}} \geq \frac{1}{\beta} - (1+\varepsilon)\alpha^{-1/2} = \frac{\alpha - 1}{\alpha} - (1+\varepsilon)\alpha^{-1/2} \quad \text{a.s.}$$

Now, letting $\alpha \to \infty$ gives $\limsup\limits_{t \to \infty} \dfrac{B_t}{\sqrt{2t \log \log t}} \geq 1$ a.s. Together with (22.4), this implies the claim of the theorem. $\square$

Corollary 22.3 *For every $s \geq 0$, a.s. we have* $\limsup\limits_{t \downarrow 0} \dfrac{B_{s+t} - B_s}{\sqrt{2t \log \log(1/t)}} = 1$.

Proof Without loss of generality, assume $s = 0$. Apply Theorem 22.1 to the Brownian motion $(t B_{1/t})$ (see Theorem 21.14). $\square$

Remark 22.4 The statement of Corollary 22.3 is about the *typical points* s of Brownian motion B. However, there might be points in which Brownian motion moves faster than $\sqrt{2t \log \log(1/t)}$. The precise statement is due to Paul Lévy [106]: Denote by $h(\delta) := \sqrt{2\delta \log(1/\delta)}$ **Lévy's modulus of continuity**. Then

$$\mathbf{P}\left[\lim_{\delta \downarrow 0} \sup_{\substack{s,t \in [0,1] \\ 0 \leq t-s \leq \delta}} |B_t - B_s|/h(\delta) = 1\right] = 1. \quad (22.7)$$

(See, e.g., [145, Theorem I.2.5] for a proof.) This implies in particular that almost surely B is not locally Hölder-$\frac{1}{2}$-continuous. $\lozenge$

Takeaways The typical fluctuation of a Brownian motion at time t is of order $\sqrt{t}$. Its maximal value by time t, however, has size $\sqrt{2t \log \log(t)}$ as $t \to \infty$. Due to the two logarithms in this formula, this statement is called law of the iterated logarithm. We have proved it by first showing it along a geometric sequence of times and then filling the gaps.

22.2   Skorohod's Embedding Theorem

In order to carry over the result of the previous section to sums of square integrable centered random variables, we use an embedding of such random variables in

a Brownian motion that is due to Skorohod. This technique also provides an
alternative proof of Donsker's invariance principle (Theorem 21.43).

Theorem 22.5 (Skorohod's embedding theorem) *Let* X *be a real random variable with* $\mathbf{E}[X] = 0$ *and* $\mathbf{Var}[X] < \infty$. *Then on a suitable probability space we
can construct a random variable* Ξ, *a Brownian motion* B *that is independent of* Ξ
and an $\mathbb{F}$-*stopping time* τ *such that*

$$B_\tau \overset{\mathcal{D}}{=} X \quad and \quad \mathbf{E}[\tau] = \mathbf{Var}[X].$$

Here the filtration $\mathbb{F}$ *is given by* $\mathcal{F}_t = \sigma(\Xi, (B_s)_{s \le t})$.

Remark 22.6 In the above theorem we can de without the additional random
variable; that is, we can choose $\mathbb{F} = \sigma(B)$. The proof is rather involved, though
(see page 579). $\Diamond$

Corollary 22.7 *Let* $X_1, X_2, \ldots$ *be i.i.d. real random variables with* $\mathbf{E}[X_1] = 0$
and $\mathbf{Var}[X_1] < \infty$. *Further, let* $S_n = X_1 + \ldots + X_n$, $n \in \mathbb{N}$. *Then on a suitable
probability space there exists a filtration* $\mathbb{F}$, *a Brownian motion* B *and* $\mathbb{F}$-*stopping
times* $0 = \tau_0 \le \tau_1 \le \tau_2 \le \ldots$ *such that* $(\tau_n - \tau_{n-1})_{n \in \mathbb{N}}$ *is i.i.d.,* $\mathbf{E}[\tau_1] = \mathbf{Var}[X_1]$
and $(B_{\tau_n})_{n \in \mathbb{N}} \overset{\mathcal{D}}{=} (S_n)_{n \in \mathbb{N}}$.

Proof (of Corollary 22.7) We only sketch the proof. The details are left to the
reader.
Choose independent triples $(B^{(n)}, \Xi^{(n)}, \tau^{(n)})$, $n \in \mathbb{N}$, as in Theorem 22.5. Let $\tau_n = \tau^{(1)} + \ldots + \tau^{(n)}$. For $t \le \tau_1$ let $B_t := B_t^{(1)}$, and define recursively

$$B_t = B_{\tau_n} + B_{t-\tau_n}^{(n+1)}, \quad \text{if } \tau_n < t \le \tau_{n+1}.$$

Using repeatedly the strong Markov property of Brownian motion, we see that B is
a Brownian motion. Now let $\mathcal{F}_t = \sigma((\Xi_n)_{n \in \mathbb{N}}, (B_s)_{s \le t})$. $\qquad\square$

We prepare for the proof of Theorem 22.5 with a lemma. In order to allow measures
as integrands, we use the following notation: If $\mu \in \mathcal{M}(E)$ is a measure and $f \in \mathcal{L}^1(\mu)$ is nonnegative, then define $\int \mu(dx) f(x) \delta_x := f\mu$, where $f\mu$ is the measure
with density f with respect to μ. This is consistent since for measurable $A \subset E$,
we then have

$$\left(\int \mu(dx) f(x) \delta_x \right)(A) = \int \mu(dx) f(x) \delta_x(A) = \int \mu(dx) f(x) \mathbb{1}_A(x) = f\mu(A).$$

Lemma 22.8 *Let* $\mu \in \mathcal{M}_1(\mathbb{R})$ *with* $\int x \, \mu(dx) = 0$ *and* $\sigma^2 := \int x^2 \, \mu(dx) < \infty$.
Then there exists a probability measure $\theta \in \mathcal{M}_1((-\infty, 0) \times [0, \infty))$ *with*

$$\mu = \int \theta(d(u, v)) \left(\frac{v}{v-u} \delta_u + \frac{-u}{v-u} \delta_v \right). \tag{22.8}$$

Furthermore, $\sigma^2 = -\int uv \, \theta(d(u, v))$.

Proof Define $m := \int_{[0,\infty)} v \, \mu(dv) = -\int_{(-\infty,0)} u \, \mu(du)$. If $m = 0$, then $\theta = \delta_{(-1,0)}$ is a possible choice. Assume now $m > 0$ and define θ by

$$\theta(d(u,v)) := m^{-1}(v-u)\,\mu(du)\mu(dv) \quad \text{for } u < 0 \text{ and } v \geq 0.$$

Then

$$\int \theta(d(u,v)) = m^{-1} \int_{(-\infty,0)} \mu(du) \int_{[0,\infty)} \mu(dv)\,(v-u)$$

$$= m^{-1} \int_{(-\infty,0)} \mu(du)\,[m - u\mu([0,\infty))]$$

$$= m^{-1}\big(m\mu((-\infty,0)) + m\mu([0,\infty))\big) = 1.$$

Hence, θ is in fact a probability measure. Furthermore,

$$\int \theta(d(u,v)) \left(\frac{v}{v-u}\delta_u + \frac{-u}{v-u}\delta_v \right)$$

$$= m^{-1} \int_{(-\infty,0)} \mu(du) \int_{[0,\infty)} \mu(dv)\,(v\delta_u - u\delta_v)$$

$$= \int_{(-\infty,0)} \mu(du)\,\delta_u + \int_{[0,\infty)} \mu(dv)\,\delta_v = \mu.$$

By (22.8), we infer

$$\sigma^2 = \int \mu(dx)\,x^2 = \int \theta(d(u,v)) \left(\frac{v}{v-u}u^2 + \frac{-u}{v-u}v^2 \right) = -\int \theta(d(u,v))\,uv. \quad \square$$

Proof (Theorem 22.5) First assume that X takes only the two values $u < 0$ and $v \geq 0$: $\mathbf{P}[X = u] = \frac{v}{v-u} = 1 - \mathbf{P}[X = v]$. Let

$$\tau_{u,v} = \inf\{t > 0 : B_t \in \{u, v\}\}.$$

By Exercise 21.2.4, we have $\mathbf{E}[B_{\tau_{u,v}}] = 0$; hence $B_{\tau_{u,v}} \overset{D}{=} X$ and $\mathbf{E}[\tau_{u,v}] = -uv$.

Now let X be arbitrary with $\mathbf{E}[X] = 0$ and $\sigma^2 := \mathbf{E}[X^2] < \infty$. Define $\mu = \mathbf{P}_X$ and $\theta = \theta_\mu$ as in Lemma 22.8. Further, let $\Xi = (\Xi_u, \Xi_v)$ be a random variable with values in $(-\infty, 0) \times [0, \infty)$ and with distribution θ.

Let $\mathbb{F} = (\mathcal{F}_t)_{t \geq 0}$ where $\mathcal{F}_t := \sigma(\Xi, B_s : s \in [0, t])$. Define $\tau := \tau_{\Xi_u, \Xi_v}$. By continuity of B, we get

$$\{\tau \leq t\} = \left\{ \sup_{s \in [0,t]} B_s \geq \Xi_v \right\} \cup \left\{ \inf_{s \in [0,t]} B_s \leq \Xi_u \right\} \in \mathcal{F}_t.$$

Hence τ is an $\mathbb{F}$-stopping time (but not a $\sigma(B)$-stopping time). For $x < 0$,

$$\mathbf{P}[X \leq x] = \int_{(-\infty,x]\times[0,\infty)} \theta(d(u,v)) \, \frac{v}{v-u}$$

$$= \int_{(-\infty,x]\times[0,\infty)} \theta(d(u,v)) \, \mathbf{P}[B_{\tau_{u,v}} = u] = \mathbf{P}[B_\tau \leq x].$$

For $x \geq 0$, we similarly get $\mathbf{P}[X > x] = \mathbf{P}[B_\tau > x]$. Summing up, we have $B_\tau \overset{\mathcal{D}}{=} X$. Furthermore,

$$\mathbf{E}[\tau] = -\mathbf{E}[\Xi_u \Xi_v] = -\int \theta(d(u,v)) \, uv = \sigma^2. \qquad \square$$

Supplement: Proof of Remark 22.6

Here we prove that in Skorohod's embedding theorem we can really do without *randomized* stopping times; that is, we can choose a stopping time with respect to the filtration generated by the Brownian motion B. In other words, the stopping time can be chosen without using additional random variables, such as the Ξ in the proof given above.

An elegant proof that is based on stochastic analysis methods can be found in Azéma and Yor; see [7] and [6]. See also [118] for a more elementary version of that proof. Here, however, we follow an elementary route whose basic idea goes back to Dubins.

For $u < 0 < v$, let $\tau_{u,v} = \inf\{t > 0 : B_t \in \{u,v\}\}$. Hence, if X is a centered random variable that takes only the values u and v, then, as shown in the proof of Theorem 22.5, $B_{\tau_{u,v}} \overset{\mathcal{D}}{=} X$ and $\mathbf{E}[\tau_{u,v}] = \mathbf{E}[X^2]$.

In a first step, we generalize this statement to binary splitting martingales. (Recall from Definition 9.42 that a binary splitting process at each time step has a choice of just two different values, which may however depend on the history of the process.) In a second step, we show that square integrable centered random variables can be expressed as limits of such martingales.

Theorem 22.9 Let $(X_n)_{n\in\mathbb{N}_0}$ be a binary splitting martingale with $X_0 = 0$. Let B be a Brownian motion and let $\mathbb{F} = \sigma(B)$ be its canonical filtration. Then there exist $\mathbb{F}$-stopping times $0 = \tau_0 \leq \tau_1 \leq \ldots$ such that

$$(X_n)_{n\in\mathbb{N}_0} \overset{\mathcal{D}}{=} (B_{\tau_n})_{n\in\mathbb{N}_0}$$

and such that $\mathbf{E}[\tau_n] = \mathbf{E}[X_n^2]$ holds for all $n \in \mathbb{N}_0$.

If $(X_n)_{n\in\mathbb{N}_0}$ is bounded in L^2 and thus converges almost surely and in L^2 to some square integrable X_∞, then $\tau := \sup_{n\in\mathbb{N}} \tau_n < \infty$ a.s., $\mathbf{E}[\tau] = \mathbf{Var}[X_\infty]$ and $X_\infty \stackrel{\mathcal{D}}{=} B_\tau$.

Proof For $n \in \mathbb{N}$, let $f_n : \mathbb{R}^{n-1} \times \{-1, +1\} \to \mathbb{R}$ and let D_n be a $\{-1, +1\}$-valued random variable such that $X_n = f_n(X_1, \ldots, X_{n-1}, D_n)$ holds (compare Definition 9.42). Without loss of generality, we may assume that f_n is monotone increasing in D_n. Let $\tau_0 := 0$ and inductively define

$$\tau_n := \inf\big\{t > \tau_{n-1} : B_t \in \{f_n(B_{\tau_1}, \ldots, B_{\tau_{n-1}}, -1), f_n(B_{\tau_1}, \ldots, B_{\tau_{n-1}}, +1)\}\big\}.$$

Let $\tilde{X}_n := B_{\tau_n}$ and

$$\tilde{D}_n := \begin{cases} 1, & \text{if } \tilde{X}_n \geq \tilde{X}_{n-1}, \\ -1, & \text{else.} \end{cases}$$

By Exercise 21.2.4 and using the strong Markov property (at τ_{n-1}), we get

$$\mathbf{P}\big[\tilde{D}_n = 1 \,\big|\, \tilde{X}_1, \ldots, \tilde{X}_{n-1}\big]$$

$$= \frac{\tilde{X}_{n-1} - f_n(\tilde{X}_1, \ldots, \tilde{X}_{n-1}, -1)}{f_n(\tilde{X}_1, \ldots, \tilde{X}_{n-1}, +1) - f_n(\tilde{X}_1, \ldots, \tilde{X}_{n-1}, -1)}$$

and $\mathbf{E}[\tau_n - \tau_{n-1}] = \mathbf{E}[(\tilde{X}_n - \tilde{X}_{n-1})^2]$. On the other hand, since $(X_n)_{n\in\mathbb{N}_0}$ is a martingale, we have

$$X_{n-1} = \mathbf{E}[X_n \,|\, X_0, \ldots, X_{n-1}]$$

$$= \sum_{i=-1,+1} \mathbf{P}[D_n = i \,|\, X_0, \ldots, X_{n-1}] \, f_n(X_1, \ldots, X_{n-1}, i).$$

Therefore,

$$\mathbf{P}\big[D_n = 1 \,\big|\, X_1, \ldots, X_{n-1}\big]$$

$$= \frac{X_{n-1} - f_n(X_1, \ldots, X_{n-1}, -1)}{f_n(X_1, \ldots, X_{n-1}, +1) - f_n(X_1, \ldots, X_{n-1}, -1)}.$$

This implies $(X_n)_{n\in\mathbb{N}_0} \stackrel{\mathcal{D}}{=} (\tilde{X}_n)_{n\in\mathbb{N}_0}$. Since $\mathbf{E}[\tau_n - \tau_{n-1}] = \mathbf{E}[(X_n - X_{n-1})^2]$, and since the martingale differences $(X_i - X_{i-1})$, $i \in \mathbb{N}$, are uncorrelated, we get $\mathbf{E}[\tau_n] = \mathbf{E}[X_n^2]$.

Finally, if (X_n) is bounded in L^2, then by the martingale convergence theorem there is a square integrable centered random variable X_∞ such that $X_n \stackrel{n\to\infty}{\longrightarrow} X_\infty$ almost surely and in L^2. In particular, we have $\mathbf{E}[X_n^2] \stackrel{n\to\infty}{\longrightarrow} \mathbf{E}[X_\infty^2]$. Clearly, $(\tau_n)_{n\in\mathbb{N}}$ is monotone increasing and thus converges to some stopping time τ. By the monotone convergence theorem, $\mathbf{E}[\tau] = \lim_{n\to\infty} \mathbf{E}[\tau_n] = \lim_{n\to\infty} \mathbf{E}[X_n^2] = \mathbf{E}[X_\infty^2] < \infty$. Hence $\tau < \infty$ a.s. As Brownian motion is continuous, we conclude

$$B_\tau = \lim_{n\to\infty} B_{\tau_n} = \lim_{n\to\infty} \tilde{X}_n \overset{D}{=} X_\infty. \qquad \Box$$

We have shown the statement of Remark 22.6 in the case where the random variable X is the limit of a binary splitting martingale. The general case is now implied by the following theorem (Figs. 22.1 and 22.2).

Theorem 22.10 *Let X be a square integrable centered random variable. Then there exists a binary splitting martingale* $(X_n)_{n\in\mathbb{N}_0}$ *with* $X_0 = 0$ *and such that* $X_n \overset{n\to\infty}{\longrightarrow} X$ *almost surely and in* L^2.

Proof We follow the idea of the proof in [118]. Let $X_0 := \mathbf{E}[X] = 0$. Inductively, for $n \in \mathbb{N}$, define

$$D_n := \begin{cases} 1, & \text{if } X \geq X_{n-1}, \\ -1, & \text{if } X < X_{n-1}, \end{cases}$$

$$\mathcal{F}_n := \sigma(D_1, \dots, D_n)$$

and

$$X_n := \mathbf{E}[X \mid \mathcal{F}_n].$$

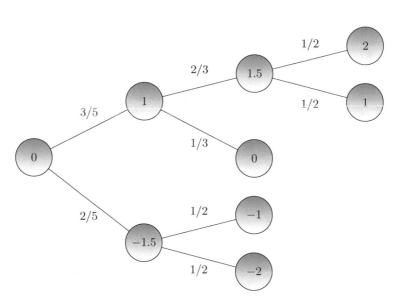

Fig. 22.1 Binary splitting martingale for the random variable X with $\mathbf{P}[X = k] = \frac{1}{5}$ for $k = -2, -1, 0, 1, 2$.

Hence there exists a map $g_n : \{-1, +1\}^n \to \mathbb{R}$ such that $g_n(D_1, \ldots, D_n) = X_n$. Clearly, $\mathbb{1}_{D_k=1} = \mathbb{1}_{X_k \geq X_{k-1}}$ almost surely for all $k \in \mathbb{N}$. Hence the $D_1, \ldots, D_k$ can be computed from the $X_1, \ldots, X_k$. Thus there exists a map $f_n : \mathbb{R}^{n-1} \times \{-1, +1\} \to \mathbb{R}$ such that $f_n(X_1, \ldots, X_{n-1}, D_n) = X_n$. Therefore, (X_n) is binary splitting.

Manifestly, $(X_n)_{n \in \mathbb{N}_0}$ is a martingale. By Jensen's inequality, we have $\mathbf{E}[X_n^2] \leq \mathbf{E}[X^2] < \infty$ for all $n \in \mathbb{N}$. Hence $(X_n)_{n \in \mathbb{N}_0}$ is bounded in L^2 and thus converges almost surely and in L^2 to some square integrable X_∞. It remains to show that $X_\infty = X$ holds almost surely. To this end, we first show

$$\lim_{n \to \infty} D_n(\omega)\big(X(\omega) - X_n(\omega)\big) = \big|X(\omega) - X_\infty(\omega)\big| \quad \text{for almost all } \omega. \quad (22.9)$$

If $X(\omega) = X_\infty(\omega)$, then (22.9) holds trivially. If $X(\omega) > X_\infty(\omega)$, then $X(\omega) > X_n(\omega)$ and thus $D_n(\omega) = 1$ for all sufficiently large n; hence (22.9) holds. Similarly, we get (22.9) if $X(\omega) < X_\infty(\omega)$.

Evidently, we have

$$\mathbf{E}\big[D_n(X - X_n)\big] = \mathbf{E}\big[D_n \, \mathbf{E}[X - X_n | \mathcal{F}_n]\big] = 0.$$

As $\big(D_n(X - X_n)\big)_{n \in \mathbb{N}}$ is bounded in L^2 (and is thus uniformly integrable), we get $\mathbf{E}[|X - X_\infty|] = \lim_{n \to \infty} \mathbf{E}[D_n(X - X_n)] = 0$; hence $X = X_\infty$ a.s. □

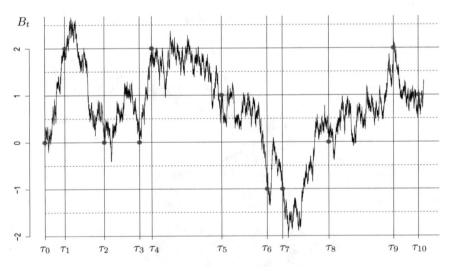

Fig. 22.2 Embedding of random variables with uniform distribution on $\{-2, -1, 0, 1, 2\}$ in a Brownian motion. After first hitting $\{-1.5, 1\}$ there are two possibilities. In the case where -1.5 is hit before 1, the Brownian motion continuous and is finally stopped upon hitting -2 or -1. In the other case, we continue until the Brownian motion hits 0 or 1.5. If it is 0, the Brownian motion stops. If it is 1.5, we continue until the motion hits 1 or 2. The random time at which the motion if finally stopped is τ_1. After τ_1, we can continue in order generate another sample of the random variable. Note that in the second period, we have to add B_{τ_1} to all numbers.

Takeaways A centred square integrable random variable can be represented by a Brownian motion evaluated at a stopping time. There is freedom in the choice of the stopping time, but there exists an optimal (minimal) stopping time in the sense that the expected value of the stopping time is the variance of the random variable. It is impossible to do better. By a repetition of the argument, the partial sums of i.i.d. centred square integrable random variables can be embedded in the path of a Brownian motion evaluated at a sequence of stopping times.

22.3 Hartman–Wintner Theorem

The goal of this section is to prove the law of the iterated logarithm for i.i.d. centered square integrable random variables X_n, $n \in \mathbb{N}$, that goes back to Hartman and Wintner (see [69]). For the special case of Rademacher random variables, the upper bound was found earlier by Khinchin in 1923 (see [97]).

Theorem 22.11 (Hartman–Wintner, law of the iterated logarithm) *Let* $X_1, X_2, \ldots$ *be i.i.d. real random variables with* $\mathbf{E}[X_1] = 0$ *and* $\mathbf{Var}[X_1] = 1$. *Let* $S_n = X_1 + \ldots + X_n$, $n \in \mathbb{N}$. *Then*

$$\limsup_{n \to \infty} \frac{S_n}{\sqrt{2n \log \log n}} = 1 \quad a.s. \tag{22.10}$$

The strategy of the proof is to embed the partial sums S_n of the random variables in a Brownian motion and then use the law of the iterated logarithm for Brownian motion. The Skorohod embedding theorem ensures that this works. We follow the exposition in [39, Section 8.8].

Proof By Corollary 22.7, on a suitable probability space there exists a filtration $\mathbb{F}$, a Brownian motion B that is an $\mathbb{F}$-martingale, and stopping times $\tau_1 \leq \tau_2 \leq \ldots$ such that $(S_n)_{n \in \mathbb{N}} \stackrel{\mathcal{D}}{=} (B_{\tau_n})_{n \in \mathbb{N}}$. Furthermore, the $(\tau_n - \tau_{n-1})_{n \in \mathbb{N}}$ are i.i.d. with $\mathbf{E}[\tau_n - \tau_{n-1}] = \mathbf{Var}[X_1] = 1$.

By the law of the iterated logarithm for Brownian motion (see Theorem 22.1), we have

$$\limsup_{t \to \infty} \frac{B_t}{\sqrt{2t \log \log t}} = 1 \quad a.s.$$

Hence, it is enough to show that

$$\limsup_{t \to \infty} \frac{|B_t - B_{\tau_{\lfloor t \rfloor}}|}{\sqrt{2t \log \log t}} = 0 \quad a.s.$$

By the strong law of large numbers (Theorem 5.17), we have $\frac{1}{n}\tau_n \overset{n\to\infty}{\longrightarrow} 1$ a.s., so let $\varepsilon > 0$ and let $t_0 = t_0(\omega)$ be large enough that

$$\frac{1}{1+\varepsilon} \le \frac{\tau_{\lfloor t \rfloor}}{t} \le 1+\varepsilon \qquad \text{for all } t \ge t_0.$$

Define

$$M_t := \sup_{s\in[t/(1+\varepsilon),\, t(1+\varepsilon)]} |B_s - B_t|.$$

It is enough to show that $\limsup\limits_{t\to\infty} \dfrac{M_t}{\sqrt{2t\log\log t}} = 0$. Consider the sequence $t_n = (1+\varepsilon)^n$, $n \in \mathbb{N}$, and define

$$M'_n := \sup_{s\in[t_{n-1},\, t_{n+2}]} |B_s - B_{t_{n-1}}|.$$

Then (by the triangle inequality), for $t \in [t_n, t_{n+1}]$,

$$M_t \le 2M'_n.$$

Let $\delta := (1+\varepsilon)^3 - 1$. Then $t_{n+2} - t_{n-1} = \delta t_{n-1}$. Brownian scaling and the reflection principle (Theorem 21.19) now yield

$$\mathbf{P}\left[M'_n > \sqrt{3\delta t_{n-1}\log\log t_{n-1}} \right]$$

$$= \mathbf{P}\left[\sup_{s\in[0,1]} |B_s| > \sqrt{3\log\log t_{n-1}} \right]$$

$$\le 2\,\mathbf{P}\left[\sup_{s\in[0,1]} B_s > \sqrt{3\log\log t_{n-1}} \right]$$

$$= 4\,\mathbf{P}\left[B_1 > \sqrt{3\log\log t_{n-1}} \right]$$

$$\le \frac{2}{\sqrt{3\log\log t_{n-1}}} \exp\left(-\frac{3}{2}\log\log t_{n-1} \right) \qquad \text{(Lemma 22.2)}$$

$$\le n^{-3/2} \quad \text{for } n \text{ sufficiently large.}$$

The probabilities can be summed over n; hence the Borel–Cantelli lemma yields

$$\limsup_{t\to\infty} \frac{M_t}{\sqrt{t\log\log t}} \le \limsup_{n\to\infty} \frac{2M'_n}{\sqrt{t_{n-1}\log\log t_{n-1}}} \le 2\sqrt{3\delta}.$$

Letting $\varepsilon \to 0$, we get $\delta = (1+\varepsilon)^3 - 1 \to 0$, and hence the proof is complete. $\square$

Takeaways The Skorohod embedding theorem allows to transfer the law of the iterated logarithm to sums of i.i.d. centred square integrable random variables.

Chapter 23
Large Deviations

Except for the law of the iterated logarithm, so far we have encountered two types of limit theorems for partial sums $S_n = X_1 + \ldots + X_n$, $n \in \mathbb{N}$, of identically distributed, real random variables $(X_i)_{i \in \mathbb{N}}$ with distribution function F:

(1) (Weak) laws of large numbers state that (under suitable assumptions on the family $(X_i)_{i \in \mathbb{N}}$), for every $x > 0$,

$$\mathbf{P}\big[\big|S_n - n\,\mathbf{E}[X_1]\big| \geq xn\big] \overset{n \to \infty}{\longrightarrow} 0. \qquad (23.1)$$

From this we get immediately that the empirical distribution functions

$$F_n : x \mapsto \frac{1}{n} \sum_{i=1}^{n} \mathbb{1}_{(-\infty, x]}(X_i)$$

converge in probability; that is, $\|F_n - F\|_\infty \overset{n \to \infty}{\longrightarrow} 0$. In other words, for any distribution function $G \neq F$ and any $\varepsilon > 0$ with $\varepsilon < \|F - G\|_\infty$, we have

$$\mathbf{P}\big[\|F_n - G\|_\infty < \varepsilon\big] \overset{n \to \infty}{\longrightarrow} 0. \qquad (23.2)$$

(2) Central limit theorems state that (under different assumptions on the family $(X_i)_{i \in \mathbb{N}}$) for every $x \in \mathbb{R}$

$$\mathbf{P}\big[S_n - n\,\mathbf{E}[X_1] \geq x\sqrt{n}\,\big] \overset{n \to \infty}{\longrightarrow} 1 - \Phi\left(\frac{x}{\sqrt{\mathrm{Var}[X_1]}}\right). \qquad (23.3)$$

Here $\Phi : t \mapsto \mathcal{N}_{0,1}((-\infty, t])$ is the distribution function of the standard normal distribution.

A. Klenke, *Probability Theory*, Universitext, https://doi.org/10.1007/978-3-030-56402-5_23

In each case, the *typical value* of S_n is $n \, \mathbf{E}[X_1]$. Equation (23.3) makes a precise statement about the average size of the deviations (which are of order $\sqrt{n}$) from the typical value. A simple consequence is of course that the probability of *large deviations* (of order n) from the typical value goes to 0; that is, (23.1) holds.

In this chapter, we compute the *speed of convergence* in (23.1) (Cramér's theorem) and in (23.2) (Sanov's theorem).

We follow in part the expositions in [31] and [74].

23.1   Cramér's Theorem

Let $X_1, X_2, \ldots$ be i.i.d. with $\mathbf{P}_{X_i} = \mathcal{N}_{0,1}$. Then, for every $x > 0$,

$$\mathbf{P}[S_n \geq xn] = \mathbf{P}\big[X_1 \geq x\sqrt{n}\,\big] = 1 - \Phi\big(x\sqrt{n}\big) = (1 + \varepsilon_n)\,\frac{1}{x\sqrt{2\pi n}}\,e^{-nx^2/2},$$

where $\varepsilon_n \overset{n\to\infty}{\longrightarrow} 0$ (by Lemma 22.2). Taking logarithms, we get

$$\lim_{n\to\infty} \frac{1}{n} \log \mathbf{P}\big[S_n \geq xn\big] = -\frac{x^2}{2} \quad \text{for every } x > 0. \tag{23.4}$$

It might be tempting to believe that a central limit theorem could be used to show (23.4) for all centered i.i.d. sequences (X_i) with finite variance. However, in general, the limit might be infinite or might be a different function of x, as we will show below. The moral is that large deviations depend more subtly on the tails of the distribution of X_i than the average-sized fluctuations do (which are determined by the variance only). The following theorem shows this for Bernoulli random variables.

Theorem 23.1 *Let $X_1, X_2, \ldots$ be i.i.d. with $\mathbf{P}[X_1 = -1] = \mathbf{P}[X_1 = 1] = \frac{1}{2}$. Then, for every $x \geq 0$,*

$$\lim_{n\to\infty} \frac{1}{n} \log \mathbf{P}[S_n \geq xn] = -I(x), \tag{23.5}$$

*where the **rate function** I is given by (see Fig. 23.1)*

$$I(z) = \begin{cases} \frac{1+z}{2} \log(1+z) + \frac{1-z}{2} \log(1-z), & \text{if } z \in [-1, 1], \\ \infty, & \text{if } |z| > 1. \end{cases} \tag{23.6}$$

Remark 23.2 Here we agree that $0 \log 0 = 0$. This makes the restriction of I to $[-1, 1]$ a continuous function with $I(-1) = I(1) = \log 2$. Note that I is strictly convex on $[-1, 1]$ with $I(0) = 0$ and I is monotone increasing on $[0, 1]$ and is monotone decreasing on $[-1, 0]$. ◊

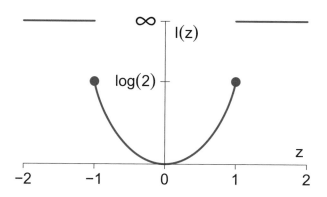

Fig. 23.1 Rate function $I(z)$ from (23.6).

Proof (of Theorem 23.1) For $x = 0$ and $x > 1$, the claim is trivial. For $x = 1$, we have $\mathbf{P}[S_n \geq n] = 2^{-n}$, and thus again (23.5) holds trivially. Hence, it is enough to consider $x \in (0, 1)$. Since $\frac{S_n + n}{2} \sim b_{n,1/2}$ is binomially distributed, we have

$$\mathbf{P}\big[S_n \geq xn\big] = 2^{-n} \sum_{k \geq (1+x)n/2} \binom{n}{k}.$$

Define $a_n(x) = \lceil n(1+x)/2 \rceil$ for $n \in \mathbb{N}$. Since $k \mapsto \binom{n}{k}$ is monotone decreasing for $k \geq \frac{n}{2}$, we get

$$Q_n(x) := \max\left\{ \binom{n}{k} : a_n(x) \leq k \leq n \right\} = \binom{n}{a_n(x)}. \qquad (23.7)$$

We make the estimate

$$2^{-n}\, Q_n(x) \ \leq \ \mathbf{P}\big[S_n \geq xn\big] \ \leq \ (n+1)\, 2^{-n}\, Q_n(x). \qquad (23.8)$$

By Stirling's formula

$$\lim_{n \to \infty} \frac{1}{n!} n^n e^{-n} \sqrt{2\pi n} \ = \ 1,$$

we obtain

$$\lim_{n\to\infty} \frac{1}{n} \log Q_n(x)$$

$$= \lim_{n\to\infty} \frac{1}{n} \log \frac{n!}{a_n(x)! \cdot (n - a_n(x))!}$$

$$= \lim_{n\to\infty} \frac{1}{n} \log \frac{n^n}{a_n(x)^{a_n(x)} \cdot (n - a_n(x))^{n - a_n(x)}}$$

$$= \lim_{n\to\infty} \left[\log(n) - \frac{a_n(x)}{n} \log\left(a_n(x)\right) - \frac{n - a_n(x)}{n} \log\left(n - a_n(x)\right) \right]$$

$$= \lim_{n\to\infty} \left[\log(n) - \frac{1 + x}{2} \left(\log\left(\frac{1 + x}{2}\right) + \log(n) \right) \right.$$

$$\left. - \frac{1 - x}{2} \left(\log\left(\frac{1 - x}{2}\right) + \log(n) \right) \right]$$

$$= -\frac{1 + x}{2} \log\left(\frac{1 + x}{2}\right) - \frac{1 - x}{2} \log\left(\frac{1 - x}{2}\right) = -I(x) + \log 2.$$

Together with (23.8), this implies (23.5).   □

Under certain assumptions on the distribution of X_1, Cramér's theorem [29] provides a general principle to compute the rate function I.

Theorem 23.3 (Cramér [29]) *Let $X_1, X_2, \ldots$ be i.i.d. real random variables with finite **logarithmic moment generating function***

$$\Lambda(t) := \log \mathbf{E}\left[e^{tX_1}\right] < \infty \quad \text{for all } t \in \mathbb{R}. \tag{23.9}$$

Let

$$\Lambda^*(x) := \sup_{t\in\mathbb{R}} \left(tx - \Lambda(t)\right) \quad \text{for } x \in \mathbb{R},$$

the Legendre transform of Λ. Then, for every $x > \mathbf{E}[X_1]$,

$$\lim_{n\to\infty} \frac{1}{n} \log \mathbf{P}\left[S_n \geq xn\right] = -I(x) := -\Lambda^*(x). \tag{23.10}$$

Proof By passing to $X_i - x$ if necessary, we may assume $\mathbf{E}[X_i] < 0$ and $x = 0$. (In fact, if $\tilde{X}_i := X_i - x$, and $\tilde{\Lambda}$ and $\tilde{\Lambda}^*$ are defined as Λ and Λ^* above but for $\tilde{X}_i$ instead of X_i, then $\tilde{\Lambda}(t) = \Lambda(t) - t \cdot x$ and thus $\tilde{\Lambda}^*(0) = \sup_{t\in\mathbb{R}}(-\tilde{\Lambda}(t)) = \Lambda^*(x)$.)

Define $\varphi(t) := e^{\Lambda(t)}$ and

$$\varrho := e^{-\Lambda^*(0)} = \inf_{t \in \mathbb{R}} \varphi(t).$$

By (23.9) and the differentiation lemma (Theorem 6.28), φ is differentiable infinitely often and the first two derivatives are

$$\varphi'(t) = \mathbf{E}\big[X_1 \, e^{t X_1}\big] \quad \text{and} \quad \varphi''(t) = \mathbf{E}\big[X_1^2 \, e^{t X_1}\big].$$

Hence φ is strictly convex and $\varphi'(0) = \mathbf{E}[X_1] < 0$.

First consider the case $\mathbf{P}[X_1 \leq 0] = 1$. Then $\varphi'(t) < 0$ for every $t \in \mathbb{R}$ and $\varrho = \lim_{t \to \infty} \varphi(t) = \mathbf{P}[X_1 = 0]$. Therefore,

$$\mathbf{P}[S_n \geq 0] = \mathbf{P}[X_1 = \ldots = X_n = 0] = \varrho^n$$

and thus the claim follows.

Now let $\mathbf{P}[X_1 < 0] > 0$ and $\mathbf{P}[X_1 > 0] > 0$. Then $\lim_{t \to \infty} \varphi(t) = \infty = \lim_{t \to -\infty} \varphi(t)$. As φ is strictly convex, there is a unique $\tau \in \mathbb{R}$ at which φ assumes its minimum; hence

$$\varphi(\tau) = \varrho \quad \text{and} \quad \varphi'(\tau) = 0.$$

Since $\varphi'(0) < 0$, we have $\tau > 0$. Using Markov's inequality (Theorem 5.11), we estimate

$$\mathbf{P}[S_n \geq 0] = \mathbf{P}\big[e^{\tau S_n} \geq 1\big] \leq \mathbf{E}\big[e^{\tau S_n}\big] = \varphi(\tau)^n = \varrho^n.$$

Thus we get the upper bound

$$\limsup_{n \to \infty} \frac{1}{n} \log \mathbf{P}[S_n \geq 0] \leq \log \varrho = -\Lambda^*(0).$$

The remaining part of the proof is dedicated to verifying the reverse inequality:

$$\liminf_{n \to \infty} \frac{1}{n} \log \mathbf{P}[S_n \geq 0] \geq \log \varrho. \tag{23.11}$$

We use the method of an exponential size-biasing of the distribution $\mu := \mathbf{P}_{X_1}$ of X_1, which turns the atypical values that are of interest here into typical values. That is, we define the **Cramér transform** $\hat{\mu} \in \mathcal{M}_1(\mathbb{R})$ of μ by

$$\hat{\mu}(dx) = \varrho^{-1} e^{\tau x} \mu(dx) \quad \text{for } x \in \mathbb{R}.$$

Let $\hat{X}_1, \hat{X}_2, \ldots$ be independent and identically distributed with $\mathbf{P}_{\hat{X}_i} = \hat{\mu}$. Then

$$\hat{\varphi}(t) := \mathbf{E}[e^{t\hat{X}_1}] = \frac{1}{\varrho} \int_{\mathbb{R}} e^{tx} e^{\tau x} \mu(dx) = \frac{1}{\varrho} \varphi(t+\tau).$$

Hence

$$\mathbf{E}[\hat{X}_1] = \hat{\varphi}'(0) = \frac{1}{\varrho} \varphi'(\tau) = 0,$$

$$\mathbf{Var}[\hat{X}_1] = \hat{\varphi}''(0) = \frac{1}{\varrho} \varphi''(\tau) \in (0, \infty).$$

Defining $\hat{S}_n = \hat{X}_1 + \ldots + \hat{X}_n$, we get

$$\begin{aligned}
\mathbf{P}[S_n \geq 0] &= \int_{\{x_1 + \ldots + x_n \geq 0\}} \mu(dx_1) \cdots \mu(dx_n) \\
&= \int_{\{x_1 + \ldots + x_n \geq 0\}} \left(\varrho \, e^{-\tau x_1}\right) \hat{\mu}(dx_1) \cdots \left(\varrho \, e^{-\tau x_n}\right) \hat{\mu}(dx_n) \\
&= \varrho^n \, \mathbf{E}\left[e^{-\tau \hat{S}_n} \mathbb{1}_{\{\hat{S}_n \geq 0\}}\right].
\end{aligned}$$

Thus, in order to show (23.11), it is enough to show

$$\liminf_{n\to\infty} \frac{1}{n} \log \mathbf{E}\left[e^{-\tau \hat{S}_n} \mathbb{1}_{\{\hat{S}_n \geq 0\}}\right] \geq 0. \qquad (23.12)$$

However, by the central limit theorem (Theorem 15.38), for every $c > 0$,

$$\begin{aligned}
\frac{1}{n} \log \mathbf{E}\left[e^{-\tau \hat{S}_n} \mathbb{1}_{\{\hat{S}_n \geq 0\}}\right] &\geq \frac{1}{n} \log \mathbf{E}\left[e^{-\tau \hat{S}_n} \mathbb{1}_{\{0 \leq \hat{S}_n \leq c\sqrt{n}\}}\right] \\
&\geq \frac{1}{n} \log\left(e^{-\tau c\sqrt{n}} \mathbf{P}\left[\frac{\hat{S}_n}{\sqrt{n}} \in [0, c]\right]\right) \\
&\xrightarrow{n\to\infty} \lim_{n\to\infty} \frac{-\tau c\sqrt{n}}{n} + \lim_{n\to\infty} \frac{1}{n} \log\left(\mathcal{N}_{0,\mathbf{Var}[\hat{X}_1]}([0, c])\right) \\
&= 0. \qquad\qquad\qquad\qquad\qquad\qquad\qquad\qquad\qquad\qquad\qquad\qquad \square
\end{aligned}$$

Example 23.4 If $\mathbf{P}_{X_1} = \mathcal{N}_{0,1}$, then

$$\Lambda(t) = \log\left(\mathbf{E}[e^{tX_1}]\right) = \log\left(\frac{1}{\sqrt{2\pi}} \int_{-\infty}^{\infty} e^{tx} e^{-x^2/2} \, dx\right) = \frac{t^2}{2}.$$

Furthermore,

$$\Lambda^*(z) = \sup_{t \in \mathbb{R}} \big(tz - \Lambda(t)\big) = \sup_{t \in \mathbb{R}} \left(tz - \frac{t^2}{2}\right) = \frac{z^2}{2}.$$

Hence the rate function coincides with that of (23.4). ◊

Example 23.5 If $\mathbf{P}_{X_1} = \frac{1}{2}\delta_{-1} + \frac{1}{2}\delta_1$, then $\Lambda(t) = \log\cosh(t)$. The maximizer $t^* = t^*(z)$ of the variational problem for Λ^* solves the equation $z = \Lambda'(t^*) = \tanh(t^*)$. Hence

$$\Lambda^*(z) = zt^* - \Lambda(t^*) = z \operatorname{arc tanh}(z) - \log\big(\cosh(\operatorname{arc tanh}(z))\big).$$

Now $\operatorname{arc tanh}(z) = \dfrac{1}{2}\log\dfrac{1+z}{1-z}$ for $z \in (-1, 1)$ and

$$\cosh\big(\operatorname{arc tanh}(z)\big) = \frac{1}{\sqrt{1-z^2}} = \frac{1}{\sqrt{(1-z)(1+z)}}.$$

Therefore,

$$\Lambda^*(z) = \frac{z}{2}\log(1+z) - \frac{z}{2}\log(1-z) + \frac{1}{2}\log(1-z) + \frac{1}{2}\log(1+z)$$

$$= \frac{1+z}{2}\log(1+z) + \frac{1-z}{2}\log(1-z).$$

However, this is the rate function from Theorem 23.1. ◊

Takeaways For random variables with exponential moments, in the weak law of large numbers, the probability for large deviations decays exponentially fast. The rate for the decay can be computed via the Legendre transform of the logarithmic moment generating function.

Exercise 23.1.1 Let X be a real random variable with density $f(x) = c^{-1}\dfrac{e^{-|x|}}{1+|x|^3}$, where $c = \displaystyle\int_{-\infty}^{\infty} \dfrac{e^{-|x|}}{1+|x|^3}\,dx$. Check if the logarithmic moment generating function Λ is continuous and sketch the graph of Λ. ♣

Exercise 23.1.2 Let X be a real random variable and let $\Lambda(t) := \log\big(\mathbf{E}[e^{tX}]\big)$, $t \in \mathbb{R}$ be its logarithmic moment generating function. Use Hölder's inequality to show that Λ is convex and is strictly convex in the interval where it is finite (if X is not almost surely constant). ♣

23.2   Large Deviations Principle

The basic idea of Cramér's theorem is to quantify the probabilities of rare events by an exponential rate and a rate function. In this section, we develop a formal framework for the quantification of probabilities of rare events in which the complete theory of large deviations can be developed. For further reading, consult, e.g., [31, 32] or [74].

Let E be a Polish space with complete metric d. Recall that

$$B_\varepsilon(x) = \{y \in E : d(x, y) < \varepsilon\}$$

denotes the open ball of radius $\varepsilon > 0$ that is centered at $x \in E$.

A map $f : E \to \bar{\mathbb{R}} = [-\infty, \infty]$ is called **lower semicontinuous** if, for every $a \in \bar{\mathbb{R}}$, the **level set** $f^{-1}([-\infty, a]) \subset E$ is closed. (In particular, continuous maps are lower semicontinuous. On the other hand, $\mathbb{1}_{(0,1)} : \mathbb{R} \to \mathbb{R}$ is lower semicontinuous but not continuous.) An equivalent condition for lower semicontinuity is that $\lim_{\varepsilon \downarrow 0} \inf f(B_\varepsilon(x)) = f(x)$ for all $x \in E$. (Recall that $\inf f(A) = \inf\{f(x) : x \in A\}$.) If $K \subset E$ is compact and nonempty, then f assumes its infimum on K. Indeed, for the case where $f(x) = \infty$ for all $x \in K$, the statement is trivial. Now assume $\inf f(K) < \infty$. If $a_n \downarrow \inf f(K)$ is strictly monotone decreasing, then $K \cap f^{-1}([-\infty, a_n]) \neq \emptyset$ is compact for every $n \in \mathbb{N}$ and hence the infinite intersection also is nonempty:

$$f^{-1}(\inf f(K)) = K \cap \bigcap_{n=1}^{\infty} f^{-1}([-\infty, a_n]) \neq \emptyset.$$

Reflection Check that suprema of lower semicontinuous functions are lower semicontinuous. In particular, suprema of continuous functions are lower semicontinuous. Which of the functions $f(x) = \mathbb{1}_{\{0\}}(x)$, $g(x) = \mathbb{1}_{\mathbb{R}\setminus\mathbb{Z}}(x)$, $h(x) = \sin(1/x)$ for $x \neq 0$ and $h(0) = -1$, are lower semicontinuous? ♠

Definition 23.6 (Rate function) *A lower semicontinuous function* $I : E \to [0, \infty]$ *is called a* **rate function***. If all level sets* $I^{-1}([-\infty, a])$, $a \in [0, \infty)$, *are compact, then* I *is called a* **good rate function***.*

Definition 23.7 (Large deviations principle) *Let* I *be a rate function and* $(\mu_\varepsilon)_{\varepsilon>0}$ *be a family of probability measures on* E. *We say that* $(\mu_\varepsilon)_{\varepsilon>0}$ *satisfies a* **large deviations principle** *(LDP) with rate function* I *if*

(LDP 1) $\liminf\limits_{\varepsilon \to 0} \varepsilon \log(\mu_\varepsilon(U)) \geq -\inf I(U)$ *for every open* $U \subset E$,
(LDP 2) $\limsup\limits_{\varepsilon \to 0} \varepsilon \log(\mu_\varepsilon(C)) \leq -\inf I(C)$ *for every closed* $C \subset E$.

We say that a family $(P_n)_{n\in\mathbb{N}}$ *of probability measures on* E *satisfies an LDP with rate* $r_n \uparrow \infty$ *and rate function* I *if (LDP 1) and (LDP 2) hold with* $\varepsilon_n = 1/r_n$ *and* $\mu_{1/r_n} = P_n$.

Often (LDP 1) and (LDP 2) are referred to as *lower bound* and *upper bound*. In many cases, the lower bound is a lot easier to show than the upper bound.

Before we show that Cramér's theorem is essentially an LDP, we make two technical statements.

Theorem 23.8 *The rate function in an LDP is unique.*

Proof Assume that $(\mu_\varepsilon)_{\varepsilon>0}$ satisfies an LDP with rate functions I and J. Then, for every $x \in E$ and $\delta > 0$,

$$I(x) \geq \inf I(B_\delta(x))$$

$$\geq -\liminf_{\varepsilon\to0} \varepsilon \log \left(\mu_\varepsilon(B_\delta(x))\right)$$

$$\geq -\limsup_{\varepsilon\to0} \varepsilon \log \left(\mu_\varepsilon\left(\overline{B_\delta(x)}\right)\right)$$

$$\geq \inf J\left(\overline{B_\delta(x)}\right) \xrightarrow{\delta\to0} J(x).$$

Hence $I(x) \geq J(x)$. Similarly, we get $J(x) \geq I(x)$.   □

Lemma 23.9 *Let $N \in \mathbb{N}$ and let a_ε^i, $i = 1, \ldots, N$, $\varepsilon > 0$, be nonnegative numbers. Then*

$$\limsup_{\varepsilon\to0} \varepsilon \log \sum_{i=1}^{N} a_\varepsilon^i = \max_{i=1,\ldots,N} \limsup_{\varepsilon\to0} \varepsilon \log(a_\varepsilon^i).$$

Proof The sum and maximum differ at most by a factor N:

$$\max_{i=1,\ldots,N} \varepsilon \log(a_\varepsilon^i) \leq \varepsilon \log \sum_{i=1}^{N} a_\varepsilon^i \leq \varepsilon \log(N) + \max_{i=1,\ldots,N} \varepsilon \log(a_\varepsilon^i).$$

The maximum and limit (superior) can be interchanged and hence

$$\max_{i=1,\ldots,N} \limsup_{\varepsilon\to0} \varepsilon \log(a_\varepsilon^i) = \limsup_{\varepsilon\to0} \varepsilon \log \left(\max_{i=1,\ldots,N} a_\varepsilon^i\right)$$

$$\leq \limsup_{\varepsilon\to0} \varepsilon \log \left(\sum_{i=1}^{N} a_\varepsilon^i\right)$$

$$\leq \limsup_{\varepsilon\to0} \varepsilon \log(N) + \max_{i=1,\ldots,N} \limsup_{\varepsilon\to0} \varepsilon \log(a_\varepsilon^i)$$

$$= \max_{i=1,\ldots,N} \limsup_{\varepsilon\to0} \varepsilon \log(a_\varepsilon^i).$$ □

Example 23.10 Let $X_1, X_2, \ldots$ be i.i.d. real random variables that satisfy the condition of Cramér's theorem (Theorem 23.3); i.e., $\Lambda(t) = \log(\mathbf{E}[e^{tX_1}]) < \infty$

for every $t \in \mathbb{R}$. Furthermore, let $S_n = X_1 + \ldots + X_n$ for every n. We will show that Cramér's theorem implies that $P_n := \mathbf{P}_{S_n/n}$ satisfies an LDP with rate n and with good rate function $I(x) = \Lambda^*(x) := \sup_{t \in \mathbb{R}}(tx - \Lambda(t))$. Without loss of generality, we can assume that $\mathbf{E}[X_1] = 0$. The function I is lower semicontinuous, strictly convex (in the interval where it is finite) and has its unique minimum at $I(0) = 0$. By convexity, we have $I(y) > I(x)$ whenever $y > x \geq 0$ or $y < x \leq 0$.

Cramér's theorem says that $\lim_{n\to\infty} \frac{1}{n} \log(P_n([x, \infty))) = -I(x)$ for $x > 0$ and (by symmetry) $\lim_{n\to\infty} \frac{1}{n} \log(P_n((-\infty, x])) = -I(x)$ for $x < 0$. Clearly, for $x > 0$,

$$-I(x) \geq \liminf_{n\to\infty} \frac{1}{n} \log P_n((x, \infty))$$

$$\geq \sup_{y>x} \liminf_{n\to\infty} \frac{1}{n} \log P_n([y, \infty)) = - \inf_{y>x} I(y)$$

Similarly, $\liminf_{n\to\infty} \frac{1}{n} \log P_n((-\infty, x)) \geq - \inf_{y<x} I(y)$ for $x < 0$. Furthermore, by the law of large numbers, for any $x > 0$, we have

$$\lim_{n\to\infty} \frac{1}{n} \log P_n((-x, \infty)) = \lim_{n\to\infty} \frac{1}{n} \log P_n([-x, \infty))$$

$$= \lim_{n\to\infty} \frac{1}{n} \log P_n((-\infty, x)) = \lim_{n\to\infty} \frac{1}{n} \log P_n((-\infty, x])) = 0 = -I(0).$$

The main work has been done by showing that the family $(P_n)_{n \in \mathbb{N}}$ satisfies conditions (LDP 1) and (LDP 2) at least for unbounded intervals. It remains to show by some standard arguments (LDP 1) and (LDP 2) for *arbitrary* open and closed sets, respectively.

First assume that $C \subset \mathbb{R}$ is closed. Define $x_+ := \inf(C \cap [0, \infty))$ as well as $x_- := \sup(C \cap (-\infty, 0])$. By monotonicity of I, on $(-\infty, 0]$ and $[0, \infty)$, we get $\inf I(C) = I(x_-) \wedge I(x_+)$ (with the convention $I(-\infty) = I(\infty) = \infty$). If $x_- = 0$ or $x_+ = 0$, then $\inf(I(C)) = 0$, and (LDP 2) holds trivially. Now let $x_- < 0 < x_+$.

Using Lemma 23.9, we get

$$\limsup_{n\to\infty} \frac{1}{n} \log P_n(C)$$

$$\leq \limsup_{n\to\infty} \frac{1}{n} \log \left(P_n((-\infty, x_-]) + P_n([x_+, \infty)) \right)$$

$$= \max \left\{ \limsup_{n\to\infty} \frac{1}{n} \log P_n((-\infty, x_-]), \; \limsup_{n\to\infty} \frac{1}{n} \log P_n([x_+, \infty)) \right\}$$

$$= \max \left\{ -I(x_-), -I(x_+) \right\} = - \inf I(C).$$

This shows (LDP 2).

Now let $U \subset \mathbb{R}$ be open. Let $x \in U \cap [0, \infty)$ with $I(x) < \infty$ (if such an x exists). Then there exists an $\varepsilon > 0$ with $(x - \varepsilon, x + \varepsilon) \subset U$. Now

$$\liminf_{n \to \infty} \frac{1}{n} \log P_n\big((x - \varepsilon, \infty)\big) \geq -I(x) > -I(x + \varepsilon)$$

$$= \lim_{n \to \infty} \frac{1}{n} \log P_n\big([x + \varepsilon, \infty)\big).$$

Therefore,

$$\liminf_{n \to \infty} \frac{1}{n} \log P_n(U) \geq \liminf_{n \to \infty} \frac{1}{n} \log P_n((x - \varepsilon, x + \varepsilon))$$

$$= \liminf_{n \to \infty} \frac{1}{n} \log \big(P_n\big((x - \varepsilon, \infty)\big) - P_n\big([x + \varepsilon, \infty)\big)\big)$$

$$= \liminf_{n \to \infty} \frac{1}{n} \log \big(P_n\big((x - \varepsilon, \infty)\big)\big) \geq -I(x).$$

Similarly, this also holds for $x \in U \cap (-\infty, 0)$ with $I(x) < \infty$; hence

$$\liminf_{n \to \infty} \frac{1}{n} \log P_n(U) \geq -\inf I(U).$$

This shows the lower bound (LDP 1). ◊

In fact, the condition $\Lambda(t) < \infty$ for all $t \in \mathbb{R}$ can be dropped. Since $\Lambda(0) = 0$, we have $\Lambda^*(x) \geq 0$ for every $x \in \mathbb{R}$. The map Λ^* is a convex rate function but is, in general, not a good rate function. We quote the following strengthening of Cramér's Theorem(see [31, Theorem 2.2.3]).

Theorem 23.11 (Cramér) *If* $X_1, X_2, \ldots$ *are i.i.d. real random variables, then* $(P_{S_n/n})_{n \in \mathbb{N}}$ *satisfies an LDP with rate function* Λ^*.

Takeaways The distributions of the arithmetic means of a growing number of i.i.d. random variables concentrate more and more around the expected value (under certain regularity assumptions, that is). This behaviour has been quantified in Sect. 23.1. Here we have developed the abstract framework (principle of large deviations) for the description of the speed of concentration for a sequence of probability measures. Finally, we have shown that the results for i.i.d. random variables that we had already fit in this framework.

Exercise 23.2.1 Let $E = \mathbb{R}$. Show that $\mu_\varepsilon := \mathcal{N}_{0,\varepsilon}$ satisfies an LDP with good rate function $I(x) = x^2/2$. Further, show that strict inequality can hold in the *upper bound* (LDP 2). ♣

Exercise 23.2.2 Let $E = \mathbb{R}$. Show that $\mu_\varepsilon := \mathcal{N}_{0,\varepsilon^2}$ satisfies an LDP with good rate function $I(x) = \infty \cdot \mathbb{1}_{\mathbb{R}\setminus\{0\}}(x)$. Further, show that strict inequality can hold in the *lower bound* (LDP 1). ♣

Exercise 23.2.3 Let $E = \mathbb{R}$. Show that $\mu_\varepsilon := \frac{1}{2}\mathcal{N}_{-1,\varepsilon} + \frac{1}{2}\mathcal{N}_{1,\varepsilon}$ satisfies an LDP with good rate function $I(x) = \frac{1}{2}\min((x+1)^2, (x-1)^2)$. ♣

Exercise 23.2.4 Compute Λ and Λ^* in the case $X_1 \sim \exp_\theta$ for $\theta > 0$. Interpret the statement of Theorem 23.11 in this case. Check that Λ^* has its unique zero at $\mathbb{E}[X_1]$. (Result: $\Lambda^*(x) = \theta x - \log(\theta x) - 1$ if $x > 0$ and $= \infty$ otherwise.) ♣

Exercise 23.2.5 Compute Λ and Λ^* for the case where X_1 is Cauchy distributed and interpret the statement of Theorem 23.11. ♣

Exercise 23.2.6 Let $X_\lambda \sim \text{Poi}_\lambda$ for every $\lambda > 0$. Show that $\mu_\varepsilon := \mathbf{P}_{\varepsilon X_{\lambda/\varepsilon}}$ satisfies an LDP with good rate function $I(x) = x\log(x/\lambda) + \lambda - x$ for $x \geq 0$ (and $= \infty$ otherwise). ♣

Exercise 23.2.7 Let $(X_t)_{t\geq0}$ be a random walk on $\mathbb{Z}$ in continuous time that makes a jump to the right with rate $\frac{1}{2}$ and a jump to the left also with rate $\frac{1}{2}$. Show that $(\mathbf{P}_{\varepsilon X_{1/\varepsilon}})_{\varepsilon>0}$ satisfies an LDP with convex good rate function

$$I(x) = 1 + x\,\text{arc}\sinh(x) - \sqrt{1+x^2}. \quad ♣$$

23.3   Sanov's Theorem

This section is close to the exposition in [31].

We present a large deviations principle that, unlike Cramér's theorem, is not based on a linear space. Rather, we consider empirical distributions of independent random variables with values in a finite set Σ, which often is called an *alphabet*.

Let μ be a probability measure on Σ with $\mu(\{x\}) > 0$ for any $x \in \Sigma$. Further, let $X_1, X_2, \ldots$ be i.i.d. random variables with values in Σ and with distribution $\mathbf{P}_{X_1} = \mu$. We will derive a large deviations principle for the empirical measures

$$\xi_n(X) := \frac{1}{n}\sum_{i=1}^n \delta_{X_i}.$$

Note that by the law of large numbers, $\mathbf{P}$-almost surely $\xi_n(X) \xrightarrow{n\to\infty} \mu$. Hence, as the state space we get $E = \mathcal{M}_1(\Sigma)$, equipped with the metric of total variation $d(\mu, \nu) = \|\mu - \nu\|_{TV}$. (As Σ is finite, in E vague convergence, weak convergence and convergence in total variation coincide.) Further, let

$$E_n := \left\{\mu \in \mathcal{M}_1(\Sigma) : n\mu(\{x\}) \in \mathbb{N}_0 \text{ for every } x \in \Sigma\right\}$$

be the range of the random variables $\xi_n(X)$.

Recall that the **entropy** of μ is defined by

$$H(\mu) := - \int \log\left(\mu(\{x\})\right) \mu(dx).$$

If $\nu \in \mathcal{M}_1(\Sigma)$, then we define the **relative entropy** (or **Kullback–Leibler information**, see [104]) of ν given μ by

$$H(\nu|\mu) := \int \log\left(\frac{\nu(\{x\})}{\mu(\{x\})}\right) \nu(dx). \tag{23.13}$$

Since $\mu(\{x\}) > 0$ for all $x \in \Sigma$, the integrand ν-a.s. is finite and hence the integral also is finite. A simple application of Jensen's inequality yields $H(\mu) \geq 0$ and $H(\nu|\mu) \geq 0$ (see Lemma 5.26 and Exercise 5.3.3). Furthermore, $H(\nu|\mu) = 0$ if and only if $\nu = \mu$. In addition, clearly,

$$H(\nu|\mu) + H(\nu) = - \int \log\left(\mu(\{x\})\right) \nu(dx). \tag{23.14}$$

Since the map $\nu \mapsto I_\mu(\nu) := H(\nu|\mu)$ is continuous, I_μ is a rate function.

Lemma 23.12 *For every $n \in \mathbb{N}$ and $\nu \in E_n$, we have*

$$(n+1)^{-\#\Sigma} e^{-n\,H(\nu|\mu)} \leq \mathbf{P}[\xi_n(X) = \nu] \leq e^{-n\,H(\nu|\mu)}. \tag{23.15}$$

Proof We consider the set of possible values for the n-tuple $(X_1, \ldots, X_n)$ such that $\xi_n(X) = \nu$:

$$A_n(\nu) := \left\{ k = (k_1, \ldots, k_n) \in \Sigma^n : \frac{1}{n}\sum_{i=1}^{n} \delta_{k_i} = \nu \right\}.$$

For every $k \in A_n(\nu)$, we have (compare (23.14))

$$\mathbf{P}[\xi_n(X) = \nu] = \#A_n(\nu)\,\mathbf{P}[X_1 = k_1, \ldots, X_n = k_n]$$

$$= \#A_n(\nu) \prod_{x \in \Sigma} \mu(\{x\})^{n\nu(\{x\})}$$

$$= \#A_n(\nu) \exp\left(n \int \nu(dx)\,\log\mu(\{x\})\right)$$

$$= \#A_n(\nu) \exp\left(-n[H(\nu) + H(\nu|\mu)]\right).$$

Now let $Y_1, Y_2, \ldots$ be i.i.d. random variables with values in Σ and with distribution $\mathbf{P}_{Y_1} = \nu$. As in the calculation for X, we obtain (since $H(\nu|\nu) = 0$)

$$1 \geq \mathbf{P}[\xi_n(Y) = \nu] = \#A_n(\nu)\,e^{-nH(\nu)};$$

hence $\#A_n(\nu) \leq e^{nH(\nu)}$. This implies the second inequality in (23.15).

The random variable $n\,\xi_n(Y)$ has the multinomial distribution with parameters $(n\nu(\{x\}))_{x\in\Sigma}$. Hence the map $E_n \to [0,1]$, $\nu' \mapsto \mathbf{P}[\xi_n(Y) = \nu']$ is maximal at $\nu' = \nu$. Therefore,

$$\#A_n(\nu) = e^{nH(\nu)}\,\mathbf{P}[\xi_n(Y) = \nu] \geq \frac{e^{nH(\nu)}}{\#E_n} \geq (n+1)^{-\#\Sigma}\,e^{nH(\nu)}.$$

This implies the first inequality in (23.15).   □

We come to the main theorem of this section, Sanov's theorem (see [150] and [151]).

Theorem 23.13 (Sanov [150]) *Let $X_1, X_2, \dots$ be i.i.d. random variables with values in the finite set Σ and with distribution μ. Then the family $(\mathbf{P}_{\xi_n(X)})_{n\in\mathbb{N}}$ of distributions of empirical measures satisfies an LDP with rate n and rate function $I_\mu := H(\cdot\,|\mu)$.*

Proof By Lemma 23.12, for every $A \subset E$,

$$\mathbf{P}\big[\xi_n(X) \in A\big] = \sum_{\nu\in A\cap E_n} \mathbf{P}[\xi_n(X) = \nu]$$

$$\leq \sum_{\nu\in A\cap E_n} e^{-nH(\nu|\mu)}$$

$$\leq \#(A \cap E_n)\exp\big(-n\,\inf I_\mu(A \cap E_n)\big)$$

$$\leq (n+1)^{\#\Sigma}\exp\big(-n\,\inf I_\mu(A)\big).$$

Therefore,

$$\limsup_{n\to\infty} \frac{1}{n}\log\mathbf{P}[\xi_n(X) \in A] \leq -\inf I_\mu(A).$$

Hence the upper bound in the LDP holds (even for arbitrary A).

Similarly, we can use the first inequality in Lemma 23.12 to get

$$\mathbf{P}\big[\xi_n(X) \in A\big] \geq (n+1)^{-\#\Sigma}\exp\big(-n\,\inf I_\mu(A \cap E_n)\big)$$

and thus

$$\liminf_{n\to\infty} \frac{1}{n}\log\mathbf{P}\big[\xi_n(X) \in A\big] \geq -\limsup_{n\to\infty}\inf I_\mu(A \cap E_n). \qquad (23.16)$$

Note that, in this inequality, in the infimum we cannot simply replace $A \cap E_n$ by A. However, we show that, for open A this can be done at least asymptotically. Hence, let $A \subset E$ be open. For $\nu \in A$, there is an $\varepsilon > 0$ with $B_\varepsilon(\nu) \subset A$. For $n \geq (2\#\Sigma)/\varepsilon$, we have $E_n \cap B_\varepsilon(\nu) \neq \emptyset$ and hence there exists a sequence $\nu_n \xrightarrow{n\to\infty} \nu$ with $\nu_n \in$

$E_n \cap A$ for large $n \in \mathbb{N}$. As I_μ is continuous, we have

$$\limsup_{n\to\infty} \inf I_\mu(A \cap E_n) \leq \lim_{n\to\infty} I_\mu(\nu_n) = I_\mu(\nu).$$

Since $\nu \in A$ is arbitrary, we get $\limsup_{n\to\infty} \inf I_\mu(A \cap E_n) = \inf I_\mu(A)$. $\square$

Example 23.14 Let $\Sigma = \{-1, 1\}$ and let $\mu = \frac{1}{2}\delta_{-1} + \frac{1}{2}\delta_1$ be the uniform distribution on Σ. Define $m = m(\nu) := \nu(\{1\}) - \nu(\{-1\})$. Then the relative entropy of $\nu \in \mathcal{M}_1(\Sigma)$ is

$$H(\nu|\mu) = \frac{1+m}{2} \log(1+m) + \frac{1-m}{2} \log(1-m).$$

Note that this is the rate function from Theorem 23.1. $\Diamond$

Next we describe formally the connection between the LDPs of Sanov and Cramér that was indicated in the previous example. To this end, we use Sanov's theorem to derive a version of Cramér's theorem for $\mathbb{R}^d$-valued random variables taking only finitely many different values.

Example 23.15 Let $\Sigma \subset \mathbb{R}^d$ be finite and let μ be a probability measure on Σ. Further, let $X_1, X_2, \ldots$ be i.i.d. random variables with values in Σ and distribution $\mathbf{P}_{X_1} = \mu$. Define $S_n = X_1 + \ldots + X_n$ for every $n \in \mathbb{N}$. Let $\Lambda(t) = \log \mathbf{E}[e^{\langle t, X_1 \rangle}]$ for $t \in \mathbb{R}^d$ (which is finite since Σ is finite) and $\Lambda^*(x) = \sup_{t \in \mathbb{R}^d} (\langle t, x \rangle - \Lambda(t))$ for $x \in \mathbb{R}^d$.

We show that $(\mathbf{P}_{S_n/n})_{n \in \mathbb{N}}$ satisfies an LDP with rate n and rate function Λ^*.

Let $\xi_n(X)$ be the empirical measure of $X_1, \ldots, X_n$. Let $E := \mathcal{M}_1(\Sigma)$. Define the map

$$m : E \to \mathbb{R}^d, \qquad \nu \mapsto \int x\, \nu(dx) = \sum_{x \in \Sigma} x\, \nu(\{x\}).$$

That is, m maps ν to its first moment. Clearly, $\frac{1}{n} S_n = m(\xi_n(X))$. For $x \in \mathbb{R}^d$ and $A \subset \mathbb{R}^d$, define

$$E_x := m^{-1}(\{x\}) = \{\nu \in E : m(\nu) = x\}$$

and

$$E_A = m^{-1}(A) = \{\nu \in E : m(\nu) \in A\}.$$

The map $\nu \mapsto m(\nu)$ is continuous; hence E_A is open (respectively closed) if A is open (respectively closed). Let $\tilde{I}(x) := \inf I_\mu(E_x)$ (where $I_\mu(\nu) = H(\nu|\mu)$ is the relative entropy). Then, by Sanov's theorem for open $U \subset \mathbb{R}^d$,

$$\liminf_{n \to \infty} \frac{1}{n} \log \mathbf{P}_{S_n/n}(U) = \liminf_{n \to \infty} \frac{1}{n} \log \mathbf{P}_{\xi_n(X)}\big(m^{-1}(U)\big)$$

$$\geq -\inf I_\mu\big(m^{-1}(U)\big) = -\inf \tilde{I}(U).$$

Similarly, for closed $C \subset \mathbb{R}^d$, we have

$$\limsup_{n \to \infty} \frac{1}{n} \log \mathbf{P}_{S_n/n}(C) \leq -\inf \tilde{I}(C).$$

In other words, $(\mathbf{P}_{S_n/n})_{n \in \mathbb{N}}$ satisfies an LDP with rate n and rate function $\tilde{I}$. Hence, it only remains to show that $\tilde{I} = \Lambda^*$.

Note that $t \mapsto \Lambda(t)$ is differentiable (with derivative Λ') and is strictly convex. Hence the variational problem for $\Lambda^*(x)$ admits a unique maximizer $t^*(x)$. More precisely,

$$\Lambda^*(x) = \langle t^*(x), x \rangle - \Lambda(t^*(x)),$$

$\Lambda^*(x) > \langle t, x \rangle - \Lambda(t)$ for all $t \neq t^*(x)$, and $\Lambda'(t^*(x)) = x$. By Jensen's inequality, for every $\nu \in \mathcal{M}_1(\Sigma)$,

$$\Lambda(t) = \log \int e^{\langle t, y \rangle} \, \mu(dy)$$

$$= \log \int \left(e^{\langle t, y \rangle} \frac{\mu(\{y\})}{\nu(\{y\})} \right) \nu(dy)$$

$$\geq \int \log \left(e^{\langle t, y \rangle} \frac{\mu(\{y\})}{\nu(\{y\})} \right) \nu(dy)$$

$$= \langle t, m(\nu) \rangle - H(\nu \,|\, \mu)$$

with equality if and only if $\nu = \nu_t$, where $\nu_t(\{y\}) = \mu(\{y\}) e^{\langle t, y \rangle - \Lambda(t)}$. Hence,

$$\langle t, x \rangle - \Lambda(t) \leq \inf_{\nu \in E_x} H(\nu \,|\, \mu)$$

with equality if $\nu_t \in E_x$. However, we now know that $m(\nu_t) = \Lambda'(t)$; hence we have $\nu_{t^*(x)} \in E_x$ and thus

$$\Lambda^*(x) = \langle t^*(x), x \rangle - \Lambda(t^*(x)) = \inf_{\nu \in E_x} H(\nu \,|\, \mu) = \tilde{I}(x). \quad \Diamond$$

The method of the proof that we applied in the last example to derive the LDP with rate function $\tilde{I}$ is called a **contraction principle**. We formulate this principle as a theorem.

Theorem 23.16 (Contraction principle) *Assume the family* $(\mu_\varepsilon)_{\varepsilon>0}$ *of probability measures on E satisfies an LDP with rate function I. If F is a topological space and* $m : E \to F$ *is continuous, then the image measures* $(\mu_\varepsilon \circ m^{-1})_{\varepsilon>0}$ *satisfy an LDP with rate function* $\tilde{I}(x) = \inf I(m^{-1}(\{x\}))$.

> **Takeaways** Consider an i.i.d. sequence of random variables with values in a finite set. The empirical distributions converge to the distribution of the random variables. The speed of convergence is exponential and the exponential rate function is the relative entropy. Using the contraction principle this large deviations principle can be reduced to functions of the random variables. In particular, we can recover Cramér's theorem for random variables that take only finitely many d-dimensional values.

23.4   Varadhan's Lemma and Free Energy

Assume that $(\mu_\varepsilon)_{\varepsilon>0}$ is a family of probability measures that satisfies an LDP with rate function I. In particular, we know that, for small $\varepsilon > 0$, the mass of μ_ε is concentrated around the zeros of I. In statistical physics, one is often interested in integrating with respect to μ_ε (where $1/\varepsilon$ is interpreted as "size of the system") functions that attain their maximal values away from the zeros of I. In addition, these functions are exponentially scaled with $1/\varepsilon$. Hence the aim is to study the asymptotics of $Z_\varepsilon^\phi := \int e^{\phi(x)/\varepsilon} \mu_\varepsilon(dx)$ as $\varepsilon \to 0$. Under some mild conditions on the continuity of ϕ, the main contribution to the integral comes from those points x that are not too unlikely (for μ_ε) and for which at the same time $\phi(x)$ is large. That is, those x for which $\phi(x) - I(x)$ is close to its maximum. These contributions are quantified in terms of the *tilted* probability measures $\mu_\varepsilon^\phi(dx) = (Z_\varepsilon^\phi)^{-1} e^{\phi(x)/\varepsilon} \mu_\varepsilon(dx)$, $\varepsilon > 0$, for which we derive an LDP. As an application, we get the statistical physics principle of minimising the free energy. As an example, we analyze the Weiss ferromagnet.

We start with a lemma that is due to Varadhan [167].

Theorem 23.17 (Varadhan's Lemma [167]) *Let I be a good rate function and let* $(\mu_\varepsilon)_{\varepsilon>0}$ *be a family of probability measures on E that satisfies an LDP with rate function I. Further, let* $\phi : E \to \mathbb{R}$ *be continuous and assume that*

$$\inf_{M>0} \limsup_{\varepsilon \to 0} \varepsilon \log \int e^{\phi(x)/\varepsilon} \mathbb{1}_{\{\phi(x) \geq M\}} \mu_\varepsilon(dx) = -\infty. \tag{23.17}$$

Then

$$\lim_{\varepsilon \to 0} \varepsilon \log \int e^{\phi(x)/\varepsilon} \mu_\varepsilon(dx) = \sup_{x \in E} \left(\phi(x) - I(x) \right). \tag{23.18}$$

Remark 23.18 (Moment condition) The tail condition (23.17) holds if there exists an $\alpha > 1$ such that

$$\limsup_{\varepsilon \to 0} \varepsilon \log \int e^{\alpha\phi/\varepsilon} d\mu_\varepsilon < \infty. \tag{23.19}$$

Indeed, for every $M \in \mathbb{R}$, we have

$$\varepsilon \log \int e^{\phi(x)/\varepsilon} \mathbb{1}_{\{\phi(x) \geq M\}} \mu_\varepsilon(dx) = M + \varepsilon \log \int e^{(\phi(x)-M)/\varepsilon} \mathbb{1}_{\{\phi(x) \geq M\}} \mu_\varepsilon(dx)$$

$$\leq M + \varepsilon \log \int e^{\alpha(\phi(x)-M)/\varepsilon} \mu_\varepsilon(dx)$$

$$= -(\alpha - 1)M + \varepsilon \log \int e^{\alpha\phi(x)/\varepsilon} \mu_\varepsilon(dx).$$

Together with (23.19), this implies (23.17). $\Diamond$

Proof We use different arguments to show that the right-hand side of (23.18) is a lower and an upper bound for the left-hand side.

Lower bound For any $x \in E$ and $r > 0$, we have

$$\liminf_{\varepsilon \to 0} \varepsilon \log \int e^{\phi/\varepsilon} d\mu_\varepsilon \geq \liminf_{\varepsilon \to 0} \varepsilon \log \int_{B_r(x)} e^{\phi/\varepsilon} d\mu_\varepsilon$$

$$\geq \inf \phi(B_r(x)) - I(x) \xrightarrow{r \to 0} \phi(x) - I(x).$$

Upper bound For $M > 0$ and $\varepsilon > 0$, define

$$F_M^\varepsilon := \int_{\{\phi \geq M\}} e^{\phi(x)/\varepsilon} \mu_\varepsilon(dx) \quad \text{and} \quad G_M^\varepsilon := \int_{\{\phi < M\}} e^{\phi(x)/\varepsilon} \mu_\varepsilon(dx).$$

Define

$$F_M := \limsup_{\varepsilon \to 0} \varepsilon \log F_M^\varepsilon \quad \text{and} \quad G_M := \limsup_{\varepsilon \to 0} \varepsilon \log G_M^\varepsilon.$$

By Lemma 23.9, for any $M > 0$,

$$\limsup_{\varepsilon \to 0} \varepsilon \log \int e^{\phi(x)/\varepsilon} \mu_\varepsilon(dx) = F_M \vee G_M.$$

As by assumption $\inf_{M>0} F_M = -\infty$, it is enough to show that

$$\sup_{M>0} G_M \leq \sup_{x \in E} \left(\phi(x) - I(x) \right). \tag{23.20}$$

Let $\delta > 0$. For any $x \in E$ there is an $r(x) > 0$ with

$$\inf I\left(B_{2r(x)}(x) \right) \geq I(x) - \delta \quad \text{and} \quad \sup \phi\left(B_{2r(x)}(x) \right) \leq \phi(x) + \delta.$$

Let $a \geq 0$. Since I is a *good* rate function, the level set $K := I^{-1}([0, a])$ is compact. Thus we can find finitely many $x_1, \ldots, x_N \in I^{-1}([0, a])$ such that $\bigcup_{i=1}^{N} B_{r(x_i)}(x_i) \supset K$. Therefore,

$$G_M^\varepsilon \leq \int_{\{\phi < M\} \cap K^c} e^{\phi(x)/\varepsilon} \mu_\varepsilon(dx) + \sum_{i=1}^{N} \int_{\{\phi < M\} \cap B_{r(x_i)}(x_i)} e^{\phi(x)/\varepsilon} \mu_\varepsilon(dx)$$

$$\leq e^{M/\varepsilon} \mu_\varepsilon(K^c) + \sum_{i=1}^{N} e^{(\phi(x_i) \wedge M + \delta)/\varepsilon} \mu_\varepsilon\left(B_{r(x_i)}(x_i) \right)$$

$$= e^{(M + \varepsilon \log(\mu_\varepsilon(K^c)))/\varepsilon} + \sum_{i=1}^{N} e^{(\phi(x_i) \wedge M + \delta + \varepsilon \log(\mu_\varepsilon(B_{r(x_i)}(x_i))))/\varepsilon}.$$

Using Lemma 23.9 and the LDP, we infer

$$G_M \leq (M - a) \vee \max_{i=1,\ldots,N} \left(\phi(x_i) - I(x_i) + 2\delta \right)$$

$$\leq (M - a) \vee \sup_{x \in E} \left(\phi(x) - I(x) \right) + 2\delta.$$

By letting first $\delta \downarrow 0$ and then $a \uparrow \infty$, we obtain (23.20). $\qquad \square$

Theorem 23.19 (Tilted LDP) *Assume that $(\mu_\varepsilon)_{\varepsilon>0}$ satisfies an LDP with good rate function I. Further, let $\phi : E \to \mathbb{R}$ be a continuous function that satisfies condition (23.17). Define $Z_\varepsilon^\phi := \int e^{\phi/\varepsilon} d\mu_\varepsilon$ and $\mu_\varepsilon^\phi \in \mathcal{M}_1(E)$ by*

$$\mu_\varepsilon^\phi(dx) = (Z_\varepsilon^\phi)^{-1} e^{\phi(x)/\varepsilon} \mu_\varepsilon(dx).$$

Further, define $I^\phi : E \to [0, \infty]$ by

$$I^\phi(x) = \sup_{z \in E} \left(\phi(z) - I(z) \right) - \left(\phi(x) - I(x) \right). \tag{23.21}$$

Then $(\mu_\varepsilon^\phi)_{\varepsilon>0}$ satisfies an LDP with rate function I^ϕ.

Proof This is left as an exercise. (Compare [32, Exercise 2.1.24], see also [43, Section II.7].)   □

Varadhan's lemma has various applications in statistical physics. Consider a Polish space Σ that is interpreted as the space of possible states of a particle. Further, let $\lambda \in \mathcal{M}_1(\Sigma)$ be a distribution that is understood as the *a priori* distribution of this particle if the influence of energy could be neglected. If Σ is finite or is a bounded subset of an $\mathbb{R}^d$, then by symmetry, typically λ is the uniform distribution on Σ. If we place n indistinguishable particles independently according to λ on the random positions $z_1, \ldots, z_n \in \Sigma$, then the *state* of this ensemble can be described by $x := \frac{1}{n} \sum_{i=1}^{n} \delta_{z_i}$. Denote by $\mu_n^0 \in \mathcal{M}_1(\mathcal{M}_1(\Sigma))$ the corresponding *a priori* distribution of x; that is, of the n-particle system.

Now we introduce the hypothesis that the energy $U_n(x)$ of a state has the form $U_n(x) = nU(x)$, where $U(x)$ is the average energy of one particle of the ensemble in state x.

Let $T > 0$ be the temperature of the system and let $\beta := 1/T$ be the so-called **inverse temperature**. In statistical physics, a key quantity is the so-called **partition function**

$$Z_n^\beta := \int e^{-\beta U_n} \, d\mu_n^0.$$

A postulate of statistical physics is that the distribution of the state x is the **Boltzmann distribution**:

$$\mu_n^\beta(dx) = (Z_n^\beta)^{-1} \, e^{-\beta U_n(x)} \, \mu_n^0(dx). \tag{23.22}$$

Varadhan's lemma (more precisely, the tilted LDP) and Sanov's theorem are the keys to building a connection to the variational principle for the free energy. For simplicity, assume that Σ is a finite set and $\lambda = \mathcal{U}_\Sigma$ is the uniform distribution on Σ. By Sanov's theorem, $(\mu_n^0)_{n \in \mathbb{N}}$ satisfies an LDP with rate n and rate function $I(x) = H(x|\lambda)$, where $H(x|\lambda)$ is the relative entropy of x with respect to λ. By (23.14), we have $H(x|\lambda) = \log(\#\Sigma) - H(x)$, where $H(x)$ is the entropy of x.

Define the **free energy** (or **Helmholtz potential**) per particle as

$$F^\beta(x) := U(x) - \beta^{-1} H(x).$$

The theorem on the tilted LDP yields that the sequence of Boltzmann distributions $(\mu_n^\beta)_{n \in \mathbb{N}}$ satisfies an LDP with rate n and rate function

$$I^\beta(x) = \beta \cdot \left(F^\beta(x) - \inf_{y \in \mathcal{M}_1(\Sigma)} F^\beta(y) \right).$$

Thus, for large n, the Boltzmann distribution is concentrated on those x that minimize the free energy. For different temperatures (that is, for different values of β) these can be very different states. This is the reason for *phase transitions* at critical temperatures (e.g., melting ice).

Example 23.20 We consider the **Weiss ferromagnet**. This is a microscopic model for a magnet that assumes that each of n indistinguishable magnetic particles has one of two possible orientations $\sigma_i \in \Sigma = \{-1, +1\}$. The mean magnetization $m = \frac{1}{n} \sum_{i=1}^{n} \sigma_i$ describes the state of the system completely (as the particles are indistinguishable). Macroscopically, this is the quantity that can be measured. The basic idea is that it is energetically favorable for particles to be oriented in the same direction. We ignore the spatial structure and assume that any particle interacts with any other particle in the same way. This is often called the **mean field** assumption. In addition, we assume that there is an exterior magnetic field of strength h. Thus up to constants the average energy of a particle is

$$U(m) = -\frac{1}{2}m^2 - hm.$$

The entropy of the state m is

$$H(m) = -\frac{1+m}{2} \log\left(\frac{1+m}{2}\right) - \frac{1-m}{2} \log\left(\frac{1-m}{2}\right).$$

Hence the average free energy of a particle is

$$F^\beta(m) = -\frac{1}{2}m^2 - hm + \beta^{-1}\left[\frac{1+m}{2} \log\left(\frac{1+m}{2}\right) + \frac{1-m}{2} \log\left(\frac{1-m}{2}\right)\right].$$

In order to obtain the minima of F^β, we compute the derivative

$$0 \stackrel{!}{=} \frac{d}{dm} F^\beta(m) = -m - h + \beta^{-1} \operatorname{arc tanh}(m).$$

Hence, m solves the equation

$$m = \tanh(\beta(m + h)). \tag{23.23}$$

In the case $h = 0$, $m = 0$ is a solution of (23.23) for any β. If $\beta \le 1$, then this is the only solution and F^β attains its global minimum at $m = 0$. If $\beta > 1$, then (23.23) has two other solutions, $m_-^{\beta,0} \in (-1, 0)$ and $m_+^{\beta,0} = -m_-^{\beta,0}$, whose values can only be computed numerically (Fig. 23.2).

In this case, F^β has a local maximum at 0 and has global minima $m_\pm^{\beta,0}$. For large n, only those values of m for which F^β is close to its minimal value can be attained and thus the distribution is concentrated around 0 if $\beta \le 1$ and around $m_\pm^{\beta,0}$ if $\beta > 1$. In the latter case, the absolute value of the mean magnetization is $|m_\pm^{\beta,0}| = m_+^{\beta,0} > 0$. Hence, there is a **phase transition** between the high temperature phase ($\beta \le 1$) without magnetization and the low temperature phase

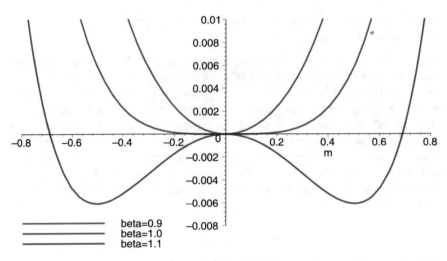

Fig. 23.2 The shifted free energy $F^\beta(m) - F^\beta(0)$ of the Weiss ferromagnet without exterior field $(h = 0)$.

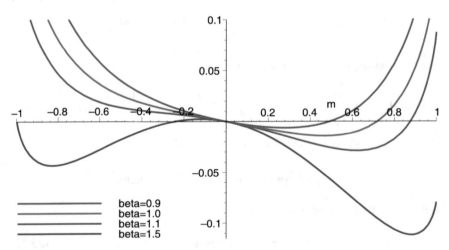

Fig. 23.3 Shifted free energy $F^\beta(m) - F^\beta(0)$ of the Weiss ferromagnet with exterior field $h = 0.04$.

$(\beta > 1)$ where so-called spontaneous magnetization occurs (that is, magnetization without an exterior field).

If $h \neq 0$, then F^β does not have a minimum at $m = 0$. Rather, F^β is asymmetric and has a global minimum $m^{\beta,h}$ with the same sign as h. Furthermore, for large β, there is another minimum with the opposite sign (Fig. 23.3). Again, the exact values can

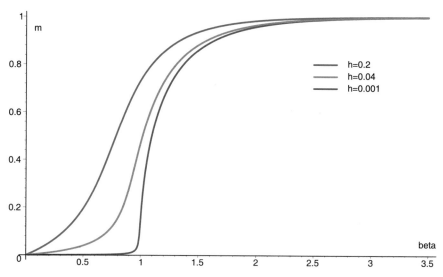

Fig. 23.4   Weiss ferromagnet: magnetization $m^{\beta,h}$ as a function of β.

only be computed numerically (Fig. 23.4). However, for high temperatures (small β), we can approximate $m^{\beta,h}$ using the approximation $\tanh(\beta(m+h)) \approx \beta(m+h)$. Hence we get

$$m^{\beta,h} \approx \frac{h}{\beta^{-1}-1} = \frac{h}{T-T_c} \quad \text{for } T \to \infty, \tag{23.24}$$

where the *Curie temperature* $T_c = 1$ is the critical temperature for spontaneous magnetization. The relation (23.24) is called the **Curie–Weiss law**. ◊

Takeaways In the context of statistical mechanics, substantial contributions to an observable are not only due to the most frequent observations but also due to rare but very large observations. In order to compute the mean values and to identify the states that yield significant contributions, we have established a so-called tilted large deviations principle. The rate function that shows up here is the analogue to the free energy of thermodynamics. Applying this formalism we have been able to describe the phase transition of the Weiss ferromagnet.

Chapter 24
The Poisson Point Process

Poisson point processes can be used as a cornerstone in the construction of very different stochastic objects such as, for example, infinitely divisible distributions, Markov processes with complex dynamics, objects of stochastic geometry and so forth.

In this chapter, we briefly develop the general framework of random measures and construct the Poisson point process and characterize it in terms of its Laplace transform. As an application we construct a certain subordinator and show that the Poisson point process is the invariant measure of systems of independent random walks. Via the connection with subordinators, in the third section, we construct two distributions that play prominent roles in population genetics: the Poisson–Dirichlet distribution and the GEM distribution.

For a nice exposition including many examples, see also [99].

24.1 Random Measures

In the following, let E be a locally compact Polish space (for example, $E = \mathbb{R}^d$ or $E = \mathbb{Z}^d$) with Borel σ-algebra $\mathcal{B}(E)$. Let

$$\mathcal{B}_b(E) = \{B \in \mathcal{B}(E) : B \text{ is relatively compact}\}$$

be the system of *bounded* Borel sets and $\mathcal{M}(E)$ the space of Radon measures on E (see Definition 13.3).

A. Klenke, *Probability Theory*, Universitext, https://doi.org/10.1007/978-3-030-56402-5_24

Definition 24.1 *Denote by* $\mathbb{M} = \sigma(I_A \;:\; A \in \mathcal{B}_b(E))$ *the smallest σ-algebra on* $\mathcal{M}(E)$ *with respect to which all maps*

$$I_A \;:\; \mu \mapsto \mu(A), \quad A \in \mathcal{B}_b(E),$$

are measurable.

Denote by $\mathcal{B}^+(E)$ the set of measurable maps $E \to [0, \infty]$ and by $\mathcal{B}_b^{\mathbb{R}}(E)$ the set of bounded measurable maps $E \to \mathbb{R}$ with compact support. For every $f \in \mathcal{B}^+(E)$, the integral $I_f(\mu) := \int f \, d\mu$ is well-defined and for every $f \in \mathcal{B}_b^{\mathbb{R}}(E)$, $I_f(\mu)$ is well-defined and finite.

Theorem 24.2 *Let τ_v be the vague topology on $\mathcal{M}(E)$. Then*

$$\mathbb{M} = \mathcal{B}(\tau_v) = \sigma\big(I_f \;:\; f \in C_c(E)\big) = \sigma\big(I_f \;:\; f \in C_c^+(E)\big).$$

Proof This is left as an exercise. (See [82, Lemma 4.1].)   □

Let $\widetilde{\mathcal{M}}(E)$ be the space of *all* measures on E endowed with the σ-algebra

$$\widetilde{\mathbb{M}} = \sigma\big(I_A \;:\; A \in \mathcal{B}_b(E)\big).$$

Choose a countable basis $\mathcal{U}$ of the topology consisting of relatively compact sets. Then we get (compare Exercise 13.1.8)

$$\mathcal{M}(E) = \bigcap_{U \in \mathcal{U}} \big\{\mu \in \widetilde{\mathcal{M}}(E) \;:\; \mu(U) < \infty\big\}.$$

Hence $\mathcal{M}(E) \in \widetilde{\mathbb{M}}$. Clearly, $\mathbb{M} = \widetilde{\mathbb{M}}\big|_{\mathcal{M}(E)}$ is the trace σ-algebra of $\widetilde{\mathbb{M}}$ on $\mathcal{M}(E)$. Here we need the slightly larger space in order to define random measures in such a way that all almost surely well-defined operations on random measures again yield random measures.

Definition 24.3 *A **random measure** on E is a random variable X on some probability space $(\Omega, \mathcal{A}, \mathbf{P})$ with values in $(\widetilde{\mathcal{M}}(E), \widetilde{\mathbb{M}})$ and with $\mathbf{P}[X \in \mathcal{M}(E)] = 1$.*

Theorem 24.4 *Let X be a random measure on E. Then the set function $\mathbf{E}[X] \;:\; \mathcal{B}(E) \to [0, \infty]$, $A \mapsto \mathbf{E}[X(A)]$ is a measure. We call $\mathbf{E}[X]$ the **intensity measure** of X. We say that X is integrable if $\mathbf{E}[X] \in \mathcal{M}(E)$.*

Proof Clearly, $\mathbf{E}[X]$ is finitely additive. Let $A, A_1, A_2, \ldots \in \mathcal{B}(E)$ with $A_n \uparrow A$. Consider the random variables $Y_n := X(A_n)$ and $Y = X(A)$. Then $Y_n \uparrow Y$ and hence, by monotone convergence, $\mathbf{E}[X](A_n) = \mathbf{E}[Y_n] \stackrel{n \to \infty}{\longrightarrow} \mathbf{E}[Y] = \mathbf{E}[X](A)$. Hence $\mathbf{E}[X]$ is lower semicontinuous and is thus a measure (by Theorem 1.36).   □

Theorem 24.5 *Let* $\mathbf{P}_X$ *be the distribution of a random measure* X. *Then* $\mathbf{P}_X$ *is uniquely determined by the distributions of either of the families*

$$((I_{f_1}, \ldots, I_{f_n}) : n \in \mathbb{N}; \ f_1, \ldots, f_n \in C_c^+(E)) \tag{24.1}$$

or

$$((I_{A_1}, \ldots, I_{A_n}) : n \in \mathbb{N}; \ A_1, \ldots, A_n \in \mathcal{B}_b(E) \ \text{pairwise disjoint}). \tag{24.2}$$

Proof The class of sets

$$\mathcal{I} = \{(I_{f_1}, \ldots, I_{f_n})^{-1}(A) : n \in \mathbb{N}; \ f_1, \ldots, f_n \in C_c^+(E), \ A \in \mathcal{B}([0, \infty)^n)\}$$

is a π-system and by Theorem 24.2 it generates $\mathbb{M}$. Hence the measure $\mathbf{P}_X$ is characterized by its values on $\mathcal{I}$.

Similarly, the claim follows for

$$((I_{A_1}, \ldots, I_{A_n}) : n \in \mathbb{N}; \ A_1, \ldots, A_n \in \mathcal{B}_b(E)).$$

If $A_1, \ldots, A_n \in \mathcal{B}_b(E)$ are arbitrary, then there exist $2^n - 1$ pairwise disjoint sets $B_1, \ldots, B_{2^n-1}$ with $A_i = \bigcup_{k: B_k \subset A_i} B_k$ for all $i = 1, \ldots, n$. The distribution of $(I_{A_1}, \ldots, I_{A_n})$ can be computed from that of $(I_{B_1}, \ldots, I_{B_{2^n-1}})$. $\square$

In the following, let $i = \sqrt{-1}$ be the imaginary unit.

Definition 24.6 *Let* X *be a random measure on* E. *Denote by*

$$\mathcal{L}_X(f) = \mathbf{E}\left[\exp\left(-\int f \, dX\right)\right], \quad f \in \mathcal{B}^+(E),$$

*the **Laplace transform** of* X *and by*

$$\varphi_X(f) = \mathbf{E}\left[\exp\left(i\int f \, dX\right)\right], \quad f \in \mathcal{B}_b^{\mathbb{R}}(E),$$

*the **characteristic function** of* X.

Theorem 24.7 *The distribution* $\mathbf{P}_X$ *of a random measure* X *is characterized by its Laplace transform* $\mathcal{L}_X(f)$, $f \in C_c^+(E)$, *as well as by its characteristic function* $\varphi_X(f)$, $f \in C_c(E)$.

Proof This is a consequence of Theorem 24.5 and the uniqueness theorem for characteristic functions (Theorem 15.9) and for Laplace transforms (Exercise 15.1.2) of random variables on $[0, \infty)^n$. $\square$

Definition 24.8 *We say that a random measure* X *on* E *has **independent increments** if, for any choice of finitely many pairwise disjoint measurable sets* $A_1, \ldots, A_n$, *the random variables* $X(A_1), \ldots, X(A_n)$ *are independent.*

Corollary 24.9 *The distribution of a random measure X on E with independent increments is uniquely determined by the family* $(\mathbf{P}_{X(A)},\ A \in \mathcal{B}_b(E))$.

Proof This is an immediate consequence of Theorem 24.5. $\square$

Definition 24.10 *Let $\mu \in \mathcal{M}(E)$. A random measure X with independent increments is called a **Poisson point process (PPP)** with intensity measure μ if, for any $A \in \mathcal{B}_b(E)$, we have $\mathbf{P}_{X(A)} = \mathrm{Poi}_{\mu(A)}$. In this case, we write $\mathrm{PPP}_\mu := \mathbf{P}_X \in \mathcal{M}_1(\mathcal{M}(E))$ and say that X is a PPP_μ.*

See Fig. 24.1 for a simulation of a Poisson point process on the unit square.

Remark 24.11 The definition of the PPP (and its construction in the following theorem) still works if $(E, \mathcal{E}, \mu)$ is only assumed to be a σ-finite measure space. However, the characterization in terms of Laplace transforms is a bit simpler in the case of locally compact Polish spaces considered here. $\Diamond$

Theorem 24.12 *For every $\mu \in \mathcal{M}(E)$, there exists a Poisson point process X with intensity measure μ.*

Proof μ is σ-finite since $\mu \in \mathcal{M}(E)$. Hence there exist $E_n \uparrow E$ with $\mu(E_n) < \infty$ for every $n \in \mathbb{N}$. Define $\mu_1 = \mu(E_1 \cap \cdot)$ and $\mu_n = \mu((E_n \setminus E_{n-1}) \cap \cdot)$ for $n \geq 2$. If $X_1, X_2, \ldots$ are independent Poisson point processes with intensity measures $\mu_1, \mu_2, \ldots$, then $X = \sum_{n=1}^{\infty} X_n$ has intensity measure $\mathbf{E}[X] = \mu$ and hence X is a random measure (see Exercise 24.1.1). Furthermore, it is easy to see that X has

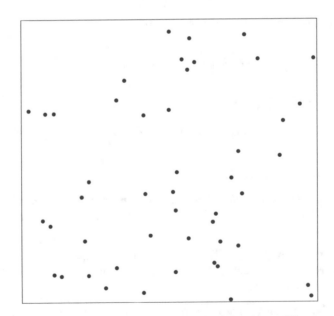

Fig. 24.1 Poisson point process on the unit square with intensity measure 50λ.

independent increments and that

$$\mathbf{P}_{X(A)} = \mathbf{P}_{X_1(A)} * \mathbf{P}_{X_2(A)} * \ldots = \mathrm{Poi}_{\mu_1(A)} * \mathrm{Poi}_{\mu_2(A)} * \ldots = \mathrm{Poi}_{\mu(A)}.$$

Hence we have $X \sim \mathrm{PPP}_\mu$.

Therefore, it is enough to consider the case $\mu(E) \in (0, \infty)$. Define

$$\nu = \frac{\mu(\cdot)}{\mu(E)} \in \mathcal{M}_1(E).$$

Let $N, Y_1, Y_2, \ldots$ be independent random variables with $N \sim \mathrm{Poi}_{\mu(E)}$ and $\mathbf{P}_{Y_i} = \nu$ for all $i \in \mathbb{N}$. Define

$$X(A) = \sum_{n=1}^{N} \mathbb{1}_A(Y_n) \quad \text{for } A \in \mathcal{B}(E).$$

The random variables $\mathbb{1}_A(Y_1), \mathbb{1}_A(Y_2), \ldots$ are independent and $\mathrm{Ber}_{\nu(A)}$-distributed; hence we have $X(A) \sim \mathrm{Poi}_{\mu(A)}$ (see Theorem 15.15(iii)). Let $n \in \mathbb{N}$ and let $A_1, \ldots, A_n \in \mathcal{B}(E)$ be pairwise disjoint. Then

$$\psi(t) := \mathbf{E}\left[\exp\left(i \sum_{l=1}^{n} t_l \mathbb{1}_{A_l}(Y_1)\right)\right] = 1 + \sum_{l=1}^{n} \nu(A_l)\left(e^{i t_l} - 1\right), \quad t \in \mathbb{R}^n,$$

is the characteristic function of $(\mathbb{1}_{A_1}(Y_1), \ldots, \mathbb{1}_{A_n}(Y_1))$. Further, let φ be the characteristic function of $(X(A_1), \ldots, X(A_n))$ and let φ_l be the characteristic function of $X(A_l)$ for $l = 1, \ldots, n$. Hence $\varphi_l(t_l) = \exp(\mu(A_l)(e^{i t_l} - 1))$. By Theorem 15.15(iii), we have

$$\varphi(t) = \mathbf{E}\left[\exp\left(i \sum_{l=1}^{n} t_l X(A_l)\right)\right]$$

$$= \exp\left(\mu(E)(\psi(t) - 1)\right)$$

$$= \exp\left(\sum_{l=1}^{n} \mu(A_l)\left(e^{i t_l} - 1\right)\right) = \prod_{l=1}^{n} \varphi_l(t_l).$$

Thus $X(A_1), \ldots, X(A_n)$ are independent. This implies $X \sim \mathrm{PPP}_\mu$. $\qquad \square$

Takeaways A random measure is a random variable taking values in the space of Radon measures on a set E. For measurable $A \subset E$, the intensity measure yields the expected value of this random variable. The distribution of a random measure is characterised by its characteristic function as well as by its Laplace transform. The Poisson point process is a specific random measure taking only integer values and whose values on disjoint sets are independent and Poisson distributed. An example are the first rain drops you see on the side walk.

Exercise 24.1.1 Let $X_1, X_2, \ldots$ be random measures and $\lambda_1, \lambda_2, \ldots \in [0, \infty)$. Define $X := \sum_{n=1}^{\infty} \lambda_n X_n$. Show that X is a random measure if and only if we have $\mathbf{P}[X(B) < \infty] = 1$ for all $B \in \mathcal{B}_b(E)$. Infer that if X is a random variable with values in $(\widetilde{\mathcal{M}}(E), \widetilde{\mathbb{M}}(E))$ and $\mathbf{E}[X] \in \mathcal{M}(E)$, then X is a random measure. ♣

Exercise 24.1.2 Let τ_w be the topology of weak convergence on $\mathcal{M}_1(E)$ and let $\sigma(\tau_w)$ be the Borel σ-algebra on $\mathcal{M}_1(E)$. Show that $\mathbb{M}\big|_{\mathcal{M}_1(E)} = \sigma(\tau_w)$. ♣

24.2   Properties of the Poisson Point Process

Rényi's Theorem

A measure μ is called atom-free, if $\mu(\{x\}) = 0$ for all $x \in E$. An integer valued measure ν is said to have no **double points**, if $\nu(\{x\}) \in \{0, 1\}$ for all $x \in E$. For a random integer valued measure the property to have no double points is in fact an *event* (i.e., it is measurable). This will be shown in the subsequent considerations.

In order to illustrate the main idea, first assume that ν is a finite integer valued measure on $(0, 1]$. Then ν has no double points if and only if

$$\lim_{n \to \infty} \sup_{k=0,\ldots,2^n-1} \nu\left(\left(\frac{k}{2^n}, \frac{k+1}{2^n}\right]\right) \leq 1. \tag{24.3}$$

Note that the condition $\nu(\{x\}) \in \{0, 1\}$ has to be fulfilled for uncountably many $x \in (0, 1]$ while condition (24.3) is a limit of conditions each involving only finitely many subsets of $(0, 1]$. In particular, for a finite integer valued random measure X on $(0, 1]$, the property to have no double points

$$\left\{ X(\{x\}) \leq 1 \forall x \in (0, 1] \right\} = \left\{ \lim_{n \to \infty} \sup_{k=0,\ldots,2^n-1} X\left(\left(\frac{k}{2^n}, \frac{k+1}{2^n}\right]\right) \leq 1 \right\} \tag{24.4}$$

is an event. In order to show the same statement for E a locally compact Polish space (instead of $(0, 1]$) we have to spend a little work. Hence, let X be a random integer valued measure on E. Since E is σ-compact, there exists a sequence $(E_n)_{n \in \mathbb{N}}$ of compacts with $E_n \uparrow E$. It is clearly enough to show that X has almost surely no double points on any E_n. Hence, without loss of generality we may and will assume that E is compact. Since X is locally finite almost surely, as a consequence X is finite almost surely. As E is compact, it is totally bounded. That is, for any $n \in \mathbb{N}$, we can cover E by finitely many open sets $B_1^n, \ldots, B_{N_n}^n$ with diameter less than 2^{-n} (see Lemma 13.2). By building intersections for any choice of $k_1 \in \{1, \ldots, N_1\}, \ldots, k_n \in \{1, \ldots, N_n\}$, we get the open sets

$$\tilde{B}_{k_1,\ldots,k_n} := \bigcap_{m=1}^{n} B_{k_m}^m.$$

By a suitable rearranging of the enumeration, we call these sets also B_k^n, $k = 1, \ldots, N_n$. By construction, for any B_k^{n+1}, there is an l such that $B_k^{n+1} \subset B_l^n$. We construct a measurable partition of E by letting $D_1^n := B_1^n$ and defining successively

$$D_k^n := B_k^n \setminus \bigcup_{l=1}^{k-1} B_l^n.$$

In fact, the sets $D_1^n, \ldots, D_{N_n}^n \in \mathcal{B}(E)$ are pairwise disjoint, have diameters at most 2^{-n}, and cover E. Note that the partition $\mathcal{D}^{n+1} := \{D_k^{n+1} : k = 1, \ldots, N_{n+1}\}$ is a refinement of $\mathcal{D}^n$ for any n. That is, for any $D = D_k^{n+1} \in \mathcal{D}^{n+1}$, there is a $D' = D_l^n \in \mathcal{D}^n$ such that $D \subset D'$. By construction, for any sequence (D^n) with $D^n \in \mathcal{D}^n$, $n \in \mathbb{N}$, we either have $\bigcap_{n=1}^\infty D^n = \emptyset$ or $\bigcap_{n=1}^\infty D^n = \{x\}$ for some $x \in E$. If $\mu \in \mathcal{M}_f(E)$ then by upper continuity, we have

$$\lim_{n \to \infty} \sup_{k=1,\ldots,N_n} \mu(D_k^n) = \sup_{x \in E} \mu(\{x\}). \tag{24.5}$$

Hence, the integer valued random measure X has almost surely no double points if and only if

$$\lim_{n \to \infty} \sup_{k=1,\ldots,N_n} X(D_k^n) \le 1 \quad \text{a.s.} \tag{24.6}$$

For the case $E = \mathbb{R}$, the following theorem goes back to Rényi [142, Theorem 2].

Theorem 24.13 (Rényi's theorem [142]) *Let $\mu \in \mathcal{M}(E)$ be atom-free and let X be an integer valued random measure on E. Then the following two statements are equivalent.*

(i) $X \sim \mathrm{PPP}_\mu$.
(ii) X almost surely has no double points and

$$\mathbf{P}[X(A) = 0] = e^{-\mu(A)} \quad \text{for all } A \in \mathcal{B}_b(E). \tag{24.7}$$

Proof (i) $\Longrightarrow$ (ii) This is obvious.
(ii) $\Longrightarrow$ (i) As E is locally compact and Polish, there is a sequence (E_n) of compacts such that $E_n \uparrow E$ and $\mu(E_n) < \infty$ for all $n \in \mathbb{N}$. Hence without loss of generality we may and will assume that E is compact and μ is a finite measure on E. Furthermore, as X is almost surely a Radon measure, it is almost surely finite on the compact E.

If $A_1, \ldots, A_n \in \mathcal{B}_b(E)$ are pairwise disjoint, then

$$\mathbf{P}\big[X(A_1) = 0, \ldots, X(A_n) = 0\big] = \mathbf{P}\big[X(A_1 \cup \ldots \cup A_n) = 0\big]$$
$$= e^{-\mu(A_1 \cup \ldots \cup A_n)}$$
$$= \prod_{l=1}^{n} e^{-\mu(A_l)} = \prod_{l=1}^{n} \mathbf{P}[X(A_l) = 0].$$

If we define $\widetilde{X}(A) := X(A) \wedge 1$ for $A \in \mathcal{B}(E)$, then the random variables $\widetilde{X}(A_l)$, $l = 1, \ldots, n$, are independent.

Now let $(\mathcal{D}^n)_{n \in \mathbb{N}}$ be a sequence of measurable partitions of E consisting of sets of diameter at most 2^{-n} and such that $\mathcal{D}^{n+1}$ is a refinement of $\mathcal{D}^n$ for any $n \in \mathbb{N}$. Such a sequence was constructed in the course of the discussion preceding Theorem 24.13.

Let

$$\varepsilon_n := \sup_{D \in \mathcal{D}^n} \mu(D).$$

As μ is atom-free, by (24.5), we have

$$\varepsilon_n \overset{n \to \infty}{\longrightarrow} 0. \tag{24.8}$$

As X almost surely has no double points and is finite, for any $A \in \mathcal{B}(E)$ by (24.5), we have

$$\lim_{n \to \infty} \sum_{D \in \mathcal{D}^n} \widetilde{X}(D \cap A) = \lim_{n \to \infty} \sum_{D \in \mathcal{D}^n} X(D \cap A) = X(A) \quad \text{a.s.} \tag{24.9}$$

In particular, we infer that also the random variables $X(A_l)$, $l = 1, \ldots, n$ are independent. In other words, X is a random measure with independent increments.

Now let $A \in \mathcal{B}(E)$ and note that $\sum_{D \in \mathcal{D}^n} \widetilde{X}(D \cap A)$ is a sum of independent Bernoulli random variables with success probabilities $p_D^n(A) := 1 - e^{-\mu(D \cap A)} \leq \varepsilon_n \overset{n \to \infty}{\longrightarrow} 0$. We have

$$\sum_{D \in \mathcal{D}^n} p_D^n(A) \overset{n \to \infty}{\longrightarrow} \mu(A)$$

and

$$\sum_{D \in \mathcal{D}^n} (p_D^n(A))^2 \leq \varepsilon_n \sum_{D \in \mathcal{D}^n} p_D^n(A) \overset{n \to \infty}{\longrightarrow} 0.$$

By the theorem on Poisson approximation (Theorem 3.7), we get $X(A) \sim \mathrm{Poi}_{\mu(A)}$. Together with the property of independent increments this shows that X is a Poisson point process with intensity measure μ. $\square$

Reflection What is the connection between Rényi's theorem and the waiting times in the construction of the Poisson process on $[0, \infty)$ in Theorem 5.36? ♠

Laplace Transform, Characteristic Function and Moments

Theorem 24.14 *Let $\mu \in \mathcal{M}(E)$ and let X be a Poisson point process with intensity measure μ. Then X has Laplace transform*

$$\mathcal{L}_X(f) = \exp\left(\int \mu(dx)\left(e^{-f(x)} - 1\right)\right), \qquad f \in \mathcal{B}^+(E),$$

and characteristic function

$$\varphi_X(f) = \exp\left(\int \mu(dx)\left(e^{if(x)} - 1\right)\right), \qquad f \in \mathcal{B}_b^{\mathbb{R}}(E).$$

Proof It is enough to show the claim for simple functions $f = \sum_{l=1}^n \alpha_l \mathbb{1}_{A_l}$ with complex numbers $\alpha_1, \ldots, \alpha_n$ and with pairwise disjoint sets $A_1, \ldots, A_n \in \mathcal{B}_b(E)$. (For general f, the claim follows by the usual approximation arguments.) For such f, however,

$$\mathbf{E}\left[\exp\left(-I_f(X)\right)\right] = \mathbf{E}\left[\prod_{l=1}^n e^{-\alpha_l X(A_l)}\right] = \prod_{l=1}^n \mathbf{E}\left[e^{-\alpha_l X(A_l)}\right]$$

$$= \prod_{l=1}^n \exp\left(\mu(A_l)\left(e^{-\alpha_l} - 1\right)\right)$$

$$= \exp\left(\sum_{l=1}^n \mu(A_l)\left(e^{-\alpha_l} - 1\right)\right)$$

$$= \exp\left(\int \mu(dx)\left(e^{-f(x)} - 1\right)\right). \qquad \square$$

Corollary 24.15 (Moments of the PPP) *Let $\mu \in \mathcal{M}(E)$ and $X \sim \mathrm{PPP}_\mu$.*

(i) *If $f \in \mathcal{L}^1(\mu)$, then $\mathbf{E}[\int f\, dX] = \int f\, d\mu$.*
(ii) *If $f \in \mathcal{L}^2(\mu) \cap \mathcal{L}^1(\mu)$, then $\mathbf{Var}[\int f\, dX] = \int f^2\, d\mu$.*

Recall that only for finite μ, we have the inclusion $\mathcal{L}^2(\mu) \subset \mathcal{L}^1(\mu)$.

Proof If $f = f^+ - f^- \in \mathcal{L}^1(\mu)$, then for the characteristic function, integral and differentiation interchange, $\frac{d}{dt}\varphi_X(tf^+) = i\varphi_X(tf^+) \int f(x) e^{itf^+(x)} \mu(dx)$ and hence (by Exercise 15.4.4(iii))

$$\mathbf{E}[I_{f^+}(X)] = \frac{1}{i}\frac{d}{dt}\varphi_X(tf^+)\Big|_{t=0} = \int f^+ d\mu.$$

Arguing similarly with f^- and adding up, we get (i).

If $f \in \mathcal{L}^1(\mu) \cap \mathcal{L}^2(\mu)$, then the argument can be iterated (using Theorem 15.35)

$$\frac{d^2}{dt^2}\varphi_X(tf) = -\varphi_X(tf)\left[\int f^2(x) e^{itf(x)} \mu(dx) + \left(\int f(x) e^{itf(x)} \mu(dx)\right)^2\right],$$

hence we have $\mathbf{E}[I_f(X)^2] = -\frac{d^2}{dt^2}\varphi_X(tf)\Big|_{t=0} = I_{f^2}(\mu) + I_f(\mu)^2.$ □

Theorem 24.16 (Mapping theorem) *Let E and F be locally compact Polish spaces and let $\phi : E \to F$ be a measurable map. Let $\mu \in \mathcal{M}(E)$ with $\mu \circ \phi^{-1} \in \mathcal{M}(F)$ and let X be a PPP on E with intensity measure μ. Then $X \circ \phi^{-1}$ is a PPP on F with intensity measure $\mu \circ \phi^{-1}$.*

Proof For $f \in \mathcal{B}^+(F)$,

$$\mathcal{L}_{X\circ\phi^{-1}}(f) = \mathcal{L}_X(f \circ \phi) = \exp\left(\int \left(e^{-f(\phi(x))} - 1\right) \mu(dx)\right)$$

$$= \exp\left(\int \left(e^{-f(y)} - 1\right) \left(\mu \circ \phi^{-1}\right)(dy)\right).$$

Now, Theorems 24.14 and 24.7 yield the claim.   □

Infinitely Divisible Random Variables

Theorem 24.17 *Let $\nu \in \mathcal{M}((0, \infty))$ and let $X \sim \mathrm{PPP}_\nu$ on $(0, \infty)$. Further, define $Y := \int x\, X(dx)$. Then the following are equivalent.*

(i) $\mathbf{P}[Y < \infty] > 0.$
(ii) $\mathbf{P}[Y < \infty] = 1.$
(iii) $\int \nu(dx)(1 \wedge x) < \infty.$

If (i)–(iii) hold, then Y is an infinitely divisible nonnegative random variable with Lévy measure ν.

Proof Let $Y_\infty = \int_{[1,\infty)} x\, X(dx)$ and $Y_t := \int_{(t,1)} x\, X(dx)$ for $t \in [0,1)$. Evidently, $Y = Y_0 + Y_\infty$. Furthermore, it is clear that

$$\mathbf{P}[Y_\infty < \infty] > 0 \iff \mathbf{P}[Y_\infty < \infty] = 1 \iff \nu([1,\infty)) < \infty. \qquad (24.10)$$

If (iii) holds, then $\mathbf{E}[Y_0] = \int_{(0,1)} x\, \nu(dx) < \infty$; hence $Y_0 < \infty$ a.s. (and thus $Y < \infty$ a.s. by (24.10)). On the other hand, if (iii) does not hold, then $Y_\infty = \infty$ a.s. or $\mathbf{E}[Y_0] = \infty$. While Y_∞ can have infinite expectation even if $Y_\infty < \infty$ a.s., for Y_0 this is impossible since, in contrast with Y_∞, Y_0 is composed not of a few large contributions but many small ones so that a law of large numbers is in force. More precisely, by Corollary 24.15, we have

$$\mathbf{Var}[Y_t] = \int_{(t,1)} x^2\, \nu(dx) \le \int_{(t,1)} x\, \nu(dx) = \mathbf{E}[Y_t] < \infty \qquad \text{for all } t \in (0,1).$$

Hence, by Chebyshev's inequality,

$$\mathbf{P}\!\left[Y_t < \frac{\mathbf{E}[Y_t]}{2}\right] \le \frac{4\,\mathbf{Var}[Y_t]}{\mathbf{E}[Y_t]^2} \xrightarrow{t \to 0} 0.$$

Thus $Y_0 = \sup_{t \in (0,1)} Y_t \ge \mathbf{E}[Y_0]/2 = \infty$ almost surely.

Now assume that (i)–(iii) hold. By Theorem 24.14, Y has the Laplace transform

$$\mathbf{E}\!\left[e^{-tY}\right] = \exp\!\left(\int \nu(dx)\big(e^{-tx} - 1\big)\right).$$

By the Lévy–Khinchin formula (Theorem 16.14), Y is infinitely divisible with Lévy measure ν. $\square$

Corollary 24.18 *Let $\mu_i \in \mathcal{M}_1([0,\infty))$, $i = 1,2$, be infinitely divisible distributions with canonical measures $\nu_i \in \mathcal{M}((0,\infty))$ and deterministic parts $\alpha_i \ge 0$ (compare Theorem 16.14). If we have*

$$\alpha_1 \le \alpha_2 \quad \text{and} \quad \nu_1([x,\infty)) \le \nu_2([x,\infty)) \quad \text{for all } x > 0, \qquad (24.11)$$

then μ_1 is stochastically smaller than μ_2; i.e., $\mu_1 \le_{\mathrm{st}} \mu_2$.

Proof *(The proof follows [100, Proof of Lemma 6.1])* The idea is to use a coupling argument where based on one Poisson point process we construct the two random variables Y_1, Y_2 with $Y_i \sim \mu_i$, $i = 1,2$, such that $Y_1 \le Y_2$ almost surely. By Theorem 17.59, this yields the claim.

Let $G_i(x) := \nu_i([x,\infty))$, $i = 1,2$, $x > 0$, and

$$\phi_i(y) := G_i^{-1}(y) = \inf\{x \ge 0 : G_i(x) \le y\} \quad \text{for } y > 0.$$

If ν_i is finite, then $\phi_i(y) = 0$ for $y \geq \nu_i((0, \infty))$. Let λ denote the Lebesgue measure on $[0, \infty)$. By construction, for the image measure restricted to the positive reals, we have

$$\left(\lambda \circ \phi_i^{-1}\right)\big|_{(0,\infty)} = \nu_i, \qquad i = 1, 2.$$

Now assume that X is a PPP on $(0, \infty)$ with intensity measure λ. By Theorem 24.16, the random measures

$$X_i := \left(\int \delta_{\phi_i(x)} X(dx)\right)\big|_{(0,\infty)} = \left(X \circ \phi_i^{-1}\right)\big|_{(0,\infty)}$$

are PPPs with intensity measures ν_i, $i = 1, 2$. By Theorem 24.17, we thus have

$$Y_i := \alpha_i + \int \phi_i(x) X(dx) \sim \mu_i \quad \text{for } i = 1, 2.$$

However, by assumption, we have $G_1 \leq G_2$ which implies $\phi_1 \leq \phi_2$ and thus $Y_1 \leq Y_2$ a.s. $\square$

Example 24.19 By Corollary 16.10, for every nonnegative infinitely divisible distribution μ with Lévy measure ν, there exists a stochastic process $(Y_t)_{t \geq 0}$ with independent stationary increments and $Y_t \sim \mu^{*t}$ (hence with Lévy measure $t\nu$). Here we give a direct construction of this process. Let X be a PPP on $(0, \infty) \times [0, \infty)$ with intensity measure $\nu \otimes \lambda$ (here λ is the Lebesgue measure). Define $Y_0 = 0$ and

$$Y_t := \int_{(0,\infty) \times (0,t]} x \, X(d(x, s)).$$

By the mapping theorem, we have $X(\,\cdot\, \times (s, t]) \sim \text{PPP}_{(t-s)\nu}$; hence $Y_t - Y_s$ is infinitely divisible with Lévy measure $(t - s)\nu$. The independence of the increments is evident. Note that $t \mapsto Y_t$ is right continuous and monotone increasing.

The process Y that we have just constructed is called a **subordinator** with Lévy measure ν. $\Diamond$

The procedure in the previous example can be generalized by allowing time sets more general than $[0, \infty)$.

Definition 24.20 *A random measure Y is called infinitely divisible if, for any $n \in \mathbb{N}$, there exist i.i.d. random measures $Y_1, \ldots, Y_n$ with $Y = Y_1 + \ldots + Y_n$.*

Theorem 24.21 *Let $\nu \in \mathcal{M}((0, \infty) \times E)$ with*

$$\int \mathbb{1}_A(t) \, (1 \wedge x) \, \nu(d(x, t)) < \infty \quad \text{for all } A \in \mathcal{B}_b(E),$$

and let $\alpha \in \mathcal{M}(E)$. Let X be a PPP_ν *and*

$$Y(A) := \alpha(A) + \int x \, \mathbb{1}_A(t) \, X(d(x, t)) \quad \text{for } A \in \mathcal{B}(E).$$

Then Y is an infinitely divisible random measure with independent increments. For $A \in \mathcal{B}(E)$, $Y(A)$ has the Lévy measure $\nu(\cdot \times A)$.

We call ν the canonical measure and α the deterministic part of Y.

Proof This is a direct consequence of Theorems 24.16 and 24.17. □

Remark 24.22 We can write Y as $Y = \alpha + \int x \delta_t \, X(d(x, t))$, where δ_t is the Dirac measure at $t \in E$. If instead of $x \, \delta_t$, we allow more general measures $\chi \in \mathcal{M}(E)$, then we get a representation

$$Y = \alpha + \int_{\mathcal{M}(E)} \chi \, X(d\chi),$$

where $X \sim PPP_\nu$ on $\mathcal{M}(E)$ and $\nu \in \mathcal{M}(\mathcal{M}(E))$ with

$$\int \nu(d\chi)(\chi(A) \wedge 1) < \infty$$

for any $A \in \mathcal{B}_b(E)$. It can be shown that this is the most general form of an infinitely divisible measure on E. We call ν the canonical measure of Y and α the deterministic part. Y is characterized by its Laplace transform which obeys the Lévy–Khinchin formula:

$$\mathcal{L}_Y(f) = \exp\left(-\int f \, d\alpha + \int \nu(d\chi)\left(e^{-\int f \, d\chi} - 1\right)\right). \quad \Diamond$$

Random Colorings

Let N be a Poisson random variable with parameter $\lambda > 0$ and let $Z_1, Z_2, \ldots$ independent (and independent of N) Bernoulli random variables with probability of success $p \in [0, 1]$. Then

$$N_0 := \sum_{i=1}^{N} \mathbb{1}_{\{0\}}(Z_i) \quad \text{and} \quad N_1 := \sum_{i=1}^{N} \mathbb{1}_{\{1\}}(Z_i)$$

are independent Poisson random variables with parameters $\lambda_0 = (1 - p)\lambda$ and $\lambda_1 = p\lambda$. Similarly, we can allow the Z_i to take values in $\{0, \ldots, k\}$ (for some

$k \in \mathbb{N}$) with probabilities $p_0, \ldots, p_k$. Then the random variables

$$N_m := \sum_{i=1}^{N} \mathbb{1}_{\{m\}}(Z_i), \qquad m = 0, \ldots, k,$$

are independent and Poisson distributed with parameters $\lambda_m = p_m \lambda$, $m = 0, \ldots, k$. Often, the Z_i are interpreted as colors or marks that are a given independently to the single points of the Poisson random variable N.

While the above statements can be shown easily using elementary methods (this remains as an exercise), we will now turn to a more general situation. We assume that the possible colors are drawn from a locally compact Polish space F (which includes finite or countable F and $F = \mathbb{R}$) with Borel σ-algebra $\mathcal{B}(F)$ and that X is a Poisson point process on a locally compact Polish space E with intensity measure $\mu \in \mathcal{M}(E)$. Let $\nu \in \mathcal{M}_1(F)$ and let $Z_1, Z_2, \ldots$ be i.i.d. F-valued random variables with distribution ν. We assume that the Z_i are independent of X. The Z_i are the colors that we want to attach to the points of X. However, in order to do so, we need an enumeration of the points of X. Any enumeration will do the trick and since the Z_i are i.i.d., the resulting will not depend on the specific enumeration we choose (in distribution).

We proceed similarly as in the proof of existence of the Poisson point process (Theorem 24.12). We decompose μ into a sum of finite measures $\mu = \sum_{n=1}^{\infty} \mu_n$ with $\mu_n \in \mathcal{M}_f(E)$, $\mu_n \neq 0$ for any $n \in \mathbb{N}$. Let $(N_n)_{n \in \mathbb{N}}$ be independent random variables with $N_n \sim \mathrm{Poi}_{\mu_n(E)}$. Define $\widetilde{\mu}_n := \mu_n / \mu_n(E) \in \mathcal{M}_1(E)$. Let $(Y_{n,i})_{n,i \in \mathbb{N}}$ and $(Z_{n,i})_{n,i \in \mathbb{N}}$ be independent (and independent of (N_n)) random variables with $Y_{n,i} \sim \widetilde{\mu}_n$ and $Z_{n,i} \sim \nu$, $n, i \in \mathbb{N}$. Let

$$X(A) := \sum_{n=1}^{\infty} \sum_{i=1}^{N_n} \mathbb{1}_A(Y_{n,i}), \qquad A \in \mathcal{B}_b(E).$$

In the proof of Theorem 24.12, we showed that $X \sim \mathrm{PPP}_\mu$.

We attach random colors $Z_{n,i}$ to the points $Y_{n,i}$ be defining

$$\widetilde{X}(A \times B) := \sum_{n=1}^{\infty} \sum_{i=1}^{N_n} \mathbb{1}_A(Y_{n,i}) \, \mathbb{1}_B(Z_{n,i}), \qquad A \in \mathcal{B}_b(E), \ B \in \mathcal{B}(F).$$

Clearly, this defines a random measure

$$\widetilde{X}(C) = \sum_{n=1}^{\infty} \sum_{i=1}^{N_n} \mathbb{1}_C((Y_{n,i}, Z_{n,i})), \qquad C \in \mathcal{B}(E \times F).$$

By construction, we have $\widetilde{X} \sim \mathrm{PPP}_{\mu \otimes \nu}$.

The next generalization we focus on is that we want to allow the distribution of the random color of a point to depend on its position but not on the positions of the other points. More precisely, let κ be a Markov kernel from E to F. If $x \in E$ is a point of X, that is, if $X(\{x\}) = 1$, then the color of that point will be chosen at random with distribution $\kappa(x, \cdot)$. The proper way to do this is to use the construction outlined above but with $Z_{n,i}$ depending on $Y_{n,i}$. More precisely, we assume that the bivariate random variables $(Y_{n,i}, Z_{n,i})$, $n, i \in \mathbb{N}$, are independent with distribution $\widetilde{\mu}_n \otimes \kappa$. That is,

$$\mathbf{P}[Y_{n,i} \in A, \ Z_{n,i} \in B] = \int \mathbb{1}_A(x)\kappa(x, B)\, \widetilde{\mu}_n(dx) \quad \text{for } A \in \mathcal{B}(E), \ B \in \mathcal{B}(F).$$

Clearly, we then have that

$$\widetilde{X}_n(C) := \sum_{i=1}^{N_n} \mathbb{1}_C((Y_{n,i}, Z_{n,i})), \qquad C \in \mathcal{B}(E \times F)$$

is a $\mathrm{PPP}_{\mu_n \otimes \kappa}$.
Finally, we define

$$\widetilde{X}_n(C) := \sum_{n=1}^{\infty} \widetilde{X}_n(C) = \sum_{n=1}^{\infty}\sum_{i=1}^{N_n} \mathbb{1}_C((Y_{n,i}, Z_{n,i})), \qquad C \in \mathcal{B}(E \times F).$$

If $x \in E$ and $X(\{x\}) = 1$, then we have $Y_{n,i} = x$ for one of the pairs (n, i) that show up in the sum. Since $\mathcal{L}(Z_{n,i} \mid Y_{n,i} = x) = \kappa(x, \cdot)$, the color of x is distributed according to $\kappa(x, \cdot)$ (and is independent of everything else).
We observe that $\mu \otimes \kappa = \sum_{n=1}^{\infty}(\mu_n \otimes \kappa)$ to conclude:

Theorem 24.23 (Coloring theorem) $\widetilde{X}$ *is a* $\mathrm{PPP}_{\mu \otimes \kappa}$.

In some situations we are interested only in the colors and not in the positions of the underlying points. In order to formalize this, we define the projection $\pi : E \times F \to F$, $(x, y) \mapsto y$ and let $X^\kappa := \widetilde{X} \circ \pi^{-1}$ be the image measure of $\widetilde{X}$. More explicitly, this means

$$X^\kappa(B) = \widetilde{X}(E \times B) \quad \text{for } B \in \mathcal{B}_b(F).$$

In order for X^κ to be a random measure, we need to assume $\mu\kappa = (\mu \otimes \kappa) \circ \pi^{-1} \in \mathcal{M}(F)$. This implies that by the projection, the colors do not concentrate to much to violate local boundedness of X^κ.

Theorem 24.24 X^κ *is a random measure with* $\mathbf{P}_{X^\kappa} = \mathrm{PPP}_{\mu\kappa}$.

Proof In the above notation, we have

$$X^\kappa(B) = \tilde{X}(E \times B) = \sum_{n=1}^{\infty} \sum_{i=1}^{N_n} \mathbb{1}_B(Z_{n,i}) \quad \text{for } B \in \mathcal{B}_b(F).$$

Now $Z_{n,i} \sim \tilde{\mu}_n\kappa$ and

$$\mu\kappa = \sum_{n=1}^{\infty} \mu_n\kappa = \sum_{n=1}^{\infty} \mu_n(E)\tilde{\mu}_n\kappa.$$

Following the construction in the proof of Theorem 24.12 we infer the claim.   □

We come back to the initial problem of a finite coloring of Poisson random variable. Since we are interested in the colors but not in the positions of the points, we take an arbitrary but sufficiently rich space E. For definiteness, let us assume $E = [0, 1]$. Let $\lambda > 0$, and let μ be λ-times the Lebesgue measure on $[0, 1]$. Let $k \in \mathbb{N}$ and $F = \{0, \ldots, k\}$ and let $(p_m)_{m=0,\ldots,k}$ be probability weights on F. We define $\kappa(x, \{m\}) = p_m$ for all $x \in E$. Then $\mu\kappa(\{m\}) = \lambda p_m$ and the randomly colored point process X^κ has the property that $N_m := X^\kappa(\{m\})$, $m = 0, \ldots, k$ are independent Poisson random variables with parameters λp_m. However, $N := X(E) \sim \text{Poi}_\lambda$.

Example 24.25 (PPP as invariant distribution) As an application of the previous theorem, consider a stochastic process on $E = \mathbb{Z}^d$ or $E = \mathbb{R}^d$ that consists of a system of independent random walks. Hence assume that we are given i.i.d. random variables Z_n^i, $i, n \in \mathbb{N}$ with distribution $\nu \in \mathcal{M}_1(E)$. Further, assume that, at time n, the position of the ith particle of our system of random walks is $S_n^i := S_0^i + \sum_{l=1}^n Z_l^i$, where S_0^i is an arbitrary, possibly random, starting point. Assume that the particles are indistinguishable. Hence we simply add the particles at each site:

$$X_n(A) := \sum_{i=1}^{\infty} \mathbb{1}_A(S_n^i) \quad \text{for } A \subset E.$$

Each X_n is a measure on E and, if at the beginning the particles are not too concentrated locally, it is a locally finite measure and hence a random measure. Assume that $X_0 \sim \text{PPP}_\mu$ for some $\mu \in \mathcal{M}(E)$. Define $\kappa(x, \cdot) = \delta_x * \nu$, and write κ^n for the n-fold application of κ; that is, $\kappa^n(x, \cdot) = \delta_x * \nu^{*n}$. We thus get $X_0^\kappa \overset{\mathcal{D}}{=} X_1$. Indeed, independence of the motions of the individual particles in the definition of X_0^κ is exactly independence of the random walks. As X_1 is also a PPP, we get inductively $X_n^\kappa \overset{\mathcal{D}}{=} X_{n+1}$ and thus $X_n \sim \text{PPP}_{\mu\kappa^n} = \text{PPP}_{\mu*\nu^{*n}}$. In particular, $X_0 \overset{\mathcal{D}}{=} X_n$ if and only if $\mu * \nu = \mu$. Clearly, this is true if we have $E = \mathbb{Z}^d$ and μ the counting measure or if $E = \mathbb{R}^d$ and μ is the Lebesgue measure. For example, if $E = \mathbb{Z}^d$, then under rather mild assumptions on ν one can show that the counting measure $\mu = \lambda$ is the *unique* (up to multiples) solution of $\mu * \nu = \mu$. In this case, every invariant measure is a convex combination of PPPs with different intensity measures $\theta\lambda$. $\Diamond$

> **Takeaways** For non-atomic intensity measure, the Poisson point process can be characterised by the probabilities for sets to be vacant (Rényi's theorem). Images of Poisson point process are again Poisson point processes. If we assign an independent random colour to each Poisson point, we get a Poisson point process on the original space enhanced by the space of colours. Infinitely divisible nonnegative random variables can be represented as the weighted sum of points of a Poisson point process with the canonical measure as intensity measure. Poisson point processes can pop up as invariant states of large particle systems, in particular in grand canonical ensembles.

Exercise 24.2.1 Use an approximation with simple functions in order to show the claim of Corollary 24.15 without using characteristic functions. ♣

Exercise 24.2.2 Let $p_1, p_2 \in (0, 1]$ and $r_1, r_2 > 0$. Show the following statement about the stochastic order of negative binomial distributions: $b^-_{r_1,p_1} \leq_{st} b^-_{r_2,p_2}$ if and only if

$$p_1 \geq p_2 \quad \text{and} \quad p_1^{r_1} \geq p_2^{r_2}. \quad ♣$$

24.3 The Poisson–Dirichlet Distribution*

The goal of this section is to solve the following problem. Take a stick of length 1. Choose a point of the stick uniformly at random and break the stick at this point. Put the left part of the stick (with length, say, W_1) aside. With the remaining part of the stick proceed just as with the original stick. Break it in two and put the left part (of length W_2) aside. Successively, we thus collect fractions of the stick of lengths $W_1, W_2, W_3, \ldots$. What is the joint distribution of $(W_1, W_2, \ldots)$? Furthermore, if we order the numbers $W_1, W_2, \ldots$ in decreasing order $W_{(1)} \geq W_{(2)} \geq \ldots$, what is the distribution of $(W_{(1)}, W_{(2)}, \ldots)$? And finally, why do we ask these questions in a chapter on Poisson point processes?

Answering these questions requires some preparation. We saw that the Beta distribution occurs naturally in Pólya's urn model as the limiting distribution of the fraction of balls of a given color. Clearly, Pólya's urn model can be considered for any number $n \geq 2$ of colors. The limiting distribution is then the n-dimensional generalization of the Beta distribution, namely the so-called Dirichlet distribution.

Define the $(n-1)$-dimensional simplex

$$\Delta_n := \{(x_1, \ldots, x_n) \in [0, 1]^n : x_1 + \ldots + x_n = 1\}.$$

Definition 24.26 *Let* $n \in \{2, 3, \ldots\}$ *and* $\theta_1, \ldots, \theta_n > 0$. *The* **Dirichlet distribution** $\text{Dir}_{\theta_1,\ldots,\theta_n}$ *is the distribution on* Δ_n *that is defined for measurable* $A \subset \Delta_n$ *by*

$$\text{Dir}_{\theta_1,\ldots,\theta_n}(A) = \int \mathbb{1}_A(x_1, \ldots, x_n) \, f_{\theta_1,\ldots,\theta_n}(x_1, \ldots, x_n) \, dx_1 \cdots dx_{n-1}.$$

Here

$$f_{\theta_1,\ldots,\theta_n}(x_1, \ldots, x_n) = \frac{\Gamma(\theta_1 + \ldots + \theta_n)}{\Gamma(\theta_1) \cdots \Gamma(\theta_n)} x_1^{\theta_1-1} \cdots x_n^{\theta_n-1}.$$

If the parameters $\theta_1, \ldots, \theta_n$ are integer-valued, they correspond to the numbers of balls of the different colors that are originally in the urn. Assume that the colors $n-1$ and n are light green and green and that in the dim light we cannot distinguish them. Then we should still end up with a Dirichlet distribution in the limit but with $n - 1$ instead of n and with $\theta_{n-1} + \theta_n$ instead of θ_{n-1} and θ_n; that is, $\text{Dir}_{\theta_1,\ldots,\theta_{n-2},\theta_{n-1}+\theta_n}$. Let $(M_t)_{t \geq 0}$ be the **Moran Gamma subordinator**, the stochastic process with right continuous, monotone increasing paths $t \mapsto M_t$ and independent, stationary, Gamma-distributed increments: $M_t - M_s \sim \Gamma_{1,t-s}$ for $t > s \geq 0$. An important connection between M and the Dirichlet distribution is revealed by the corollaries of the following theorem and by Theorem 24.32.

Theorem 24.27 *Let* $n \in \mathbb{N}$, $\theta_1, \ldots, \theta_n > 0$ *and* $\Theta := \theta_1 + \ldots + \theta_n$. *Let* $X \sim \text{Dir}_{\theta_1,\ldots,\theta_n}$ *and let* $Z \sim \Gamma_{1,\Theta}$ *be independent random variables. Then the random variables* $S_i := Z \cdot X_i$, $i = 1, \ldots, n$ *are independent and* $S_i \sim \Gamma_{1,\theta_i}$.

Proof In the following, always let $x_n := 1 - \sum_{i=1}^{n-1} x_i$ and $s = \sum_{j=1}^{n} s_j$. Let

$$\Delta_n' := \left\{ (x_1, \ldots, x_{n-1}) \in (0, 1)^{n-1} : \sum_{i=1}^{n-1} x_i < 1 \right\}.$$

For $x \in \Delta_n'$ and $z \geq 0$, the distribution of $(X_1, \ldots, X_{n-1}, Z)$ has the density

$$f(x_1, \ldots, x_{n-1}, z) = \prod_{j=1}^{n} \left(x_j^{\theta_j-1} \Big/ \Gamma(\theta_j) \right) z^{\Theta-1} e^{-z}.$$

Consider the map

$$F : \Delta_n' \times (0, \infty) \to (0, \infty)^n, \quad (x_1, \ldots, x_{n-1}, z) \mapsto (zx_1, \ldots, zx_n).$$

This map is invertible with inverse map

$$F^{-1} : (s_1, \ldots, s_n) \mapsto (s_1/s, \ldots, s_{n-1}/s, s).$$

The Jacobian determinant of F is $\det(F'(x_1, \ldots, x_{n-1}, z)) = z^{n-1}$. By the transformation formula for densities (Theorem 1.101), $(S_1, \ldots, S_n)$ has density

$$
g(s_1, \ldots, s_n) = \frac{f(F^{-1}(s_1, \ldots, s_n))}{|\det(F'(F^{-1}(s_1, \ldots, s_n)))|}
$$

$$
= \frac{s^{\Theta-1} e^{-s}}{s^{n-1}} \prod_{j=1}^{n} \left((s_j/s)^{\theta_j - 1} / \Gamma(\theta_j) \right)
$$

$$
= \prod_{j=1}^{n} \left(s_j^{\theta_j - 1} e^{-s_j} / \Gamma(\theta_j) \right).
$$

However, this is the density for independent Gamma distributions. $\square$

Corollary 24.28 *If $t_i := \sum_{j=1}^{i} \theta_j$ for $i = 0, \ldots, n$, then the random variables $X = ((M_{t_i} - M_{t_{i-1}})/M_{t_n},\ i = 1, \ldots, n)$ and $S := M_{t_n}$ are independent with distributions $X \sim \mathrm{Dir}_{\theta_1, \ldots, \theta_n}$ and $S \sim \Gamma_{1, t_n}$.*

Corollary 24.29 *Let $(X_1, \ldots, X_n) \sim \mathrm{Dir}_{\theta_1, \ldots, \theta_n}$. Then $X_1 \sim \beta_{\theta_1, \sum_{i=2}^{n} \theta_i}$ and $(X_2/(1-X_1), \ldots, X_n/(1-X_1)) \sim \mathrm{Dir}_{\theta_2, \ldots, \theta_n}$ are independent.*

Proof Let M be as in Corollary 24.28. Then $X_1 = M_{t_1}/M_{t_n} \sim \beta_{\theta_1, t_n - \theta_1}$. Since $X_1 = \left(\frac{M_{t_n} - M_{t_1}}{M_{t_1}} + 1 \right)^{-1}$, we see that X_1 depend only on M_{t_1} and $M_{t_n} - M_{t_1}$. On the other hand,

$$
\left(\frac{X_2}{1 - X_1}, \ldots, \frac{X_n}{1 - X_1} \right) = \left(\frac{M_{t_2} - M_{t_1}}{M_{t_n} - M_{t_1}}, \ldots, \frac{M_{t_n} - M_{t_{n-1}}}{M_{t_n} - M_{t_1}} \right)
$$

is independent of M_{t_1}. By Corollary 24.28, it is also independent of $M_{t_n} - M_{t_1}$ and is $\mathrm{Dir}_{\theta_2, \ldots, \theta_n}$-distributed. $\square$

Corollary 24.30 *Let $V_1, \ldots, V_{n-1}$ be independent, $V_i \sim \beta_{\theta_i, \theta_{i+1} + \ldots + \theta_n}$ and $V_n = 1$. Then*

$$
\left(V_1, (1 - V_1)V_2, (1 - V_1)(1 - V_2)V_3, \ldots, \left(\prod_{i=1}^{n-1} (1 - V_i) \right) V_n \right) \sim \mathrm{Dir}_{\theta_1, \ldots, \theta_n}.
$$

Proof This follows by iterating the claim of Corollary 24.29. $\square$

It is natural to ask what happens if we distinguish more and more colors (instead of pooling them). For simplicity, consider a symmetric situation where we have $\theta_1 = \ldots = \theta_n = \theta/n$ for some $\theta > 0$. Hence we consider

$$
\mathrm{Dir}_{\theta;n} := \mathrm{Dir}_{\theta, \ldots, \theta} \quad \text{for } \theta > 0.
$$

If $X^n = (X_1^n, \ldots, X_n^n) \sim \mathrm{Dir}_{\theta/n;n}$, then, by symmetry, we have $\mathbf{E}[X_i^n] = 1/n$ for every $n \in \mathbb{N}$ and $i = 1, \ldots, n$. Hence, clearly $(X_1^n, \ldots, X_k^n) \overset{n\to\infty}{\Longrightarrow} 0$ for any $k \in \mathbb{N}$. In order to obtain a nontrivial limit, one possibility is to reorder the values by decreasing size: $X_{(1)}^n \geq X_{(2)}^n \geq \ldots$.

Definition 24.31 *Let $\theta > 0$ and let $(M_t)_{t\in[0,\theta]}$ be a Moran Gamma subordinator. Let $m_1 \geq m_2 \geq \ldots \geq 0$ be the jump sizes of M in decreasing order and let $\tilde{m}_i = m_i/M_\theta$, $i = 1, 2, \ldots$. The distribution of the random variables $(\tilde{m}_1, \tilde{m}_2, \ldots)$ on $S := \{(x_1 \geq x_2 \geq \ldots \geq 0) : x_1 + x_2 + \ldots = 1\}$ is called the **Poisson–Dirichlet distribution** PD_θ with parameter $\theta > 0$.*

To be honest, we still have to show that $\sum_{i=1}^\infty \tilde{m}_i = 1$. To this end, let Y be a PPP on $(0, \infty) \times (0, \theta]$ with intensity measure $\nu \otimes \lambda$, where λ is the Lebesgue measure and $\nu(dx) = e^{-x} x^{-1}\, dx$ is the Lévy measure of the $\Gamma_{1,1}$ distribution. We can define M by $M_t := \sum_{(x,s):\, Y(\{x,s\})=1,\, s\leq t} x$. Now we have $m_1 = \sup\{x \in (0, \infty) : Y(\{x\} \times (0, \theta]) = 1\}$. Inductively, we get $m_n = \sup\{x < m_{n-1} : Y(\{x\} \times (0, \theta]) = 1\}$ for $n \geq 2$. Interchanging the order of summations, we obtain $M_\theta = \sum_{n=1}^\infty m_n$.

Theorem 24.32 *If $X^n \sim \mathrm{Dir}_{\theta/n;n}$ for $n \in \mathbb{N}$, then $\mathbf{P}_{(X_{(1)}^n, X_{(2)}^n, \ldots)} \overset{n\to\infty}{\Longrightarrow} \mathrm{PD}_\theta$.*

Proof The idea is to express the random variables X^n, $n \in \mathbb{N}$, in terms of the increments of the Moran Gamma subordinator $(M_t)_{t\in[0,\theta]}$ in such a way that convergence of distributions implies almost sure convergence. Hence, let $X_i^n = (M_{\theta i/n} - M_{\theta(i-1)/n})/M_\theta$. By Corollary 24.28, we have $X^n \sim \mathrm{Dir}_{\theta/n;n}$. Let $t_1, t_2, \ldots \in (0, \theta]$ be the positions of the jumps $m_1 \geq m_2 \geq \ldots$. Evidently, $X_{(1)}^n \geq \tilde{m}_1$ for every n. If n is large enough that $|t_1 - t_2| > \theta/n$, then $X_{(2)}^n \geq \tilde{m}_2$. Inductively, we get $\liminf_{n\to\infty} X_{(i)}^n \geq \tilde{m}_i$ almost surely. Using the convention $X_{(i)}^n = 0$ for $i > n$, we have $\sum_{i=1}^\infty X_{(i)}^n = 1$ for every $n \in \mathbb{N}$. By Fatou's lemma, we thus get

$$1 = \sum_{i=1}^\infty \tilde{m}_i \leq \sum_{i=1}^\infty \liminf_{n\to\infty} X_{(i)}^n \leq \liminf_{n\to\infty} \sum_{i=1}^\infty X_{(i)}^n = 1.$$

Therefore, $\lim_{n\to\infty} X_{(i)}^n = \tilde{m}_i$ almost surely. $\qquad\square$

Instead of *ordering* the values of X^n by their sizes, there is a different way of arranging the terms so that the distributions converge. Think of a biological population in which a certain phenotypical property can be measured with different levels of precision. If we distinguish n different values of this property, then we write X_i^n for the proportion of the population that has type $i \in \{1, \ldots, n\}$.

Now successively choose individuals from the population at random. Let I_1^n be the type of the first individual. Denote by I_2^n the type of the first individual that is not of type I_1^n. That is, I_2^n is the second *type* that we see. Now inductively define I_k^n as the kth type that we see; that is, the type of the first individual that has none of the types $I_1^n, \ldots, I_{k-1}^n$. Consider the vector $\hat{X}^n = (\hat{X}_1^n, \ldots, \hat{X}_n^n)$, where $\hat{X}_k^n =$

$X_{I_k^n}^n$. Since the probability of the event $\{I_1^n = i\}$ is proportional to the size of the subpopulation of type i, we say that $\hat{X}^n$ is the successively size-biased vector.

The distribution of $\hat{X}^n$ does not change if we change the order of the $X_1^n, \ldots, X_n^n$. For example, instead of $X_1^n, \ldots, X_n^n$, we can use the order statistics (in decreasing order) $(X_{(1)}^n, \ldots, X_{(n)}^n)$ and again end up with $\hat{X}^n$ as the successively size-biased vector. Hence we can define the successively size-biased vector $\hat{X}$ for the infinite vector $X \sim PD_\theta$. If $X^n \sim Dir_{\theta/n;n}$, then by Theorem 24.32, we have $\hat{X}^n \overset{n\to\infty}{\Longrightarrow} \hat{X}$. Hence we can compute the distribution of $\hat{X}$.

Theorem 24.33 *Let* $\theta > 0$ *and* $X^n \sim Dir_{\theta/n;n}$, $n \in \mathbb{N}$. *Let* $X \sim PD_\theta$. *Further, let* $V_1, V_2, \ldots$ *be i.i.d. random variables on* $[0, 1]$ *with density* $x \mapsto \theta(1 - x)^{\theta-1}$. *Define* $Z_1 = V_1$ *and* $Z_k = \left(\prod_{i=1}^{k-1}(1 - V_i) \right) V_k$ *for* $k \geq 2$. *Then:*

(i) $\hat{X}^n \overset{n\to\infty}{\Longrightarrow} \hat{X}$.

(ii) $\hat{X} \overset{D}{=} Z$.

The distribution of Z *is called the* **GEM$_\theta$ distribution** *(Griffiths–Engen–McCloskey).*

Proof Statement (i) was shown in the discussion preceding the theorem. In order to show (ii), we compute the distribution of $\hat{X}^n$ and show that it converges to the distribution of Z.

Let $\hat{X}^{n,1}$ be the vector $\hat{X}^{n,1} = (X_{I_1^n}^n, X_1^n, X_2^n, \ldots, X_{I_1^n-1}^n, X_{I_1^n+1}^n, \ldots, X_n^n)$, in which only the first coordinate is sampled size-biasedly. We show that

$$\hat{X}^{n,1} \sim Dir_{(\theta/n)+1,\theta/n,\ldots,\theta/n}. \tag{24.12}$$

Let $f(x) = \left(\Gamma(\theta)/\Gamma(\theta/n)^n \right) \cdot \prod_{k=1}^n x_k^{(\theta/n)-1}$ be the density of $Dir_{\theta/n;n}$. We compute the density $\hat{f}^{n,1}$ of $\hat{X}^{n,1}$ by decomposing according to the value i of I_1^n:

$$\hat{f}^{n,1}(x) = \sum_{i=1}^n x_1 f(x_2, \ldots, x_i, x_1, x_{i+1}, \ldots, x_n) = n x_1 f(x)$$

$$= \frac{n\Gamma(\theta)}{\Gamma(\theta/n)^n} x_1^{\theta/n} \prod_{i=2}^n x_i^{(\theta/n)-1}$$

$$= \frac{\Gamma(\theta + 1)}{\Gamma((\theta/n) + 1)\,\Gamma(\theta/n)^{n-1}} x_1^{\theta/n} \prod_{i=2}^n x_i^{(\theta/n)-1}.$$

However, this is the density of $Dir_{(\theta/n)+1,\theta/n,\ldots,\theta/n}$. By Corollary 24.29, we have

$$\hat{X}^{n,1} \overset{D}{=} \left(V_1^n, (1 - V_1^n)Y_1, \ldots, (1 - V_1^n)Y_{n-1} \right),$$

where

$$V_1^n \sim \beta_{(\theta/n)+1,\theta(n-1)/n} \quad \text{and} \quad Y = (Y_1, \ldots, Y_{n-1}) \sim \mathrm{Dir}_{\theta/n;n-1}$$

are independent. Applying this to Y, we get inductively

$$\hat{X}^n \overset{D}{=} Z^n, \tag{24.13}$$

where

$$Z_1^n = V_1^n \quad \text{and} \quad Z_k^n = \left(\prod_{i=1}^{k-1} (1 - V_i^n) \right) V_k^n \quad \text{for } k \geq 2$$

and where $V_1^n, \ldots, V_{n-1}^n$ are independent and $V_i^n \sim \beta_{(\theta/n)+1,\theta(n-i)/n}$. Now it is easy to check that $\beta_{(\theta/n)+1,\theta(n-i)/n} \overset{n\to\infty}{\longrightarrow} \beta_{1,\theta}$ for every $i \in \mathbb{N}$. Recall that $\beta_{1,\theta}$ has the density $x \mapsto \theta(1-x)^{\theta-1}$. Hence $V_i^n \overset{n\to\infty}{\Longrightarrow} V_i$ for every i and thus $Z^n \overset{n\to\infty}{\Longrightarrow} Z$ and $\hat{X}^n \overset{n\to\infty}{\Longrightarrow} Z$. Together with (i), this proves claim (ii). $\square$

At the beginning of this chapter, we raised the question of how the sizes $W_1, W_2, \ldots$ of the stick lengths are distributed if at each step, we break the remaining part of the stick at a point chosen uniformly at random. The preceding theorem gives the answer: The vector $(W_{(1)}, W_{(2)}, \ldots)$ has distribution PD_1, and $(W_1, W_2, \ldots)$ has distribution GEM_1.

The Chinese Restaurant Process

We will study a further situation in which the Poisson–Dirichlet distribution arises naturally. As the technical details get a bit tricky, we content ourselves with the description of the problem and with stating (but not proving) two fundamental theorems. An excellent reference for this type of problem is [130].

Consider a Chinese restaurant with countably many enumerated round tables. At each table, there is enough space for arbitrarily many guests. Initially, the restaurant is empty. One by one an infinite number of guests arrive. The first guest sits down at table number one. If there are already n guests sitting at k tables, then the $(n+1)$th guest can choose between sitting down at any of the k occupied tables or at the free table with the smallest number (that is, $k+1$). Assume that the guest makes his choice at random (and independently of the previous choices of the other guests). For $l \leq k$, denote by N_l^n the number of guests at the lth table and assume that the probability of choosing the lth table is $(N_l^n - \alpha)/(n+\theta)$. Then the probability of choosing the first free table is $(\theta + k\alpha)/(n+\theta)$. Here $\alpha \in [0, 1]$ and $\theta > -\alpha$ are parameters. We say that $(N^n)_{n\in\mathbb{N}} = (N_1^n, N_2^n, \ldots)_{n\in\mathbb{N}}$ is the **Chinese restaurant process** with parameters (α, θ).

In the special case $\alpha = 0$, there is a nice interpretation: Assume that the new guest can also choose his seating position at the table (that is, his neighbor to the right). Then, for any of the present guests, the probability of being chosen as a right neighbor is $1/(n + \theta)$. The probability of starting a new table is $\theta/(n + \theta)$.

In order to study the large n behavior of $N^n/n = (N_1^n/n, N_2^n/n, \ldots)$, we extend the Poisson–Dirichlet distribution and the GEM distribution by a further parameter.

Definition 24.34 *Let $\alpha \in [0, 1)$ and $\theta > -\alpha$. Let $V_1, V_2, \ldots$ be independent and $V_i \sim \beta_{1-\alpha,\theta+i\alpha}$. Define $Z = (Z_1, Z_2, \ldots)$ by $Z_1 = V_1$ and*

$$Z_k = V_k \prod_{i=1}^{k-1} \left(1 - V_i\right) \quad \textit{for } k \geq 2.$$

*Then $\mathrm{GEM}_{\alpha,\theta} := \mathbf{P}_Z$ is called the **GEM distribution** with parameters (α, θ). The distribution of the size-biased vector $(Z_{(1)}, Z_{(2)}, \ldots)$ is called the **Poisson–Dirichlet distribution** with parameters (α, θ), or briefly $\mathrm{PD}_{\alpha,\theta}$.*

Explicit formulas for the densities of the finite-dimensional marginals of $\mathrm{PD}_{\alpha,\theta}$ can be found in [132]. Note that, for $\alpha = 0$, we recover the classical distributions $\mathrm{GEM}_\theta = \mathrm{GEM}_{0,\theta}$ and $\mathrm{PD}_\theta = \mathrm{PD}_{0,\theta}$.

Theorem 24.35 *Let $\alpha \in [0, 1)$, $\theta > -\alpha$ and let $(N^n)_{n\in\mathbb{N}}$ be the Chinese restaurant process with parameters (α, θ). Then $\mathbf{P}_{N^n/n} \xrightarrow{n\to\infty} \mathrm{PD}_{\alpha,\theta}$.*

Proof See [129] or [130, Theorem 25]. $\qquad\qquad\qquad\qquad\qquad\qquad\qquad\qquad\qquad\Box$

Reflection The Hoppe urn is a variation of the Pólya Urne. Initially, there is only one ball in the urn and this one has a unique label. Whenever this ball is drawn it will be returned together with a ball of a completely new colour. Whenever one of the coloured balls is drawn, it will be returned together with a second ball of the same colour. The unique ball is sometimes called the mutator. What is the connection between the Hoppe urn and the Chinese restaurant process? ♠

As for the one-parameter Poisson–Dirichlet distribution, there is a representation of $\mathrm{PD}_{\alpha,\theta}$ in terms of the size-ordered jumps of a certain subordinator. In the following, let $\alpha \in (0, 1)$ and let $(M_t)_{t\in[0,1]}$ be an α-stable subordinator; that is, a subordinator with Lévy measure $\nu(dx) = x^{-\alpha-1}\, dx$. Further, let $m_1 \geq m_2 \geq \cdots \geq 0$ be the jumps of M, $\tilde{m}_i = m_i/M_1$ for $i \in \mathbb{N}$, and $\tilde{m} = (\tilde{m}_1, \tilde{m}_2, \ldots)$. We quote the following theorem from [130, Section 4.2].

Theorem 24.36 *Let $\alpha \in (0, 1)$.*

(i) $\tilde{m} \sim \mathrm{PD}_{\alpha,0}$.
(ii) If $\theta > -\alpha$, then $\mathrm{PD}_{\alpha,\theta} \ll \mathrm{PD}_{\alpha,0} = \mathbf{P}[\tilde{m} \in \cdot]$ and the density is

$$\mathrm{PD}_{\alpha,\theta}(dx) = \frac{M_1^{-\theta}}{\mathbf{E}[M_1^{-\theta}]}\, \mathbf{P}[\tilde{m} \in dx].$$

Takeaways The Dirichlet distribution generalises the beta distribution from two to n types. After sorting the values of a Dirichlet random variable $(X_1, \ldots, X_n)$ in decreasing order, we can take the limit as $n \to \infty$ and get the so-called Poisson-Dirichlet distribution. Drawing the values of a given realisation of a Poisson-Dirichlet random variable successively with size biased probabilities, we get a GEM distributed vector. The GEM distribution plays an important role in the mathematical descriptions of samples in ecology and evolutionary genetics.

Exercise 24.3.1 Let $(X, 1 - X) \sim \mathrm{Dir}_{\theta_1, \theta_2}$. Show that $X \sim \beta_{\theta_1, \theta_2}$ is Beta-distributed. ♣

Exercise 24.3.2 Let $X = (X_1, \ldots, X_n) \sim \mathrm{Dir}_{\theta_1, \ldots, \theta_n}$. Show the following.

(i) For any permutation σ on $\{1, \ldots, n\}$, we have

$$(X_{\sigma(1)}, \ldots, X_{\sigma(n)}) \sim \mathrm{Dir}_{\theta_{\sigma(1)}, \ldots, \theta_{\sigma(n)}}.$$

(ii) $(X_1, \ldots, X_{n-2}, X_{n-1} + X_n) \sim \mathrm{Dir}_{\theta_1, \ldots, \theta_{n-2}, \theta_{n-1}+\theta_n}$. ♣

Exercise 24.3.3 Let $(N^n)_{n \in \mathbb{N}}$ be the Chinese restaurant process with parameters $(0, \theta)$.

(i) Let $\theta = 1$.

   (a) Show that $\mathbf{P}[N_1^n = k] = 1/n$ for any $k = 1, \ldots, n$,

   (b) Show that, for $k_l = 1, \ldots, n - (k_1 + \ldots + k_{l-1})$,

$$\mathbf{P}\big[N_l^n = k_l \,\big|\, N_1^n = k_1, \ldots, N_{l-1}^n = k_{l-1}\big] = \frac{1}{n - (k_1 + \ldots + k_{l-1})}.$$

   (c) Infer the claim of Theorem 24.35 in the case $\alpha = 0$ and $\theta = 1$.

(ii) Let $\theta > 0$.

   (a) Show that $n\,\mathbf{P}[N_1^n = \lfloor nx \rfloor] \overset{n \to \infty}{\longrightarrow} \theta(1 - x)^{\theta-1}$ for $x \in (0, 1)$.

   (b) Show that

$$n\,\mathbf{P}\big[N_l^n = \lfloor nx_l \rfloor \,\big|\, N_1^n = \lfloor nx_1 \rfloor, \ldots, N_{l-1}^n = \lfloor nx_{l-1} \rfloor\big]$$
$$\overset{n \to \infty}{\longrightarrow} (\theta/y_l)\big(1 - x_l/y_l\big)^{\theta-1}$$

   for $x_1, \ldots, x_l \in (0, 1)$ with $y_l = 1 - (x_1 + \ldots + x_{l-1}) > x_l$.

   (c) As in (i), infer the claim of Theorem 24.35 for $\alpha = 0$ and $\theta > 0$. ♣

Chapter 25
The Itô Integral

The Itô integral allows us to integrate stochastic processes with respect to the increments of a Brownian motion or a somewhat more general stochastic process. We develop the Itô integral first for Brownian motion and then for generalized diffusion processes (so-called Itô processes). In the third section, we derive the celebrated Itô formula. This is the chain rule for the Itô integral that enables us to do explicit calculations with the Itô integral. In the fourth section, we use the Itô formula to obtain a stochastic solution of the classical Dirichlet problem. This in turn is used in the fifth section in order to show that like symmetric simple random walk, Brownian motion is recurrent in low dimensions and transient in high dimensions.

25.1 Itô Integral with Respect to Brownian Motion

Let $W = (W_t)_{t \geq 0}$ be a Brownian motion on the space $(\Omega, \mathcal{F}, \mathbf{P})$ with respect to the filtration $\mathbb{F}$ that satisfies the usual conditions (see Definition 21.22). That is, W is a Brownian motion and an $\mathbb{F}$-martingale. The aim of this section is to construct an integral

$$I_t^W(H) = \int_0^t H_s \, dW_s$$

for a large class of integrands $H : \Omega \times [0, \infty) \to \mathbb{R}$, $(\omega, t) \mapsto H_t(\omega)$ in such a way that $(I_t^W(H))_{t \geq 0}$ is a continuous $\mathbb{F}$-martingale. Since almost all paths $s \mapsto W_s(\omega)$ of Brownian motion are of locally infinite variation, $W(\omega)$ is not the distribution function of a signed Lebesgue–Stieltjes measure on $[0, \infty)$. Hence $I_t^W(H)$ cannot be defined in the framework of classical integration theory. The basic new idea is to establish the integral as an L^2-limit. We start with an elementary example to illustrate this.

A. Klenke, *Probability Theory*, Universitext, https://doi.org/10.1007/978-3-030-56402-5_25

Example 25.1 Assume that $X_1, X_2, \ldots$ are i.i.d. $\mathrm{Rad}_{1/2}$ random variables; that is, $\mathbf{P}[X_n = 1] = \mathbf{P}[X_n = -1] = \frac{1}{2}$. Let $(h_n)_{n \in \mathbb{N}}$ be a sequence of real numbers. Under which assumptions on $(h_n)_{n \in \mathbb{N}}$ is the series

$$R := \sum_{n \in \mathbb{N}} h_n X_n \tag{25.1}$$

well-defined? If $\sum_{n \in \mathbb{N}} |h_n| < \infty$, then the series converges absolutely for *every* ω. In this case, there is no problem. Now assume that only the weaker condition $\sum_{n \in \mathbb{N}} h_n^2 < \infty$ holds. In this case, the series (25.1) does not necessarily converge any more for every ω. However, we have $\mathbf{E}[h_n X_n] = 0$ for each $n \in \mathbb{N}$ and $\sum_{n=1}^{\infty} \mathbf{Var}[h_n X_n] = \sum_{n=1}^{\infty} h_n^2 < \infty$. Hence $R_N := \sum_{k=1}^{N} h_k X_k$ converges in L^2 (for $N \to \infty$). We can thus define the series R in (25.1) as the L^2-limit of the partial sums R_N. Note that (at least formally) for the approximating sums the order of summation matters. In a sense, we have constructed $\sum_{n=1}^{\infty}$ instead of $\sum_{n \in \mathbb{N}}$.

An equivalent formulation that gives a flavor of what is to come is the following. Denote by ℓ^2 the Hilbert space of square summable sequences of real numbers with inner product $\langle h, g \rangle = \sum_{n=1}^{\infty} h_n g_n$ and norm $\|g\| = \langle g, g \rangle^{1/2}$. Let ℓ^f be the subspace of those sequences with only finitely many nonzero entries. Then $R(h) = \sum_{n \in \mathbb{N}} h_n X_n$ for $h \in \ell^f$ is well-defined (since it is a finite sum). Since

$$\mathbf{E}[R(h)^2] = \mathbf{Var}[R(h)] = \sum_{n \in \mathbb{N}} \mathbf{Var}[h_n X_n] = \sum_{n \in \mathbb{N}} h_n^2 = \|h\|^2,$$

the map $R : \ell^f \to \mathcal{L}^2(\mathbf{P})$ is an isometry. As $\ell^f \subset \ell^2$ is dense, there is a unique continuous extension of R to ℓ^2. Hence, if $h \in \ell^2$ and $(h^N)_{N \in \mathbb{N}}$ is a sequence in ℓ^f with $\|h^N - h\| \overset{N \to \infty}{\longrightarrow} 0$, then $R(h^N) \overset{N \to \infty}{\longrightarrow} R(h)$ in the L^2 sense. In particular, $h_n^N := h_n \mathbb{1}_{\{n \le N\}}$, $n \in \mathbb{N}$, $N \in \mathbb{N}$, is an approximating sequence for h, and we have $R(h^N) = \sum_{n=1}^{N} h_n X_n$. Thus the approximation of R with the partial sums R_N that we described above is a special case of this construction. $\lozenge$

The programme for the construction of the Itô integral $I_t^W(H)$ is the following. First consider *simple functions* as integrands H; that is, the map $t \mapsto H_t(\omega)$ is a step function. For these H, the integral can easily be defined as a finite sum. The next step is to extend the integral, as in Example 25.1, to integrands that can be approximated in a certain L^2-space by simple integrands.

Definition 25.2 *Denote by $\mathcal{E}$ the vector space of maps $H : \Omega \times [0, \infty) \to \mathbb{R}$ of the form*

$$H_t(\omega) = \sum_{i=1}^{n} h_{i-1}(\omega)\, \mathbb{1}_{(t_{i-1}, t_i]},$$

where $n \in \mathbb{N}$, $0 = t_0 < t_1 < \ldots < t_n$ and h_{i-1} is bounded and $\mathcal{F}_{t_{i-1}}$-measurable for every $i = 1, \ldots, n$. $\mathcal{E}$ is called the vector space of predictable simple processes.

We equip $\mathcal{E}$ with a (pseudo) norm $\| \cdot \|_{\mathcal{E}}$ by defining

$$\|H\|_{\mathcal{E}}^2 = \sum_{i=1}^{n} \mathbf{E}[h_{i-1}^2] (t_i - t_{i-1}) = \mathbf{E}\left[\int_0^{\infty} H_s^2 \, ds\right].$$

Definition 25.3 *For $H \in \mathcal{E}$ and $t \geq 0$, define*

$$I_t^W(H) = \sum_{i=1}^{n} h_{i-1} \left(W_{t_i \wedge t} - W_{t_{i-1} \wedge t}\right)$$

and

$$I_{\infty}^W(H) = \sum_{i=1}^{n} h_{i-1} \left(W_{t_i} - W_{t_{i-1}}\right).$$

Clearly, for every bounded stopping time τ,

$$\mathbf{E}[I_{\tau}^W(H)] = \sum_{i=1}^{n} \mathbf{E}[h_{i-1} \left(W_{t_i}^{\tau} - W_{t_{i-1}}^{\tau}\right)]$$

$$= \sum_{i=1}^{n} \mathbf{E}[h_{i-1} \, \mathbf{E}[W_{t_i}^{\tau} - W_{t_{i-1}}^{\tau} \,|\, \mathcal{F}_{t_{i-1}}]] \; = \; 0$$

since, by the optional stopping theorem (OST), the stopped Brownian motion W^{τ} is an $\mathbb{F}$-martingale. Hence (again by the OST) $(I_t^W(H))_{t \geq 0}$ is an $\mathbb{F}$-martingale. In particular, we have $\mathbf{E}[(I_{t_{i+1}}^W(H) - I_{t_i}^W(H))(I_{t_{j+1}}^W(H) - I_{t_j}^W(H))] = 0$ for $i \neq j$. Therefore,

$$\mathbf{E}[I_{\infty}^W(H)^2] = \sum_{i=1}^{n} \mathbf{E}\left[\left(I_{t_i}^W(H) - I_{t_{i-1}}^W(H)\right)^2\right]$$

$$= \sum_{i=1}^{n} \mathbf{E}\left[h_{i-1}^2 \left(W_{t_i} - W_{t_{i-1}}\right)^2\right] \tag{25.2}$$

$$= \sum_{i=1}^{n} \mathbf{E}[h_{i-1}^2] (t_i - t_{i-1}) \; = \; \|H\|_{\mathcal{E}}^2.$$

From these considerations, the following statement is immediate.

Theorem 25.4

(i) *The map $I_{\infty}^W : \mathcal{E} \to L^2(\Omega, \mathcal{F}, \mathbf{P})$ is an isometric linear map (with respect to $\| \cdot \|_{\mathcal{E}}$ and $\| \cdot \|_2$).*

(ii) *The process $(I_t^W(H))_{t \geq 0}$ is an L^2-bounded continuous $\mathbb{F}$-martingale.*

Proof Only the linearity remains to be shown. However, this is trivial.   □

The idea is to extend the map I_∞^W continuously from $\mathcal{E}$ to a suitable closure $\overline{\mathcal{E}}$ of $\mathcal{E}$. Now as a subspace of what space should we close $\mathcal{E}$? A minimal requirement is that $(\omega, t) \mapsto H_t(\omega)$ be measurable (with respect to $\mathcal{F} \otimes \mathcal{B}([0, \infty)))$ and that H be adapted.

Definition 25.5 *A stochastic process* $X = (X_t)_{t \geq 0}$ *with values in a Polish space* E *is called*

(i) **product measurable** *if* $(\omega, t) \mapsto X_t(\omega)$ *is measurable with respect to* $\mathcal{F} \otimes \mathcal{B}([0, \infty)) - \mathcal{B}(E),$

(ii) **progressively measurable** *if, for every* $t \geq 0$, *the map* $\Omega \times [0, t] \to E$, $(\omega, s) \mapsto X_s(\omega)$ *is measurable with respect to* $\mathcal{F}_t \otimes \mathcal{B}([0, t]) - \mathcal{B}(E),$

(iii) **predictable** (*or* **previsible**) *if* $(\omega, t) \mapsto X_t(\omega)$ *is measurable with respect to the predictable* σ-*algebra* $\mathcal{P}$ *on* $\Omega \times [0, \infty)$:

$$\mathcal{P} := \sigma\big(X : X \text{ is a left continuous adapted process}\big).$$

Remark 25.6 Any $H \in \mathcal{E}$ is predictable. This property ensures that $I^M(H)$ is a martingale for every (even discontinuous) martingale M. The notion of predictability is important only for integration with respect to discontinuous martingales. As we will not develop that calculus in this book, predictability will not be central for us. ◇

Remark 25.7 If H is progressively measurable, then H is evidently also product measurable and adapted. With a little work, the converse can also be shown: If H is adapted and product measurable, then there is a progressively measurable modification of H (see, e.g., [115, pages 68ff]). ◇

Theorem 25.8 *If H is adapted and right continuous or left continuous, then H is progressively measurable. If H is adapted and a.s. right continuous or left continuous, then there exists a version of H that is progressively measurable.*

In particular, every predictable process is progressively measurable.

Proof See Exercise 21.1.4.   □

We consider $\mathcal{E}$ as a subspace of

$$\mathcal{E}_0 := \Big\{ H : \text{ product measurable, adapted and } \|H\|^2 := \mathbf{E}\Big[\int_0^\infty H_t^2 \, dt\Big] < \infty \Big\}.$$

Let $\overline{\mathcal{E}}$ denote the closure of $\mathcal{E}$ in $\mathcal{E}_0$.

Theorem 25.9 *If H is progressively measurable (for instance, left continuous or right continuous and adapted) and $\mathbf{E}\Big[\int_0^\infty H_t^2 \, dt\Big] < \infty$, then $H \in \overline{\mathcal{E}}$.*

Proof Let H be progressively measurable and $\mathbf{E}\left[\int_0^\infty H_t^2\, dt\right] < \infty$. It is enough to show that, for any $T > 0$, there exists a sequence $(H^n)_{n \in \mathbb{N}}$ in $\mathcal{E}$ such that

$$\mathbf{E}\left[\int_0^T (H_s - H_s^n)^2\, ds\right] \overset{n \to \infty}{\longrightarrow} 0. \tag{25.3}$$

Step 1. First assume that H is continuous and bounded. Define $H_0^n = 0$ and

$$H_t^n = H_{i2^{-n} T} \quad \text{if } i2^{-n} T < t \le (i+1)2^{-n} T \quad \text{for some } i = 0, \ldots, 2^n - 1$$

and $H_t^n = 0$ for $t > T$. Then $H^n \in \mathcal{E}$, and we have $H_t^n(\omega) \overset{n \to \infty}{\longrightarrow} H_t(\omega)$ for all $t > 0$ and $\omega \in \Omega$. By the dominated convergence theorem, we get (25.3).

Step 2. Now let H be progressively measurable and bounded. It is enough to show that there exist continuous adapted processes H^n, $n \in \mathbb{N}$, for which (25.3) holds. Let

$$H_t^n := n \int_{(t-1/n)\vee 0}^{t \wedge T} H_s\, ds \quad \text{for } t \ge 0,\ n \in \mathbb{N}.$$

Then H^n is continuous, adapted and bounded by $\|H\|_\infty$. By the fundamental theorem of calculus (see Exercise 13.1.7), we have

$$H_t^n(\omega) \overset{n \to \infty}{\longrightarrow} H_t(\omega) \quad \text{for } \lambda\text{-almost all } t \in [0, T] \text{ and for all } \omega \in \Omega. \tag{25.4}$$

By Fubini's theorem and the dominated convergence theorem, we thus conclude that

$$\mathbf{E}\left[\int_0^T (H_s - H_s^n)^2\, ds\right] = \int_{\Omega \times [0,T]} \left(H_s(\omega) - H_s^n(\omega)\right)^2 (\mathbf{P} \otimes \lambda)(d(\omega, s)) \overset{n \to \infty}{\longrightarrow} 0.$$

Step 3. Now let H be progressively measurable, and assume $\mathbf{E}\left[\int_0^\infty H_t^2\, dt\right] < \infty$. It is enough to show that there exists a sequence $(H^n)_{n \in \mathbb{N}}$ of bounded, progressively measurable processes such that (25.3) holds. Manifestly, we can choose $H_t^n = H_t \mathbb{1}_{\{|H_t| < n\}}$. $\qquad\square$

Definition 25.10 (Itô integral) *For $H \in \overline{\mathcal{E}}$, define the **Itô integral***

$$\int_0^\infty H_s\, dW_s := I_\infty^W(H)$$

as the continuous extension of the map $I_\infty^W : \mathcal{E} \to \mathcal{L}^2(\mathbf{P})$ to the closure $\overline{\mathcal{E}}$ of $\mathcal{E}$. In other words, if $(H^n)_{n \in \mathbb{N}}$ is a sequence in $\mathcal{E}$ with $\|H - H^n\| \overset{n \to \infty}{\longrightarrow} 0$, then we define $I_\infty^W(H)$ by

$$I_\infty^W(H) := \lim_{n\to\infty} I_\infty^W(H^n) \quad in \ L^2.$$

If τ is a stopping time, then in the following we use the abbreviation

$$H_t^{(\tau)} := H_t \, \mathbb{1}_{\{t\le\tau\}} \quad for \ t \ge 0.$$

(Note that this is not the stopped process $H_t^\tau = H_{\tau\wedge t}$.)

Theorem 25.11

(i) *The map* $I_\infty^W : \overline{\mathcal{E}} \to \mathcal{L}^2(\Omega, \mathcal{F}, \mathbf{P})$ *is linear and*

$$\mathbf{E}\big[I_\infty^W(H)^2\big] = \mathbf{E}\bigg[\int_0^\infty H_s^2 \, ds\bigg].$$

(ii) *For every* $H \in \overline{\mathcal{E}}$, *the process* $\tilde{I}^W(H)$ *defined by* $\tilde{I}_t^W(H) := I_\infty^W(H^{(t)})$ *is an* L^2-*bounded* $\mathbb{F}$-*martingale that has a continuous modification* $I^W(H)$.

Definition 25.12 (Itô integral as a process) *Let* $I^W(H)$ *be the continuous version of the martingale* $(I_\infty^W(H^{(t)}))_{t\ge 0}$ *(see Theorem 25.11(ii)). Denote by*

$$\int_s^t H_r \, dW_r := I_t^W(H) - I_s^W(H) \quad for \ 0 \le s \le t \le \infty$$

the Itô integral of H with respect to Brownian motion W on the interval $[s, t]$.

Proof (of Theorem 25.11)

(i) This is a direct consequence of the definition of $I_\infty^W(H)$.

(ii) Let $(H^n)_{n\in\mathbb{N}}$ be a sequence in $\mathcal{E}$ with $\|H^n - H\| \overset{n\to\infty}{\longrightarrow} 0$. By Theorem 25.4(ii), we have

$$I_\infty^W\big((H^n)^{(t)}\big) = I_t^W(H^n) = \mathbf{E}\big[I_\infty^W(H^n)\,\big|\,\mathcal{F}_t\big] \quad for \ all \ t \ge 0, \ n \in \mathbb{N}.$$

Since $\|(H^n)^{(t)} - H^{(t)}\| \le \|H^n - H\| \overset{n\to\infty}{\longrightarrow} 0$, this implies (using Corollary 8.21)

$$\tilde{I}_t^W(H) = \lim_{n\to\infty} I_t^W(H^n) = \lim_{n\to\infty} \mathbf{E}\big[I_\infty^W(H^n)\,\big|\,\mathcal{F}_t\big] = \mathbf{E}\big[I_\infty^W(H)\,\big|\,\mathcal{F}_t\big].$$

Hence $\tilde{I}^W(H)$ is an L^2-bounded martingale and $I_t^W(H^n) \overset{n\to\infty}{\longrightarrow} \tilde{I}_t^W(H)$ in L^2 for every $t \ge 0$. By Theorem 25.4(ii), $I^W(H^n)$ is continuous for every $n \in \mathbb{N}$. Thus, by Exercise 21.4.3, there exists a continuous modification $I^W(H)$ of $\tilde{I}^W(H)$.   □

The last step in the construction of the Itô integral is to weaken the strong integrability condition $\mathbf{E}\big[\int_0^\infty H_s^2 \, ds\big] < \infty$. We start with a simple observation.

Let τ be a stopping time and recall that $\int_0^\tau H_s \, dW_s$ denotes the random variable that for any ω assumes the value $\left(\int_0^{\tau(\omega)} H_s \, dW_s \right)(\omega)$.

Lemma 25.13 *Let τ be a stopping time and let $H \in \overline{\mathcal{E}}$.*

(i) *We have*

$$\int_0^\tau H_s \, dW_s = \int_0^\infty H_s^{(\tau)} \, dW_s := \int_0^\infty H_s \mathbb{1}_{\{s \le \tau\}} \, dW_s \quad a.s.$$

(ii) *In particular, for any $t \ge 0$, on the event $\{\tau \ge t\}$ we have*

$$\int_0^t H_s \, dW_s = \int_0^t H_s^{(\tau)} \, dW_s \quad a.s.$$

(iii) *Let $G \in \overline{\mathcal{E}}$ be such that $H_s = G_s$ for all $s \le \tau$. Then*

$$\int_0^\tau H_s \, dW_s = \int_0^\tau G_s \, dW_s \quad a.s.$$

Proof

(i) Assume first that τ takes values in $\{k/2^n : k \in \mathbb{N}_0\} \cup \{\infty\}$ for some $n \in \mathbb{N}$. Then $\mathbb{1}_{\{k/2^n \le \tau\}} \mathbb{1}_{\{t \in ((k-1)/2^n, k/2^n]\}} \in \mathcal{E}$ for all $k \in \mathbb{N}$. If, in addition, $H \in \mathcal{E}$, then also $H^{(\tau)} \in \mathcal{E}$ and the claim follows directly from the definition of the Itô integral (Definition 25.3). Now let $H \in \overline{\mathcal{E}}$ and let $(H^k)_{k \in \mathbb{N}}$ be a sequence in $\mathcal{E}$ such that $\| H^k - H \|_{\mathcal{E}} \overset{k \to \infty}{\longrightarrow} 0$. Writing $H_t^{k,(\tau)} := H_t^k \mathbb{1}_{\{t \le \tau\}}$ we get that $\| H^{k,(\tau)} - H^{(\tau)} \|_{\mathcal{E}} \overset{k \to \infty}{\longrightarrow} 0$. By choosing a suitable sequence $k_m \uparrow \infty$, we obtain

$$\int_0^\tau H_s \, dW_s = \lim_{m \to \infty} \int_0^\tau H_s^{k_m} \, dW_s$$

$$= \lim_{m \to \infty} \int_0^\infty H_s^{k_m,(\tau)} \, dW_s = \int_0^\infty H_s^{(\tau)} \, dW_s \quad a.s.$$

Finally, assume that τ is an arbitrary stopping time and define $\tau_n := 2^{-n} \lceil 2^n \tau \rceil$ for $n \in \mathbb{N}$. Then (τ_n) is a sequence of stopping times with $\tau_n \downarrow \tau$. Recall that $I^W(H)$ is continuous and note that $\| H^{(\tau_n)} - H^{(\tau)} \|_{\mathcal{E}} \overset{n \to \infty}{\longrightarrow} 0$.

Hence by taking a suitable sequence $n(m) \uparrow \infty$, we get

$$
\int_0^\tau H_s \, dW_s = \lim_{m \to \infty} \int_0^{\tau_{n(m)}} H_s \, dW_s
$$

$$
= \lim_{m \to \infty} \int_0^\infty H_s^{(\tau_{n(m)})} \, dW_s = \int_0^\infty H_s^{(\tau)} \, dW_s \quad \text{a.s.}
$$

(ii), (iii) These statements are direct consequences of (i). □

Definition 25.14 Let $\mathcal{E}_{loc}$ be the space of progressively measurable stochastic processes H with

$$
\int_0^T H_s^2 \, ds < \infty \quad \text{a.s.} \quad \text{for all } T > 0.
$$

Lemma 25.15 For every $H \in \mathcal{E}_{loc}$, there exists a sequence $(\tau_n)_{n \in \mathbb{N}}$ of stopping times with $\tau_n \uparrow \infty$ almost surely and $\mathbf{E}\left[\int_0^{\tau_n} H_s^2 \, ds \right] < \infty$ and hence such that $H^{(\tau_n)} \in \overline{\mathcal{E}}$ for every $n \in \mathbb{N}$.

Proof Define

$$
\tau_n := \inf \left\{ t \geq 0 : \int_0^t H_s^2 \, ds \geq n \right\}.
$$

By the definition of $\mathcal{E}_{loc}$, we have $\tau_n \uparrow \infty$ almost surely. By construction, we have $\| H^{(\tau_n)} \|^2 = \mathbf{E}\left[\int_0^{\tau_n} H_s^2 \, ds \right] \leq n$. □

Definition 25.16 Let $H \in \mathcal{E}_{loc}$ and let $(\tau_n)_{n \in \mathbb{N}}$ be as in Lemma 25.15. For $t \geq 0$, define the Itô integral as the almost sure limit

$$
\int_0^t H_s \, dW_s := \lim_{n \to \infty} \int_0^t H_s^{(\tau_n)} \, dW_s. \tag{25.5}
$$

Theorem 25.17 Let $H \in \mathcal{E}_{loc}$.

(i) The limit in (25.5) is well-defined and continuous at t. Up to a.s. equality, it is independent of the choice of the sequence $(\tau_n)_{n \in \mathbb{N}}$.

(ii) If τ is a stopping time with $\mathbf{E}\left[\int_0^\tau H_s^2 \, ds \right] < \infty$, then the stopped Itô integral $\left(\int_0^{\tau \wedge t} H_s \, dW_s \right)_{t \geq 0}$ is an L^2-bounded, continuous martingale.

(iii) If $\mathbf{E}\left[\int_0^T H_s^2 \, ds \right] < \infty$ for all $T > 0$, then $\left(\int_0^t H_s \, dW_s \right)_{t \geq 0}$ is a square integrable continuous martingale.

Proof

(i) By Lemma 25.13(ii), on the event $\{\tau_n \geq t\}$, we have

$$\int_0^t H_s \, dW_s = \int_0^t H_s^{(\tau_n)} \, dW_s.$$

Hence the limit exists, is continuous and is independent of the choice of the sequence $(\tau_n)_{n \in \mathbb{N}}$.

(ii) This is immediate by Theorem 25.11.

(iii) As we can choose $\tau_n = n$, this follows from (ii). □

Theorem 25.18 Let H be progressively measurable and $\mathbf{E}\left[\int_0^T H_s^2 \, ds\right] < \infty$ for all $T > 0$. Then

$$M_t := \int_0^t H_s \, dW_s, \quad t \geq 0,$$

defines a square integrable continuous martingale, and

$$(N_t)_{t \geq 0} := \left(M_t^2 - \int_0^t H_s^2 \, ds\right)_{t \geq 0}$$

is a continuous martingale with $N_0 = 0$.

Proof It is enough to show that N is a martingale. Clearly, N is adapted. Let τ be a bounded stopping time. Then

$$\mathbf{E}[N_\tau] = \mathbf{E}\left[M_\tau^2 - \int_0^\tau H_s^2 \, ds\right]$$

$$= \mathbf{E}\left[\left(\int_0^\infty H_s^{(\tau)} \, dW_s\right)^2\right] - \mathbf{E}\left[\int_0^\infty \left(H_s^{(\tau)}\right)^2 \, ds\right] = 0.$$

Thus, by the optional stopping theorem (see Exercise 21.1.3(iii)), N is a martingale. □

Recall the notions of local martingales and square variation from Sect. 21.10.

Corollary 25.19 If $H \in \mathcal{E}_{loc}$, then the Itô integral $M_t = \int_0^t H_s \, dW_s$ is a continuous local martingale with square variation process $\langle M \rangle_t = \int_0^t H_s^2 \, ds$.

Example 25.20

(i) $W_t = \int_0^t 1 \, dW_s$ is a square integrable martingale, and $(W_t^2 - t)_{t \geq 0}$ is a continuous martingale.

(ii) Since $\mathbf{E}\left[\int_0^T W_s^2 \, ds\right] = \frac{T^2}{2} < \infty$ for all $T \geq 0$, $M_t := \int_0^t W_s \, dW_s$ is a continuous, square integrable martingale, and $\left(M_t^2 - \int_0^t W_s^2 \, ds\right)_{t \geq 0}$ is a continuous martingale.

(iii) Assume that H is progressively measurable and bounded, and let $M_t := \int_0^t H_s \, dW_s$. Then M is progressively measurable (since it is continuous and adapted) and

$$\mathbf{E}\left[\int_0^T M_s^2 \, ds\right] = \int_0^T \left(\int_0^s \mathbf{E}[H_r^2] \, dr\right)^2 ds \leq \frac{T^2 \|H\|_\infty^2}{2}.$$

Hence $\tilde{M}_t := \int_0^t M_s \, dW_s$ is a square integrable, continuous martingale and $\left(\tilde{M}_t^2 - \int_0^t M_s^2 \, dW_s\right)_{t \geq 0}$ is a continuous martingale. $\Diamond$

Takeaways We cannot define an integral with respect to Brownian motion as integrator as a Stieltjes integral. This is due to the infinite variation of paths. The Itô integral is therefore fundamentally different and is constructed in two steps. For piecewise constant adapted integrands, it is just the weighted sum of Brownian increments. By a clever choice of an L^2 norm, the operator that maps the integrand to the integral is an isometry. Hence, the integral is obtained by continuous extension of the operator. It is defined for locally square integrable adapted integrands and is itself a continuous local martingale.

25.2   Itô Integral with Respect to Diffusions

If

$$H = \sum_{i=1}^n h_{i-1} \mathbb{1}_{(t_{i-1}, t_i]} \in \mathcal{E}, \tag{25.6}$$

then the elementary integral

$$I_t^M(H) = \sum_{i=1}^n h_{i-1}\left(M_{t_i \wedge t} - M_{t_{i-1} \wedge t}\right)$$

is a martingale (respectively local martingale) if M is a martingale (respectively local martingale). Furthermore,

$$\mathbf{E}\left[(I_\infty^M(H))^2\right] = \sum_{i=1}^n \mathbf{E}\left[h_{i-1}^2(M_{t_i} - M_{t_{i-1}})^2\right] = \sum_{i=1}^n \mathbf{E}\left[h_{i-1}^2(\langle M\rangle_{t_i} - \langle M\rangle_{t_{i-1}})\right]$$

$$= \mathbf{E}\left[\int_0^\infty H_t^2 \, d\langle M\rangle_t\right]$$

if the expression on the right-hand side is finite. Roughly speaking, the procedure in Sect. 25.1 by which we defined the Itô integral for Brownian motion and integrands $H \in \bar{\mathcal{E}}$ can be repeated to construct a stochastic integral with respect to M for a large class of integrands H. Essentially, in the definition of the norm on $\mathcal{E}$ we have to replace dt (that is, the square variation of Brownian motion) by the square variation $d\langle M \rangle_t$ of M:

$$\|H\|_M^2 := \mathbf{E}\left[\int_0^\infty H_t^2 \, d\langle M \rangle_t \right].$$

Extending the integral to the closure $\bar{\mathcal{E}}$ works just as for Brownian motion. The tricky point is to check whether a given integrand is in $\bar{\mathcal{E}}$. For example, for discontinuous martingales M the integrands have to be predictable in order for the stochastic integral to be a martingale (not to mention the difficulty of establishing for such M, the existence of the square variation process). For the case of discrete time processes, we saw this in Sect. 9.3. Now if M is a continuous martingale with continuous square variation $\langle M \rangle$, then the following problem occurs. In the proof of Theorem 25.9 in Step 2, in order to show that progressively measurable processes H are in $\bar{E}$, we used the fact that $H_t^n(\omega) \overset{n \to \infty}{\longrightarrow} H_t(\omega)$ for Lebesgue-almost all t and all ω. Now if $d\langle M \rangle_t$ is not *absolutely* continuous with respect to the Lebesgue measure, then this is not sufficient to infer convergence of the integrals with respect to $d\langle M \rangle_t$. In the case of absolutely continuous square variation, however, that proof works without change. As in Sect. 25.1, we obtain the following theorem.

Theorem 25.21 *Let M be a continuous local martingale with absolutely continuous square variation $\langle M \rangle$ and let H be a progressively measurable process with $\int_0^T H_s^2 \, d\langle M \rangle_s < \infty$ a.s. for all $T \geq 0$. Then the Itô integral $N_t := \int_0^t H_s \, dM_s$ is well-defined and is a continuous local martingale with square variation $\langle N \rangle_t = \int_0^t H_s^2 \, d\langle M \rangle_s$. For any sequence $(\tau_n)_{n\in\mathbb{N}}$ with $\tau_n \uparrow \infty$ and $\left\| H^{(\tau_n)} \right\|_M < \infty$, and for any family $(H^{n,m}, \, n, m \in \mathbb{N}) \subset \mathcal{E}$ with $\left\| H^{n,m} - H^{(\tau_n)} \right\|_M \overset{m\to\infty}{\longrightarrow} 0$, we have*

$$\int_0^t H_s \, dM_s = \lim_{n\to\infty} \lim_{m\to\infty} I_t^M(H^{m,n}) \quad \text{in probability for all } t \geq 0.$$

The following theorem formulates a certain generalization.

Theorem 25.22 *Let M^1 and M^2 be continuous local martingales with absolutely continuous square variation. Let H^i be progressively measurable processes with $\int_0^T (H_s^i)^2 \, d\langle M^i \rangle_s < \infty$ for $i = 1, 2$ and $T < \infty$. Let $N_t^i := \int_0^t H_s^i \, dM_s^i$ for $i = 1, 2$. Then N^1 and N^2 are continuous local martingales with quadratic covariation $\langle N^i, N^j \rangle_t = \int_0^t H_s^i H_s^j \, d\langle M^i, M^j \rangle_s$ for $i, j \in \{1, 2\}$. If M^1 and M^2 are independent, then $\langle N^1, N^2 \rangle \equiv 0$.*

Proof First assume $H^1, H^2 \in \mathcal{E}$. Then there are numbers $0 = t_0 < t_1 < \ldots < t_n$ and $\mathcal{F}_{t_k}$-measurable bounded maps h_k^i, $i = 1, 2$, $k = 0, \ldots, n - 1$ such that

$$H_t^i(\omega) = \sum_{k=1}^{n} h_{k-1}^i(\omega)\, \mathbb{1}_{(t_{k-1}, t_k]}(t).$$

Therefore,

$$N_t^i N_t^j = \sum_{k,l=1}^{n} h_{k-1}^i h_{l-1}^j \big(M_{t_k \wedge t}^i - M_{t_{k-1} \wedge t}^i\big)\big(M_{t_l \wedge t}^j - M_{t_{l-1} \wedge t}^j\big).$$

Those summands with $k \neq l$ are local martingales. For any of the summands with $k = l$,

$$\left(h_{k-1}^i h_{k-1}^j \Big(\big(M_{t_k \wedge t}^i - M_{t_{k-1} \wedge t}^i\big)\big(M_{t_k \wedge t}^j - M_{t_{k-1} \wedge t}^j\big) \right.$$
$$\left. - \big(\langle M^i, M^j\rangle_{t_k \wedge t} - \langle M^i, M^j\rangle_{t_{k-1} \wedge t}\big)\Big)\right)_{t \geq 0}$$

is a local martingale. Since

$$\sum_{k=1}^{n} h_{k-1}^i h_{k-1}^j \big(\langle M^i, M^j\rangle_{t_k \wedge t} - \langle M^i, M^j\rangle_{t_{k-1} \wedge t}\big) = \int_0^t H_s^i H_s^j \, d\langle M^i, M^j\rangle_s,$$

$\big(N_t^i N_t^j - \int_0^t H_s^i H_s^j \, d\langle M^i, M^j\rangle_s\big)_{t \geq 0}$ is a continuous local martingale.

The case of general progressively measurable H^1, H^2 that satisfy an integrability condition follows by the usual L^2-approximation arguments.

If M^1 and M^2 are independent, then $\langle M^1, M^2\rangle \equiv 0$. $\square$

In the following, we consider processes that can be expressed as Itô integrals with respect to a Brownian motion. For these processes, we give a different and more detailed proof of Theorem 25.21.

Definition 25.23 _Let W be a Brownian motion and let σ and b be progressively measurable stochastic processes with $\int_0^t \sigma_s^2 + |b_s| \, ds < \infty$ almost surely for all $t \geq 0$. Then we say that the process X defined by_

$$X_t = \int_0^t \sigma_s \, dW_s + \int_0^t b_s \, ds \quad \text{for } t \geq 0$$

is a generalized **diffusion process** (or, briefly, generalized diffusion) with **diffusion coefficient** σ and **drift** b. Often X is called an **Itô process**.

In particular, if σ and b are of the form $\sigma_s = \tilde{\sigma}(X_s)$ and $b_s = \tilde{b}(X_s)$ for certain maps $\tilde{\sigma} : \mathbb{R} \to [0, \infty)$ and $\tilde{b} : \mathbb{R} \to \mathbb{R}$, then X is called a diffusion (in the proper sense).

In contrast with generalized diffusions, we will see that under certain regularity assumptions on the coefficients, diffusions in the proper sense are Markov processes (compare Theorems 26.8, 26.10 and 26.26).

A diffusion X can always be decomposed as $X = M + A$, where $M_t = \int_0^t \sigma_s \, dW_s$ is a continuous local martingale with square variation $\langle M \rangle_t = \int_0^t \sigma_s^2 \, ds$ (by Corollary 25.19) and $A_t = \int_0^t b_s \, ds$ is a continuous process of locally finite variation.

Clearly, for the H in (25.6), we have

$$\int_0^t H_s \, dM_s = \sum_{i=1}^n h_{i-1}\left(M_{t_i \wedge t} - M_{t_{i-1} \wedge t}\right)$$

$$= \sum_{i=1}^n h_{i-1} \int_{t_{i-1} \wedge t}^{t_i \wedge t} \sigma_s \, dW_s = \int_0^t (H_s \, \sigma_s) \, dW_s.$$

For progressively measurable H with $\int_0^T H_s^2 \, d\langle M \rangle_s = \int_0^T (H_s \sigma_s)^2 \, ds < \infty$ for all $T \geq 0$, we thus define the Itô integral as

$$\int_0^t H_s \, dM_s := \int_0^t (H_s \sigma_s) \, dW_s.$$

Without further work, in particular, without relying on Theorem 25.21, we get the following theorem.

Theorem 25.24 *Let $X = M + A$ be a generalized diffusion with σ and let b be as in Definition 25.23. Let H be progressively measurable with*

$$\int_0^T H_s^2 \sigma_s^2 \, ds < \infty \quad a.s. \quad for \ all \ T \geq 0 \tag{25.7}$$

and

$$\int_0^T |H_s b_s| \, ds < \infty \quad a.s. \quad for \ all \ T \geq 0. \tag{25.8}$$

Then the process Y defined by

$$Y_t := \int_0^t H_s \, dX_s := \int_0^t H_s \, dM_s + \int_0^t H_s \, dA_s := \int_0^t H_s \sigma_s \, dW_s + \int_0^t H_s b_s \, ds$$

is a generalized diffusion with diffusion coefficient $(H_s \sigma_s)_{s \geq 0}$ *and drift* $(H_s b_s)_{s \geq 0}$. *In particular,* $N_t := \int_0^t H_s \, dM_s$ *is a continuous local martingale with square variation process* $\langle N \rangle_t = \int_0^t H_s^2 \, d\langle M \rangle_s = \int_0^t H_s^2 \sigma_s^2 \, ds$.

> **Takeaways** Diffusions are stochastic processes that are a sum of a process of bounded variation and a stochastic integral with respect to a Brownian motion. We have defined the Itô integral for diffusions as integrators in a procedure similar to the one for Brownian motion. The stochastic integrals are again diffusions.

Exercise 25.2.1 Let M be a continuous local martingale with absolutely continuous square variation $\langle M \rangle$ (e.g., a generalized diffusion), and let H be progressively measurable and continuous with $\int_0^T H_s^2 \, d\langle M \rangle_s < \infty$ for all $T \geq 0$. Further, assume that $\mathcal{P} = (\mathcal{P}^{(n)})_{n \in \mathbb{N}}$ is an admissible sequence of partitions (see Definition 21.56).

(i) Show that for all $T \geq 0$, in the sense of stochastic convergence, we have

$$\int_0^T H_s \, dM_s = \lim_{n \to \infty} \sum_{t \in \mathcal{P}_T^n} H_t (M_{t'} - M_t). \tag{25.9}$$

(ii) Show that there exists a subsequence of $\mathcal{P}$ such that almost surely, we have (25.9) for all $T \geq 0$. ♣

25.3   The Itô Formula

This and the following two sections are based on lecture notes of Hans Föllmer.

If $t \mapsto X_t$ is a differentiable map with derivative X' and $F \in C^1(\mathbb{R})$ with derivative F', then we have the classical substitution rule

$$F(X_t) - F(X_0) = \int_0^t F'(X_s) \, dX_s = \int_0^t F'(X_s) X_s' \, ds. \tag{25.10}$$

This remains true even if X is continuous and has locally finite variation (see Sect. 21.10); that is, if X is the distribution function of an absolutely continuous signed measure on $[0, \infty)$. In this case, the derivative X' exists as a Radon–Nikodym derivative almost everywhere, and it is easy to show that (25.10) also holds in this case.

The paths of Brownian motion W are nowhere differentiable (Theorem 21.17 due to Paley, Wiener and Zygmund) and thus have everywhere locally infinite variation. Hence a substitution rule as simple as (25.10) cannot be expected. Indeed, it is easy to see that such a rule *must* be false: Choose $F(x) = x^2$. Then the right-hand side in (25.10) (with X replaced by W) is $\int_0^t 2W_s\, dW_s$ and is hence a martingale. The left-hand side, however, equals W_t^2, which is a submartingale that only becomes a martingale by subtracting t. Indeed, this t is the additional term that shows up in the substitution rule for Itô integrals, the so-called Itô formula. A somewhat bold heuristic puts us on the right track: For small t, W_t is of order $\sqrt{t}$. If we formally write $dW_t = \sqrt{dt}$ and carry out a Taylor expansion of $F \in C^2(\mathbb{R})$ up to second order, then we obtain

$$dF(W_t) = F'(W_t)\, dW_t + \frac{1}{2} F''(W_t)\, (dW_t)^2 = F'(W_t)\, dW_t + \frac{1}{2} F''(W_t)\, dt.$$

Rewriting this as an integral yields

$$F(W_t) - F(W_0) = \int_0^t F'(W_s)\, dW_s + \int_0^t \frac{1}{2} F''(W_s)\, ds. \tag{25.11}$$

(For certain discrete martingales, we derived a similar formula in Example 10.9.) The main goal of this section is to show that this so-called **Itô formula** is indeed correct.

The subsequent discussion in this section does not explicitly rely on the assumption that we integrate with respect to Brownian motion. All that is needed is that the function with respect to which we integrate have continuous square variation (along a suitable admissible sequence of partitions $\mathcal{P} = (\mathcal{P}^n)_{n\in\mathbb{N}}$). In particular, for Brownian motion, $\langle W \rangle_t = t$.

In the following, let $\mathcal{P} = (\mathcal{P}^n)_{n\in\mathbb{N}}$ be an admissible sequence of partitions (recall the definition of $C_{qv} = C_{qv}^{\mathcal{P}}$, $\mathcal{P}_T^n$, $\mathcal{P}_{S,T}^n$, t' and so on from Definitions 21.56 and 21.58). Let $X \in C([0,\infty))$ with continuous square variation (along $\mathcal{P}$)

$$T \mapsto \langle X \rangle_T = V_T^2(X) = \lim_{n\to\infty} \sum_{t\in\mathcal{P}_T^n} (X_{t'} - X_t)^2.$$

For Brownian motion, we have $W \in C_{qv}^{\mathcal{P}}$ almost surely for any admissible sequence of partitions (Theorem 21.64) and $\langle W \rangle_T = T$. For continuous local martingales M passing to a suitable subsequence $\mathcal{P}'$ of $\mathcal{P}$ ensures that $M \in C_{qv}^{\mathcal{P}'}$ almost surely (Theorem 21.70).

Now fix $\mathcal{P}$ and let $X \in C_{qv}$ be a (deterministic) function.

Theorem 25.25 (Pathwise Itô formula) *Let $X \in C_{qv}$ and $F \in C^2(\mathbb{R})$. Then, for all $T \geq 0$, there exists the limit*

$$\int_0^T F'(X_s)\, dX_s := \lim_{n\to\infty} \sum_{t\in\mathcal{P}_T^n} F'(X_t)(X_{t'} - X_t). \tag{25.12}$$

Furthermore, the Itô formula holds:

$$F(X_T) - F(X_0) = \int_0^T F'(X_s)\,dX_s + \frac{1}{2}\int_0^T F''(X_s)\,d\langle X\rangle_s. \qquad (25.13)$$

Here the right integral in (25.13) is understood as a classical (Lebesgue–Stieltjes) integral.

Remark 25.26 If M is a continuous local martingale, then, by Exercise 25.2.1, the Itô integral $\int_0^T F'(M_s)\,dM_s$ is the stochastic limit of $\sum\limits_{t\in\mathcal{P}_T^n} F'(M_t)(M_{t'} - M_t)$ as $n \to \infty$. Thus, in fact, for $X = M(\omega)$, the pathwise integral in (25.12) coincides with the Itô integral (a.s.). In particular, for the Itô integral of Brownian motion, the Itô formula (25.11) holds. ◊

Proof (of Theorem 25.25) We have to show that the limit in (25.12) exists and that (25.13) holds.

For $n \in \mathbb{N}$ and $t \in \mathcal{P}_T^n$ (with successor $t' \in \mathcal{P}_T^n$), the Taylor formula yields

$$F(X_{t'}) - F(X_t) = F'(X_t)(X_{t'} - X_t) + \frac{1}{2}F''(X_t)\cdot(X_{t'} - X_t)^2 + R_t^n, \qquad (25.14)$$

where the remainder

$$R_t^n = \big(F''(\xi) - F''(X_t)\big)\cdot\frac{1}{2}(X_{t'} - X_t)^2$$

(for a suitable ξ between X_t and $X_{t'}$) can be bounded as follows. As X is continuous, $C := \{X_t : t \in [0, T]\}$ is compact and $F''\big|_C$ is uniformly continuous. Thus, for every $\varepsilon > 0$, there exists a $\delta > 0$ with

$$|F''(X_r) - F''(X_s)| < \varepsilon \quad \text{for all } r, s \in [0, T] \text{ with } |X_r - X_s| < \delta.$$

Since X is uniformly continuous on $[0, T]$ and since the mesh size $|\mathcal{P}^n|$ of the partition goes to 0 as $n \to \infty$, for every $\delta > 0$, there exists an N_δ such that

$$\sup_{n\geq N_\delta}\ \sup_{t\in\mathcal{P}_T^n} |X_{t'} - X_t| < \delta.$$

Hence, for $n \geq N_\delta$ and $t \in \mathcal{P}_T^n$,

$$|R_t^n| \leq \frac{1}{2}\varepsilon\,(X_{t'} - X_t)^2.$$

Summing over $t \in \mathcal{P}_T^n$ in (25.14) yields

$$\sum_{t\in\mathcal{P}_T^n}\big(F(X_{t'}) - F(X_t)\big) = F(X_T) - F(X_0)$$

and

$$\sum_{t \in \mathcal{P}_T^n} |R_t^n| \leq \varepsilon \sum_{t \in \mathcal{P}_T^n} (X_{t'} - X_t)^2 \overset{n \to \infty}{\longrightarrow} \varepsilon \langle X \rangle_T < \infty.$$

As $\varepsilon > 0$ was arbitrary, we get $\sum_{t \in \mathcal{P}_T^n} |R_t^n| \overset{n \to \infty}{\longrightarrow} 0$. We have (see Exercise 21.10.2)

$$\sum_{t \in \mathcal{P}_T^n} \frac{1}{2} F''(X_t)(X_{t'} - X_t)^2 \overset{n \to \infty}{\longrightarrow} \frac{1}{2} \int_0^T F''(X_s) \, d\langle X \rangle_s.$$

Hence, in (25.14) the sum of the remaining terms also has to converge. That is, the limit in (25.12) exists.   □

As a direct consequence, we obtain the Itô formula for the Itô integral with respect to diffusions.

Theorem 25.27 (Itô formula for diffusions) *Let* $Y = M + A$ *be a (generalized) diffusion (see Definition 25.23), where* $M_t = \int_0^t \sigma_s \, dW_s$ *and* $A_t = \int_0^t b_s \, ds$. *Let* $F \in C^2(\mathbb{R})$. *Then we have the Itô formula*

$$F(Y_t) - F(Y_0) = \int_0^t F'(Y_s) \, dM_s + \int_0^t F'(Y_s) \, dA_s + \frac{1}{2} \int_0^t F''(Y_s) \, d\langle M \rangle_s$$

$$= \int_0^t F'(Y_s)\sigma_s \, dW_s + \int_0^t \left(F'(Y_s)b_s + \frac{1}{2} F''(Y_s)\sigma_s^2 \right) ds.$$

$$(25.15)$$

In particular, for Brownian motion,

$$F(W_t) - F(W_0) = \int_0^t F'(W_s) \, dW_s + \frac{1}{2} \int_0^t F''(W_s) \, ds. \qquad (25.16)$$

As an application of the Itô formula, we characterize Brownian motion as a continuous local martingale with a certain square variation process.

Theorem 25.28 (Lévy's characterization of Brownian motion) *Let* $X \in \mathcal{M}_{loc,c}$ *with* $X_0 = 0$. *Then the following are equivalent.*

(i) $(X_t^2 - t)_{t \geq 0}$ *is a local martingale.*
(ii) $\langle X \rangle_t = t$ *for all* $t \geq 0$.
(iii) X *is a Brownian motion.*

Proof **(iii)** $\Longrightarrow$ **(i)**   This is obvious.
(i) $\Longleftrightarrow$ **(ii)**   This is clear since the square variation process is unique.
(ii) $\Longrightarrow$ **(iii)**   It is enough to show that $X_t - X_s \sim \mathcal{N}_{0,t-s}$ given $\mathcal{F}_s$ for $t > s \geq 0$. Employing the uniqueness theorem for characteristic functions, it is enough to show

that (with $i = \sqrt{-1}$) for $A \in \mathcal{F}_s$ and $\lambda \in \mathbb{R}$, we have

$$\varphi_{A,\lambda}(t) := \mathbf{E}\big[e^{i\lambda(X_t-X_s)}\,\mathbb{1}_A\big] = \mathbf{P}[A]\,e^{-\lambda^2(t-s)/2}.$$

Applying Itô's formula separately to the real and the imaginary parts, we obtain

$$e^{i\lambda X_t} - e^{i\lambda X_s} = \int_s^t i\,\lambda\,e^{i\lambda X_r}\,dX_r - \frac{1}{2}\int_s^t \lambda^2\,e^{i\lambda X_r}\,dr.$$

Therefore,

$$\mathbf{E}\big[e^{i\lambda(X_t-X_s)}\,\big|\,\mathcal{F}_s\big] - 1$$
$$= \mathbf{E}\left[\int_s^t i\,\lambda\,e^{i\lambda(X_r-X_s)}\,dX_r\,\bigg|\,\mathcal{F}_s\right] - \frac{1}{2}\lambda^2\,\mathbf{E}\left[\int_s^t e^{i\lambda(X_r-X_s)}\,dr\,\bigg|\,\mathcal{F}_s\right].$$

Now $M_t := \operatorname{Re}\int_s^t i\,\lambda\,e^{i\lambda(X_r-X_s)}\,dX_r$ and $N_t := \operatorname{Im}\int_s^t i\,\lambda\,e^{i\lambda(X_r-X_s)}\,dX_r,\, t \geq s$ are continuous local martingales with

$$\langle M\rangle_t = \int_s^t \lambda^2 \sin\big(\lambda(X_r - X_s)\big)^2\,dr \;\leq\; \lambda^2(t-s)$$

and

$$\langle N\rangle_t = \int_s^t \lambda^2 \cos\big(\lambda(X_r - X_s)\big)^2\,dr \;\leq\; \lambda^2(t-s).$$

Thus, by Corollary 21.76, M and N are martingales. Hence we have

$$\mathbf{E}\left[\int_s^t i\,\lambda\,e^{i\lambda(X_r-X_s)}\,dX_r\,\bigg|\,\mathcal{F}_s\right] = 0.$$

Since $A \in \mathcal{F}_s$, Fubini's theorem yields

$$\varphi_{A,\lambda}(t) - \varphi_{A,\lambda}(s) = \mathbf{E}\big[e^{i\lambda(X_t-X_s)}\,\mathbb{1}_A\big] - \mathbf{P}[A]$$
$$= -\frac{1}{2}\lambda^2\int_s^t \mathbf{E}\big[e^{i\lambda(X_r-X_s)}\,\mathbb{1}_A\big]\,dr = -\frac{1}{2}\lambda^2\int_s^t \varphi_{A,\lambda}(r)\,dr.$$

That is, $\varphi_{A,\lambda}$ is the solution of the linear differential equation

$$\varphi_{A,\lambda}(s) = \mathbf{P}[A] \quad\text{and}\quad \frac{d}{dt}\varphi_{A,\lambda}(t) = -\frac{1}{2}\lambda^2\,\varphi_{A,\lambda}(t).$$

The unique solution is $\varphi_{A,\lambda}(t) = \mathbf{P}[A]\,e^{-\lambda^2(t-s)/2}$. $\square$

As a consequence of this theorem, we get that any continuous local martingale whose square variation process is absolutely continuous (as a function of time) can be expressed as an Itô integral with respect to some Brownian motion.

Theorem 25.29 (Itô's martingale representation theorem) *Let M be a continuous local martingale with $M_0 = 0$ and absolutely continuous square variation $t \mapsto \langle M \rangle_t$. Then, on a suitable extension of the underlying probability space, there exists a Brownian motion W with*

$$M_t = \int_0^t \sqrt{\frac{d\langle M \rangle_s}{ds}} \, dW_s \quad \text{for all } t \geq 0.$$

Proof Assume that the probability space is rich enough to carry a Brownian motion $\widetilde{W}$ that is independent of M. Let

$$f_t := \lim_{n \to \infty} n\left(\langle M \rangle_t - \langle M \rangle_{t-1/n}\right) \quad \text{for } t > 0.$$

Then f is a progressively measurable version of the Radon–Nikodym derivative $\frac{d\langle M \rangle_t}{dt}$. Clearly, $\int_0^T \mathbb{1}_{\{f_t > 0\}} f_t^{-1} d\langle M \rangle_t \leq T < \infty$ for all $T > 0$. Hence the following integrals are well-defined, and furthermore, as a sum of continuous martingales,

$$W_t := \int_0^t \mathbb{1}_{\{f_s > 0\}} f_s^{-1/2} \, dM_s + \int_0^t \mathbb{1}_{\{f_s = 0\}} \, d\widetilde{W}_s$$

is a continuous local martingale. By Theorem 25.22, we have

$$\langle W \rangle_t = \int_0^t \mathbb{1}_{\{f_s > 0\}} f_s^{-1} \, d\langle M \rangle_s + \int_0^t \mathbb{1}_{\{f_s = 0\}} \, ds$$

$$= \int_0^t \mathbb{1}_{\{f_s > 0\}} f_s^{-1} f_s \, ds + \int_0^t \mathbb{1}_{\{f_s = 0\}} \, ds = t.$$

Hence, by Theorem 25.28, W is a Brownian motion. On the other hand, we have

$$\int_0^t f_s^{1/2} \, dW_s = \int_0^t \mathbb{1}_{\{f_s > 0\}} f_s^{1/2} f_s^{-1/2} \, dM_s + \int_0^t \mathbb{1}_{\{f_s = 0\}} f_s^{1/2} \, d\widetilde{W}_s$$

$$= \int_0^t \mathbb{1}_{\{f_s > 0\}} \, dM_s.$$

However,

$$M_t - \int_0^t \mathbb{1}_{\{f_s > 0\}} \, dM_s = \int_0^t \mathbb{1}_{\{f_s = 0\}} \, dM_s$$

is a continuous local martingale with square variation $\int_0^t \mathbb{1}_{\{f_s=0\}}\, d\langle M\rangle_s = 0$ and hence it almost surely equals zero. Therefore, we have $M_t = \int_0^t f_s^{1/2}\, dW_s$, as claimed. $\square$

Reflection In Theorem 25.29, why is it not sufficient to postulate continuity of the map $t \mapsto \langle M\rangle_t$ instead of absolute continuity? Find a counterexample! ♠

We come next to a multidimensional version of the pathwise Itô formula. To this end, let $\mathcal{C}_{qv}^d$ be the space of continuous maps $X : [0, \infty) \to \mathbb{R}^d$, $t \mapsto X_t = (X_t^1, \ldots, X_t^d)$ such that, for $k, l = 1, \ldots, d$, the quadratic covariation (see Definition 21.58) $\langle X^k, X^l\rangle$ exists and is continuous. Further, let $C^2(\mathbb{R}^d)$ be the space of twice continuously differentiable functions F on $\mathbb{R}^d$ with partial derivatives $\partial_k F$ and $\partial_k \partial_l F$, $k, l = 1, \ldots, d$. Denote by ∇ the gradient and by $\Delta = (\partial_1^2 + \ldots + \partial_d^2)$ the **Laplace operator**.

Theorem 25.30 (Multidimensional pathwise Itô formula) *Let* $X \in \mathcal{C}_{qv}^d$ *and* $F \in C^2(\mathbb{R}^d)$. *Then*

$$F(X_T) - F(X_0) = \int_0^T \nabla F(X_s)\, dX_s + \frac{1}{2} \sum_{k,l=1}^d \int_0^T \partial_k \partial_l F(X_s)\, d\langle X^k, X^l\rangle_s,$$

where

$$\int_0^T \nabla F(X_s)\, dX_s := \lim_{n\to\infty} \sum_{t\in\mathcal{P}_T^n} \sum_{k=1}^d \partial_k F(X_t)(X_{t'}^k - X_t^k).$$

Proof This works just as in the one-dimensional case. We leave the details as an exercise. $\square$

Remark 25.31 If each of the integrals $\int_0^T \partial_k F(X_s)\, dX_s^k$ exists, then

$$\int_0^T \nabla F(X_s)\, dX_s = \sum_{k=1}^d \int_0^T \partial_k F(X_s)\, dX_s^k.$$

Note that existence of the individual integrals does not follow from the existence of the integral $\int_0^T \nabla F(X_s)\, dX_s$. ◊

Corollary 25.32 (Product rule) *If* $X, Y, X - Y, X + Y \in \mathcal{C}_{qv}$, *then*

$$X_T Y_T = X_0 Y_0 + \int_0^T Y_s\, dX_s + \int_0^T X_s\, dY_s + \langle X, Y\rangle_T \quad \text{for all } T \geq 0$$

if both integrals exist. In particular, the product rule holds if X and Y are continuous local martingales.

Proof By assumption (and using the polarization formula), the covariation $\langle X, Y \rangle$ exists. Applying Theorem 25.30 with $F(x, y) = xy$, the claim follows.

For continuous local martingales, by Exercise 25.2.1, there exists a suitable sequence of partitions $\mathcal{P}$ such that the integrals exist (pathwise). □

Now let $Y = M + A$ be a d-dimensional generalized diffusion. Hence

$$M_t^k = \sum_{l=1}^{d} \int_0^t \sigma_s^{k,l} \, dW_s^l \quad \text{and} \quad A_t^k = \int_0^t b_s^k \, ds \quad \text{for } t \geq 0, \, k = 1, \ldots, d.$$

Here $W = (W^1, \ldots, W^d)$ is a d-dimensional Brownian motion and $\sigma^{k,l}$ (respectively b^k) are progressively measurable, locally square integrable (respectively locally integrable) stochastic processes for every $k, l = 1, \ldots, d$. Since $\langle W^k, W^l \rangle_t = t \cdot \mathbb{1}_{\{k=l\}}$, we have $\langle Y^k, Y^l \rangle_t = \langle M^k, M^l \rangle_t = \int_0^t a_s^{k,l} \, ds$, where

$$a_s^{k,l} := \sum_{i=1}^{d} \sigma_s^{k,i} \sigma_s^{l,i}$$

is the covariance matrix of the diffusion M. In particular, we have $M \in \mathcal{C}_{qv}^d$ almost surely. Note that (by Exercise 25.2.1), there exists a partition sequence $\mathcal{P}$ such that the integrals $\int_0^t \sigma_s^{k,l} \, \partial_k F(Y_s) \, dW_s^l$ in (25.17) exist in the pathwise sense. As a corollary of the multidimensional pathwise Itô formula (Theorem 25.30 and Remark 25.31), we thus get the following theorem.

Theorem 25.33 (Multidimensional Itô formula) *Let Y be as above and let $F \in C^2(\mathbb{R}^d)$. Then*

$$F(Y_T) - F(Y_0) = \int_0^T \nabla F(Y_s) \, dY_s + \frac{1}{2} \sum_{k,l=1}^{d} \int_0^T \partial_k \partial_l F(Y_s) \, d\langle M^k, M^l \rangle_s$$

$$= \sum_{k,l=1}^{d} \int_0^T \sigma_s^{k,l} \, \partial_k F(Y_s) \, dW_s^l + \sum_{k=1}^{d} \int_0^T b_s^k \, \partial_k F(Y_s) \, ds \quad (25.17)$$

$$+ \frac{1}{2} \sum_{k,l=1}^{d} \int_0^T a_s^{k,l} \, \partial_k \partial_l F(Y_s) \, ds.$$

In particular, for Brownian motion, we have

$$F(W_t) - F(W_0) = \sum_{k=1}^{d} \int_0^t \partial_k F(W_s) \, dW_s^k + \frac{1}{2} \int_0^t \triangle F(W_s) \, ds. \quad (25.18)$$

Corollary 25.34 *The process* $(F(W_t))_{t \geq 0}$ *is a continuous local martingale if and only if* F *is harmonic (that is,* $\triangle F \equiv 0$*).*

Proof If F is harmonic, then as a sum of Itô integrals, $F(W_t) = F(W_0) + \sum_{k=1}^{d} \int_0^t \partial_k F(W_s) \, dW_s^k$ is a continuous local martingale.

On the other hand, if $(F(W_t))_{t \geq 0}$ is a continuous local martingale, then as a difference of continuous local martingales, $\int_0^t \triangle F(W_s) \, ds$ is also a continuous local martingale. As $t \mapsto \int_0^t \triangle F(W_s) \, ds$ has finite variation, we have $\int_0^t \triangle F(W_s) \, ds = 0$ for all $t \geq 0$ almost surely (by Corollary 21.72). Hence $\triangle F \equiv 0$.   □

Corollary 25.35 (Time-dependent Itô formula) *If* $F \in C^{2,1}(\mathbb{R}^d \times \mathbb{R})$*, then*

$$F(W_T, T) - F(W_0, 0)$$

$$= \sum_{k=1}^{d} \int_0^T \partial_k F(W_s, s) \, dW_s^k + \int_0^T \left(\partial_{d+1} + \frac{1}{2}(\partial_1^2 + \ldots + \partial_d^2) \right) F(W_s, s) \, ds.$$

Proof Apply Theorem 25.33 to $Y = (W_t^1, \ldots, W_t^d, t)_{t \geq 0}$.   □

Takeaways The Itô formula is the probabilistic analogue to the substitution rule for Riemann integrals. Any continuous local martingale with absolutely continuous variance process can be represented as an Itô integral with respect to some Brownian motion. In particular, it is a diffusion. Compare with the representation theorem in discrete time that we derived in Theorem 9.43.

Exercise 25.3.1 (Fubini's theorem for Itô integrals) Let $X \in \mathcal{C}_{qv}$ and assume that $g : [0, \infty)^2 \to \mathbb{R}$ is continuous and (in the interior) twice continuously differentiable in the second coordinate with derivative $\partial_2 g$. Use the product rule (Corollary 25.32) to show that

$$\int_0^s \left(\int_0^t g(u, v) \, du \right) dX_v = \int_0^t \left(\int_0^s g(u, v) \, dX_v \right) du$$

and

$$\int_0^s \left(\int_0^v g(u, v) \, du \right) dX_v = \int_0^s \left(\int_u^s g(u, v) \, dX_v \right) du. \quad \clubsuit$$

Exercise 25.3.2 (Stratonovich integral) Let $\mathcal{P}$ be an admissible sequence of partitions, $X \in \mathcal{C}_{qv}^{\mathcal{P}}$ and $F \in C^2(\mathbb{R})$ with derivative $f = F'$. Show that, for every

$t \geq 0$, the **Stratonovich integral**

$$\int_0^T f(X_t) \circ dX_t := \lim_{n \to \infty} \sum_{t \in \mathcal{P}_T^n} f\left(\frac{X_{t'} + X_t}{2}\right)(X_{t'} - X_t)$$

is well-defined, and that the classical substitution rule

$$F(X_T) - F(X_0) = \int_0^T F'(X_t) \circ dX_t$$

holds. Show that, in contrast with the Itô integral, the Stratonovich integral with respect to a continuous local martingale is, in general, not a local martingale. ♣

25.4   Dirichlet Problem and Brownian Motion

As for discrete Markov chains (compare Sect. 19.1), the solutions of the Dirichlet problem in a domain $G \subset \mathbb{R}^d$ can be expressed in terms of a d-dimensional Brownian motion that is stopped upon hitting the boundary G.

In the following, let $G \subset \mathbb{R}^d$ be a bounded open set.

Definition 25.36 (Dirichlet problem) *Let $f : \partial G \to \mathbb{R}$ be continuous. A function $u : \overline{G} \to \mathbb{R}$ is called a solution of the Dirichlet problem on G with boundary value f if u is continuous, twice differentiable in G and*

$$\begin{aligned} \triangle u(x) &= 0 &&\text{for } x \in G, \\ u(x) &= f(x) &&\text{for } x \in \partial G. \end{aligned} \tag{25.19}$$

For sufficiently smooth domains, there always exists a solution of the Dirichlet problem (see, e.g., [79, Section 4.4]). If there is a solution, then as a consequence of Theorem 25.38, it is unique.

In the following, let $W = (W^1, \ldots, W^d)$ be a d-dimensional Brownian motion with respect to a filtration $\mathbb{F}$ that satisfies the usual conditions. We write $\mathbf{P}_x$ and $\mathbf{E}_x$ for probabilities and expectations if W is started at $W_0 = x = (x^1, \ldots, x^d) \in \mathbb{R}^d$. If $A \subset \mathbb{R}^d$ is open, then

$$\tau_{A^c} := \inf\{t > 0 : W_t \in A^c\}$$

is an $\mathbb{F}$-stopping time (see Exercise 21.4.4). Since G is bounded, we have $G \subset (-a, a) \times \mathbb{R}^{d-1}$ for some $a > 0$. Thus $\tau_{G^c} \leq \tau_{((-a,a) \times \mathbb{R}^{d-1})^c}$. By Exercise 21.2.4 (applied to W^1), for $x \in G$,

$$\mathbf{E}_x[\tau_{G^c}] \leq \mathbf{E}_x[\tau_{((-a,a) \times \mathbb{R}^{d-1})^c}] = (a - x^1)(a + x^1) < \infty. \tag{25.20}$$

In particular, $\tau_{G^c} < \infty$ $\mathbf{P}_x$-almost surely. Hence $W_{\tau_{G^c}}$ is a $\mathbf{P}_x$-almost surely well-defined random variable with values in ∂G.

Definition 25.37 *For* $x \in G$, *denote by*

$$\mu_{x,G} = \mathbf{P}_x \circ W_{\tau_{G^c}}^{-1}$$

the **harmonic measure** *on* ∂G.

Theorem 25.38 *If* u *is a solution of the Dirichlet problem on* G *with boundary value* f, *then*

$$u(x) = \mathbf{E}_x\big[f(W_{\tau_{G^c}})\big] = \int_{\partial G} f(y)\,\mu_{x,G}(dy) \quad \text{for } x \in G. \tag{25.21}$$

In particular, the solution of the Dirichlet problem is always unique.

Proof Let $G_1 \subset G_2 \subset \ldots$ be a sequence of open sets with $x \in G_1$, $G_n \uparrow G$ and $\overline{G}_n \subset G$ for every $n \in \mathbb{N}$. Hence, in particular, every $\overline{G}_n$ is compact and thus ∇u is bounded on $\overline{G}_n$. We abbreviate $\tau := \tau_{G^c}$ and $\tau_n := \tau_{G_n^c}$.

As u is harmonic (that is, $\triangle u = 0$), by the Itô formula, for $t < \tau$,

$$u(W_t) = u(W_0) + \int_0^t \nabla u(W_s)\,dW_s = u(W_0) + \sum_{k=1}^d \int_0^t \partial_k u(W_s)\,dW_s^k. \tag{25.22}$$

In particular, $M := (u(W_t))_{t \in [0,\tau)}$ is a local martingale up to τ (however, in general, it is not a martingale). For $t < \tau_n$, we have

$$(\partial_k u(W_s))^2 \leq C_n := \sup_{y \in \overline{G}_n} \|\nabla u(y)\|_2^2 < \infty \quad \text{for all } k = 1, \ldots, d.$$

Hence, by (25.20),

$$\mathbf{E}\left[\int_0^{\tau_n} (\partial_k u(W_s))^2\,ds\right] \leq C_n\,\mathbf{E}_x[\tau_n] \leq C_n\,\mathbf{E}[\tau] < \infty.$$

Thus, by Theorem 25.17(ii), for every $n \in \mathbb{N}$, the stopped process M^{τ_n} is a martingale. Therefore,

$$\mathbf{E}_x[u(W_{\tau_n})] = \mathbf{E}_x[M_{\tau_n}] = \mathbf{E}_x[M_0] = u(x). \tag{25.23}$$

As W is continuous and $\tau_n \uparrow \tau$, we have $W_{\tau_n} \xrightarrow{n\to\infty} W_\tau \in \partial G$. Since u is continuous, we get

$$u(W_{\tau_n}) \xrightarrow{n\to\infty} u(W_\tau) = f(W_\tau). \tag{25.24}$$

Again, since u is continuous and $\overline{G}$ is compact, u is bounded. By the dominated convergence theorem, (25.24) implies convergence of the expectations; that is (also using (25.23)),

$$u(x) = \lim_{n\to\infty} \mathbf{E}_x\big[u(W_{\tau_n})\big] = \mathbf{E}_x\big[f(W_\tau)\big]. \qquad \square$$

Reflection A Dirichlet problem need not have a solution. Find an example based on the punctured disc $G := \{(x, y) \in \mathbb{R}^2 : x^2 + y^2 \in (0, 1)\}$ that does not have a solution. ♠

Takeaways If the solution of a Dirichlet problem exists, then it is the expected value of the boundary value at the random point where Brownian motion first leaves the domain. As similar statement for discrete Markov chains was shown in equation (19.1).

Exercise 25.4.1 Let $G = \mathbb{R} \times (0, \infty)$ be the open upper half plane of $\mathbb{R}^2$ and $x = (x_1, x_2) \in G$. Show that $\tau_{G^c} < \infty$ almost surely and that the harmonic measure $\mu_{x,G}$ on $\mathbb{R} \cong \partial G$ is the Cauchy distribution with scale parameter x_2 that is shifted by x_1: $\mu_{x,G} = \delta_{x_1} * \mathrm{Cau}_{x_2}$. ♣

Exercise 25.4.2 Let $d \geq 3$ and let $G = \mathbb{R}^{d-1} \times (0, \infty)$ be an open half space of $\mathbb{R}^d$. Let $x = (x_1, \ldots, x_d) \in G$. Show that $\tau_{G^c} < \infty$ almost surely and that the harmonic measure $\mu_{x,G}$ on $\mathbb{R}^{d-1} \cong \partial G$ has the density

$$\frac{\mu_{x,G}(dy)}{dy} = \frac{\Gamma(d/2)}{\pi^{d/2}} \frac{x_d}{\sqrt{(x_1 - y_1)^2 + \ldots + (x_{d-1} - y_{d-1})^2 + x_d^2}}. \qquad ♣$$

Exercise 25.4.3 Let $r > 0$ and let $B_r(0) \subset \mathbb{R}^d$ be the open ball with radius r centered at the origin. For $x \in B_r(0)$, determine the harmonic measure $\mu_{x,B_r(0)}$. ♣

25.5   Recurrence and Transience of Brownian Motion

By Pólya's theorem (Theorem 17.40), symmetric simple random walk $(X_n)_{n\in\mathbb{N}}$ on $\mathbb{Z}^d$ is recurrent (that is, it visits every point infinitely often) if and only if $d \leq 2$. If $d > 2$, then the random walk is transient and eventually leaves every bounded set $A \subset \mathbb{Z}^d$. To give a slightly different (though equivalent) formulation of this,

$$\liminf_{n\to\infty} \|X_n\| = 0 \text{ a.s.} \qquad \Longleftrightarrow \qquad d \leq 2$$

and

$$\lim_{n\to\infty} \|X_n\| = \infty \quad \text{a.s.} \qquad \Longleftrightarrow \qquad d > 2.$$

The main result of this section is that a similar dichotomy also holds for Brownian motion.

Theorem 25.39 *Let* $W = (W^1, \ldots, W^d)$ *be a d-dimensional Brownian motion.*

(i) *If* $d \le 2$, *then* W *is recurrent in the sense that*

$$\liminf_{t\to\infty} \|W_t - y\| = 0 \quad \text{a.s.} \quad \text{for every } y \in \mathbb{R}^d.$$

In particular, almost surely the path $\{W_t : t \ge 0\}$ *is dense in* $\mathbb{R}^d$.

(ii) *If* $d > 2$, *then* W *is transient in the sense that*

$$\lim_{t\to\infty} \|W_t\| = \infty \quad \text{a.s.},$$

and for any $y \in \mathbb{R}^d \setminus \{0\}$, *we have* $\inf\{\|W_t - y\| : t \ge 0\} > 0$ *almost surely.*

The basic idea of the proof is to use a suitable Dirichlet problem (and the result of Sect. 25.4) to compute the probabilities for W to hit certain balls,

$$B_R(x) := \{y \in \mathbb{R}^d : \|x - y\| < R\}.$$

Let $0 < r < R < \infty$ and let $G_{r,R}$ be the annulus

$$G_{r,R} := B_R(0) \setminus \overline{B}_r(0) = \{x \in \mathbb{R}^d : r < \|x\| < R\}.$$

Recall that, for closed $A \subset \mathbb{R}^d$, we write $\tau_A = \inf\{t > 0 : W_t \in A\}$ for the stopping time of first entrance into A. We further write

$$\tau_s := \inf\{t > 0 : \|W_t\| = s\} \quad \text{and} \quad \tau_{r,R} = \inf\{t > 0 : W_t \notin G_{r,R}\}.$$

If we start W at $W_0 \in G_{r,R}$, then clearly $\tau_{r,R} = \tau_r \wedge \tau_R$. On the boundary of $G_{r,R}$, define the function f by

$$f(x) = \begin{cases} 1, & \text{if } \|x\| = r, \\ 0, & \text{if } \|x\| = R. \end{cases} \tag{25.25}$$

Define $u_{r,R} : \overline{G}_{r,R} \to \mathbb{R}$ by

$$u_{r,R}(x) = \frac{V(\|x\|) - V(R)}{V(r) - V(R)},$$

where $V : (0, \infty) \to \mathbb{R}$ is Newton's potential function

$$V(s) = V_d(s) = \begin{cases} s, & \text{if } d = 1, \\ \log(s), & \text{if } d = 2, \\ -s^{2-d}, & \text{if } d > 2. \end{cases} \qquad (25.26)$$

It is easy to check that $\varphi : \mathbb{R}^d \setminus \{0\} \to \mathbb{R}$, $x \mapsto V_d(\|x\|)$ is harmonic (that is, $\triangle \varphi \equiv 0$). Hence $u_{r,R}$ is the solution of the Dirichlet problem on $G_{r,R}$ with boundary value f. By Theorem 25.38, for $x \in G_{r,R}$,

$$\mathbf{P}_x[\tau_{r,R} = \tau_r] = \mathbf{P}_x[\|W_{\tau_{r,R}}\| = r] = \mathbf{E}_x[f(W_{\tau_{r,R}})] = u_{r,R}(x). \qquad (25.27)$$

Theorem 25.40 *For $r > 0$ and $x, y \in \mathbb{R}^d$ with $\|x - y\| > r$, we have*

$$\mathbf{P}_x[W_t \in B_r(y) \text{ for some } t > 0] = \begin{cases} 1, & \text{if } d \leq 2, \\ \left(\frac{\|x-y\|}{r}\right)^{2-d}, & \text{if } d > 2. \end{cases}$$

Proof Without loss of generality, assume $y = 0$. Then

$$\mathbf{P}_x[\tau_r < \infty] = \lim_{R \to \infty} \mathbf{P}_x[\tau_{r,R} = \tau_r] = \lim_{R \to \infty} \frac{V(\|x\|) - V(R)}{V(r) - V(R)}$$

$$= \begin{cases} 1, & \text{if } d \leq 2, \\ \frac{V_d(\|x\|)}{V_d(r)}, & \text{if } d > 2, \end{cases}$$

since $\lim_{R \to \infty} V_d(R) = \infty$ if $d \leq 2$ and $= 0$ if $d > 2$. $\qquad \square$

Proof (of Theorem 25.39) Using the strong Markov property of Brownian motion, we get for $r > 0$

$$\mathbf{P}_x\left[\liminf_{t \to \infty} \|W_t\| < r\right] = \mathbf{P}_x\left[\bigcup_{s \in (0,r)} \bigcap_{R > \|x\|} \{\|W_t\| \leq s \text{ for some } t > \tau_R\}\right]$$

$$= \sup_{s \in (0,r)} \inf_{R > \|x\|} \mathbf{P}_x\left[\|W_t\| \leq s \text{ for some } t > \tau_R\right]$$

$$= \sup_{s \in (0,r)} \inf_{R > \|x\|} \mathbf{P}_x\left[\mathbf{P}_{W_{\tau_R}}[\tau_s < \infty]\right].$$

However, by Theorem 25.40 (since $\|W_{\tau_R}\| = R$ for $R > \|x\|$), we have

$$\mathbf{P}_{W_{\tau_R}}[\tau_s < \infty] = \begin{cases} 1, & \text{if } d \le 2, \\ (s/R)^{d-2}, & \text{if } d > 2. \end{cases}$$

Therefore,

$$\mathbf{P}\left[\liminf_{t \to \infty} \|W_t\| < r\right] = \begin{cases} 1, & \text{if } d \le 2, \\ 0, & \text{if } d > 2. \end{cases}$$

This implies the claim.   □

Definition 25.41 (Polar set) *A set* $A \subset \mathbb{R}^d$ *is called **polar** if*

$$\mathbf{P}_x\big[W_t \notin A \text{ for all } t > 0\big] = 1 \quad \text{for all } x \in \mathbb{R}^d.$$

Theorem 25.42 *If* $d = 1$, *then only the empty set is polar. If* $d \ge 2$, *then* $\{y\}$ *is polar for every* $y \in \mathbb{R}^d$.

Proof For $d = 1$, the statement is obvious since

$$\limsup_{t \to \infty} W_t = \infty \quad \text{and} \quad \liminf_{t \to \infty} W_t = -\infty \quad \text{a.s.}$$

Hence, due to the continuity of W, every point $y \in \mathbb{R}$ will be hit (infinitely often).

Now let $d \ge 2$. Without loss of generality, assume $y = 0$. If $x \ne 0$, then

$$\begin{aligned} \mathbf{P}_x[\tau_{\{0\}} < \infty] &= \lim_{R \to \infty} \mathbf{P}_x[\tau_{\{0\}} < \tau_R] \\ &= \lim_{R \to \infty} \inf_{r > 0} \mathbf{P}_x[\tau_{r,R} = \tau_r] \qquad (25.28) \\ &= \lim_{R \to \infty} \inf_{r > 0} u_{r,R}(x) = 0 \end{aligned}$$

since $V_d(r) \overset{r \to 0}{\longrightarrow} -\infty$ if $d \ge 2$.

On the other hand, if $x = 0$, then the strong Markov property of Brownian motion (and the fact that $\mathbf{P}_0[W_t = 0] = 0$ for all $t > 0$) implies

$$\begin{aligned} \mathbf{P}_0[\tau_{\{0\}} < \infty] &= \sup_{t > 0} \mathbf{P}_0[W_s = 0 \text{ for some } s \ge t] \\ &= \sup_{t > 0} \mathbf{P}_0[\mathbf{P}_{W_t}[\tau_{\{0\}} < \infty]] = 0. \end{aligned}$$

Note that in the last step, we used (25.28).   □

Takeaways Using the well known analytic solution of a specific Dirichlet problem (the Newton potential), we compute the probability for d-dimensional Brownian motion to ever hit a certain ball depending on d, the size and the distance of the ball. We conclude that Brownian motion misses single points if $d \geq 2$ and is transient if $d > 2$.

Chapter 26
Stochastic Differential Equations

Stochastic differential equations describe the time evolution of certain continuous Markov processes with values in $\mathbb{R}^n$. In contrast with classical differential equations, in addition to the derivative of the function, there is a term that describes the random fluctuations that are coded as an Itô integral with respect to a Brownian motion. Depending on how seriously we take the concrete Brownian motion as the driving force of the noise, we speak of strong and weak solutions. In the first section, we develop the theory of strong solutions under Lipschitz conditions for the coefficients. In the second section, we develop the so-called (local) martingale problem as a method of establishing weak solutions. In the third section, we present some examples in which the method of duality can be used to prove weak uniqueness.

As stochastic differential equations are a very broad subject, and since things quickly become very technical, we only excursively touch some of the most important results, partly without proofs, and illustrate them with examples.

26.1 Strong Solutions

Consider a stochastic differential equation (SDE) of the type

$$X_0 = \xi,$$
$$dX_t = \sigma(t, X_t)\, dW_t + b(t, X_t)\, dt. \tag{26.1}$$

Here $W = (W^1, \ldots, W^m)$ is an m-dimensional Brownian motion, ξ is an $\mathbb{R}^n$-valued random variable with distribution μ that is independent of W, $\sigma(t, x) = \left(\sigma_{ij}(t, x)\right)_{\substack{i=1,\ldots,n \\ j=1,\ldots,m}}$ is a real $n \times m$ matrix and $b(t, x) = \left(b_i(t, x)\right)_{i=1,\ldots,n}$ is an n-

A. Klenke, *Probability Theory*, Universitext,
https://doi.org/10.1007/978-3-030-56402-5_26

dimensional vector. Assume the maps $(t, x) \mapsto \sigma_{ij}(t, x)$ and $(t, x) \mapsto b_i(t, x)$ are measurable.

By a solution of (26.1) we understand a continuous adapted stochastic process X with values in $\mathbb{R}^n$ that satisfies the integral equation

$$X_t = \xi + \int_0^t \sigma(s, X_s)\, dW_s + \int_0^t b(s, X_s)\, ds \quad \textbf{P}\text{-a.s. for all } t \geq 0. \qquad (26.2)$$

Written in full, this is

$$X_t^i = \xi^i + \sum_{j=1}^m \int_0^t \sigma_{ij}(s, X_s)\, dW_s^j + \int_0^t b_i(s, X_s)\, ds \qquad \text{for all } i = 1, \ldots, n.$$

Now the following problem arises: To which filtration $\mathbb{F}$ do we wish X to be adapted? Should it be the filtration that is generated by ξ and W, or do we allow $\mathbb{F}$ to be larger? Already for ordinary differential equations, depending on the equation, uniqueness of the solution may fail (although existence is usually not a problem); for example, for $f' = |f|^{1/3}$. If $\mathbb{F}$ is larger than the filtration generated by W, then we can define further random variables that select one out of a variety of possible solutions. We thus have more possibilities for solutions than if $\mathbb{F} = \sigma(W)$. Indeed, it will turn out that in some situations for the existence of a solution, it is necessary to allow a larger filtration. Roughly speaking, X is a strong solution of (26.1) if (26.2) holds and if X is adapted to $\mathbb{F} = \sigma(W)$. On the other hand, X is a weak solution if X is adapted to a larger filtration $\mathbb{F}$ with respect to which W is still a martingale. Weak solutions will be dealt with in Sect. 26.2.

Definition 26.1 (Strong solution) *We say that the stochastic differential equation (SDE) (26.1) has a **strong solution** X if there exists a map $F : \mathbb{R}^n \times C([0, \infty); \mathbb{R}^m) \to C([0, \infty); \mathbb{R}^n)$ with the following properties:*

(i) *For every $t \geq 0$, the map $(x, w) \mapsto F(x, w)$ is measurable with respect to $\mathcal{B}(\mathbb{R}^n) \otimes \mathcal{G}_t^m - \mathcal{G}_t^n$, where (for $k = m$ or $k = n$) $\mathcal{G}_t^k := \sigma(\pi_s : s \in [0, t])$ is the σ-algebra generated by the coordinate maps $\pi_s : C([0, \infty); \mathbb{R}^k) \to \mathbb{R}$, $w \mapsto w(s)$, $0 \leq s \leq t$.*

(ii) *The process $X = F(\xi, W)$ satisfies (26.2).*

Condition (i) says that the path $(X_s)_{s \in [0,t]}$ depends only on ξ and $(W_s)_{s \in [0,t]}$ but not on further information. In particular, X is adapted to $\mathcal{F}_t = \sigma(\xi, W_s : s \in [0, t])$ and is progressively measurable; hence the Itô integral in (26.2) is well-defined if σ and b do not grow too quickly for large x.

Remark 26.2 Clearly, a strong solution of an SDE is a generalized n-dimensional diffusion. If the coefficients σ and b are independent of t, then the solution is an n-dimensional diffusion. $\Diamond$

Remark 26.3 Let X be a strong solution and let F be as in Definition 26.1. If W' is an m-dimensional Brownian motion on a space $(\Omega', \mathcal{F}', \mathcal{P}')$ with filtration $\mathbb{F}'$, and if ξ' is independent of W' and is $\mathcal{F}'_0$-measurable, then $X' = F(\xi', W')$ satisfies the integral equation (26.2). Hence, it is a strong solution of (26.1) with W replaced by W'. Thus the existence of a strong solution does not depend on the actual realization of the Brownian motion or on the filtration $\mathbb{F}$. $\Diamond$

Definition 26.4 *We say that the SDE (26.1) has a unique strong solution if there exists an F as in Definition 26.1 such that:*

(i) *If W is an m-dimensional Brownian motion on some probability space $(\Omega, \mathcal{F}, \mathbf{P})$ with filtration $\mathbb{F}$ and if ξ is an $\mathcal{F}_0$-measurable random variable that is independent of W and such that $\mathbf{P} \circ \xi^{-1} = \mu$, then $X := F(\xi, W)$ is a solution of (26.2).*

(ii) *For every solution (X, W) of (26.2), we have $X = F(\xi, W)$.*

Example 26.5 Let $m = n = 1$, $b \in \mathbb{R}$ and $\sigma > 0$. The **Ornstein–Uhlenbeck process**

$$X_t := e^{bt}\xi + \sigma \int_0^t e^{(t-s)b} \, dW_s, \quad t \geq 0, \tag{26.3}$$

is a strong solution of the SDE $X_0 = \xi$ and

$$dX_t = \sigma \, dW_t + b \, X_t \, dt.$$

In the language of Definition 26.1, we have (in the sense of the pathwise Itô integral with respect to w)

$$F(x, w) = \left(t \mapsto e^{bt}x + \int_0^t e^{(t-s)b} \, dw(s) \right)$$

for all $w \in C_{qv}$ (that is, with continuous square variation). Since $\mathbf{P}[W \in C_{qv}] = 1$, we can define $F(x, w) = 0$ for $w \in C([0, \infty); \mathbb{R}) \setminus C_{qv}$.

Indeed, by Fubini's theorem for Itô integrals, we have (Exercise 25.3.1)

$$\xi + \int_0^t \sigma \, dW_s + \int_0^t b \, X_s \, ds$$

$$= \xi + \sigma W_t + \int_0^t b \, e^{bs}\xi \, ds + \int_0^t \sigma b \left(\int_0^s e^{b(s-r)} \, dW_r \right) ds$$

$$= \xi + \sigma W_t + (e^{bt} - 1)\xi + \int_0^t \sigma \left(\int_r^t b \, e^{b(s-r)} \, ds \right) dW_r$$

$$= e^{bt}\xi + \int_0^t \left(\sigma + (e^{b(t-r)} - 1)\sigma \right) dW_r$$

$$= X_t.$$

It can be shown (see Theorem 26.8) that the solution is also (strongly) unique. $\lozenge$

Example 26.6 Let $\alpha, \beta \in \mathbb{R}$. The one-dimensional SDE $X_0 = \xi$ and

$$d X_t = \alpha \, X_t \, d W_t + \beta \, X_t \, dt \tag{26.4}$$

has the strong solution

$$X_t = \xi \, \exp\left(\alpha \, W_t + \left(\beta - \frac{\alpha^2}{2}\right) t\right).$$

In the language of Definition 26.1, we have $\sigma(t, x) = \alpha x$, $b(t, x) = \beta x$ and

$$F(x, w) = \left(t \mapsto x \exp\left(\alpha \, w(t) + \left(\beta - \frac{\alpha^2}{2}\right) t\right)\right)$$

for all $w \in C([0, \infty); \mathbb{R})$ and $x \in \mathbb{R}$. Indeed, by the time-dependent Itô formula (Corollary 25.35),

$$X_t = \xi + \int_0^t \alpha X_s \, d W_s + \int_0^t \left(\left(\beta - \frac{\alpha^2}{2}\right) + \frac{1}{2}\alpha^2\right) X_s \, ds.$$

Also in this case, we have strong uniqueness of the solution (see Theorem 26.8). The process X is called a **geometric Brownian motion** and, for example, serves in the so-called **Black–Scholes model** as the process of stock prices. $\lozenge$

We give a simple criterion for existence and uniqueness of strong solutions. For an $n \times m$ matrix A, define the **Hilbert–Schmidt norm**

$$\|A\| = \sqrt{\text{trace}\left(A \, A^T\right)} = \sqrt{\sum_{i=1}^{n} \sum_{j=1}^{m} A_{i,j}^2}. \tag{26.5}$$

For $b \in \mathbb{R}^n$, we use the Euclidean norm $\|b\|$. Since all norms on finite-dimensional vector spaces are equivalent, it is not important exactly which norm we use. However, the Hilbert–Schmidt norm simplifies the computations, as the following lemma shows.

Lemma 26.7 *Let $t \mapsto H(t) = (H_{ij}(t))_{i=1,\dots,n,\, j=1,\dots,m}$ be progressively measurable and $\mathbf{E}\left[\int_0^T H_{ij}^2(t) \, dt\right] < \infty$ for all i, j. Then*

$$\mathbf{E}\left[\left\|\int_0^T H(t) \, d W_t\right\|^2\right] = \mathbf{E}\left[\int_0^T \|H(t)\|^2 \, dt\right], \tag{26.6}$$

where $\|H\|$ is the Hilbert–Schmidt norm from (26.5).

Proof For $i = 1, \ldots, n$, the process $I_i(t) := \sum_{j=1}^m \int_0^t H_{ij}(s) \, dW_s^j$ is a continuous martingale with square variation process $\langle I_i \rangle_t = \int_0^t \sum_{j=1}^m H_{ij}^2(s) \, ds$. Hence

$$\mathbf{E}\big[(I_i(T))^2\big] = \mathbf{E}\left[\int_0^T \sum_{j=1}^m H_{ij}^2(s) \, ds\right].$$

The left-hand side in (26.6) equals

$$\sum_{i=1}^n \mathbf{E}\big[(I_i(T))^2\big] = \mathbf{E}\left[\int_0^T \sum_{i=1}^n \sum_{j=1}^m H_{ij}^2(s) \, ds\right].$$

Hence the claim follows by the definition of $\|H(s)\|^2$. $\qquad\square$

Theorem 26.8 *Let b and σ be Lipschitz continuous in the second coordinate. That is, we assume that there exists a $K > 0$ such that, for all $x, x' \in \mathbb{R}^n$ and $t \geq 0$,*

$$\|\sigma(t, x) - \sigma(t, x')\| + \|b(t, x) - b(t, x')\| \leq K \|x - x'\|. \tag{26.7}$$

Further, assume the growth condition

$$\|\sigma(t, x)\|^2 + \|b(t, x)\|^2 \leq K^2 (1 + \|x\|^2) \quad \text{for all } x \in \mathbb{R}^n, \, t \geq 0. \tag{26.8}$$

Then, for every initial point $X_0 = x \in \mathbb{R}^n$, there exists a unique strong solution X of the SDE (26.1). This solution is a Markov process and in the case where σ and b do not depend on t, it is a strong Markov process.

As the main tool, we need the following lemma.

Lemma 26.9 (Gronwall) *Let $f, g : [0, T] \to \mathbb{R}$ be integrable and let $C > 0$ such that*

$$f(t) \leq g(t) + C \int_0^t f(s) \, ds \quad \text{for all } t \in [0, T]. \tag{26.9}$$

Then

$$f(t) \leq g(t) + C \int_0^t e^{C(t-s)} g(s) \, ds \quad \text{for all } t \in [0, T].$$

In particular, if $g(t) \equiv G$ is constant, then $f(t) \leq G e^{Ct}$ for all $t \in [0, T]$.

Proof Let $F(t) = \int_0^t f(s) \, ds$ and $h(t) = F(t) \, e^{-Ct}$. Then, by (26.9),

$$\frac{d}{dt} h(t) = f(t) \, e^{-Ct} - C F(t) \, e^{-Ct} \leq g(t) \, e^{-Ct}.$$

Integration yields

$$F(t) = e^{Ct} h(t) \le \int_0^t e^{C(t-s)} g(s) \, ds.$$

Substituting this into (26.9) gives

$$f(t) \le g(t) + CF(t) \le g(t) + C \int_0^t g(s) e^{C(t-s)} \, ds. \qquad \square$$

Proof (of Theorem 26.8) It is enough to show that, for every $T < \infty$, there exists a unique strong solution up to time T.

Uniqueness We first show uniqueness of the solution. Let X and X' be two solutions of (26.2). Then

$$X_t - X'_t = \int_0^t \big(b(s, X_s) - b(s, X'_s)\big) \, ds + \int_0^t \big(\sigma(s, X_s) - \sigma(s, X'_s)\big) \, dW_s.$$

Hence

$$\|X_t - X'_t\|^2 \le 2 \left\| \int_0^t \big(b(s, X_s) - b(s, X'_s)\big) \, ds \right\|^2 \tag{26.10}$$
$$+ 2 \left\| \int_0^t \big(\sigma(s, X_s) - \sigma(s, X'_s)\big) \, dW_s \right\|^2.$$

For the first summand in (26.10), use the Cauchy–Schwarz inequality, and for the second one use Lemma 26.7 to obtain

$$\mathbf{E}\big[\|X_t - X'_t\|^2\big] \le 2t \int_0^t \mathbf{E}\Big[\|b(s, X_s) - b(s, X'_s)\|^2\Big] \, ds$$
$$+ 2 \int_0^t \mathbf{E}\Big[\|\sigma(s, X_s) - \sigma(s, X'_s)\|^2\Big] \, ds.$$

Write $f(t) = \mathbf{E}\big[\|X_t - X'_t\|^2\big]$ and $C := 2(T+1)K^2$. Then $f(t) \le C \int_0^t f(s) \, ds$. Hence Gronwall's lemma (with $g \equiv 0$) yields $f \equiv 0$.

Existence We use a version of the Picard iteration scheme. For $N \in \mathbb{N}_0$, recursively define processes X^N by $X_t^0 \equiv x$ and

$$X_t^N := x + \int_0^t b\big(s, X_s^{N-1}\big) \, ds + \int_0^t \sigma\big(s, X_s^{N-1}\big) \, dW_s \quad \text{for } N \in \mathbb{N}. \tag{26.11}$$

Using the growth condition (26.8), it can be shown inductively that

$$\int_0^T \mathbf{E}\left[\|X_t^N\|^2\right] dt \leq 2(T+1) K^2 \left(T + \int_0^T \mathbf{E}\left[\|X_t^{N-1}\|^2\right] dt\right)$$
$$\leq \left(2T(T+1) K^2\right)^N \left(1 + \|x\|^2\right) < \infty, \qquad N \in \mathbb{N}.$$

Hence, at each step, the Itô integral is well-defined.

Consider now the differences

$$X_t^{N+1} - X_t^N = I_t + J_t,$$

where

$$I_t := \int_0^t \left(\sigma(s, X_s^N) - \sigma(s, X_s^{N-1})\right) dW_s$$

and

$$J_t := \int_0^t \left(b(s, X_s^N) - b(s, X_s^{N-1})\right) ds.$$

By applying Doob's L^2-inequality to the nonnegative submartingale $(\|I_t\|^2)_{t \geq 0}$, using Lemma 26.7 and (26.7), we obtain

$$\mathbf{E}\left[\sup_{s \leq t} \|I_s\|^2\right] \leq 4\mathbf{E}\left[\|I_t\|^2\right]$$
$$= 4\mathbf{E}\left[\int_0^t \|\sigma(s, X_s^N) - \sigma(s, X_s^{N-1})\|^2 ds\right] \qquad (26.12)$$
$$\leq 4K^2 \int_0^t \mathbf{E}\left[\|X_s^N - X_s^{N-1}\|^2\right] ds.$$

For J_t, by the Cauchy–Schwarz inequality, we get

$$\|J_t\|^2 \leq t \int_0^t \|b(s, X_s^N) - b(s, X_s^{N-1})\|^2 ds.$$

Hence

$$\mathbf{E}\left[\sup_{s \leq t} \|J_s\|^2\right] \leq t\mathbf{E}\left[\int_0^t \|b(s, X_s^N) - b(s, X_s^{N-1})\|^2 ds\right]$$
$$\leq tK^2 \int_0^t \mathbf{E}\left[\|X_s^N - X_s^{N-1}\|^2\right] ds. \qquad (26.13)$$

Defining

$$\Delta^N(t) := \mathbf{E}\left[\sup_{s \le t} \|X_s^N - X_s^{N-1}\|^2\right],$$

and $C := 2K^2(4 + T) \vee 2(T + 1)K^2(1 + \|x\|^2)$, we obtain (using the growth condition (26.8))

$$\Delta^{N+1}(t) \le C \int_0^t \Delta^N(s)\, ds \quad \text{for } N \ge 1$$

and

$$\Delta^1(t) \le 2t \int_0^t \|b(s, x)\|^2\, ds + 2 \int_0^t \|\sigma(s, x)\|^2\, ds$$

$$\le 2(T + 1)K^2\big(1 + \|x\|^2\big) \cdot t \le C t.$$

Inductively, we get $\Delta^N(t) \le \frac{(Ct)^N}{N!}$. Thus, by Markov's inequality,

$$\sum_{N=1}^{\infty} \mathbf{P}\left[\sup_{s \le t} \|X_s^N - X_s^{N-1}\|^2 > 2^{-N}\right] \le \sum_{N=1}^{\infty} 2^N \Delta^N(t)$$

$$\le \sum_{N=1}^{\infty} \frac{(2Ct)^N}{N!} \le e^{2Ct} < \infty.$$

Using the Borel–Cantelli lemma, we infer $\sup_{s \le t} \|X_s^N - X_s^{N-1}\|^2 \overset{N \to \infty}{\longrightarrow} 0$ a.s. Hence a.s. $(X^N)_{N \in \mathbb{N}}$ is a Cauchy sequence in the Banach space $(C([0, T]), \|\cdot\|_\infty)$. Therefore, X^N converges a.s. uniformly to some X. As uniform convergence implies convergence of the integrals, X is a strong solution of (26.2).

Markov property The strong Markov property follows from the strong Markov property of the Brownian motion that drives the SDE.   □

We have already seen some important examples of this theorem. Many interesting problems, however, lead to stochastic differential equations with coefficients that are not Lipschitz continuous. In the one-dimensional case, using special comparison methods, one can show that it is sufficient that σ is Hölder-continuous of order $\frac{1}{2}$ in the space variable.

Theorem 26.10 (Yamada–Watanabe) *Consider the one-dimensional situation where $m = n = 1$. Assume that there exist $K < \infty$ and $\alpha \in \left[\frac{1}{2}, 1\right]$ such that, for all* *t ≥ 0 and x, x′ ∈ ℝ, we have*

$$\left| b(t, x) - b(t, x') \right| \leq K \left| x - x' \right| \quad and \quad \left| \sigma(t, x) - \sigma(t, x') \right| \leq K \left| x - x' \right|^{\alpha}.$$

Then, for every $X_0 \in \mathbb{R}$, the SDE (26.1) has a unique strong solution X and X is a strong Markov process.

Proof See [173] or [85, Proposition 5.2.13] and [49, Theorem 5.3.11] for existence and uniqueness. The strong Markov property follows from Theorem 26.26.   □

Example 26.11 Consider the one-dimensional SDE

$$dX_t = \sqrt{\gamma\, X_t^+}\, dW_t + a\left(b - X_t^+ \right) dt \tag{26.14}$$

with initial point $X_0 = x \geq 0$, where $\gamma > 0$ and $a, b \geq 0$ are parameters. The conditions of Theorem 26.10 are fulfilled with $\alpha = \frac{1}{2}$ and $K = \sqrt{\gamma} + a$. Obviously, the unique strong solution X remains nonnegative if $X_0 \geq 0$. (In fact, it can be shown that $X_t > 0$ for all $t > 0$ if $2ab/\gamma \geq 1$, and that X_t hits zero arbitrarily often with probability 1 if $2ab/\gamma < 1$. See, e.g., [78, Example IV.8.2, page 237]. Compare Example 26.16. See Figs. 26.1 and 26.2 for computer simulations.)

Depending on the context, this process is sometimes called **Feller's branching diffusion with immigration** or the **Cox–Ingersoll–Ross model** for the time evolution of interest rates.

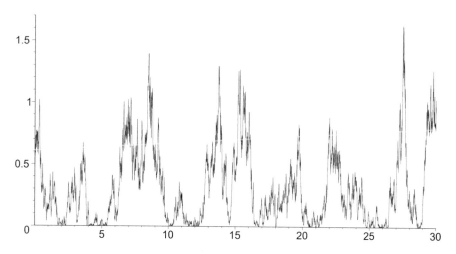

Fig. 26.1 Cox–Ingersoll–Ross diffusion with parameters $\gamma = 1, b = 1$ and $a = 0.3$. The path hits zero again and again since $2ab/\gamma = 0.6 < 1$.

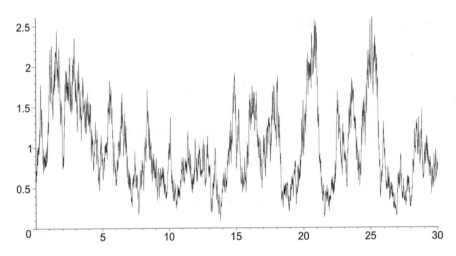

Fig. 26.2 Cox–Ingersoll–Ross diffusion with parameters $\gamma = 1, b = 1$ and $a = 2$. The path never hits zero since $2ab/\gamma = 4 \geq 1$.

For the case $a = b = 0$, use the Itô formula to compute that

$$e^{-\lambda X_t} - e^{-\lambda x} - \gamma \frac{\lambda^2}{2} \int_0^t e^{-\lambda X_s} X_s \, ds \ = \ \lambda \int_0^t e^{-\lambda X_s} \sqrt{\gamma X_s} \, dW_s$$

is a martingale. Take expectations for the Laplace transform $\varphi(t, \lambda, x) = \mathbf{E}_x[e^{-\lambda X_t}]$ to get the differential equation

$$\frac{d}{dt}\varphi(t, \lambda, x) \ = \ \gamma \frac{\lambda^2}{2} \mathbf{E}[X_t \, e^{-\lambda X_t}] \ = \ -\frac{\gamma \lambda^2}{2} \frac{d}{d\lambda}\varphi(t, \lambda, x).$$

With initial value $\varphi(0, \lambda, x) = e^{-\lambda x}$, the unique solution is

$$\varphi(t, \lambda, x) = \exp\left(-\frac{\lambda}{(\gamma/2)\lambda t + 1} x\right).$$

However (for $\gamma = 2$), this is exactly the Laplace transform of the transition probabilities of the Markov process that we defined in Theorem 21.48 and that in Lindvall's theorem (Theorem 21.51) we encountered as the limit of rescaled Galton–Watson branching processes. ◇

Takeaways A stochastic differential equation is a standard way of describing a continuous stochastic process. If the coefficients are Lipschitz continuous, to (almost) every path of the Brownian motion we can assign uniquely a path of the solution of the differential equation. This is the so-called strong solution. For a one-dimensional stochastic differential equation, we can relax the Lipschitz condition on the diffusion coefficient: it is enough that the diffusion coefficient be Hölder-$\frac{1}{2}$-continuous.

Exercise 26.1.1 Let $a, b \in \mathbb{R}$. Show that the stochastic differential equation

$$dX_t = \frac{b - X_t}{1 - t} \, dt + dW_t$$

with initial value $X_0 = a$ has a unique strong solution for $t \in [0, 1)$ and that $X_1 := \lim_{t \uparrow 1} X_1 = b$ almost surely. Furthermore, show that the process $Y = (X_t - a - t(b - a))_{t \in [0,1]}$ can be described by the Itô integral

$$Y_t = (1 - t) \int_0^t (1 - s)^{-1} \, dW_s, \qquad t \in [0, 1),$$

and is hence a Brownian bridge (compare Exercise 21.5.3). ♣

26.2   Weak Solutions and the Martingale Problem

In the last section, we studied strong solutions of the stochastic differential equation

$$dX_t = \sigma(t, X_t) \, dW_t + b(t, X_t) \, dt. \tag{26.15}$$

A strong solution is a solution where any path of the Brownian motion W gets mapped onto a path of the solution X. In this section, we will study the notion of a weak solution where additional information (or additional noise) can be used to construct the solution.

Definition 26.12 (Weak solution of an SDE) *A **weak solution** of (26.15) with initial distribution $\mu \in \mathcal{M}_1(\mathbb{R}^n)$ is a triple*

$$L = \big((X, W), (\Omega, \mathcal{F}, \mathbf{P}), \mathbb{F}\big),$$

where

- *$(\Omega, \mathcal{F}, \mathbf{P})$ is a probability space,*
- *$\mathbb{F} = (\mathcal{F}_t)_{t \geq 0}$ is a filtration on $(\Omega, \mathcal{F}, \mathbf{P})$ that satisfies the usual conditions,*

- W is a Brownian motion on $(\Omega, \mathcal{F}, \mathbf{P})$ and is a martingale with respect to $\mathbb{F}$,
- X is continuous and adapted (hence progressively measurable),
- $\mathbf{P} \circ (X_0)^{-1} = \mu$, and
- (X, W) satisfies

$$X_t = X_0 + \int_0^t \sigma(s, X_s)\, dW_s + \int_0^t b(s, X_s)\, ds \quad \textbf{P}\text{-a.s.} \tag{26.16}$$

A weak solution L is called (weakly) unique if, for any further solution L' with initial distribution μ, we have $\mathbf{P}' \circ (X')^{-1} = \mathbf{P} \circ X^{-1}$.

Remark 26.13 Clearly, a weak solution of an SDE is a generalized n-dimensional diffusion. If the coefficients σ and b do not depend on t, then the solution is an n-dimensional diffusion. $\Diamond$

Remark 26.14 Clearly, every strong solution of (26.15) is a weak solution. The converse is false, as the following example shows. $\Diamond$

Example 26.15 Consider the SDE (with initial value $X_0 = 0$)

$$dX_t = \text{sign}(X_t)\, dW_t, \tag{26.17}$$

where $\text{sign} = \mathbb{1}_{[0,\infty)} - \mathbb{1}_{(-\infty,0)}$ is the sign function. Then

$$X_t = X_0 + \int_0^t \text{sign}(X_s)\, dW_s \quad \text{for all } t \geq 0 \tag{26.18}$$

if and only if

$$W_t = \int_0^t dW_s = \int_0^t \text{sign}(X_s)\, dX_s \quad \text{for all } t \geq 0. \tag{26.19}$$

A weak solution of (26.17) is obtained as follows. Let X be a Brownian motion on a probability space $(\Omega, \mathcal{F}, \mathbf{P})$ and $\mathbb{F} = \sigma(X)$. If we define W by (26.19), then W is a continuous $\mathbb{F}$-martingale with square variation

$$\langle W \rangle_t = \int_0^t (\text{sign}(X_s))^2\, ds = t.$$

Thus, by Lévy's characterization (Theorem 25.28), W is a Brownian motion. Hence $((X, W), (\Omega, \mathcal{F}, \mathbf{P}), \mathbb{F})$ is a weak solution of (26.17).

In order to show that a strong solution does not exist, take any weak solution and show that X is not adapted to $\sigma(W)$. Since, by (26.18), X is a continuous martingale with square variation $\langle X \rangle_t = t$, X is a Brownian motion.

Let $F_n \in C^2(\mathbb{R})$ be a convex even function with derivatives F_n' and F_n'' such that

$$\sup_{x \in \mathbb{R}} |F_n(x) - |x|| \overset{n \to \infty}{\longrightarrow} 0,$$

$|F_n'(x)| \le 1$ for all $x \in \mathbb{R}$ and $F_n'(x) = \text{sign}(x)$ for $|x| > \frac{1}{n}$. In particular, we have

$$\int_0^t \left(F_n'(X_s) - \text{sign}(X_s) \right)^2 ds \overset{n \to \infty}{\longrightarrow} 0 \quad \text{a.s.}$$

and thus

$$\int_0^t F_n'(X_s) \, dX_s \overset{n \to \infty}{\longrightarrow} \int_0^t \text{sign}(X_s) \, dX_s \quad \text{in } L^2. \tag{26.20}$$

By passing to a subsequence, if necessary, we may assume that almost sure convergence holds in (26.20).

Since F_n'' is even, we have

$$W_t = \int_0^t \text{sign}(X_s) \, dX_s = \lim_{n \to \infty} \int_0^t F_n'(X_s) \, dX_s$$

$$= \lim_{n \to \infty} \left(F_n(X_t) - F_n(0) - \frac{1}{2} \int_0^t F_n''(X_s) \, ds \right)$$

$$= |X_t| - \lim_{n \to \infty} \frac{1}{2} \int_0^t F_n''(|X_s|) \, ds.$$

As the right-hand side depends only on $|X_s|$, $s \in [0, t]$, W is adapted to $\mathbb{G} := (\sigma(|X_s| : s \in [0, t]))$. Hence $\sigma(W) \subset \mathbb{G} \subsetneq \sigma(X)$, and thus X is not adapted to $\sigma(W)$. ◇

Example 26.16 Let $B = (B^1, \ldots, B^n)$ be an n-dimensional Brownian motion started at $y \in \mathbb{R}^n$. Let $x := \|y\|^2$, $X_t := \|B_t\|^2 = (B_t^1)^2 + \ldots + (B_t^n)^2$ and

$$W_t := \sum_{i=1}^n \int_0^t \frac{1}{\sqrt{X_s}} B_s^i \, dB_s^i.$$

Then W is a continuous local martingale with $\langle W \rangle_t = t$ for every $t \ge 0$ and

$$X_t = x + nt + \int_0^t \sqrt{X_s} \, dW_s.$$

That is, (X, W) is a weak solution of the SDE $dX_t = \sqrt{2X_t} \, dW_t + n \, dt$. X is called an n-dimensional **Bessel process**. By Theorem 25.42, B (and thus X) hits the origin

for some $t > 0$ if and only if $n = 1$. Clearly, we can define X also for noninteger $n \geq 0$. One can show that X hits zero if and only if $n \leq 1$. Compare Example 26.11.
◊

For the connection between existence and uniqueness of weak solutions and strong solutions, we only quote here the theorem of Yamada and Watanabe.

Definition 26.17 (Pathwise uniqueness) *A solution of the SDE* (26.15) *with initial distribution* μ *is said to be **pathwise unique** if, for every* $\mu \in \mathcal{M}_1(\mathbb{R}^n)$ *and for any two weak solutions* (X, W) *and* (X', W) *on the same space* $(\Omega, \mathcal{F}, \mathbf{P})$ *with the same filtration* $\mathbb{F}$, *we have* $\mathbf{P}[X_t = X'_t \text{ for all } t \geq 0] = 1$.

Theorem 26.18 (Yamada and Watanabe) *The following are equivalent.*

(i) *The SDE* (26.15) *has a unique strong solution.*
(ii) *For any* $\mu \in \mathcal{M}_1(\mathbb{R}^n)$, (26.15) *has a weak solution, and pathwise uniqueness holds.*

If (i) and (ii) hold, then the solution is weakly unique.

Proof See [173], [148, pages 151ff] or [78, pages 163ff].   □

Example 26.19 Let X be a weak solution of (26.17). Then $-X$ is also a weak solution; that is, pathwise uniqueness does not hold (although it can be shown that the solution is weakly unique; see Theorem 26.25). ◊

Consider the one-dimensional case $m = n = 1$. If X is a solution (strong or weak) of (26.15), then

$$M_t := X_t - \int_0^t b(s, X_s)\, ds$$

is a continuous local martingale with square variation

$$\langle M \rangle_t = \int_0^t \sigma^2(s, X_s)\, ds.$$

We will see that this characterizes a weak solution of (26.15) (under some mild growth conditions on σ and b).

Now assume that, for all $t \geq 0$ and $x \in \mathbb{R}^n$, the $n \times n$ matrix $a(t, x)$ is symmetric and nonnegative definite, and let $(t, x) \mapsto a(t, x)$ be measurable.

Definition 26.20 *An n-dimensional continuous process X is called a solution of the **local martingale problem** for a and b with initial condition* $\mu \in \mathcal{M}_1(\mathbb{R}^n)$ *(briefly, LMP(a, b, μ)) if* $\mathbf{P} \circ X_0^{-1} = \mu$ *and if, for every* $i = 1, \ldots, n$,

$$M_t^i := X_t^i - \int_0^t b_i(s, X_s)\, ds, \qquad t \geq 0,$$

is a continuous local martingale with quadratic covariation

$$\langle M^i, M^j \rangle_t = \int_0^t a_{ij}(s, X_s) \, ds \quad \text{for all } t \geq 0, \, i, j = 1, \ldots, n.$$

We say that the solution of LMP(a, b, μ) is unique if, for any two solutions X and X', we have $\mathbf{P} \circ X^{-1} = \mathbf{P} \circ (X')^{-1}$.

Denote by σ^T the transposed matrix of σ. Clearly, $a = \sigma\sigma^T$ is a nonnegative semidefinite symmetric $n \times n$ matrix.

Theorem 26.21 *X is a solution of LMP$(\sigma\sigma^T, b, \mu)$ if and only if (on a suitable extension of the probability space) there exists a Brownian motion W such that (X, W) is a weak solution of (26.15).*

In particular, there exists a unique weak solution of the SDE (26.15) with initial distribution μ if LMP$(\sigma\sigma^T, b, \mu)$ is uniquely solvable.

Proof We show the statement only for the case $m = n = 1$. The general case needs some consideration on the roots of nonnegative semidefinite symmetric matrices, which, however, do not yield any further insight into the stochastics of the problem. For this we refer to [85, Proposition 5.4.6].
" $\Longleftarrow$ " If (X, W) is a weak solution, then, by Corollary 25.19, X solves the local martingale problem.
" $\Longrightarrow$ " Let X be a solution of LMP(σ^2, b, μ). By Theorem 25.29, on an extension of the probability space there exists a Brownian motion $\tilde{W}$ such that $M_t = \int_0^t |\sigma(s, X_s)| \, d\tilde{W}_s$. If we define

$$W_t := \int_0^t \text{sign}(\sigma(s, X_s)) \, d\tilde{W}_s,$$

then $M_t = \int_0^t \sigma(s, X_s) \, dW_s$ and hence (X, W) is a weak solution of (26.15). $\qquad \square$

In some sense, a local martingale problem is a very natural way of writing a stochastic differential equation; that is:

X locally has derivative (drift) b and additionally has random normally distributed fluctuations of size σ.

Here, a concrete Brownian motion does not appear. In fact, in most problems its occurrence is rather artificial. Just as Markov chains are described by their transition probabilities and not by a concrete realization of the random transitions (as in Theorem 17.17), many continuous (space and time) processes are most naturally described by the drift and the *size* of the fluctuations but not by the concrete realization of the random fluctuations.

From a technical point of view, the formulation of a stochastic differential equation as a local martingale problem is very convenient since it makes SDEs accessible to techniques such as martingale inequalities and approximation theorems that can be used to establish existence and uniqueness of solutions. Here we simply quote two important results.

Theorem 26.22 (Existence of solutions) *Let* $(t, x) \mapsto b(t, x)$ *and* $(t, x) \mapsto a(t, x)$ *be continuous and bounded. Then, for every* $\mu \in \mathcal{M}_1(\mathbb{R}^n)$, *there exists a solution* X *of the LMP*(a, b, μ).

Proof See [148, Theorem V.23.5]. $\qquad\qquad\qquad\qquad\qquad\qquad\qquad\qquad\qquad\qquad\qquad\square$

Definition 26.23 *The LMP*(a, b) *is said to be* **well-posed** *if, for every* $x \in \mathbb{R}^n$, *there exists a unique solution* X *of LMP*(a, b, δ_x).

Remark 26.24 If σ and b satisfy the Lipschitz conditions of Theorem 26.8, then the LMP$(\sigma \sigma^T, b)$ is well-posed. This follows by Theorems 26.8, 26.18 and 26.21. $\lozenge$

In the following, we assume

$$(t, x) \mapsto \sigma(t, x) \text{ resp. } (t, x) \mapsto a(t, x) \text{ is bounded on compact sets.} \qquad (26.21)$$

This condition ensures the equivalence of the local martingale problems to the somewhat more common martingale problem (see [85, Proposition 5.4.11]).

Theorem 26.25 (Uniqueness in the martingale problem) *Assume* (26.21) *and that, for any* $x \in \mathbb{R}^n$, *there exists a solution* X^x *of LMP*(a, b, δ_x). *The distribution of* X^x *will be denoted by* $\mathbf{P}_x := \mathbf{P} \circ (X^x)^{-1}$.

Assume that, for any two solutions X^x *and* Y^x *of LMP*(a, b, δ_x), *we have*

$$\mathbf{P} \circ (X_T^x)^{-1} = \mathbf{P} \circ (Y_T^x)^{-1} \quad \text{for any } T \geq 0. \qquad (26.22)$$

Then LMP(a, b) *is well-posed, and the canonical process* X *is a strong Markov process with respect to* $(\mathbf{P}_x, x \in \mathbb{R}^n)$. *If* $a = \sigma \sigma^T$, *then under* $\mathbf{P}_x$, *the process* X *is the unique weak solution of the SDE* (26.15).

Proof See [49, Theorem 4.4.2 and Problem 49] and [85, Proposition 5.4.11]. $\quad\square$

A fundamental strength of this theorem is that we do not need to check the uniqueness of the whole process but only have to check in (26.22) the one-dimensional marginal distributions. We will use this in Sect. 26.3 in some examples.

The existence of solutions of a stochastic differential equation (or equivalently of a local martingale problem) is often easier to show than the uniqueness of solutions. We know already that Lipschitz conditions for the coefficients b and σ (not $\sigma \sigma^T$!) ensure uniqueness (Theorems 26.8 and 26.18), as here strong uniqueness of the solution holds.

At first glance, it might seem confusing that random fluctuations have a stabilising effect on the solution. That is, there are deterministic differential equations whose solution is unique only after adding random noise terms. For example, consider the following equation:

$$dX_t = \text{sign}(X_t) |X_t|^{1/3} dt + \sigma dW_t, \qquad X_0 = 0. \qquad (26.23)$$

If $\sigma = 0$, then the deterministic differential equation has a continuum of solutions that can be parameterized by $v \in \{-1, +1\}$ and $T \geq 0$, namely $X_t = v\,2\sqrt{2}\,(t - T)^{3/2}\mathbb{1}_{\{t>T\}}$. If $\sigma > 0$, then the noise eliminates the instability of (26.23) at $x = 0$. We quote the following theorem for the time-independent case from [148, Theorem V.24.1] (see also [162, Chapter 10]).

Theorem 26.26 (Stroock–Varadhan) *Let $a_{ij} : \mathbb{R}^n \to \mathbb{R}$ be continuous and let $b_i : \mathbb{R}^n \to \mathbb{R}$ be measurable for $i, j = 1, \ldots, n$. Assume*

(i) $a(x) = (a_{ij}(x))$ is symmetric and strictly positive definite for every $x \in \mathbb{R}^n$,
(ii) there exists a $C < \infty$ such that, for all $x \in \mathbb{R}^n$ and $i, j = 1, \ldots, n$, we have

$$\left|a_{ij}(x)\right| \leq C\left(1 + \|x\|^2\right) \quad \text{and} \quad \left|b_i(x)\right| \leq C\left(1 + \|x\|\right).$$

Then the LMP(a, b) is well-posed and the SDE (26.15) has a unique strong solution that is a strong Markov process. The solution X has the Feller property: For every $t > 0$ and every bounded measurable $f : \mathbb{R}^n \to \mathbb{R}$, the map $x \mapsto \mathbf{E}_x[f(X_t)]$ is continuous.

We will present explicit examples in Sect. 26.3. Here we just remark that we have developed a particular method in order to construct Markov processes, namely as the solution of a stochastic differential equation or of a local martingale problem. In the framework of models in discrete time, in Sect. 17.2 and especially in Exercise 17.2.1, we characterized certain Markov chains as solutions of martingale problems. In order for drift and square variation to be sufficient for uniqueness of the Markov chain described by the martingale problem, it was essential that, for any step of the chain, we only allowed three possibilities. Here, however, the decisive restriction is the continuity of the processes.

Takeaways For a weak solution of a stochastic differential equation, the Brownian motion does not exist *a priori*. Rather it will be constructed together with the paths of the solution. Hence, for weak solutions the focus lies on quantitative properties such as strength of noise and drift. The explicit dependence of the solution path on a specific Brownian motion path is typically of no interest in this situation. Weak solutions can be constructed via martingale problems under rather weak assumptions. However, uniqueness of weak solutions is often an issue and requires tailor-made methods.

Exercise 26.2.1 Consider the time-homogeneous one-dimensional case ($m = n = 1$). Let σ and b be such that, for every $X_0 \in \mathbb{R}$, there exists a unique weak solution of

$$dX_t = \sigma(X_t)\,dW_t + b(X_t)\,dt$$

that is a strong Markov process. Further, assume that there exists an $x_0 \in \mathbb{R}$ with

$$C := \int_{-\infty}^{\infty} \frac{1}{\sigma^2(x)} \exp\left(\int_{x_0}^{x} \frac{2b(r)}{\sigma^2(r)}\, dr\right) dr < \infty.$$

(i) Show that the measure $\pi \in \mathcal{M}_1(\mathbb{R})$ with density

$$\frac{\pi(dx)}{dx} = C^{-1} \frac{1}{\sigma^2(x)} \exp\left(\int_{x_0}^{x} \frac{2b(r)}{\sigma^2(r)}\, dr\right)$$

is an invariant distribution for X.

(ii) For which values of b does the Ornstein–Uhlenbeck process $dX_t = \sigma\, dW_t + bX_t\, dt$ have an invariant distribution? Determine this distribution and compare the result with what could be expected by an explicit computation using the representation in (26.3).

(iii) Compute the invariant distribution of the Cox–Ingersoll–Ross SDE (26.14) (i.e., Feller's branching diffusion).

(iv) Let $\gamma, c > 0$ and $\theta \in (0, 1)$. Show that the invariant distribution of the solution X of the SDE on $[0, 1]$,

$$dX_t = \sqrt{\gamma X_t(1 - X_t)}\, dW_t + c(\theta - X_t)\, dt$$

is the Beta distribution $\beta_{2c\gamma/\theta,\, 2c\gamma/(1-\theta)}$. ♣

Exercise 26.2.2 Let $\gamma > 0$. Let X^1 and X^2 be solutions of $dX_t^i = \sqrt{\gamma X_t^i}\, dW_t^i$, where W^1 and W^2 are two independent Brownian motions with initial values $X_0^1 = x_0^1 > 0$ and $X_0^2 = x_0^2 > 0$. Show that $Z := X^1 + X^2$ is a weak solution of $Z_0 = 0$ and $dZ_t = \sqrt{\gamma Z_t}\, dW_t$. ♣

26.3 Weak Uniqueness via Duality

The Stroock–Varadhan theorem provides a strong criterion for existence and uniqueness of solutions of stochastic differential equations. However, in many cases, the condition of locally uniform ellipticity of a (Condition (i) in Theorem 26.26) is not fulfilled. This is the case, in particular, if the solutions are defined only on subsets of $\mathbb{R}^n$.

Here we will study a powerful tool that in many special cases can yield weak uniqueness of solutions.

Definition 26.27 (Duality) *Let* $X = (X^x,\, x \in E)$ *and* $Y = (Y^y,\, y \in E')$ *be families of stochastic processes with values in the spaces E and E', respectively, and such that $X_0^x = x$ a.s. and $Y_0^y = y$ a.s. for all $x \in E$ and $y \in E'$. We say that X and Y are **dual** to each other with **duality function** $H : E \times E' \to \mathbb{C}$ if, for all*

$x \in E$, $y \in E'$ and $t \geq 0$, the expectations $\mathbf{E}\big[H(X_t^x, y)\big]$ and $\mathbf{E}\big[H(x, Y_t^y)\big]$ exist and are equal:

$$\mathbf{E}\big[H(X_t^x, y)\big] = \mathbf{E}\big[H(x, Y_t^y)\big].$$

In the following, we assume that $\sigma_{ij} : \mathbb{R}^n \to \mathbb{R}$ and $b_i : \mathbb{R}^n \to \mathbb{R}$ are bounded on compact sets for all $i = 1, \ldots, n$, $j = 1, \ldots, m$. Consider the time-homogeneous stochastic differential equation

$$dX_t = \sigma(X_t)\, dW_t + b(X_t)\, dt. \tag{26.24}$$

Theorem 26.28 (Uniqueness via duality) *Assume that, for every $x \in \mathbb{R}^n$, there exists a solution of the local martingale problem for $(\sigma \sigma^T, b, \delta_x)$. Further, assume that there exists a family $(Y^y, \ y \in E')$ of Markov processes with values in the measurable space $(E', \mathcal{E}')$ and a measurable map $H : \mathbb{R}^n \times E' \to \mathbb{C}$ such that, for every $y \in E'$, $x \in \mathbb{R}^n$ and $t \geq 0$, the expectation $\mathbf{E}[H(x, Y_t^y)]$ exists and is finite. Further, let $(H(\,\cdot\,, y),\ y \in E')$ be a separating class of functions for $\mathcal{M}_1(\mathbb{R}^n)$ (see Definition 13.9).*

For every $x \in \mathbb{R}^n$ and every solution X^x of $LMP(\sigma \sigma^T, b, \delta_x)$, assume that the duality equation holds:

$$\mathbf{E}[H(X_t^x, y)] = \mathbf{E}[H(x, Y_t^y)] \quad \text{for all } y \in E', \ t \geq 0. \tag{26.25}$$

Then the local martingale problem of $(\sigma \sigma^T, b)$ is well-posed and hence (26.24) has a unique weak solution that is a strong Markov process.

Proof By Theorem 26.25, it is enough to check that, for every $x \in \mathbb{R}^n$, every solution X^x of $LMP(\sigma \sigma^T, b, \delta_x)$ and every $t \geq 0$, the distribution $\mathbf{P} \circ (X_t^x)^{-1}$ is unique. Since $(H(\,\cdot\,, y),\ y \in E')$ is a separating class of functions, this follows from (26.16). $\qquad\square$

Example 26.29 (Wright–Fisher diffusion) Consider the Wright–Fisher SDE

$$dX_t = \mathbb{1}_{[0,1]}(X_t)\sqrt{\gamma\, X_t(1 - X_t)}\, dW_t, \tag{26.26}$$

where $\gamma > 0$ is a parameter. See Fig. 26.3 for a computer simulation. By Theorem 26.22, for every $x \in \mathbb{R}$, there exists a weak solution $(\tilde{X}, W)$ of (26.26). $\tilde{X}$ is a continuous local martingale with square variation

$$\langle \tilde{X} \rangle_t = \int_0^t \gamma \tilde{X}_s(1 - \tilde{X}_s)\mathbb{1}_{[0,1]}(\tilde{X}_s)\, ds.$$

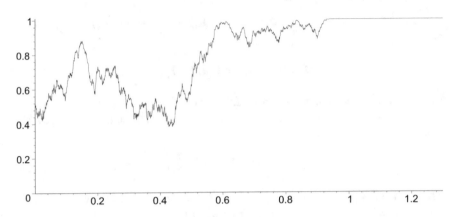

Fig. 26.3 Simulation of a Wright–Fisher diffusion with parameter $\gamma = 1$.

Let $\tau := \inf\{t > 0 : \tilde{X}_t \notin [0, 1]\}$ and let $X := \tilde{X}^\tau$ be the process stopped at τ. Then X is a continuous bounded martingale with

$$\langle X \rangle_t = \int_0^t \gamma X_s (1 - X_s) \mathbb{1}_{[0,1]}(X_s)\, ds.$$

Hence, (X, W) is a solution of (26.26). By construction, $X_t \in [0, 1]$ for all $t \geq 0$ if $X_0 = \tilde{X}_0 \in [0, 1]$.

Let $\tau' := \inf\{t > 0 : \tilde{X}_t \in [0, 1]\}$. If $\tilde{X}_0 \notin [0, 1]$, then $\tau' > 0$ since $\tilde{X}$ is continuous. Since $\tilde{X}^{\tau'}$ is a continuous local martingale with $\langle \tilde{X}^{\tau'} \rangle \equiv 0$, we have $\tilde{X}_t^{\tau'} = \tilde{X}_0$ for all $t \geq 0$. However, this implies $\tilde{X}_t = \tilde{X}_0$ for all $t < \tau'$. Again, by continuity of $\tilde{X}$, we get $\tau' = \infty$ and $\tilde{X}_t = \tilde{X}_0$ for all $t \geq 0$.

Hence, it is enough to show uniqueness of the solution for $\tilde{X}_0 = x \in [0, 1]$. To this end, let $Y = (Y_t)_{t \geq 0}$ be the Markov process on $\mathbb{N}$ with Q-matrix

$$q(m, n) = \begin{cases} \gamma \binom{m}{2}, & \text{if } n = m - 1, \\ -\gamma \binom{m}{2}, & \text{if } n = m, \\ 0, & \text{else.} \end{cases}$$

We show duality of X and Y with respect to $H(x, n) = x^n$:

$$\mathbf{E}_x\left[X_t^n\right] = \mathbf{E}_n\left[x^{N_t}\right] \quad \text{for all } t \geq 0,\ x \in [0, 1],\ n \in \mathbb{N}. \tag{26.27}$$

Define $m^{x,n}(t) = \mathbf{E}_x\left[X_t^n\right]$ and $g^{x,n}(t) = \mathbf{E}_n\left[x^{N_t}\right]$. By the Itô formula,

$$X_t^n - x^n - \int_0^t \gamma \binom{n}{2} X_s^{n-1}(1 - X_s)\, ds = \int_0^t n X_s^{n-1} \sqrt{\gamma X_s(1 - X_s)}\, dW_s$$

is a martingale.

Taking expectations, we obtain the following recursive equations for the moments of X:

$$m^{x,1}(t) = x,$$

$$m^{x,n}(t) = x^n + \gamma \binom{n}{2} \int_0^t \left(m^{x,n-1}(s) - m^{x,n}(s) \right) ds \quad \text{for } n \geq 2.$$

(26.28)

Clearly, this system of linear differential equations can be uniquely solved recursively in n.

Due to the Markov property of Y, for $h > 0$ and $t \geq 0$, we have

$$g^{x,n}(t+h) = \mathbf{E}_n\left[x^{Y_{t+h}}\right] = \mathbf{E}_n\left[\mathbf{E}_{Y_h}\left[x^{Y_t}\right]\right]$$

$$= \sum_{m=1}^n \mathbf{P}_n[Y_h = m]\,\mathbf{E}_m\left[x^{Y_t}\right]$$

$$= \sum_{m=1}^n \mathbf{P}_n[Y_h = m]\,g^{x,m}(t).$$

This implies

$$\frac{d}{dt}g^{x,n}(t) = \lim_{h\downarrow 0} h^{-1}\left[g^{x,n}(t+h) - g^{x,n}(t)\right]$$

$$= \lim_{h\downarrow 0} h^{-1} \sum_{m=1}^n \mathbf{P}_n[Y_h = m]\left(g^{x,m}(t) - g^{x,n}(t)\right)$$

$$= \sum_{m=1}^n q(n,m)\,g^{x,m}(t)$$

(26.29)

$$= \gamma\binom{n}{2}\left(g^{x,n-1}(t) - g^{x,n}(t)\right).$$

Evidently, $g^{x,1}(t) = x$ for all $x \in [0, 1]$ and $t \geq 0$ and $g^{x,n}(0) = x^n$. That is, $g^{x,n}$ solves (26.28), and thus (26.27) holds.

By Theorem 15.4, the family $(H(\cdot, n),\, n \in \mathbb{N}) \subset C([0, 1])$ is separating for $\mathcal{M}_1([0, 1])$; hence the conditions of Theorem 26.28 are fulfilled. Therefore, X is the unique weak solution of (26.26) and is a strong Markov process. $\Diamond$

Remark 26.30 The martingale problem for the Wright–Fisher diffusion is almost identical to the martingale problem for the Moran model (see Example 17.22) $M^N = (M_n^N)_{n\in\mathbb{N}_0}$ with population size N: M^N is a martingale with values in the set $\{0, 1/N, \ldots, (N-1)/N, 1\}$ and with square variation process

$$\langle M^N \rangle_n = \frac{2}{N^2} \sum_{k=0}^{n-1} M_k^N (1 - M_k^N).$$

At each step, M^N can either stay put or increase or decrease by $1/N$. In Exercise 17.2.1, we saw that this determines the process M^N uniquely. Similarly as in Theorem 21.51 for branching processes, it can be shown that the time-rescaled Moran processes $\tilde{M}_t^N = M_{\lfloor N^2 t \rfloor}^N$ converge to the Wright–Fisher diffusion with $\gamma = 2$. The Wright–Fisher diffusion thus occurs as the limiting model of a genealogical model and describes the gene frequency (that is, the fraction) of a certain allele in a population that fluctuates randomly due to resampling. $\Diamond$

Example 26.31 (Feller's branching diffusion) Let $(Z_n^N)_{n \in \mathbb{N}_0}$ be a Galton–Watson branching process with critical geometric offspring distribution $p_k = 2^{-k-1}, k \in \mathbb{N}_0$ and $Z_0^N = N$ for any $N \in \mathbb{N}$. Then Z^N is a discrete martingale and we have

$$\mathbf{E}\left[(Z_n^N - Z_{n-1}^N)^2 \,\middle|\, Z_{n-1}^N \right] = Z_{n-1}^N \left(\sum_{k=0}^{\infty} p_k \, k^2 - 1 \right) = 2 \, Z_{n-1}^N.$$

Hence Z^N has square variation

$$\langle Z^N \rangle_n = \sum_{k=0}^{n-1} 2 Z_k^N.$$

Define the linearly interpolated version

$$Z_t^N := \left(t - N^{-1} \lfloor t N \rfloor \right) \left(Z_{\lfloor t N \rfloor + 1}^N - Z_{\lfloor t N \rfloor}^N \right) + \frac{1}{n} Z_{\lfloor t N \rfloor}^N$$

of $N^{-1} Z_{\lfloor t N \rfloor}^N$. By Lindvall's theorem (Theorem 21.51), there is a continuous Markov process Z such that $Z^N \xrightarrow{N \to \infty} Z$ in distribution. See Fig. 26.4 for a computer simulation of Z. Since it can be shown that the moments also converge, we have that Z is a continuous martingale with square variation

$$\langle Z \rangle_t = \int_0^t 2 Z_s \, ds.$$

In fact, in Example 26.11, we have already shown that Z is the unique solution of the SDE

$$dZ_t = \sqrt{2 Z_t} \, dW_t \tag{26.30}$$

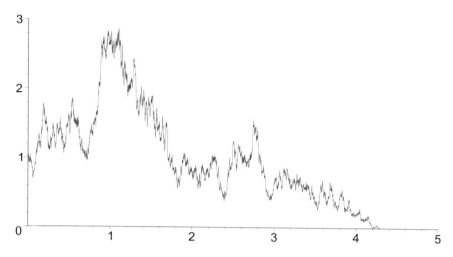

Fig. 26.4 Simulation of Feller's branching diffusion with parameter $\gamma = 1$.

with initial value $Z_0 = 1$. There we also showed that Z is dual to $Y_t^y = \left(\frac{t\gamma}{2} + \frac{1}{y}\right)^{-1}$ with $H(x, y) = e^{-xy}$. This implies uniqueness of the solution of (26.30) and the strong Markov property of Z. ◊

It could be objected that in Examples 26.29 and 26.31, we considered only one-dimensional problems for which the Yamada–Watanabe theorem (Theorem 26.10) yields uniqueness (indeed of a strong solution) anyway. The full strength of the method of duality is displayed only in higher-dimensional problems. As an example, we consider an extension of Example 26.29.

Example 26.32 (Interacting Wright–Fisher diffusions) The Wright–Fisher diffusion from Example 26.29 describes the fluctuations of the gene frequency of an allele in *one* large population. Now we consider more populations, which live at the points $i \in S := \{1, \ldots, N\}$ and interact with each other by a migration that is quantified by migration rates $r(i, j) \geq 0$. As a model for the gene frequencies $X_t(i)$ at site i at time t we use the following N-dimensional SDE for $X = (X(1), \ldots, X(N))$:

$$dX_t(i) = \sqrt{\gamma\, X_t(i)(1 - X_t(i))}\, dW_t^i + \sum_{j=1}^{N} r(i, j)\big(X_t(j) - X_t(i)\big)\, dt. \qquad (26.31)$$

Here $W = (W^1, \ldots, W^N)$ is an N-dimensional Brownian motion. By Theorem 26.22, this SDE has weak solutions; however, none of our general criteria for weak uniqueness apply. We will thus show weak uniqueness by virtue of duality.

As in Example 26.29, it is not hard to show that solutions of (26.31), started at $X_0 = x \in E := [0, 1]^S$, remain in $[0, 1]^S$. The diagonal terms $r(i, i)$ do not appear in (26.31). We use our freedom and define these terms as $r(i, i) = -\sum_{j \neq i} r(i, j)$. Let $Y = (Y_t)_{t \geq 0}$ be the Markov process on $E' := (\mathbb{N}_0)^S$ with the following Q-matrix:

$$
q(\varphi, \eta) = \begin{cases}
\varphi(i)\, r(i, j), & \text{if } \eta = \varphi - \mathbb{1}_{\{i\}} + \mathbb{1}_{\{j\}} \text{ for} \\
 & \text{some } i, j \in S, \ i \neq j, \\[2mm]
\gamma\binom{\varphi(i)}{2}, & \text{if } \eta = \varphi - \mathbb{1}_{\{i\}} \text{ for some } i \in S, \\[2mm]
\sum_{i \in S}\left(\varphi(i) r(i, i) - \gamma\binom{\varphi(i)}{2}\right), & \text{if } \eta = \varphi, \\[2mm]
0, & \text{else.}
\end{cases}
$$

Here $\varphi \in E'$ denotes a generic state with $\varphi(i)$ particles at site $i \in S$, and $\mathbb{1}_{\{i\}} \in E'$ denotes the state with exactly one particle at site i. The process Y describes a system of particles that independently with rate $r(i, j)$ jump from site i to site j. If there is more than one particle at the same site i, then any of the $\binom{\varphi(i)}{2}$ pairs of particles coalesce with the same rate γ to one particle. The common genealogical interpretation of this process is that (in reversed time) it describes the lines of descent of samples of $Y_0(i)$ individuals at each site $i \in S$. By migration, the lines change sites. If two individuals have the same common ancestor, then their lines coalesce. Clearly, for two particles to have the same ancestor at a given time, it is necessary but not sufficient for them to be at the same site.

For $x \in \mathbb{R}^n$ and $\varphi \in E'$, we denote $x^\varphi := \prod_{i \in S} x(i)^{\varphi(i)}$. We show that X and Y are dual to each other with the duality function $H(x, \varphi) = x^\varphi$:

$$
\mathbf{E}_x[X_t^\varphi] = \mathbf{E}_\varphi[x^{Y_t}] \quad \text{for all } \varphi \in S^{\mathbb{N}_0}, \ x \in [0, 1]^S, \ t \geq 0. \tag{26.32}
$$

Let $m^{x,\varphi}(t) := \mathbf{E}_x[X_t^\varphi]$ and $g^{x,\varphi}(t) := \mathbf{E}_\varphi[x^{Y_t}]$. Clearly, H has the derivatives $\partial_i H(\cdot, \varphi)(x) = \varphi(i) x^{\varphi - \mathbb{1}_{\{i\}}}$ and $\partial_i \partial_i H(\cdot, \varphi)(x) = 2\binom{\varphi(i)}{2} x^{\varphi - 2\mathbb{1}_{\{i\}}}$.

By the Itô formula,

$$
X_t^\varphi - X_0^\varphi - \int_0^t \sum_{i,j \in S} \varphi(i) r(i, j)\bigl(X_s(j) - X_s(i)\bigr) X_t^{\varphi - \mathbb{1}_{\{i\}}} \, ds
$$

$$
- \sum_{i \in S} \int_0^t \gamma\binom{\varphi(i)}{2}\bigl(X_s(i)(1 - X_s(i))\bigr) X_s^{\varphi - 2\mathbb{1}_{\{i\}}} \, ds
$$

is a martingale. Taking expectations, we get a system of linear integral equations

$$m^{x,0}(t) = 1,$$

$$m^{x,\varphi}(t) = x^\varphi + \int_0^t \sum_{i,j \in S} \varphi(i) r(i,j) \left(m^{x,\varphi + 1_{\{j\}} - 1_{\{i\}}}(s) - m^{x,\varphi}(s) \right) ds$$

$$+ \int_0^t \gamma \sum_{i \in S} \binom{\varphi(i)}{2} \left(m^{x,\varphi - 1_{\{i\}}}(s) - m^{x,\varphi}(s) \right) ds. \tag{26.33}$$

This system of equations can be solved uniquely by induction on $n = \sum_{i \in I} \varphi(i)$. However, we do not intend to compute this solution explicitly. We show only that it coincides with $g^{x,\varphi}(t)$ by showing that g solves an equivalent system of differential equations.

For g as in (26.29), we obtain

$$\frac{d}{dt} g^{x,\varphi}(t) = \sum_{\eta \in E'} q(\varphi, \eta) g^{x,\varphi}(t)$$

$$= \sum_{i,j \in S} r(i,j) \left(g^{x,\varphi + 1_{\{j\}} - 1_{\{i\}}}(t) - g^{x,\varphi}(t) \right) \tag{26.34}$$

$$+ \sum_{i \in S} \gamma \binom{\varphi(i)}{2} \left(g^{x,\varphi - 1_{\{i\}}}(t) - g^{x,\varphi}(t) \right).$$

Together with the initial values $g^{x,0}(t) = 1$ and $g^{x,\varphi}(0) = x^\varphi$, the system (26.34) of differential equations is equivalent to (26.33). Hence the duality (26.32) holds, and thus the SDE (26.31) has a unique weak solution. (In fact, it can be shown that there exists a unique strong solution, even if S is countably infinite, as long as r then satisfies certain regularity conditions such as if it is the Q-matrix of a random walk on $S = \mathbb{Z}^d$; see [154].) ◊

Takeaways One tool to establish uniqueness of a weak solution of a stochastic differential equation is a duality of the solution to a second process (either deterministic or random). For Feller's branching diffusion, we have a deterministic dual. Interacting Wright-Fisher diffusions are dual to a system of (delayed) coalescing random walks. In many cases the dual process can be used to compute quantitative properties such as extinction probabilities, fixation probabilities and so on.

Exercise 26.3.1 (Extinction probability of Feller's branching diffusion) Let $\gamma > 0$ and let Z be the solution of $dZ_t := \sqrt{\gamma Z_t}\, dW_t$ with initial value $Z_0 = z > 0$. Use the duality to show

$$\mathbf{P}_z[Z_t = 0] = \exp\left(-\frac{2z}{\gamma t}\right). \tag{26.35}$$

Use Lemma 21.44 to compute the probability that a Galton–Watson branching process X with critical geometric offspring distribution and with $X_0 = N \in \mathbb{N}$ is extinct by time $n \in \mathbb{N}$. Compare the result with (26.35). ♣

References

1. R.J. Adler, *An Introduction to Continuity, Extrema, and Related Topics for General Gaussian Processes*. Institute of Mathematical Statistics Lecture Notes-Monograph Series, vol. 12 (Institute of Mathematical Statistics, Hayward, CA, 1990)
2. M. Aizenman, H. Kesten, C.M. Newman, Uniqueness of the infinite cluster and continuity of connectivity functions for short and long range percolation. Commun. Math. Phys. **111**(4), 505–531 (1987)
3. M. Aizenman, H. Kesten, C.M. Newman, Uniqueness of the infinite cluster and related results in percolation, in *Percolation theory and ergodic theory of infinite particle systems (Minneapolis, Minn., 1984–1985). IMA Volumes in Mathematics and Its Applications*, vol. 8 (Springer, New York, 1987), pp. 13–20
4. D.J. Aldous, Exchangeability and related topics, in *École d'été de probabilités de Saint-Flour, XIII—1983*. Lecture Notes in Mathematics, vol. 1117 (Springer, Berlin, 1985), pp. 1–198
5. K.B. Athreya, P.E. Ney, *Branching Processes* (Springer, Berlin, 1972)
6. J. Azéma, M. Yor, Le problème de Skorokhod: compléments à "Une solution simple au problème de Skorokhod", in *Séminaire de Probabilités, XIII (Univ. Strasbourg, Strasbourg, 1977/78)*. Lecture Notes in Mathematics, vol. 721 (Springer, Berlin, 1979), pp. 625–633
7. J. Azéma, M. Yor, Une solution simple au problème de Skorokhod, in *Séminaire de Probabilités, XIII (Univ. Strasbourg, Strasbourg, 1977/78). Lecture Notes in Mathematics*, vol. 721 (Springer, Berlin, 1979), pp. 90–115
8. L.E. Baum, M. Katz, Convergence rates in the law of large numbers. Trans. Am. Math. Soc. **120**, 108–123 (1965)
9. M. Baxter, R. Rennie, *Financial Calculus* (Cambridge University Press, Cambridge, 1997)
10. A.C. Berry, The accuracy of the Gaussian approximation to the sum of independent variates. Trans. Am. Math. Soc. **49**, 122–136 (1941)
11. A. Beutelspacher, *Kryptologie* (Vieweg + Teubner, Wiesbaden, 9th edn, 2009)
12. P. Billingsley, *Convergence of Probability Measures* (Wiley, New York, 1968)
13. P. Billingsley, *Weak Convergence of Measures: Applications in Probability*. Conference Board of the Mathematical Sciences Regional Conference Series in Applied Mathematics, No. 5 (Society for Industrial and Applied Mathematics, Philadelphia, PA, 1971)
14. P. Billingsley, *Convergence of Probability Measures*, 2nd edn. Wiley Series in Probability and Statistics: Probability and Statistics (Wiley, New York, 1999). A Wiley-Interscience Publication
15. K. Binder, D.W. Heermann, *Monte Carlo Simulation in Statistical Physics: An Introduction*, 3rd edn. *Springer Series in Solid-State Sciences*, vol. 80 (Springer, Berlin, 1997)

© The Editor(s) (if applicable) and The Author(s), under exclusive license
to Springer Nature Switzerland AG 2020
A. Klenke, *Probability Theory*, Universitext,
https://doi.org/10.1007/978-3-030-56402-5

16. G.D. Birkhoff, Proof of the ergodic theorem. Proc. Nat. Acad. Sci. **17**, 656–660 (1931)
17. D. Blackwell, D, Kendall, The Martin boundary of Pólya's urn scheme, and an application to stochastic population growth. J. Appl. Probab. **1**, 284–296 (1964)
18. R.M. Blumenthal, An extended Markov property. Trans. Amer. Math. Soc. **85**, 52–72 (1957)
19. S. Bochner, *Vorlesungen über Fouriersche Integrale*. (Chelsea Publishing Company, New York, 1932) Reprinted 1948
20. L. Breiman, *Probability* (Addison-Wesley Publishing Company, Reading, MA, 1968)
21. P. Brémaud, *Markov Chains*. Texts in Applied Mathematics, vol. 31. Gibbs fields, Monte Carlo simulation, and queues (Springer, New York, 1999)
22. D. Brüggemann. Starke Gesetze der großen Zahlen bei blockweisen Unabhängigkeits-bedingungen. PhD thesis, Universität zu Köln, 2002
23. R.M. Burton, M. Keane, Density and uniqueness in percolation. Commun. Math. Phys. **121**(3), 501–505 (1989)
24. G. Choquet, J. Deny, Sur l'équation de convolution $\mu = \mu * \sigma$. C. R. Acad. Sci. Paris **250**, 799–801 (1960)
25. Y.S. Chow, H. Teicher, *Probability Theory: Independence, Interchangeability, Martingales*, 3rd edn. Springer Texts in Statistics (Springer, New York, 1997)
26. K.L. Chung, *Markov Chains with Stationary Transition Probabilities*. Die Grundlehren der mathematischen Wissenschaften, Bd. 104 (Springer, Berlin, 1960)
27. K.L. Chung, W.H.J. Fuchs, On the distribution of values of sums of random variables. Mem. Am. Math. Soc. **6**, 1–12 (1951)
28. P. Clifford, A. Sudbury, A model for spatial conflict. Biometrika **60**, 581–588 (1973)
29. H. Cramér, Sur un nouveau théorème-limite de la théorie des probabilités. Actualités Scientifiques et Industrielles **763**, 5–23 (1938). Colloque consacré à la théorie des probabilités
30. F. Delbaen, W. Schachermayer, A general version of the fundamental theorem of asset pricing. Math. Ann. **300**(3), 463–520 (1994)
31. A. Dembo, O. Zeitouni, *Large Deviations Techniques and Applications*. Stochastic Modelling and Applied Probability, vol. 38 (Springer, Berlin, 2010). Korrigierter Nachdruck der zweiten Auflage von 1998
32. J.-D. Deuschel, D.W. Stroock, *Large Deviations*. Pure and Applied Mathematics, vol. 137 (Academic, Boston, MA, 1989)
33. P. Diaconis, D. Freedman, Finite exchangeable sequences. Ann. Probab. **8**(4), 745–764 (1980)
34. J. Dieudonné, *Foundations of Modern Analysis*. Pure and Applied Mathematics, vol. X (Academic, New York/London, 1960)
35. M.D. Donsker, An invariance principle for certain probability limit theorems. Mem. Am. Math. Soc. **6**, 1–12 (1951)
36. P.G. Doyle, J.L. Snell, *Random Walks and Electric Networks*. Carus Mathematical Mono-graphs, vol. 22 (Mathematical Association of America, Washington, DC, 1984)
37. R.M. Dudley, *Real Analysis and Probability*. Cambridge Studies in Advanced Mathematics, vol. 74 (Cambridge University Press, Cambridge, 2002). Revised reprint of the 1989 original
38. N. Dunford, J.T. Schwartz, *Linear Operators. I. General Theory*. With the assistance of W. G. Bade and R. G. Bartle. Pure and Applied Mathematics, vol. 7 (Interscience Publishers, Inc., New York, 1958)
39. R. Durrett, *Probability: Theory and Examples*, 4th edn. Cambridge Series in Statistical and Probabilistic Mathematics. (Cambridge University Press, Cambridge 2010)
40. A. Dvoretzky, P. Erdős, S. Kakutani, Nonincrease everywhere of the Brownian motion process, in *Proceedings of the 4th Berkeley Symposium on Mathematics, Statistics and Probability*, vol. II (University of California Press, Berkeley, CA, 1961), pp. 103–116
41. D. Egoroff, Sur les suites des fonctions measurables. C. R. Acad. Sci. Paris **152**, 135–157 (1911)
42. R.J. Elliott, P.E. Kopp, *Mathematics of Financial Markets*, 2nd edn. Springer Finance (Springer, New York, 2005)
43. R.S. Ellis, *Entropy, Large Deviations, and Statistical Mechanics*. Grundlehren der Mathema-tischen Wissenschaften, vol. 271 (Springer, New York, 1985)

44. J. Elstrodt, *Maß- und Integrationstheorie*, 8th edn. (Springer, New York, 2018)
45. P. Erdős, R.L. Graham, On a linear diophantine problem of Frobenius. Acta Arith. **21**, 399–408 (1972)
46. C.-G. Esseen, On the Liapounoff limit of error in the theory of probability. Ark. Mat. Astr. Fys. **28A**(9), 1–19 (1942)
47. N. Etemadi, An elementary proof of the strong law of large numbers. Z. Wahrsch. Verw. Gebiete **55**(1), 119–122 (1981)
48. A. Etheridge, *A Course in Financial Calculus* (Cambridge University Press, Cambridge, 2002)
49. S.N. Ethier, T.G. Kurtz, *Markov Processes: Characterization and Convergence*. Wiley Series in Probability and Mathematical Statistics: Probability and Mathematical Statistics (Wiley, New York, 1986)
50. S.N. Evans, X. Zhou, Identifiability of exchangeable sequences with identically distributed partial sums. Electron. Comm. Probab. **4**, 9–13 (electronic) (1999)
51. W. Feller, Über den zentralen Grenzwertsatz der Wahrscheinlichkeitstheorie I. Math. Zeit. **40**, 521–559 (1935)
52. W. Feller, Über den zentralen Grenzwertsatz der Wahrscheinlichkeitstheorie II. Math. Zeit. **42**, 301–312 (1937)
53. W. Feller, *An Introduction to Probability Theory and its Applications*, vol. I, 3rd edn. (Wiley, New York, 1968)
54. W. Feller, *An Introduction to Probability Theory and its Applications*, vol. II, 2nd edn. (Wiley, New York, 1971)
55. J.A. Fill, An interruptible algorithm for perfect sampling via Markov chains. Ann. Appl. Probab. **8**(1), 131–162 (1998)
56. J.A. Fill, M. Machida, D.J. Murdoch, J.S. Rosenthal, Extension of Fill's perfect rejection sampling algorithm to general chains. Random Struct. Algoritm. **17**(3–4), 290–316 (2000). Proceedings of the Ninth International Conference "Random Structures and Algorithms" (Poznan, 1999)
57. H. Föllmer, A. Schied, *Stochastic Finance. de Gruyter Studies in Mathematics*, vol. 27, 2nd edn. (Walter de Gruyter & Co., Berlin, 2004)
58. D.A. Freedman, Bernard Friedman's urn. Ann. Math. Statist **36**, 956–970 (1965)
59. H.-O. Georgii, *Stochastics: Introduction to Probability Theory and Statistics*. de Gruyter Lehrbuch, 2nd edn. (Walter de Gruyter & Co., Berlin, 2012)
60. A.L. Gibbs, F.E. Su, On choosing and bounding probability metrics. Int. Stat. Rev. **70**(3), 419–435 (2002)
61. M.L. Glasser, I.J. Zucker, Extended Watson integrals for the cubic lattices. Proc. Nat. Acad. Sci. U.S.A. **74**(5), 1800–1801 (1977)
62. B.V. Gnedenko, A.N. Kolmogorov, *Limit Distributions for Sums of Independent Random Variables* (Addison-Wesley Publishing Co., Reading, MA/London/Don Mills, ON, 1968)
63. G. Grimmett, *Percolation*. Grundlehren der Mathematischen Wissenschaften, vol. 321, 2nd edn. (Springer, Berlin, 1999)
64. G.R. Grimmett, D.R. Stirzaker, *Probability and Random Processes*, 3rd edn. (Oxford University Press, New York, 2001)
65. E. Grosswald, The Student *t*-distribution of any degree of freedom is infinitely divisible. Z. Wahrsch. Verw. Gebiete **36**(2), 103–109 (1976)
66. O. Häggström, *Finite Markov Chains and Algorithmic Applications*. London Mathematical Society Student Texts, vol. 52 (Cambridge University Press, Cambridge, 2002)
67. T. Hara, G. Slade, Mean-field critical behaviour for percolation in high dimensions. Commun. Math. Phys. **128**(2), 333–391 (1990)
68. J.M. Harrison, S.R. Pliska, Martingales and stochastic integrals in the theory of continuous trading. Stoch. Process. Appl. **11**(3), 215–260 (1981)
69. P. Hartman, A. Wintner, On the law of the iterated logarithm. Am. J. Math. **63**, 169–176 (1941)

70. W.K. Hastings, Monte Carlo sampling methods using Markov chains and their applications. Biometrika **57**, 97–109 (1970)

71. E. Hewitt, K.A. Ross, *Abstract Harmonic Analysis. Vol. II: Structure and analysis for compact groups. Analysis on locally compact Abelian groups.* Die Grundlehren der mathematischen Wissenschaften, Band 152 (Springer, New York, 1970)

72. E. Hewitt, L.J. Savage, Symmetric measures on Cartesian products. Trans. Math. Soc. **80**, 470–501 (1955)

73. C.C. Heyde, On a property of the lognormal distribution. J. R. Stat. Soc. B **29**, 392–393 (1963)

74. F. den Hollander, *Large Deviations.* Fields Institute Monographs, vol. 14 (American Mathematical Society, Providence, RI, 2000)

75. R.A. Holley, T.M. Liggett, Ergodic theorems for weakly interacting infinite systems and the voter model. Ann. Probab. **3**(4), 643–663 (1975)

76. B.D. Hughes, *Random Walks and Random Environments*, vol. 1. Oxford Science Publications (The Clarendon Press/Oxford University Press, New York, 1995). Random walks

77. B.D. Hughes, *Random Walks and Random Environments*, vol. 2. Oxford Science Publications (The Clarendon Press/Oxford University Press, New York, 1996). Random environments

78. N. Ikeda, S. Watanabe, *Stochastic Differential Equations and Diffusion Processes.* North-Holland Mathematical Library, vol. 24, 2nd edn. (North-Holland Publishing Co., Amsterdam, 1989)

79. J. Jost, *Partial Differential Equations.* Graduate Texts in Mathematics, vol. 214, 3rd edn. (Springer, New York, 2013)

80. G.S. Joyce, Singular behaviour of the lattice Green function for the d-dimensional hypercubic lattice. J. Phys. A **36**(4), 911–921 (2003)

81. S. Kakutani, Examples of ergodic measure preserving transformations which are weakly mising but not strongly mixing, in *Recent Advances in Topological Dynamics (Proceedings of the Conference at Yale University, New Haven, CT, 1972, in honor of Gustav Arnold Hedlund).* Lecture Notes in Mathematics, vol. 318 (Springer, Berlin, 1973), pp. 143–149

82. O. Kallenberg, *Random Measures*, 4th edn. (Akademie-Verlag, Berlin, 1986)

83. O. Kallenberg, *Foundations of Modern Probability*, 2nd edn. Probability and Its Applications (Springer, New York/Berlin, 2002)

84. L.V. Kantorovič, G.Š. Rubinštein, On a space of completely additive functions. Vestnik Leningrad Univ. **13**(7), 52–59 (1958)

85. I. Karatzas, S.E. Shreve, *Brownian Motion and Stochastic Calculus.* Graduate Texts in Mathematics, vol. 113, 2nd edn. (Springer, New York, 1991)

86. I. Karatzas, S.E. Shreve, *Methods of Mathematical Finance.* Applications of Mathematics, vol. 39 (Springer, New York, 1998)

87. T. Kato, *Perturbation Theory for Linear Operators*, 2nd edn. Grundlehren der Mathematischen Wissenschaften, Band 132. (Springer, Berlin, 1976)

88. G. Keller, *Equilibrium States in Ergodic Theory.* London Mathematical Society Student Texts, vol. 42 (Cambridge University Press, Cambridge, 1998)

89. G. Keller, *Wahrscheinlichkeitstheorie.* Lecture Notes (German) (Universität Erlangen, 2003)

90. J.L. Kelley, *General Topology.* Graduate Texts in Mathematics, vol. 27 (Springer, New York, 1975). Reprint of the 1955 edition [Van Nostrand, Toronto, Ontario]

91. J.G. Kemeny, J.L. Snell, *Finite Markov Chains.* Undergraduate Texts in Mathematics (Springer, New York, 1976). Reprinting of the 1960 original

92. R.W. Kenyon, J.G. Propp, D.B. Wilson, Trees and matchings. Electron. J. Combin. **7**, Research Paper 25, 34 pp. (electronic) (2000)

93. H. Kesten, Sums of stationary sequences cannot grow slower than linearly. Proc. Am. Math. Soc. **49**, 205–211 (1975)

94. H. Kesten, The critical probability of bond percolation on the square lattice equals $\frac{1}{2}$. Commun. Math. Phys. **74**(1), 41–59 (1980)

95. H. Kesten, B.P. Stigum, A limit theorem for multidimensional Galton-Watson processes. Ann. Math. Statist. **37**, 1211–1223 (1966)

96. H. Kesten, M.V. Kozlov, F. Spitzer, A limit law for random walk in a random environment. Compos. Math. **30**, 145–168 (1975)
97. A. Khintchine, Über dyadische Brüche. Math. Z. **18**, 109–116 (1923)
98. J.F.C. Kingman, Uses of exchangeability. Ann. Probab. **6**(2), 183–197 (1978)
99. J.F.C. Kingman, *Poisson Processes*. Oxford Studies in Probability, vol. 3 (The Clarendon Press/Oxford University Press, New York, 1993). Oxford Science Publications
100. A. Klenke, L. Mattner, Stochastic ordering of classical discrete distributions. Adv. in Appl. Probab. **42**(2), 392–410 (2010)
101. A.N. Kolmogorov, Sulla determinazione empirica di una legge di distibuzione. Giornale Istituto Italiano degli Attuari **4**, 83–91 (1933)
102. R. Korn, E. Korn, *Option Pricing and Portfolio Optimization*. Graduate Studies in Mathematics, vol. 31 (American Mathematical Society, Providence, RI, 2001). Modern methods of financial mathematics, translated from the 1999 German original by the authors
103. U. Krengel, *Ergodic Theorems*. de Gruyter Studies in Mathematics, vol. 6 (Walter de Gruyter & Co., Berlin, 1985)
104. S. Kullback, R.A. Leibler, On information and sufficiency. Ann. Math. Stat. **22**, 79–86 (1951)
105. S.L. Lauritzen, *Extremal Families and Systems of Sufficient Statistics*. Lecture Notes in Statistics, vol. 49 (Springer, New York, 1988)
106. P. Lévy, *Théorie de l'Addition des Variables Aléatoires* (Gauthier-Villars, Paris, 1937)
107. P. Lévy, *Processus Stochastiques et Mouvement Brownien. Suivi d'une note de M. Loève* (Gauthier-Villars, Paris, 1948)
108. J.W. Lindeberg, Eine neue Herleitung des Exponentialgesetzes in der Wahrscheinlichkeitsrechnung. Math. Zeit. **15**, 211–225 (1922)
109. T. Lindvall, Convergence of critical Galton-Watson branching processes. J. Appl. Probab. **9**, 445–450 (1972)
110. R. Lyons, Y. Peres, *Probability on Trees and Networks*. Cambridge Series in Statistical and Probabilistic Mathematics, vol. 42 (Cambridge University Press, New York, 2016)
111. R. Lyons, R. Pemantle, Y. Peres, Conceptual proofs of $L \log L$ criteria for mean behavior of branching processes. Ann. Probab. **23**(3), 1125–1138 (1995)
112. N. Madras, *Lectures on Monte Carlo Methods*. Fields Institute Monographs, vol. 16 (American Mathematical Society, Providence, RI, 2002)
113. D.E. Menchoff, Sur les séries des fonctions orthogonales (première partie). Fund. Math. **4**, 92–105 (1923)
114. N. Metropolis, A.W. Rosenbluth, M.N. Rosenbluth, A.H. Teller, E. Teller, Equation of state calculations by fast computing machines. J. Chem. Phys. **21**, 1087–1092 (1953)
115. P.-A. Meyer, *Probability and Potentials* (Blaisdell Publishing Co. Ginn and Co., Waltham, MA/Toronto, ON/London, 1966)
116. S.P. Meyn, R.L. Tweedie, *Markov Chains and Stochastic Stability*. Communications and Control Engineering Series (Springer, London, 1993)
117. F. Móricz, K. Tandori, An improved Menshov–Rademacher theorem. Proc. Am. Math. Soc. **124**(3), 877–885 (1996)
118. P. Mörters, Y. Peres, *Brownian Motion*. Cambridge Series in Statistical and Probabilistic Mathematics (Cambridge University Press, Cambridge, 2010). With an appendix by Oded Schramm and Wendelin Werner
119. R. Motwani, P. Raghavan, *Randomized Algorithms* (Cambridge University Press, Cambridge, 1995)
120. A. Müller, D. Stoyan, *Comparison Methods for Stochastic Models and Risks*. Wiley Series in Probability and Statistics (Wiley, Chichester, 2002)
121. M. Musiela, M. Rutkowski, *Martingale Methods in Financial Modelling*. Stochastic Modelling and Applied Probability, vol. 36, 2nd edn. (Springer, Berlin, 2005)
122. J. von Neumann, Proof of the quasi-ergodic hypothesis. Proc. Nat. Acad. Sci. **18**, 70–82 (1932)
123. J.R. Norris, *Markov Chains*. Cambridge Series in Statistical and Probabilistic Mathematics (Cambridge University Press, Cambridge, 1998). Reprint of the 1997 edition

124. E. Nummelin, *General Irreducible Markov Chains and Nonnegative Operators*. Cambridge Tracts in Mathematics, vol. 83 (Cambridge University Press, Cambridge, 1984)

125. R.E.A.C. Paley, N. Wiener, *Fourier Transforms in the Complex Domain*. American Mathematical Society Colloquium Publications, vol. 19 (American Mathematical Society, Providence, RI, 1987). Reprint of the 1934 original

126. R.E.A.C. Paley, N. Wiener, A. Zygmund, Note on random functions. Math. Zeit. **37**, 647–668 (1933)

127. R.F. Peierls, On Ising's model of ferromagnetism. Proc. Camb. Philol. Soc. **32**, 477–481 (1936)

128. V.V. Petrov, *Sums of Independent Random Variables*. Ergebnisse der Mathematik und ihrer Grenzgebiete, vol. 82 (Springer, New York, 1975)

129. J. Pitman, Exchangeable and partially exchangeable random partitions. Probab. Theory Related Fields **102**(2), 145–158 (1995)

130. J. Pitman, *Combinatorial Stochastic Processes*. Lecture Notes in Mathematics, vol. 1875 (Springer, Berlin, 2006). Lectures from the 32nd Summer School on Probability Theory held in Saint-Flour, July 7–24, 2002, with a foreword by Jean Picard

131. J. Pitman, M. Yor, Bessel processes and infinitely divisible laws, in *Stochastic integrals (Proceedings of the Symposium at the University of Durham, Durham, 1980)*. Lecture Notes in Mathematics, vol. 851 (Springer, Berlin, 1981), pp. 285–370

132. J. Pitman, M. Yor, The two-parameter Poisson-Dirichlet distribution derived from a stable subordinator. Ann. Probab. **25**(2), 855–900 (1997)

133. J. Pitman, M. Yor, On the distribution of ranked heights of excursions of a Brownian bridge. Ann. Probab. **29**(1), 361–384 (2001)

134. G. Pólya, Über eine Aufgabe der Wahrscheinlichkeitsrechnung betreffend die Irrfahrt im Straßennetz. Math. Ann. **84**, 149–160 (1921)

135. G. Pólya, Sur quelques points de la théorie de probabilités. Ann. Inst. H. Poincaré **1**, 117–161 (1931)

136. Y.V. Prohorov, Convergence of random processes and limit theorems in probability theory. Teor. Veroyatnost. i Primenen. **1**, 177–238 (1956). Russian with English summary

137. J. Propp, D. Wilson, Coupling from the past: a user's guide, in *Microsurveys in Discrete Probability (Princeton, NJ, 1997)*. DIMACS Series in Discrete Mathematics and Theoretical Computer Science, vol. 41 (American Mathematical Society, Providence, RI, 1998), pp. 181–192

138. J.G. Propp, D.B. Wilson, Exact sampling with coupled Markov chains and applications to statistical mechanics. Random Struct. Algoritm. **9**(1–2), 223–252 (1996)

139. J.G. Propp, D.B. Wilson, How to get a perfectly random sample from a generic Markov chain and generate a random spanning tree of a directed graph. J. Algorithms **27**(2), 170–217 (1998). 7th Annual ACM-SIAM Symposium on Discrete Algorithms (Atlanta, GA, 1996)

140. P.E. Protter, *Stochastic Integration and Differential Equations*. Applications of Mathematics (New York), vol. 21, 2nd edn. (Springer, Berlin, 2004). Stochastic Modelling and Applied Probability

141. H. Rademacher, Einige Sätze über Reihen von allgemeinen Orthogonalfunktionen. Math. Ann. **87**, 112–138 (1922)

142. A. Rényi, Remarks on the Poisson process. Studia Sci. Math. Hungar **2**, 119–123 (1967)

143. P. Révész, *Random Walk in Random and Non-random Environments*, 2nd edn. (World Scientific Publishing Co. Pte. Ltd., Hackensack, NJ, 2005)

144. D. Revuz, *Markov Chains*. North-Holland Mathematical Library, vol. 11, 2nd edn. (North-Holland Publishing Co., Amsterdam, 1984)

145. D. Revuz, M. Yor, *Continuous Martingales and Brownian Motion. Grundlehren der Mathematischen Wissenschaften*, vol. 293, 3rd edn. (Springer, Berlin, 1999)

146. R.T. Rockafellar, *Convex Analysis*. Princeton Mathematical Series, No. 28 (Princeton University Press, Princeton, NJ, 1970)

147. L.C.G. Rogers, D. Williams, *Diffusions, Markov Processes, and Martingales. Vol. 1: Foundations*. Cambridge Mathematical Library (Cambridge University Press, Cambridge, 2000). Reprint of the 2nd edition from 1994

148. L.C.G. Rogers, D. Williams, *Diffusions, Markov Processes, and Martingales. Vol. 2: Itô Calculus*. Cambridge Mathematical Library (Cambridge University Press, Cambridge, 2000). Reprint of the 2nd edition from 1994

149. W. Rudin, *Principles of Mathematical Analysis*. International Series in Pure and Applied Mathematics, 3rd edn. (McGraw-Hill, New York, 1976)

150. I.N. Sanov, On the probability of large deviations of random magnitudes (Russian). Mat. Sb. N. S. **42**(84), 11–44 (1957)

151. I.N. Sanov, On the probability of large deviations of random variables. Sel. Transl. Math. Stat. Probab. **1**, 213–244 (1961)

152. R.L. Schilling, L. Partzsch, *Brownian Motion* (De Gruyter, Berlin, 2012). An introduction to stochastic processes, With a chapter on simulation by Björn Böttcher.

153. E. Seneta, *Non-negative Matrices and Markov Chains*. Springer Series in Statistics (Springer, New York, 2006). Revised reprint of the second (1981) edition

154. T. Shiga, A. Shimizu, Infinite-dimensional stochastic differential equations and their applications. J. Math. Kyoto Univ. **20**(3), 395–416 (1980)

155. A.N. Shiryaev, *Probability*. Graduate Texts in Mathematics, vol. 95, 2nd edn. (Springer, New York, 1996). Translation of the Russian edition from 1980

156. J. Sinaï, On the concept of entropy for a dynamic system. Dokl. Akad. Nauk SSSR **124**, 768–771 (1959)

157. N.V. Smirnov, Sur les écarts de la courbe de distribution empirique. Matematicheskij Sbornik, Rossijskaya Akademiya Nauk, Moscow **2**, 3–16 (1939). Russian with French summary

158. F. Solomon, Random walks in a random environment. Ann. Probab. **3**, 1–31 (1975)

159. F. Spitzer, *Principles of Random Walks*. Graduate Texts in Mathematics, vol. 34, 2nd edn. (Springer, New York, 1976)

160. J.M. Steele, *Stochastic Calculus and Financial Applications*. Applications of Mathematics (New York), vol. 45 (Springer, New York, 2001)

161. V. Strassen, The existence of probability measures with given marginals. Ann. Math. Statist. **36**, 423–439 (1965)

162. D.W. Stroock, S.R.S. Varadhan, *Multidimensional Diffusion Processes*. Grundlehren der Mathematischen Wissenschaften, vol. 233 (Springer, Berlin, 1979)

163. J.J. Sylvester, Mathematical questions with their solutions. Educ. Times **41**, 171–178 (1884)

164. K. Tandori, Über die orthogonalen Funktionen. I. Acta Sci. Math. Szeged **18**, 57–130 (1957)

165. K. Tandori, Über die Divergenz der Orthogonalreihen. Publ. Math. Debrecen **8**, 291–307 (1961)

166. K. Tandori, Bemerkung über die paarweise unabhängigen zufälligen Größen. Acta Math. Hungar. **48**(3–4), 357–359 (1986)

167. S.R.S. Varadhan, Asymptotic probabilities and differential equations. Commun. Pure Appl. Math. **19**, 261–286 (1966)

168. P. Walters, *An Introduction to Ergodic Theory*. Graduate Texts in Mathematics, vol. 79 (Springer, New York, 1982)

169. G.N. Watson, Three triple integrals. Q. J. Math. Oxford Ser. **10**, 266–276 (1939)

170. D. Williams, *Probability with Martingales*. Cambridge Mathematical Textbooks (Cambridge University Press, Cambridge, 1991)

171. D.B. Wilson, J.G. Propp, How to get an exact sample from a generic Markov chain and sample a random spanning tree from a directed graph, both within the cover time, in *Proceedings of the Seventh Annual ACM-SIAM Symposium on Discrete Algorithms (Atlanta, GA, 1996)* (ACM, New York, 1996), pp. 448–457

172. S. Wright, Evolution in Mendelian populations. Genetics **16**, 97–159 (1931)

173. T. Yamada, S. Watanabe, On the uniqueness of solutions of stochastic differential equations. J. Math. Kyoto Univ. **11**, 155–167 (1971)

174. K. Yosida, *Functional Analysis*. Classics in Mathematics. (Springer, Berlin, 1995). Reprint of the sixth edition from 1980
175. O. Zeitouni, Random walks in random environment, in *Lectures on Probability Theory and Statistics*. Lecture Notes in Mathematics, vol. 1837 (Springer, Berlin, 2004), pp. 189–312

Notation Index

© The Editor(s) (if applicable) and The Author(s), under exclusive license
to Springer Nature Switzerland AG 2020
A. Klenke, *Probability Theory*, Universitext,
https://doi.org/10.1007/978-3-030-56402-5

Name Index

B

Banach, Stefan, 1892 (Kraków, now Poland) – 1945 (Lvov, now Ukraine), 171

Bayes, Thomas, 1702 (London) – 1761 (Tunbridge Wells, England), 192

Bernoulli, Jakob, 1654 (Basel, Switzerland) – 1705 (Basel), 19

Bienaymé, Irénée-Jules, 1796 (Paris) – 1878 (Paris), 116

Blackwell, David, 1919 (Centralia, Illinois) – 2010 (Berkeley, California), 119

Bochner, Salomon, 1899 (Kraków, now Poland) – 1982 (Houston, Texas), 348

Boltzmann, Ludwig, 1844 (Vienna, Austria) – 1906 (Duino, near Trieste, Italy), 447

Borel, Emile, 1871 (Saint-Affrique, France) – 1956 (Paris), 8

Brown, Robert, 1773 (Montrose, Scotland) – 1858 (London), 522

C

Cantelli, Francesco Paolo, 1875 (Palermo, Italy) – 1966 (Rome, Italy), 58

Carathéodory, Constantin, 1873 (Berlin) – 1950 (Munich, Germany), 20

Cauchy, Augustin Louis, 1789 (Paris) – 1857 (near Paris), 117

Cesàro, Ernesto, 1859 (Naples, Italy) – 1906 (Torre Annunziata, Italy), 70

Chebyshev, Pafnutij Lvovich (Чебышёв, Пафнутий Львович), 1821 (Okatavo, Russia) – 1894 (Saint Petersburg), 121

Cramér, Harald, 1893 (Stockholm) – 1985 (Stockholm), 365

Curie, Pierre, 1859 (Paris) – 1906 (Paris), 609

D

Dieudonné, Jean Alexandre 1906 (Lille, France) – 1992 (Paris), 328

Dirac, Paul Adrien Maurice, 1902 (Bristol, England) – 1984 (Tallahassee, Florida), 12

Dirichlet, Lejeune, 1805 (Düren, Germany) – 1859 (Göttingen, Germany), 463

Doob, Joseph Leo, 1910 (Cincinnati, Ohio) – 2004 (Urbana, Illinois), 229

Dynkin, Eugene, 1924 (Petrograd, now Saint Petersburg) – 2014 (Ithaca, New York), 3

E

Egorov, Dmitrij Fedorovich (Егоров, Дмитрий Фёдорович), 1869 (Moscow) – 1931 (Kazan, Russia), 152

Esseen, Carl-Gustav, 1918 (Linköping, Sweden) – 2001 (Uppsala, Sweden ?), 363

© The Editor(s) (if applicable) and The Author(s), under exclusive license to Springer Nature Switzerland AG 2020
A. Klenke, *Probability Theory*, Universitext,
https://doi.org/10.1007/978-3-030-56402-5

Subject Index

© The Editor(s) (if applicable) and The Author(s), under exclusive license
to Springer Nature Switzerland AG 2020
A. Klenke, *Probability Theory*, Universitext,
https://doi.org/10.1007/978-3-030-56402-5